W9-AEJ-729

SALEM HEALTH
GENETICS
& INHERITED CONDITIONS

SALEM HEALTH
GENETICS
& INHERITED CONDITIONS

Volume 3

Palmoplantar keratoderma — Zellweger syndrome

Appendixes

Indexes

Editor

Jeffrey A. Knight, Ph.D.

Mount Holyoke College

SALEM PRESS

Pasadena, California Hackensack, New Jersey

Some of the updated and revised essays in this work originally appeared in the *Encyclopedia of Genetics, Revised Edition* (2004), edited by Bryan Ness, Ph.D. Substantial new material has been added.

∞ The paper used in these volumes conforms to the American National Standard for Permanence of Paper for Printed Library Materials, Z39.48-1992 (R1997).

Note to Readers
The material presented in *Salem Health: Genetics and Inherited Conditions* is intended for broad informational and educational purposes. Readers who suspect that they or someone whom they know or provide caregiving for suffers from any disorder, disease, or condition described in this set should contact a physician without delay; this work should not be used as a substitute for professional medical diagnosis. Readers who are undergoing or about to undergo any treatment or procedure described in this set should refer to their physicians and other health care team members for guidance concerning preparation and possible effects. This set is not to be considered definitive on the covered topics, and readers should remember that the field of health care is characterized by a diversity of medical opinions and constant expansion in knowledge and understanding.

Library of Congress Cataloging-in-Publication Data

Genetics and inherited conditions / editor, Jeffrey A. Knight.
p. cm. — (Salem health)
Includes bibliographical references and index.
ISBN 978-1-58765-650-7 (set : alk. paper) — ISBN 978-1-58765-651-4 (v. 1 : alk. paper) —
ISBN 978-1-58765-652-1 (v. 2 : alk. paper) — ISBN 978-1-58765-653-8 (v. 3 : alk. paper)
1. Genetic disorders. 2. Genetics. I. Knight, Jeffrey A., 1948-
RB155.5.G4616 2010
616'.042—dc22

2010005289

First Printing

PRINTED IN THE UNITED STATES OF AMERICA

Contents

Complete List of Contents

Volume 1

Volume 2

Volume 3

SALEM HEALTH
GENETICS
& INHERITED CONDITIONS

P

Palmoplantar keratoderma

CATEGORY: Diseases and syndromes
ALSO KNOWN AS: PPK; keratosis palmaris et plantaris; Palmoplantar keratosis; Hyperkeratosis of palms and soles; palmar plantar keratodermas

DEFINITION

Palmoplantar keratoderma (PPK) is a generic term for an heterogenous group of mostly hereditary disorders characterized by unusual thickening of the skin in soles of the feet and palms of the hands. Excessive or abnormal keratin production leads to hypertrophy of the stratum corneum (hyperkeratosis of the top layer of the skin). Their initial classification is based on whether they are inherited (congenital) or acquired. Hereditary variants result from several gene abnormalities that cause abnormal skin protein.

RISK FACTORS

Hereditary predisposition is associated with the genetic PPK forms. Therefore, familial association is a common risk factor for patients. Mutations in up to sixteen different chromosomes have been identified as responsible for the inherited variants of PPK in an autosomal dominant or recessive manner. In certain populations, it is common to see a higher frequency of a particular inherited skin disorder, as seen with Naxos disease (the Greek island of Naxos), diffuse PPK (Norrbotten, Sweden), or Mal de Meleda (the island of Meleda, Yugoslavia).

ETIOLOGY AND GENETICS

Keratin is a tough fibrous structural protein and a component of the top layer of the skin, the epidermis. The keratinocyte is the epidermal cell that produces keratin. The normal keratinized skin provides waterproofing and works as a natural barrier to infection or harmful chemicals. Keratoderma is a disease of the skin marked by overgrowth of horny tissue or keratosis. In hyperkeratosis, there is an excess of keratin production in the skin.

Although PPK can manifest itself as an acquired condition, in the inherited form it is frequently present as the result of various genetic mutations. In the acquired type, the condition is not the result of genetic abnormalities and appears because of a shift in the patient's health or environmental factors. The most evident causes in the acquired variants are mechanical, vascular, endocrine, infective, or drug-induced.

The genetic PPK disorders are inherited from one or both parent(s) by their children. Generally, PPKs result from genetic abnormalities that affect the keratin protein expression, causing atypical thickening of the skin. Inheritance may occur either by an autosomal dominant or autosomal recessive pattern. In autosomal dominant keratodermas, each affected patient has at least one parent phenotypically affected and there is a transmission of the gene defect from generation to generation. There is a 50 percent chance that each offspring will be affected. For the autosomal recessive form, the affected patient has phenotypical unaffected parents and there is no transmission of the abnormal gene from generation to generation. People that carry the abnormal gene are referred to as carriers of the disease. Therefore, the carriers pass on the genetic defect to the next generation, but the children can only exhibit the disease if their other parent also is a carrier of the same abnormal gene and successfully passes it forward. There is a 25 percent chance that each offspring will be affected by the autosomal recessive disorder. Recessive disorders are frequently found in families with increased consanguinity (descendent from a common ancestor).

More than eighty PPK disorders are indexed in the Online Mendelian Inheritance in Man (OMIM) catalog. These mutated genes code for structural skin proteins such as keratins or subcellular structures such as desmosomes. Additional mutations affect protein products related to keratinocyte cytoskeleton structure, cell-to-cell adhesion, connexins, and gap junctions. The physiologic role of other molecules related to PPK pathogenesis is not well understood. What is known is that all these affected protein structures interact in a highly regulated and integrated manner to keep epithelia and skin integrity. Keratin gene mutations are responsible for several types of hereditary PPK. Some of the main PPKs, with their corresponding protein product of its mutated gene in parenthesis, are Mal de Meleda (SLURP-1), Thost-Unna keratoderma (keratin-1), Vörner keratoderma (keratin-1 and 9), Vohwinkel mutilating keratoderma (Loricrin and Connexin-26), Papillon-Lefévre syndrome (Cathepsin C), Carrajo syndrome (Desmoplakin), Naxos disease (Plakoglobin), and Richner-Hanhart syndrome (tyrosine amino transferase).

When these PPK mutations alter the normal skin surface functions, the epidermal protective properties are compromised, giving the chance for bacterial and fungal infections, inflammatory skin reactions (eczema or psoriasis), or other severe conditions to arise that have an impact on quality of life or might worsen morbidity and mortality for the affected patient.

Only one pertinent PPK clinical study is listed and completed in ClinicalTrials.gov, and none are ongoing.

Symptoms

In this cutaneous hereditary disorder, there is marked hyperkeratosis present over the palms and the soles of the patient. The three main clinical patterns of epidermal involvement are diffuse, focal, and punctate. In the diffuse form there is uniform hyperkeratosis of the palmoplantar surface. The focal pattern mainly affects pressure points or sites of constant friction. The punctate pattern consists of multiple, hyperkeratotic nodules or tiny bumps on the palms and soles. Generally, the abnormal thickening of the skin is seen only on the surface of the palms and soles (nontransgradient), but sometimes there is transmigration to close areas outside the palmoplantar surface (transgradient). The understanding of the pathogenesis is critical in narrowing down the target genes for each individual case. More apparent symptoms could be present such as knuckle pads, oral lesions, atrophy, blisters, deafness, mental retardation, starfish keratoses, nail changes, constricting bands, redness, malignancies, cardiomyopathies, or sclerodactyly. For some rare PPK types, organs in the body may be implicated so the keratoderma could be a marker of some internal abnormality.

Screening and Diagnosis

The rare and large group of hyperkeratotic disorders that inherited PPKs comprise are phenotypically characterized as extensively heterogeneous. Overcoming the diagnostic challenges posed by phenotypic and genotypic heterogeneity, several classification strategies have been designed using morphology, associated symptoms, distribution, and inheritance mode to improve screening between different genetic variants. Traditionally, diagnosis has been clinically based, but in recent years the molecular genetic classification is slowly being adopted and replacing the historical descriptive system. Clinical reference laboratories for genetic diagnostics are available, and information can be found through OMIM. Screening includes evaluation of the main genetic mutations that affect protein products associated to keratinocyte cytoskeleton structure, gap junctions, connexins, keratins, and subcellular structures such as desmosomes. Differential diagnosis might include blood tests, skin biopsy, histology, PCR, and mutation, microsatellite, and haplotype analysis. Knowledge of genetic mutations associated with distinct clinical findings together with a systematic and comprehensive analysis makes the exact diagnosis possible by their molecular pathology. This integrative approach leads to the discovery of the real underlying genetic defect. A full personal and family history, regular medical examinations, and appropriate cancer screenings are crucial for early detection of any associated internal malignancies or cardiomyophathies for some of these patients or whether other skin findings are evident and/or other organs are affected. Dermatopathology plays in increasing role in the diagnosis.

Treatment and Therapy

Currently, there is no effective treatment for any PPK types. Common therapy alternatives mainly cause short-term improvements that are often aggravated by unacceptable adverse side effects. Treatment is tailored to the severity of the symptoms, the age of the patient, the affected areas, and the degree of hyperkeratosis. Generally, treatment will either focus on softening the thickened skin or making it less noticeable. Therapies may range from saltwater soaks and paring to emollients, topical keratolytics, topical or systemic retinoids, topical vitamin D ointment, potent topical steroids, careful footwear selection, or even reconstructive surgery in severe cases to excise the hyperkeratotoic skin, followed by grafting. When the PPK syndrome is also showcased as an associated feature with other diseases (such as deafness, corneal dystrophy, internal malignancy, cardiomyopathy, alopecia, and severe periodontitis), treatment and therapy are particularly adapted to the special needs of each case. Future studies on genotype/phenotype correlations in PPK and their treatment response are vital to improve the patient's quality of life.

Prevention and Outcomes

Since genetic factors play an important role in PPK inherited disorders and there is no effective way of prevention, genetic counseling should be provided. Disease management is difficult considering there is no effective treatment for any PPK variants. Early diagnosis is essential for disease management purposes, to mitigate some symptoms, and to define possible outcomes. The PPK inherited variant and the severity of the condition defines disease outcome. Outcome strategies for PPK with associated diseases, as with cardiomyopathies, internal malignancies, and systemic conditions, add another layer of variability, complexity, and challenge to disease management.

Ana Maria Rodriguez-Rojas, M.S.

Further Reading

Bergman, Reuven. "Dermatopathology and Molecular Genetics." *Journal of the American Academy of Dermatology* 58, no. 3 (2008): 452-457.

Itin, Peter H., and Stephen Lautenschlager. "Palmoplantar Keratoderma and Associated Syndromes." *Seminars in Dermatology* 14, no. 2 (1995): 152-161.

Kimyai-Asadi, A., L. B. Kotcher, and M. H. Jih. "The Molecular Basis of Hereditary Palmoplantar Keratodermas." *Journal of the American Academy of Dermatology* 47, no. 3 (September, 2002): 327-343.

Lucker, G. P., P. C. Van de Kerkhof, and P. M. Steijlen. "The Hereditary Palmoplantar Keratoses: An Updated Review and Classification." *British Journal of Dermatology* 131, no. 1 (July, 1994): 1-14.

Torres, Gisela, et al. "'I Forgot to Shave My Hands': A Case of Spiny Keratoderma." *Journal of the American Academy of Dermatology* 58, no. 2 (2008): 344-348.

Web Sites of Interest

DermWeb
http://www.dermweb.com

Foundation for Ichthyosis & Related Skin Types (F.I.R.S.T.)
http://www.scalyskin.org

Genetic Alliance
http://www.geneticalliance.org

Keratodermas
http://www.thedoctorsdoctor.com/diseases/keratodermas.htm

Madisons Foundation
http://www.madisonsfoundation.org/

National Foundation for Ectodermal Dysplasias
http://nfed.org

National Institute of Arthritis and Musculoskeletal and Skin Diseases
http://www.niams.nih.gov

National Organization for Rare Disorders (NORD)
http://www.rarediseases.org/

Palmoplantar Keratoderma, Genetic and Rare Diseases Information Center (GARD), National Institutes of Health
http://rarediseases.info.nih.gov/GARD/Condition/8167/Palmoplantar_Keratoderma.aspx

Online Mendelian Inheritance in Man (OMIM)
http://www.ncbi.nlm.nih.gov/omim/

See also: Albinism; Chediak-Higashi syndrome; Epidermolytic hyperkeratosis; Hermansky-Pudlak syndrome; Ichthyosis; Melanoma.

Pancreatic cancer

Category: Diseases and syndromes
Also known as: Cancer of the pancreas

Definition

Pancreatic cancer is the growth of cancer cells in the pancreas. The pancreas is a long, flattened pear-shaped organ in the abdomen. It makes digestive enzymes and hormones, including insulin.

Cancer occurs when cells in the body divide without control or order. If cells keep dividing uncontrollably, a mass of tissue forms. This is called a growth or tumor. The term "cancer" refers to malignant tumors. They can invade nearby tissue and spread to other parts of the body.

Risk Factors

Males and people who are forty years of age or older are at an increased risk for pancreatic cancer. Other risk factors include smoking and using smokeless tobacco, such as chewing tobacco; having diabetes; having chronic pancreatitis, hereditary pancreatitis, or familial nonpolyposis colon cancer syndrome; having a family or personal history of certain types of colon polyps or colon cancer; having a family history of pancreatic cancer, especially in Ashkenazi Jews with the *BRCA2* (breast cancer associated) gene; and eating a high-fat diet.

Etiology and Genetics

While there is a substantial genetic component to the development of pancreatic cancer, the majority of new cases of the disease occur in families where there is no previous family history. Several genes, however, have been identified that are associated with an increased risk of pancreatic cancer. For example, mutations in the *PRSS1* gene, found on the long arm of chromosome 7 at position 7q35, can cause hereditary pancreatitis, a condition that involves inflammation and swelling of the pancreas and severe abdominal pain. Affected individuals also have about a 40 percent risk of developing pancreatic cancer at some point in their adult years.

Most other cases of inherited pancreatic cancer result from mutations in any of several genes called tumor-suppressor genes, and these mutations can be inherited from either the male or female parent. Tumor-suppressor genes encode proteins that normally function in a variety of ways to limit or prevent cell growth and division. Mutations in these genes can lead to a loss in the ability to restrict tumor formation due to uncontrolled cell growth. When mutations occur in tumor-suppressor genes, it is not unusual to find that there is an increased risk for several different types of cancer to develop.

Peutz-Jeghers syndrome (PJS) is a rare condition in which affected individuals have multiple polyps in the digestive tract. People with PJS have about a 35 percent risk of developing pancreatic cancer, as well as an increased risk for skin, uterine, ovarian, breast, and lung cancers. Mutations in a single gene called *STK11* (at position 19p13.3) are responsible for this condition.

The two genes most commonly associated with an increased risk of breast and ovarian cancer are *BRCA1*, found on the long arm of chromosome 17 at position 17q21, and *BRCA2* (at position 13q12.3), yet mutations in these same two genes have also been linked to an increased risk of pancreatic cancer—about ten times greater than the average risk. These mutations are inherited in an autosomal dominant fashion, meaning that a single copy of the mutation is sufficient to cause the increased cancer risk. An affected individual has a 50 percent chance of transmitting the mutation to each of his or her children.

Individuals carrying a mutation that predisposes them to familial nonpolyposis colon cancer syndrome (Lynch syndrome) have been shown to have about an 8 to 10 percent risk of developing pancreatic cancer, as well as a 20 to 50 percent risk of developing uterine cancer and a 10 percent increased risk for ovarian cancer. Four different genes have been identified in which such mutations might occur: *MLH1* (at position 3p21.3), *MSH2* (at position 2p22-p21), *MSH6* (at position 2p16), and *PMS2* (at position 7p22).

Finally, there is a condition known as familial atypical multiple mole melanoma and pancreatic cancer (FAMMM-PC), which is associated with mutations in the *CDKN2A* gene (at position 9p21). Affected individuals have a 15 to 17 percent risk of developing pancreatic cancer and as much as a 70 percent risk of developing melanoma skin cancer.

Symptoms

Pancreatic cancer does not cause symptoms in its early stages. The cancer may grow for some time before it causes symptoms. When symptoms do appear, they may be very vague. In many cases, the cancer has spread outside the pancreas by the time it is discovered.

Symptoms will vary depending on the location

and size of the tumor. Symptoms include nausea; loss of appetite; unexplained weight loss; pain in the upper abdomen, sometimes spreading to the back (a result of the cancer growing and spreading); jaundice—yellowness of skin and whites of the eyes; dark urine (if the tumor blocks the common bile duct); tan stool or stool that floats to the top of the bowl; and weakness, dizziness, chills, muscle spasms, and diarrhea (especially if the cancer involves the islet cells that make insulin and other hormones). These symptoms may also be caused by other, less serious health conditions. Anyone experiencing these symptoms should see a doctor.

SCREENING AND DIAGNOSIS

The doctor will ask about a patient's symptoms and medical history, and a physical exam may be done. The doctor may order blood and urine tests and may also check for hidden blood in bowel movements.

Tests may include an upper GI (gastrointestinal) series, a series of X rays of the upper digestive system taken after drinking a barium solution; a computed tomography (CT) scan, a type of X ray that uses a computer to make pictures of structures inside the abdomen; a magnetic resonance imaging (MRI) scan, a test that uses magnetic waves to make pictures of structures inside the abdomen; and ultrasonography, a test that uses sound waves to find tumors. Other tests may include an endoscopic retrograde cholangiopancreatography (ERCP), a type of X ray that shows the pancreatic ductal system after dye has been sent through a tube down the throat and into the pancreas; percutaneous transhepatic cholangiography (PTC), a type of X-ray test that shows blockages in the bile ducts of the liver; angiography, X rays of blood vessels taken after an injection of dye that makes the blood vessels show up on the X rays; and a biopsy, the removal of a sample of pancreatic tissue to test for cancer cells.

TREATMENT AND THERAPY

Once cancer of the pancreas is found, staging tests are performed. These tests help to find out if the cancer has spread and, if so, to what extent. Treatments for pancreatic cancer depend on the stage of the cancer.

Treatments include surgery to remove the cancerous tumor and nearby tissue. Nearby lymph nodes may also need to be removed. Surgery may also be performed to relieve symptoms of pancreatic cancer. Surgeries include the Whipple procedure, which is the removal of the head of the pancreas, part of the small intestine, and some of the tissues around it; a total pancreatectomy, which is the removal of the whole pancreas, part of the small intestine, part of the stomach, the bile duct, the gallbladder, the spleen, and most of the lymph nodes in the area; and a distal pancreatectomy, which is the removal of the body and tail of the pancreas.

Radiation therapy (radiotherapy) uses radiation to kill cancer cells and shrink tumors. In external radiation therapy, radiation is directed at the tumor from a source outside the body; in internal radiation therapy, radioactive materials are placed into the body in or near the cancer cells.

Chemotherapy is the use of drugs to kill cancer cells. It may be given in many forms, including pill, injection, and via a catheter. The drugs enter the bloodstream and travel through the body, killing mostly cancer cells. Some healthy cells are also killed.

Biological therapy is the use of medications or substances made by the body in order to increase or restore the body's natural defenses against cancer. It is also called biologic response modifier (BRM) therapy.

Most times, pancreatic cancer is discovered at an advanced stage. Surgery may not be appropriate in this case. If surgery cannot be done, then chemotherapy and radiation are offered together to prolong survival.

Surgery would be appropriate in only 25 percent of patients with this disease in the early stage. In these cases, the patient would benefit from surgery. After surgery, follow-up chemotherapy and radiation therapy have been found to prolong survival in some cases.

PREVENTION AND OUTCOMES

There are no guidelines for preventing this disease. Individuals who think they are at risk for pancreatic cancer should talk to their doctors about ways to reduce their risk factors. Patients and doctors can work together to make an appropriate schedule for checkups.

Laurie LaRusso, M.S., ELS;
reviewed by Igor Puzanov, M.D.
"Etiology and Genetics" by Jeffrey A. Knight, Ph.D.

FURTHER READING

Boffetta, P., et al. "Smokeless Tobacco and Cancer." *Lancet Oncology* 9, no. 7 (July, 2008): 667-675.

Cameron, John L. *Pancreatic Cancer.* Hamilton, Ont.: B. C. Decker, 2001.

EBSCO Publishing. *Health Library: Pancreatic Cancer.* Ipswich, Mass.: Author, 2009. Available through http://www.ebscohost.com.

O'Reilly, Eileen, and Joanne Frankel Kelvin. *One Hundred Questions and Answers About Pancreatic Cancer.* 2d ed. Sudbury, Mass.: Jones and Bartlett, 2010.

WEB SITES OF INTEREST

American Cancer Society
http://www.cancer.org

Canadian Cancer Society
http://www.cancer.ca

National Cancer Institute
http://www.cancer.gov

Pancreatic Cancer Action Network
http://www.pancan.org

See also: Cancer; Colon cancer; Hereditary diffuse gastric cancer; Hereditary leiomyomatosis and renal cell cancer; Hereditary mixed polyposis syndrome; Hereditary non-VHL clear cell renal cell carcinomas; Hereditary papillary renal cancer; Mutagenesis and cancer; Mutation and mutagenesis; Oncogenes; Ovarian cancer; Tumor-suppressor genes.

Pancreatitis

CATEGORY: Diseases and syndromes
ALSO KNOWN AS: Acute pancreatitis; chronic pancreatitis; hereditary pancreatitis

DEFINITION

Pancreatitis is an inflammation of the pancreas, a large gland located near the liver and gallbladder that secretes insulin and enzymes that help to digest food. These enzymes normally join bile and other substances flowing through the digestive system, becoming active only when they reach the intestines. When the pancreas is unable to release these enzymes, they become active before they leave the pancreas, causing inflammation and damage to pancreatic tissue. Pancreatitis can be hereditary or may be related to other diseases that are genetic in nature.

RISK FACTORS

Pancreatitis is often caused by gallstones that block the pancreatic duct and prevent enzymes from flowing out of the pancreas. Other factors are heavy use of alcohol, trauma, certain medications, and genetic abnormalities of the pancreas. Men are affected by pancreatitis more than women. Chronic pancreatitis usually affects people sometime in their thirties or forties.

ETIOLOGY AND GENETICS

Two major gene mutations, known as R122H and N29I, to the cationic tryspinogen gene (*PRSS1*) are associated with hereditary pancreatitis. Family members may carry one, but usually not more than one, mutation. However, even families with a strong history of pancreatitis have none of the known mutations, so scientists believe that other gene mutations for this disease exist. Hereditary pancreatitis follows an autosomal dominant inheritance pattern. It is a possible diagnosis if a person has two or more family members in one generation with pancreatitis and has pancreatic problems before they are thirty; however, most often this condition is not diagnosed for several years as the symptoms (abdominal pain and diarrhea) come and go seemingly in a random fashion.

Some disorders of the pancreas are hereditary, such as a genetic abnormality in structure. For example, pancreas divisum is an inherited condition where two pancreatic ducts form rather than one. Another inherited disorder is a genetic mutation that causes pancreatic enzymes to become active when they are produced. Other hereditary conditions are related to pancreatitis, but the reasons for the relationship are unclear. Hereditary conditions that seem to be related to pancreatitis include cystic fibrosis, certain autoimmune conditions, porphyria, hypercalcemia (high levels of calcium in the blood, which may be related to gallstone production), and hyperlipidemia (high levels of fat in the blood).

SYMPTOMS

Abdominal pain, either acute or chronic, is the main symptom of pancreatitis. Pain may extend into the back or be worse after eating. A distended abdomen, nausea, oily stool, or fever may also be present. Weight loss (from malabsorption) may be a symptom. In severe cases, bleeding and infection may be symptoms.

Screening and Diagnosis

Genetic testing may help diagnose hereditary pancreatitis. When symptoms are present, a blood test for amylase and lipase, the enzymes normally found in the pancreas, is usually the first test to determine pancreatitis. Ultrasound, CT, and MRI testing may also be helpful. Blood, urine, and stool tests may be used to confirm diagnosis and monitor treatment. Glucose tolerance testing can help determine whether the pancreas is still releasing insulin.

Treatment and Therapy

No cure currently exists for pancreatitis. Pain relief is generally the first step. In acute cases, one may be hospitalized to receive narcotics and possibly intravenous or tube-feeding to give the pancreas time to rest and heal. Once the acute symptoms are under control, treatment of the underlying cause can begin. Treatment may involve surgery to remove any gallstones blocking the pancreatic duct or draining any cysts or removing any scar tissue that may be present. Other strategies include sphincterotomy, an enlargement of the sphincter muscle that keeps the pancreatic duct closed, or placing a stent, a small piece of material that keeps the duct open. Replacement of pancreatic enzymes may be necessary in cases where the pancreas has impaired function. Removing the pancreas is not generally recommended, as this procedure results in a type of diabetes that is extremely difficult to manage. However, autologous islet cell transplantation is a therapy that may be helpful for patients with hereditary pancreatitis.

Prevention and Outcomes

Heavy alcohol use is a prime risk factor for developing pancreatitis; avoiding alcohol is the best way to avoid this disease. Outcomes are better for patients who make dietary and lifestyle changes, such as not smoking or drinking alcohol, drinking more water, and eating a low-fat, healthy diet. Management of this condition may involve treatment for alcohol or tobacco addiction. Some alternative therapies, such as acupuncture or meditation, may help manage the pain associated with this condition. Keeping the pancreas active producing enzymes and releasing insulin is important; otherwise, it may begin to calcify or die and portions may need to be removed surgically. In advanced stages, malabsorption, diabetes, impairment of lung function, and kidney failure can occur. Long-term damage to the pancreas is also a risk factor for pancreatic cancer.

Marianne M. Madsen, M.S.

Further Reading

Buechler, M. W., et al. *Chronic Pancreatitis.* New York: Wiley-Blackwell, 2002.

Etemad, B., and D. C. Whitcomb. "Chronic Pancreatitis: Diagnosis, Classification, and New Genetic Developments." *Gastroenterology* 120 (2001): 682-707.

Forsmark, C. E., ed. *Pancreatitis and Its Complications (Clinical Gastroenterology).* Totowa. N.J.: Humana Press, 2004.

Howard, J. M., and W. Hess. *History of the Pancreas: Mysteries of a Hidden Organ.* New York: Springer, 2002.

John, C. D., and C. W. Imrie, eds. *Pancreatic Disease.* New York: Springer, 2004.

Neoptolemos, J. P., and S. B. Manoop. *Diseases of the Pancreas and Biliary Tract.* Abingdon, Oxfordshire, England: Health Press, 2006.

Parker, James, ed. *The Official Patient's Sourcebook on Pancreatitis: A Revised and Updated Directory for the Internet Age.* San Diego: Icon Health, 2002.

Whitcomb, D. C. "Hereditary Pancreatitis: New Insights into Acute and Chronic Pancreatitis." *Gut* 45 (1999): 317-322.

Web Sites of Interest

American Pancreatic Association
http://www.american-pancreatic-association.org

National Pancreas Foundation
http://www.pancreasfoundation.org

Pancreas.org
http://www.pancreas.org

Pancreatitis Association
http://pancassociation.org

See also: Adrenomyelopathy; Androgen insensitivity syndrome; Autoimmune polyglandular syndrome; Congenital hypothyroidism; Diabetes; Diabetes insipidus; Graves' disease; Obesity; Pancreatic cancer; Steroid hormones.

Parkinson disease

CATEGORY: Diseases and syndromes
ALSO KNOWN AS: Parkinson's disease; PD; paralysis agitans; shaking palsy

DEFINITION

Parkinson disease (PD) is a progressive movement disorder that causes muscle rigidity, tremor at rest, slowing down of movements (bradykinesia), difficulty moving, and gait instability. The disorder is caused by a loss of nerve cells in the brain, including loss in an area called the substantia nigra.This loss decreases the amount of dopamine in the brain. Low dopamine results in PD symptoms.

RISK FACTORS

Individuals at increased risk for Parkinson disease include those who are fifty years old or older, have a history of polio, have family members with PD, and are nonsmokers. Men are slightly more likely than women to develop PD. Other risk factors are exposure to toxins, drugs, or the conditions listed above.

ETIOLOGY AND GENETICS

The causes of Parkinson disease are complex and poorly understood, but it is clear that both genetic and environmental factors are involved. In about 85 percent of new cases, the disease appears in people with no family history of the disorder. In the remaining 15 percent of cases, mutations in any one of five different genes (*PARK2, PARK7, PINK1, LRRK2,* and *SNCA*) might be responsible.

The *PARK2* gene, found on chromosome 6 at position 6q25.2-q27, is a very large gene that specifies the protein called parkin. This protein is an important component of a cellular system designed to rid the cell of damaged or excess proteins. Loss of parkin function because of a mutation in the gene could result in an accumulation of defective proteins in the nerve cells of the brain that can lead to cell death and the characteristic symptoms of Parkinson disease. The *PARK7* gene on chromosome 1 (at position 1p36.33-p36.12) encodes a protein called DJ-1, which has several functions, one of which is similar to parkin. The *PINK1* gene, also found at location 1p36, encodes the protein PTEN induced putative kinase 1. This protein is important for proper mitochondrial function, but it is unclear why the lack of this protein as a result of a mutation in the *PINK1* gene causes the death of nerve cells that lead to Parkinson disease.

Parkinson disease that results from mutations in any of these three genes is inherited in an autosomal recessive manner, which means that both copies of the gene must be deficient in order for the individual to be afflicted. Typically, an affected child is born to two unaffected parents, both of whom are carriers of the recessive mutant allele. The probable outcomes for children whose parents are both carriers are 75 percent unaffected and 25 percent affected.

The *LRRK2* gene (at position 12q12) specifies a protein called dardarin, which is a complex protein with at least three different enzyme activities. It is unclear why loss of this protein causes Parkinson disease symptoms to occur. Finally, the *SNCA* gene (at position 4q21) encodes the protein alpha-synuclein, which is abundant in the presynaptic vesicles in brain neurons. Most mutations in this gene which cause Parkinson disease result in an excess production of alpha-synuclein, although it is unclear how this excess impairs proper nerve cell function.

Parkinson disease that results from mutations in either of these two genes is inherited in an autosomal dominant fashion, meaning that a single copy of the mutation is sufficient to cause full expression of the disease. An affected individual has a 50 percent chance of transmitting the mutation to each of his or her children.

SYMPTOMS

Symptoms of PD begin mildly, and they will worsen over time. Symptoms include a "pill-rolling" tremor in the hands. Tremors are present at rest, improve with movement, and are absent during sleep.

Other symptoms include stiffness and rigidity of muscles, usually beginning on one side of the body; difficulty and shuffling when walking; short steps; slowness of purposeful movements; trouble performing usual tasks due to shaking in the hands; trouble speaking; a flat, monotonous voice; stuttering; a shaky, spidery handwriting; poor balance; difficulty rising from a sitting position; and "freezing." Anxiety; seborrhea (a skin problem that causes a red rash and white scales); a tendency to fall;

stooped posture; an increasingly masklike face, with little variation in expression; trouble chewing and swallowing; depression; dementia; difficulty thinking; problems with memory; loss of or decreased sense of smell; and sleep problems, such as rapid eye movement (REM) behavior disorder, are also symptoms.

Screening and Diagnosis

The doctor will ask about a patient's symptoms and medical history and a physical exam will be done. There are no tests to definitively diagnose PD. The doctor will ask many questions, which will help to rule out other causes of a patient's symptoms.

Tests to rule out other conditions may include blood tests; urine tests; a computed tomography (CT) scan, a type of X ray that uses a computer to make pictures of structures inside the head; a magnetic resonance imaging (MRI) scan, a test that uses magnetic waves to make pictures of structures inside the head; and a positron emission tomography (PET) scan, a test that makes images that show the amount of activity in the brain.

Treatment and Therapy

Currently, there are no treatments to cure PD. There are also no proven treatments to slow or stop its progression.

Some medications are used to improve symptoms. Over time the side effects may become troublesome, and the medications may lose their effectiveness. Medications include levodopa/carbidopa (Sinemet); amantadine (Symmetrel); anticholinergics, such as benztropine (Cogentin) and biperiden (Akineton); selegiline (Eldepryl); and dopamine agonists, such as bromocriptine (Parlodel), pergolide (Permax), pramipexole (Mirapex), cabergoline (Dostinex), and ropinirole (Requip). Pergolide (Permax) was withdrawn in March, 2007. This medication had a high risk of serious heart valve damage. Cabergoline (Dostinex) has also been linked to this risk. Additional medications for PD include apomorphine (Apokyn) and COMT (catechol-O-methyltransferase) inhibitors, such as entacapone (Comtan) and tolcapone (Tasmar).

Medicine may also be given to relieve the depression or hallucinations that may also occur with PD. These medications may include selective serotonin reuptake inhibitors (SSRIs) and antipsychotic clozapine. These drugs can worsen other symptoms. Patients who take them will need to be closely followed.

Different brain operations are available, and many more are being researched. Deep brain stimulation (DBS) involves implanting a device to stimulate certain parts of the brain. DBS can decrease tremor and rigidity. Thalamotomy and pallidotomy destroy certain areas of the brain to improve tremor when medication does not work; these operations are not as common as DBS. Research is being conducted about the possible use of nerve-cell transplants to increase the amount of dopamine made in the brain.

Physical therapy can improve muscle tone, strength, and balance; it will include exercises and stretches. Patients with PD can also join a support group with other people who have the disorder. Participation in the support group will help patients learn how others are learning to live with the challenges of PD.

Prevention and Outcomes

There are no guidelines for preventing PD.

Rosalyn Carson-DeWitt, M.D.;
reviewed by Rimas Lukas, M.D.
"Etiology and Genetics" by Jeffrey A. Knight, Ph.D.

Further Reading

EBSCO Publishing. *DynaMed: Parkinson's Disease.* Ipswich, Mass.: Author, 2009. Available through http://www.ebscohost.com/dynamed.

_______. *Health Library: Parkinson Disease.* Ipswich, Mass.: Author, 2009. Available through http://www.ebscohost.com.

Goetz, Christopher G., ed. *Textbook of Clinical Neurology.* 3d ed. Philadelphia: Saunders Elsevier, 2007.

Rowland, Lewis P., ed. *Merritt's Neurology.* 11th ed. Philadelphia: Lippincott Williams & Wilkins, 2005.

Samii, A., J. G. Nutt, and B. R. Ransom. "Parkinson's Disease." *Lancet* 363, no. 9423 (May 29, 2004): 1783-1793.

Samuels, Martin A., and Steven K. Feske, eds. *Office Practice of Neurology.* 2d ed. Philadelphia: Churchill Livingstone, 2003.

Siderowf, A., and M. Stern. "Update on Parkinson's Disease." *Annals of Internal Medicine* 138, no. 8 (April 15, 2003): 651-658.

Wider, C., and C. K. Wszolek. "Movement Disorders:

Insights into Mechanisms and Hopes for Treatments. *Lancet Neurology* 8, no. 1 (January, 2009): 8-10.

WEB SITES OF INTEREST

American Academy of Neurology
http://www.aan.com

Genetics Home Reference
http://ghr.nlm.nih.gov

Health Canada
http://www.hc-sc.gc.ca/index-eng.php

National Institute of Neurological Disorders and Stroke: NINDS Parkinson's Disease Information Page
http://www.ninds.nih.gov/disorders/parkinsons_disease/parkinsons_disease.htm

National Parkinson Foundation
http://www.parkinson.org/Page.aspx?pid=201

Parkinson Society Canada
http://www.parkinson.ca

Parkinson's Disease Foundation, Inc.
http://www.pdf.org

U.S. Food and Drug Administration: "FDA Announces Voluntary Withdrawal of Pergolide Products"
http://www.fda.gov/NewsEvents/Newsroom/PressAnnouncements/2007/ucm108877.htm

See also: Adrenoleukodystrophy; Alexander disease; Alzheimer's disease; Amyotrophic lateral sclerosis; Arnold-Chiari syndrome; Ataxia telangiectasia; Canavan disease; Cerebrotendinous xanthomatosis; Charcot-Marie-Tooth syndrome; Chediak-Higashi syndrome; Dandy-Walker syndrome; Deafness; Epilepsy; Essential tremor; Friedreich ataxia; Huntington's disease; Jansky-Bielschowsky disease; Joubert syndrome; Kennedy disease; Krabbé disease; Leigh syndrome; Leukodystrophy; Limb girdle muscular dystrophy; Maple syrup urine disease; Metachromatic leukodystrophy; Myoclonic epilepsy associated with ragged red fibers (MERRF); Narcolepsy; Nemaline myopathy; Neural tube defects; Neurofibromatosis; Prion diseases: Kuru and Creutzfeldt-Jakob syndrome; Spinal muscular atrophy; Vanishing white matter disease.

Paroxysmal nocturnal hemoglobinuria

CATEGORY: Diseases and syndromes
ALSO KNOWN AS: PNH

DEFINITION

Paroxysmal nocturnal hemoglobinuria (PNH) is a rare noninherited polyclonal genetic disorder caused by somatic mutations of the hematopoietic stem cells, which normally give rise to the trilineage cell lines including white blood cells, red blood cells, and platelets. The abnormal stem cells in PNH give rise to cells that are deficient in an anchor protein that normally bind certain proteins to cell membranes. Resultant complications can lead to hemolysis, aplastic anemia, venous thrombosis, and acute leukemia.

RISK FACTORS

PNH is a rare disorder, with an incidence of 1 to 2 per million per year. It frequently affects those who are middle-aged, and affects both sexes at all age ranges. It is usually diagnosed in those with an existing diagnosis of myelodysplastic syndrome or aplastic anemia, the latter of which is a risk factor for the development of PNH in about 5 percent of cases.

ETIOLOGY AND GENETICS

PNH is an acquired disorder due to somatic mutations of the phosphatidylinositol glycan class A (*pig*-A) gene on the X chromosome of hematopoietic stem cells. This defect leads to an absent or defective protein product that normally is involved in the synthesis of the glycosyl-phosphatidylinositol (GPI) anchor protein that is required for binding of cell membrane surface proteins. The majority of *pig*-A defects are due to frame-shift mutations that end with an early stop codon, leading to a lack of the protein product. Substitution mutations are also possible that may lead to protein products that are less effective in the synthesis of the GPI anchor. In most cases, PNH patients will have more than one type of mutation, leading to a polyclonal population of stem cells admixed with a population of normal stem cells. Most of these mutations are unique to each patient.

Because this defect occurs in hematopoietic stem

cells, all cell lines are affected, including red blood cells, white blood cells, and platelets. The critical cell membrane surface proteins that are lacking on red blood cells include the decay-accelerating factor (also known as CD55), membrane inhibitor of reactive lysis (also known as CD59), homologous restriction factor (HRF), and C8 binding protein. These surface proteins normally interact and slow down the complement process, notably through proteins C3b and C4b. Their absence therefore leads to the unhampered destruction of the red cell membrane by the complement system, leading to intravascular hemolysis.

Though the *pig-A* gene defect appears to be a necessary component for the development of PNH, it is not clear if this is sufficient, since *pig-A* mutations have been found in small numbers in many, if not most, normal individuals. The abnormal population increases in aplastic anemia and becomes overwhelmingly large in PNH. The Luzzatto-Young hypothesis on the development of PNH notes that the processes that lead to aplastic anemia tend to suppress the proliferation of normal cells and not defective cells, therefore leading to the preferential selection of defective cells characteristic of PNH.

SYMPTOMS

One of the defining manifestations of PNH is dark urine in the morning, which is due to the leakage of hemoglobin from breakdown of red blood cells. Other symptoms vary depending on the degree of disease progression. Yellowing of the skin can occur with prolonged red blood cell breakdown; bleeding can occur with an abnormally low platelet count; risks for infections can occur with low white blood cell count; and headache, abdominal pain, and skin findings can occur due to vessel thrombosis.

SCREENING AND DIAGNOSIS

The principal diagnostic test of PNH is flow cytometry, which is able to detect the CD55 and CD59 glycoproteins on the surface of red blood cells. The absence or the reduced expression of both such glycoproteins, along with clinical and laboratory features, is diagnostic of PNH. Another more specific method is using fluorescently labeled inactive bacterial toxin aerolysin, which normally initiates hemolysis by binding to GPI anchors on RBCs. The lack of such binding suggests an absence of GPI and is also diagnostic of PNH.

TREATMENT AND THERAPY

Therapy for PNH includes addressing clinical issues related to anemia, leukopenia, thrombocytopenia, and thrombosis. Blood or platelet transfusions are indicated in clinically indicated anemia or thrombocytopenia. Other therapies such as folic acid, glucocorticoids, and eculizumab (an agent that inhibits terminal complement pathway activation) have been used. Treatment of thrombosis includes thrombolysis to break up blood clots in life-threatening thrombosis, or with anticoagulation agents such as heparin or warfarin for preventive measures. Hematopoietic stem cell transplants and medical agents such as cyclosporine or antithymocyte globulin have been used to address the issue of aplastic anemia. Although gene therapy is theoretically possible since the *pig-A* gene has been cloned, it is still currently under further evaluation.

PREVENTION AND OUTCOMES

Although PNH is a chronic disorder, many patients are able to live for extended periods of time, and spontaneous recovery is possible. The median survival after the onset of disease is found to be approximately ten to fifteen years. The majority of deaths and morbidity are related to complications from venous thrombosis.

Andrew Ren, M.D.

FURTHER READING

Lichtman, Marshall, et al. *Williams Hematology.* 6th ed. McGraw-Hill Professional, 2000.

Omime, M. *Paroxysmal Nocturnal Hemoglobinuria and Related Disorders.* New York: Springer, 2003.

Young, Neal S. *PNH and the GPI-Linked Proteins.* New York: Academic Press, 2000.

WEB SITES OF INTEREST

PNH Research and Support Home
http://www.pnhfoundation.org

PNH Source
http://www.pnhsource.org

PNH Support Group
http://www.pnhdisease.org

See also: ABO blood types; Chronic myeloid leukemia; Fanconi anemia; Hemophilia; Hereditary spherocytosis; Infantile agranulocytosis; Myelodysplastic syndromes; Rh incompatibility and isoimmunization; Sickle-cell disease.

Parthenogenesis

CATEGORY: Genetic engineering and biotechnology

SIGNIFICANCE: Parthenogenesis is the development of unfertilized eggs, which produces individuals that are genetically alike and allows rapid expansion of a population of well-adapted individuals into a rich environment. This clonal reproduction strategy is used by a number of species for rapid reproduction under very favorable conditions, and it appears to offer a selective advantage to individuals living in disturbed habitats.

KEY TERMS

adaptive advantage: increased fertility in offspring as a result of passing on favorable genetic information

diploid: having two sets of homologous chromosomes

fertilization: the fusion of two cells (egg and sperm) in sexual reproduction

haploid: having one set of chromosomes

meiosis: nuclear division that reduces the chromosome number from diploid to haploid in the production of the sperm and the egg

zygote: the product of fertilization in sexually reproducing organisms

THE NATURE OF PARTHENOGENESIS

Parthenogenesis is derived from two Greek words that mean "virgin" (*parthenos*) and "origin" (*genesis*) and describes a form of reproduction in which females lay diploid eggs (containing two sets of chromosomes) that develop into new individuals without fertilization—there is no fusion of a sperm nucleus with the egg nucleus to produce the new diploid individual. This is a form of clonal reproduction because all of the individuals are genetically identical to the mother and to each other. The mechanisms of parthenogenesis do not show any single pattern and have evolved independently in different groups of organisms. In some organisms, such as rotifers and aphids, parthenogenesis alternates with normal sexual reproduction. When there is a rich food source, such as new rose bushes emerging in the early spring, aphids reproduce by parthenogenesis; late in the summer, however, as the food source is decreasing, sexually reproducing females appear. The same pattern has been observed in rotifers, in which a decrease in the quality of the food supply leads to the appearance of females that produce haploid eggs by normal meiosis that require fertilization for development. The strategy appears to involve the clonal production of large numbers of genetically identical individuals that are well suited to the environment when the conditions are favorable and the production of a variety of different types, by the recombination that occurs during normal meiosis and the mixing of alleles from two individuals in sexual reproduction, when the conditions are less favorable. In social insects, such as bees, wasps, and ants, parthenogenesis is a major factor in sex determination, although it may not be the only factor. In these insects, eggs that develop by parthenogenesis remain haploid and develop into males, while fertilized eggs develop into diploid, sexually reproducing females.

In algae and some forms of plants, parthenogenesis also allows rapid reproduction when conditions are favorable. In citrus, seed development by parthenogenesis maintains the favorable characteristics of each plant. For this reason, most commercial citrus plants are propagated by asexual means, such as grafting. Parthenogenesis has also been induced in organisms that do not show the process in natural populations. In sea urchins, for example, development can be induced by mechanical stimulation of the egg or by changes in the chemistry of the medium. Even some vertebrate eggs have shown signs of early development when artificially stimulated, but haploid vertebrate cells lack all of the information required for normal development, so such "zygotes" cease development very early.

PARTHENOGENESIS IN VERTEBRATES

Parthenogenesis has been observed in vertebrates such as fish, frogs, and lizards. In these parthenogenetic populations, all the individuals are females, so reproduction of the clone is restricted to parthenogenesis. Parthenogenetic fish often occur in populations along with sexually reproducing individuals. The parthenogenetic forms produce diploid eggs that develop without fertilization; in rare cases, however, fertilization of a parthenogenetic egg gives rise to a triploid individual that has three sets of chromosomes rather than the normal two sets (two from the diploid egg and one from the sperm). In some groups, penetration of a sperm is necessary to activate development of the zygote, but the sperm nucleus is not incorporated into the zygote.

Evidence indicates that in each of these vertebrate situations, the parthenogenetic populations have resulted from a hybridization between two different species. The parthenogenetic forms always occur in regions where the two parental species overlap in their distribution, often an area that is not the most favorable habitat for either species. The hybrid origin has been confirmed by the demonstration that the animals have two different forms of an enzyme that have been derived from the two different species in the region. Genetic identity has also been confirmed using skin graft studies. In unrelated organisms, skin grafts are quickly rejected because of genetic incompatibilities; clonal animals, on the other hand, readily accept grafts from related donors. Parthenogenetic fish from the same clone accept grafts that confirm their genetic identity, but rejection of grafts by other parthenogenetic forms from different populations shows that they are different clones and must have a different origin. This makes it possible to better understand the structure of the populations and helps in the study of the origins of parthenogenesis within those populations. Comparisons using nuclear and mitochondrial DNA also allow the determination of species origin and the maternal species of the parthenogenetic form since the mitochondria are almost exclusively transmitted through the vertebrate egg. Within the hybrid, a mechanism has originated that allows the egg to develop without fertilization, although, as already noted, penetration by a sperm may be required to activate development in some of the species.

The advantage of parthenogenesis appears to be the production of individuals that are genetically identical. Since the parthenogenetic form may, at least in vertebrates, be a hybrid, it is heterozygous at most of its genetic loci. This provides greater variation that may provide the animal with a greater range of responses to the environment. Maintaining this heterozygous genotype may give the animals an advantage in environments where the parental species are not able to reproduce successfully and may be a major reason for the persistence of this form of reproduction. Many vertebrate parthenogenetic populations are found in disturbed habitats, so their unique genetic composition may allow for adaptation to these unusual conditions.

Mechanisms of Development

The mechanisms of diploid egg development are as diverse as the organisms in which this form of reproduction is found. In normal meiosis, the like chromosomes of each pair separate at the first division and the copies of each chromosome separate at the second division (producing four haploid cells). During the meiotic process in the egg, three small cells (the polar bodies), each with one set of chromosomes, are produced, and one set of chromosomes remains as the egg nucleus. In parthenogenetic organisms, some modification of this process occurs that results in an egg nucleus with two sets of chromosomes—the diploid state. In some forms, the first meiotic division does not occur, so two chromosome sets remain in the egg following the second division. In other forms, one of the polar bodies fuses back into the cell so that there are two sets of chromosomes in the final egg. In another variation, there is a replication of chromosomes after the first division, but no second division takes place in the egg, so the chromosome number is again diploid. In all of these mechanisms, the genetic content of the egg is derived from the mother's genetic content, and there is no contribution to the genetic content from male material.

The situation may be even more complex, however, because some hybrid individuals may retain the chromosomal identity of one species by a selective loss of the chromosomes of the other species during meiosis. The eggs may carry the chromosomes of one species but the mitochondria of the other species. The haploid eggs must be fertilized, so these individuals are not parthenogenetic, but their presence in the population shows how complex reproductive strategies can be and how important it is to study the entire population in order to understand its dynamics fully: A single population may contain individuals of the two sexual species, true parthenogenetic individuals, and triploid individuals resulting from fertilization of a diploid egg.

D. B. Benner, Ph.D.

FURTHER READING

Beatty, Richard Alan. *Parthenogenesis and Polyploidy in Mammalian Development.* Cambridge, England: Cambridge University Press, 1957. An early but still useful study.

Kaufman, Matthew H. *Early Mammalian Development: Parthenogenetic Studies.* New York: Cambridge University Press, 1983. Written by a well-known expert in mouse studies.

Lim, Hwa A. *Multiplicity Yours: Cloning, Stem Cell Research, and Regenerative Medicine.* Hackensack, N.J.: World Scientific, 2006. This overview of reproduction, cloning, stem cell research, and regenerative medicine lists several references to parthenogenesis in its index. Includes discussions of parthenogenesis versus cloning and parthenogenesis in humans, sea urchins, rabbits, and monkeys.

Schon, Isa, Koen Martens, and Peter van Dijk, eds. *Lost Sex: The Evolutionary Biology of Parthenogenesis.* New York: Springer, 2009. Focuses on the fate of animal and plant groups in which sex is lost. Discusses the theory behind asexual reproduction, the disadvantages confronted by asexual groups, and the genetic and ecological consequences of asexuality.

WEB SITES OF INTEREST

Kimball's Biology Pages
http://users.rcn.com/jkimball.ma.ultranet/BiologyPages/A/AsexualReproduction.html

John Kimball, a retired Harvard University biology professor, includes a page about asexual reproduction, with information about parthenogenesis, in his online cell biology text.

Nova Science.now: An Alternative to Cloning
http://www.pbs.org/wgbh/nova/sciencenow/3209/04-alternative.html

The Nova Science.now site, maintained by the Public Broadcasting Service (PBS), features a discussion of parthenogenesis as an alternative to cloning.

See also: Totipotency.

Patau syndrome

CATEGORY: Diseases and syndromes
ALSO KNOWN AS: Trisomy 13; Patau's syndrome

DEFINITION

Patau syndrome is a severe systemic disorder that affects many essential body systems and functions. Typically, each cell in the body will have an extra copy of chromosome 13, thus yielding a total of 47 chromosomes per cell instead of the usual 46.

RISK FACTORS

The only consistently reported risk factor is advanced maternal age, since the extra copy of chromosome 13 most commonly arises from an error in meiosis during egg cell maturation. There is a slightly higher incidence of affected females reported at birth, but this is most probably a result of slightly decreased survival of affected male fetuses.

ETIOLOGY AND GENETICS

The presence of an extra chromosome 13 in the cells of a developing fetus results from a type of error called nondisjunction, which can occur during gamete (sperm or egg) production in either parent. Either the failure of homologous chromosomes to separate from each other during the first meiotic division or the failure of sister chromatids to separate from each other during the second meiotic division will result in mature sperm cells or egg cells that have either one extra or one missing chromosome. Since each chromosome contains thousands of genes, it is not surprising that individuals with extra or missing chromosomes in all cells would have a severe imbalance of genetic information and suffer from multiple developmental anomalies. In fact, only three autosomal trisomies (conditions in which each cell has three copies of a nonsex chromosome) are generally known to be consistent with full-term delivery, and Patau syndrome is the least common and most severe of these. The most common and least severe is Down syndrome (trisomy 21). Edwards syndrome (trisomy 18), like Patau syndrome, results in affected newborns with multiple structural and developmental problems, and survival beyond the first year is rare.

Very occasionally, a case of trisomy 13 occurs in which the extra copy of chromosome 13 does not appear as a separate chromosome but rather is physically attached onto the end of another chromosome; such cases are known as translocation Patau syndrome. While the clinical features of the affected newborn do not differ from the usual form of the syndrome, it is particularly important to identify this variety of Patau, since it may be transmitted with high frequency by a normal-appearing parent who carries the translocation chromosome.

One additional variant that is infrequently encountered is known as mosaic trisomy 13. The bodies of mosaic individuals are composed of two distinctly different cell lines, in which only some of the cells have the extra chromosome 13, while the remainder have a normal chromosome complement. The severity of the clinical presentation in these cases depends on the type and number of cells that carry the extra chromosome, but in almost all cases a less severe form of the syndrome is manifested.

Symptoms

The most consistent symptoms present at birth include microcephaly (small head), cleft lip and/or palate, and polydactyly (extra fingers or toes). Ears are often low-set and malformed, and the nose can be oddly shaped or occasionally altogether absent. Most affected individuals are presumed deaf, and many are blind as well. Other neurological problems are common, including profound mental retardation and failure of the brain to divide into its proper hemispheres during gestation. About 80 percent of affected newborns are reported to have moderate to severe heart defects.

Screening and Diagnosis

Patau syndrome occurs in about 1 out of 12,000 live births, and diagnosis is most often immediately apparent, although there is some overlap of symptoms with Edwards syndrome. Genetic studies should be performed to confirm the diagnosis. Ultrasound examinations and imaging studies should be done to check for more extensive developmental problems. They include brain, heart, and kidney defects as well as an extra spleen, rotated intestines, and defects of the liver and pancreas. Males may have undescended testes, while females frequently have a divided uterus.

Treatment and Therapy

Because of the heterogeneous nature of each clinical presentation, treatment is usually specifically directed to the particular physical problems with which each affected child is born. About 80 percent of affected newborns die within the first month, most from serious heart defects or severe neurological problems. For many, medical treatment may focus primarily on patient comfort and noninvasive symptom treatment rather than on prolonging life. Surgery may be performed to repair heart defects or cleft lip and palate. In those rare cases where survival extends beyond one or two years, additional surgeries and physical therapy are often undertaken to allow the affected child to reach his or her full developmental potential.

Prevention and Outcomes

Except for the rare translocation form of Patau syndrome, there is no effective means of prevention. Genetic counseling should always be available for parents of an affected child, and amniocentesis is an option for older at-risk mothers. Only about 5 percent of affected newborns survive the first year of life, and survival into the teenage years is exceedingly rare.

Jeffrey A. Knight, Ph.D.

Further Reading

Cummings, Michael. *Human Heredity: Principles and Issues.* 8th ed. Pacific Grove, Calif.: Brooks/Cole, 2008. A comprehensive yet accessible introduction to all aspects of human genetics.

Lewis, Ricki. *Human Genetics.* 8th ed. New York: McGraw-Hill, 2007. A basic human genetics reference text written by a practicing genetic counselor.

Nussbaum, Robert L., Roderick R. McInnes, and Huntington F. Willard. *Thompson and Thompson Genetics in Medicine.* 7th ed. New York: Saunders, 2007. A classic and complete medical school textbook that is nevertheless understandable to nonprofessionals.

Web Sites of Interest

Daily Strength Trisomy 13 (Patau Syndrome) Support Group
http://www.dailystrength.org/c/Trisomy-13-Patau-Syndrome/support-group

Organized Wisdom
http://organizedwisdom.com/Patau_Syndrome_Support_Groups

Support Organization for Trisomy 18, 13, and Related Disorders (SOFT)
http://www.trisomy.org

See also: Apert syndrome; Brachydactyly; Carpenter syndrome; Cleft lip and palate; Congenital defects;

Cornelia de Lange syndrome; Cri du chat syndrome; Crouzon syndrome; Down syndrome; Edwards syndrome; Ellis-van Creveld syndrome; Holt-Oram syndrome; Ivemark syndrome; Meacham syndrome; Opitz-Frias syndrome; Polydactyly; Robert syndrome; Rubinstein-Taybi syndrome.

Patents on life-forms

CATEGORY: Bioethics; Human genetics and social issues

SIGNIFICANCE: In 1980, the U.S. Supreme Court upheld the right to patent a live, genetically altered organism. The decision was opposed by many scientists and theologians who believed that such organisms would pose a threat to the future of humanity. Although "legally" settled, the debate has continued, opponents arguing that patenting life-forms and DNA sequences imposes too great a cost and greatly inconveniences genetic research.

KEY TERM

patent: a grant made by the government that gives the creator or inventor the sole right to make, use, or sell that invention for a specific period of time, usually twenty years in the United States

PATENT ON LIFE-FORM UPHELD

On June 16, 1980, the U.S. Supreme Court voted 5 to 4 that living organisms could be patented under federal law. The case involved Ananda M. Chakrabarty, a scientist who, while working for General Electric in 1972, had created a new form of bacteria, *Pseudomona originosa*, which could break down crude oil, and, therefore, could be used to clean up oil spills. Chakrabarty filed for a patent, but an examiner for the Patent Office rejected the application on the ground that living things are not patentable subject matter under existing patent law. Commissioner of Patents and Trademarks Sidney A. Diamond supported this view. Federal patent law provided that a patent could be issued only to a person who invented or discovered any new and useful "manufacture" or "composition of matter." The U.S. Court of Customs and Patent Appeals reversed that decision in 1979, concluding that the fact that microorganisms are alive has no legal significance. It held that a live, human-made bacterium is a patentable item since the microorganism was manufactured by crossbreeding four existing strains of bacteria and had never existed in nature.

Writing for the majority, Supreme Court Chief Justice Warren Burger upheld the patent appeals court judgment, making a distinction between the new bacterium and "laws of nature, physical phenomena and abstract ideas," which are not patentable. In the Court majority's view, Chakrabarty had invented a form of life that did not exist in the natural world, so it could not be considered part of nature. Instead, it was a product of human "ingenuity and research" that deserved patent protection. Items not patentable include new minerals that are discovered in the earth or a new species of plant found in a distant forest. These things occur naturally and are not created by humans. Burger also stressed that physicist Albert Einstein could not have patented his formula $E = mc^2$, since it is a law of nature, nor could Sir Isaac Newton have received a patent for the law of gravity. Discoveries such as these are part of the natural world and cannot be owned by a single individual.

Chakrabarty, on the other hand, had not found an unknown, natural species, nor had he discovered a law of nature. His new bacterium had a distinctive name and was developed in the laboratory for a specific purpose. None of the characteristics of the new organism could be found in nature. His discovery, Burger reemphasized, was patentable because he had created it.

OPPOSITION TO THE RULING

The Court majority refused to consider arguments made in friend-of-the-court briefs filed by opponents of genetic engineering. The briefs were presented by groups representing scientists, including several Nobel Prize winners, and religious organizations. One brief suggested that genetic research posed a dangerous and serious threat to the future of humanity and should, therefore, be prohibited. Possible dangers included the spread of pollution and disease by newly created bacteria, none of which would have any natural enemies. Other threats involved the possible loss of genetic diversity, if, for instance, only the "best" form of laboratory-created plant seeds were grown. Research into human genetics could lead to newly designed gene material that could be used to build a "master race," thereby devaluing other human lives. Chief Justice Burger concluded, however, that humans could be trusted not to create such horrible things. Quoting William Shakespeare's *Hamlet*, the chief justice asserted that it is sometimes better "to bear those ills we have than to fly to others that we know not of." People can try to guess what genetic manipulation could lead to, but it would also be a good idea to expect good things from science rather than "a gruesome parade of horribles." Besides, he then said, it did not matter whether a patent was granted in this case; in either case, scientific research would continue into the nature of genes.

The People's Business Commission, a nonprofit educational foundation, had argued that granting General Electric and Chakrabarty a patent would give corporations the right "to own the processes of life in the centuries to come" through genetic manipulation. Chief Justice Burger wrote that the Court was "without competence to entertain these arguments." They did not have enough information available to determine whether to ignore such fears "as fantasies generated by fear of the unknown" or accept them. Such a determination was not the responsibility of the Court, however. Questions of the morality of genetic research and manipulation were better left to Congress and the political process. How to proceed in these matters could only be resolved "after the kind of investigation, examination, and study that legislative bodies can provide and courts cannot."

Supreme Court Justice William J. Brennan, Jr., presented a brief dissenting opinion. He noted that Congress had twice, in 1930 and 1970, permitted new types of plants to be patented. However, those laws made no mention of bacteria. Thus, Brennan argued, Congress had indicated that only plants could receive patents and that the legislators had thus clearly indicated that other life-forms were excluded from the patent process. The Court majority rejected this view, arguing that Congress had not specifically excluded other life-forms.

The U.S. Patent and Trademark Office (USPTO)

Those opposed to patents on DNA sequences have a wide variety of contentions, ranging from the concern that sole privileges will impede research to the notion that genes represent the very basis of life and thus no one should have exclusive rights to them.

In 2001, the USPTO thoughtfully considered whether genetic discoveries were patentable by evaluating opposition comments from thirty-five individuals and seventeen organizations. Their decisions were published in the *Federal Register*. To address the contention that a gene is not a new invention because it exists in nature, the USPTO emphasized that only DNA in an unnatural form, excised and purified from its chromosome or synthesized in a laboratory, is patentable. To tackle the notion that no one person or company should own a human gene sequence because it inherently belongs to all humans, the USPTO asserted that progress is promoted and secrecy reduced when a patent gives an inventor purely the legal right to exclude others from making, using, selling, or importing the gene for twenty years, not ownership. To avoid reckless patenting of any gene sequence found, the USPTO stipulated that the utility of a gene or expressed sequence tag (EST), not just its sequence, must be known when filing for sole rights. Examples of a gene's utility include being involved in the cell regulation, coding for a useful protein, or flagging a disease.

The USPTO established these four criteria when considering a patent grant on a gene, gene fragment, single nucleotide polymorphism (SNP), gene test, protein, or stem cell: novelty, usefulness, nonobviousness, and enablement, whereby the life-form could be reproduced by someone skilled in the biochemistry field. Patentees must deposit a sample of the unique life-form in one of twenty-six worldwide culture depositories.

Sharing Knowledge, Licensing, and Commercializing Life-Forms

Measures have been taken by key institutions to even the genomics playing field for all researchers. The Human Genome Project publically lists the genome for free on the Internet. In April, 1999, the U.K. Wellcome Trust vowed unrestricted access to the 1.5 million SNPs they identified and patented in order to prevent others from gaining exclusive rights to them first.

As one of the United States' primary financial supporters of scientific research, the National Institutes of Health (NIH) weighed in on the patenting debate in 2004 by developing its "Best Practices for the Licensing of Genomic Inventions." These guidelines leave room for researchers to protect their work with patents, especially when there is potential to commercialize their product through the private sector for society's health benefit, but the NIH also strongly encourages investigators to propagate information by granting nonexclusive licenses to other universities receiving NIH funding.

In 1980, Congress passed the Bahy-Dole Act, which permits universities to commercialize via private industry their federally funded discoveries. This prompted colleges to establish a campus Technology Transfer Office and implement a material transfer agreement (MTA). Academic researchers now need to obtain an MTA before they disclose information to other investigators, adding another potential hurdle to the flow of knowledge. A 2005 survey of ninety-three agricultural biology departments found that MTAs have taken the lead over patents as the hindrance of conveying scientific information. This study and others have found that patents are not aggressively enforced and not truly encumbering academic research.

Leslie V. Tischauser, Ph.D., Bryan Ness, Ph.D.; updated by Cherie Dewar

FURTHER READING

Barfield, Claude, and John E. Calfee. *Biotechnology and the Patent System: Balancing Innovation and Property Rights.* Washington, D.C.: AEI Press, 2007. Explores ways the current patent system promotes additional research and venture capitalism, but also obstructs new developments. Concludes with reform recommendations to Congress.

Chapman, Audrey R., ed. *Perspectives on Genetic Patenting: Religion, Science, and Industry in Dialogue.* Washington, D.C.: American Association for the Advancement of Science, 1999. Discusses questions such as, should products of nature be patentable? Are genes or gene fragments discoveries or inventions? Should patenting of genes, cell lines, or genetically modified organisms be equated with ownership of them? Is the DNA in genes just a complex molecule or is it sacred? Does patenting human DNA and tissue demean human life and human dignity?

Diamond, Commissioner of Patents and Trademarks v. Chakrabarty (1980), 447 U.S. 303. The official citation of the Supreme Court decision.

Doll, John. "Talking Gene Patents." *Scientific American,* August, 2001. A brief interview with the director of biotechnology for the U.S. Patent and Trademark Office on what makes a gene eligible for a chemical compound patent and the number of genes patented.

Hanson, Mark J. "Religious Voices in Biotechnology: The Case of Gene Patenting." *Hastings Center Law Report,* November/December, 1997. Discusses the religious, legal, moral, and scientific concerns about patenting human genetic material, DNA and patents, and the biotechnology industry.

Hope, Janet. *Biobazaar, the Open Source Revolution and Biotechnology.* Cambridge, Mass.: Harvard University Press, 2008. With insight from Nobel Prize winners, licensing experts, and others, the author makes a case why the next wave of progress in biotechnology should be open sourcing.

Resnick, David B. *Owning the Genome: A Moral Analysis of DNA Patenting.* Albany: State University of New York Press, 2004. The author examines the morals, consequences, and main arguments for and against DNA patenting in the areas of human dignity, scientific progress, medicine, and agriculture.

U.S. Patent and Trademark Office. *Federal Register* 66, no. 4 (2001): 1092-1099. The decision of the USPTO regarding the patenting of life-forms.

Vogel, Fredrich, and Reinhard Grunwald, eds. *Patenting of Human Genes and Living Organisms.* New York: Springer, 1994. Provides an overview of patent acquisition and legal concerns. Illustrated.

WEB SITES OF INTEREST

Human Genome Project Genetics and Patents
http://www.ornl.gov/sci/techresources/Human_Genome/elsi/patents.shtml

NIH Office of Science Policy, Patents and Access
http://oba.od.nih.gov/SACGHS/sacghs_focus_patents.html

NIH Office of Technology Transfer
http://www.ott.nih.gov

U.S. Patent and Trademark Office
http://www.uspto.gov

See also: Genetic engineering: Historical development; Genetic engineering: Industrial applications; Genetic engineering: Social and ethical issues; Human genetics; Human Genome Project; Hybridization and introgression; Model organism: *Mus musculus*; Transgenic organisms.

Paternity tests

Category: Human genetics and social issues

Significance: Establishing paternity can be important for establishing legal responsibility for child support, health insurance, veterans' and social security benefits, and legal access to medical records. It may also affect a child's future as it relates to inherited diseases.

Key terms

forensic genetics: the use of genetic tests and principles to resolve legal questions

human leukocyte antigens (HLA): antigens produced by a cluster of genes that play a critical role in the outcome of transplants; because they are made up of a large number of genes, they are used in individual identification and the matching of parents and offspring

paternity exclusion: the indication, through genetic testing, that a particular man is not the biological father of a particular child

Genetic Principles of Paternity Testing

The basic genetic principles utilized in paternity testing have remained the same from the first applications of ABO blood groups to applications of DNA fingerprinting. Available tests may positively exclude a man from being a child's biological father. Evidence supporting paternity, however, cannot be considered conclusive. Ultimately, a court must decide whether a man is determined to be the legal father based on all lines of evidence.

The genetic principles can be illustrated with a very simple example that uses ABO blood types. The four blood groups (A, B, AB, and O) are controlled by three pairs of genes. In the example, however, only three of the blood groups will be used to demonstrate the range of matings with the possible children for each of them (see the table headed "Blood Types, Genes, and Possible Offspring").

Example 1: A man is not excluded.
Mother: A
Child: A
Putative Father: AB

It can be seen that the mothers in matings 1 and 4 satisfy the condition of the mother being A and possibly having a child being A. Mating 4 satisfies the condition of a father being AB, the mother A, and a possible child being A. Results indicate that the putative father could be the father. He is not excluded.

Example 2: A man is excluded.
Mother: A
Child: A
Putative Father: B

Blood Types, Genes, and Possible Offspring

Mating Number	Genes of Parents		Blood Type of Parents		Possible Children	
	Father	Mother	Father	Mother	Genes	Blood Type
1	AA	AA	A	A	AA	A
2	AA	AB	A	AB	AA or AB	A or AB
3	AA	BB	A	B	AB	AB
4	AB	AA	AB	A	AA or AB	A or AB
5	AB	AB	AB	AB	AA, AB, or BB	A, AB, or B
6	AB	BB	AB	B	AB or BB	AB or B
7	BB	AA	B	A	AB	AB
8	BB	AB	B	AB	AB or BB	AB or B
9	BB	BB	B	B	BB	B

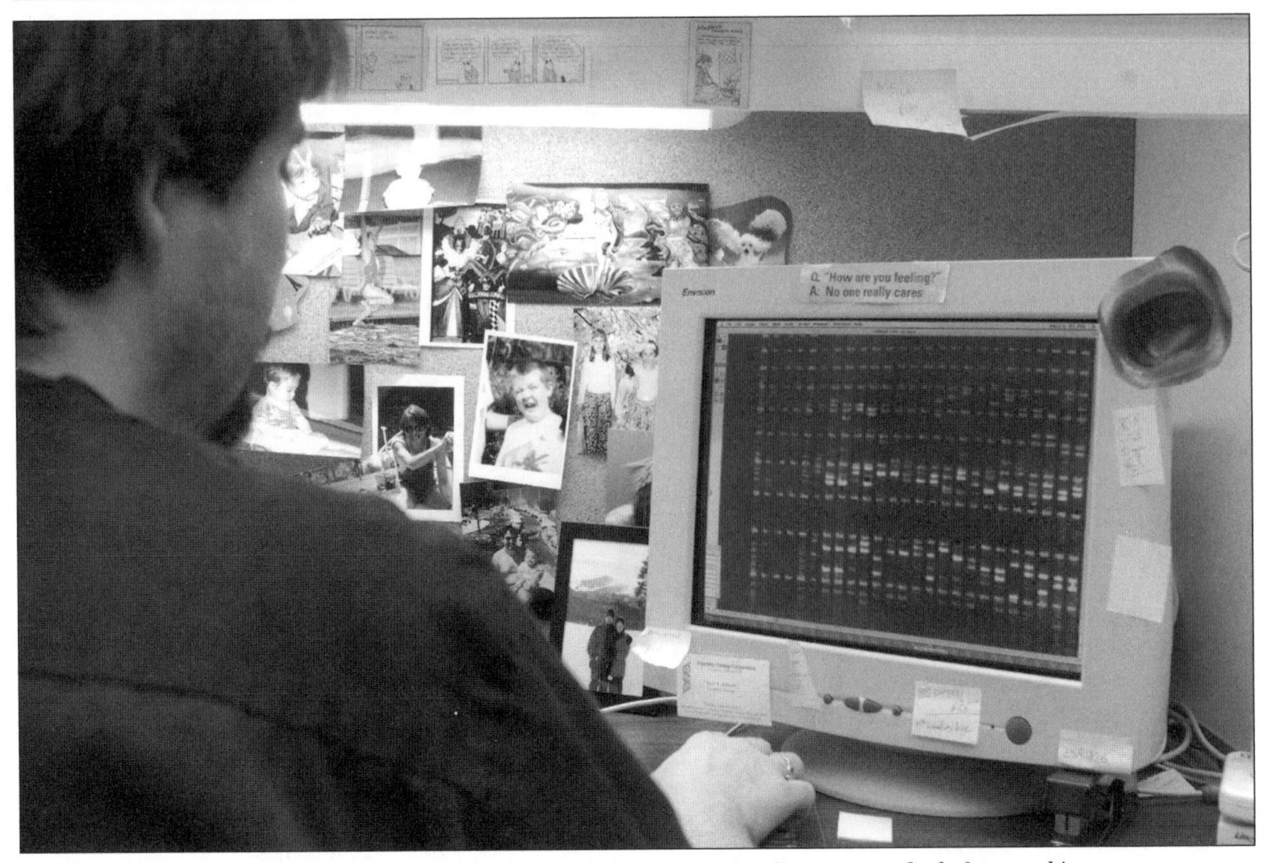

A technician at the Paternity Testing Corporation performs an autolanding test to check for matching genes among a mother, child, and alleged father. (AP/Wide World Photos)

Again, it is seen that the mothers in matings 1, 4, and 7 satisfy the condition of the mother being A and possibly having a child being A. Mating 7 satisfies the condition of a father being B and the mother A, but mating 7 cannot produce a child being A. The putative father cannot be the father, and he is excluded.

DNA Fingerprinting

After the initial use of ABO blood groups in paternity testing, it became apparent that there were many cases in which the ABO phenotypes did not permit exclusion. Other blood group systems have also been used, including the MN and Rh groups. As more blood groups are utilized, the probability of exclusion (or nonexclusion) increases. Paternity tests have not been restricted to blood groups alone; tissue types and serum enzymes have also been used.

The most powerful tool developed has been DNA testing. DNA fingerprinting was developed in England by Sir Alec Jeffreys. DNA is extracted from white blood cells and broken down into fragments by bacterial enzymes (restriction endonucleases). The fragments are separated by size, and specific fragments are identified. Each individual has a different DNA profile, but the profiles of parents and children have similarities in greater proportion than those between unrelated people. Also, frequencies of different fragments tend to vary among ethnic groups. It is possible not only to exclude someone who is not the biological father but also to determine actual paternity with a probability approaching 100 percent.

Impact and Applications

The personal, social, and economic implications involved in paternity testing have far-reaching consequences. Blood-group analysis is cheaper but less consistent than DNA testing. Paternity can often be excluded but rarely proven with the same degree of accuracy that DNA testing provides. Human leukocyte antigen (HLA) testing can also be used but suffers from many of the same problems as blood-group analysis. The development of DNA testing after 1984 revolutionized the field of paternity testing. DNA fingerprinting has made decisions on paternity assignments virtually 100 percent accurate. The same technique has also been applied in cases of individual identification, and results have helped to release people who have been falsely imprisoned as well as convict other people with the analysis of trace evidence.

Donald J. Nash, Ph.D.

FURTHER READING

Anderlik, Mary R., and Mark A. Rothstein. "DNA-Based Identity Testing and the Future of the Family: A Research Agenda." *American Journal of Law and Medicine* 28, nos. 2/3 (2002): 215. Covers DNA-based identity testing, misattributed paternity, legal issues, and more.

Cohen, Warren. "Kid Looks Like the Mailman? Genetic Labs Boom as the Nation Wonders Who's Daddy." *U.S. News and World Report* 122, no. 3 (January 27, 1997): 62. Discusses paternity testing at genetic laboratories.

Goodman, Christi. *Paternity, Marriage, and DNA.* Denver, Colo.: National Conference of State Legislatures, 2001. A concise discussion of critical policy issues surrounding paternity, DNA, and marriage.

Lasarow, Avi. *Who Is Really Who? The Comprehensive Guide to DNA Paternity Testing.* London: John Blake, 2006. A consumer primer on paternity testing. Includes explanations of the DNA test process, the test's accuracy, emotional and legal aspects of paternity testing, and extended family testing of siblings and grandparents.

Rothstein, Mark A., et al., eds. *Genetic Ties and the Family: The Impact of Paternity Testing on Parents and Children.* Baltimore: Johns Hopkins University Press, 2005. Collection of essays that examine the tensions between biological and social conceptions of parentage. Some of the contributors discuss the ethical, legal, and social implications of paternity testing and the effect of this testing on family relationships and children's well-being.

Sonenstein, Freya L., Pamela A. Holcomb, and Kristin S. Seefeldt. *Promising Approaches to Improving Paternity Establishment Rates at the Local Level.* Washington, D.C.: Urban Institute, 1993. Reviews paternity establishment procedures.

Weir, Bruce S. *Human Identification: The Use of DNA Markers.* New York: Kluwer Academic, 1995. Discussion includes the debates over using DNA profiles to identify paternity. Bibliography.

WEB SITES OF INTEREST

American Pregnancy Association, Paternity Testing
http://www.americanpregnancy.org/prenataltesting/paternitytesting.html

Answers consumers' questions about paternity testing, including its cost, why it is important, and the various types of tests that are available.

DNA Diagnostics Center
http://www.dnacenter.com/index.html

This company boasts it is the world's largest provider of private DNA paternity and other DNA tests. Its Web site explains the type of DNA tests that are available to consumers.

National Newborn Screening and Genetics Resource Center
http://genes-r-us.uthscsa.edu

Site serves as a resource for information on genetic screening, including paternity testing.

See also: DNA fingerprinting; Forensic genetics; Repetitive DNA.

Pattern baldness

CATEGORY: Diseases and syndromes
ALSO KNOWN AS: Male-pattern baldness; MPB; androgenetic alopecia

DEFINITION

Pattern baldness, or androgenetic alopecia, refers to the most common form of scalp hair loss. Male pattern baldness follows a typical progression that begins with mild frontal hairline recession, developing to more noticeable frontal hair loss and vertex loss. Female hair loss does not typically follow this same pattern and may have a different underlying etiology.

RISK FACTORS

The three main risk factors for pattern baldness are male gender, age, and family history. Early-onset baldness is almost entirely dependent on genetic predisposition. By middle age, more than half of the male population will have some degree of pattern baldness. By age eighty, more than 80 percent of males will have pattern baldness. The incidence is much lower in females.

ETIOLOGY AND GENETICS

Anecdotally, pattern baldness is often described as being autosomal dominant or X-linked in inheritance. In reality, population studies and genetic research suggest that male pattern baldness is a complex, multifactorial condition. A combination of both genetic predisposition and environmental factors leads to an individual's risk for baldness.

The environmental factors involved in baldness development are poorly defined and appear to be significantly less influential than genetic factors. Preliminary studies of multiple different environmental exposures and influences have yet to find a significant association to male-pattern baldness.

While genetic predisposition to baldness appears to be the strongest risk factor, researchers are still in the early stages of identifying the genes involved. Pattern baldness is androgen-dependent, and therefore, preliminary investigations into the genetics of baldness have relied on the candidate gene approach, focusing on genes involved in the sex steroid pathway. The 5 alpha-reductase, aromatase, and androgen receptor genes have all been investigated, with only the androgen receptor (*AR*) gene showing a clear association with pattern baldness to date.

The androgen receptor gene is approximately 90 kilobases (kb) long and is located on the X chromosome. Mutations in this gene have been associated with other genetic conditions, but research to date has not found a specific mutation or functional alteration in males with pattern baldness. Different polymorphisms in the *AR* gene have been investigated but not confirmed. Current research is focusing on the regulatory regions of the gene, splicing variants, or other epigenetic factors in an attempt to identify specific genetic changes involved in baldness predisposition. *AR* has been postulated to be the primary genetic factor in the development of early-onset pattern baldness, possibly accounting for up to 40 percent of the total genetic risk.

While other genes in the sex-steroid pathway have not shown a significant association to pattern baldness in previous studies, new technologies and considerations warrant further investigation into these candidate genes. Male-pattern baldness has been associated with coronary heart disease, disorders of insulin resistance, and prostate cancer. Investigation of genes involved in these conditions may lead to further identification of predisposition genes.

Genome-wide association studies (GWAS) have yielded new susceptibility variants for pattern baldness. A locus on chromosome 20p11 has shown significant association with male-pattern baldness, and was identified by this method. This locus appears to have a strong influence on the development of early-onset hair loss, and does not show an association to the androgen pathway. Additionally, another locus at chromosome 3q26 has been identified and is being investigated. In discovering more about these loci, additional information about the molecular basis for hair loss will be gained. Ultimately, this may lead to improved treatments, and advance the molecular understanding of the complex diseases that have been associated with pattern baldness.

SYMPTOMS

The degree of male-pattern baldness is defined by the Hamilton-Norwood baldness scale, which contains seven distinct categories of hair loss. As baldness progresses, the hair in the affected region becomes shorter, finer, and less pigmented. Ultimately, the hair follicle is incapable of producing a noticeable hair, rendering the area bald.

SCREENING AND DIAGNOSIS

The diagnosis of male-pattern baldness is made by clinical assessment of scalp hair loss. As female scalp hair loss does not typically follow the same pattern as males, the diagnosis cannot be made by the degrees defined by the Hamilton-Norwood baldness scale. Due to the limited information currently available about the genetic risk factors, predisposition genetic testing is not available.

TREATMENT AND THERAPY

Current treatments for hair loss mainly involve medications that were originally intended for other uses but found to have hair growth as a side effect. The best-known treatment, minoxidil, was originally prescribed for hypertension. Its mechanism for stimulation of hair growth is unknown, but as with other similar medications, it is not curative. Other medications that inhibit the actions of 5 alpha-reductase have also been effective and are the basis of future studies.

Ideally, a more complete understanding of an individual's personal predisposition to hair loss, and the molecular pathways involved, will allow for a more targeted approach to treatment.

Prevention and Outcomes

There are no known preventive measures for male-pattern baldness. A small study has suggested a link between increased alcohol consumption and an increased risk for baldness, but further evaluation is necessary before confirming a link.

Trudy McKanna, M.S.

Further Reading

Ellis, Justine A., and Stephen B. Harrap. "The Genetics of Androgenetic Alopecia." *Clinical Dermatology* 19 (2001): 149-154.

Ellis, Justine A., and Rodney D. Sinclair. "Male Pattern Baldness: Current Treatments, Future Prospects." *Drug Discovery Today* 13 (2008): 791-797.

Hillmer, Axel M., et al. "Genetic Variation in the Human Androgen Receptor Gene Is the Major Determinant of Common Early-Onset Androgenetic Alopecia." *The American Journal of Human Genetics* 77 (2005): 140-148.

Web Sites of Interest

Genetics Home Reference: Androgenetic Alopecia
http://ghr.nlm.nih.gov/condition=androgeneticalopecia

Hair Loss Heaven
http://www.hairlossheaven.com

See also: Albinism; Chediak-Higashi syndrome; Epidermolytic hyperkeratosis; Hermansky-Pudlak syndrome; Ichthyosis; Melanoma; Palmoplantar keratoderma.

Pearson syndrome

Category: Diseases and syndromes

Also known as: Pearson's syndrome; Pearson's disease; Pearson's marrow-pancreas syndrome

Definition

Pearson syndrome—named for Howard Pearson, a pediatric hematologist oncologist—is a rare, generally fatal disorder of infancy that affects the hematopoietic system and exocrine pancreas, with variable kidney, liver, and endocrine failure. Pearson syndrome is commonly caused by mitochondrial DNA (mtDNA) deletions or duplications, with high levels of heteroplasmy (proportion of mutant mitochondrial genomes) in the affected tissues.

Risk Factors

There are no specific risk factors for mtDNA deletions, which usually arise de novo, either in the oocyte or early in embryogenesis. There is no race or sex specific association for Pearson syndrome. Siblings of affected individuals are rarely affected. Males do not transmit mitochondrial disease, while affected females occasionally transmit mtDNA deletions.

Etiology and Genetics

Mitochondria are the cell's energy-producing organelles. They contain hundreds to thousands of copies of mtDNA, a small double-stranded circular genome that encodes thirteen subunits of respiratory chain enzymes and part of the protein synthetic machinery necessary to translate these mitochondrial transcripts, two ribosomal RNA genes, and twenty-two transfer RNA genes. Pearson syndrome is caused by deletions or, less commonly, duplications/rearrangements of mtDNA. Depending on their size, mtDNA deletions can disrupt protein-coding sequences or decrease their expression by removing rRNA and tRNA genes. The proportion of deleted mitochondrial genomes varies between tissues and can change over time. When heteroplasmy levels reach a critical threshold, oxidative metabolism is impaired and cells experience an energy deficit. Abnormal iron metabolism also occurs in Pearson syndrome, which may contribute to cellular damage.

An important feature of mtDNA mutations is

that, depending on the level of heteroplasmy and/or tissue-specific factors, a single mutation can cause different multisystemic disorders. MtDNA deletions are associated with three overlapping syndromes, with a general correspondence between the level of heteroplasmy and the affected tissue. Pearson syndrome primarily involves bone marrow and exocrine pancreas, but rare survivors may accumulate mtDNA deletions as the mitochondrial genome replicates in nerve and muscle, leading to Kearns-Sayre syndrome or progressive external ophthalmoplegia (paralysis of extraocular eye muscle).

Most mitochondrial diseases show maternal inheritance because mtDNA is transmitted exclusively in oocytes. However, mitochondrial DNA deletion syndromes usually occur sporadically. In rare inherited cases, mtDNA deletions are transmitted maternally, but the mother is usually affected with a later onset syndrome that involves other tissues, such as Kearns-Sayre syndrome.

Symptoms

The major symptoms of Pearson syndrome are bone marrow failure resulting in pancytopenia (reduced number of red and white blood cells and platelets) and exocrine pancreas dysfunction (problems with digestive enzymes, leading to chronic diarrhea and malabsorption). Typically sideroblastic anemia occurs despite the presence of sufficient iron (sideroblasts are nucleated erythrocytes with cytoplasmic iron granules because they cannot incorporate iron into hemoglobin). Often the patient is transfusion-dependent. Defective oxidative phosphorylation leads to lactic acidemia, which can be persistent or intermittent. The liver and kidneys are variably affected.

Symptoms begin during infancy and early childhood. Parents notice paleness due to anemia, chronic diarrhea and fatty stools, and failure to thrive. Death is usually caused from sepsis, hepatic failure, or metabolic crisis.

Screening and Diagnosis

In Pearson syndrome, mtDNA deletions are usually more common in blood than in muscle. However, bone marrow is the most reliable tissue for diagnosis, because less-involved tissues may have lower levels of heteroplasmy that can reduce the chance of detection. Although the junction-fragment created by an mtDNA deletion can be detected by polymerase chain reaction (PCR), deletions occur in normal individuals as part of the aging process, and PCR tests can thus give false positive results; Southern blot analysis is preferred. Molecular testing should be performed on blood isolated before transfusions to avoid false negative results caused by dilution of the patient's abnormal blood with transfused donor blood.

Treatment and Therapy

There is no specific therapy, but individual symptoms have corresponding treatments: infection (antibiotics), metabolic acidosis (bicarbonate supplements or dichloracetic acid, DCA, which can be neurotoxic), pancytopenia (transfusions, with erythropoietin to possibly decrease their frequency), neutropenia (granulocyte colony-stimulating factor), and malabsorption (pancreatic enzyme replacement and vitamins). Other endocrine imbalances are treated with the appropriate hormones. One study suggests that high carbohydrate diets should be avoided, as those diets can stress the liver. Transplantation is not effective due to the multisystemic nature of the disease.

Prevention and Outcomes

Pearson syndrome is usually sporadic. Prenatal screening for females with mtDNA deletions is problematic because of unknown levels of heteroplasmy in untested fetal tissues. Pearson syndrome is often fatal in childhood, but survivors may develop Kearns-Sayre syndrome and should be monitored for cardiac and muscle function.

Toni R. Prezant, Ph.D.

Further Reading

Bernes, S. M., et al. "Identical Mitochondrial DNA Deletion in Mother with Progressive External Ophthalmoplegia and Son with Pearson Marrow-Pancreas Syndrome." *Journal of Pediatrics* 123 (1993): 598-602. Demonstration of identical mtDNA deletion with different clinical outcomes.

Rotig, A., et al. "Pearson's Marrow-Pancreas Syndrome: A Multisystem Mitochondrial Disorder in Infancy." *The Journal of Clinical Investigation* 86 (1990): 1601-1608. A technical investigation that identifies the pathogenic mechanism in five infants with Pearson syndrome.

Van den Ouweland, J. M. W., et al. "Characterization of a Novel Mitochondrial DNA Deletion in a

Patient with a Variant of the Pearson Marrow-Pancreas Syndrome." *European Journal of Human Genetics* 8 (2000): 195-203. A patient with multisystemic disease that resembles Pearson syndrome, but without pancreas involvement, has a similar genetic cause.

Web Sites of Interest

eMedicine Pediatrics: Pearson Syndrome (Charles Quinn)
http://emedicine.medscape.com/article/957186-print

GeneReviews: Mitochondrial DNA Deletion Syndromes
http://www.ncbi.nlm.nih.gov/bookshelf/br.fcgi?book=gene&part=kss

Online Mendelian Inheritance in Man: Pearson Marrow-Pancreas Syndrome
http://www.ncbi.nlm.nih.gov/entrez/dispomim.cgi?id=557000

See also: Adrenomyelopathy; Androgen insensitivity syndrome; Autoimmune polyglandular syndrome; Congenital hypothyroidism; Diabetes insipidus; Graves' disease; Obesity; Pancreatic cancer; Pancreatitis; Steroid hormones.

Pedigree analysis

Category: Population genetics; Techniques and methodologies

Significance: Charts called pedigrees are used to represent the members of a family and to indicate which individuals have particular inherited traits. A pedigree is built of shapes connected by lines. Pedigrees are used by genetic counselors to help families determine the risk of genetic disease and are used by research scientists in determining how traits are inherited.

Key terms

alleles: alternate forms of a gene locus, some of which may cause disease

autosomal trait: a trait that typically appears just as frequently in either sex because an autosomal chromosome, rather than a sex chromosome, carries the gene

dominant allele: an allele that is expressed even when only one copy (instead of two) is present

hemizygous: the human male is considered to be hemizygous for X-linked traits, because he has only one copy of X-linked genes

heterozygous carriers: individuals who have one copy of a particular recessive allele that is expressed only when present in two copies

homozygote: an organism that has identical alleles at the same locus

recessive allele: an allele that is expressed only when there are two copies present

X-linked trait: a trait caused by a gene carried on the X chromosome, which has different patterns of inheritance in females and males because females have two X chromosomes while males have only one

Overview and Definition

Pedigree analysis involves the construction of family trees that can be used to trace inheritance of a trait over several generations. It is a graphical representation of the appearance of a particular trait or disease in related individuals along with the nature of the relationships.

Standardized symbols are used in pedigree charts. Males are designated by squares, females by circles. Symbols for individuals affected by a trait are shaded, while symbols for unaffected individuals are not. Heterozygous carriers are indicated by shading of half of the symbol, while carriers of X-linked recessive traits have a dot in the middle of the symbol. Matings are indicated by horizontal lines linking the mated individuals. The symbols of the individuals who are offspring of the mated individuals are linked to their parents by a vertical line intersecting with the horizontal mating line.

The classic way to determine the mode of inheritance of a trait is to conduct experimental matings of large numbers of individuals. Such experimental matings between humans are not possible, so it is necessary to infer the mode of inheritance of traits in humans through the use of pedigrees. Large families with good historical records are the easiest to analyze. Once a pedigree is established, it can be used to determine the likely mode of inheritance of a particular trait and, if the mode of inheritance can be determined with certainty, to determine the risk of the trait's appearing in offspring.

TYPICAL PEDIGREES

There are four common modes of inheritance detected using pedigree analysis: autosomal dominant, autosomal recessive, X-linked dominant, and X-linked recessive. Autosomal traits are governed by genes found on one of the autosomes (chromosomes 1-22), while the genes that cause X-linked traits are found on the X chromosome. Males and females are equally likely to be affected by autosomal traits, whereas X-linked traits are never passed on from father to son and all affected males in a family received the mutant allele from their mothers.

The pattern of autosomal dominant inheritance is perhaps the easiest type of Mendelian inheritance to recognize in a pedigree. A trait that appears in successive generations, and is found only among offspring where at least one of the parents is affected, is normally due to a dominant allele.

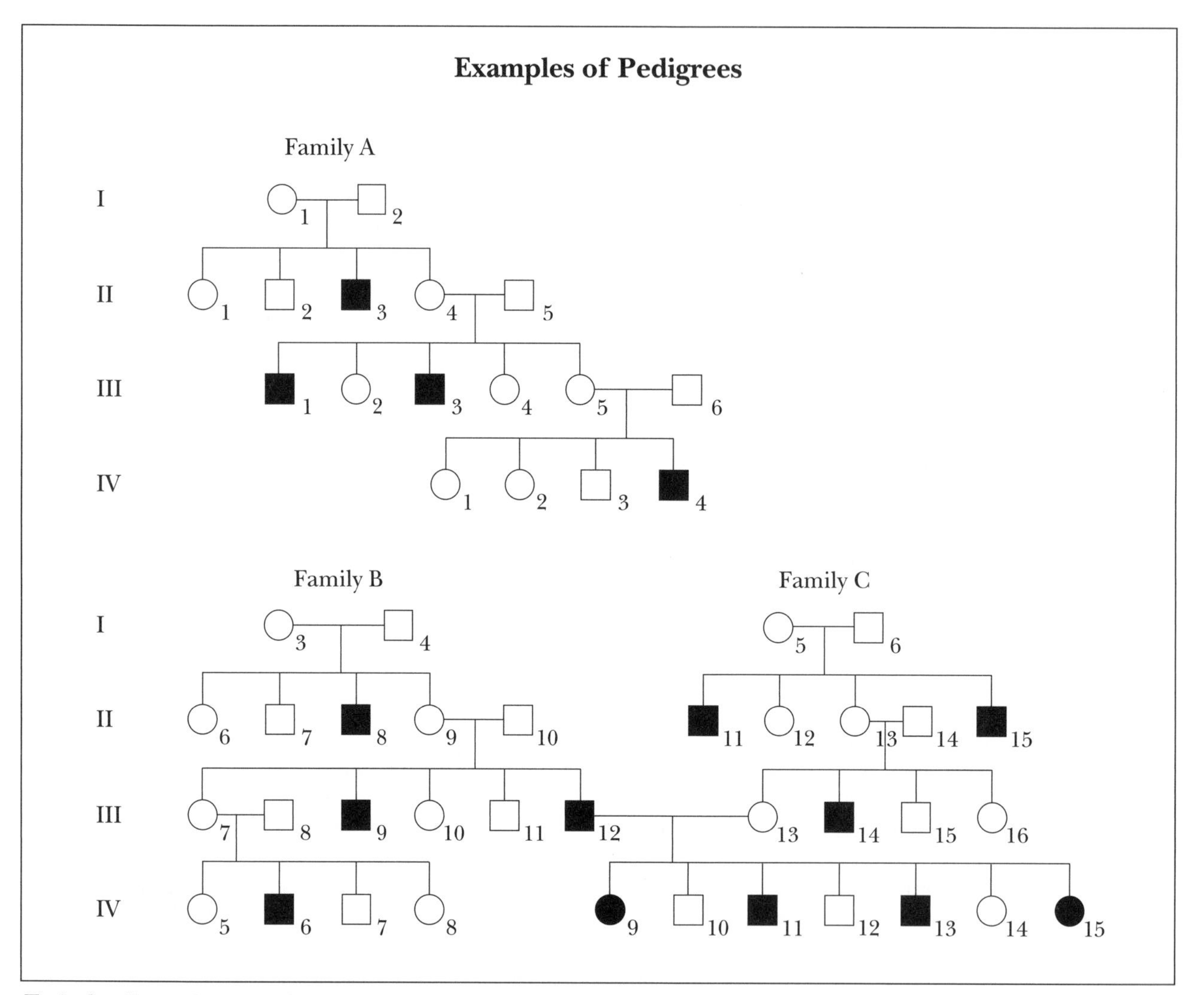

Typical pedigree charts for three families: Roman numerals indicate generations. Squares denote male individuals; circles, female individuals; white or blank individuals are "normal" phenotype; black denotes "affected" phenotype. The charts read like a family tree, with "mother" and "father" at the top and vertical lines denoting offspring; individuals connected only by horizontal lines are mates that have entered the genetic line from outside ("in-laws" in the case of humans). Family A provides an example of a sex-linked recessive trait. Families B and C (joined at 12 and 13) give examples of autosomal traits and how they can resemble sex-linked recessive traits sometimes—and hence the reason for using large families when constructing pedigrees. (Bryan Ness)

If neither parent has the characteristic phenotype displayed by the child, the trait is recessive. For recessive traits, on average, the recurrence risk to the unborn sibling of an affected individual is one in four. The majority of X-linked traits are recessive. The hallmark of X-linked recessive inheritance is that males are much more likely to be affected than females, because males are hemizygous, that is, they possess only one X chromosome, while females have two X chromosomes. Therefore, a recessive trait on the X chromosome will be expressed in all males who possess that X chromosome, while females with one affected X chromosome will be asymptomatic carriers unless their other X chromosome also carries the recessive trait. The trait or disease is typically passed from an affected grandfather through his carrier daughters to half of his grandsons.

X-linked dominant traits are rare but distinctive. All daughters of an affected male and a normal female are affected, while all sons of an affected male and a normal female are normal. For matings between affected females and normal males, the risk of having an affected child is one in two, regardless of the sex of the child. Males are usually more severely affected than females. The trait may be lethal in males. In the general population, females are more likely to be affected than males, even if the disease is not lethal in males.

USEFULNESS

Pedigrees are important both for helping families identify the risk of transmitting an inherited disease and as starting points for searching for the genes responsible for inherited diseases. Mendelian ratios do not apply in individual human families because of the small size. Pooling of families is possible; in the United States, the Mormons and the Amish have kept good records that have aided genetic studies.

However, even using large, carefully constructed records, pedigrees can be difficult to construct and interpret for several reasons. Tracing family relationships can be complicated by adoption, children born out of wedlock, blended families, and assisted reproductive technologies that result in children who may not be genetically related to their parents. Additionally, people are sometimes hesitant to supply information because they are embarrassed by genetic conditions that affect behavior or mental stability.

Many traits do not follow clear-cut Mendelian ratios. Extensions and exceptions to Mendel's laws that can confound efforts to develop a useful pedigree are numerous. In diseases with variable expressivity, some of the symptoms of the disease are always expressed but may range from very mild to severe. In autosomal dominant diseases with incomplete penetrance, some individuals who possess the dominant allele may not express the disease phenotype at all. Some traits have a high recurrent mutation rate. An example is achondroplasia (a type of dwarfism), in which 85 percent of cases are due to new mutations, where both parents have a normal phenotype. Traits due to multifactorial inheritance have variable expression as a result of interactions of the genes involved with the environment. Early-acting lethal alleles can lead to embryonic death and a resulting dearth of expected affected individuals. Pleiotropy is the situation in which a single gene controls several functions and therefore has several effects; it can result in different symptoms in different affected individuals. Finally, one trait can have a different basis of inheritance in different families. For example, mutations in any one of more than four hundred different genes can result in hereditary deafness.

MODERN APPLICATIONS

Genetic counseling is one of the key areas in which pedigrees are employed. A genetic counseling session usually begins with the counselor taking a family history and sketching a pedigree with paper and pencil, followed by use of a computer program to create an accurate pedigree. The Human Genome Project has accelerated the number of genetic disorders that can be detected by heterozygote and prenatal screening. A large part of the genetic counselor's job is to determine for whom specific genetic tests are appropriate.

Although genetic tests for many disorders are now available, the genes involved in many other disorders have yet to be identified. Therefore most human gene mapping utilizes molecular DNA markers, which reflect variation at noncoding regions of the DNA near the affected gene, rather than biochemical, morphological, or behavioral traits. A DNA marker is a piece of DNA of known size, representing a specific locus, that comes in identifiable variations. These allelic variations segregate according to Mendel's laws, which means it is possible to follow their transmission as one would any gene's transmission. If a particular allelic variant of the DNA marker is found in individuals with a particular phenotype, the DNA marker can be used to develop a pedigree. The DNA from all available family members is examined and the pedigree is constructed using the presence of the DNA marker rather than phenotypic categories. This method is particularly useful for late-onset diseases such as Huntington's disease, whose victims may not know they carry the deleterious allele until they are in their forties or fifties, well past reproductive years. Although using DNA markers is a powerful method, crossover in the chromosome between the marker and the gene can cause an individual to be normal but still have the marker that suggests presence of the mutant allele. Thus, for all genetic tests there is a small percentage of false positive and false negative results, which must be factored into the advice given during genetics counseling.

Lisa M. Sardinia, Ph.D.

FURTHER READING

Bennett, Robin L. *The Practical Guide to the Genetic Family History.* New York: Wiley-Liss, 1999. Designed for primary care physicians, this practical book provides the foundation in human genetics necessary to recognize inherited disorders and familial disease susceptibility. Shows how to create a family pedigree.

Bennett, Robin L., et al. "Recommendations for Standardized Human Pedigree Nomenclature." *American Journal of Human Genetics* 56, no. 3 (1995): 745-752. A report from the Pedigree Standardization Task Force that addresses current usage, consistency among symbols, computer compatibility, and the adaptability of symbols to reflect the rapid technical advances in human genetics.

Cummings, Michael R. "Pedigree Analysis in Human Genetics." In *Human Heredity: Principles and Issues.* 8th ed. Florence, Ky.: Brooks/Cole/Cengage Learning, 2009. Textbook designed for an introductory human genetics course for nonscience majors. This chapter contains many useful diagrams and pictures, ending with several case studies and numerous problems.

Hartl, D. L., and Elizabeth W. Jones. "Human Pedigree Analysis." In *Genetics: Analysis of Genes and Genomes.* 7th ed. Sudbury, Mass.: Jones and Bartlett, 2009. This excellent introductory genetics textbook devotes a section of chapter 3 to a discussion of human pedigree analysis within the broader context of transmission genetics.

Thompson, James N., Jr., et al. "Pedigree Analysis." In *Primer of Genetic Analysis: A Problems Approach.* 3d ed. New York: Cambridge University Press, 2007. A textbook providing guided instruction about the analysis and interpretation of genetic data.

Wolff, G., T. F. Wienker, and H. Sander. "On the Genetics of Mandibular Prognathism: Analysis of Large European Noble Families." *Journal of Medical Genetics* 30, no. 2 (1993): 112-116. Good, not overly technical, example of the use of human pedigrees to determine modes of inheritance.

WEB SITES OF INTEREST

Biology Web, Pedigree Analysis
http://faculty.clintoncc.suny.edu/faculty/michael.gregory/files/Bio%20101/Bio%20101%20Laboratory/Pedigree%20Analysis/Pedigree.htm

Biology Web, a site containing information for biology courses taught at Clinton Community College, devotes a page to an illustrated explanation of pedigree analysis.

Pedigree Analysis
http://www.ndsu.nodak.edu/instruct/mcclean/plsc431/mendel/mendel9.htm

Philip McClean, a professor in the department of plant science at North Dakota State University, provides a section about pedigree analysis in his online explanation of Mendelian genetics.

See also: Artificial selection; Classical transmission genetics; Complete dominance; Eugenics; Genetic counseling; Homosexuality; Incomplete dominance; Multiple alleles.

Pelizaeus-Merzbacher disease

CATEGORY: Diseases and syndromes

ALSO KNOWN AS: Pelizaeus Merzbacher brain sclerosis; sclerosis; diffuse familial brain; sudanophilic leukodystrophy; Pelizaeus-Merzbacher type; PMD

DEFINITION

Pelizaeus-Merzbacher disease (PMD) is an X-linked neurological disorder caused by mutation and/or duplication of the proteolipid protein gene (*PLP1*) and characterized by dysmyelination, resulting in permanent hypomyelination or lack of the myelin sheath, the fatty covering of nerve cells.

RISK FACTORS

PMD occurs in 1 out of every 200,000 to 500,000 births in the United States. It is an X-linked disorder, affecting only males who inherit it from their mothers, who are carriers. For females who carry the *PLP1* gene, there is a 50 percent risk of passing it on with every pregnancy—sons inherit the gene and have PMD, while daughters become carriers. Genetic counseling and in utero testing are advised for those with a family history.

ETIOLOGY AND GENETICS

PMD is one of the leukodystrophies, a group of inherited and progressive metabolic diseases affecting myelination of the nervous system and development of white matter in the brain. Each disorder has a separate gene abnormality that affects a different enzyme (protein). In PMD, the defect is in the *PLP1* gene, usually a point mutation (substitution of a single AT or GC base), which results in misfolding of the proteolipid protein or a duplication of the entire gene, causing overexpression of the protein. These mutant proteins are toxic to the oligodendrocyte cells that make myelin.

Myelin constitutes the myelin sheath, which is a fatty covering surrounding axons in the central and peripheral nervous systems and acts as an electrical insulator, allowing impulses to be transmitted quickly along the nerve cells. Without myelin, impulses leak out and nerves cannot function normally. Normal myelination is a step-by-step, ordered process that begins at about five months gestation and continues until a child is two to three years old. In PMD, myelin simply never develops, resulting in permanent hypomyelination and axonal degeneration, primarily in the subcortical region of the cerebrum, cerebellum, and/or brain stem. This prevents impulses from being transmitted from neuron to neuron and causes a range of neurological and motor dysfunctions. It is now known that duplication of the *PLP1* gene accounts for 50 to 75 percent of PMD cases.

The gene encoding the PLP protein is located on the long arm of the X chromosome at band Xq22 and is about 17 kilobases in size, consisting of seven exons and six introns. Two transcript variants encoding distinct isoforms (Isoform 1 and Isoform DM-20) have been identified. The normal PLP protein is a four-transmembrane domain structure that correlates well with one exon of the gene, except at the C terminal end, and binds strongly to other copies of itself.

Many mutations in the *PLP1* gene have been reported. Molecular analysis of the gene revealed a variety of mutations, deletions, and duplications, including two mutations in the 5 untranslated region, missense mutations in exon 2, and an A-to-T transition in exon 4 leading to an Asp-to-Val substitution at residue 202. Exonic mutations tend to be more severe than simple point mutations. Forms of the disorder include the classical X-linked PMD, a severe acute infantile (connate) PMD, and an autosomal dominant late-onset PMD.

SYMPTOMS

Symptoms of PMD are typically slowly progressive, but in the case of connate PMD, occur in early childhood. The first symptom in infants is usually involuntary oscillatory movements of the eyes (nystagmus) and may be concomitant with labored and noisy breathing (stridor) and lack of muscle tone/floppiness (hypotonia). Involuntary muscle spasms (spasticity) and associated muscle and joint stiffness develop. With time, other symptoms become evident, such as impaired ability to coordinate movement (ataxia), developmental delays, loss of motor function and head/trunk control, and deterioration of intellectual abilities.

SCREENING AND DIAGNOSIS

DNA-based testing can be used to diagnose PMD in symptomatic patients, as well as in utero, and to determine carrier status in family members. Identification of pathologic mutations and copies of the *PLP1* gene is the definitive test, using sequence analysis and quantitative polymerase chain reaction (PCR) or fluorescence in situ hybridization (FISH) methods, respectively. Pathological signs of dysmyelination can be examined using magnetic resonance imaging (MRI), once a child is one to two years old when white matter pathways in the brain are maturing and hypomyelination can be detected.

Treatment and Therapy

PMD cannot be cured and there is no effective treatment. Currently, treatment is symptomatic and supportive, but medications are available to alleviate stiffness or spasticity and control seizures. Cell-based therapies are being investigated, including transplantation of a functioning neuregulin gene into unmyelinated nerve cells, which may reprogram them to produce myelin and the use of human adult-derived glial progenitor cells as vectors. A separate study using purified human neural stem cells to treat PMD is currently in Phase I of clinical trials.

Prevention and Outcomes

There are no means of preventing PMD, but genetic counseling/testing is available for couples who have the *PLP1* gene mutation. The prognosis for patients with PMD varies by severity of mutation and form of PMD, with survival as short as early childhood and as long as into the sixties.

Barbara Woldin

Further Reading

Hannigan, Steve, and National Information Centre for Metabolic Diseases. *Inherited Metabolic Diseases: A Guide to 100 Conditions.* Abingdon, England: Radcliffe, 2007. Reader-friendly information on metabolic diseases and their genetics.

Martenson, Russell. *Myelin—Biology and Chemistry.* New York: CRC Press, 1992. In-depth reference discussing role of myelin in disease, with chapter on PMD.

Vinken, Pierre, G. W. Bruyn, Christopher Goetz, et al. *Neurodystrophies and Neurolipidoses.* Amsterdam: Elsevier Science, 1992. Compendium on neuromuscular diseases, myelination, and clinical/pathological manifestations.

Web Sites of Interest

eMedicine: Pelizaeus-Merzbacher Disease
http://emedicine.medscape.com/article/1153103-overview

Genetics Home Reference: Pelizaeus-Merzbacher Disease
http://ghr.nlm.nih.gov/condition=pelizaeusmerzbacherdisease

The Myelin Project: Pelizaeus-Merzbacher disease
http://www.myelin.org/en/cms/291

The Pelizaeus-Merzbacher Disease Foundation
http://www.pmdfoundation.org

United Leukodystrophy Foundation
http://www.ulf.org/types/pelizaeus.html

See also: Adrenoleukodystrophy; Alexander disease; Alzheimer's disease; Amyotrophic lateral sclerosis; Arnold-Chiari syndrome; Ataxia telangiectasia; Canavan disease; Cerebrotendinous xanthomatosis; Charcot-Marie-Tooth syndrome; Chediak-Higashi syndrome; Dandy-Walker syndrome; Deafness; Epilepsy; Essential tremor; Friedreich ataxia; Huntington's disease; Jansky-Bielschowsky disease; Joubert syndrome; Kennedy disease; Krabbé disease; Leigh syndrome; Leukodystrophy; Limb girdle muscular dystrophy; Maple syrup urine disease; Metachromatic leukodystrophy; Myoclonic epilepsy associated with ragged red fibers (MERRF); Narcolepsy; Nemaline myopathy; Neural tube defects; Neurofibromatosis; Parkinson disease; Prion diseases: Kuru and Creutzfeldt-Jakob syndrome; Spinal muscular atrophy; Vanishing white matter disease.

Pendred syndrome

CATEGORY: Diseases and syndromes
ALSO KNOWN AS: Autosomal recessive sensorineural hearing impairment and goiter; deafness with goiter; goiter-deafness syndrome; Pendred's syndrome

Definition

Pendred syndrome (PS), first described by Vaughan Pendred in 1896, is a genetic condition caused by mutations in the *SLC26A4* gene. PS is one of the most common forms of syndromic hearing loss associated with developmental abnormalities of the inner ear, ranging from enlarged vestibular aqueducts (EVA) to cochlea malformations (Mondini dysplasia), goiter (enlarged thyroid), and severe to profound sensorineural hearing loss.

Risk Factors

There are no reported factors associated with an increased risk for having a child with PS; it is diagnosed in both males and females and in all ethnicities. The exact prevalence for PS is unknown, however, reports state that it accounts for approximately 4.3 to 7.5 percent of all congenital deafness.

Etiology and Genetics 50 Percent

PS is associated with mutations in the *SLC26A4* gene located on the long arm of chromosome 7 (7q21-34) and is inherited in an autosomal recessive manner, in which a condition has to be inherited by both parents. A person who has one working gene and one nonworking gene is referred to as a carrier. A carrier is unaffected by the condition. However, when two carriers of the same nonworking gene have children, they have a 25 percent chance of both passing on the nonworking gene and thus, of having a child affected with the condition. This is also a 75 percent chance with each pregnancy that the child will not have PS. If one parent has PS, the chance of having an affected child depends upon the carrier status of the other parent. If both parents have PS, every child born will also have PS.*SLC26A4,* a member of the solute carrier 26 gene family, codes for the protein pendrin. Pendrin is involved in the transport of chloride, iodide, and bicarbonate ions into and out of cells, which is important for the normal function of the inner ear and thyroid. Ion transport is disrupted when *SLC26A4* mutations alter the function or structure of pendrin. Currently more than seventy mutations have been reported. Three mutations are recurrently found in individuals of Northern European ancestry, accounting for up to 50 percent of mutations in this population. Mutations in *SLC26A4* are also associated with nonsyndromic enlarged vestibular aqueduct (DFNB4). DFNB4 is similar to PS, but thyroid abnormalities are not associated with it.

Genetic testing is available for individuals suspected of having PS. However only 50 percent of individuals from families where multiple people are affected, have a mutation identified. Families with only one individual affected have approximately a 20 percent chance of gene detection. Therefore PS is likely a genetically heterogeneous condition (caused by more than one gene).

Although a large number of individuals with a clinical diagnosis of PS lack mutations or only have one mutation detected in the *SLC26A4* gene, other possible genes are being discovered. Recently there is evidence that the *FOXI1* gene, a transcriptional regulator of *SLC26A4,* may be involved in some individuals with PS. Recently an individual has been reported who is a double heterozygote: having a heterozygous mutation in the SLC26A4 gene and a heterozygous mutation in the *FOXI1* gene.

Symptoms

PS is characterized by bilateral, severe to profound sensorineural hearing loss. Although mild-to-moderate and progressive hearing loss has been reported, hearing loss is usually nonprogressive and congenital. Individuals with PS have inner ear abnormalities of the temporal bones. In fact, more than 60 percent of individuals have bilateral enlarged vestibular aqueducts (EVA). Approximately 75 percent of individuals have evidence of goiter on clinical examination. In approximately 40 percent the goiter develops in late childhood or early puberty and for 60 percent in early adult life. There is significant interfamilial and intrafamilial variability.

Screening and Diagnosis

A clinical diagnosis is given to individuals with sensorineural hearing loss, bilateral enlarged vestibular aqueducts, and either a goiter or an abnormal perchlorate discharge test (a test to determine if the thyroid is working properly). Molecular genetic testing is available clinically to confirm the diagnosis if PS is suspected, to clarify risks for family members, and for prenatal diagnosis. Population screening for PS is not available.

Treatment and Therapy

There is no cure for PS. However, benefits are gained from early detection and treatment with hearing aids or cochlear implants and speech and language therapy. EVA may cause increased intracranial pressure that can cause a decline in hearing; therefore activities such as weightlifting, contact sports, scuba diving should be avoided, and head protection for activities such as bicycling should be encouraged. The abnormal thyroid function should be treated in the standard manner. For optimal care, patients should see a variety of specialists, including a clinical geneticist, genetic counselors, otolaryngologists, audiologists, speech-language pathologists, and an endocrinologist.

Prevention and Outcomes

A consultation with a genetic counselor should be made available for individuals with personal or family histories. Prenatal or preimplantation genetic diagnosis is available if the cause is known. Most individuals with PS have normal intelligence and life expectancies.

Amber M. Mathiesen, M.S.

Further Reading

Kochhar, A., M. S. Hildebrand, and R. J. H. Smith. "Clinical Aspects of Hereditary Hearing Loss." *Genetics in Medicine* 9 (2007): 393-409.

Maciaszczyk, K., and A. Lewinski. Phenotypes of SLC26A4 Gene Mutations: "Pendred Syndrome and Hypoacusis with Enlarged Vestibular Aqueduct." *Neuroendocrinology Letters* 29, no. 1 (2008): 29-36.

Web Sites of Interest

American Society for Deaf Children
www.deafchildren.org

National Association of the Deaf
www.nad.org

See also: Adrenoleukodystrophy; Alexander disease; Alzheimer's disease; Amyotrophic lateral sclerosis; Arnold-Chiari syndrome; Ataxia telangiectasia; Canavan disease; Cerebrotendinous xanthomatosis; Charcot-Marie-Tooth syndrome; Chediak-Higashi syndrome; Dandy-Walker syndrome; Deafness; Epilepsy; Essential tremor; Friedreich ataxia; Huntington's disease; Jansky-Bielschowsky disease; Joubert syndrome; Kennedy disease; Krabbé disease; Leigh syndrome; Leukodystrophy; Limb girdle muscular dystrophy; Maple syrup urine disease; Metachromatic leukodystrophy; Myoclonic epilepsy associated with ragged red fibers (MERRF); Narcolepsy; Nemaline myopathy; Neural tube defects; Neurofibromatosis; Parkinson disease; Pelizaeus-Merzbacher disease; Prion diseases: Kuru and Creutzfeldt-Jakob syndrome; Spinal muscular atrophy; Vanishing white matter disease.

Penetrance

Category: Population genetics

Significance: Penetrance is a measure of how frequently a specific genotype results in the same, predictable phenotype. Such variable expression of the same genotype is the result of different genetic backgrounds and the effects of variations in the environment. Geneticists desire 100 percent penetrance for desirable genes that offer disease resistance but reduced penetrance and low expressivity for others that may contribute to human diseases.

Key terms

expressivity: the degree to which a phenotype is expressed, or the extent of expression of a phenotype

phenotype: the physical appearance or biochemical and physiological characteristics of an individual, which is determined by both heredity and environment

Gene Expression and Environment

Gene expression results in a chemical product (protein) with a specific function. The genotype (genetic makeup, or gene) and environmental conditions determine the phenotype of an individual.

Penetrance and Expressivity

Gene expression is dependent upon environmental factors and may be modified, enhanced, silenced, and/or timed by the regulatory mechanisms of the cell in response to internal and external forces. A range of phenotypes can result from a genotype in response to different environments; the phenomenon is called "norms of reaction" or "phenotypic plasticity." Norms of reaction represent the expression of phenotypic variability in individuals of a single genotype.

The question of which is more important in the formation of an organism, nature (genotype) or nurture (environment), has been debated for centuries. The answer is that it depends. The genotype defines phenotypic potential. The environment works on the plasticity of expression to produce different phenotypes from similar genotypes.

Penetrance is the proportion of individuals with a specific genotype who display a defined phenotype. Some individuals may not express a gene if modifiers, epistatic genes, or suppressors are also present in the genome. Penetrance is the likelihood, or probability, that a condition or disease phenotype will, in fact, appear when a given genotype is present. If every person carrying a gene for a domi-

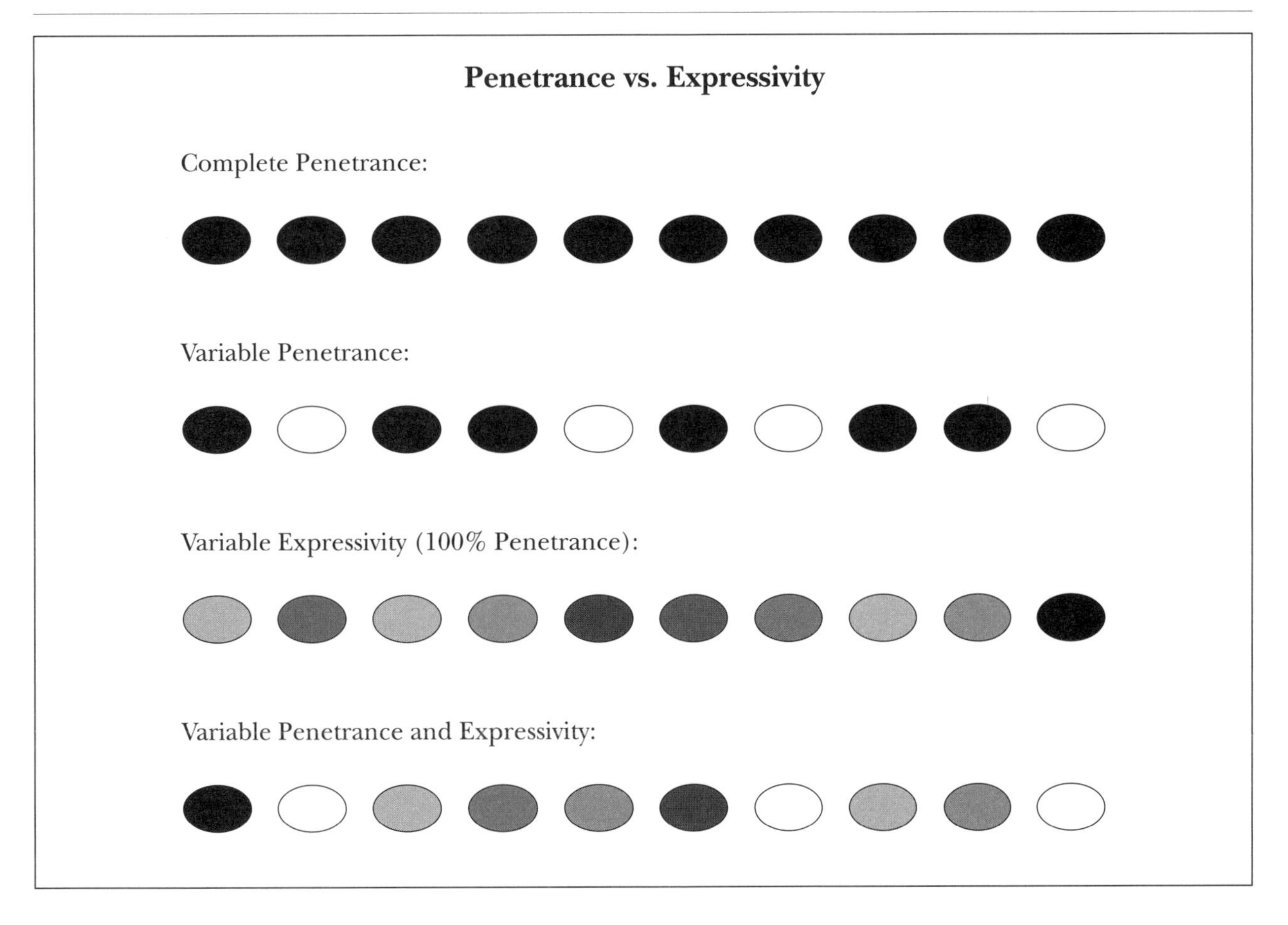

nantly inherited disorder has the mutant phenotype, then the gene is said to have 100 percent penetrance. If only 30 percent of those carrying the mutant allele exhibit the mutant phenotype, the penetrance is 30 percent. Sometimes an individual with a certain genotype fails to express the expected phenotype, and then the allele is said to be nonpenetrant in the individual. If the phenotype is expressed to any degree, the genotype is penetrant.

Given a particular phenotypic trait and a genotype, penetrance can be expressed as the probability of the phenotype given the genotype. For example, penetrance can be the probability of round seeds, a phenotype, given the genotype *G*; it can also be the probability of wrinkled seeds, another phenotype, given the genotype *G*. One could label the specific phenotype of interest as P_i (P_i might refer to either the round or wrinkled seeds) and the specific genotype among many possibilities as G_j. The penetrance would then be the probability of P_i given G_j. These penetrances can all be expressed using the mathematical notation of conditional probabilities as follows:

Case 1: $\Pr(\text{round} | G)$
Case 2: $\Pr(\text{wrinkled} | G)$
Case 3: $\Pr(P_i | Gj)$

A 100 percent penetrance means that all individuals who possess a particular genotype express the phenotype (common in all homozygous lethal genes). Tay-Sachs disease shows complete, or 100 percent, penetrance, as all homozygotes for this allele develop the disease and die.

An allele, *Fu*, in mice causes fusion in the tail in heterozygotes, *Fufu*, and extremely fused and abnormal tails in the homozygotes, *FuFu*. From testcross matings of *Fufu* × *fufu*, 87 fused-tailed mice and 129 nonfused-tailed mice resulted. Genetic analyses of the 129 nonfused-tailed mice revealed that 22 were genotypically *Fufu*. The number of fused-tailed mice was 87 and the number of mice with the *Fufu* genotype but nonfused tails was 22.

The total number of fused-tailed mice expected was (87 + 22) = 109. Therefore, penetrance was calculated at 87/109 = 0.798

Expressivity

Whereas penetrance describes the frequency with which a genotype is expressed as a specified phenotype, expressivity describes the range of variation in the phenotype when expression is observed. Expressivity is variation in allelic expression when the allele is penetrant. Not all traits are expressed 100 percent of the time even though the allele is present. Expressivity is the range of variation in a phenotype; it refers to the degree of expression of a given trait or combination of traits that is associated with a gene. Affected individuals may have severe or mild symptoms; they may have symptoms that show up in one organ or combination of organs in one individual but not in the same locations in other individuals.

Phenotype may be altered by heterogeneity of other genes that affect the expression of a particular locus in question, or by environmental influence. Variable expressivity is a common feature of a variety of cancers. The lower the penetrance, the fewer individuals will be affected. In humans, the dominant allele *P* produces polydactyly—extra toes and/or fingers. Matings between two normal appearing parents sometimes produce offspring with polydactyly. The parent with the *Pp* genotype exhibits reduced penetrance for the *P* allele.

Manjit S. Kang, Ph.D.

Further Reading

Fairbanks, Daniel J., and W. Ralph Anderson. *Genetics: The Continuity of Life.* New York: Brooks/Cole, 1999. This is one of the rare books that contains a good discussion, in Chapter 13, of the concepts of penetrance and expressivity.

Kang, Manjit S. "Using Genotype-by-Environment Interaction for Crop Cultivar Development." *Advances in Agronomy* 62 (November, 1997): 199-252. Discusses environmental influences on heredity.

Kang, Manjit S., and Hugh G. Gauch, Jr. *Genotype-by-Environment Interaction.* Boca Raton, Fla.: CRC Press, 1996. For those interested in in-depth treatment of the interactions between genotypes and environments.

Snustad, D. Peter, and Michael J. Simmons. "Gene Action: From Genotype to Phenotype." In *Principles of Genetics.* 5th ed. Hoboken, N.J.: John Wiley and Sons, 2009. Discusses the concepts of penetrance and expressivity.

Strachan, Tom, and Andrew P. Read. "Complications to the Basic Mendelian Pedigree Patterns." In *Human Molecular Genetics 3.* 3d ed. New York: Garland Press, 2004. This section of the textbook includes a discussion of penetrance, nonpenetrance, and variable expression.

Web Sites of Interest

Cancer Institute of New Jersey
http://www.cinj-genetics.org/health_pro/health_pro.htm#patterns_inherit

Provides information about cancer genetics, including a discussion of patterns of inheritance that features explanations of reduced penetrance and variable expressivity.

Genetics Home Reference
http://ghr.nlm.nih.gov/handbook/inheritance/penetranceexpressivity

Offers explanations of reduced penetrance and variable expressivity for the general reader.

Scitable
http://www.nature.com/scitable/topicpage/Phenotype-Variability-Penetrance-and-Expressivity-573

Scitable, a library of science-related articles compiled by the Nature Publishing Group, contains the article "Phenotype Variability: Penetrance and Expressivity," with links to additional information.

See also: Hereditary diseases; Pedigree analysis.

Periodic paralysis syndrome

CATEGORY: Diseases and syndromes

ALSO KNOWN AS: Familial periodic paralysis; hypokalemic periodic paralysis; Andersen-Tawil syndrome; paraneoplastic periodic paralysis

Definition

Periodic paralysis is a rare inherited condition that causes occasional episodes of severe muscle weakness. The two most common types of periodic paralysis are hypokalemic and hyperkalemic.

Risk Factors

Individuals should tell their doctors if they have any of the risk factors for periodic paralysis. Risk factors include having a family history of the condition and thyroid disorder; the latter factor is a particular risk for Asian males.

Etiology and Genetics

"Periodic paralysis syndrome" is a general term that refers to any of several rare genetic diseases, all of which are inherited as autosomal dominant disorders. This means that a single copy of the defective gene is sufficient to cause the disease, although not all family members who carry the gene are affected to the same extent. An affected individual has a 50 percent chance of transmitting the mutation to each of his or her children. Many cases of periodic paralysis syndrome, however, result from a spontaneous new mutation, so in these instances affected individuals will have unaffected parents.

Mutations in at least four different genes are known to cause this syndrome. All these genes encode ion channel proteins, which form bridges across cell membranes to facilitate the movement of electrically charged ions, such as potassium, sodium, and calcium. Mutations can result in nonfunctional or partially functioning ion channels that allow ions to leak in or out of muscle cells and result in the symptoms associated with the syndrome.

In hypokalemic periodic paralysis, potassium leaks into muscle cells from the blood. This condition is caused by mutations in the genes *KCNE3* (at chromosomal position 11q13-q14), *CACNL1A3* (at position 1q32), and *SCN4A* (at position 17q23.1-25.3). Different mutations in *SCN4A* can cause hyperkalemic periodic paralysis, in which potassium leaks out of muscle cells.

A fourth gene, *KCNJ2* (at position 17q23.1-q24.2), is associated with Andersen-Tawil syndrome, a condition that involves changes in heart rhythms and other developmental abnormalities in addition to the characteristic periodic muscle weakness. It encodes an ion channel protein that is particularly prevalent in cardiac muscle, as well as skeletal muscle.

Symptoms

While muscle strength returns to normal between attacks, repeated bouts of weakness may lead to chronic muscle weakness later in life. An individual remains alert and aware during attacks, and there is no accompanying loss of sensation.

Episodic bouts of severe weakness in the arms and legs are the most prominent symptom of periodic paralysis. Typically, these bouts occur during sleep, especially after strenuous activity. Cold, stress, and alcohol may also produce attacks. Other, less common symptoms may include weakness in the eyelids and face muscles, muscle pain, and irregular heartbeats (arrhythmias). Another symptom may be difficulty breathing or swallowing, which requires emergency care.

Some features are specific to the type of periodic paralysis. In the hypokalemic type, potassium levels are low during attacks and the frequency of attacks varies from daily to yearly. Attacks usually last between four and twenty-four hours, but can last for several days. Attacks usually begin in adolescence, but they can occur before age ten.

In the hyperkalemic type, potassium levels are high or normal during attacks. Attacks are usually shorter (lasting one to two hours), more frequent, and less severe than in the hypokalemic form; breathing and swallowing difficulties are extremely rare. Between attacks, patients often experience muscle spasms or difficulty relaxing their muscles, a condition known as myotonia.

Individuals with some types of periodic paralysis are at risk for a condition known as malignant hyperthermia, which can occur during the use of general anesthesia. Anyone with a family history of periodic paralysis needs to notify the anesthesiologist of this history prior to any surgery.

Screening and Diagnosis

Because this primarily is an inherited condition, the most important aspect of diagnosis is obtaining a family history. In addition to asking about symptoms and a patient's medical history, the doctor will perform a physical exam.

Attacks do not usually occur during an office visit, so the doctor may prescribe several blood tests to check potassium levels during an attack. The doctor may wish to bring on an attack during an office visit. This should be done only under careful monitoring by an experienced neurologist.

If an attack is triggered, several tests may be done, including blood tests to look for the gene mutation or to look for antibodies that may cause these types of symptoms; an electrocardiogram (ECG), a test that records the heart's activity by measuring electrical currents through the heart muscle; and electromyography (EMG) to test the functioning of nerves and muscles. If the diagnosis is in question, the doctor may do a muscle biopsy.

Treatment and Therapy

Since there is no cure for periodic paralysis, lifelong treatment is usually required. Treatment focuses on preventing attacks and relieving symptoms. Individuals can adopt a few behaviors to reduce the frequency and severity of attacks. For individuals with hypokalemic period paralysis, these lifestyle changes include eating a low-carbohydrate, low-sodium diet and avoiding strenuous exercise. Individuals with the hyperkalemic type can eat a low-potassium diet; stay warm; and avoid fasting, alcohol, and heavy exercise.

Medications for both hypokalemic and hyperkalemic periodic paralysis include acetazolamide (Diamox), which may prevent an attack by reducing the flow of potassium from the bloodstream into the cells of the body. For patients with the hypokalemic type, potassium in pill or liquid form may stop an attack; intravenous potassium may be prescribed for severe weakness. Avoiding certain commonly prescribed medications may help reduce the onset of attacks. If these patients also have thyroid conditions, they should be sure to get treatment for these conditions.

Medications for patients with hyperkalemic periodic paralysis include thiazide diuretics, or water pills, which may be prescribed to prevent an attack; and glucose, glucose and insulin, or calcium carbonate, which may be prescribed to slow or stop an attack.

Prevention and Outcomes

Familial periodic paralysis cannot be prevented. Because it can be inherited, genetic counseling may be advised for couples at risk of passing on the disorder.

For the hypokalemic type, patients may reduce attacks by avoiding corticosteroids and glucose infusions and following a diet low in carbohydrates and sodium and rich in potassium. For the hyperkalemic type, patients may reduce attacks by avoiding high-potassium foods, fasting, and drugs known to increase potassium levels. Patients may also decrease attacks by engaging in regular, mild exercise.

Jill Buchanan;
reviewed by J. Thomas Megerian, M.D., Ph.D., F.A.A.P.
"Etiology and Genetics" by Jeffrey A. Knight, Ph.D.

Further Reading

EBSCO Publishing. *DynaMed: Hyperkalemic Period Paralysis.* Ipswich, Mass.: Author, 2009. Available through http://www.ebscohost.com/dynamed.

_______. *Health Library: Periodic Paralysis Syndrome.* Ipswich, Mass.: Author, 2009. Available through http://www.ebscohost.com.

Jurkat-Ratt, K., and F. Lehmann-Horn. "Paroxysmal Muscle Weakness: The Periodic Paralyses." *Journal of Neurology* 253, no. 11 (November, 2006): 1391-1398.

Levitt, J., P. Cochran, and J. Jankowiak. "Patient Page: Attacks of Immobility Caused by Diet or Exercise? The Mystery of Periodic Paralyses." *Neurology* 63, no. 9 (November 9, 2004): E17-18.

Miller, T. M., et al. "Correlating Phenotype and Genotype in the Periodic Paralyses." *Neurology* 63, no. 9 (November 9, 2004): 1647-1655.

Ropper, Allan H., and Martin A. Samuels. *Adams and Victor's Principles of Neurology.* 9th ed. New York: McGraw-Hill Medical, 2009.

Web Sites of Interest

Genetics Home Reference
http://ghr.nlm.nih.gov

Health Canada
http://www.hc-sc.gc.ca/index-eng.php

Muscular Dystrophy Association
http://www.mdausa.org

Muscular Dystrophy Canada
http://www.mdac.ca

National Institute of Neurological Disorders and Stroke: NINDS Familial Periodic Paralyses Information Page
http://www.ninds.nih.gov/disorders/periodic_paralysis/periodic_paralysis.htm

National Organization for Rare Disorders
http://www.rarediseases.org

Periodic Paralysis Resource Center, Periodic Paralysis Association
http://www.periodicparalysis.org

See also: Adrenoleukodystrophy; Alexander disease; Alzheimer's disease; Amyotrophic lateral sclerosis; Arnold-Chiari syndrome; Ataxia telangiectasia; Canavan disease; Cerebrotendinous xanthomatosis; Charcot-Marie-Tooth syndrome; Chediak-Higashi syndrome; Dandy-Walker syndrome; Deafness; Epilepsy; Essential tremor; Friedreich ataxia; Huntington's disease; Jansky-Bielschowsky disease; Joubert syndrome; Kennedy disease; Krabbé disease; Leigh syndrome; Leukodystrophy; Limb girdle muscular dystrophy; Maple syrup urine disease; Metachromatic leukodystrophy; Myoclonic epilepsy associated with ragged red fibers (MERRF); Narcolepsy; Nemaline myopathy; Neural tube defects; Neurofibromatosis; Parkinson disease; Pelizaeus-Merzbacher disease; Pendred syndrome; Prion diseases: Kuru and Creutzfeldt-Jakob syndrome; Spinal muscular atrophy; Vanishing white matter disease.

Phenylketonuria (PKU)

Category: Diseases and syndromes

Definition

Phenylketonuria is a relatively common genetic disease affecting about one in every ten thousand newborn babies. Phenylketonuria, or PKU, was discovered in 1934 by Asbjørn Følling in Norway. Følling discovered that the urine of retarded children turned green when ferric chloride, a chemical used to detect ketones in the urine of diabetics, was added. The urine of diabetics normally turns purple or burgundy with the addition of ferric chloride. Følling conducted further investigations and discovered that the substance responsible for turning urine green upon addition of ferric chloride was phenylpyruvic acid. Følling discovered that the origin of phenylpyruvic acid was the amino acid phenylalanine.

Risk Factors

Both of a child's parents must pass along the mutated *PAH* gene in order for the child to develop this condition. Children who have only one parent with this gene are not at risk. PKU primarily affects white people of Northern European ancestry and is much less common in African Americans.

Etiology and Genetics

PKU can be caused by a mutation in the *PAH* gene. This gene encodes the information for the liver enzyme phenylalanine hydroxylase (PAH), which catalyzes the conversion of phenylalanine to tyrosine. In people with normal metabolisms, phenylalanine, an essential amino acid, must be consumed in the diet. Phenylalanine is either incorporated into the body's proteins or converted by the enzyme phenylalanine hydroxylase into tyrosine, another amino acid. Tyrosine is either incorporated into protein or converted into other important biological molecules, such as dopamine, epinephrine, norepinephrine, and melanin. Alternatively, tyrosine can be completely metabolized and eliminated from the body.

People with PKU cannot metabolize phenylalanine into tyrosine at normal rates. The disease-causing mutant PKU gene is recessive. Thus, in order for a person to have PKU, he or she must inherit two copies of the mutant gene. Approximately one in every fifty people in the United States is a heterozygous carrier for the disease. About one in every ten thousand newborn babies has the disease.

The PKU gene was isolated in 1992, and soon afterward it was discovered that there is no one type of PKU mutation. Instead, the disease can be caused by a variety of defects affecting the PKU gene. Many of these defects are "point" mutations resulting in single base-pair changes in the DNA which lead to amino acid substitutions in the *PAH* gene. Other defects include base-pair changes leading to splicing defects in *PAH* messenger RNA (mRNA), deletions resulting in one or more missing amino acids in PAH, and insertions resulting in mRNA reading frame shifts. More than four hundred mutations have been found in the PKU gene. The variety of different defects in the PKU gene leads to variability in the activity of PAH and the severity of the disease.

Symptoms

Common characteristics of untreated patients with PKU are mental retardation, light-colored skin, hyperactivity, schizophrenia, tremors, and eczema. PKU also has major metabolic effects. Normally, blood phenylalanine concentrations are between 2 and 6 milligrams per deciliter (mg/dl), but in PKU phenylalanine accumulates to 20 mg/dl or more. Since phenylalanine cannot be properly converted into tyrosine, melanin, dopamine, norepinephrine, and epinephrine, there is a deficiency of those important compounds, which probably contributes to the development of symptoms characteristic of the disease.

The high levels of phenylalanine may also interfere with the transport of other important amino acids into the brain. Since several amino acids use the same transport system as phenylalanine, phenylalanine is preferentially transported at the expense of the others. This may also contribute to the development of symptoms characteristic of PKU. Additionally, since phenylalanine cannot be metabolized normally, it is metabolized into abnormal compounds such as phenylpyruvic acid, which further contributes to the development of PKU symptoms.

Screening and Diagnosis

In 1957 Willard Centerwall introduced ferric chloride as a screening technique by impregnating babies' diapers with ferric chloride. If the babies' urine contained phenylpyruvic acid, the diaper would turn green. Since the test was reliable only after the baby was several weeks old and after brain damage may already have occurred, a new, more reliable and more sensitive test was needed.

Robert Guthrie developed a more sensitive test. In the Guthrie test, bacteria are grown on an agar medium that contains an inhibitor of growth that can be overcome by exogenously added phenylalanine. If a small piece of filter paper containing blood is placed on the agar medium with the bacteria, the phenylalanine in the blood leaches out of the filter paper and stimulates growth of the bacteria. The extent of the growth around the filter paper is directly proportional to the amount of phenylalanine in the blood. Guthrie published his procedure in 1961. In 1963 Massachusetts became the first state to legislate mandatory PKU screening of all newborns. It is now mandatory in all fifty states.

Chorionic villus sampling can detect PKU in a developing fetus. This test can be performed by inserting a needle through a pregnant woman's lower abdomen, or by inserting a catheter up through the cervix into the chorionic villi, which forms the lining of the placenta. The needle removes a small sample of cells for genetic testing. A doctor or genetics counselor can help a pregnant woman determine whether these tests are accurate and how she should respond to the results.

Treatment and Therapy

The treatment of choice for PKU is dietary or nutritional intervention. PKU babies placed on very low phenylalanine diets show normal cognitive development. The PKU diet eliminates high-protein foods, which are replaced with low-phenylalanine foods and supplemented with a nutritional formula. In 1954, Horst Bickel was the first to treat PKU with diet therapy.

It is recommended that dietary intervention begin as soon as possible after birth and continue for life. It is especially important that pregnant PKU women adhere closely to the diet, or their babies will be mentally retarded. Studies have shown that if children or adults are taken off the diet, some PKU symptoms may develop.

The U.S. Food and Drug Administration (FDA) in December, 2007, approved the drug sapropterin dihydrochloride (Kuvan) for treatment of some people with PKU, to be used in combination with a PKU diet. The FDA directed that studies continue to be conducted in order to determine this drug's efficacy.

Prevention and Outcomes

If PKU is not detected and treatment is not begun within the first few weeks of life, the child will develop various neurological symptoms, including retardation. If the disease is detected shortly after birth and dietary treatment is instituted, symptoms characteristic of the disease usually will not develop.

Women with PKU can prevent birth defects by maintaining a low-phenylalanine diet before they become pregnant. Individuals with PKU, or those who have a close relative or child with the condition, may benefit from genetic counseling before deciding to have a child.

Charles L. Vigue, Ph.D.; updated by Rebecca Kuzins

Further Reading

Kaufman, Seymour. *Overcoming a Bad Gene: The Story of the Discovery and Successful Treatment of Phenylketonuria, a Genetic Disease That Causes Mental Retardation.* Bloomington, Ind.: AuthorHouse, 2004. Kaufman, who was chief of the laboratory of neurochemistry at the National Institutes of Health, recounts the history of PKU's discovery and the subsequent research to learn more about the disease. He also provides a thorough explanation of the disease's symptoms, treatment, diag-

nosis, and how it can cause mental retardation.
Koch, Jean Holt. *Robert Guthrie, the PKU Story: A Crusade Against Mental Retardation.* Pasadena, Calif.: Hope, 1997. A longtime friend profiles the scientific work and personal life work of Robert Guthrie.
National PKU News. This newsletter, published in Seattle, Washington, three times a year, provides information about PKU.
Parker, James N. *The Official Parent's Sourcebook on Phenylketonuria.* San Diego: Icon Health, 2002. This resource, created for parents with PKU children, tells parents how and where to look for information about PKU.
Surendran, Sankar, et al. "Neurochemical Changes and Therapeutical Targets in Phenylketonuria (PKU)." In *Neurochemistry of Metabolic Diseases: Lysosomal Storage Diseases, Phenylketonuria, and Canavan Disease, 2007,* edited by Surendran. Kerala, India: Transworld Research Network, 2007. Focuses on PKU and other health conditions in which a single gene defect hampers normal metabolic activity, resulting in pathophysiological abnormalities.

Web Sites of Interest

Genetic and Rare Diseases Information Center, Phenylketonuria
http://rarediseases.info.nih.gov/GARD/Disease.aspx?PageID=4&DiseaseID=7383
Links to online resources that explain the disease and provide management guidelines and information about newborn screening.

Genetics Home Reference
http://ghr.nlm.nih.gov/condition=phenylketonuria
Describes the disease and explains how it results from a mutation in the *PAH* gene. Links to other online resources.

Medline Plus
http://www.nlm.nih.gov/medlineplus/phenylketonuria.html
Contains a brief description of PKU and numerous links to additional information.

National Organization for Rare Disorders
http://www.rarediseases.org
This site contains an index of rare diseases and a rare disease database that enables users to retrieve information about phenylketonura, as well as a list of other names for the disease and a list of related organizations.

See also: Biochemical mutations; Genetic screening; Genetic testing; Hereditary diseases; Inborn errors of metabolism; Model organism: *Mus musculus.*

Plasmids

Category: Molecular genetics

Significance: Plasmids are DNA molecules that exist separately from the chromosome. Plasmids exist in a commensal relationship with their host and may provide the host with new abilities. They are used in genetic research as vehicles for carrying genes. In the wild, they promote the exchange of genes and contribute to the problem of antibiotic resistance.

Key terms

commensalism: a relationship in which two organisms rely on each other for survival
gene: a region of DNA containing instructions for the manufacture of a protein
transposon: a piece of DNA that can copy itself from one location to another

Plasmid Structure

The structure of plasmids is usually circular, although linear forms do exist. Their size ranges from a few thousand base pairs to hundreds of thousands of base pairs. They are found primarily in bacteria but have also been found in fungi, plants, and even humans.

In its commensal relationship with its host, the plasmid can be thought of as a molecular parasite whose primary function is to maintain itself within its host and to spread itself as widely as possible to other hosts. The majority of genes that are present on a plasmid will be dedicated to this function. Researchers have discovered that despite the great diversity of plasmids, most of them have similar genes, dedicated to this function. This relative simplicity of plasmids makes them ideal models of gene function, as well as useful tools for molecular biology.

Genes of interest can be placed on a plasmid, which can easily be moved in and out of cells. Using plasmids isolated from the wild, molecular biologists have designed many varieties of artificial plasmids, which have greatly facilitated research in molecular biology.

PLASMID REPLICATION

To survive and propagate, a plasmid must be able to copy itself, or replicate. The genes that direct this process are known as the replication genes. These genes do not carry out all the functions of replication, but instead coopt the host's replication machinery to replicate the plasmid. Replication allows the plasmid to propagate by creating copies of itself that can be passed to each daughter cell when the host divides. In this manner, the plasmid propagates along with the host.

A second function of the replication genes is to control the copy number of the plasmid. The number of copies of a plasmid that exist inside a host can vary considerably. Plasmids can exist at a very low copy number (one or two copies per cell) or at a higher copy number, with dozens of copies per cell. Adjusting the copy number is an important consideration for a plasmid. Plasmid replication is an expensive process that consumes energy and resources of the host cell. A plasmid with a high copy number can place a significant energy drain on its host cell. In environments where the nutrient supply is low, a plasmid-bearing cell may not be able to compete successfully with other, non-plasmid-containing cells. Wild plasmids often exist at a low copy number, or create a high copy number for only a brief period of time.

PLASMID PARTITIONING

Because the presence of a plasmid is expensive in terms of energy, a cell harboring a plasmid will grow more slowly than a similar cell with no plasmid. This can cause a problem for a plasmid if it fails to partition properly during its host's division. If the plasmid does not partition properly, then one of the host's daughter cells will not contain a plasmid. Since this cell does not have to spend energy replicating a plasmid, it will gain an ability to grow faster, as will all of its offspring. In such a situation, the population of non-plasmid-containing cells could outgrow the population of plasmid-containing cells and use up all the nutrients in the environment. To avoid this problem, plasmids have evolved strategies to prevent improper partitioning. One strategy is for the plasmid to contain partitioning genes. Partitioning genes encode proteins that actively partition plasmids into each daughter cell during the host cell's division. Active partitioning greatly reduces the errors in partitioning that might occur if partitioning were left to chance.

A second strategy that plasmids use to prevent partitioning errors is the plasmid addition system. In this strategy, genes on the plasmid direct the production of both a toxin and an antidote. The antidote protein is very unstable and degrades quickly, but the toxin is quite stable. As long as the plasmid is present, the cytoplasm of the cell will be full of toxin and antidote. Should a daughter cell fail to receive a plasmid during division, the residual antidote and toxin present in the cytoplasm from the mother cell will begin to degrade, since there is no longer a plasmid present to direct the synthesis of either toxin or antidote. Since the antidote is very unstable, it will degrade first, leaving only toxin, which will kill the cell.

PLASMID TRANSFER BETWEEN CELLS

Propagation of plasmids can occur through the spread of plasmids from parent cells to their offspring (referred to as vertical transfer), but propagation can also occur between two different cells (referred to as horizontal transfer). Many plasmids are able to transfer themselves from one host to another through the process of conjugation. Conjugal plasmids contain a collection of genes that direct the host cell that contains them to attach to other cells and transfer a copy of the plasmid. In this manner, the plasmid can spread itself to other hosts and is not limited to spreading itself only to the descendants of the original host cell.

One of the first plasmids to be identified was discovered because of its ability to conjugate. This plasmid, known as the F plasmid, or F factor, is a plasmid found in the bacterium *Escherichia coli*. Cells harboring the F plasmid are designated F^+ cells and can transfer their plasmid to other *E. coli* cells that do not contain the F plasmid (called F^- cells).

Conjugal plasmids can be very specific and transfer only between closely related members of the same species (such as the F plasmid), or they can be very promiscuous and allow transfer between unrelated species. An extreme example of cross-species transfer is the Ti plasmid of the bacterial species *Agrobacterium tumefaciens.* The Ti plasmid is capable of transferring part of itself from *A. tumefaciens* into the cells of dicotyledonous plants. Plant cells that receive parts of the Ti plasmid are induced to grow and form a tumorlike structure, called a gall, that provides a hospitable environment for *A. tumefaciens.*

Host Benefits from Plasmids

In most commensal relationships, there is an exchange of benefits between the two partners. The same is true for plasmids and their hosts. In many cases, plasmids provide their host cells with a collection of genes that enhance the ability of the host cell to survive. Enhancements include the ability to metabolize a wider range of materials for food and the ability to survive in hostile environments. One particular hostile environment in which plasmids can provide the ability to survive is the human body. A number of pathogenic microorganisms gain their ability to inhabit the human body, and thus cause disease, from genes contained on plasmids. An example of this is *Bacillus anthracis,* the agent that causes anthrax. Many of the genes that allow this organism to cause disease are contained on one of two plasmids, called pXO1 and pXO2. *Yersinia pestis,* the causative agent of bubonic plague, also gains its disease-causing ability from plasmids.

R Factors

Another example of plasmids conferring on their hosts the ability to survive in a hostile environment is antibiotic resistance. Plasmids known as R factors contain genes that make their bacterial hosts resistant to antibiotics. These R factors are usually conjugal plasmids, so they can move easily from cell to cell. Because the antibiotic resistance genes they carry are usually parts of transposons, they can readily copy themselves from one piece of DNA to another. Two different R factors that happened to be together in one cell could exchange copies of each other's antibiotic resistance genes. A number of R factors exist that contain multiple antibiotic resistance genes. Such plasmids can result in the formation of "multidrug resistant" (MDR) strains of pathogenic bacteria, which are difficult to treat. There is much evidence to suggest that the widespread use of antibiotics has contributed to the development of MDR pathogens, which are emerging as an important health concern.

Role of Plasmids in Evolution

Through conjugation, plasmids can transfer genetic information from one species of bacterial cell to another. During its stay in a particular host, a plasmid may acquire some of the chromosomal genes of the host, which it then carries to a new host by conjugation. These genes can then be transferred from the plasmid to the chromosome of the new host. If the new host and the old host are different species, this gene transfer can result in the introduction of new genes, and thus new traits, into a cell. Bacteria, being asexual, produce daughter cells that are genetically identical to their parent. The existence of conjugal plasmids, which allow for the transfer of genes between bacterial species, may represent an important mechanism by which bacteria generate diversity and create new species.

Genetic Engineering of Plasmids

The identification of restriction endonuclease sites within plasmids allowed scientists to manipulate the organization and makeup of these molecules. Researchers were now able to insert genes of interest into these restriction sites within the plasmids and have the recombined plasmid vector taken up by bacteria through the process of transformation. Transcription of the inserted gene by the host bacterium is dependent on an upstream promoter that is active in that bacterial strain. The use of genetic engineering and the creation of artificial plasmid vectors have revolutionized basic research and have led to the creation of modern industrial microbiology. Because bacteria containing the vector are able to express proteins encoded by the inserted gene(s) at a high level, mass quantities of desired proteins can be produced on an industrial scale. Vectors have been used to express medically important proteins such as insulin, human growth hormone, and human factor IX, a blood-clotting factor. Vectors have also been used to express proteins within eukaryotic cells. Additional DNA elements are necessary for the optimal production of proteins in these cells. Viral or mammalian promoters recognized by host RNA polymerases, as well as enhancer sequences, allow proteins to be expressed from the vector. Poly-adenylation sequences are added downstream of the inserted gene for proper messenger RNA (mRNA) expression.

The ability to express proteins within mammalian cells has enabled the development of DNA vaccines in which antigenic proteins are expressed from plasmids. These plasmids are taken up by cells following introduction into the animal or human subject by injection or alternate means of inoculation. Expression of these proteins in the cell then allows for an immunological response by the vaccine. Expression of cytokine or other immunomodulatory molecules from the same plasmid has been explored as a means of enhancing the immune response to the coexpressed antigenic protein using DNA vaccines. Plasmid DNA itself can sometimes have immunostimulatory effects due to the presence of CpG dinucleotides. Clinical trials using DNA vaccines targeting cancer have demonstrated the immunogenicity of these vaccines in humans. Veterinary applications of DNA vaccines have been approved for use.

Plasmids have also been used as delivery vehicles for the expression of double-stranded RNAs (dsRNAs) in order to suppress specific mRNAs by RNA interference (RNAi). Because of the abbreviated length (about 21 nucleotides) and need for a distinction termination point of these dsRNAs, RNA polymerase III promoters and the accompanying termination signals have often been included on the plasmids to express these transcripts within cells. Delivery of the RNAi expression plasmids to the desired location within the body remains a challenge for the development of this technology.

Douglas H. Brown, Ph.D.;
updated by Daniel E. McCallus, Ph.D.

FURTHER READING

Bower, D. M., and K. L. J. Prather. "Engineering of Bacterial Strains and Vectors for the Production of Plasmid DNA." *Applied Microbiology and Biotechnology* 82, no. 5 (2009): 805-813. Discusses innovations in both plasmid vectors and the bacterial strains that produce them in current biotechnology practice.

Levy, Stuart B. "The Challenge of Antibiotic Resistance." *Scientific American* 278 (1998): 46-53. A discussion on the growing problem of antibiotic resistance. Written by one of the experts in the field.

Summers, David K. *The Biology of Plasmids.* Malden, Mass.: Blackwell, 1996. A comprehensive book on plasmid biology written for college undergraduates.

Thomas, Christopher M. "Paradigms of Plasmid Organization." *Molecular Microbiology* 37, no. 3 (2000): 485-491. A review that clearly discusses the evolution and organization of plasmid genes.

Van Gaal, E. V. B., W. E. Hennink, D. J. A. Crommelin, and E. Mastrobattista. "Plasmid Engineering for Controlled and Sustained Gene Expression for Nonviral Gene Therapy." *Pharmaceutical Research* 23, no. 6 (2006): 1053-1074. In-depth review describing the factors and genetic regions necessary for optimal expression of proteins from plasmid vectors.

WEB SITES OF INTEREST

Plasmid Sequencing
http://www.sanger.ac.uk/Projects/Plasmids

Plasmid.org
http://www.plasmid.org

Plasmids: History of a Concept
http://histmicro.yale.edu/mainfram.htm

See also: Anthrax; Antisense RNA; Archaea; Bacterial genetics and cell structure; Bacterial resistance and super bacteria; Biopesticides; Biopharmaceuticals; Blotting: Southern, Northern, and Western; Cloning; Cloning vectors; DNA sequencing technology; Emerging and remerging infectious diseases; Extrachromosomal inheritance; Gene regulation: Bacteria; Genetic engineering; Genetic engineering: Agricultural applications; Genetic engineering: Historical development; Genetic engineering: Industrial applications; Genome size; Genomics; High-yield crops; Human growth hormone; Immunogenetics; Model organism: *Chlamydomonas reinhardtii*; Model organism: *Escherichia coli*; Model organism: *Saccharomyces cerevisiae*; Model organism: *Xenopus laevis*; Noncoding RNA molecules; Polymerase chain reaction; Proteomics; Shotgun cloning; Transgenic organisms; Transposable elements.

PMS genes

CATEGORY: Molecular genetics

SIGNIFICANCE: *PMS* genes function in DNA mismatch repair, where they help correct errors that arise during replication. If these errors go uncorrected, mutations will accumulate in the genome, eventually leading to cancer. Deficiency in the

PMS2 gene can cause hereditary nonpolyposis colon cancer as well as Turcot syndrome.

KEY TERMS

apoptosis: programmed cell death

DNA mismatch repair (MMR): cellular mechanism for correcting mismatches in DNA that occur during replication; also suppresses aberrant recombination and signals cell death from chemical damage

hereditary nonpolyposis colon cancer (HNPCC): also called Lynch syndrome; cancer caused by a deficiency in one copy of a mismatch repair gene; patients with HNPCC have an 80 percent lifetime risk of developing colorectal cancer and women with HNPCC have a 20 to 60 percent lifetime risk of developing endometrial cancer

heterodimer: two different proteins binding together to form a complex

methylation damage: certain chemicals, termed methylators, cause methyl (CH_3) groups to incorporate into DNA; these lesions are toxic to cells

Turcot syndrome: cancer syndrome defined by early onset hematological and CNS malignancies as well as colorectal cancer

GENETICS

There are two genes in the *PMS* family, *PMS1* and *PMS2*. *PMS* stands for postmeiotic segregation, the name given to the yeast homologue due to its role in regulating recombination during meiosis. In humans, *PMS1* is located on chromosome 2q31-33, and *PMS2* is located on 7p22.

MMR AND CANCER

The main cellular role for the *PMS* genes is in mismatch repair. When cells divide, the DNA is replicated by enzymes called polymerases that use the existing strand as a copy for the new strand. The polymerases occasionally make an error, creating a mismatch between the template strand and the newly replicated strand. MMR functions to correct these errors before they become mutations. The *PMS* genes are part of the MutL heterodimer. *PMS2* is in MutLalpha, where it binds *MLH1*. *PMS1* also binds MLH1, as part of MutLbeta. The MutL heterodimer gets recruited to mismatched bases in the DNA by the MutS heterodimer, typically MutSalpha, which is composed of *MSH2* and *MSH6*. Once bound to DNA, MutS and MutL recruit downstream factors that excise the newly replicated strand in the area of the mismatch and fill it in with the correct base. If MMR is compromised, then the mismatched bases will go uncorrected and become mutations in the next round of replication. An increased rate of mutation accumulation can eventually lead to cancer.

If a mismatch repair gene is defective, it causes hereditary nonpolyposis colon cancer, or Lynch syndrome. The majority of HNPCC cases are caused by mutations in either *MLH1* or *MSH2*; less than 5 percent of HNPCC cases are due to inherited mutations in *PMS2* and no HNPCC-causing mutations have been definitively attributed to *PMS1*. However, *PMS2* has a number of pseudogenes that have made detecting mutations in *PMS2* difficult. HNPCC is an autosomal dominant disorder; only one copy of *PMS2* must be mutated to cause the disease. Eventually, the remaining copy of *PMS2* is lost in a subset of cells. These cells have a high mutation rate, increasing the likelihood of activating genes that promote cancer as well as altering genes that prevent it.

In addition to its role in repairing replication errors, *PMS2* also responds to damage caused by methylating agents. Methylating agents damage DNA by creating lesions that cannot be repaired by normal cellular repair mechanisms. The cells then signal apoptosis in order to prevent the damaged cells from continuing to proliferate. *PMS2* is necessary for cells to signal apoptosis in response to methylation. Cells deficient in *PMS2* will continue to grow in the presence of methylating agents.

TURCOT SYNDROME

Turcot syndrome, also known as mismatch repair cancer syndrome, is caused by mutation in both alleles of a mismatch repair gene, including *PMS2*. It differs from HNPCC in that it causes early-onset malignancies of the central nervous system as well as colorectal tumors. In addition, patients often present with light brown spots on the skin, termed café-au-lait spots. Due to having mutations in both copies of *PMS2*, the severity of Turcot syndrome is increased compared to HNPCC, with cancers presenting much earlier.

IMPACT

The *PMS* genes are an essential component of mismatch repair, which is necessary to maintain genomic stability in the cell. Mutation in one copy of

PMS2 can lead to HNPCC, while mutation in both copies can cause Turcot syndrome. Furthermore, *PMS2* was the first MMR gene to be identified as causing Turcot syndrome. It was later shown that the other MMR genes could cause Turcot syndrome as well. *PMS2* also responds to DNA damage caused by methylating agents, signaling cell death. Such methylating agents are often used in chemotherapeutic settings; for example, 6-mercaptopurine, which is used to treat childhood leukemias. Functioning MMR is necessary for these drugs to be effective. It has been demonstrated that tumors can become resistant to these drugs by mutating MMR.

Jennifer Johnson

FURTHER READING

Alberts, Bruce, et al. *Molecular Biology of the Cell.* New York: Garland Science, 2002. General background on DNA replication and repair.

Kufe, Donald W., et al., eds. *Cancer Medicine.* Hamilton, Ont.: BC Decker, 2003. In-depth information on cancer formation, progression, and treatment.

Nussbaum, Robert L., Roderick McInnes, and Huntington F. Willard. *Genetics in Medicine.* 7th ed. New York: Saunders, 2007. Thorough yet basic background on genetic principles that includes a spotlight on HNPCC.

WEB SITES OF INTEREST

International Society for Gastrointestinal Hereditary Tumors
www.insight-group.org

Online Mendelian Inheritance in Man (OMIM): Postmeiotic Segregation Increased, S. cerevisiae, 1; PMS1
http://www.ncbi.nlm.nih.gov/entrez/dispomim.cgi?id=600258

Online Mendelian Inheritance in Man (OMIM): Postmeiotic Segregation Increased, S. cerevisiae, 2; PMS2
http://www.ncbi.nlm.nih.gov/entrez/dispomim.cgi?id=600259

See also: *BRAF* gene; *BRCA1* and *BRCA2* genes; *MLH1* gene; *RB1* gene; *RhoGD12* gene; *SCLC1* gene.

Polycystic kidney disease

CATEGORY: Diseases and syndromes
ALSO KNOWN AS: PKD; autosomal dominant polycystic kidney disease; AKPKD; adult polycystic disease; polycystic kidney disease type 2

DEFINITION

The word polycystic means many cysts. Polycystic kidney disease (PKD) is an inherited disease that causes many cysts to form in the kidneys.

Cysts, which are sacs filled with fluid, grow in both kidneys causing them to become enlarged. The number of cysts can range from a few to a great number. The size of the cysts can vary from too small to detect to cysts larger than the kidney itself.

PKD can be painful and interfere with the normal functioning of the kidney, resulting in infection, kidney stones, high blood pressure, and, eventually, kidney failure. PKD is a potentially serious condition that requires care from a doctor.

RISK FACTORS

The primary risk factor for PKD is having a parent with the disease. Fifty percent of children born to a parent with the PKD gene develop the disease. In about 10 percent of cases, the gene for the disease was not inherited, but mutated. PKD affects men and women equally.

ETIOLOGY AND GENETICS

The majority of cases of adult-onset polycystic kidney disease result from a genetic disorder, but some cases are known in which the disease has been acquired, usually after years of dialysis for kidney failure of unrelated origin. The genetic forms result from mutations in either of two genes, *PKD1* or *PKD2*. About 85 percent of cases result from mutations in *PKD1*, found on the short arm of chromosome 16 at position 16p13.3. This gene encodes a large protein called polycystin-1, which normally spans the membrane of kidney cells and receives chemical signals from outside the cell that help the cell respond properly under varying environmental conditions. The protein interacts with polycystin-2 (the product of the *PKD2* gene, found on chromosome 4 at position 4q21-q23) to trigger a cascade of cellular reactions. A missing or altered polycystin-1 protein disrupts this signaling, and cells in the renal tissue may grow and divide abnormally, resulting in the cyst development that is symptomatic of the disease. Mutations in *PKD2*, which account for the remaining 15 percent of adult cases, result in abnormal polycystin-2 proteins that lead to similar physiological malfunctions. Mutations in either gene are inherited in an autosomal dominant fashion, meaning that a single copy of the mutation is sufficient to cause full expression of the disease. An affected individual has a 50 percent chance of transmitting the mutation to each of his or her children. About 10 percent of cases of polycystic kidney disease, however, result from a spontaneous new mutation, so in these instances affected individuals will have unaffected parents.

A rare severe form of polycystic kidney disease that usually is diagnosed at birth or in early infancy results from mutations in the *PKHD1* gene (at position 6p12.2). The protein encoded by this gene is fibrocystin, a membrane-spanning protein with functions similar to polycystin-1. This often lethal form of the disease is inherited in an autosomal recessive manner, which means that both copies of the *PKHD1* gene must be deficient in order for the individual to be afflicted. Typically, an affected child is born to two unaffected parents, both of whom are carriers of the recessive mutant allele. The probable outcomes for children whose parents are both carriers are 75 percent unaffected and 25 percent affected.

Symptoms

During the early stages of PKD, there are often no symptoms. Some people are never diagnosed because their symptoms are mild. Most symptoms appear in middle age.

Frequently, the first symptom is pain in the back or flank area. Other signs of PKD include high blood pressure, blood in the urine, urinary tract infection, and kidney stones. Additional, less common symptoms may include nail abnormalities, painful menstruation, joint pain, and drowsiness.

Individuals who experience any of these symptoms should not assume it is due to PKD. These symptoms may be caused by other, less serious health conditions. Individuals who experience any one of the symptoms should see their physicians.

Screening and Diagnosis

The doctor will ask about a patient's symptoms and medical history and will perform a physical exam. When diagnosing PKD, the doctor may begin by looking for signs of the disease, including high blood pressure, enlarged or tender kidneys, enlarged liver, and protein or blood in the urine.

An abdominal ultrasound is usually the test of first choice to detect the presence of cysts on the kidneys. If cysts are too small to be detected by ultrasound and the diagnosis is still in question, an abdominal computed tomography (CT) scan or a magnetic resonance imaging (MRI) scan may be performed.

If the diagnosis still remains unconfirmed, additional tests may be ordered, including a gene linkage study, a blood test that tests the deoxyribonucleic acid (DNA) of the patient and family members with and without PKD; and direct DNA sequencing, a blood sample of a patient's DNA to look for presence of the PKD gene.

Between 10 and 40 percent of patients with PKD also have an aneurysm (a weakness in the wall of a blood vessel) in the brain. If a patient is diagnosed with PKD and there is a family history of a brain aneurysm, the doctor may recommend an arteriogram to detect the presence of an aneurysm.

Treatment and Therapy

Most treatments for PKD treat the disease symptoms or prevent complications. Some of these treatment options may include high blood pressure medication. Since high blood pressure is common with PKD, antihypertensive medications are often prescribed to control blood pressure.

Pain medications must be used cautiously, since some of them can cause further damage to the kidneys. In the event of a urinary tract infection, aggressive treatment with antibiotics is required to avoid further damage to the kidneys.

Cysts may be drained through surgery to relieve pain, blockage, infection, or bleeding. Cyst drainage may also temporarily lower blood pressure. Sometimes, one or both kidneys may be removed, a procedure called a nephrectomy, if pain is severe.

A low-protein diet may reduce stress on the kidney. Avoiding salt can help maintain normal blood pressure, and drinking lots of water can help reduce the risk of kidney stones.

More than half of PKD patients develop kidney failure and require dialysis. Dialysis is used to remove wastes from the blood, since the kidneys cannot. At this stage, dialysis will be a lifelong requirement unless a kidney transplant from a donor can be arranged and performed successfully.

Recent research has led to the development of several drugs that may prevent cysts from developing. An example is somatostatin, which has been studied in humans and may one day be available to prevent polycystic kidneys from developing.

Prevention and Outcomes

PKD is an inherited disease and is not preventable. Individuals who have a family history of PKD may want to talk to their doctors about genetic testing.

Patricia Griffin Kellicker, B.S.N.;
reviewed by Adrienne Carmack, M.D.
"Etiology and Genetics" by Jeffrey A. Knight, Ph.D.

Further Reading

Chang, M. Y., and A. C. Ong. "Autosomal Dominant Polycystic Kidney Disease: Recent Advances in Pathogenesis and Treatment." *Nephron: Physiology* 108, no. 1 (2008): 1-7.

EBSCO Publishing. *DynaMed: Autosomal Dominant Polycystic Kidney Disease.* Ipswich, Mass.: Author, 2009. Available through http://www.ebscohost.com/dynamed.

_______. *Health Library: Polycystic Kidney Disease.* Ipswich, Mass.: Author, 2009. Available through http://www.ebscohost.com.

McPhee, Stephen J., and Maxine A. Papadakis., eds. *Current Medical Diagnosis and Treatment 2009.* 48th ed. New York: McGraw-Hill Medical, 2008.

Web Sites of Interest

American Academy of Family Physicians: Polycystic Kidney Disease
http://familydoctor.org/online/famdocen/home/common/kidney/142.html

American Urological Association: Renal Dysplasia and Cystic Disease
http://www.urologyhealth.org/auafhome.asp

Genetics Home Reference
http://ghr.nlm.nih.gov

HealthLink B. C. (British Columbia)
http://www.healthlinkbc.ca/kbaltindex.asp

The Kidney Foundation of Canada
http://www.kidney.ab.ca

MedLine Plus: Polycystic Kidney Disease
http://www.nlm.nih.gov/medlineplus/ency/article/000502.htm

PKD Foundation
http://www.pkdcure.org

See also: Alport syndrome; Bartter syndrome; Hereditary leiomyomatosis and renal cell cancer; Hereditary non-VHL clear cell renal cell carcinomas; Hereditary papillary renal cancer.

Polydactyly

CATEGORY: Diseases and syndromes
ALSO KNOWN AS: Polydactylia; polydactylism; hyperdactyly; extra digits; supernumerary digits

Definition

Polydactyly is a relatively common congenital deformity that occurs due to errors during fetal development and results in the presence of one or more extra digits on the hand and/or foot. It occurs as follows: preaxial duplication—thumb side of hand/hallux (big toe) side of foot; central duplication—middle digit area of hand/foot; postaxial duplication—little digit side of hand/foot.

Extra digits are usually smaller in size and abnormal in appearance. However, they range from barely noticeable and consisting solely of soft tissue to fully developed and possibly functional.

Risk Factors

Polydactyly can occur spontaneously (familial polydactyly) or in conjunction with a number of genetic disorders. Asphyxiating thoracic dystrophy (Jeune syndrome) is a bone growth disorder. Carpenter syndrome is an acrocephalopolysyndactyly (ACPS) disorder that affects bone growth. Ellis-van Creveld syndrome (chondroectodermal dysplasia) is a bone growth disorder involving growth hormone derficiency; the highest incidence rate of this syndrome occurs in Old Order Amish from Lancaster County, Pennsylvania. Laurence-Moon-Biedl syndrome is a mitochondrial myopathic disorder characterized by mental retardation and possibly short stature.

Rubenstein-Taybi syndrome is a rare disorder characterized by mental retardation and short stature. Smith-Lemli-Opitz syndrome is a disorder characterized by microcephaly (small head circumference), hypotonia (weak muscle tonus), and possible organ malformations; the highest incidence of this syndrome occurs in Caucasians of Central European ancestry. Trisomy 13 (Patau syndrome) is a usually fatal disorder characterized by multiple structural and developmental abnormalities.

Familial polydactyly may occur independent of any other symptoms or disease. The highest incidence of sixth digit inheritance occurs in African Americans. In those rare instances where polydactylism causality cannot be definitively attributed to genetic abnormality, it is hypothesized that womb abnormality or exposure to toxins may be contributing factors.

Etiology and Genetics

Although it can occur independent of genetic factors, polydactyly is most commonly a heritable, autosomal dominant trait involving a single gene that is capable of causing several variations in expression. Therefore, inheritance is not gender linked and does not require that both parents have the trait. It most likely occurs as the result of duplication of a single embryologic bud.

Incidence rate for polydactyly of the hand is approximately 1 per 1,000 births. It is the most common hand anomaly and occurs most frequently as preaxial (thumb) polydactyly in those of Asian ancestry who have the trait and as postaxial (little finger) polydactyly in those of African ancestry who have the trait. It does not usually occur bilaterally.

Incidence rate for polydactyly of the foot is approximately 1 to 2 per 1,000 births. In approximately 50 percent of all cases, it occurs bilaterally although not necessarily symmetrically. In approximately 33 percent of all cases, it occurs in conjunction with polydactyly of the hand.

Incidence of fused extra digits indicates the concomitant occurrence of polydactyly and syndactyly (fusion of digits) and is termed polysyndactyly.

Symptoms

The only symptom of polydactyly is the presence of one or more extra digits. The presence of other symptoms may indicate the possibility of a concomitant genetic disorder.

Screening and Diagnosis

Screening consists of obtaining a comprehensive medical history and performing chromosome studies. Diagnosis is possibly via fetal sonogram and is immediately apparent at birth. Radiographic evaluation may be necessary to determine the extent of possible skeletal involvement and to confirm that there is no underlying deformity. Other tests that may be used to confirm the diagnosis are enzyme tests and metabolic studies.

Treatment and Therapy

Treatment most often involves surgery, the extent of which is dependent upon the degree of bone, ligament, and tendon involvement. In order to reduce the risk of anesthesia yet allow for the maximum potential for remodeling, surgical excision usually occurs when a baby is approximately one year old.

Standard practice for correction of a "floppy digit," one attached only by soft tissue, is application of suture ligature while the baby is in the hospital nursery. This practice is not recommended, however, if the extra digit has metacarpal/metatarsal duplication and/or residual cartilage, due to the risk of future deformity.

Extensive surgical intervention may require subsequent stabilization of the joint area via short-term casting and maximization of function via physical therapy. Additional surgery may be required during childhood to prevent or correct growth deformity.

Prevention and Outcomes

There is no known means of prevention for spontaneously occurring polydactyly. Possible prevention of polydactyly that occurs in conjunction with genetic disorders would require genetic screening and counseling of prospective parents. The prognosis for isolated polydactyly is extremely favorable with surgical excision.

Cynthia L. De Vine

Further Reading

Barnhill, Raymond L., and A. Neil Crowson. *Textbook of Dermatopathology*. 2d ed. New York: McGraw-Hill, 2004. A comprehensive, illustrated reference written for professionals and scholars yet comprehensible for nonprofessionals; includes photomicrographs.

McGlamry, Dalton, et al. *McGlamry's Comprehensive*

Textbook of Foot and Ankle Surgery. 3d ed. Philadelphia: Lippincott, Williams & Wilkins, 2001. A standard, core podiatric textbook written for physicians, researchers, and students.

Tickle, C. "Embryology." In *The Growing Hand: Diagnosis and Management of the Upper Extremity in Children,* edited by A. Gupta, S. P. J. Kay, and L. R. Scheker. London: Mosby, 2000.

WEB SITES OF INTEREST

Children's Craniofacial Association
http://www.ccakids.com/syn.asp#Carpenter_Syndrome

Children's Hospital Boston
http://www.childrenshospital.org/az/Site1073/mainpageS1073P0.html

Genetics Home Reference; US National Library of Medicine
http://ghr.nlm.nih.gov/

March of Dimes Pregnancy & Newborn Health Education Center
http://www.marchofdimes.com/pnhec/4439_4136.asp

MedlinePlus Medical Encyclopedia; U.S. National Library of Medicine and the National Institutes of Health
http://www.nlm.nih.gov/medlineplus/ency/article/003176.htm

Penn State Milton Hershey Medical Center College of Medicine
http://www.hmc.psu.edu/healthinfo/pq/poly.htm

University of Maryland Medical Center
http://www.umm.edu/news/

Wheeless' Textbook of Orthopaedics; Duke University Medical Center's Division of Orthopaedic Surgery
http://www.wheelessonline.com

See also: Apert syndrome; Brachydactyly; Carpenter syndrome; Cleft lip and palate; Congenital defects; Cornelia de Lange syndrome; Cri du chat syndrome; Crouzon syndrome; Down syndrome; Edwards syndrome; Ellis-van Creveld syndrome; Holt-Oram syndrome; Ivemark syndrome; Meacham syndrome; Opitz-Frias syndrome; Patau syndrome; Robert syndrome; Rubinstein-Taybi syndrome.

Polygenic inheritance

CATEGORY: Classical transmission genetics

SIGNIFICANCE: Polygenically inherited traits—characterized by the amount of some attribute that they possess but not by their presence or absence—are central to plant and animal breeding, medicine, and evolutionary biology. Most of the economically important traits in plants and animals—for example, yield and meat production—are polygenic in nature. Quantitative genetic principles are applied to improve such traits.

KEY TERMS

heritability: the proportion of the total observed variation for a trait attributable to heredity or genes

meristic trait: traits that are counted, such as number of trichomes or bristles

quantitative trait loci (QTLs): genomic regions that condition a quantitative trait, generally identified via DNA-based markers

quantitative trait: a trait, such as human height or weight, that shows continuous variation in a population and can be measured; also called a metric trait

threshold traits: characterized by discrete classes at an outer scale but exhibiting continuous variation at an underlying scale; for example, diabetes, schizophrenia, and cancer

DISCOVERY OF POLYGENIC INHERITANCE

Soon after the rediscovery of Gregor Mendel's laws of inheritance in 1900, Herman Nilsson-Ehle, a Swedish geneticist, showed in 1909 how multiple genes with small effects could collectively affect a continuously varying character. He crossed dark, red-grained wheat with white-grained wheat and found the progeny with an intermediate shade of red. Upon crossing the progeny among themselves, he obtained grain colors ranging from dark red to white. He could classify the grains into five groups in a symmetric ratio of 1:4:6:4:1, with the extreme phenotypes being one-sixteenth dark red and one-sixteenth white. This suggested two-gene segregation. For a two-gene ($n = 2$) model, the number and frequency of phenotypic classes ($2n + 1 = 5$) can be determined by expanding the binomial $(a + b)4$, where a represents the number of favorable alleles and b represents the number of nonfavorable alleles.

Subsequently, Nilsson-Ehle crossed a different variety of red-grained wheat with white-grained wheat. He found that one-sixty-fourth of the plants produced dark red kernels and one-sixty-fourth produced white kernels. There were a total of seven phenotypic (color) classes instead of five. The segregation ratio corresponded to three genes: $(a + b)^6 = 1a^6 + 6a^5b^1 + 15a^4b^2 + 20a^3b^3 + 15a^2b^4 + 6a^1b^5 + 1b^6$. Here, a^6 means that one of sixty-four individuals possessed six favorable alleles, $20a^3b^3$ means that twenty of sixty-four individuals had three favorable and three nonfavorable alleles, and b^6 means that one individual had six nonfavorable alleles. An assumption was that each of the alleles had an equal, additive effect. These experiments led to what is known as the multiple-factor hypothesis, or polygenic inheritance (Kenneth Mather coined the terms "polygenes" and "polygenic traits"). Around 1920, Ronald Aylmer Fisher, Sewall Green Wright, and John Burdon Sanderson Haldane developed methods of quantitative analysis of genetic effects.

Polygenic traits are characterized by the amount of some attribute that they possess but not by presence or absence, as is the case with qualitative traits that are controlled by one or two major genes. Environmental factors generally have little or no effect on the expression of a gene or genes controlling a qualitative trait, whereas quantitative traits are highly influenced by the environment and genotype is poorly represented by phenotype. Genes controlling polygenic traits are sometimes called minor genes.

EXAMPLES AND CHARACTERISTICS OF POLYGENIC TRAITS

Quantitative genetics encompasses analyses of traits that exhibit continuous variation caused by polygenes and their interactions among themselves and with environmental factors. Such traits include height, weight, and some genetic defects.

Diabetes and cancer are considered to be threshold traits because all individuals can be classified as affected or unaffected (qualitative). They are also continuous traits because severity varies from nearly undetectable to extremely severe (quantitative). Because it is virtually impossible to determine the exact genotype for such traits, it is difficult to control defects with a polygenic mode of inheritance.

DETECTION OF GENES CONTROLLING POLYGENIC TRAITS

The detection of genes controlling polygenic traits is challenging and complex for the following reasons.

(1) The expression of genes controlling such traits is modified by fluctuations in environmental and/or management factors.
(2) A quantitative trait is usually a composite of many other traits, each influenced by many genes with variable effects.
(3) Effects of allele substitution are small because many genes control the trait.
(4) Expression of an individual gene may be modified by the expression of other genes and environment.

Polygenic traits are best analyzed with statistical methods, the simplest of which are estimation of arithmetic mean, standard error, variance, and standard deviation. Two populations can have the same mean, but their distribution may be different. Thus, one needs information on variances for describing the two populations more fully. From variances, effects of genes can be ascertained in the aggregate rather than as individual genes.

The issues in quantitative genetics are not only how many and which genes control a trait but also how much of what is observed (phenotype) is attributable to genes (heritability) and how much to the environment. The concept of heritability in the broad sense is useful for quantitative traits, but heritability itself does not give any clues to the total number of genes involved. If heritability is close to 1.0, the variance for a trait is attributable entirely to genetics, and when it is close to zero, the population's phenotype is due entirely to the variation in the underlying environment. Environmental effects mask or modify genetic effects.

Distribution or frequency of different classes in segregating populations—for example, F_2—may provide an idea about the number of genes, particularly if the gene number is small (say, three to four). Formulas have been devised to estimate the number of genes conditioning a trait, but these estimates are not highly reliable. Genes controlling quantitative traits can be estimated via use of chromosomal translocations or other cytogenetic procedures. The advent of molecular markers, such as restriction fragment length polymorphisms, has made it easier and more reliable to pinpoint the location of genes on chromosomes of a species of interest. With much work in a well-characterized organism, these polygenes can be mapped to chromosomes as quantitative trait loci.

Manjit S. Kang, Ph.D.

Further Reading

Jurmain, Robert, et al. "Heredity and Evolution." In *Introduction to Physical Anthropology.* 12th ed. Florence, Ky.: Wadsworth Cengage Learning, 2009. Textbook includes a discussion of polygenic inheritance.

Kang, Manjit S. *Quantitative Genetics, Genomics, and Plant Breeding.* Wallingford, Oxon, England: CABI, 2002. Provides various methods of studying metric or quantitative traits, especially with DNA-based markers.

Lynch, Michael, and Bruce Walsh. *Genetics and Analysis of Quantitative Traits.* Sunderland, Mass.: Sinauer Associates, 1998. Gives an overview of the history of quantitative genetics and covers evolutionary genetics.

Young, Ian D. "Polygenic Inheritance and Complex Diseases." In *Medical Genetics.* New York: Oxford University Press, 2005. Explains polygenic inheritance and its relationship to multifactorial disorders, such as Alzheimer's disease and schizophrenia.

Web Sites of Interest

Online Biology Book
http://www.emc.maricopa.edu/faculty/farabee/BIOBK/BioBookgeninteract.html#Polygenic%20inheritance

Michael J. Farabee, a professor at the Maricopa Community Colleges, includes a chapter on gene interactions, including information about polygenic inheritance, in his online book.

Scitable
http://www.nature.com/scitable/topicpage/Polygenic-Inheritance-and-Gene-Mapping-915

Scitable, a library of science-related articles compiled by the Nature Publishing Group, contains the article "Polygenic Inheritance and Gene Mapping," with links to additional information.

The Virtually Biology Course: Polygenic Inheritance
http://staff.jccc.net/pdecell/evolution/polygen.html

Paul Decelles, a professor at Johnson Community College in Overland Park, Kansas, has included a page about polygenic inheritance in his "virtually biology" course.

See also: Congenital defects; Genetic engineering; Hereditary diseases; Neural tube defects; Pedigree analysis; Plasmids; Quantitative inheritance.

Polymerase chain reaction

CATEGORY: Genetic engineering and biotechnology; Molecular genetics; Techniques and methodologies

SIGNIFICANCE: Polymerase chain reaction (PCR) is the in vitro (in the test tube) amplification of specific nucleic acid sequences. In a few hours, a single piece of DNA can be copied one billion times. Because this technique is simple, rapid, and very sensitive, it is used in a very wide range of applications, including forensics, disease diagnosis, molecular genetics, and nucleic acid sequencing.

Key terms

DNA polymerase: an enzyme that copies or replicates DNA; it uses a single-stranded DNA as a template for synthesis of a complementary new strand and requires an RNA primer or a small section of double-stranded DNA to initiate synthesis

molecular cloning: the process of splicing a piece of DNA into a plasmid, virus, or phage vector to obtain many identical copies of that DNA

The Development of the Polymerase Chain Reaction

The polymerase chain reaction (PCR) was developed by Kary B. Mullis in the mid-1980's. The technique revolutionized molecular genetics and the study of genes. One of the difficulties in studying genes is that a specific gene can be one of approximately twenty-one thousand genes in a complex genome. To obtain the number of copies of a specific gene needed for accurate analysis required the time-consuming techniques of molecular cloning and detection of specific DNA sequences. The polymerase chain reaction changed the science of molecular genetics by allowing huge numbers of copies of a specific DNA sequence to be produced without the use of molecular cloning. The tremendous significance of this discovery was recognized by the awarding of the 1993 Nobel Prize in Chemistry to Mullis for the invention of the PCR method. (The 1993 prize was also awarded to Michael Smith, for work on oligonucleotide-based, site-directed mutagenesis and its development for protein studies.)

HOW POLYMERASE CHAIN REACTION WORKS

PCR begins with the creation of a single-stranded DNA template to be copied. This is done by heating double-stranded DNA to temperatures near boiling (about 94 to 99 degrees Celsius, or about 210 degrees Fahrenheit). This is followed by the annealing (binding of a complementary sequence) of pairs of oligonucleotides (short nucleic acid molecules about ten to twenty nucleotides long) called primers. Because DNA polymerase requires a double-stranded region to prime (initiate) DNA synthesis, the starting point for DNA synthesis is specified by the location at which the primer anneals to the template. The primers are chosen to flank the DNA to be amplified. This annealing is done at a lower temperature (about 30-65 degrees Celsius, or about 86-149 degrees Fahrenheit). The final step is the synthesis by DNA polymerase of a new strand of DNA complementary to the template starting from the primers. This step is carried out at temperatures about 65-75 degrees Celsius (149-167 degrees Fahrenheit). These three steps are repeated many times (for many cycles) to amplify the template DNA. The time for each of the three steps is typically one to two minutes. If, in each cycle, one copy is made of

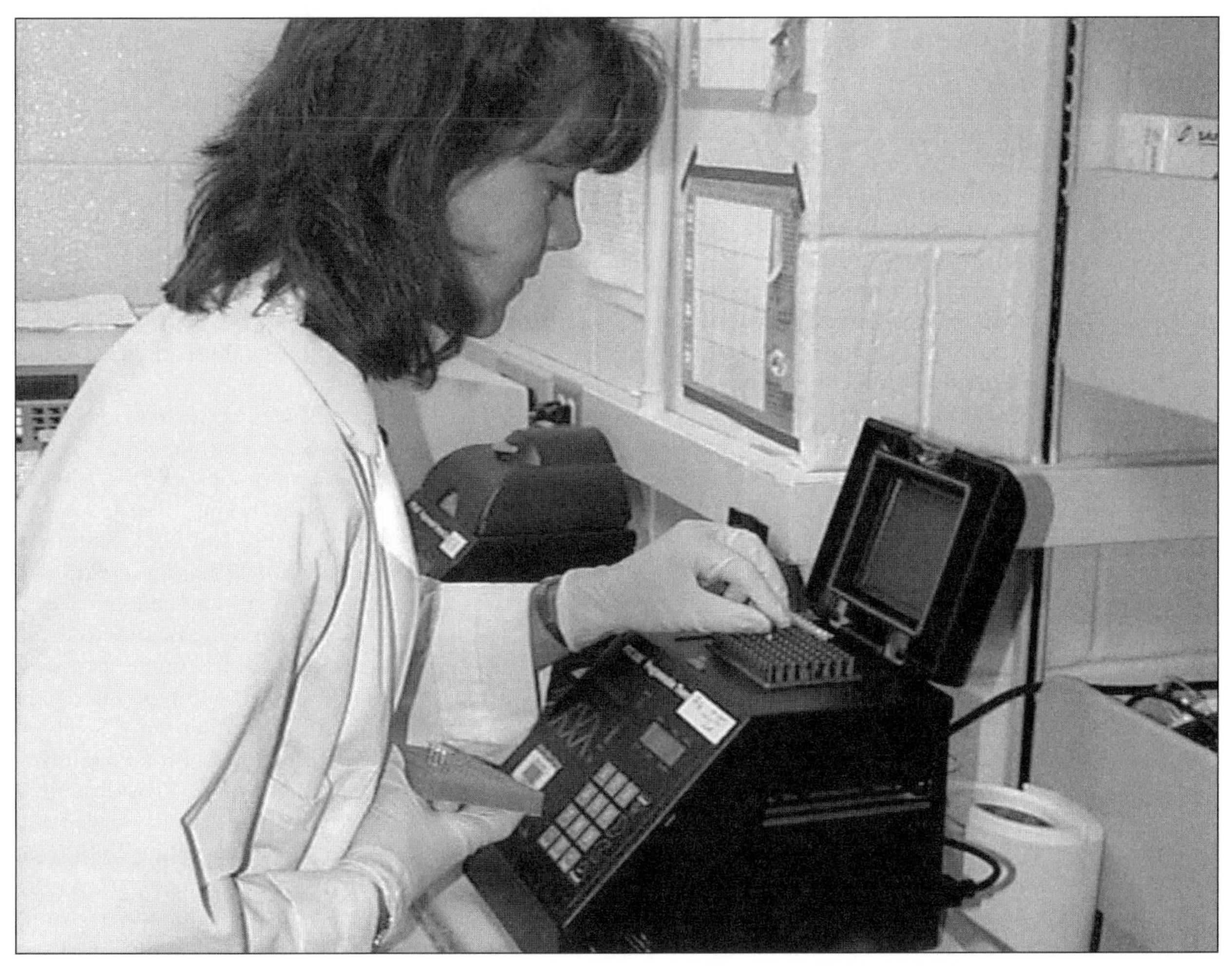

A technician performs polymerase chain reaction testing of anthrax samples. (AP/Wide World Photos)

each of the strands of the template, the number of DNA molecules produced doubles each cycle. Because of this doubling, more than one million copies of the template DNA are made at the end of twenty cycles.

The PCR reaction is made more efficient by the use of heat-stable DNA polymerases, isolated from bacteria that live at very high temperatures in hot springs or deep-sea vents, and by the use of a programmable water bath (called a thermal cycler) to change the temperatures of samples quickly to each of the temperatures needed in each of the steps of a cycle.

Impact and Applications

PCR is extremely rapid. One billion copies of a specific DNA can be made in a few hours. It is also extremely sensitive. It is possible to copy a single DNA molecule. Great care must be taken to avoid contamination, however, for even trace contaminants can readily be amplified by this method.

PCR is a useful tool for many different applications. It is used in basic research to obtain DNA for sequencing and other analyses. PCR is used in disease diagnosis, in prenatal diagnosis, and to match donor and recipient tissues for organ transplants. Because a specific sequence can be amplified greatly, much less clinical material is needed to make a diagnosis. The assay is also rapid, so results are available sooner. PCR is used to detect pathogens, such as the causative agents for Lyme disease or for acquired immunodeficiency syndrome (AIDS), that are difficult to culture. PCR can even be used to amplify DNA from ancient sources such as mummies, bones, and other museum specimens. PCR is an important tool in forensic investigations. Target DNA from trace amounts of biological material such as semen, blood, and hair roots can be amplified. There are probes for regions of human DNA that show hypervariability in the population and therefore make good markers to identify the source of the DNA. PCR can therefore be used to evaluate evidence at the scene of a crime, help identify missing people, and resolve paternity cases.

Susan J. Karcher, Ph.D.

Further Reading

Budowle, Bruce, et al. *DNA Typing Protocols: Molecular Biology and Forensic Analysis.* Natick, Mass.: Eaton, 2000. Discussion includes DNA extraction and PCR-based analyses. Illustrations, bibliography, index.

Chen, Bing-Yuan, and Harry W. Janes, eds. *PCR Cloning Protocols.* Rev. 2d ed. Totowa, N.J.: Humana Press, 2002. Presents helpful introductory chapters with each section and guidelines for PCR cloning. Illustrations, bibliographies, index.

Dorak, M. Tevfik, ed. *Real-Time PCR.* New York: Taylor and Francis, 2006. Focuses on the practical aspects of PCR techniques, emphasizing how these methods can be used in the laboratory.

Guyer, Ruth L., and Daniel E. Koshland, Jr. "The Molecule of the Year." *Science* 246, no. 4937 (December 22, 1989): 1543-1546. Reviews the "major scientific development of the year," the polymerase chain reaction, noting that the technique, although introduced earlier, "truly burgeoned" in 1989.

Innis, Michael A., David H. Gelfand, and John J. Sninsky, eds. *PCR Applications: Protocols for Functional Genomics.* San Diego: Academic Press, 1999. Discusses gene discovery, genomics, and DNA array technology. Includes entries on nomenclature, expression, sequence analysis, structure and function, electrophysiology, and information retrieval. Illustrations, bibliography, index.

Kochanowski, Bernd, and Udo Reischl, eds. *Quantitative PCR Protocols.* Totowa, N.J.: Humana Press, 1999. Outlines protocols and includes methodological and process notes. Illustrations, bibliography, index.

Lloyd, Ricardo V., ed. *Morphology Methods: Cell and Molecular Biology Techniques.* Totowa, N.J.: Humana Press, 2001. Includes an overview of PCR. Illustrations, bibliography, index.

Logan, Julie, Kirstin Edwards, and Nick Saunders, eds. *Real-Time PCR: Current Technology and Applications.* Norfolk, England: Caister Academic Press, 2009. Discusses PCR technologies and tools, as well as applications for PCR in gene expression, mutation detection, diagnosis of fungal infections, and determination of food authenticity.

McPherson, M. J., and S. G. Møller. *PCR.* 2d ed. New York: Taylor and Francis, 2006. Provides introductory information about PCR theory, background, and protocols. Illustrations, bibliography, index.

Mullis, Kary B. "The Unusual Origin of the Polymerase Chain Reaction." *Scientific American* 262, no. 4 (April, 1990): 56. Nobel laureate Mullis describes the initial development of the technique for the general audience.

Watson, James D., et al. *Recombinant DNA—Genes and Genomes: A Short Course.* 3d ed. New York: W. H. Freeman, 2007. Summarizes polymerase chain reaction and its applications. Full-color illustrations, diagrams, bibliography, index.

WEB SITES OF INTEREST

Dolan DNA Learning Center, Biology Animation Library
http://www.dnalc.org/resources/animations/pcr.html

Viewers can watch an animated demonstration of PCR.

Library, University of California, Berkeley, PCR Project
http://sunsite3.berkeley.edu/PCR

An introductory overview to PCR prepared by professors and librarians at the university, including Paul Rainbow, an anthropology professor who wrote the book *Making PCR: A Story of Biotechnology* (1996). Provides access to several papers discussing the fundamentals of PCR, applications of PCR basics, and technical variations in basic PCR methods.

University of Utah, Genetic Science Learning Center, PCR Virtual Lab
http://learn.genetics.utah.edu/content/labs/pcr

The virtual lab provides a demonstration of PCR.

See also: Ancient DNA; Anthrax; Bioinformatics; Blotting: Southern, Northern, and Western; Central dogma of molecular biology; Cloning vectors; DNA fingerprinting; DNA sequencing technology; Forensic genetics; Genetic engineering: Historical development; Human Genome Project; In vitro fertilization and embryo transfer; Mitochondrial diseases; Molecular genetics; Paternity tests; Repetitive DNA; Reverse transcription polymerase chain reaction (RT-PCR); RFLP analysis; RNA isolation.

Polyploidy

CATEGORY: Population genetics

SIGNIFICANCE: Polyploids have three or more complete sets of chromosomes in their nuclei instead of the two sets found in diploids. Polyploids are especially common in plants, with some examples also existing in animals, and have a prominent role in the evolution of species. Some tissues of diploid organisms are polyploid, while the remaining cells in the organism are diploid.

KEY TERMS

allopolyploid: a type of polyploid species that contains genomes from more than one ancestral species

aneuploid: a cell or an organism with one or more missing or extra chromosomes; the opposite is "euploid," a cell with the normal chromosome number

autopolyploid: a type of polyploid species that contains more than two sets of chromosomes from the same species

homologous chromosomes: chromosomes that are structurally the same and have the same gene loci, although they may have different alleles (alternative forms of a gene) at many of their shared loci

THE FORMATION OF POLYPLOIDY

Most animals are diploid, meaning that they have two homologous sets of chromosomes in their cells; and their gametes (eggs and sperm) are haploid, that is, having one set of chromosomes. Plants, a variety of single-celled eukaryotes, and some insects have individual or parts of an individual's life cycle when they are haploid. In any case, when there are more than two sets of homologous chromosomes, the cell or organism is considered polyploid. A triploid organism has three sets of homologous chromosomes, a tetraploid has four sets, a dodecaploid has twelve sets, and there are organisms known to have many more than a dozen sets of homologous chromosomes.

How polyploids are formed in nature is still debated. Regardless of what theory is accepted, the first step certainly involves a failure during cell division, in either meiosis or mitosis. For example, if cytokinesis (division of the cytoplasm) fails at the conclusion of meiosis II, the daughter cells will be diploid. If, by chance, a diploid sperm fertilizes a diploid egg, the resulting zygote will be tetraploid. Although polyploidy might occur this way, biologists have proposed an alternative model involving a triploid intermediate stage.

The triploid intermediate model has been applied primarily to plants, in which polyploidy is better studied. Hybrids between two species are often sterile, but occasionally a diploid gamete from one of the species joins with a normal haploid gamete from the other species, which produces a triploid hybrid. Triploids are also sterile, for the most part, but do produce a small number of gametes, many of which are diploid. This makes the probability that two diploid gametes will join, to form a tetraploid, much higher. This hypothesis is supported by the discovery of triploid hybrid plants that do produce a small number of viable gametes. This type of polyploid, formed as a result of hybridization between two species, is called an allopolyploid. Allopolyploids are typically fertile and represent a new species.

Polyploidy can also occur within a single species, without hybridization, in which case it is called an autopolyploid. Autopolyploids can form in the same way as allopolyploids, but they can also occur as the result of a failure in cell division in a bud. If a cell in the meristematic region (a rapidly dividing group of cells at the tip of a bud) completes mitosis but not cytokinesis, it will be a tetraploid cell. All daughter cells from this cell will also be tetraploid, so that any flowers borne on this branch will produce diploid gametes. If the plant is self-compatible, it can then produce tetraploid offspring from these flowers. Autopolyploids are often a little larger and more robust than the diploids that produce them, but they are often so similar they cannot be easily distinguished. An autopolyploid, when formed, represents a new species but is not generally recognized as such unless it looks different enough physically from diploids.

Wheat is one of many important polyploid crops. (©Tammy Mcallister/Dreamstime.com)

The Genetics of Polyploids

A polyploid has more copies of each gene than a diploid. For example, a tetraploid has four alleles at each locus, which means tetraploids can contain much more individual variability than diploids. This has led some evolutionists to suggest that polyploids should have higher fitness than the diploids from which they came. With more variation, the individual would be preadapted to a much wider range of conditions. Because there are so many extra copies of genes, a certain amount of gene silencing (loss of genes through mutation or other processes) occurs, with no apparent detriment to the plant.

The pairing behavior of chromosomes in polyploids is also unique. In a diploid, during meiosis, homologous chromosomes associate in pairs. In an autotetraploid there are four homologous chromosomes of each type which associate together in groups of four. In an allotetraploid, the chromosomes from the two species from which they are derived are commonly not completely homologous and do not associate together. Consequently, the pairs of homologous chromosomes from one parent species associate together in pairs, as do the chromosomes from the other parent species. For this reason, sometimes allopolyploids are referred to as amphidiploids, because their pairing behavior looks the same as it does in a diploid. This is also why an allopolyploid is fertile (because meiosis occurs normally), but a hybrid between two diploids commonly is not, because the chromosomes from the two species are unable to pair properly.

POLYPLOID PLANTS AND ANIMALS

In the plant kingdom, it is estimated by some that 95 percent of pteridophytes (plants, including ferns, that reproduce by spores) and perhaps as many as 80 percent of angiosperms (flowering plants that form seeds inside an ovary) are polyploid, although there is high variability in its occurrence among families of angiosperms. In contrast, polyploidy is uncommon in gymnosperms (plants that have naked seeds that are not within specialized structures). Extensive polyploidy is observed in chrysanthemums, in which chromosome numbers range from 18 to 198. The basic chromosome number (haploid or gamete number of chromosomes) is 9. Polyploids from triploids (with 27 chromosomes) to 22-ploids (198 chromosomes) are observed. The stonecrop *Sedum suaveolens*, which has the highest chromosome number of any angiosperm, is believed to be about 80-ploid (720 chromosomes). Many important agricultural crops, including wheat, corn, sugarcane, potatoes, coffee, apples, and cotton, are polyploid.

Polyploid animals are less common than polyploid plants but are found among some groups, including crustaceans, earthworms, flatworms, and insects such as weevils, sawflies, and moths. Polyploidy has also been observed in some vertebrates, including tree frogs, lizards, salamanders, and fish. It has been suggested that the genetic redundancy observed in vertebrates may be caused by ancestral polyploidy.

POLYPLOIDY IN TISSUES

Most plants and animals contain particular tissues that are polyploid or polytene, while the rest of the organism is diploid. Polyploidy is observed in multinucleate cells and in cells that have undergone endomitosis, in which the chromosomes condense but the cell does not undergo nuclear or cellular division. For example, in vertebrates, liver cells are binucleate and therefore tetraploid. In addition, in humans, megakaryocytes can have polyploidy levels of up to sixty-four. A megakaryocyte is a giant bone-marrow cell with a large, irregularly lobed nucleus that is the precursor to blood platelets. A megakaryocyte does not circulate, but forms platelets by budding. A single megakaryocyte can produce three thousand to four thousand platelets. A platelet is an enucleated, disk-shaped cell in the blood that has a role in blood coagulation. In polytene cells, the replicated copies of the chromosomal DNA remain associated to produce giant chromosomes that have a continuously visible banding pattern. The trophoblast cells of the mammalian placenta are polytene.

IMPORTANCE OF POLYPLOIDS TO HUMANS

Most human polyploids die as embryos or fetuses. In a few rare cases, a polyploid infant is born that lives for a few days. In fact, polyploidy is not tolerated in most animal systems. Plants, on the other hand, show none of these problems with polyploidy. Some crop plants are much more productive because they are polyploid. For example, wheat (*Triticum aestivum*) is an allohexaploid and contains chromosome sets that are derived from three different ancient types. Compared to the species from which it evolved, *T. aestivum* is far more productive and produces larger grains of wheat. *Triticum aestivum* was not developed by humans but appears to have arisen by a series of chance events in the past, humans simply recognizing the better qualities of *T. aestivum*. Another fortuitous example involves three species of mustard that have given rise to black mustard, turnips, cabbage, broccoli, and several other related crops, all of which are allotetraploids.

Polyploids may be induced by the use of drugs such as colchicine, which halts cell division. Because of the advantages of the natural polyploids used in agriculture, many geneticists have experimented with artificially producing polyploids to improve crop yields. One prime example of this approach is *Triticale*, which represents an allopolyploid produced by hybridizing wheat and rye. Producing artificial polyploids often produces a new variety that has unexpected negative characteristics, so that only a few such polyploids have been successful. Nevertheless, research on polyploidy continues.

Susan J. Karcher, Ph.D.; updated by Bryan Ness, Ph.D.

FURTHER READING

Adams, Keith L., et al. "Genes Duplicated by Polyploidy Show Unequal Contributions to the Transcriptome and Organ-Specific Reciprocal Silencing." *Proceedings of the National Academy of Sciences* 100 (April 15, 2003): 4649-4654. This article shows that with multiple copies of a gene due to polypoidy, some of the copies are silenced.

Gregory, T. Ryan, ed. *The Evolution of the Genome.* Burlington, Mass.: Elsevier Academic, 2005. Two chapters focus on polyploidy, one dealing with polyploidy in plants, the other with polyploidy in animals.

Hunter, Kimberley L., et al. "Investigating Polyploidy: Using Marigold Stomates and Fingernail Polish." *American Biology Teacher* 64, no. 5 (May, 2002): 364. A guide to exploring polyploidy through hands-on learning. Experiment supports National Science Education Standards.

Leitch, Illia J., and Michael D. Bennett. "Polyploidy in Angiosperms." *Trends in Plant Science* 2, no. 12 (December, 1997): 470-476. Describes the role of polyploidy in the evolution of higher plants.

Lewis, Ricki. *Human Genetics: Concepts and Applications.* 9th ed. Dubuque, Iowa: McGraw-Hill, 2009. Gives an overview of polyploidy and aneuploidy in humans. Color ilustrations, and maps.

Miller, Orlando J., and Eeva Therman. *Human Chromosomes.* 4th ed. New York: Springer, 2001. A textbook about the function and dysfunction of human chromosomes, with information about nondisjunction in meiosis and gametes and the origin of diploid gametes.

Sumner, Adrian T. *Chromosomes: Organization and Function.* Malden, Mass.: Blackwell, 2003. Discusses the origins of polyploidy and polyploidy and evolution, gene expression, and disease.

WEB SITES OF INTEREST

Kimball's Biology Pages
http://users.rcn.com/jkimball.ma.ultranet/BiologyPages/P/Polyploidy.html

John Kimball, a retired Harvard University biology professor, includes a page about polyploidy in plants and animals and polyploidy in speciation in his online cell biology text.

The Polyploidy Portal
http://www.polyploidy.org/index.php/Main_Page

This site calls itself a "Web entry point to information about polyploidy," and it contains both basic and advanced information about polyploidy, descriptions of polyploidy-related research projects, educational activities, and a bibliography.

See also: Cell division; Cytokinesis; Gene families; Genome size; Hereditary diseases; High-yield crops; Nondisjunction and aneuploidy.

Pompe disease

CATEGORY: Diseases and syndromes
ALSO KNOWN AS: Lysosomal storage disease; acid maltase deficiency disease; glycogen storage disease type II

DEFINITION

Pompe disease is a metabolic disorder caused by mutations in the acid alpha-glucosidase (*GAA*) gene. GAA, an enzyme responsible for breaking down glycogen in the cells, is either absent (resulting in the rapidly progressive infantile form of Pompe disease) or deficient (resulting in the late-onset juvenile or adult form). As a result, glycogen builds up in the lysosomes of cells and tissues, primarily in cardiac and skeletal muscles, affecting their function and causing progressive weakness and organ failure.

RISK FACTORS

This is an autosomal recessive disorder; therefore, each parent must carry a defective *GAA* gene, both of which are inherited by the affected child. The incidence is estimated at 1 in 40,000 people worldwide. About one-third of patients have the infantile-onset form. Both sexes are equally affected, although the incidence does vary by geography and ethnic group.

ETIOLOGY AND GENETICS

The *GAA* gene, located on the long arm of chromosome 17, is the only gene associated with Pompe disease. More than two hundred mutations have been identified throughout the gene. Some defects are more common than others. For example, more than half of Caucasians with late-onset Pompe disease share a common splice-site mutation. Some infantile-onset mutations are observed more frequently in certain geographic (such as the Netherlands) or ethnic (such as African Americans and persons of Chinese descent) populations.

In general, the type and combination of mutations inherited determine the residual level of GAA activity and thus the severity of the disease. If both chromosomes are fully compromised, GAA activity is nonexistent. Combinations of one severely mutated allele and one mildly affected allele usually preserve some GAA activity, meaning a slower disease progression, although the age of onset can vary. Researchers are cautious about correlating genotype with clinical features, however, because both infantile and late-onset forms have been observed in the same family.

Several factors explain how glycogen buildup in the lysosomes likely disrupts muscle function. As the lysosomes become bloated, they can displace myofibrils in neighboring cells, disrupting the muscle's ability to contract and transmit force. In late-onset Pompe disease, swollen lysosomes can rupture or release other enzymes into surrounding tissues, damaging muscles. Disuse and oxidative stress may also play a role in muscle wasting.

Symptoms

Manifestations of Pompe disease vary depending on age of onset and level of residual GAA activity. In the classic infantile-onset form, symptoms are observed shortly after birth and include an enlarged heart, poor muscle tone (inability to hold the head up, roll over), feeding problems (difficulty swallowing, enlarged tongue), and respiratory distress (frequent lung infections). In the nonclassic infantile form, cardiac involvement is moderate and muscle weakness is delayed. In late-onset Pompe disease, symptoms can appear from two to seventy years. Muscle weakness and pain, primarily in the legs and trunk (difficulty climbing stairs or playing sports, frequent falls) and respiratory distress (shortness of breath, sleep apnea) are typical. In all cases, early diagnosis is critical for disease management.

Screening and Diagnosis

Pompe disease shares many symptoms with other muscle disorders, complicating diagnosis. Initial clinical studies include chest radiography and electrocardiograms, as well as muscle tests, electromyography, and nerve conduction tests in adults. The diagnosis is confirmed through tests of GAA activity (blood tests, skin fibroblasts cultures, and/or muscle biopsy in adults) or through DNA analysis. DNA analysis is also useful for identifying familial mutations and carriers and for newborn screening.

Treatment and Therapy

Historically, patients with Pompe disease were given supportive care only. Today, enzyme replacement therapy (ERT) using recombinant human GAA is a promising treatment, especially in infants younger than six months who do not yet require ventilatory assistance. Clinical trials of ERT in late-onset Pompe disease are ongoing. Other treatment is multidisciplinary and aimed at preventing secondary complications such as infections, treating symptoms, and maintaining function as long as possible. These treatments include frequent cardiac evaluations, use of bronchodilators, steroids, and mechanical ventilation, and special diets and tube feeding. Physical, occupational, and speech therapies and current immunizations are also advised.

Prevention and Outcomes

Before ERT, patients with infantile-onset Pompe disease typically died of cardiac and/or respiratory complications by one year of age. ERT has enhanced ventilator-free survival for many young patients; reduced heart size and improvements in cardiac and skeletal muscle function have also been seen. In late-onset Pompe disease, juvenile patients are usually more severely affected than adults and rarely survive past the second or third decade of life due to respiratory failure. They often require mechanical ventilation and wheelchairs. Older patients may also experience steadily progressive debilitation and premature mortality. However, the outlook is improving. Newer screening techniques have aided in early diagnosis, ERT has shown promise for treatment, and gene therapies and pharmacologic chaperones are being aggressively investigated.

Judy Majewski

Further Reading

ACMG Work Group on Management of Pompe Disease. "Pompe Disease Diagnosis and Management Guideline." *Genetics in Medicine* 8, no. 5 (2006): 267-288. A discussion of best practices for diagnosis and treatment.

Anand, G. *The Cure: How a Father Raised $100 Million—And Bucked the Medical Establishment—In a Quest to Save His Children.* New York: William Morrow, 2006. A firsthand account of nonclassic infantile-onset Pompe disease.

Hirschhorn, R., and A. J. Reuser. "Glycogen Storage Disease Type II: Acid Alpha-glucosidase (Acid Maltase) Deficiency." In *The Metabolic and Molecular Bases of Inherited Disease,* edited by Charles Scriver et al. 8th ed. New York: McGraw-Hill, 2001. A comprehensive scientific discussion.

WEB SITES OF INTEREST

Acid Maltase Deficiency Association (AMDA)
http://www.amda-pompe.org/

Association for Glycogen Storage Disease
http://www.agsdus.org/html/typeiipompe.htm

Pompe Disease
http://ghr.nlm.nih.gov/condition=pompedisease

Pompe Registry
http://www.lsdregistry.net/pomperegistry/

See also: Andersen's disease; Forbes disease; Galactokinase deficiency; Galactosemia; Gaucher disease; Glucose galactose malabsorption; Glucose-6-phosphate dehydrogenase deficiency; Glycogen storage diseases; Hereditary diseases; Hers disease; Inborn errors of metabolism; McArdle's disease; Tarui's disease.

Population genetics

CATEGORY: Population genetics

SIGNIFICANCE: Population genetics is the study of how genes behave in populations. It is concerned with both theoretical and experimental investigations of changes in genetic variation caused by various forces; therefore, the field has close ties to evolutionary biology. Population genetics models can be used to explore the evolutionary histories of species, make predictions about future evolution, and predict the behavior of genetic diseases in human populations.

KEY TERMS

allele: one of the different forms of a particular gene (locus)

fitness: a measure of the ability of a genotype or individual to survive and reproduce compared to other genotypes or individuals

gene pool: all of the alleles in all the gametes of all the individuals in a population

genetic drift: random changes in genetic variation caused by sampling error in small populations

genotype: the pair of alleles carried by an individual for a specific gene locus

Hardy-Weinberg law: a mathematical model that predicts, under particular conditions, that allele frequencies will remain constant over time, with genotypes in specific predictable proportions

modern synthesis: the merging of the Darwinian mechanisms for evolution with Mendelian genetics to form the modern fields of population genetics and evolutionary biology

neutral theory of evolution: Motoo Kimura's theory that nucleotide substitutions in the DNA often have no effect on fitness, and thus changes in allele frequencies in populations are caused primarily by genetic drift

THE HARDY-WEINBERG LAW

The branch of genetics called population genetics is based on the application of nineteenth century Austrian botanist Gregor Mendel's principles of inheritance to genes in a population. (Although, for some species, "population" can be difficult to define, the term generally refers to a geographic group of interbreeding individuals of the same species.) Mendel's principles can be used to predict the expected proportions of offspring in a cross between two individuals of known genotypes, where the genotype describes the genetic content of an individual for one or more genes. An individual carries two copies of all chromosomes (except perhaps for the sex chromosomes, as in human males) and therefore has two copies of each gene. These two copies may be identical or somewhat different. Different forms of the same gene are called alleles. A genotype in which both alleles are the same is called a homozygote, while one in which the two alleles are different is a heterozygote. Although a single individual can carry no more than two alleles for a particular gene, there may be many alleles of a gene present in a population.

It would be essentially impossible to track the inheritance patterns of every single mating pair in a population, in essence tracking all the alleles in the gene pool. However, by making some simplifying assumptions about a population, it is possible to predict what will happen to the gene pool over time. Working independently in 1908, the British mathematician Godfrey Hardy and the German physiologist Wilhelm Weinberg were the first to formulate a simple mathematical model describing the behavior of a gene (locus) with two alleles in a population. In this model, the numbers of each allele and of each genotype are not represented as actual numbers but as proportions (known as allele frequencies and genotype frequencies, respectively) so that the model can be applied to any population regardless of its size. By assuming Mendelian inheritance of alleles, Hardy and Weinberg showed that allele frequencies in a population do not change over time and that genotype frequencies will change to specific proportions, determined by the allele frequencies, within one generation and remain at those proportions in future generations. This result is known as the Hardy-Weinberg law, and the stable genotype proportions predicted by the law are known as Hardy-Weinberg equilibrium. It was shown in subsequent work by others that the Hardy-Weinberg law remains true in more complex models with more than two alleles and more than one locus.

In order for the Hardy-Weinberg law to work, certain assumptions about a population must be true:

(1) The gene pool must be infinite in size.
(2) Mating among individuals (or the fusion of gametes) must be completely random.
(3) There must be no new mutations.
(4) There must be no gene flow (that is, no alleles should enter or leave the population.
(5) There should be no natural selection.

Since real populations cannot meet these conditions, it may seem that the Hardy-Weinberg model is too unrealistic to be useful, but, in fact, it can be useful. First, the conditions of a natural population may be very close to Hardy-Weinberg assumptions, so the Hardy-Weinberg law may be approximately true for at least some populations. Second, if genotypes in a population are not in Hardy-Weinberg equilibrium, it is an indication that one or more of these assumptions is not met. The Hardy-Weinberg law has been broadly expanded, using sophisticated mathematical modeling, and with adequate data can be used to determine why a population's allele and genotype frequencies are out of Hardy-Weinberg equilibrium.

Genetic Variation and Mathematical Modeling

Sampling and genetic analyses of real populations of many different types of organisms reveal that there is usually a substantial amount of genetic variation, meaning that for a fairly large proportion of genes (loci) that are analyzed, there are multiple alleles, and therefore multiple genotypes, within populations. For example, in the common fruit fly *Drosophila melanogaster* (an organism that has been well studied genetically since the very early 1900's), between one-third and two-thirds of the genes that have been examined by protein electrophoresis have been found to be variable. Genetic variation can be measured as allele frequencies (allelic variation) or genotype frequencies (genotypic variation). A major task of population geneticists has been to describe such variation, to try to explain why it exists, and to predict its behavior over time.

The Hardy-Weinberg law predicts that if genetic variation exists in a population, it will remain constant over time, with genotypes in specific proportions. However, the law cannot begin to explain natural variation, since genotypes are not always found in Hardy-Weinberg proportions, and studies that involve sampling populations over time often show that genetic variation can be changing. The historical approach to explaining these observations has been to formulate more complex mathematical models based on the simple Hardy-Weinberg model that violate one or more of the implicit Hardy-Weinberg conditions.

Beginning in the 1920's and 1930's, a group of population geneticists, working independently, began exploring the effects of violating Hardy-Weinberg assumptions on genetic variation in populations. In what has become known as the "modern synthesis," Ronald A. Fisher, J. B. S. Haldane, and Sewall Wright merged Darwin's theory of natural selection with Mendel's theory of genetic inheritance to create a field of population genetics that allows for genetic change. They applied mathematics to the problem of variation in populations and were eventually able to incorporate what happens when each, or combinations, of the Hardy-Weinberg assumptions are violated.

Assortative Mating and Inbreeding

One of the implicit conditions of the Hardy-Weinberg model is that genotypes form mating pairs at random. In most cases mates are not selected based on genotype. Unless the gene in question has some direct effect on mate choice, mating with respect to that gene is random. However, there are conditions in natural populations in which mating is not random. For example, if a gene controls fur color and mates are chosen by appropriate fur color, then the genotype of an individual with respect to that gene will determine mating success. For this gene, then, mating is not random but rather "assortative." Positive assortative mating means that individuals tend to choose mates with genotypes like their own, while negative assortative mating means that individuals tend to choose genotypes different than their own.

Variation in a population for a gene subject to assortative mating is altered from Hardy-Weinberg expectations. Although allele frequencies do not change, genotype frequencies are altered. With positive assortative mating, the result is higher proportions of homozygotes and fewer heterozygotes, while the opposite is true when assortative mating is negative. Sometimes random mating in a population is not possible because of the geographic organization of the population or general mating habits. Truly random mating would mean that any individual can mate with any other, but this is nearly impossible because of gender differences and practical limitations. In natural populations, it is often the case that mates are somewhat related, even closely related, because the population is organized into extended family groups whose members do not (or cannot, as in plants) disperse to mate with members of other groups. Mating between relatives is called inbreeding. Because related individuals tend to have similar genotypes for many genes, the effects of inbreeding are much like those of positive assortative mating for many genes. The proportions of homozygotes for many genes tend to increase. Again, this situation has no effect on allelic variation, only genotypic variation. Clearly, the presence of nonrandom mating patterns cannot by itself explain the majority of patterns of genetic variation in natural populations but can contribute to the action of other forces, such as natural selection.

Migration and Mutation

In the theoretical Hardy-Weinberg population, there are no sources of new genetic variation. In real populations, alleles may enter or leave the population, a process called migration or "gene flow" (a more accurate term, since migration in this context means not only movement between populations but also successful reproduction to introduce alleles in the new population). Also, new alleles may be introduced by mutation, the change in the DNA sequence of an existing allele to create a new one, as a result of errors during DNA replication or the inexact repair of DNA damage from environmental influences such as radiation or mutagenic chemicals. Both of these processes can change both genotype frequencies and allele frequencies in a population. If the tendency to migrate is associated with particular genotypes, a long period of continued migration tends to push genotype and allele frequencies toward higher proportions of one type (in general, more homozygotes) so that the overall effect is to reduce genetic variation. However, in the short term, migration may enhance genetic variation by allowing new alleles and genotypes to enter. The importance of migration depends on the particular population. Some populations may be relatively isolated from others so that migration is a relatively weak force affecting genetic variation, or there may be frequent migration among geographic populations. There are many factors involved, not the least of which is the ability of members of the particular species to move over some distance.

Mutation, because it introduces new alleles into a population, acts to increase genetic variation. Before the modern synthesis, one school of thought was that mutation might be the driving force of evolution, since genetic change over time coming about from continual introduction of new forms of genes seemed possible. In fact, it is possible to develop simple mathematical models of mutation that show resulting patterns of genetic variation that resemble those found in nature. However, to account for the rates of evolution that are commonly observed, very high rates of mutation are required. In general, mutation tends to be quite rare, making the hypothesis of evolution by mutation alone unsatisfactory.

The action of mutation in conjunction with other forces, such as selection, may account for the low-frequency persistence of clearly harmful alleles in populations. For example, one might expect that alleles that can result in genetic diseases (such as cystic fibrosis) would be quickly eliminated from human populations by natural selection. However, low rates of mutation can continually introduce these alleles into populations. In this "mutation-selection balance," mutation tends to introduce alleles while selection tends to eliminate them, with a net result of continuing low frequencies in the population.

Genetic Drift

Real populations are not, of course, infinite in size, though some are large enough that this Hardy-Weinberg condition is a useful approximation.

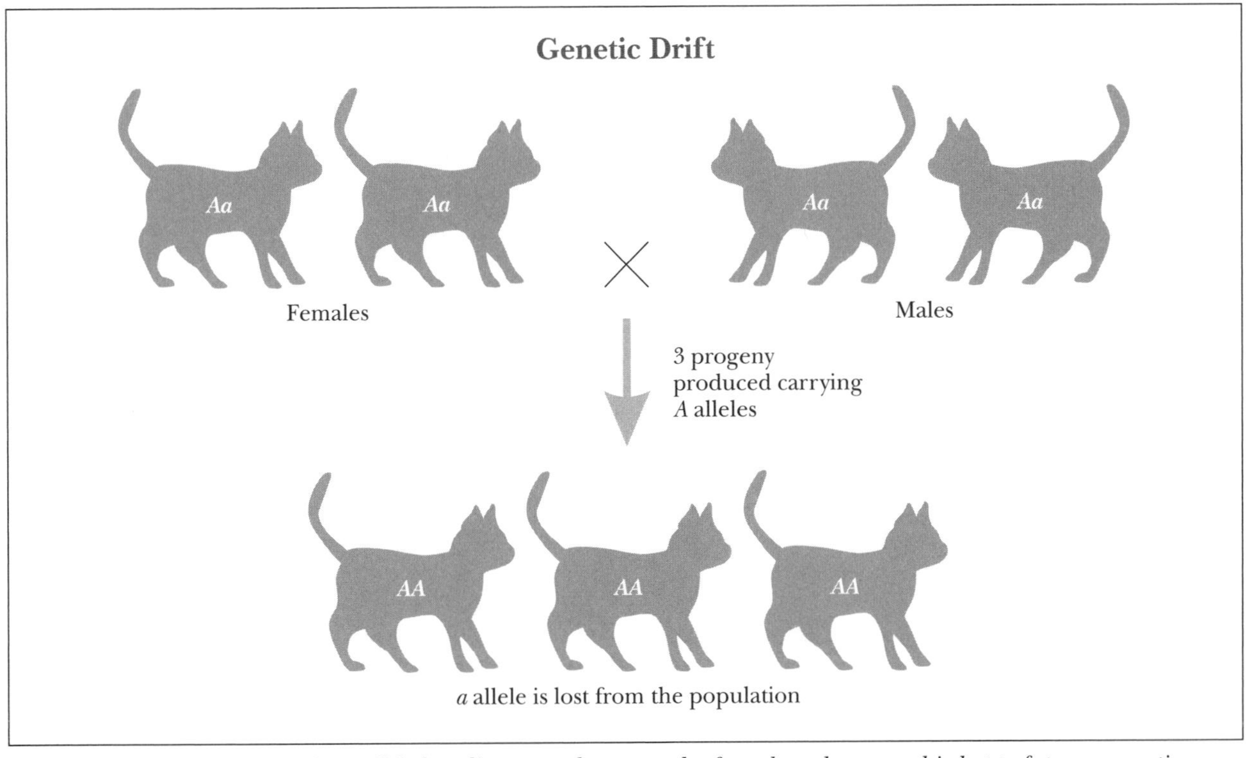

In this small population, the a *allele has disappeared as a result of random chance and is lost to future generations.*

However, many natural populations are small, and any population with less than about one thousand individuals will vary randomly in the pattern of genetic variation from generation to generation. These random changes in allele and genotype frequencies are called genetic drift. The situation is analogous to coin tossing. With a fair coin, the expectation is that half of the tosses will result in heads and half in tails. On average, this will be true, but in practice a small sample will not show the expectation. For example, if a coin is tossed ten times, it is unlikely that the result will be exactly five heads and five tails. On the other hand, with a thousand tosses, the results will be closer to half and half. This higher deviation from the expected result in small samples is called a sampling error. In a small population, there is an expectation of the pattern of genetic variation based on the Hardy-Weinberg law, but sampling error during the union of sex cells to form offspring genotypes will result in random deviations from that expectation. The effect is that allele frequencies increase or decrease randomly, with corresponding changes in genotype frequencies. The smaller the population, the greater the sampling error and the more pronounced genetic drift will be.

Genetic drift has an effect on genetic variation that is similar to that of other factors. Over the long term, allele frequencies will drift until all alleles have been eliminated but one, eliminating variation. (For the moment, ignore the action of other forces that increase variation.) Over a period of dozens of generations, however, drift can allow variation to be maintained, especially in larger populations in which drift is minimal.

In the early days of population genetics, the possibility of genetic drift was recognized but often considered to be a minor consideration, with natural selection as a dominant force. Fisher in particular dismissed the importance of genetic drift, engaging over a number of years in a published debate with Wright, who always felt that drift would be important in small populations. Beginning in the 1960's with the acquisition of data on DNA-level population variation, the role of drift in natural populations became more recognized. It appears to be an especially strong force in cases in which a small number of individuals leave the population and migrate to a new area where they establish a new population. Large changes can occur, especially if the number of migrants is only ten or twenty. This type of situation is now referred to as a founder effect.

Natural Selection

Natural selection in a simple model of a gene with two alleles in a population can be easily represented by assuming that genotypes differ in their ability to survive and produce offspring. This ability is called fitness. In applying natural selection to a theoretical population, each genotype is assigned a fitness value between zero and one. Typically, the genotype in a population that is best able to survive and can, on average, produce more offspring than other genotypes is assigned a fitness value of one, and genotypes with lower fitness are assigned fitnesses with fractional values relative to the high-fitness genotype.

The study of this simple model of natural selection has revealed that it can alter genetic variation in different ways, depending on which genotype has the highest fitness. In the simple one-gene, two-allele model, there are three possible genotypes: two homozygotes and one heterozygote. If one homozygote has the highest fitness, it will be favored, and the genetic composition of the population will gradually shift toward more of that genotype (and its corresponding allele). This is called directional selection. If both homozygotes have higher fitness than the heterozygote (disruptive selection), one or the other will be favored, depending on the starting conditions. Both of these situations will decrease genetic variation in the population, because eventually one allele will prevail. Although each of these types of selection (particularly directional) may be found for genes in natural populations, they cannot explain why genetic variation is present, and is perhaps increasing, in nature.

Heterozygote advantage, in which the heterozygote has higher fitness than either homozygote, is the other possible situation in this model. In this case, because the heterozygote carries both alleles, both are expected to be favored together and therefore maintained. This is the only condition in this simple model in which genetic variation may be maintained or increased over time. Although this seems like a plausible explanation for the observed levels of natural variation, studies in which fitness values are measured almost never show heterozygote advantage in genes from natural populations. As a general explanation for the presence of genetic variation, this simple model of selection is unsatisfactory.

Studies of more complex theoretical models of selection (for example, those with many genes and different forms of selection) have revealed conditions that allow patterns of variation very similar to those observed in natural populations, and in some cases it seems clear that natural selection is a major factor determining patterns of genetic change. However, in many cases, selection does not seem to be the most important factor or even a factor at all.

Experimental Population Genetics and the Neutral Theory

Population genetics has always been a field in which the understanding of theory is ahead of empirical observation and experimental testing, but these have not been neglected. Although Fisher, Haldane, and Wright were mainly theorists, there were other architects of the modern synthesis who concentrated on testing theoretical predictions in natural populations. Beginning in the 1940's, for example, Theodosius Dobzhansky showed in natural and experimental populations of *Drosophila* species that frequency changes and geographic patterns of variation in chromosome variants are consistent with the effects of natural selection.

Natural selection was the dominant hypothesis for genetic changes in natural populations for the first several decades of the modern synthesis. In the 1960's, new techniques of molecular biology allowed population geneticists to examine molecular variation, first in proteins and later, with the use of restriction enzymes in the 1970's and DNA sequencing in the 1980's and 1990's, in DNA sequences. These types of studies only confirmed that there is a large amount of genetic variation in natural populations, much more than can be attributed only to natural selection. As a result, Motoo Kimura proposed the "neutral theory of evolution," the idea that most DNA sequence differences do not have fitness differences and that population changes in DNA sequences are governed mainly by genetic drift, with selection playing a minor role. This view, although still debated by some, was mostly accepted by the 1990's, although it was recognized that evolution of proteins and physical traits may be governed by selection to a greater extent.

Impact and Applications

The field of population genetics is a fundamental part of the current field of evolutionary biology. One possible definition of evolution would be "genetic change in a population over time," and population geneticists try to describe patterns of genetic variation, document changes in variation, determine their theoretical causes, and predict future patterns. These types of research have been valuable in studying the evolutionary histories of organisms for which there are living representatives, including humans.

In addition to the scientific value of understanding evolutionary history better, there are more immediate applications of such work. In conservation biology, data about genetic variation in a population can help to assess its ability to survive in the future. Data on genetic similarities between populations can aid in decisions about whether they can be considered as the same species or are unique enough to merit preservation.

Population genetics has had an influence on medicine, particularly in understanding why "disease genes," while clearly harmful, persist in human populations. The field has also affected the planning of vaccination protocols to maximize their effectiveness against parasites, since a vaccine-resistant strain is a result of a rare allele in the parasite population. In the 1990's it began to be recognized that effective treatments for medical conditions would need to take into account genetic variation in human populations, since different individuals might respond differently to the same treatment.

Stephen T. Kilpatrick, Ph.D.; updated by Bryan Ness, Ph.D.

Further Reading

Christiansen, Freddy B. *Population Genetics of Multiple Loci.* New York: Wiley, 2000. Reinterprets classical population genetics to include the mixture of genes not only from one generation to the next but also within existing populations. Illustrations, map, bibliography, index.

Dobzhansky, Theodosius. *Genetics and the Origin of Species.* 3d ed. New York: Columbia University Press, 1951. A classic treatment of population genetics and evolution.

Gillespie, John H. *Population Genetics: A Concise Guide.* 2d ed. Baltimore: Johns Hopkins University Press, 2004. Boils down the basics to about two hundred pages.

Hamilton, Matthew B. *Population Genetics.* Hoboken, N.J.: Wiley-Blackwell, 2009. Basic textbook presenting the major concepts in population genetics, such as genotype frequency, genetic drift and population size, mutation, quantitative trait variation, and natural selection. Each chapter concludes with a review and suggestions for further reading.

Hartl, Daniel L. *A Primer of Population Genetics.* Rev. 3d ed. Sunderland, Mass.: Sinauer Associates, 2000. Sections cover genetic variation, the causes of evolution, molecular population genetics, and the genetic architecture of complex traits. Illustrations, bibliography, index.

Hedrick, Philip W. *Genetics of Populations.* 3d ed. Boston: Jones and Bartlett, A quantitative analysis of population genetics. Illustrations, bibliography.

Landweber, Laura F., and Andrew P. Dobson, eds. *Genetics and the Extinction of Species: DNA and the Conservation of Biodiversity.* Princeton, N.J.: Princeton University Press, 1999. Offers theories on and methods for maintaining biodiversity and for preventing species' extinction. Illustratations, bibliography, index.

Lewontin, Richard C. *The Genetic Basis of Evolutionary Change.* New York: Columbia University Press, 1974. Discusses genetic variation in populations. Bibliography.

Papiha, Surinder S., Ranjan Deka, and Ranajit Chakraborty, eds. *Genomic Diversity: Applications in Human Population Genetics.* New York: Kluwer Academic/Plenum, 1999. Emphasis is on genetic variation and the application of molecular markers. Illustrations (some in color), bibliography, index.

Provine, William B. *The Origins of Theoretical Population Genetics.* 2d ed. Chicago: University of Chicago Press, 2001. An account of the early history of the field. Illustrated, bibliography, index.

Slatkin, Montgomery, and Michel Veuille, eds. *Modern Developments in Theoretical Population Genetics: The Legacy of Gustave Malécot.* New York: Oxford University Press, 2002. Discusses the work of the late cofounder of population genetics. Focuses on the theory of coalescents. Illustrations, bibliography, index.

Templeton, Alan R. *Population Genetics and Microevolutionary Theory.* Hoboken, N.J.: Wiley-Liss, 2006. Textbook divided into three sections: population structure and history, genotype and phenotype interactions, and natural selection and adaptation. Includes review questions, bibliography, and list of Web resources.

WEB SITES OF INTEREST

The Genographic Project
https://genographic.nationalgeographic.com/genographic/index.html

In April, 2005, the National Geographic Society, IBM, and a team of international scientists embarked on a five-year project that aimed to uncover new knowledge about the migratory history of human beings. The team planned to collect and analyze at least 100,000 DNA samples from indigenous and traditional people throughout the world. The project's Web site provides information about the team's activities, as well as basic genetic information.

Population and Evolutionary Genetics
http://www.ndsu.nodak.edu/instruct/mcclean/plsc431/popgen/index.htm

Philip McClean, a professor in the department of plant science at North Dakota State University, provides several pages on population and evolutionary genetics in his course-related Web site. Some of the pages discuss population variability, genotypic and allelic frequencies, the Hardy-Weinberg Law, evolutionary genetics, and natural selection.

See also: Artificial selection; Behavior; Consanguinity and genetic disease; Emerging and reemerging infectious diseases; Evolutionary biology; Genetic load; Genetics: Historical development; Hardy-Weinberg law; Heredity and environment; Hybridization and introgression; Inbreeding and assortative mating; Lateral gene transfer; Natural selection; Polyploidy; Punctuated equilibrium; Quantitative inheritance; Sociobiology; Speciation.

Porphyria

CATEGORY: Diseases and syndromes

DEFINITION

The term "porphyria" refers to a group of disorders. They do differ in some ways, but all share the same problem: a buildup of porphyrins in the body. Porphyrins help to make heme, a part of the red blood cell. However, a buildup of the porphyrins in the body causes damage. It most often affects the nervous system and skin.

Some porphyria disorders include acute intermittent porphyria, porphyria cutanea tarda, and erythropoietic protoporphyria. Another disorder, congenital erythropoietic protoporphyria, is present from birth. Some types of porphyria start in early childhood, some at puberty, and others during adulthood. Attacks may be separated by long periods of time. The attacks can be triggered by drugs, infections, alcohol consumption, and dieting.

RISK FACTORS

The most common risk factor for porphyria is having a family member with this disease. Caucasians are at greater risk than blacks or Asians, and females also have an increased risk (related to their menstrual cycles). Most onsets happen between the ages of twenty and forty.

ETIOLOGY AND GENETICS

"Porphyria" is a general term used to describe a group of disorders caused by faulty heme production. Mutations in at least eight different genes are known to be responsible. The protein products of these genes are all enzymes necessary for the biosynthesis of heme, so a mutation which results in a missing or altered form of any of the enzymes could severely limit the amount of heme produced in various body tissues.

The *CPOX* gene (found on chromosome 3 at position 3q12) specifies an enzyme known as coproporphyrinogen oxidase; *HMBS* (at position 11q23.3) encodes the enzyme hydroxymethylbilane synthase; *PPOX* (at position 1q22) specifies protoporphyrinogen oxidase; and *UROD* (at position 1p34) encodes uroporphyrinogen decarboxylase. Mutations in any of these four genes can result in an autosomal dominant form of porphyria, meaning that a single copy of the mutation is sufficient to cause full expression of the disease. An affected individual has a 50 percent chance of transmitting the mutation to each of his or her children. Many cases of autosomal dominant porphyria, however, result from a spontaneous new mutation, so in these instances affected individuals will have unaffected parents.

The *ALAD* gene (at position 9q34) specifies the

enzyme delta-aminolevulinate dehydratase, and the *UROS* gene (at position 10q25.2-q26.3) encodes uroporphyrinogen III synthase. Mutations in either of these two genes cause autosomal recessive forms of porphyria, which means that both copies of the gene must be deficient in order for the individual to be afflicted. Typically, an affected child is born to two unaffected parents, both of whom are carriers of the recessive mutant allele. The probable outcomes for children whose parents are both carriers are 75 percent unaffected and 25 percent affected.

The *FECH* gene (at position 18q21.3), which encodes the enzyme ferrochelatase, is unique in that it is associated with both autosomal dominant and autosomal recessive porphyria, depending on the specific molecular nature of the mutation. Finally, the *ALAS2* gene (at position Xp11.21), which specifies the enzyme aminolevulinate,delta-,synthase 2, is associated with an X-linked dominant form of porphyria. A single copy of the mutation is sufficient for disease expression, but while an affected woman will transmit the mutation to one-half of her children, regardless of sex, an affected father will transmit the disease to all of his daughters and none of his sons.

Symptoms

Porphyria can cause skin or nervous system problems. Urine from some types of the disorder may be reddish in color due to the presence of excess porphyrins. The urine may darken after standing in the light. Specific symptoms depend on the type.

In acute intermittent porphyria (AIP), nervous system symptoms occur most often after puberty. Nerves of the intestines can cause gastrointestinal problems. Attacks can last from days to weeks. Symptoms of future attacks resemble the initial episode and may include abdominal pain and cramping; nausea and vomiting; constipation; painful urination or urinary retention; pain in the limbs, head, neck, or chest; muscle weakness; loss of sensation; tremors; sweating; rapid heart rate; high blood pressure; breathing problems; heart arrhythmia; and seizures. Mental symptoms of AIP include hallucinations, restlessness, depression, anxiety, insomnia, confusion, and paranoia.

Porphyria cutanea tarda (PCT) is the most common porphyria. Most are not inherited; they are acquired at some point. Symptoms of PCT may include sun sensitivity, and sun-exposed skin may be fragile. Minor injury may damage the skin. Additional symptoms may include blisters on the face, hands, arms, feet, and legs; skin that heals slowly; skin susceptible to infection; skin that thickens and scars; skin color changes; excess hair growth; and reddish urine in infancy or childhood.

Symptoms of erythropoietic protoporphyria may include sun sensitivity; redness or swelling, but usually no blisters; an itching or burning sensation; long-term skin and nail changes; and gallstones.

Congenital erythropoietic protoporphyria (CEP) is extremely rare. Symptoms may include reddish urine in infancy; sun sensitivity, beginning in early infancy; fragile sun-exposed skin; blisters on sun-exposed skin; blisters that are open and are prone to infection; and changes in skin color. Additional symptoms may include skin thickening, excess hair growth, scarring, reddish-brown teeth, an enlarged spleen, and hemolytic anemia.

Screening and Diagnosis

The doctor will ask a patient about any symptoms and will take a medical and family history. A physical exam will also be done. The symptoms can be very vague. As a result, the diagnosis is often delayed.

Tests differ for the various types of porphyria and may include blood, urine, and/or stool tests. These tests check for excess porphyrin or a specific enzyme deficiency. In some cases, specific genetic testing may also be available.

Treatment and Therapy

For all types of porphyria, treatment includes avoiding known triggers and drugs that can precipitate an attack and eating a high-carbohydrate diet. Porphyria that affects the skin requires special attention to protect the skin from injury and/or infection.

Specific treatment depends on the type of porphyria. Patients with acute intermittent porphyria may need to be hospitalized during an attack. In the hospital, they may be given heme by vein (intravenous) in the form of hematin, heme albumin, or heme arginate; glucose by vein; and drugs to control symptoms, such as pain, nausea, anxiety, and insomnia.

Porphyria cutanea tarda treatment may include blood removal weekly to monthly; low doses of antimalarial drugs, such as chloroquine (Aralen

Phosphate) or hydroxychloroquine (Plaquenil); and radiological imaging to monitor for increased risk of liver cancer.

Erythropoietic protoporphyria treatment may include oral beta-carotene; maintaining normal iron levels with food or supplements; medication to aid excretion of porphyrins in the stool, such as activated charcoal, cholestyramine, blood transfusions, and heme by vein in the form of hematin, heme albumin, or heme arginate; and splenectomy, the removal of the spleen.

Treatment for congenital erythropoietic protoporphyria may include oral beta-carotene; oral charcoal, to aid excretion of porphyrins in the stool; blood transfusions; splenectomy, the removal of the spleen; and bone marrow transplantation.

Prevention and Outcomes

Genetic testing may identify individuals at risk for porphyria. An individual who has a family member with the diagnosis of porphyria may be eligible for testing. An individual whose family member has had a test that showed DNA changes can also be tested for these changes.

A genetic counselor can review family history. The counselor will help find the risks for this disorder in an individual and in his or her offspring. The counselor will also discuss appropriate testing.

Genetic mutation cannot be corrected. However, attacks can be anticipated, prevented, or controlled. Steps to avoid porphyria attacks and complications include protecting skin from injury or infection and avoiding drugs and other triggers.

Triggers for acute intermittent porphyria include drugs, such as barbiturates; sulfa drugs; seizure drugs; and steroid hormones, such as estrogen and progesterone. Other triggers include hormonal changes related to the menstrual cycle, weight-loss diets or fasting, infections, alcohol, stress, surgery, and cigarette smoke.

Triggers for porphyria cutanea tarda include alcohol, estrogens, hydrocarbons, and certain pesticides or chemicals. Sunlight and weight-loss diets or fasting can trigger erythropoietic protoporphyria. Sunlight can trigger congenital erythropoietic protoporphyria.

Debra Wood, R.N.; reviewed by Igor Puzanov, M.D.
"Etiology and Genetics" by Jeffrey A. Knight, Ph.D.

Further Reading

Beers, Mark H., ed. *The Merck Manual of Medical Information.* 2d home ed., new and rev. Whitehouse Station, N.J.: Merck Research Laboratories, 2003.

Conn, H. F., and R. E. Rakel. *Conn's Current Therapy.* 53d ed. Philadelphia: W.B. Saunders, 2001.

EBSCO Publishing. *Health Library: Porphyria.* Ipswich, Mass.: Author, 2009. Available through http://www.ebscohost.com.

Fauci, Anthony S., et al., eds. *Harrison's Principles of Internal Medicine.* 17th ed. New York: McGraw-Hill Medical, 2008.

Goldman, Lee, and Dennis Ausiello, eds. *Cecil Medicine.* 23d ed. Philadelphia: Saunders Elsevier, 2008.

Web Sites of Interest

American Liver Foundation
http://www.liverfoundation.org

The American Porphyria Foundation
http://porphyriafoundation.com

Canadian Liver Foundation
http://www.liver.ca/Home.aspx

Canadian Organization for Rare Disorders
http://www.cord.ca

Genetics Home Reference
http://ghr.nlm.nih.gov

MedLine Plus: Porphyria
http://www.nlm.nih.gov/medlineplus/porphyria.html

National Institute of Diabetes and Digestive and Kidney Diseases
http://www2.niddk.nih.gov

National Organization for Rare Disorders
http://www.rarediseases.org

See also: ABO blood types; Chronic myeloid leukemia; Fanconi anemia; Hemophilia; Hereditary spherocytosis; Infantile agranulocytosis; Myelodysplastic syndromes; Paroxysmal nocturnal hemoglobinuria; Rh incompatibility and isoimmunization; Sickle-cell disease.

Prader-Willi and Angelman syndromes

CATEGORY: Diseases and syndromes

DEFINITION

Both Prader-Willi syndrome (PWS) and Angelman syndrome (AS) are caused by errors at the same site in the long arm of chromosome 15 (15q11-q13), but the clinical outcomes of these errors are markedly different, because the chromosomes containing the errors come from different parents. Thus, these syndromes offer a striking example of the concept known as parental imprinting.

RISK FACTORS

In most cases, the defects and deletions in paternal genes that are the cause of PWS occur randomly and are not inherited. However, in a few instances a genetic mutation inherited from the father may cause this disorder.

Researchers do not know what causes the genetic deletions that result in most cases of AS. The majority of people with AS do not have a family history of the disease. However, in a small number of cases AS may be inherited from the mother.

ETIOLOGY AND GENETICS

The primary cause of both syndromes appears to be a small deletion on the long arm of chromosome 15 (del 15q11-q13). The deleted area is estimated to be about 4 million base pairs (bp), small by molecular standards but large enough to contain several genes. This area of chromosome 15 is known to contain several genes that are activated or inactivated depending on the chromosome's parent of origin (that is, a gene may be turned on in the chromosome inherited from the mother but turned off in the chromosome inherited from the father). This parent-specific activation is referred to as genetic imprinting. It is now known that the deletions causing AS appear in the chromosome inherited from the mother, while those causing PWS occur in the chromosome inherited from the father. Since the genes of only one chromosome are active at a time, any disruption (deletion) in the active chromosome will lead to the effects seen in one of these syndromes.

In 1997 a gene within the AS deletion region called *UBE3A* was found to be mutated in approximately 5 percent of AS individuals. These mutations can be as small as a single base pair. This gene codes for a protein/enzyme called a ubiquitin protein ligase, and *UBE3A* is believed to be the causative gene in AS. All mechanisms known to cause AS appear to cause inactivation or absence of this gene. *UBE3A* is an enzymatic component of a complex protein degradation system termed the ubiquitin-proteasome pathway. This pathway is located in the cytoplasm of all cells. The pathway involves a small protein molecule (ubiquitin) that can be attached to proteins, thereby causing them to be degraded. In the normal brain, *UBE3A* inherited from the father is almost completely inactive, so the maternal copy performs most of the ubiquitin-producing function. Inheritance of a *UBE3A* mutation from the mother causes AS; inheritance of the mutation from the father has no apparent effect on the child. In some families, AS caused by a *UBE3A* mutation can occur in more than one family member.

Another cause of AS (3 percent of cases) is paternal uniparental disomy (UPD). In this case a child inherits both copies of chromosome 15 from the father, with no copy inherited from the mother. Even though there is no deletion or mutation, the child is still missing the active *UBE3A* gene because the paternally derived chromosomes only have brain-inactivated *UBE3A* genes.

A fourth class of AS individuals (3-5 percent) have chromosome 15 copies inherited from both parents, but the copy inherited from the mother functions in the same way as a paternally inherited one would. This is referred to as an "imprinting defect." Some individuals may have a very small deletion of a region known as the imprinting center (IC), which regulates the activity of *UBE3A* from a distant location. The mechanism for this is not yet known.

While there are several genetic mechanisms for AS, all of them lead to the typical clinical features found in AS individuals, although minor differences in incidence of features may occur between each group.

The primary genes involved in PWS are *SNRPN*, a gene that encodes the small ribonucleotide polypeptide SmN that is found in the fetal and adult brain, and *ZFN127*, a gene that encodes a zinc-finger protein of unknown function. *SNRPN* is involved in messenger RNA (mRNA) processing, an intermediate step between DNA transcription and protein formation. A mouse model of PWS has been developed with a large deletion that includes the *SNRPN* region and the PWS imprinting center and shows a phenotype similar to that of infants with PWS.

It is probable that the hypothalamic problems (such as overeating) associated with PWS might result from a loss of *SNRPN*. The production of this protein is found mainly in the hypothalamic regions of the brain and in the olfactory cortex. Thus, disruption of hypothalamic functions such as satiety are a likely result of this defect. Prader-Willi syndrome is the most common genetic cause of obesity. In addition to its role in satiety, the hypothalamus regulates growth, sexual development, metabolism, body temperature, pigmentation, and mood—all functions that are affected in those with PWS.

PWS may also be caused by uniparental disomy, as seen in AS. However, in PWS both copies of chromosome 15 are derived from the mother instead of from the father.

As mentioned above, the imprinting center may be involved in at least some cases of both syndromes. This chromosome 15 IC is about 100 kilobase pairs (kb) long and includes exon 1 of the *SNRPN* gene. Mutations in this area appear to prevent the paternal-to-maternal imprinting switch in the AS families and prevents the maternal-to-paternal switch in PWS families. Therefore, it is possible that the IC is needed to regulate alternate RNA splicing in the *SNRPN* gene transcripts.

Few examples of known parental imprinting occur in the human, so AS and PWS provide rare opportunities for geneticists and biologists to study this important phenomenon. Examples of nonhuman parental imprinting are well known, but the genetic and biochemical mechanisms have not been established. Detailing the IC for chromosome 15 will be key to understanding how imprinting occurs and how the effects of AS and PWS are manifested.

The suggestion has been made that PWS (and therefore disruption of the IC) may also, at least in some cases, have an environmental trigger. A high association of PWS with fathers employed in hydrocarbon-related occupations (such as factory workers, lumbermen, machinists, chemists, and mechanics) at the time of conception has been reported by one investigative team. This is an area that needs further exploration.

Symptoms

Angelman syndrome was first described in 1965 by Dr. Harry Angelman, who described three children with a stiff, jerky gait, absent speech, excessive laughter, and seizures. Newer reports include severe mental retardation and a characteristic face that is small with a large mouth and prominent chin. These characteristics give rise to the alternate name for the syndrome, that being "happy puppet syndrome." The syndrome is fairly rare, with an incidence estimated to be between one in fifteen thousand to one in thirty thousand. It is usually not recognized at birth or in infancy, since the developmental problems are nonspecific during this period.

Prader-Willi syndrome, by comparison, is characterized by mental retardation, hypotonia (decreased muscle tone), skin picking, short stature, crytorchidism (small or undescended testes), and hyperphagia (overeating leading to severe obesity). Delayed motor and language development are common, as is intellectual impairment (the average IQ is about 70). The syndrome was first described by Doctors Andrea Prader, Alexis Labhart, and Heinrich Willi in 1956. Like Angelman syndrome, PWS has a fairly low incidence, estimated at one in fifteen thousand. Neither condition is race-specific, and neither is considered to be a familial disease.

Screening and Diagnosis

The usual chromosome studies carried out during prenatal diagnosis are interpreted as normal in fetuses with AS and PWS syndromes, since the small abnormalities on chromosome 15 are not detected by this type of study. Likewise, fetal ultrasound offers no help in detecting physical abnormalities related to AS or PWS, since the affected fetus is well formed. Amniotic fluid volume and alpha-feto protein levels also appear normal.

Fluorescence in situ hybridization (FISH) is an extremely sensitive assay for determining the presence of deletions on chromosomes. It uses a fluorescence-tagged segment of DNA that binds to the DNA region being studied. Specialized chromosome 15 FISH studies are needed to determine the presence of either syndrome resulting from chromosomal deletions. Testing for parent-specific methylation imprints at the 15q11-q13 locus detects more than 95 percent of cases. For cases caused by uniparental disomy, polymerase chain reaction (PCR) testing can be used.

Treatment and Therapy

Treatment for both PWS and AS focuses on managing the medical and development problems that are caused by these conditions. Most people with PWS will require specialized care throughout their lives. Many infants who have this disorder need a high-calorie formula to help them gain weight. Older children will need to maintain a reduced-calorie diet in order to control their weight and ensure proper nutrition.

Some children with PWS receive human growth hormones to increase growth and reduce body fat. However, the long-term effects of hormone treatment are not known, and parents should discuss this treatment with an endocrinologist—a specialist who treats hormonal disorders—to determine if it is right for their children. The endocrinologist may also recommend that a child receive hormone replacement therapy (testosterone for males or estrogen and progesterone for females) to replenish low levels of sex hormones.

Children with PWS may also receive physical, speech, occupational, and developmental therapies in order to improve movement and language skills, learn to perform everyday tasks, behave appropriately, and acquire social and interpersonal skills. In addition, a psychologist or psychiatrist may be needed to address a child's psychological problems, including mood or obsessive-compulsive disorders.

Treatment for AS may include antiseizure medication to control the seizures caused by the disorder. Physical therapy can help children walk and improve other mobility problems; behavioral therapy can teach children to overcome their hyperactivity and short attention spans. Because people with AS usually have limited verbal language skills, communication therapy can help them develop nonverbal language skills through sign language and picture communication.

Prevention and Outcomes

There is no cure for PWS or AS, but the proper therapy and medication can help address the developmental and medical difficulties associated with these disorders.

Kerry L. Cheesman, Ph.D.;
updated by Rebecca Kuzins

Further Reading

Butler, Merlin G., Phillip D. K. Lee, and Barbara Y. Whitman, eds. *Management of Prader-Willi Syndrome.* 3d ed. New York: Springer, 2006. Collection of articles describing the diagnosis, genetics, medical physiology, and treatment of PWS. The majority of the articles examine how to manage the lives of PWS patients, including discussions of associated speech and language disorders, the education of children and adolescents, vocational training, and behavior management.

Cassidy, S. B., and S. Schwartz. "Prader-Willi and Angelman Syndromes: Disorders of Genomic Imprinting." *Medicine* (Baltimore) 77, no. 2 (March, 1998): 140-151. A review of these syndromes written for health care professionals.

Dan, Bernard. *Angelman Syndrome.* London: Mac Keith Press, 2008. Reviews the genetic issues, research, and treatment of Angelman syndrome, as well as specific clinical problems associated with the syndrome, such as motor impairment, behavior, learning difficulties, communication, sleep, and epilepsy.

Eiholzer, Urs. *Prader-Willi Syndrome: Coping with the Disease, Living with Those Involved.* New York: Karger, 2005. Designed for family members and others who have contact with PWS-afflicted children, this book provides information about the genetic cause of PWS, the results of research about the syndrome, and treatment options.

Hall, J. G. "Genomic Imprinting: Nature and Clinical Relevance." *Annual Review of Medicine* 48, no. 1 (1997): 35-44. A well-documented review that includes discussion of both AS and PWS as examples of human genomic imprinting.

Lai, L. W., R. P. Erickson, and S. B. Cassidy. "Clinical Correlates of Chromosome Fifteen Deletions and Maternal Disomy in Prader-Willi Syndrome." *American Journal of Diseases of Children* 147, no. 11 (November, 1993): 1217-1223. Discusses the signs and symptoms of PWS, including those that are needed to make a clinical diagnosis in children.

Lalalande, M. "Parental Imprinting and Human Disease." *Annual Review of Genetics* 30, no. 1 (1996): 173-195. A well-written, well-documented review of imprinting and how it relates to both AS and PWS. Good reading for undergraduate students.

Mann, M. R., and M. S. Bartolomei. "Towards a Molecular Understanding of Prader-Willi and Angelman Syndromes." *Human Molecular Genetics* 8, no. 10 (1999): 1867-1873. Details the molecular mechanisms, rather than the clinical correlates, of AS and PWS; fairly technical but readable by students of biology.

Nicholls, R. D. "Genomic Imprinting and Uniparental Disomy in Angelman and Prader-Willi Syndromes: A Review." *American Journal of Medical Genetics* 46, no. 1 (April 1, 1993): 16-25. Not as far-reaching as Lalalande, but a well-written review of the genetic aspects of both AS and PWS.

Whittington, Joyce, and Tony Holland. *Prader-Willi Syndrome: Development and Manifestations.* New York: Cambridge University Press, 2004. The authors participated in a major PWS study at Cambridge University and the study's findings are the basis for this book, which describes how to manage the medical, nutritional, psychological, educational, social, and therapeutic needs of patients with PWS.

WEB SITES OF INTEREST

Angelman Syndrome Foundation
http://www.angelman.org

This support group's Web site features an online booklet entitled *The Facts About Angelman Syndrome* and an alphabetical list of topics of importance to patients and their families.

Genetics Home Reference
http://ghr.nlm.nih.gov

Users can search the "Genetic Disorders A to Z" section of the site to find individual pages about the Prader-Willi and Angelman syndromes. The pages describe the genetic changes and inheritance patterns related to both conditions and provide links to additional information.

National Organization for Rare Disorders
http://www.rarediseases.org

This site contains an index of rare diseases and a rare disease database that enables users to retrieve information about the Prader-Willi and Angelman syndromes, as well as a list of related organizations.

Prader-Willi Syndrome Association
http://www.pwsausa.org

Offers background information on the syndrome, a research/medical section, and links to other resources.

See also: Amniocentesis; Chorionic villus sampling; Chromosome structure; Congenital defects; Down syndrome; Fragile X syndrome; Hereditary diseases; Human growth hormone; Huntington's disease; Intelligence; Polymerase chain reaction; Prenatal diagnosis.

Prenatal diagnosis

CATEGORY: Human genetics and social issues

SIGNIFICANCE: Tests ranging from ultrasound and maternal blood tests to testing fetal cells from the amniotic fluid or placenta are performed to detect genetic disorders that the fetus may have. Although tests may show the absence of specific genetic defects, the detection of a genetic defect can produce an ethical dilemma for the parents and their physician.

KEY TERMS

amniotic fluid: the liquid that surrounds the developing fetus

neural tube: the embryonic structure that becomes the brain and spinal cord

placenta: an organ composed of both fetal and maternal tissue through which the fetus is nourished

trisomy: the presence of three copies (instead of two) of a particular chromosome in a cell

PRENATAL TESTING

Prenatal testing is administered to a large number of women, and the tests are becoming more informative. Some of the tests are only mildly invasive to the mother, but others involve obtaining fetal cells. Some are becoming routine for all pregnant women; others are offered only when an expectant mother meets a certain set of criteria. Some physicians will not offer the testing (especially the more invasive procedures) unless the parents have agreed that they will abort the fetus if the testing reveals a major developmental problem, such as Down syndrome or Tay-Sachs disease. Others will order testing without any such guarantees, believing that test results will give the parents time to prepare themselves for a special-needs baby. The test results are also used to determine whether additional medical teams should be present at the delivery to deal with a newborn who is not normal and healthy. Most often, prenatal testing is offered if the mother is age thirty-five or older, if a particular disorder is present in relatives on one or both sides of the family, or if the parents have already produced one child with a genetic disorder.

MATERNAL BLOOD TESTS AND ULTRASOUND

Screening maternal blood for the presence of alpha fetoprotein (AFP) is offered to pregnant women who are about eighteen weeks into a pregnancy. Although AFP is produced by the fetal liver, some will cross the placenta into the mother's blood. Elevated levels of AFP can indicate an open neural tube defect (such as spina bifida), although it can also indicate twins. Unusual AFP findings are usually followed up by ultrasound examination of the fetus.

Other tests of maternal blood measure the amounts of two substances that are produced by the fetal part of the placenta: hCG and UE3. Lower-than-average levels of AFP and UE3, combined with a higher-than-average amount of hCG, increases the risk that the woman is carrying a Down syndrome (trisomy 21) fetus. For example, a nineteen-year-old woman has a baseline risk of conceiving a fetus with Down syndrome of 1 in 1,193. When blood-test results show low AFP and UE3 along with high hCG, the probability of Down syndrome rises to 1 in 145.

During an ultrasound examination, harmless sound waves are bounced off the fetus from an emitter placed on the surface of the mother's abdomen or in her vagina. They are used to make a picture of the fetus on a television monitor. Measurements on the monitor can often be used to determine the overall size, the head size, and the sex of the fetus, and whether all the arms and legs are formed and of the proper length. Successive ultrasound tests will indicate if the fetus is growing normally. Certain ultrasound findings, such as shortened long bones, may indicate an increased probability for a Down syndrome baby. Because Down syndrome is a highly variable condition, normal ultrasound findings do not guarantee that the child will be born without Down syndrome. Only a chromosome analysis can determine this for certain.

AMNIOCENTESIS, KARYOTYPING, AND FISH

Amniocentesis is the process of collecting fetal cells from the amniotic fluid. Fetal cells collected by

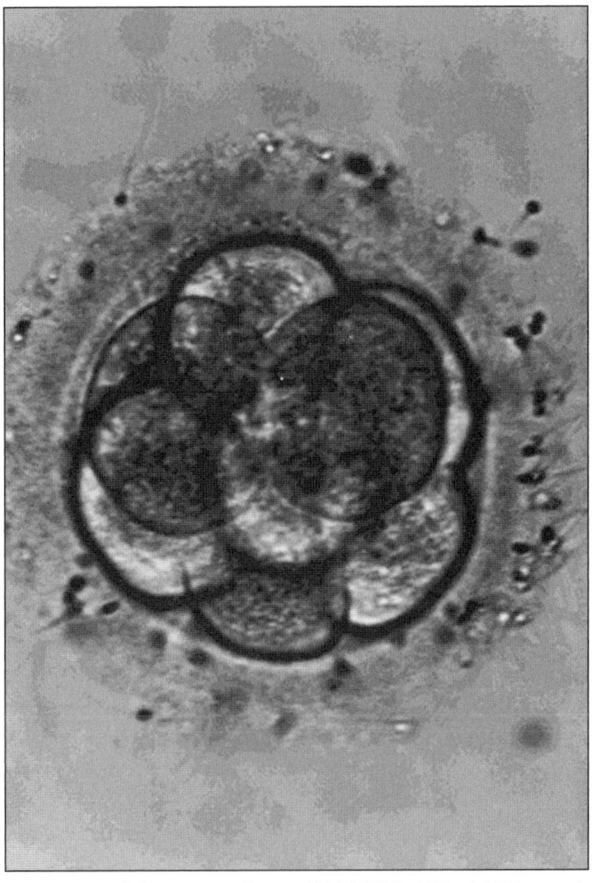

An eight-cell human embryo. (AP/Wide World Photos)

amniocentesis can be grown in culture; then the fluid around the cells is collected and analyzed for enzymes produced by the cells. If an enzyme is missing (as in the case of Tay-Sachs disease), the fetus may be diagnosed with the disorder before it is born. Because disorders such as Tay-Sachs disease are untreatable and fatal, a woman who has had one Tay-Sachs child may not wish to give birth to another. Early diagnosis of a second Tay-Sachs fetus would permit her to have a therapeutic abortion.

Chromosomes in the cells obtained by amniocentesis may be stained to produce a karyotype. In a normal karyotype, the chromosomes will be present in pairs. If the fetus has Down syndrome (trisomy 21), there will be three copies of chromosome 21. Other types of chromosome abnormalities that also appear in karyotypes are changes within a single chromosome. If a chromosome has lost a piece, it is said to contain a deletion. Large deletions will be obvious when a karyotype is analyzed because the

The Eight-Cell Stage

Preimplantation genetic diagnosis (PGD) has been used since 1988 to screen for genetic disorders. The most common type of PGD involves embryo biopsy at the 6-8 cell stage after fertilization has occurred in vitro. This early form of prenatal diagnosis is typically performed on day 3 embryos. One to two blastomeres (cells) are removed from the embryo (either by aspiration or by extrusion) using a fine glass needle. The biopsied embryo is then returned to culture, where the lost cells are replenished. Genetic testing is then carried out on the biopsied cells using either a technique known as fluorescence in situ hybridization (FISH) or a second technique known as fluorescent polymerase chain reaction (PCR). The FISH technique can be used to determine the presence of chromosomes 13, 16, 18, 21, 22, X, and Y. Aneuploidies (abnormal numbers of chromosomes) involving these chromosomes account for the majority of first-trimester miscarriages and for 95 percent of all postnatal chromosomal abnormalities. PCR involves amplification of DNA and allows diagnosis of single-gene diseases.

By enabling very early diagnosis of these abnormalities, PGD allows physicians to determine which embryos are most likely to be chromosomally normal prior to placement in the uterus. This increases the probability of a successful pregnancy and a healthy baby. Genetic testing generally takes only six to eight hours to complete, so that intrauterine transfer of the chromosomally normal embryo can take place within one day. If more normal embryos are obtained than one wishes to implant, the extra embryos may be preserved for future use by cryopreservation. The survival rate of frozen embryos is thought to be about 50 percent.

While this early form of prenatal diagnosis allows the elimination of many embryos carrying major genetic defects prior to implantation, it is still recommended that follow-up prenatal diagnosis (using chorionic villus sampling or amniocentesis) be done on resulting pregnancies. It must be realized that in order to employ preimplantation diagnostic testing, a couple must undergo in vitro fertilization even if they are fertile. This is an expensive, time-consuming process, and generally results in only a 20 percent pregnancy rate per cycle. For couples with fertility problems, this is an easy path to choose as they try to ensure implantation of normal, healthy embryos leading to healthy babies.

This technique has been used increasingly by couples who have a history of genetic disorders. In the past, couples who had a history of genetic abnormalities could decide not to have children, to become pregnant and knowingly risk and accept abnormalities, or to become pregnant and rely on chorionic villus sampling or amniocentesis to diagnose genetic problems and terminate (abort) problem pregnancies.

To date, PGD has been used to detect cases of cystic fibrosis, Tay-Sachs disease, beta-thalassemia, Huntington's disease, myotonic dystrophy, X-linked disorders, and aneuploidies such as trisomy 13, 18, or 21, Turner syndrome, or Klinefelter syndrome. The number of detectable genetic defects has greatly increased since 1988. In that time, hundreds of healthy children have been born to parents undergoing preimplantation diagnosis.

Robin Kamienny Montvilo, Ph.D.

chromosome will appear smaller than normal. Sometimes the deletion is so small that it is not visible on a karyotype.

If chromosome analysis is needed early in pregnancy before the volume of amniotic fluid is large enough to permit amniocentesis, the mother and doctor may opt for chorionic villus sampling (CVS). The embryo produces fingerlike projections (villi) into the uterine lining. Because these projections are produced by the embryo, their cells will have the same chromosome number as the rest of the embryonic cells. After growing in culture, the cells may be karyotyped in the same way as those obtained by amniocentesis. Both amniocentesis and CVS carry risks of infection and miscarriage. Normally these procedures are not offered unless the risk of having an affected child is found to be greater than the risk of complications from the procedures.

If the doctor is convinced that the fetus has a tiny chromosomal defect that is not visible on a karyotype, it will then be necessary to probe (or "FISH") the fetal chromosomes; the initials "FISH" stand for "fluorescence in situ hybridization." A chromosome probe is a piece of DNA that is complementary to DNA within a gene. Complementary pieces of DNA will stick together (hybridize) when they come in contact. The probe also has an attached molecule that will glow when viewed under fluorescent light. A probe for a particular gene will stick to the part of the chromosome where the gene is located and make a glowing spot. If the gene is not present because it has been lost, no spot will appear. Probes have been developed for many individual genes that cause developmental abnormalities when they are deleted from the chromosomes.

Cells obtained by amniocentesis can be probed in less time than it takes to grow and prepare them for karyotyping. Probes have been developed for the centromeres of the chromosomes that are frequently present in extra copies, such as 13, 18, 21, X, and Y. Y chromosomes that have been probed appear as red spots, X chromosomes as green spots, and number 18 chromosomes as aqua spots. A second set of probes attached to other cells from the same fetus will cause number 13 chromosomes to appear as green spots and number 21 chromosomes to appear as red spots. Cells from a girl with trisomy 21 would have two green spots and two aqua spots, but no red spot when the first set of probes is used. Some other cells from the same girl will show two red spots, but three green ones, when the second set of probes is used.

Tests for many more genetic defects using advanced molecular genetics tests have been developed. DNA can be isolated from fetal cells, obtained by one of the methods already described, which is then probed for single gene defects. Hundreds of potential genetic defects can be detected in this way, although only a few such tests are generally available. Another barrier to their use is their high cost. Costs will likely drop in the future as the tests are perfected and are used more widely. These same tests may be performed on the parents to determine whether they are carriers of certain genetic diseases.

IMPACT AND APPLICATIONS

Until the development of prenatal techniques, pregnant women had to wait until delivery day to find out the sex of their child and whether or not the baby was normal. Now much more information is available to both the woman and her doctor weeks

Yury Verlinsky (right), known for his cutting-edge work in prenatal testing, and Ridvan Seckin Ozen examine human chromosomes at the Reproductive Genetics Institute in Chicago. (AP/Wide World Photos)

before the baby is due. Even though tests are not available for all possible birth defects, normal blood tests, karyotypes, or FISH can be very comforting. On the other hand, abnormal test results give the parents definite information about birth defects, as opposed to the possibilities inherent in a statement of risk. The parents must decide whether to continue the pregnancy. If they do, they must then cope with the fact that they are not going to have a normal child. When properly administered, the test results are explained by a genetic counselor who is also equipped to help the parents deal with the strong emotions that bad news can produce. Genetic testing also has far-reaching implications. If insurance companies pay for the prenatal testing, they receive copies of the results. Information about genetic abnormalities could cause the insurance companies to deny claims arising from treatment of the newborn or to deny insurance to the individual later in life.

Nancy N. Shontz, Ph.D.;
updated by Bryan Ness, Ph.D.

Further Reading

Bianchi, Diana W., Timothy M. Crombleholme, and Mary E. D'Alton. *Fetology: Diagnosis and Management of the Fetal Patient.* New York: McGraw-Hill, 2000. A resource for practitioners and a guide for parents. Illustrations, bibliography, index.

De Crespigny, Lachlan, and Frank Chervenak. *Prenatal Tests: The Facts.* New York: Oxford University Press, 2006. Designed to provide prospective parents with objective information about prenatal testing. Discusses the reasons for prenatal diagnosis and describes normal and abnormal fetal development. Provides a general overview of the available prenatal tests, as well as more specific information about testing during various stages of pregnancy.

Evans, Mark I., et al., eds. *Prenatal Diagnosis.* New York: McGraw-Hill Medical, 2006. Begins with a review of genetic and reproductive risks and then discusses various prenatal diagnostic procedures, including ultrasound, amniocentesis, and fetoscopy. Covers laboratory diagnostics, management of prenatal problems, types of fetal therapy, and the ethical, legal, and social issues surrounding prenatal diagnosis.

Heyman, Bob, and Mette Henriksen. *Risk, Age, and Pregnancy: A Case Study of Prenatal Genetic Screening and Testing.* New York: Palgrave, 2001. Provides a detailed case study of a prenatal genetic screening and testing system in a British hospital, giving perspectives of pregnant women, hospital doctors, and midwives, and elucidating the communication between women and the hospital doctors who advise them.

McConkey, Edwin H. *Human Genetics: The Molecular Revolution.* Boston: Jones and Bartlett, 1993. Contains additional information on FISH. Illustrations, bibliography, index.

New, Maria I., ed. *Diagnosis and Treatment of the Unborn Child.* Reddick, Fla.: Idelson-Gnocchi, 1999. Provides an overview of prenatal testing and treatment. Illustrations (some color), bibliography.

Petrikovsky, Boris M., ed. *Fetal Disorders: Diagnosis and Management.* New York: Wiley-Liss, 1999. Discusses testing and treatment protocols, prognoses, and counseling. Illustrations (some color), bibliography, index.

Pilu, Gianluigi, and Kypros H. Nicolaides. *Diagnosis of Fetal Abnormalities: The Eighteen-Twenty-three-Week Scan.* New York: Parthenon Group, 1999. Covers midtrimester fetal ultrasound. Illustrations, index.

Rodeck, Charles H., and Martin J. Whittle, eds. *Fetal Medicine: Basic Science and Clinical Practice.* New York: Churchill Livingstone, 1999. Good general reference for residents and fellows in neonatal and maternal-fetal medicine. Illustrations (some color), bibliography, index.

Twining, Peter, Josephine M. McHugo, and David W. Pilling, eds. *Textbook of Fetal Abnormalities.* 2d ed. Edinburgh: Churchill Livingstone, Elsevier, 2007. Covers the safety of ultrasound; the routine fetal anomaly scan; disorders of amniotic fluid; cranial, spinal, cardiac, pulmonary, abdominal, skeletal, urinary tract, and chromosomal abnormalities; intrauterine therapy; and counseling. Illustrations (some color), bibliography, index.

Weaver, David D., with the assistance of Ira K. Brandt. *Catalog of Prenatally Diagnosed Conditions.* 3d ed. Baltimore: Johns Hopkins University Press, 1999. Covers about 800 conditions and has 1,221 literature references. Bibliography, index.

Web Sites of Interest

Association of Women's Health, Obstetric, and Neonatal Nurses
http://www.awhonn.org
Offers pages for education and practice resources, as well as legal policy.

Family Doctor.org, Prenatal Diagnosis: Amniocentesis and CVS
http://familydoctor.org/online/famdocen/home/women/pregnancy/fetal/144.html

Family Doctor.org, a Web site sponsored by the American Academy of Family Physicians, includes a page describing amniocentesis and chorionic villus sampling, with links to additional information.

Internet Pathology Laboratory, Prenatal Diagnosis
http://library.med.utah.edu/WebPath/TUTORIAL/PRENATAL/PRENATAL.html

The site, cosponsored by Mercer University School of Medicine and the University of Utah Eccles Health Sciences Library, is an online source of information about pathologic findings associated with human diseases. The site's page on prenatal diagnosis describes the benefits of prenatal testing, the various types of tests, and techniques for pathologic examination. It also provides an overview of fetal-placental abnormalities, with links to additional information about medical conditions.

March of Dimes
http://www.marchofdimes.com

This site is searchable by keyword and includes information on the basics of amniocentesis and chorionic villus sampling and articles on how the two procedures relate to genetics.

National Newborn Screening and Genetics Resource Center
http://genes-r-us.uthscsa.edu

A resource for information on genetic screening.

See also: Albinism; Amniocentesis; Androgen insensitivity syndrome; Burkitt's lymphoma; Chorionic villus sampling; Color blindness; Congenital defects; Consanguinity and genetic disease; Cystic fibrosis; Down syndrome; Dwarfism; Fragile X syndrome; Gender identity; Genetic counseling; Genetic screening; Genetic testing; Genetic testing: Ethical and economic issues; Heart disease; Hemophilia; Hereditary diseases; Hermaphrodites; Human genetics; Human Genome Project; Huntington's disease; In vitro fertilization and embryo transfer; Inborn errors of metabolism; Klinefelter syndrome; Metafemales; Monohybrid inheritance; Neural tube defects; Pedigree analysis; Phenylketonuria (PKU); Polymerase chain reaction; Prader-Willi and Angelman syndromes; Pseudohermaphrodites; RFLP analysis; Sickle-cell disease; Tay-Sachs disease; Thalidomide and other teratogens; Turner syndrome; XYY syndrome.

Prion diseases

Kuru and Creutzfeldt-Jakob syndrome

CATEGORY: Diseases and syndromes
SIGNIFICANCE: Kuru and Creutzfeldt-Jakob syndrome are rare, fatal diseases of the brain and spinal cord. Nerve cell death is caused by the accumulation of a protein called a "prion" that appears to be a new infectious agent that interferes with gene expression in nerve cells. Understanding these diseases has far-reaching implications for the study of other degenerative mental disorders.

KEY TERMS

dementia: mental deterioration ranging from forgetfulness and disorientation to complete unresponsiveness

prion: short for "proteinaceous infectious particle," an element consisting mainly of protein and generally lacking nucleic acid (DNA and RNA), which is often the causative agent behind various spongiform encephalopathies

CAUSES, SYMPTOMS, AND TREATMENT

Kuru and Creutzfeldt-Jakob syndrome, degenerative diseases of the human central nervous system, are among a group of diseases that also affect cattle (mad cow disease) and sheep (scrapie). They have been classified in several ways, including "slow-virus" infections (because of the extremely long incubation period between contact and illness) and "spongiform encephalopathies" (because of the large holes seen in the brain after death). However, a virus that may cause such a disease has never been found, and the body does not respond to the disease as an infection. The only clue to the cause is the accumulation of a transmissible, toxic protein known as a prion; therefore, these disorders are now known simply as "prion diseases."

Creutzfeldt-Jakob syndrome is rare: Approximately 250 people die from it yearly in the United States. It usually begins in middle age with symptoms that include rapidly progressing dementia, jerking spastic movements, and visual problems. Within one year after the symptoms begin, the patient is comatose and paralyzed, and powerful seizures affect the entire body. Death occurs shortly thereafter. The initial symptom of the illness (rapid mental deterioration) is similar to other disorders; therefore, diagnosis is difficult. No typical infectious agent (bacteria or viruses) can be found in the blood or in the fluid that surrounds the brain and spinal cord. X rays and other scans are normal. There is no inflammation, fever, or antibody production. Brain wave studies are, however, abnormal, and at autopsy, the brain is found to have large holes and massive protein deposits in it.

Kuru is found among the Fore tribe of Papua New Guinea. Until the early 1960's, more than one thousand Fore died of kuru each year. Anthropologists recording their customs described their practice of eating the brains of their dead relatives in order to gain the knowledge they contained. Clearly, some infectious agent was being transmitted during this ritual. Such cannibalism has since stopped, and kuru has declined markedly. Kuru, like Creutzfeldt-Jakob syndrome, exhibits the same spongiform changes and protein deposits in the brain after death. Similarly, early symptoms include intellectual deterioration, spastic movements, and visual problems. Within a year, the patient becomes unresponsive and dies.

The outbreak of "mad cow" disease in the mid-1990's in Great Britain led to widespread fear. Thousands of cattle were killed to prevent human consumption of contaminated beef. The cows were

The Discovery of Prions

In 1972, Stanley B. Prusiner, then a resident in neurology at the University of California School of Medicine at San Francisco, lost a patient to Creutzfeldt-Jakob disease. He resolved to learn more about the condition. He read that it and related diseases, scrapie and kuru, could be transmitted by injecting extracts from diseased brain into the brains of healthy animals. At the time, the diseases were thought to be caused by a slow-acting virus, but it had not been identified. He was intrigued by a study from the laboratory of Tikvah Alper that suggested that the scrapie agent lacked nucleic acid. When he started his own lab in 1974, Prusiner decided to pursue the nature of the infectious agent.

He and his associates determined to purify the causative agent in scrapie-infected brains and, by 1982, had a highly purified preparation. They subjected it to extensive analysis, and all of their results indicated that it indeed lacked DNA or RNA and that it consisted mainly, if not exclusively, of protein. The infectivity was lost when treated with procedures that denatured protein, but not when treated with those detrimental to nucleic acids. He named the agent a "prion," an abbreviation for "proteinaceous infectious particle." Shortly afterward, he showed that it consisted of a single protein. This was a highly unorthodox discovery because all pathogens studied to date contained nucleic acid. Skeptics were convinced that a very small amount of nucleic acid must be contaminating the prions, although the limits on detection showed that it contained fewer than one hundred nucleotides and would have to be smaller than any known virus.

Prusiner and his collaborators subsequently learned that the gene for the prion protein was found in chromosomes of hamsters, mice, humans, and all other mammals that have been examined. Furthermore, most of the time these animals make the prion protein without getting sick—a startling observation. Prusiner and his team subsequently showed that the prion protein existed in two forms, one harmless and the other leading to disease. The latter proved to be highly resistant to degradation by proteolytic enzymes and accumulated in the brain tissue of affected animals and people. In infectious disease, the harmful form of the prions appears to convert the harmless form to the harmful form, although the mechanism is not understood. In inherited disease, mutations in the prion may cause it to adopt the harmful form spontaneously or after some unknown signal, leading eventually to the disease state. While questions remain, research since the 1980's has established the involvement of prions in various spongiform encephalopathies.

In 1997, Prusiner was awarded the Nobel Prize in Physiology or Medicine for his pioneering discovery of prions and their role in various neurological diseases. The Nobel Committee also noted his perseverance in pursuing an unorthodox hypothesis in the face of major skepticism.

James L. Robinson, Ph.D.

infected by supplemental feedings tainted by infected sheep meat. Animal-to-human transmission of these diseases appears to occur, and research has shown that human-to-animal infection is possible as well.

Both kuru and Creutzfeldt-Jakob syndrome, as well as the animal forms, have no known treatment or cure. Because of the long incubation period, decades may pass before symptoms appear, but once they do, the central nervous system is rapidly destroyed, and death comes quickly. It is likely that many more people die of these disorders than is known because they are so rarely diagnosed.

Properties of Prions

Most of the research on prion diseases has focused on scrapie in sheep. It became clear that the infectious particle had novel properties: It was not a virus as had been suspected, nor did the body react to it as an invader. It was discovered that this transmissible agent was an abnormal version of a common protein, which defied medical understanding. This protein is normally secreted by nerve cells and is found on their outer membranes. Its gene is on chromosome 20 in humans. The transmissible, infectious fragment of the prion somehow disrupts the nerve cell, causing it to produce the abnormal fragment instead of the normal protein. This product accumulates to toxic levels in the tissue and fluid of the brain and spinal cord over many years, finally destroying the central nervous system.

Prion infection appears to occur from exposure to infected tissues or fluids. Transmission has occurred accidentally through nerve tissue transplants and neurosurgical instruments. Prions are not affected by standard sterilization techniques; prevention requires careful handling of infected materials and extended autoclaving of surgical instruments (for at least one hour) or thorough rinsing in chlorine bleach. The agent is not spread by casual contact or air, and isolating the patient is not necessary.

Other human degenerative nervous system diseases whose causes remain unclear also show accumulations of proteins to toxic levels. Alzheimer's disease is the best-studied example, and it is possible that a process similar to that in prion diseases is at work. The discovery of prions has far-reaching implications for genetic and cellular research. Scientists have already learned a startling fact: Substances as inert as proteins and far smaller than viruses can act as agents of infection.

Connie Rizzo, M.D., Ph.D.

Further Reading

Aguzzi, Adriano, and Charles Weismann. "Prion Research: The Next Frontiers." *Nature* 389, no. 6653 (October 23, 1997): 796. An overview of the history of research into prions and prion diseases.

Baker, Harry F., ed. *Molecular Pathology of the Prions.* Totowa, N.J.: Humana Press, 2001. Overview of research on prion diseases. Illustrated, bibliography, index.

Goldman, Lee, and Dennis Ausiello, eds. *Cecil Textbook of Medicine.* 23rd ed. Philadelphia: Saunders Elsevier, 2008. A classic medical reference text that covers prion diseases. Bibliography, index.

Groschup, Martin H., and Hans A. Kretzschmar, eds. *Prion Diseases: Diagnosis and Pathogenesis.* New York: Springer, 2000. Comprehensive collection of research articles on the pathogenesis of prion diseases in humans and other animals, pharmacology, epidemiology, and diagnosis. Illustrations (some color), bibliography.

Harris, David A., ed. *Prions: Molecular and Cellular Biology.* Portland, Oreg.: Horizon Scientific Press, 1999. Focuses on the cellular, biochemical, and genetic aspects of prion diseases. Illustrations, bibliography, index.

Klitzman, Robert. *The Trembling Mountain: A Personal Account of Kuru, Cannibals, and Mad Cow Disease.* New York: Plenum Trade, 1998. Autobiographical account of a study of kuru disease in Papua New Guinea. Illustrations, index.

Max, D. T. *The Family That Couldn't Sleep: A Medical Mystery.* New York: Random House, 2006. Traces the understanding of prion diseases beginning in 1765, when members of a Venetian family began to die of a mysterious inability to sleep.

Prusiner, Stanley B. "The Prion Diseases." *Scientific American* 272, no. 1 (January, 1995): 48. Discusses how meat consumption by humans can lead to prion diseases.

_______, ed. *Prion Biology and Diseases.* 2d ed. Cold Spring Harbor, N.Y.: Cold Spring Harbor Laboratory Press, 2004. Prusiner, who won the 1997 Nobel Prize in Medicine for his discovery of prions, and contributors provide an overview of research. Illustrations (some color), bibliography, index.

Rabenau, Holger F., Jindrich Cinatl, and Hans Wilhelm Doerr, eds. *Prions: A Challenge for Science, Medicine, and Public Health System.* 2d ed, rev. and extended. New York: Karger, 2004. Focuses on the etiological, clinical, and diagnostic aspects of prion diseases, as well as epidemiology, disease management, and how prions might be inactivated. Illustrations, bibliography, index.

Ratzan, Scott C., ed. *The Mad Cow Crisis: Health and the Public Good.* New York: New York University Press, 1998. A look at the disease from scientific, historical, political, health, preventive, and management perspectives. Illustrations, bibliography, index.

Yam, Philip. *The Pathological Protein: Mad Cow, Chronic Wasting, and Other Deadly Prion Diseases.* New York: Copernicus, 2003. Chronicles medical and scientific efforts to discover and understand prions and to devise treatments for the diseases they create.

WEB SITES OF INTEREST

Genetics Home Reference, Prion Disease
http://ghr.nlm.nih.gov/condition=priondisease
Describes the genes related to prion disease and offers links to additional sources of information.

Medline Plus, Creutzfeldt-Jakob Disease
http://www.nlm.nih.gov/medlineplus/creutzfeldtjakobdisease.html
This page in the site created by the U.S. National Library of Medicine and the National Institutes of Health provides numerous links to information about Creutzfeldt-Jakob disease.

National Institute of Neurological Disorders and Stroke, NINDS Kuru Information Page
http://www.ninds.nih.gov/disorders/kuru/kuru.htm
Offers a brief overview of the disease and links to additional information about kuru and other prion diseases.

National Organization for Rare Disorders
http://www.rarediseases.org
Searchable site by type of disorder. Includes background information on Creutzfeldt-Jakob syndrome, a list of other names for the disorder, and a list of related organizations.

See also: Alzheimer's disease; Huntington's disease.

Progressive external ophthalmoplegia

CATEGORY: Diseases and syndromes
ALSO KNOWN AS: PEO

DEFINITION

Progressive external ophthalmoplegia (PEO) is a mitochondrial myopathy disorder that is characterized by painless, slowly progressive paralysis of certain eye muscles. PEO is typically caused by a mitochondrial DNA (mtDNA) deletion in the skeletal muscle during oogenesis or embryogenesis.

RISK FACTORS

Males and females are affected equally by this mutation. Onset of symptoms tends to begin in young adulthood. Although the mutation occurs almost always as a sporadic point mutation, it can be inherited as a point mutation of maternal mitochondrial tRNA or as autosomal dominant and autosomal recessive deletions of mtDNA, with the autosomal recessive inheritance usually being more severe. However, the mother of a child with PEO is usually unaffected. The risk to siblings is extremely low as no cases of maternal transmission to more than one child have been reported. The offspring of males with an mtDNA mutation are not at risk.

ETIOLOGY AND GENETICS

Mitochondrial diseases affect the mitochondria, which are found in almost all cells and are responsible for providing the energy that cells need to survive. Since most cells rely on mitochondria for energy, a mitochondrial disease can affect many different types of cells. When a mitochondrial disease causes prominent muscular complications, it is called a mitochondrial myopathy. PEO is a mitochondrial myopathy that is characterized by slowly progressive paralysis of the six muscles that control the movement of the human eye (extraocular muscles).

PEO involves multiple mtDNA deletions in the skeletal muscles. Several types of deletions of varying lengths have been identified in PEO. Most mitochondrial DNA deletions associated with PEO typically range in size from two to ten kilobases. The same person may have variable proportions of deleted mtDNA in different tissues throughout their body which can ultimately determine how significantly the tissue is clinically affected. However, no correlation exists between the size or location of the mtDNA deletion and phenotype.

Mitochondrial DNA deletions are almost always caused by a point mutation. Specifically in PEO, multiple deletions of mtDNA have been identified by mutations in the *SLC25A4*, *PEO1*, *POLGI*, and *ECGF1* genes. When the mutation is inherited, it is always transmitted by maternal inheritance and various mtDNA point mutations have been found in people with maternally inherited PEO.

SYMPTOMS

Severity and phenotypic variability of symptoms vary greatly with PEO. This often makes an early diagnosis more difficult. The most consistent symptoms usually begin to manifest in late adolescence or young adulthood. Symptoms typically progress slowly. The defect in ocular motility is usually bilateral and symmetric so most patients do not complain of double vision. Primary symptoms include reduced superior visual field, corneal dryness and ulceration due to poor blinking reflexes, droopy eyelids, loss of Bell's reflex (eyes roll upward when lids are closed), muscle weakness and wasting, exercise intolerance, muscle cramps, and low energy.

SCREENING AND DIAGNOSIS

Diagnosis of PEO relies upon the presence of characteristic physical findings as well as through laboratory and imaging studies. When PEO is suspected, a complete ophthalmologic exam including a dilated retinal exam, a cranial nerve test, and a forced duction test should be performed. Thin, symmetrical extraocular muscles are often found during magnetic resonance imaging (MRI), computed tomography (CT), and ultrasound studies. Laboratory testing is often used to differentiate between PEO and other conditions with similar physical symptoms (such as myasthenia gravis). Serum creatine kinase is usually measured since this may be elevated in mitochondrial disease. In addition, deletions in mtDNA in PEO patients can be detected in muscle tissue by Southern blot analysis of samples taken from a muscle biopsy. Genetic tests can be performed to screen for known mutations/deletions.

TREATMENT AND THERAPY

There is no cure for PEO, but treatments do exist for the alleviation of some of the associated symptoms. Mechanical devices such as placement of eyelid slings have helped to decrease severe drooping. Surgical procedures can also be performed to lift a drooping eyelid. However, since poor blink responses are often associated with PEO, the cornea of the eye can become dry or scratched following surgical intervention. The use of artificial tears and/or eye patching at night can often help to alleviate this problem. Recently, dietary supplements such as Q10 and L-carnitine, which are involved in increasing ATP production in the cells, have shown modest results in some people. Physical and occupational therapy as well as treatments aimed at preventing/treating symptoms of depression which are often present in people suffering from a chronic disease are also frequently incorporated into the overall treatment strategy for PEO.

PREVENTION AND OUTCOMES

There is no effective means of prevention of PEO. Genetic counseling enables people to make more informed medical and personal decisions and should always be encouraged and available to patients and their families especially regarding options for at-risk mothers. PEO is not a fatal disease; however, associated disabilities may affect mortality.

Kimberly Lynch

FURTHER READING

Bau, Viktoria, and Stephan Zierz. "Update on Chronic Progressive External Ophthalmoplegia." *Strabismus* 13, no. 3 (September, 2005): 133-142. A journal article that provides a thorough review of PEO.

Lewin, Benjamin. *Genes IX.* Sudbury, Mass.: Jones and Bartlett, 2008. A popular textbook that sets the standard for teaching molecular biology and genetics with a unified approach.

Riordan-Eva, Paul, and John P. Whitcher. *Vaughan & Asbury's General Ophthalmology.* 17th ed. New York: McGraw-Hill, 2008. A widely used textbook that provides a comprehensive review of ophthalmology.

WEB SITES OF INTEREST

Eye Rounds
http://www.eyerounds.org

Gene Reviews
http://www.genetests.org

See also: Aniridia; Best disease; Choroideremia; Color blindness; Congenital muscular dystrophy; Corneal dystrophies; Duchenne muscular dystrophy; Glaucoma; Gyrate atrophy of the choroid and retina; Kennedy disease; Limb girdle muscular dystrophy; McArdle's disease; Macular degeneration; Myotonic dystrophy; Nemaline myopathy; Norrie syndrome; Retinitis pigmentosa; Retinoblastoma.

Prostate cancer

CATEGORY: Diseases and syndromes
ALSO KNOWN AS: Prostatic carcinoma

DEFINITION

Prostate cancer is a disease in which cancer cells grow in the prostate gland. The prostate is a walnut-sized gland in men. The prostate makes a fluid that is part of semen.

Cancer occurs when cells in the body (in this, case prostate cells) divide without control or order. Normally, cells divide in a regulated manner. If cells keep dividing uncontrollably when new cells are not needed, a mass of tissue forms, called a growth or tumor. The term "cancer" refers to malignant tumors, which can invade nearby tissue and spread to other parts of the body. A benign tumor does not invade or spread.

The sooner prostate cancer is treated, the better the outcome. Men should contact their doctors right away if they think they have this condition.

RISK FACTORS

Men should tell their doctors if they have any of the risk factors for prostate cancer. Men who are fifty-five years of age or older and black men are at increased risk for the disease. Other risk factors include having a family history of prostate cancer, especially in a father or brother; having a family history of prostate cancer diagnosed at a young age; and eating a high-fat diet.

ETIOLOGY AND GENETICS

Prostate cancer is a complex condition that may involve both genetic and environmental determinants. About 75 percent of cases are sporadic, with no known underlying genetic basis. An additional 15 to 20 percent of cases are termed "familial" because they run in families, and these probably result from a combination of shared genes and shared lifestyles. Only 5 to 10 percent of cases are termed "hereditary," yet mutations in a dozen or more genes have been identified that may increase an individual's susceptibility to prostate cancer. Mutations in susceptibility genes for hereditary prostate cancer probably account for about half of the cases of disease in patients aged 50 years or younger. In these cases, the vast majority show an autosomal dominant pattern of transmission, meaning that a single copy of the mutation is sufficient to cause the increased susceptibility. An affected individual has a 50 percent chance of transmitting the mutation to each of his or her children. Other cases of hereditary prostate cancer, however, result from a spontaneous new mutation, so in these instances affected individuals will have an unaffected father.

Six genes that are known to be associated with increased susceptibility to prostate cancer are *HPC1* (found on chromosome 1 at position 1q25), *HPC2* (located on chromosome 17 at position 17p11), *HPCX* (at position 11p11.22), *CAPB* (at position 1p36.1-p35), *PCA3* (at position 9q21-q22), and *TMPRSS2* (at position 21q22.3). Additionally, mutations in the *BRCA1* and *BRCA2* genes, most commonly associated with hereditary breast and ovarian cancer, also confer an increased risk for prostate cancer. One study suggests that men with *BRCA1* mutations have a slightly increased risk, while men with *BRCA2* mutations may have as much as a 20 percent risk of developing prostate cancer.

SYMPTOMS

Men who have any of these symptoms should not assume it is due to prostate cancer. These symptoms may also be caused by other, less serious health conditions, such as benign prostatic hyperplasia (BPH) or an infection.

Men should tell their doctors if they have any of the symptoms, which include a need to urinate frequently, especially at night; difficulty starting urination or holding back urine; an inability to urinate; a weak or interrupted urine flow; and painful or burning urination. Additional symptoms include difficulty having an erection, painful ejaculation, blood in urine or semen, and frequent pain or stiffness in the lower back, hips, or upper thighs.

Screening and Diagnosis

The doctor will ask about symptoms and medical history and will perform a physical exam. Tests include a digital rectal exam, in which the doctor will insert a gloved finger into the rectum in order to examine it; urine tests to check for blood or infection; and blood tests to measure prostate specific antigen (PSA) and prostatic acid phosphatase (PAP).

Other tests to learn more about the cause of a patient's symptoms include transrectal ultrasonography, a test that uses sound waves and a probe inserted into the rectum to find tumors; intravenous pyelogram, a series of X rays of the organs of the urinary tract; cystoscopy, in which the doctor looks into the urethra and bladder through a thin, lighted tube; and biopsy, the removal of a sample of prostate tissue to test for cancer cells.

Treatment and Therapy

Once prostate cancer is found, tests are done to find out if the cancer has spread and, if so, to what extent. Treatment depends on how far the cancer has spread. Patients should talk to a radiation oncologist and urologist, who can help decide the best treatment plan. Patients should also discuss the benefits and risks of each treatment option.

Standard options include watchful waiting. There is no treatment with watchful waiting. The doctor will do tests to see if the cancer is growing. Watchful waiting is used for early-stage prostate cancer that seems to be growing slowly, for older prostate cancer patients, and those with serious medical problems that may make the treatment risks outweigh the possible benefits.

Surgery involves removing the cancerous tumor, nearby tissues, and possibly the nearby lymph nodes. Surgery is offered to patients who are in good health and are younger than seventy years old.

Types of surgery include pelvic lymphadenectomy, the removal of lymph nodes in the pelvis to determine if they contain cancer. If they do, removal of the prostate and other treatment may be recommended. Radical retropubic prostatectomy is the removal of the entire prostate and nearby lymph nodes through an incision in the abdomen. Radical perineal prostatectomy is the removal of the entire prostate through an incision between the scrotum and the anus; nearby lymph nodes are sometimes removed through a separate incision in the abdomen.

Transurethral resection of the prostate (TURP) is the removal of part of the prostate with an instrument inserted through the urethra. TURP is not a cancer surgery, but it can be used to relieve the symptoms of patients who have either prostate cancer or an enlarged gland due to other reasons.

Prostate cancer surgery can cause impotence, and it can also cause urine leakage from the bladder or stool from the rectum. Nerve-sparing surgery may reduce these risks. However, this surgery may not treat very large tumors or tumors that are very close to nerves.

Robotic surgery and laparoscopic surgery may be other options. These minimally invasive techniques can reduce side effects, blood loss, and recovery time.

Radiation therapy involves the use of radiation to kill cancer cells and shrink tumors. Radiation may be external radiation therapy, in which radiation is directed at the tumor from a source outside the body, or internal radiation therapy, in which radioactive materials are placed into the body near the cancer cells. Internal radiation therapy is often used for treating earlier-stage cancers. Radiation therapy for prostate cancer may cause impotence and urinary problems. However, most studies show that impotence rates are less for radiation therapy than for standard prostatectomy and slightly less than rates for nerve-sparing procedures. Incontinence also occurs less frequently following radiation therapy than it does following prostatectomy. However, there is a slightly increased risk of cystitis due to radiation.

Hormone therapy is used for patients whose prostate cancer has spread beyond the prostate or has recurred after treatment. The goal of hormone therapy is to lower levels of the male hormones, called androgens. The main androgen is testosterone. Lowering androgen levels can cause prostate cancers to shrink or grow more slowly, but it does not cure cancer.

Methods of hormone therapy include orchiectomy, a surgical procedure to remove one or both of the testicles, which are the main source of male hormones. Orchiectomy decreases hormone production, which can shrink or slow the growth of most prostate cancers. Luteinizing hormone-releasing hormone (LHRH) agonists are injections that can decrease the amount of testosterone made by the testicles. Antiandrogens, such as flutamide or bicalutamide, are medications that can block the action of androgens. These medications are used in combination with orchiectomy or LHRH agonists, a combination called total androgen blockade. Other forms of hormone therapy include the use of drugs, such as ketoconazole or aminoglutethimide, that prevent the adrenal glands from making androgens; and estrogens, drugs that prevent the production of testosterone in the testicles. Estrogens are rarely used today because of the risk of serious side effects. Hormone therapy for prostate cancer may cause hot flashes, impaired sexual function, loss of sexual desire, and weakened bones.

Other treatments are being tested. Patients may want to consider taking part in a clinical trial when weighing their treatment options. The treatments that are currently being tested include cryosurgery, which uses an instrument to freeze and destroy prostate cancer cells.

Chemotherapy is the use of drugs to kill cancer cells. It may be given in many forms, including pill, injection, and via a catheter. The drugs enter the bloodstream and travel through the body, killing mostly cancer cells, but also some healthy cells. One type of chemotherapy is the use of docetaxel (Taxotere). This drug was found to prolong life in men with hormone refractory prostate cancer (HRPC). In HRPC, PSA levels continue to rise or the tumor continues to grow despite hormone therapy.

Biological therapy is the use of medications or substances made by the body to increase or restore the body's natural defenses against cancer. It is also called biological response modifier (BRM) therapy. High-intensity focused ultrasound is a treatment that uses an endorectal probe to makes ultrasound (high-energy sound waves). This can destroy cancer cells. Conformal radiation therapy uses three-dimensional radiation beams that are conformed into the shape of the diseased prostate. This treatment spares nearby tissue the damaging effects of radiation. Intensity-modulated radiation therapy (IMRT) uses radiation beams of different intensities to deliver higher doses of radiation therapy to the tumor and lower doses to nearby tissues at the same time.

Prevention and Outcomes

Beginning at age fifty, men should be offered a digital rectal exam and PSA blood test to screen for prostate cancer. Many, but not all, professional organizations recommend a yearly PSA blood test for men older than fifty. Black men and men with close family members who have had prostate cancer diagnosed at a young age should begin screening when they are forty-five years old. All men should discuss PSA testing with their doctors.

Krisha McCoy, M.S.; reviewed by Igor Puzanov, M.D.
"Etiology and Genetics" by Jeffrey A. Knight, Ph.D.

Further Reading

American Cancer Society. *Quick Facts Prostate Cancer: What You Need to Know—Now.* Atlanta: Author, 2007.

Berthold, D. R., et al. "Docetaxel Plus Prednisone or Mitoxantrone for Advanced Prostate Cancer: Updated Survival of the TAX 327 Study." *Journal of Clinical Oncology* 26, no. 2 (January 10, 2008): 242-245.

Burnett, Arthur. *The Johns Hopkins Patients' Guide to Prostate Cancer.* Sudbury, Mass.: Jones and Bartlett, 2010.

EBSCO Publishing. *Health Library: Prostate Cancer.* Ipswich, Mass.: Author, 2009. Available through http://www.ebscohost.com.

Klein, Eric A. *The Cleveland Clinic Guide to Prostate Cancer.* New York: Kaplan, 2009.

Tannock, I. F., et al. "Docetaxel Plus Prednisone or Mitoxantrone Plus Prednisone for Advanced Prostate Cancer." *New England Journal of Medicine* 351, no. 15 (October 7, 2004): 1502-1512.

Web Sites of Interest

American Cancer Society
http://www.cancer.org

American Urological Association
http://www.auanet.org

Hormone Refractory Prostate Cancer
http://www.hrpca.org/index.htm

National Cancer Institute
http://www.cancer.gov

Prostate Cancer Canada
http://www.prostatecancer.ca/english/home

Prostate Cancer Canada Network
http://www.cpcn.org

Zero: The Project to End Prostate Cancer
http://www.zerocancer.org/index.html

See also: *BRAF* gene; *BRCA1* and *BRCA2* genes; Breast cancer; Burkitt's lymphoma; Chemical mutagens; Chromosome mutation; Chronic myeloid leukemia; Colon cancer; Cowden syndrome; *DPC4* gene testing; Familial adenomatous polyposis; Gene therapy; Harvey *ras* oncogene; Hereditary diffuse gastric cancer; Hereditary diseases; Hereditary leiomyomatosis and renal cell cancer; Hereditary mixed polyposis syndrome; Hereditary non-VHL clear cell renal cell carcinomas; Hereditary papillary renal cancer; Homeotic genes; *HRAS* gene testing; Hybridomas and monoclonal antibodies; Li-Fraumeni syndrome; Lynch syndrome; Mutagenesis and cancer; Multiple endocrine neoplasias; Mutation and mutagenesis; Nondisjunction and aneuploidy; Oncogenes; Ovarian cancer; Pancreatic cancer; Tumor-suppressor genes; Wilms' tumor aniridia-genitourinary anomalies-mental retardation (WAGR) syndrome.

Protein structure

Category: Molecular genetics

Significance: Proteins have three-dimensional structures that determine their functions, and slight changes in overall structure may significantly alter their activity. Correlation of protein structure and function can provide insights into cellular metabolism and its many interconnected processes. Because most diseases result from improper protein function, advances in this field could lead to effective molecular-based disease treatments.

Key terms

amino acid: the basic subunit of a protein; there are twenty commonly occurring amino acids, any of which may join together by chemical bonds to form a complex protein molecule

enzymes: proteins that are able to increase the rate of chemical reactions in cells without being altered in the process

hydrogen bond: a weak bond that helps stabilize the folding of a protein

polypeptide: a chain of amino acids joined by chemical bonds

R group: a functional group that is part of an amino acid that gives each amino acid its unique properties

Protein Structure and Function

Proteins consist of strings of individual subunits called amino acids that are chemically bonded together with peptide bonds. Once amino acids are bonded, the resulting molecule is called a polypeptide. The properties and arrangement of the amino acids in the polypeptide cause it to fold into a specific shape or conformation that is required for proper protein function. Proteins have been called the "workhorses" of the cell because they perform most of the activities encoded in the genes of the cell. Proteins function by binding to other molecules, frequently to other proteins. The precise three-dimensional shape of a protein determines the specific molecules it will be able to bind to, and for many proteins binding is specific to just one other specific type of molecule.

In 1973, Christian B. Anfinsen performed experiments that showed that the three-dimensional structure of a protein is determined by the sequence of its amino acids. He used a protein called ribonuclease (RNase), an enzyme that degrades RNA in the cell. The ability of ribonuclease to degrade RNA is dependent upon its ability to fold into its proper three-dimensional shape. Anfinsen showed that if the enzyme was completely unfolded by heat and chemical treatment (at which time it would not function), it formed a linear chain of amino acids. Although there were 105 possible conformations that the enzyme could take upon refolding, it would refold into the single correct functional conformation upon removal of heat and chemicals. This established that the amino acid sequences of proteins, which are specified by the genes of the cell, carry all of the information necessary for proteins to fold into their proper three-dimensional shapes.

To understand protein conformation better, it is helpful to analyze the underlying levels of structure that determine the final three-dimensional shape. The primary structure of a polypeptide is the simplest level of structure and is, by definition, its amino acid sequence. Because the primary structure of polypeptides ultimately determines all succeeding levels of structure, knowing the primary structure should theoretically allow scientists to predict the final three-dimensional structure. Building on a detailed knowledge of the structure of many proteins, scientists can now develop computer programs that are able to predict three-dimensional shape with some degree of accuracy, but much more research will be required to increase the accuracy of these methods.

Primary Protein Structure

There are twenty naturally occurring amino acids that are commonly found in proteins, and each of these has a common structure consisting of a nitrogen-containing amino group (—NH_2), a carboxyl group (—COOH), a hydrogen atom (H), and a unique functional group referred to as an R group, all bonded to a central carbon atom (known as the alpha carbon, or Cα) as shown in the following figure:

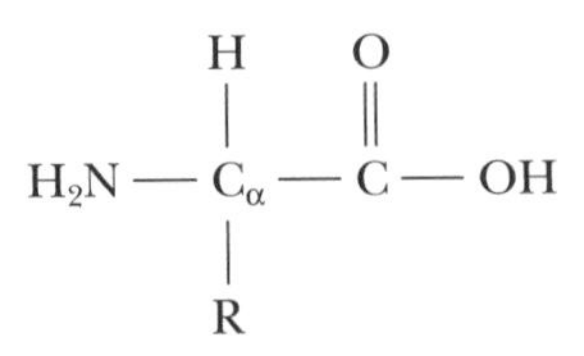

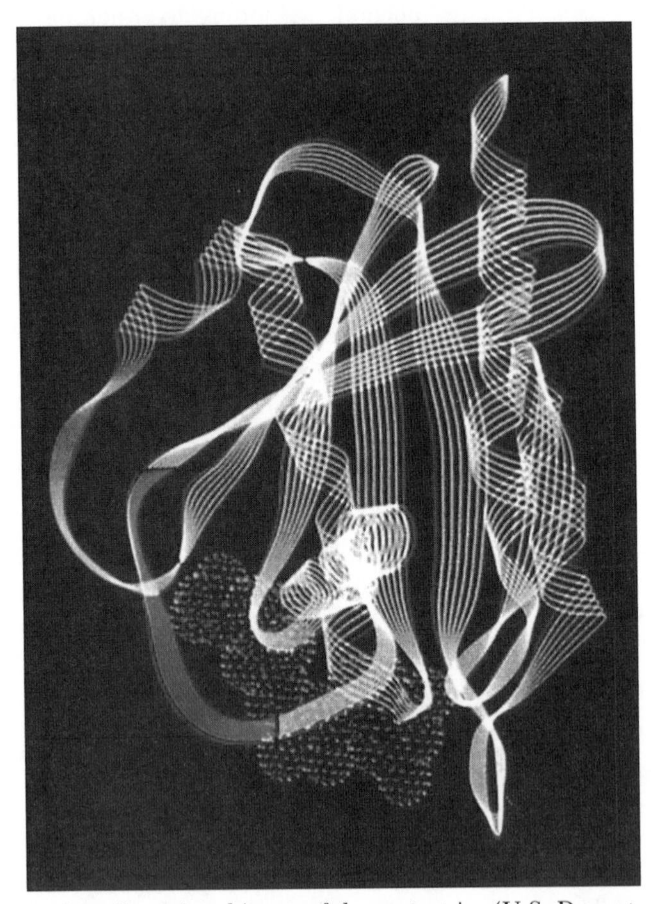
A three-dimensional image of the ras protein. (U.S. Department of Energy Human Genome Program, http://www.ornl.gov/hgmis)

The uniqueness of each of the twenty amino acids is determined by the R group. This group may be as simple as a hydrogen atom (in the case of the amino acid glycine) or as complex as a ring-shaped structure (as found in the amino acid phenylalanine). It may be charged, either positively or negatively, or it may be uncharged.

Cells join amino acids together to form peptides (strings of up to ten amino acids), polypeptides (strings of ten to one hundred amino acids), or proteins (single or multiple polypeptides folded and oriented to one another so they are functional). The amino acids are joined together by covalent bonds, called peptide bonds (in the box in the following figure), between the carbon atom of the carboxyl group of one amino acid (—COOH) and the nitrogen atom of the amino group (—NH_2) of the next adjacent amino acid:

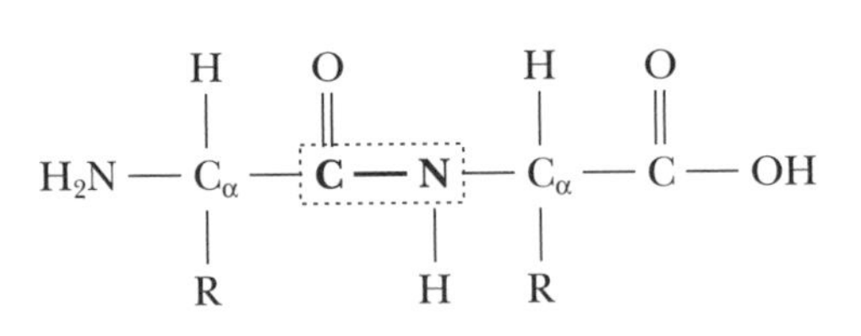

During the formation of the peptide bond, a molecule of water (H_2O) is lost (an -OH from the carboxyl group and an -H from the amino group), so this reaction is also called a dehydration synthesis. The result is a dipeptide (a peptide made of two amino acids joined by a peptide bond) that has a "backbone" of nitrogens and carbons (N—Cα—C—N—Cα—C) with other elements and R groups protruding from the backbone. An amino acid may be joined to the growing peptide chain by formation of a peptide bond between the carbon atom of the free carboxyl group (on the right of the preceding figure) and the nitrogen atom of the amino acid being added. The end of a polypeptide with an exposed carboxyl group is called the C-terminal end, and the end with an exposed amino group is called the N-terminal end.

The atoms and R groups that protrude from the backbone are capable of interacting with each other, and these interactions lead to higher-order secondary, tertiary, and quaternary structures.

SECONDARY STRUCTURE

The next level of structure is secondary structure, which involves the formation of hydrogen bonds between the oxygen atoms in carboxyl groups with the hydrogen atoms of amino groups from different parts of the polypeptide. Hydrogen bonds are weak bonds that form between atoms that have a very strong attraction for electrons (such as oxygen or nitrogen), and a hydrogen atom that is bound to another atom with a very strong attraction for electrons. Secondary structure does not involve the formation of bonds with R groups or atoms that are parts of R groups, but involves bonding just between amino and carboxyl groups that are in the peptide bonds making up the backbone of polypeptides.

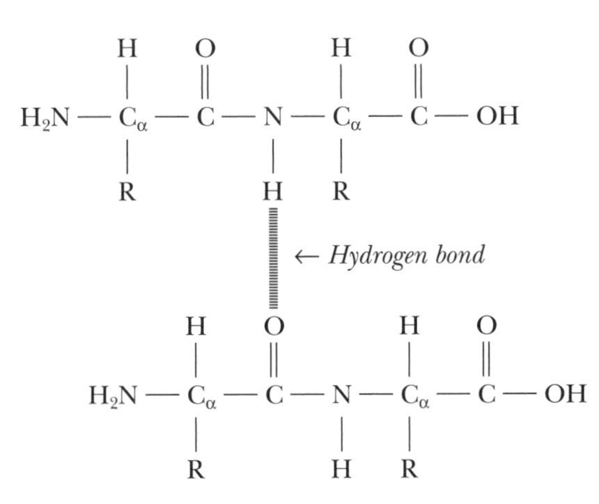

These hydrogen bonds between backbone molecules lead to the formation of two major types of structures: alpha helices and beta-pleated sheets. An alpha helix is a rigid structure shaped very much like a telephone cord; it spirals around as the oxygen of one amino acid of the chain forms a hydrogen bond with the hydrogen atom of an amino acid five amino acids away on the protein strand. The rigidity of the structure is caused by the large number of hydrogen bonds (individually weak but collectively strong) and the compactness of the helix that forms. Many alpha helices are found in proteins that function to maintain cell structure.

Beta sheets are formed by hydrogen bonding between amino acids in different regions (often very far apart on the linear strand) of a polypeptide. The shape of a beta-pleated sheet may be likened to the bellows of an accordion or a sheet of paper that has been folded multiple times to form pleats. Because of the large number of hydrogen bonds in them, beta sheets are also strong structures, and they form planar regions that are often found at the bottom of "pockets" inside proteins to which other molecules attach.

In addition to alpha helices or beta-pleated sheets, other regions of the protein may have no obvious secondary structure; these regions are said to have a "random coil" shape. It is the combinations of random coils, alpha helices, and beta sheets that form the secondary structure of the protein.

TERTIARY STRUCTURE

The final level of protein shape (for a single polypeptide or simple protein) is called tertiary structure. Tertiary structure is caused by the numerous interactions of R groups on the amino acids and of the protein with its environment, which is usually aqueous (water based). Various R groups may either be attracted to and form bonds with each other, or they may be repelled from each other. For example, if an R group has an overall positive electrical charge, it will be attracted to R groups with a negative charge but repelled from other positively charged R groups. For a polypeptide with one hundred amino acids, if amino acid number 6 is negatively charged, it could be attracted to a positively charged amino acid at position 74, thus bringing two ends of the protein that are linearly distant into close proximity. Many of these attractions lead to the formation of hydrogen, ionic, or covalent bonds. For example, sulfur is contained in the R groups of a few of the amino acids, and sometimes a disulfide bond (a covalent bond) will be formed between two of these. It is the arrangement of disulfide bonds in hair proteins that gives hair its physical properties of curly versus straight. Hair permanent treatments actually break these disulfide bonds and then reform them when the hair is arranged as desired. Many other R groups in the protein will also be attracted to or repelled from each other, leading to an overall folded shape that is most stable. In addition, because most proteins exist in an aqueous environment in the cell, most proteins are folded such that their amino acids with hydrophilic R groups (R groups attracted to water) are on the outside, while their amino acids with hydrophobic R groups (R groups repelled from water) are tucked away in the interior of the protein.

Quaternary Structure

Many polypeptides are nonfunctional until they physically associate with another polypeptide, forming a functional unit made up of two or more subunits. Proteins of this type are said to have quaternary structure. Quaternary structure is caused by interactions between the R groups of amino acids of two different polypeptides. For example, hemoglobin, the oxygen-carrying protein found in red blood cells, functions as a tetramer, with four polypeptide subunits.

Because secondary, tertiary, and quaternary interactions are caused by the R groups of the specific amino acids, the folding is ultimately dictated by the amino acid sequence of the protein. Although there may be numerous possible final conformations that a polypeptide could take, it usually assumes only one of these, and this is the conformation that leads to proper protein function. Many polypeptides are capable of folding into their final conformation spontaneously. More complex ones may need the assistance of other proteins, called chaperones, to help in the folding process.

Impact and Applications

The function of a protein may be altered by changing its shape, because proper function is dependent on proper conformation. Many genetic defects are detrimental because they represent a mutation that results in a change in protein structure. Changes in protein conformation are also an integral part of metabolic control in cells. Normal cellular processes are controlled by "turning on" and "turning off" proteins at the appropriate time. A protein's activity may be altered by attaching a molecule or ion to that protein that results in a change of shape. Because the shape is caused by R group interactions, binding of a charged ion such as calcium to the protein will alter these interactions and thus alter the shape and function of the protein. One molecular "on/off" switch that is used frequently within a cell involves the attachment or removal of a phosphate group to or from a protein. Attachment of a phosphate will significantly alter the shape of the protein by repelling negatively charged amino acids and attracting positively charged amino acids, which will either activate the protein to perform its function (turn it on) or deactivate it (turn it off).

Cancer and diseases caused by bacterial or viral infections are often the result of nonfunctional proteins that have been produced with incorrect shapes or that cannot be turned on or off by a molecular switch. The effects may be minor or major, depending upon the protein, its function, and the severity of the structural deformity. Understanding how a normal protein is shaped and how it is altered in the disease process allows for the development of drugs that may block the disease. This may be accomplished by blocking or changing the effect of the protein of interest or by generating drugs or therapies that mimic the normal functioning of the protein. Thus, understanding protein structure is essential for understanding proper protein function and for developing molecular-based disease treatments.

Sarah Lea McGuire, Ph.D.;
updated by Bryan Ness, Ph.D.

Further Reading

Banaszak, Leonard J. *Foundations of Structural Biology.* San Diego: Academic Press, 2000. Focuses on three-dimensional visualization strategies for proteins and DNA segments. Illustrations, bibliography, index.

Brändén, Carl-Ivar, and John Tooze. *Introduction to Protein Structure.* 2d ed. New York: Garland, 1999. Covers research into the structure and logic of proteins. Illustrations (mostly color), bibliography, index.

Buxbaum, Engelbert. *Fundamentals of Protein Structure and Function.* New York: Springer, 2007. Textbook provides basic information about protein structure, characterization, and function; amino acids; and enzymes.

Johnson, George B. *How Scientists Think: Twenty-one Experiments That Have Shaped Our Understanding of Genetics and Molecular Biology.* Dubuque, Iowa: Wm. C. Brown, 1996. Gives an excellent introductory account of Christian Anfinsen's experiments leading to the determination that primary sequence dictates protein shape. Illustrations and index.

Kyte, Jack. *Structure in Protein Chemistry.* 2d ed. New York: Garland, 2007. This textbook provides information about protein purification, amino acid sequences, crystallography, hydrogen bonding, and other topics relevant to protein chemistry.

Lodish, Harvey, et al. *Molecular Cell Biology.* 6th ed. New York: W. H. Freedman, 2008. Provides both summary and detailed accounts of protein structure and the chemical bonds that lead to the various levels of structure. Illustrations, bibliography, index.

McRee, Duncan Everett. *Practical Protein Crystallography.* 2d ed. San Diego: Academic Press, 1999. Introductory protein structure handbook. Illustrations (some color), bibliography, index.

Maddox, Brenda. *Rosalind Franklin: The Dark Lady of DNA.* New York: HarperCollins, 2002. Biography of the foundational but little recognized work of the physical chemist, whose photographs of DNA were critical in helping James Watson, Francis Crick, and Maurice Wilkins discover the double-helical structure of DNA, for which they won the Nobel Prize in 1962. Illustrations, bibliography, index.

Murphy, Kenneth P. *Protein Structure, Stability, and Folding.* Totowa, N.J.: Humana Press, 2001. Describes cutting-edge experimental and theoretical methodologies for investigating proteins and protein folding.

Whitford, David. *Proteins: Structure and Function.* Hoboken, N.J.: John Wiley and Sons, 2005. Introduction to the study of proteins, including information about amino acids, fibrous and membrane proteins, protein synthesis, and protein expression. Illustrations.

Web Sites of Interest

Introduction to Protein Structure
http://webhost.bridgew.edu/fgorga/proteins/default.htm

Frank R. Gorga, a professor at Bridgewater State College, has prepared this introduction as an adjunct to one of his course lectures. It provides information on proteins, amino acids, and peptides.

RCSB Protein Data Bank
http://www.rcsb.org/pdb/home/home.do

The Research Collaboratory for Structural Bioinformatics (RCSB) Protein Data Bank is an archive of information about "experimentally-determined structures of proteins, nucleic acids, and complex assemblies" that enables users to conduct searches "based on annotations relating to sequence, structure and function."

See also: Central dogma of molecular biology; DNA repair; DNA replication; DNA structure and function; Genetic code; Genetic code, cracking of; Molecular genetics; Protein synthesis; RNA structure and function; RNA transcription and mRNA processing; RNA world; Synthetic genes.

Protein synthesis

Category: Molecular genetics

Significance: Cellular proteins can be grouped into two general categories: proteins with a structural function that contribute to the three-dimensional organization of a cell, and proteins with an enzymatic function that catalyze the biochemical reactions required for cell growth and function. Understanding the process by which proteins are synthesized provides insight into how a cell organizes itself and how defects in this process can lead to disease.

Key terms

amino acid: the basic subunit of a protein; there are twenty commonly occurring amino acids, any of which may join together by chemical bonds to form a complex protein molecule

peptide bond: the chemical bond between amino acids in protein

polypeptide: a linear molecule composed of amino acids joined together by peptide bonds; all proteins are functional polypeptides

RNA: ribonucleic acid, that molecule that acts as the messenger between genes in DNA and their protein product, directing the assembly of proteins;

as an integral part of ribosomes, RNA is also involved in protein synthesis

translation: the process of forming proteins according to instructions contained in an RNA molecule

THE FLOW OF INFORMATION FROM STORED TO ACTIVE FORM

The cell can be viewed as a unit that assembles resources from the environment into biochemically functional molecules and organizes these molecules in three-dimensional space in a way that allows cellular growth and replication. In order to carry out this organizational process, a cell must have a biosynthetic means to assemble resources into useful molecules, and it must contain the information required to produce the biosynthetic and structural machinery. DNA serves as the stored form of this information, whereas protein is its active form. Although there are thousands of different proteins in cells, they either serve a structural role or are enzymes that catalyze the biosynthetic reactions of a cell. Following the discovery of the structure of DNA in 1953 by James Watson and Francis Crick, scientists began to study the process by which the information stored in this molecule is converted into protein.

Proteins are linear, functional molecules composed of a unique sequence of amino acids. Twenty different amino acids are used as the protein building blocks. Although the information for the amino acid sequence of each protein is present in DNA, protein is not synthesized directly from this source. Instead, RNA serves as the intermediate form from which proteins are synthesized. RNA plays three roles during protein synthesis. Messenger RNA (mRNA) contains the information for the amino acid sequence of a protein. Transfer RNAs (tRNAs) are small RNA molecules that serve as adapters that decipher the coded information present within an mRNA and bring the appropriate amino acid to the polypeptide as it is being synthesized. Ribosomal RNAs (rRNAs) act as the engine that carries out most of the steps during protein synthesis. Together with a specific set of proteins, rRNAs form ribosomes that bind the mRNA, serve as the platform for tRNAs to decode an mRNA, and catalyze the formation of peptide bonds between amino acids. Each ribosome is composed of two subunits: a small (or 40s) and a large (or 60s) subunit, each of which has its own function. The "s" in 40s and 60s is an abbreviation for Svedberg units, which are a measure of how quickly a large molecule or complex molecular structure sediments (or sinks) to the bottom of a centrifuge tube while being centrifuged. The larger the number, the larger the molecule.

Like all RNA, mRNA is composed of just four types of nucleotides: adenine (A), guanine (G), cytosine (C), and uracil (U). Therefore, the information in an mRNA is contained in a linear sequence of nucleotides that is converted into a protein molecule composed of a linear sequence of amino acids. This process is referred to as "translation," since it converts the "language" of nucleotides that make up an mRNA into the "language" of amino acids that make up a protein. This is achieved by a three-letter genetic code in which each amino acid in a protein is specified by a three-nucleotide sequence in the mRNA called a codon. The four possible "letters" means that there are sixty-four possible three-letter "words." As there are only twenty amino acids used to make proteins, most amino acids are encoded by several different codons. For example, there are six different codons (UCU, UCC, UCA, UCG, AGU, and AGC) that specify the amino acid serine, whereas there is only one codon (AUG) that specifies the amino acid methionine. The mRNA, therefore, is simply a linear array of codons (that is, three-nucleotide "words" that are "read" by tRNAs together with ribosomes). The region within an mRNA containing this sequence of codons is called the coding region.

Before translation can occur in eukaryotic cells, mRNAs undergo processing steps at both ends to add features that will be necessary for translation. (These processing steps do not occur in prokaryotic cells.) Nucleotides are structured such that they have two ends, a 5′ and a 3′ end, that are available to form chemical bonds with other nucleotides. Each nucleotide present in an mRNA has a 5′ to 3′ orientation that gives a directionality to the mRNA so that the RNA begins with a 5′ end and finishes in a 3′ end. The ribosome reads the coding region of an mRNA in a 5′ to 3′ direction. Following the synthesis of an mRNA from its DNA template, one guanine is added to the 5′ end of the mRNA in an inverted orientation and is the only nucleotide in the entire mRNA present in a 3′ to 5′ orientation. It is referred to as the cap. A long stretch of adenosine is added to the 3′ end of the mRNA to make what is called the poly-A tail.

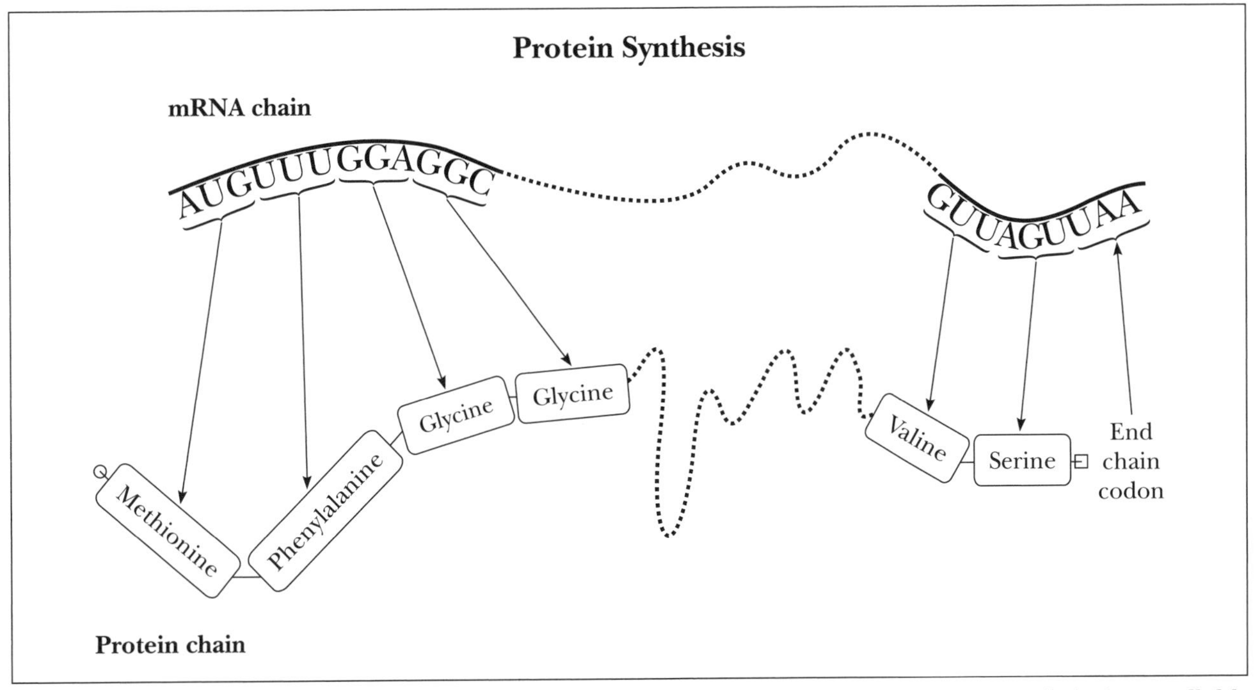

Protein synthesis is directed by messenger RNA (mRNA). The order of the amino acids in the protein chain is controlled by the order of the bases in the mRNA chain. It takes a codon of three bases to specify one amino acid.

Typically, mRNAs have a stretch of nucleotide sequence that lies between the cap and the coding region. This is referred to as the leader sequence and is not translated. Therefore, a signal is necessary to indicate where the coding region initiates. The codon AUG usually serves as this initiation codon; however, other AUG codons may be present in the coding region. Any one of three possible codons (UGA, UAG, or UAA) can serve as stop codons that signal the ribosome to terminate translation. Several accessory proteins assist ribosomes in binding mRNA and help carry out the required steps during translation.

The Translation Process: Initiation

Translation occurs in three phases: initiation, elongation, and termination. The function of the 40s ribosomal subunit is to bind to an mRNA and locate the correct AUG as the initiation codon. It does this by binding close to the cap at the 5′ end of the mRNA and scanning the nucleotide sequence in its 5′ to 3′ direction in search of the initiation codon. Marilyn Kozak identified a certain nucleotide sequence surrounding the initiator AUG of eukaryotic mRNAs that indicates to the ribosome that this AUG is the initiation codon. She found that the presence of an A or G three nucleotides prior to the AUG and a G in the position immediately following the AUG were critical in identifying the correct AUG as the initiation codon. This is referred to as the "sequence context" of the initiation codon. Therefore, as the 40s ribosomal subunit scans the leader sequence of an mRNA in a 5′ to 3′ direction, it searches for the first AUG in this context and may bypass other AUGs not in this context.

Nahum Sonenberg demonstrated that the scanning process by the 40s subunit can be impeded by the presence of stem-loop structures present in the leader sequence. These form from base pairing between complementary nucleotides present in the leader sequence. Two nucleotides are said to be complementary when they join together by hydrogen bonds. For instance, the nucleotide (or base) A is complementary to U, and these two can form what is called a "base pair." Likewise, the nucleotides C and G are complementary. Several accessory proteins, called eukaryotic initiation factors (eIFs), aid the binding and scanning of 40s subunits. The first of these, eIF4F, is composed of three subunits called eIF4E, eIF4A, and eIF4G. The protein eIF4E is the subunit responsible for recognizing and binding to the cap of the mRNA. The eIF4A subunit of eIF4F, together with another factor called eIF4B, functions to remove the presence of stem-loop structures in the leader sequence through the disruption of the base pairing between nucleotides in the stem loop. The protein eIF4G is the large subunit of eIF4F, and it serves to interact with several other proteins, one of which is eIF3. It is this latter initiation factor that the 40s subunit first associates with during its initial binding to an mRNA.

Through the combined action of eIF4G and eIF3, the 40s subunit is bound to the mRNA, and through the action of eIF4A and eIF4B, the mRNA is prepared for 40s subunit scanning. As the cellular concentration of eIF4E is very low, mRNAs must compete for this protein. Those that do not compete well for eIF4E will not be translated efficiently. This represents one means by which a cell can regulate protein synthesis. One class of mRNA that competes poorly for eIF4E encodes growth-factor proteins. Growth factors are required in small amounts to stimulate cellular growth. Sonenberg has shown that the overproduction of eIF4E in animal cells leads to a reduction in the competition for this protein, and mRNAs such as growth-factor mRNAs that were previously poorly translated when the concentration of eIF4E was low are now translated at a

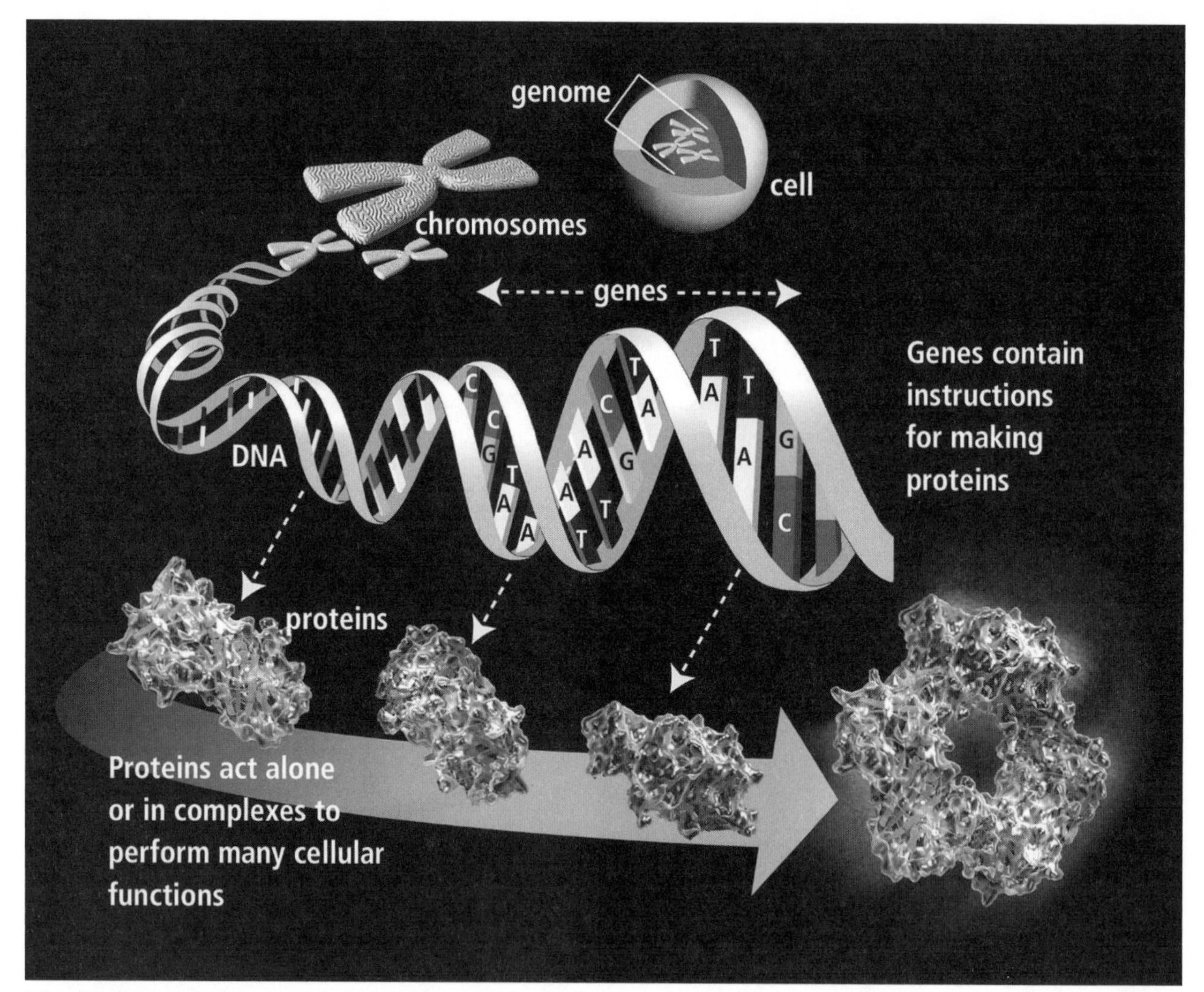

The fundamental steps in protein synthesis. (U.S. Department of Energy Human Genome Program, http://www.ornl.gov/hgmis)

higher rate when eIF4E is abundant. This in turn results in the overproduction of growth factors, which leads to uncontrolled growth, a characteristic typical of cancer cells.

A protein that specifically binds to the poly-A tail at the 3′ end of an mRNA is called the poly-A-binding protein (PABP). Discovered in the 1970's, the only function of this protein was thought to be to protect the mRNA from attack at its 3′ end by enzymes that degrade RNA. Daniel Gallie demonstrated another function for PABP by showing that the PABP-poly-A-tail complex was required for the function of the eIF4F-cap complex during translation initiation. The idea that a protein located at the 3′ end of an mRNA should participate in events occurring at the opposite end of an mRNA seemed strange initially. However, RNA is quite flexible and is rarely present in a straight, linear form in the cellular environment. Consequently, the poly-A tail can easily approach the cap at the 5′ end. Gallie showed that PABP interacts with eIF4G and eIF4B, two initiation factors that are closely associated with the cap, through protein-to-protein contacts. The consequence of this interaction is that the 3′ end of an mRNA is held in close physical proximity to its cap. The interaction between these proteins stabilizes their binding to the mRNA, which in turn promotes protein synthesis. Therefore, mRNAs can be thought of as adopting a circular form during translation that looks similar to a snake biting its own tail. This idea is now widely accepted by scientists.

One additional factor, called eIF2, is needed to bring the first tRNA to the 40s subunit. Along with the initiator tRNA (which decodes the AUG codon specifying the amino acid methionine), eIF2 aids the 40s subunit in identifying the AUG initiation. Once the 40s subunit has located the initiation codon, the 60s ribosomal subunit joins the 40s subunit to form the intact 80s ribosome. (Svedberg units are not additive; therefore, a 40s and 60s unit joined together do not make a 100s unit.) This marks the end of the initiation phase of translation.

THE TRANSLATION PROCESS: ELONGATION AND TERMINATION

During the elongation phase, tRNAs bind to the 80s ribosome as it passes over the codons of the mRNA, and the amino acids attached to the tRNAs are transferred to the growing polypeptide. Binding of the tRNAs to the ribosome is assisted by an accessory protein called eukaryotic elongation factor 1 (eEF1). A codon is decoded by the appropriate tRNA through base pairing between the three nucleotides that make up the codon in the mRNA and three complementary nucleotides within a specific region (called the anticodon) within the tRNA. The tRNA binding sites in the 80s ribosome are located in the 60s subunit. The ribosome moves over the coding region one codon at a time, or in steps of three nucleotides, in a process referred to as "translocation." When the ribosome moves to the next codon to be decoded, the tRNA containing the appropriate anticodon will bind tightly in the open site in the 60s subunit (the A site). The tRNA that bound to the previous codon is present in a second site in the 60s subunit (the P site). Once a new tRNA has bound to the A site, the ribosomal RNA itself catalyzes the formation of a peptide bond between the growing polypeptide and the new amino acid. This results in the transfer of the polypeptide attached to the tRNA present in the P site to the amino acid on the tRNA present in the A site. A second elongation factor, eEF2, catalyzes the movement of the ribosome to the next codon to be decoded. This process is repeated one codon at a time until a stop codon is reached.

The termination phase of translation begins when the ribosome reaches one of the three termination or stop codons. These are also referred to as "nonsense" codons as the cell does not produce any tRNAs that can decode them. Accessory factors, called release factors, are also required to assist this stage of translation. They bind to the empty A site in which the stop codon is present, and this triggers the cleavage of the bond between the completed protein from the last tRNA in the P site, thereby releasing the protein. The ribosome then dissociates into its 40s and 60s subunits, the latter of which diffuse away from the mRNA. The close physical proximity of the cap and poly-A tail of an mRNA maintained by the interaction between PABP and the initiation factors (eIF4G and eIF4B) is thought to assist the recycling of the 40s subunit back to the 5′ end of the mRNA to participate in a subsequent round of translation.

IMPACT AND APPLICATIONS

The elucidation of the process and control of protein synthesis provides a ready means by which scientists can manipulate these processes in cells. In addition to infectious diseases, insufficient dietary protein represents one of the greatest challenges to world health. The majority of people now living are limited to obtaining their dietary protein solely through the consumption of plant matter. Knowledge of the process of protein synthesis may allow molecular biologists to increase the amount of protein in important crop species. Moreover, most plants contain an imbalance in the amino acids needed in the human diet that can lead to disease. For example, protein from corn is poor in the amino acid lysine, whereas the protein from soybeans is poor in methionine and cysteine. Molecular biologists may be able to correct this imbalance by changing the codons present in plant genes, thus improving this source of protein for those people who rely on it for life.

Daniel R. Gallie, Ph.D.

FURTHER READING

Crick, Francis. "The Genetic Code III." *Scientific American* 215 (October, 1966): 57. The codiscoverer of DNA's helical structure provides a good summary of the code specifying the amino acids.

Lake, James. "The Ribosome." *Scientific American* 245 (August, 1981): 84-97. Summarizes information about the structure and function of ribosomes.

Lewin, Benjamin. *Genes IX.* Sudbury, Mass.: Jones and Bartlett, 2007. Details the translational process.

Liljas, Anders. *Structural Aspects of Protein Synthesis.* Hackensack, N.J.: World Scientific, 2004. Summarizes the translation process for bacterial ribosomes and explains the functions of ribosomes. Illustrations.

Rich, Alexander, and Sung Hou Kim. "The Three-Dimensional Structure of Transfer RNA." *Scientific American* 238 (January, 1978): 52-62. Presents a structural description of tRNA.

Tropp, Burton E., and David Freifelder. *Molecular Biology: Genes to Proteins.* 3d ed. Sudbury, Mass.: Jones and Bartlett, 2008. Includes two chapters about protein synthesis: "Protein Synthesis: The Genetic Code" and "Protein Synthesis: The Ribosome."

Whitford, David. "Protein Synthesis, Processing, and Turnover." In *Proteins: Structure and Function.* Hoboken, N.J.: John Wiley and Sons, 2005. This chapter of the textbook discusses the cell cycle, transcription, an outline of and the structural basis for protein synthesis, and other information about protein synthesis. Illustrated.

WEB SITES OF INTEREST

Natural Science Pages, DNA and Protein Synthesis
http://web.jjay.cuny.edu/~acarpi/NSC/12-dna.htm

One of the natural science pages created by Anthony Carpi, a professor at John Jay College in New York City, offers an explanation of DNA and protein synthesis, with links to additional information.

Online Biology Book
http://www.emc.maricopa.edu/faculty/farabee/BIOBK/BioBookPROTSYn.html#Links

Michael J. Farabee, a professor at the Maricopa Community Colleges, includes a page on protein synthesis, with links to additional information, in his online book.

Public Broadcasting Service (PBS), DNA Workshop Activity
http://www.pbs.org/wgbh/aso/tryit/dna/index.html

The workshop includes a page of information about protein synthesis and an interactive activity enabling users to become involved with the process of DNA replication and protein synthesis.

See also: Central dogma of molecular biology; DNA repair; DNA replication; DNA structure and function; Genetic code; Genetic code, cracking of; Molecular genetics; Protein structure; RNA structure and function; RNA transcription and mRNA processing; RNA world; Synthetic genes.

Proteomics

CATEGORY: Molecular genetics; Techniques and methodologies

SIGNIFICANCE: The study of proteomics and its relationship to genomics currently focuses on the vast family of gene-regulating proteins. These polypeptides and their functions affect the expression of various genetically related diseases, such as Alzheimer's and cancer. By focusing on the interrelated groups of regulator functions, geneticists are learning the connections between structure, abundance within the cell, and how each protein relates to expression.

Key terms

chromatography: a separation technique involving a mobile solvent and a stationary, adsorbent phase

mass spectroscopy: a method of analyzing molecular structure in which sample molecules are ionized and the resulting fragmented particles are passed through electric and magnetic fields to a detector

peripheral proteins: proteins of the chromosome that do not directly affect transcription

protein folding structure: the three-dimensional structure of proteins created by the folding of linked amino acids upon each other; this structure is held together by intermolecular forces, such as hydrogen bonds and ionic attractions

protein marker: a sequence of DNA that chemically attracts a particular regulatory protein sequence or structure

regulators: proteins that control the transcription of a gene

senile plaques: protein sections that are no longer functional and clutter the intercellular space of the brain, disrupting proper processes

transcription: the process by which mRNA is formed using DNA as a template

translation: the process of building a protein by bonding amino acids according to the mRNA marker present

What Is "Proteomics"?

Historically, much of the focus in genetic research has been on genes and completion of the Human Genome Project. More recently, the focus has shifted to a new and related topic, the proteome. Proteins are known to perform most of the important functions of cells. Therefore, proteomics is, essentially, the study of proteins in an organism and, most important, their function. There are many aspects to the understanding of protein function, including where a particular protein is located in the cell, what modifications occur during its activity, what ligands may bind to it, and its activity. Researchers are seeking to identify all the proteins made in a given cell, tissue, or organism and determine how those proteins interact with metobolites, with themselves, and with nucleic acids. By studying proteomics, scientists hope to uncover underlying causes of disease at the cellular level, invent better methods of diagnosis, and discover new, more efficient medicines for the treatment of disease.

Proteomics has moved to the forefront of molecular research, especially in the area of drug research. Neither the structure nor the function of a protein can be predicted from the DNA sequence alone. Although genes code for proteins, there is a large difference between the number of messenger (mRNA) molecules transcribed from DNA and the number of proteins in a cell. In addition, two hundred known modifications occur during the stages between transcription and post-translation, including phosphorylation, glycosylation, proteolytic processing, deamidation, sulfation, and nitration. Other factors that affect the expression of proteins include aging, stress, environmental forces, and medications. In addition, changes to the sequence of amino acids may occur during or after translation.

Methods of Proteomic Research

In order to study the functions of a protein, it must be separated from other proteins or contaminants, purified, and structurally characterized. These are the major tasks facing researchers in the field.

In order to obtain a sufficient quantity of a particular protein for study, the coding plasmid can be injected into *Escherichia coli* bacteria and the cells will translate the protein multiple times. Alternatively, it must be extracted from biological tissues. The desired polypeptide must then be separated from cells or tissues that may contain thousands of unique proteins. This can be accomplished by homogenizing the tissue, extracting the proteins with solvents or by centrifugation, and further purifying the protein by various means, including high-pressure liquid chromatography (HPLC, separation by solubility differences) and two-dimensional (2-D) gel electrophoresis (separation of molecules by charge and molecular mass). A relatively recent development in laboratory technique research is three-dimensional (3-D) gel electrophoresis, allowing for further separation and identification of proteins.

Structural characterization begins with establishing the order of linked amino acids in the protein. This can be accomplished by the classical techniques of using proteases to fragment the protein chemically and then analyzing the fragments by separation and spectroscopic analysis. The molecular mass of small polypeptides can be investigated by employing several techniques involving mass spectrometry (MS). Sequentially coupled mass spectrometers (the "tandem" MS/MS techniques) are being used to analyze the amino acid sequence and molecular masses of isolated larger polypeptides. These MS/MS analyses are sometimes added to a separation method, such as HPLC, to analyze mixtures of polypeptides.

Historically, Linus Pauling used analytical data from X-ray diffraction (or crystallography) to determine the three-dimensional, helical structure of proteins. The method is still being used to investigate the structures of proteins and ligand-protein complexes. Such studies may lead to significant improvements in the design of medicinal drugs. One significant drawback to analyzing protein structure by X-ray diffraction, however, is that the method requires a significant quantity (approximately 1 milligram) of the protein. Transmission electron microscopy (TEM), which uses electron beams to produce images and diffraction patterns from extremely small samples or regions of a sample, is therefore often preferrable. The TEM method may involve auxiliary techniques to analyze data, including enhancement of images by means of computer software.

Although such methods provide valuable information in analyzing the structure of proteins, they suffer from the loss of spatial information that occurs when tissues are homogenized, when the protein is obtained from a manufactured, bacterial environment, or when it is otherwise isolated. Matrix-assisted laser desorption/ionization time-of-flight (MALDI-TOF) mass spectrometry is a complementary method of analysis that does not yield structural information but provides protein profiles from intact tissue, allowing comparison of diseased versus normal tissue.

Large databases of mass spectroscopic data are being assembled to assist in future identification of known proteins. Further databases of proteome information include particular molecular masses, charges, and, in some cases, connections to the genes regulated or the parent gene of the peptide in question. Scientists hope to relate regulators and the complex web of peripheral proteins that affect the function of each gene.

Challenges and Limitations of Current Methods

The amount of data being obtained by proteomics research poses a problem in organizing and processing the information obtained on proteins. The Human Proteome Organization (HUPO) and the European Bioinformatics Institute (EBI) are two organizations whose purposes include the management and organization of proteomics information and databases, and the facilitation of the advancement of this scientific endeavor.

Analyzing MS data from proteins and relating the complex array of proteins within a single cell to the linear genetic material of DNA present challenges to researchers that they are tackling through computer algorithms, programs, and databases. The SWISS-PROT database, for example, is an annotated protein-sequence database maintained by the Swiss Bioinformatics Institute.

Other obstacles to relating proteins to parent genes include the loss of quaternary structure during separation and the presence of post-translation processing, which can alter the amino acid sequence to the extent that it becomes almost unrecognizable from the parent gene. A lack of protein amplification methods—techniques that would produce more copies of a protein to aid in study—requires sensitive analysis methods and increasingly strong detectors. Currently new methods are being developed, but the limit of study is as large as 1 nanometer.

Disease

Proteins often act as markers for disease. As researchers study proteins, they have found that disease may be characterized by some proteins that are being overproduced, not being produced at all, or being produced at inappropriate times. As the correlation of proteins to disease becomes clearer, better diagnostic tests and drugs are being explored. For example, Alzheimer's disease and Down syndrome are associated with a common protein fragment as the major extracellular protein component of senile plaques.

Researchers are investigating changes in protein expression in heart disease and heart failure, and several hundred cardiac proteins have already been identified. The study of proteomics in immunological diseases has revealed that there is a connection between the human neutrophil α-defensins (HNPs) and human immunodeficiency virus, HIV-1. HNPs are small, cysteine-rich, cationic antimicrobial proteins that are stored in the azurophilic granules of neutrophils and released during phagocytosis to kill ingested foreign microbes. To date, the three most abundant forms of the protein have been implicated in suppressing HIV-1 in vivo.

Similarly, cancer is being studied to find a roster of proteins that are present in cancerous cells but not in normal cells. The Clinical Proteomics Program, a joint effort of the National Cancer Institute and the Food and Drug Administration is searching for the differences between cancerous and normal cells, and also for protein "markers."

Possible Future Directions

Although proteomics is a relatively new area of genetic research, there are three fields that are gaining attention, glycomics, metabolomics, and metabonomics. Glycomics addresses the importance of the sugar coatings of proteins and cells and is gaining attention. This area of study has arisen because of the many roles of sugar coatings in important cellular functions, including the immunological recognition sites, barriers, and sites for attack by pathogens. Metabolomics is the study of the proteins left behind as the cell performs its processes. This field primarily looks at small proteins produced as byproducts. Metabonomics is often used interchangeably with metabolomics but differs in that it examines the change that proteins produce when the cell responds to stresses, such as disease.

Audrey Krumbach, Kayla Williams, and Massimo D. Bezoari, M.D.; updated by Kecia Brown, M.P.H.

Further Reading

Hopker, Hans-Rudolf, et al. *Proteomics in Practice: Guide for Successful Research Design*, 2d ed. Weinham, Germany: Wiley-VCH Verlag GmbH, 2008. A combination textbook and lab manual geared toward academic and industry researchers.

Liebler, David G. *Introduction to Proteomics: Tools for the New Biology.* Totowa, N.J.: Humana Press, 2001. Basics of protein and proteome analysis, key concepts of proteomics, workings of the analytical instrumentation, overview of software tools, and applications of protein and peptide separation techniques, mass spectrometry, and more.

Link, Andrew J., ed. *2-D Proteome Analysis Protocols.* Totowa, N.J.: Humana Press, 1999. Practical proteomics, presenting techniques with step-by-step instructions for laboratory researchers. Fifty-five chapters prepared by more than seventy specialists.

Modern Drug Discovery (October, 2002). The entire issue is devoted to proteomics, with many interesting articles on methods of research, Web sites, and computer-assisted methods of data analysis.

Twyman, Richard M. *Principles of Proteomics.* Oxon, England: Garland Science/BIOS Scientific, 2009. This book is designed for students in advanced-level courses.

Web Sites of Interest

Biotechnology News and Information Portal
http://www.bioexchange.com/news/proteomics.html

Provides information about new developments in the biotechnology business.

Cambridge Healthtech Institute
http://www.genomicglossaries.com/content/proteomics.asp

Provides a useful glossary of technical terms used in proteomics, as well as many links to related sites.

Clinical Proteomics Program
http://home.ccr.cancer.gov/ncifdaproteomics

Serves as an entrance to information on the joint program between the National Cancer Institute of the National Institutes of Health and the U.S. Food and Drug Administration to support the development of proteomics-based technologies.

Human Proteomics Organization
http://www.hupo.org

HUPO works to consolidate regional proteome organizations into a worldwide group, conduct scientific and educational activities, and disseminate knowledge about both the human proteome and model organisms.

Introduction to Proteomics
http://web1.tch.harvard.edu/cfapps/research/data_admin/Site602/mainpageS602P0.html

A Web site was developed by the Proteomics Center at Children's Hospital, Boston. Has an interactive component, which serves as an excellent basic educational tool for understanding proteomics.

See also: Bioinformatics; Genomics; Human genetics; Human Genome Project; Protein structure; Protein synthesis.

Pseudogenes

CATEGORY: Molecular genetics

SIGNIFICANCE: Pseudogenes are DNA sequences derived from partial copies, mutated complete copies of functional genes, or normal copies of a gene that has lost its control sequences and therefore cannot be transcribed. They may originate by gene duplication or retrotransposition. They are apparently nonfunctional regions of the genome that may evolve at a maximum rate, free from the evolutionary constraints of natural selection.

KEY TERMS

introns: noncoding segments of DNA within a gene that are removed from pre-messenger RNA (pre-mRNA) as a part of the process of producing mature mRNA

long interspersed sequences (LINES): long repeats of DNA sequences scattered throughout a genome

neutral theory of molecular evolution: the theory that most DNA sequence evolution is a result of mutations that are neutral with respect to the fitness of the organism

retrotransposon (retroposon): a DNA sequence that is transcribed to RNA and reverse transcribed to a DNA copy able to insert itself at another location in the genome

reverse transcriptase: an enzyme, isolated from retroviruses, that synthesizes a DNA strand from an RNA template

short interspersed sequences (SINES): short repeats of DNA sequences scattered throughout a genome

DEFINITION AND ORIGIN

Pseudogenes are DNA sequences that resemble genes but are not correctly transcribed or translated to a functional polypeptide. If a functional gene is duplicated so that there are two nonhomologous copies of it in the genome, one of the copies can retain the code for the original polypeptide product, while the other is free from such constraints, since one copy of the gene is sufficient to produce the protein. Because mutations in one copy do not destroy the gene's function, they may be retained, and the unneeded copy can evolve more quickly. It may change to produce a different, functional polypeptide (and effectively become a new gene), or it may remain nonfunctional as a pseudogene. There are two types of pseudogenes, defined by how they were produced: nonprocessed and processed.

NONPROCESSED PSEUDOGENES

Nonprocessed (or duplicated) pseudogenes arise when a portion of the original gene is duplicated, with portions necessary for proper functioning missing or altered or when the complete original gene is duplicated. They can be identified by the presence of introns and may have mutations in the promoter that prevent transcription or the correct removal of introns, or they may have other mutations (such as premature stop codons) within exons that result in translation of a nonfunctional polypeptide. A series of tandem duplications of a gene can result in clustered gene families, which can include expressed genes, expressed pseudogenes (which are transcribed but produce no functional polypeptide), and nonexpressed pseudogenes that are not transcribed. The alpha-globulin and beta-globulin clusters are examples of such gene families. Other examples of nonprocessed pseudogenes include members of the immunoglobulin (Ig) and major histocompatibility complex (MHC) gene families.

PROCESSED PSEUDOGENES

Processed pseudogenes originate from transcribed RNA copies of genes that are copied back to DNA by the enzyme reverse transcriptase. Processed pseudogenes are usually integrated into the genome in a different location from the original gene. Reverse transcriptase is an enzyme produced by retroviruses, which have RNA genomes that are reverse transcribed to DNA when the viruses infect host cells. Retrotransposons, which are related to retroviruses, are DNA sequences that transpose or duplicate themselves by reverse transcription of a transcribed RNA copy of the sequence.

Often, retrotransposons will carry along a copy of the surrounding host DNA, resulting in the duplication of that sequence—a processed pseudogene. Because the introns of a gene are removed from the RNA transcript, processed pseudogenes are not exact copies of the original DNA sequence; the introns are missing. Copies of protein-coding genes copied by this mechanism are members of a type of repetitive DNA called LINES (for long interspersed sequences) and exist in multiple copies scattered around the genome, each up to several thousand base pairs in length. Short processed pseudogenes are members of another class of repetitive DNA called SINES (short interspersed sequences of up to several hundred base pairs in length) and result from the retrotransposon-mediated copying of tRNA or rRNA genes. SINES of this type are sometimes very abundant in genomes because they may have internal promoters, so that they are more easily transcribed, and therefore transposed. The most prominent of SINES are those that are members of the *Alu* family, which occur an average of once every six thousand base pairs in the human genome.

Pseudogenes and Neutral Evolution

In 2003, Japanese researchers reported the discovery of a mouse pseudogene that is involved in regulating the expression of its related "functional" gene. This discovery suggested that at least some pseudogenes may have important functions. Although pseudogenes are very commonly found across genomes, most do not appear to serve any function, and until further research uncovers more functional pseudogenes this assumption appears warranted. Their abundance can be explained by the tendency of duplicated sequences to be further copied. Retrotransposition increases the number of copies of processed pseudogenes, and gene duplication leading to unprocessed pseudogenes favors mechanisms that generate additional copies, leading to clustered gene families. Natural selection does not tend to eliminate these additional copies because their presence does not harm the organism as long as there is at least one functional copy of the original gene. In other words, pseudogenes are selectively neutral.

Because of their selective neutrality, pseudogenes are especially useful for estimating neutral mutation rates in genomes. The neutral theory of evolution predicts that, because of the constraints of selection, functional regions of the genome (such as the exons, or coding sequences, of genes) will evolve more slowly than less critical sequences, such as introns, or nonfunctional sequences like pseudogenes. The number of nucleotide differences between homologous sequences of related species can be used to calculate estimates of evolutionary rates, and such estimates support the neutral theory: the greatest rates of divergence occur within pseudogenes. Using comparisons from several pseudogenes, researchers can establish the baseline neutral mutation rate for a group of species.

Stephen T. Kilpatrick, Ph.D.

Further Reading

Alberts, Bruce, et al. "DNA, Chromosomes, and Genomes." In *Molecular Biology of the Cell.* 5th ed. New York: Garland Science, 2008. Includes information about pseudogenes.

Gerstein, Mark, and Deyou Zheng. "The Real Life of Pseudogenes." *Scientific American* 295, no. 2 (August, 2006): 48-55. An overview of pseudogenes, which describes how they may provide scientists with information about mapping genomes and understanding the evolution of the human genome.

Graur, Dan, and Wen-Hsiung Li. *Fundamentals of Molecular Evolution.* 2d ed. Sunderland, Mass.: Sinauer Associates, 1999. A detailed review of the topic, including the importance of pseudogenes in determining genomic rates of neutral evolution.

Lewin, Benjamin. *Genes IX.* Sudbury, Mass.: Jones and Bartlett, 2007. Provides several examples of clustered gene families that include pseudogenes and describes the mechanism for the origin of processed pseudogenes.

Li, Wen-Hsiung, Takashi Gojobori, and Masatoshi Nei. "Pseudogenes as a Paradigm of Neutral Evolution." *Nature* 292, no. 5820 (1981): 237-239. Discusses the predictions of the neutral theory, how rates of nucleotide substitution may be calculated, and how data from pseudogenes support the neutral theory.

Zhang, Zhaolei, and Mark Gerstein. "The Human Genome Has Forty-nine Cytochrome C Pseudogenes, Including a Relic of a Primordial Gene That Still Functions in a Mouse." *Gene* 312 (July, 2003): 61. The authors explain how they identified forty-nine cytochrome c (cyc) pseudogenes in the human genome, which are full-length and originated primarily from independent retrotransposition events.

Web Sites of Interest

Pseudogene.org
http://www.pseudogene.org

Developed and maintained by the Gerstein Laboratory at Yale University, this site contains information about pseudogenes, including a comprehensive database of identified pseudogenes.

University of Iowa Pseudogenes
http://genome.uiowa.edu/pseudogenes
Provides statistics about pseudogenes, a database of pseudogenes in the human genome, and a ranked list of human pseudogenes that have been identified as candidates for gene conversion.

See also: Gene families; Repetitive DNA.

Pseudohermaphrodites

CATEGORY: Diseases and syndromes

SIGNIFICANCE: Pseudohermaphrodites are individuals born with either ambiguous genitalia or external genitalia that are the opposite of their chromosomal sex. These individuals need a thorough medical evaluation and appropriate medical intervention to help ensure a healthy, well-adjusted life.

KEY TERMS

ambiguous genitalia: external sexual organs that are not clearly male or female

genotype: the sum total of the genes present in an individual

gonads: organs that produce reproductive cells and sex hormones, for example, testes in males and ovaries in females

karyotype: the number and kind of chromosomes present in every cell of the body (normal female karyotype is 46,XX and normal male karyotype is 46,XY)

phenotype: the physical appearance and physiological characteristics of an individual, which depend on the interaction of genotype and environment

NORMAL FETAL DEVELOPMENT

Prior to nine weeks in gestational age, a male and a female fetus have identical external genitalia (sexual organs) consisting of a phallus and labioscrotal folds. The phallus develops into a penis in males and a clitoris in females; labioscrotal folds become the scrotum in males and the labial folds in females. Early in development, the gonads can develop into either testes or ovaries. In a fetus with a normal male karyotype (46,XY), the primitive gonads become testes, which produce testosterone. Testosterone in turn causes enlargement of the primitive phallus into a penis. It is the presence of the Y chromosome, and in particular a small, sex-determining region of the Y chromosome termed the *SRY* locus, that drives the formation of the testes. The presence of the *SRY* locus appears to be essential for development of a normal male.

PSEUDOHERMAPHRODITISM

A true hermaphrodite is born with both ovarian and testicular tissue. A male pseudohermaphrodite has a 46,XY karyotype with either female genitalia or ambiguous genitalia (but only testicular tissue); a female pseudohermaphrodite has a 46,XX karyotype with either male genitalia or ambiguous genitalia (but only ovarian tissue). Ambiguous genitalia typically consist of a small, abnormally shaped, phalluslike structure, often with hypospadias (in which urine comes from the base of the penis instead of the tip) and abnormal development of the labioscrotal folds (not clearly a scrotum or labia). A vaginal opening may be present.

Most cases of pseudohermaphroditism result from abnormal exposure to increased or decreased amounts of sex hormones during embryonic development. The most common cause of female pseudohermaphroditism is exposure of a female fetus to increased levels of testosterone during the first half of pregnancy. Maternal use of anabolic steroids can cause this condition, but the most common genetic cause of increased testosterone exposure is congenital adrenal hyperplasia (CAH). CAH results from an abnormality in the enzymatic pathways of the fetus that make both cortisol (a stress hormone) and the sex steroids (such as testosterone). At several points in these pathways, there may be a nonfunctioning enzyme that results in too little production of cortisol and too much production of the sex steroids. This will result in partial masculinization of the external genitalia of a female embryo. Females with CAH are usually born with an enlarged clitoris (often mistakenly thought to be a penis) and partial fusion of the labia. Males can also have CAH, but the excess testosterone does not affect their genital development since a relatively high level of testosterone exposure is a normal part of their development.

The most common causes of male pseudohermaphroditism are abnormalities of testosterone production or abnormalities in the testosterone receptor at the cellular level. One example is a deficiency in 5-alpha-reductase, the enzyme that converts testosterone to dihydrotestosterone (DHT). When there is a deficiency of this enzyme, there will be a deficiency of DHT, which is the hormone primarily responsible for masculinization of external genitalia. A male who lacks DHT will have female-appearing external genitalia or ambiguous genitalia at birth. Often these individuals are reared as females, but at puberty they will masculinize because of greatly increased production of testosterone. These individuals may actually develop into nearly normal-appearing males. Abnormalities of the testosterone receptor can also result in a range of different conditions in affected males, from normal female appearance (a totally defective receptor) to ambiguous genitalia (partially defective receptor) in a 46,XY male. These individuals will not masculinize at puberty because no matter how much testosterone or DHT they produce, their bodies cannot respond to the hormones.

Both male and female pseudohermaphroditism can result from chromosomal abnormalities. The absence or dysfunction of the *SRY* locus produces an individual with normal female genitalia but a 46,XY karyotype. Individuals with a 46,XX karyotype who have the *SRY* locus transposed to one of their X chromosomes will have a normal male appearance.

Impact and Applications

Some forms of pseudohermaphroditism are life threatening, and so early diagnosis is imperative. Both males and females with CAH are at risk for sudden death caused by low cortisol levels and other hormone deficiencies. Early diagnosis is relatively easy in affected females since their genital abnormalities are noticeable at birth. Affected males are often not recognized until they have a life-threatening event, which usually occurs in the first two weeks of life. Treatment of CAH consists of appropriate hormone supplementation that, if instituted early in life, can help prevent serious problems. CAH is inherited in an autosomal recessive manner, so parents of an affected individual have a 25 percent chance of having another affected child with each pregnancy.

The sex of rearing of a child with ambiguous genitalia is usually determined by the child's type of pseudohermaphroditism. Typically, sex of rearing will be based on the chromosomal sex of the child. These children may need sex hormone supplementation or surgery to assist in developing gender-appropriate genitalia. Children with pseudohermaphroditism with normal-appearing genitalia at birth may not be recognized until puberty, when abnormal masculinization or feminization may occur. These individuals need medical evaluation and karyotype determination to guide the proper medical treatment.

Patricia G. Wheeler, M.D.

Further Reading

Hunter, R. H. F. *Sex Determination, Differentiation, and Intersexuality in Placental Mammals.* New York: Cambridge University Press, 1995. Discusses the genetic determination of sex in mammals. Illustrations (some color), bibliography, index.

Meyer-Bahlburg, Heino. "Intersexuality and the Diagnosis of Gender Identity Disorder." *Archives of Sexual Behavior* 23, no. 1 (February, 1994): 21. Addresses gender identity and its relation to pseudohermaphroditism.

Moore, Keith L., and T. V. N. Persaud. "The Urogential System." In *The Developing Human: Clinically Oriented Embryology.* 8th ed. Philadelphia: Saunders/Elsevier, 2008. Details embryology from a clinical perspective, providing discussions of the stages of organs and genital system development, discussing pseudohermaphrodites.

Preves, Sharon E. "Medical Sex Assignment." In *Intersex and Identity: The Contested Self.* New Brunswick, N.J.: Rutgers University Press, 2003. Includes a discussion of pseudohermaphrodites.

Simpson, J. L. "Disorders of the Gonads, Genital Tract, and Genitalia." In *Emery and Rimoin's Principles and Practice of Medical Genetics*, edited by David L. Rimoin et al. 5th ed. 3 vols. Philadelphia: Churchill Livingstone Elsevier, 2007. A detailed account of male and female pseudohermaphroditism. Illustrations, bibliography, index.

Speiser, Phyllis W., ed. *Congenital Adrenal Hyperplasia.* Philadelphia: W. B. Saunders, 2001. Contents address prenatal treatment, newborn screening, new treatments, surgery, gender, sexuality, cognitive function, and pregnancy outcomes. Illustrated.

WEB SITES OF INTEREST

Intersex Society of North America
http://www.isna.org
The society is "devoted to systemic change to end shame, secrecy, and unwanted genital surgeries for people born with an anatomy that someone decided is not standard for male or female." Its Web site offers information on such conditions as clitoromegaly, micropenis, hypospadias, ambiguous genitals, early genital surgery, adrenal hyperplasia, Klinefelter syndrome, and androgen insensitivity syndrome.

Johns Hopkins University, Division of Pediatric Endocrinology, Syndromes of Abnormal Sex Differentiation
http://www.hopkinschildrens.org/intersex
A guide for parents and their families providing information about syndromes of abnormal sex differentiation.

Medline Plus, Intersex
http://www.nlm.nih.gov/medlineplus/ency/article/001669.htm
Features information on 46,XX Intersex and other intersex conditions, including the causes, diagnosis, symptoms, and treatment of these conditions. Additional information can be retrieved by using the search engine to search on the word "pseudohermaphroditism."

National Organization for Rare Disorders (NORD)
http://www.rarediseases.org
Offers information and articles about rare genetic conditions and diseases, including XYY syndrome, in several searchable databases.

See also: Androgen insensitivity syndrome; Biological clocks; Gender identity; Hermaphrodites; Homosexuality; Human genetics; Metafemales; RNA transcription and mRNA processing; Steroid hormones; X chromosome inactivation; XYY syndrome.

Pseudohypoparathyroidism

CATEGORY: Diseases and syndromes
ALSO KNOWN AS: Albright's hereditary osteodystrophy; Types 1A and 1B pseudohypoparathyroidism

DEFINITION

Pseudohypoparathyroidism denotes a group of inherited disorders characterized by deficiency in the end-organ response to parathyroid hormone (PTH) and in some instances because of abnormal PTH receptors. The term "pseudohypoparathyroidism" is used because there is evidence of clinical hypoparathyroidism but serum PTH levels are normal.

RISK FACTORS

A family history of pseudohypoparathyroidism is the primary risk factor for the disease.

ETIOLOGY AND GENETICS

Parathyroid glands assist in the control of calcium use and removal by the body by producing parathyroid hormone, or PTH. PTH helps control calcium, phosphorus, and vitamin D levels within the blood and bone. Individuals with pseudohypoparathyroidism produce the required amount of PTH, but the body is "resistant" to its effect. This causes low blood calcium levels and high blood phosphate levels. Pseudohypoparathyroidism is very rare and is caused by abnormal genes. Type 1A, also called Albright's hereditary osteodystrophy, is an inherited autosomal dominant trait, and therefore only one parent could potentially pass the defective gene to cause the condition. The condition results in short stature, round face, abnormally developed metacarpal and metatarsal bones, and subcutaneous bone ossification. Type 1B is less understood than type 1A and involves resistance to PTH only in the kidneys. Type 2 is very similar to type 1 in its clinical features, but the events that take place in the kidneys are different.

Type 2 pseudohypoparathyroidism also involves low blood calcium and high blood phosphate levels, but persons with this form do not develop the physical characteristics seen in those with Type 1A. Recent studies report the *GNAS* gene encodes the alpha-subunit of the stimulatory G proteins. These proteins are important in intracellular signal transduction of peptide and neurotransmitter receptors. Heterozygous inactivating maternally inherited mutations of *GNAS* lead to a phenotype in which Albright hereditary osteodystrophy is associated with pseudohypoparathyroidism type 1A. Such mutations include translation initiation mutations, amino acid substitutions, nonsense mutations, splice-site mutations, and small insertions or deletions. E. Fernandezrebollo and colleagues found the *GNAS* gene of a five-year-old boy with pseudohypoparathyroidism to have a heterozygous inversion of exon 2 and part of intron 1 of de novo origin. This report demonstrated the first evidence for an inversion at the *GNAS* gene responsible for pseudohypoparathyroidism type 1.

Symptoms

Symptoms of pseudohypoparathyroidism are related to low levels of calcium known as hypocalcemia. Hypocalcemia causes increased irritability of nerves, leading to numbness and tingling of the hands, feet, and lips. Tetany, which is manifested clinically as muscular spasms first affecting the hands and feet, is another symptom of pseudohypoparathyroidism. Laryngeal spasms can occur, leading to obstruction of the respiratory system. Severe hypocalcemia can lead to generalized convulsions in children. Other complications of hypocalcemia associated with pseudohypoparathyroidism may include other endocrine problems, leading to lowered sexual drive and lowered sexual development, lowered energy levels, and increased weight. Individuals with Albright's hereditary osteodystrophy may have symptoms of calcium deposits under the skin and the physical characteristics of a round face and short neck and short stature in children.

Screening and Diagnosis

Blood tests are evaluated to determine calcium, phosphorus, and PTH levels. Urine tests are also utilized to determine key endocrine levels. Genetic testing is utilized to determine positive family history of the abnormalities. Magnetic resonance imaging (MRI) creates a detailed picture of the brain and surrounding nerve tissues and is helpful in diagnosing decreased levels of calcium causing increased irritability of nerve tissue.

Treatment and Therapy

Calcium and vitamin D supplements are prescribed to maintain proper calcium levels. A low-phosphorus diet may be necessary if blood phosphate levels remain higher than normal. Emergency treatment with intravenous calcium is needed if an individual presents an acute hypocalcemia crisis such as tetany, laryngeal spasm, and convulsions.

Prevention and Outcomes

Low blood calcium in pseudohypoparathyroidism is usually milder than in other forms of hypoparathyroidism. Proper medical attention and follow-up are important in prevention and treatment of other endocrine problems, as patients with type 1A pseudohypoparathyroidism have an increased occurrence of other endocrine abnormalities to include hypothyroidism and hypogonadism.

Jeffrey P. Larson, P.T., A.T.C.

Further Reading

"Albright's Hereditary Osteodystrophy." In *A Dictionary of Nursing*. Oxford, England: Oxford University Press. 2008.

Bringhurst, F. R., M. B. Demay, and H. M. Kronenberg. "Disorders of Mineral Metabolism." In *Williams Textbook of Endocrinology*, edited by H. M. Kronenberg, M. Schlomo, K. S. Polansky, and P. R. Larsen. 11th ed. St. Louis: W. B. Saunders, 2008.

Chandrasoma, P., and C. Taylor. *Concise Pathology*. Norwalk, Conn.: Appleton and Lange, 1991.

Copstead, Lee-Ellen. *Perspectives on Pathophyiology*. Philadelphia: W. B. Saunders, 1994.

Wysolmerski, J. J. and K. L. Insogna. "The Parathyroid Glands, Hypercalcemia, and Hypocalcemia." In *Williams Textbook of Endocrinology*, edited by H. M. Kronenberg, M. Schlomo, K. S. Polansky, and P. R. Larsen. 11th ed. St. Louis: W. B. Saunders, 2008.

Web Sites of Interest

Endocrine Associates
www.endocrine-associates.com

Endocrine Society
http://www.endo-society.org

Endocrineweb
http://www.endocrineweb.com

HighBeam Research
http://www.highbeam.com

See also: Adrenomyelopathy; Androgen insensitivity syndrome; Autoimmune polyglandular syndrome; Congenital hypothyroidism; Diabetes insipidus; Graves' disease; Obesity; Pancreatic cancer; Pancreatitis; Pearson syndrome; Steroid hormones.

Punctuated equilibrium

CATEGORY: Evolutionary biology; Population genetics

SIGNIFICANCE: Punctuated equilibrium is a model of evolutionary change in which new species originate abruptly and then exist through a long period of stasis. This model is important as an explanation of the stepwise pattern of species change seen in the fossil record.

KEY TERMS

allopatric speciation: a theory that suggests that small parts of a population may become genetically isolated and develop differences that would lead to the development of a new species

heterochrony: a change in the timing or rate of development of characters in an organism relative to those same events in its evolutionary ancestors

phyletic gradualism: the idea that evolutionary change proceeds by a progression of tiny changes, adding up to produce new species over immense periods of time

EVOLUTIONARY PATTERNS

Nineteenth century English naturalist Charles Darwin viewed the development of new species as occurring slowly by a shift of characteristics within populations, so that a gradual transition from one species to another took place. This is now generally referred to as phyletic gradualism. A number of examples from the fossil record were put forward to support this view, particularly that of the horse, in which changes to the feet, jaws, and teeth seem to have progressed in one direction over a long period of time. Peter Sheldon in 1987 documented gradual change in eight lineages of trilobites over a three-million-year period in the Ordovician period of Wales. Despite these and other examples (some of which have been reinterpreted), it is clear that the fossil record more commonly shows a picture of populations that are stable through time but are separated by abrupt morphological breaks. This pattern was recognized by Darwin but was attributed by him to the sketchy and incomplete nature of the fossil record. So few animals become fossilized, and conditions for fossilization are so rare, that he felt only a fragmentary sampling of gradual transitions was present, giving the appearance of abrupt change.

One hundred years later, the incompleteness of the fossil record no longer seemed convincing as an explanation. In 1972, Niles Eldredge and Stephen Jay Gould published their theory of the evolutionary process, called by them "punctuated equilibrium." This model explains the lack of intermediates by suggesting that evolutionary change occurs only in short-lived bursts in which a new species arises abruptly from a parent species, often with relatively large morphological changes, and thereafter remains more or less stable until its extinction.

THE PROCESS OF PUNCTUATED EQUILIBRIUM

A number of explanations have been put forward to show how this process might take place. One of these, termed allopatric speciation, was first proposed by Ernst Mayr in 1963. He pointed out that a reproductive isolating mechanism is needed to provide a barrier to gene flow and that this could be provided by geographic isolation. Allopatric or geographical isolation could result when the normal range of a population of organisms is reduced or fragmented. Parts of the population become separated in peripheral isolates, and if the population is small, it may become modified rapidly by natural selection or genetic drift, particularly if it is adapting to a new environment. This type of process is commonly called the founder effect, because it is the characteristics of the small group of individuals that will overwhelmingly determine the possible characteristics of their descendants. As the initial members of the peripheral isolate may be few in number, it might take only a few generations for the population to have changed enough to become reproductively isolated from the parent population. In the fossil record, this will be seen as a period of stasis representing the parent population, followed by a rapid morphological change as the peripheral population is isolated from it and then replaces it, either competitively or because it has become extinct or has moved to follow a shifting habitat. Because this is thought to take place rapidly in small populations, fossilization potential is low, and unequivocal examples are not common in the fossil record. However, in 1981, Peter Williamson published a well-documented example from the Tertiary period of Lake Turkana in Kenya, which showed episodes of stasis and rapid change in populations of freshwater mollusks. The increases in evolutionary rate were apparently driven by severe environmental change that caused parts of the lake to dry up.

Punctuated changes may also have taken place because of heterochrony, which is a change in the rate of development or timing of appearance of ancestral characters. Paedomorphosis, for example, would result in the retention of juvenile characters in the adult, while its opposite, peramorphosis, would result in an adult morphologically more advanced than its ancestor. Rates of development could be affected by a mutation, perhaps resulting in the descendant growing for much longer than the ancestral form, thus producing a giant version. These changes would be essentially instantaneous and thus would show as abrupt changes of species in the fossil record.

Stephen Jay Gould, who (with Niles Eldredge) developed the theory of punctuated equilibrium to explain gaps in the fossil record. (Author)

Impact and Applications

The publication of the idea of punctuated equilibrium ignited a storm of controversy that still persists. It predicts that speciation can be very rapid, but more important, it is consistent with the prevalence of stasis over long periods of time so often observed in the fossil record. Species had long been viewed as flexible and responsive to the environment, but fossil species showed no change over long periods despite a changing environment. Biologists have thus had to review their ideas about the concept of species and the processes that operate on them. Species are now seen as real entities that have characteristics that are more than the sum of their component populations. Thus the tendency of a group to evolve rapidly or slowly may be intrinsic to the group as a whole and not dependent on the individuals that compose it. This debate has helped show that the fossil record can be important in detecting phenomena that are too large in scale for biologists to observe.

David K. Elliott, Ph.D.

Further Reading

Ayala, Francisco J. "Punctuated Equilibrium and Species Selection." In *Back to Darwin: A Richer Account of Evolution*, edited by John B. Cobb, Jr. Grand Rapids, Mich.: William B. Eerdmans, 2008. Contests Gould's theories about species selection, maintaining that "evolutionary biologists have largely, if not completely, lost interest in PE (punctuated equilibrium) as a scientific theory."

Eldredge, Niles. *Time Frames: The Rethinking of Darwinian Evolution and the Theory of Punctuated Equilibria.* New York: Simon & Schuster, 1985. The theory's coauthor explains how punctuated equilibria complements Darwin's thesis. Illustrations, bibliography, index.

Eldredge, Niles, and Stephen Jay Gould. "Punctuated Equilibria: An Alternative to Phyletic Gradualism." In *Models in Paleobiology*, edited by Thomas J. M. Schopf. San Francisco: Freeman, Cooper, 1972. The 1972 paper that introduced the theory of punctuated equilibrium to the scientific community. Illustrations, bibliography.

Gould, Stephen Jay. "The Meaning of Punctuated Equilibria and Its Role in Validating a Hierarchical Approach to Macroevolution." In *Perspectives on Evolution*, edited by Roger Milkman. Sunderland, Mass.: Sinauer Associates, 1982. The founder of the theory expands on its implications for evolution.

Gould, Stephen Jay, and Niles Eldredge. "Punctuated Equilibrium: The Tempo and Mode of Evolution Reconsidered." *Paleobiology* 3 (1977): 115-151. A follow-up to the original exposition of the theory.

Prothero, Donald R. *Bringing Fossils to Life: An Introduction to Paleobiology.* 2d ed. Boston: McGraw-Hill Higher Education, 2004. Includes an introductory discussion of punctuated equilibrium. Bibliography, index.

Sepkoski, David, and Michael Ruse, eds. *The Paleobiological Revolution: Essays on the Growth of Modern Paleontology.* Chicago: University of Chicago Press, 2009. Includes three essays: "Punctuated Equilibria and Speciation: What Does It Mean to Be a Darwinian?" by Patricia Princehouse, "'Radical' or 'Conservative'? The Origin and Early Reception of Punctuated Equilibrium" by David Sepkoski, and "Punctuated Equilibrium Versus Community Evolution" by Arthur J. Boucot.

Somit, Albert, and Steven A. Peterson, eds. *The Dynamics of Evolution: The Punctuated Equilibrium Debate in the Natural and Social Sciences.* Ithaca, N.Y.: Cornell University Press, 1992. Provides an overview of the punctuated equilibrium debate, including updates by Gould and Eldredge. Bibliography, index.

Web Sites of Interest

Evolution 101, More on Punctuated Equilibrium
http://evolution.berkeley.edu/evosite/evo101/VIIA1bPunctuated.shtml

This site, created by the University of California Museum of Paleontology, provides information about evolution for teachers and their students. It devotes a page to punctuated equilibrium and another page to competing hypotheses.

The Talk Origins Archive, Punctuated Equilibrium
http://www.talkorigins.org/faqs/punc-eq.html

The archive, which examines the evolution versus creationism debate, includes an explanation of punctated equilibrium by Wesley R. Elsberry, a marine biologist and a critic of the antievolution movement.

See also: Artificial selection; Consanguinity and genetic disease; Evolutionary biology; Genetic load; Hardy-Weinberg law; Inbreeding and assortative mating; Molecular clock hypothesis; Natural selection; Population genetics; Speciation.

Purine nucleoside phosphorylase deficiency

CATEGORY: Diseases and syndromes
ALSO KNOWN AS: PNP-deficiency

Definition

Purine nucleoside phosphorylase (PNP) deficiency is a rare and severe, inherited primary immunodeficiency disease. PNP-deficiency destroys T-cell lymphocytes in the immune system, severely weakening the immune system and decreasing life expectancy.

Risk Factors

Family history is the only known risk factor for this disease. It is an autosomal recessive disorder, and symptomatic individuals inherit defective PNP genes on band 14q13 from each parent. Symptoms occur only in those with homozygous mutations of the *PNP* gene. Carrier parents have a 25 percent chance of passing this genetic defect to their offspring. Neurological symptoms often accompany this disease, but the route by which they develop is still undiscovered. Risk is evenly distributed between males and females.

Etiology and Genetics

This inherited genetic disease causes the absence or mutation of the nucleoside phosphorylase gene NP(14q13.1). PNP deficiency is one of two severe combined immunodeficiency (SCID) diseases, adenosine deaminase (ADA) deficiency and purine nucleoside phosphorylase (PNP) deficiency. PNP deficiency accounts for only about 4 percent of all combined immunodeficiency diseases (CID) and is differentiated by recurring symptoms of chronic and unusual infections resistant to drug therapies, neurological symptoms, lymphopenia, and increased incidences of lymphomas and autoimmune diseases at a young age.

Purines are components necessary in cellular energy systems and for the production of DNA and RNA. Purines are often recycled during catabolism using the purine salvage pathway. When the PNP enzyme is deficient along this pathway, elevated levels of deoxy-guanosine triphosphate (dGTP) begins to accumulate in lymph tissues, whose primary function is immunity. Subsequently, the production of ribonucleotide reductase necessary for the synthesis of deoxynucleotides, which are used in the genetic code, and mitochondrial DNA repair is inhibited. As a result, T-cell lymphocyte and thymocyte sensitivity increases, leading to T-cell destruction and a decreased ability to fight infection. Unlike ADA deficiency, B-cell lymphocytes in the immune system are not affected.

Symptoms

PNP deficiency is often difficult to diagnosis because of recurring sinopulmonary and urinary tract infections that may also indicate other diseases, such as B-cell immunodeficiency, vitamin B_{12} deficiency, or interstitial cystitis. But PNP deficiency is distinguished from other diseases by neurological and autoimmune diseases coupled with unusual, recurring, and drug-resistant bacterial, fungal, mycobacterial, protozoal, and viral infections. Drug-resistant pneumonia and oral thrush before one year of age are usually the first indications of a PNP deficiency. Often neurological symptoms precede infections because T-cell immunity declines gradually, thus delaying onset of infections. Ataxia, autoimmune hemolytic anemia, autoimmune neutropenia, behavioral issues, bronchiectasis, central nervous vasculitis, developmental and motor delays, diarrhea, failure to thrive, herpes infections, hypertonia, idiopathic thrombocytopenia, lupus, malabsorption, mental retardation, neurological symptoms, nodular lymphoid hyperplasia of the gastrointestinal tract, spasticity, recurrent sinopulmonary and urinary tract infections, thyroiditis, and tremors are all symptoms that have been associated with a PNP deficiency.

Screening and Diagnosis

When an immunodeficiency disease is suspected, family and patient history, physical exam, blood screening tests, and skin tests are used to make an accurate diagnosis. A physical exam usually finds absent or underdeveloped lymph glands and tissue. Blood screening tests include complete blood count and manual differential, quantitative serum immunoglobulin levels, antibody response measurement, and complement response.

These blood tests usually show a marked decrease in circulating lymphocytes and serum antibodies. Sixty-seven percent of PNP deficient individuals present with neurological symptoms, including muscle spasticity and mental retardation. Autoimmune hemolytic anemia, immune thrombocytopenia, neutropenia, thyroiditis, and lupus diseases are often found. Chest X rays usually show an underdeveloped thymus gland. Average age of diagnosis is six and a half months.

Treatment and Therapy

PNP deficiency is fatal, but recent treatments can extend life span. Treatment first involves treating any infections with appropriate antibiotic therapies. If failure to thrive is present, then appropriate nutrition interventions are also prescribed. In the case of anemia, red blood cell transfusions improve symptoms for a limited time. Successful bone marrow and stem cell transplantation from a compatible donor cure the underlying immunodeficiency and improve immune system function. However, accompanying neurological disorders or autoimmune diseases are not cured by bone marrow transplantation, and treatment of these disorders depends on the specific disease. Prophylactic precautions include avoiding situations where germs may spread and cause infections, use of good hygiene and nutrition practices, avoidance of live viral immunizations, and treatment with long-term, low-dose antibiotics to prevent and control infections. In vitro studies show gene and enzyme replacement therapies as well as infusing PNP fusion proteins may resolve PNP deficiency and offer promising treatments in the future. Transplants of stem cells from umbilical cord blood may also resolve neurological symptoms.

Prevention and Outcomes

Genetic counseling to assess risk is recommended when there is a family history of PNP deficiency. Prenatal diagnosis is possible in families with a previously affected child.

However, prognosis is generally poor for PNP-deficiency, and life expectancy is usually not more than ten years of age at this time.

Alice C. Richer, RD

Further Reading

Dror, Yigal, et al. "Purine Nucleoside Phosphorylase Deficiency Associated with a Dysplastic Marrow Morphology." *Pediatric Research* 55, no. 3 (March, 2004): 472-477.

Gates, Robert H. *Infectious Disease Secrets.* 2d ed. St. Louis: Elsevier Health Sciences, 2003.

Janeway, Charles A., Paul Travers, Mark Walport, and Mark Shlomchik. *Immunobiology: The Immune System in Health and Disease.* 5th ed. New York: Garland, 2001.

WEB SITES OF INTEREST

eMedicine from WebMD
http://emedicine.medscape.com

Immune Deficiency Foundation
http://www.primaryimmune.org

Jeffrey Modell Foundation
http://www.jmfworld.com

National Institute of Child Health and Human Development
http://www.nichd.nih.gov/publications/pubs/primary_immuno.cfm

National Institutes of Health, Office of Rare Diseases Research
http://rarediseases.info.nih.gov/GARD

Purine Research Society
http://www.purineresearchsociety.org

See also: Agammaglobulinemia; Allergies; Antibodies; Anthrax; Ataxia telangiectasia; Autoimmune disorders; Autoimmune polyglandular syndrome; Chediak-Higashi syndrome; Chronic granulomatous disease; Hybridomas and monoclonal antibodies; Immunodeficiency with hyper-IgM; Immunogenetics; Infantile agranulocytosis; Myeloperoxidase deficiency.

Pyloric stenosis

CATEGORY: Diseases and syndromes
ALSO KNOWN AS: Infantile hypertrophic pyloric stenosis

DEFINITION

Food passes from the stomach to the small intestine. In pyloric stenosis, food cannot pass freely because the entrance between the stomach and the small intestine narrows. The narrowing is caused by the enlargement of the pylorus (the muscle at the entrance to the stomach). Almost all cases of pyloric stenosis happen in very young babies (usually three to twelve weeks old). This problem happens about two to four times out of every one thousand births. It is much more common in males than in females. The sooner pyloric stenosis is treated, the fewer problems will result and the healthier a baby will be, so parents who think their child has this condition should contact their doctors immediately.

RISK FACTORS

The following risk factors increase a baby's chance of developing pyloric stenosis; parents whose child has any of these risk factors should tell their doctor. Risk factors include prematurity and a family history of pyloric stenosis. Pyloric stenosis is more common in male babies (particularly firstborn males) and more common in Caucasian than in Latino, Asian, or African American babies.

ETIOLOGY AND GENETICS

The etiology of pyloric stenosis is multifactorial, suggesting that there is a broad range of genetic and environmental factors that may contribute to the appearance of this birth defect. While there is no single gene that when mutated will always cause this condition, researchers have identified at least five genes in which mutations may significantly increase the risk of developing this disorder. Since input from several genes is probably involved, most cases do not conform to a predictable pattern of inheritance. Scattered reports in the literature note isolated families in which pyloric stenosis is inherited as an autosomal dominant condition, but these are exceptions to the norm. In autosomal dominant inheritance, a single copy of the mutation is sufficient to cause full expression of the trait. An affected individual has a 50 percent chance of transmitting the mutation to each of his or her children.

The five genes that are known to be associated with increased susceptibility to the development of pyloric stenosis are known as *IHPS1-5*, with the gene letters standing for infantile hypertrophic pyloric stenosis. *IHPS1* is located on the long arm of chromosome 12, at position 12q24.2-q24.31, and *IHPS2* is located on chromosome 16 at position 16p13-p12. The remaining genes are *IHPS3* (at position 11q14-q22), *IHPS4* (at position Xq23), and *IHPS5* (at position 16q24.3). Little is known about the cellular action or interaction of the protein products of these genes or the molecular mechanism by which susceptibility may be increased.

Symptoms

Symptoms of pyloric stenosis usually begin when babies are three to five weeks old. They include forceful vomiting of formula or milk; acting hungry most of the time; weight loss; signs of dehydration, such as less urination, dry mouth, and crying without tears; tiredness; fewer bowel movements; and blood-tinged vomit (which happens when repeated vomiting irritates the stomach, causing mild stomach bleeding).

Screening and Diagnosis

The doctor will ask about the symptoms a child is experiencing and about his or her medical history. The doctor will also perform a physical examination. An olive-shaped knot caused by the presence of pyloric stenosis is often felt by the experienced examiner. If a baby is diagnosed with pyloric stenosis, parents and their families will be referred to a pediatric surgeon (a doctor specializing in surgery in children).

Tests may include an abdominal ultrasound, a procedure that uses sound waves to make detailed computer pictures of the inside of the abdomen. Tests may also include a barium upper gastrointestinal X-ray series. A medicine (barium) is swallowed to outline the esophagus and stomach. X-ray pictures of the abdomen can then tell if food is moving normally through the stomach.

Treatment and Therapy

Pyloric stenosis is treated with a surgery called a pyloromyotomy, with a baby asleep under anesthesia. In a pyloromyotomy, the outside of the pylorus muscle is cut to relieve the blockage. Prior to surgery, fluids and electrolytes will be given intravenously to correct the dehydration and electrolyte imbalances that are common in babies with pyloric stenosis. After the operation, fluids are given by vein until the baby can take all of his or her normal feedings by mouth.

Prevention and Outcomes

There are no known ways of preventing pyloric stenosis, although it is possible that breast-feeding might reduce the risk.

Nathalie Smith, M.S.N, R.N.;
reviewed by Daus Mahnke, M.D.
"Etiology and Genetics" by Jeffrey A. Knight, Ph.D.

Further Reading

EBSCO Publishing. *Health Library: Pyloric Stenosis.* Ipswich, Mass.: Author, 2009. Available through http://www.ebscohost.com.

Kim, S. S., et al. "Pyloromyotomy: A Comparison of Laparoscopic, Circumumbilical, and Right Upper Quadrant Operative Techniques." *Journal of the American College of Surgeons* 201, no. 1 (July, 2005): 66-70.

Pisacane, A., et al. "Breast Feeding and Hypertrophic Pyloric Stenosis: Population-Based Case-Control Study." *British Medical Journal* 312, no. 7033 (March 23, 1996): 745-746.

Udassin, R. "New Insights in Infantile Hypertrophic Pyloric Stenosis." *Israel Medical Association Journal* 6, no. 3 (March, 2004): 160-161.

White, J. S., et al. "Treatment of Infantile Hypertrophic Pyloric Stenosis in a District General Hospital: A Review of 160 Cases." *Journal of Pediatric Surgery* 38, no. 9 (2003): 1333-1336.

Web Sites of Interest

American Association of Pediatrics
http://www.aap.org

The American Pediatric Surgical Association
http://www.eapsa.org/parents/pyloric.htm

Caring for Kids, Canadian Paediatric Society
http://www.caringforkids.cps.ca

The Montreal Children's Hospital
http://www.thechildren.com

See also: Celiac disease; Colon cancer; Crohn disease; Familial adenomatous polyposis; Hereditary diffuse gastric cancer (HDGC); Hereditary mixed polyposis syndrome; Hirschsprung's disease; Lynch syndrome.

Q

Quantitative inheritance

CATEGORY: Population genetics

SIGNIFICANCE: Quantitative inheritance involves metric traits. These traits are generally associated with adaptation, reproduction, yield, form, and function. They are thus of great importance to evolution, conservation biology, psychology, and especially to the improvement of agricultural organisms.

KEY TERMS

genotype: the genetic makeup of an organism at all loci that affect a quantitative trait

heritability: the proportion of phenotypic differences among individuals that are a result of genetic differences

metric traits: traits controlled by multiple genes with small individual effects and continuously varying environmental effects, resulting in continuous variation in a population

phenotype: the observed expression of a genotype that results from the combined effects of the genotype and the environment to which the organism has been exposed

THE GENETICS UNDERLYING METRIC TRAITS

An understanding of the genetics affecting metric traits came with the unification of the Mendelian and biometrical schools of genetics early in the 1900's. The statistical relationships involved in inheritance of metric traits such as height of humans were well known in the late 1800's. Soon after that, Gregor Mendel's breakthrough on particulate inheritance, obtained from work utilizing traits such as colors and shapes of peas, was rediscovered. However, some traits did not follow Mendelian inheritance patterns. As an example, Francis Galton crossed pea plants having uniformly large seeds with those having uniformly small seeds. The seed size of the progeny was intermediate. However, when the progeny were mated among themselves, seed size formed a distribution from small to large with many intermediate sizes.

How could particulate genetic factors explain a continuous distribution? The solution was described early in the twentieth century when Swedish plant breeder Herman Nilsson-Ehle crossed red and white wheat. The resulting progeny were light red in color. When matings were made within the progeny, the resulting kernels of wheat ranged in color from white to red. He was able to categorize the wheat into five colors: red, intermediate red, light red, pink, and white. Intermediate colors occurred with greater frequency than extreme colors. Nilsson-Ehle deduced that particulate genetic factors (now known as alleles) were involved, with red wheat inheriting four red alleles, intermediate red inheriting three red alleles, light red inheriting two red alleles, pink inheriting one red allele, and white inheriting no red alleles. These results were consistent with Mendel's findings, except that two sets of factors (now known as loci) were controlling this trait rather than the single locus observed for the traits considered by Mendel. Further, these results could be generalized to account for additional inheritance patterns controlled by more than two loci. Quantitative inheritance was mathematically described by British statistician and geneticist Ronald A. Fisher.

Under many circumstances, the environment also modifies the expression of traits. A combination of many loci with individually small effects alone would produce a rough bell-shaped distribution for a quantitative trait. Environmental effects are continuous and are independent of genetic effects. Environmental effects blur the boundaries of the genetic categories and can make it difficult or impossible to identify the effects of individual loci for many quantitative traits. The distribution of phenotypes, reflecting combined genetic and environmental effects, is typically a smooth, bell-shaped curve.

Genetic and environmental effects jointly influence the value of most metric traits. The relative magnitudes of genetic and environmental effects are measured using heritability statistics. Although essentially equivalent, heritability has several practical definitions. One definition states that heritability is equal to the proportion of observed differences among organisms for a trait due to genetic differences. For example, if one-quarter of the differences among cows for the amount of milk they produce are caused by differences among their genotypes, the heritability of milk production is 25 percent. The remaining 75 percent of differences among the animals are attributed to environmental effects. An alternative definition is that heritability is equal to the proportion of differences among sets of parents that are passed on to their progeny. For example, if the average height of a pair of parents is 8 inches (20 centimeters) more than the mean of their population and the heritability of height is 50 percent, their progeny would be expected to average 4 inches (10 centimeters) taller than their peers in the population.

FUNDAMENTAL RELATIONSHIPS OF QUANTITATIVE GENETICS

Two relationships are fundamental to the understanding and application of quantitative genetics. First, there is a tendency for likeness among related individuals. Although similarities of human stature and facial appearance within families are familiar to most people, similar relationships hold for such traits in all organisms. Correlation among relatives exists for such diverse traits as blood pressure, plant height, grain yield, and egg production. These correlations are caused by relatives sharing a portion of genes in common. The more closely the individuals are related, the greater the proportion of genes that are shared. Identical twins share all their genes, and full brothers and sisters or parent and offspring are expected to share one-half their genes. This relationship is commonly utilized in the improvement of agricultural organisms. Individuals are chosen to be parents based on the performance of their relatives. For example, bulls of dairy breeds are chosen to become widely used as sires based on the milk-producing ability of their sisters and daughters.

The second fundamental relationship is that, in organisms that do not normally self-fertilize, vigor is depressed in progeny that result from the mating of closely related individuals. This effect is known as inbreeding depression. It may be the basis of the social taboos regarding incestuous relationships in humans and for the dispersal systems for some other species of mammals such as wolves. Physiological barriers have evolved to prevent fertilization between close relatives in many species of plants. Some mechanisms function as an anatomical inhibitor to prevent union of pollen and ova from the same plant; in maize, for example, the male and female flower are widely separated on the plant. Indeed, in some species such as asparagus and holly trees, the sexes are separated in different individuals; thus all seeds must consequently result from cross-pollination. In other systems, cross-pollination is required for fertile seeds to result. The pollen must originate from a plant genetically different from the seed parent. These phenomena are known as self-incompatibility and are present in species such as broccoli, radishes, some clovers, and many fruit trees.

The corollary to inbreeding depression is hybrid vigor, a phenomenon of improved fitness that is often evident in progeny resulting from the mating of individuals less related than the average in a population. Hybrid vigor has been utilized in breeding programs to achieve remarkable productivity of hybrid seed corn as well as crossbred poultry and livestock. Hybrid vigor results in increased reproduction and efficiency of nutrient utilization. The mule, which results from mating a male donkey to a female horse, is a well-known example of a hybrid that has remarkable strength and hardiness compared to the parent species, but which is, unfortunately, sterile.

QUANTITATIVE TRAITS OF HUMANS

Like other organisms, many traits of humans are quantitatively inherited. Psychological characteristics, intelligence quotient (IQ), and birth weight have been studied extensively. The heritability of IQ has been reported to be high. Other personality characteristics such as incidence of depression, introversion, and enthusiasm have been reported to be highly heritable. Musical ability is another characteristic under some degree of genetic control. These results have been consistent across replicated studies and are thus expected to be reliable; however, some caution must be exercised when considering the reliability of results from individual studies. Most studies of heritability in humans have involved likeness of twins reared together and apart. The difficulty in obtaining such data results in a relatively small sample size, at least relative to similar experiments in animals. An unfortunate response to studies of quantitative inheritance in humans was the eugenics movement.

Birth weight of humans is of interest because it is both under genetic control and subject to influence by well-known environmental factors, such as smoking by the mother. Birth weight is subject to stabilizing selection, in which individuals with intermediate values have the highest rates of survival. This results in genetic pressure to maintain the average birth weight at a relatively constant value.

Quantitative Characters in Agricultural Improvement

The ability to meet the demand for food by a growing world population is dependent upon continuously increasing agricultural productivity. Reserves of high-quality farmland have nearly all been brought into production, and a sustainable increase in the harvest of fish is likely impossible. Many countries that struggle to meet the food demands of their populations are too poor to increase agricultural yields through increased inputs of fertilizer and chemicals. Increased food production will, therefore, largely depend on genetic improvement of the organisms produced by farmers worldwide.

Most characteristics of economic value in agriculturally important organisms are quantitatively inherited. Traits such as grain yield, baking quality, milk and meat production, and efficiency of nutrient utilization are under the influence of many genes as well as the production environment. The task of breeders is not only to identify organisms with superior genetic characteristics but also to identify those breeds and varieties well adapted to the specific environmental conditions in which they will be produced. The type of dairy cattle that most efficiently produces milk under the normal production circumstances in the United States, which includes high health status, unlimited access to high-quality grain rations, and protection from extremes of heat and cold, may not be ideal under conditions in New Zealand in which cattle are required to compete with herdmates for high-quality pasture forage. Neither of these animals may be ideal under tropical conditions where extremely high temperatures, disease, and parasites are common.

Remarkable progress has been made in many important food crops. Grain yield has responded to improvement programs. Development of hybrid corn increased yield several-fold over the last few decades of the twentieth century. Development of improved varieties of small grains resulted in an increased ability of many developing countries to be self-sufficient in food production. Grain breeder Norman Borlaug won the Nobel Peace Prize in 1970 for his role in developing grain varieties that contributed to the Green Revolution.

Can breeders continue to make improvements in the genetic potential for crops, livestock, and fish to yield enough food to support a growing human population? Tools of biotechnology are expected to increase the rate at which breeders can make genetic change. Ultimately, the answer depends upon the genetic variation available in the global populations of food-producing organisms and their wild relatives. The potential for genetic improvement of some species has been relatively untapped. Domestication of fish for use in aquaculture and utilization of potential crop species such as amaranth are possible food reserves. Wheat, corn, and rice provide a large proportion of the calories supporting the world population. The yields of these three crop species have already benefited from many generations of selective breeding. For continued genetic improvement, it is critical that variation not be lost through the extinction of indigenous strains and wild relatives of important food-producing organisms.

Impact and Applications

Molecular genetics and biotechnology have also added new tools for analyzing the genetics of quantitative traits. In any organism that has had its genome adequately mapped, genetic markers can be used to determine the number of loci involved in a particular trait. In carefully constructed crosses geneticists look for statistical correlations between markers and the trait of interest. When a high correlation is found, the marker is said to represent a quantitative trait locus (QTL). Often a percentage effect for each QTL can be determined and because the location of markers is typically known, the potential location of the gene can also be inferred (that is, somewhere near the marker). A good understanding of the QTLs involved in the expression of a quantitative trait can help determine the best way to improve the organism.

Although QTLs are much easier to discover in organisms where controlled crosses are possible, studies have also been carried out in humans. In humans, geneticists must rely on whatever matings have happened, and due to ethical limitations, cannot set up specific crosses. Studies in humans have attempted to quantify the number of QTLs responsible for such things as IQ and various physical traits. One study even purported to show that homosexuality is genetically based. Although there is some support for such studies, much controversy surrounds them, and ethicists continue to worry that conclusions from such research will be used in a new wave of eugenics. In spite of the risk of misusing an improved understanding of human quantitative traits, human biology and medicine stand to benefit.

William R. Lamberson, Ph.D.; updated by Bryan Ness, Ph.D.

Further Reading

Falconer, D. S., and Trudy F. MacKay. *Introduction to Quantitative Genetics.* 4th ed. Reading, Mass.: Addison-Wesley, 1996. The standard text on the subject, outlining the genetics of differences in quantitative phenotypes and their applications to animal breeding, plant improvement, and evolution.

Gillespie, John H. "Quantitative Genetics." In *Population Genetics: A Concise Guide.* 2d ed. Baltimore: Johns Hopkins University Press, 2004. This chapter provides a brief but comprehensive overview of one of the aspects of population genetics.

Hamilton, Matthew B. *Population Genetics.* Hoboken, N.J.: Wiley-Blackwell, 2009. Basic textbook presenting the major concepts in population genetics. Two of the chapters focus on quantitative genetics: chapter 9, "Quantitative Trait Variation and Evolution," and chapter 10, "The Mendelian Basis of Quantitative Trait Variation." Each chapter concludes with a review and suggestions for further reading.

Templeton, Alan R. *Population Genetics and Microevolutionary Theory.* Hoboken, N.J.: Wiley-Liss, 2006. Part 2, "Genotype and Phenotype," focuses on quantitative genetics, with one chapter providing basic quantitative genetic definitions and theory and two other chapters examining measured and unmeasured genotypes. Includes review questions, bibliography, and list of Web resources.

Web Sites of Interest

Scitable
http://www.nature.com/scitable/topicpage/Quantitative-Genetics-Growing-Transgenic-Tomatoes-1123

Scitable, a library of science-related articles compiled by the Nature Publishing Group, contains the article "Quantitative Genetics: Growing Transgenic Tomatoes," which discusses the importance of quantitative trait loci in agriculture.

Tutor Vista.com, Quantitative Inheritance
http://www.tutorvista.com/content/biology/biology-iii/heredity-and-variation/quantitative-inheritance.php

TutorVista.com, an online tutorial for students, includes a page discussing quantitative inheritance.

See also: Artificial selection; Biofertilizers; Consanguinity and genetic disease; Epistasis; Genetic load; Hardy-Weinberg law; Hybridization and introgression; Inbreeding and assortative mating; Mendelian genetics; Polygenic inheritance; Population genetics; Speciation; Twin studies.

R

Race

Category: Human genetics and social issues

Significance: Humans typically have been categorized into a small number of races based on common traits, ancestry, and geography. Knowledge of human genomic diversity has increased awareness of ambiguities associated with traditional racial groups. The sociopolitical consequences of using genetics to devalue certain races are profound, and based on the available data, are completely baseless.

Key terms

eugenics: a movement concerned with the improvement of human genetic traits, predominantly by the regulation of mating

Human Genome Diversity Project: an extension of the Human Genome Project in which DNA of native people around the world is collected for study

population: a group of geographically localized, interbreeding individuals

race: a collection of geographically localized populations with well-defined genetic traits

Conflicting Definitions of Race

Few ideas have had such a contentious history as the use of the term "race." Categorization relied on consideration of salient traits such as skin color, body form and hair texture to classify humans into distinct subcategories. The term "race" is currently believed to have little biological meaning, in great part because of advances in genetic research. Studies have revealed that a person's genes cannot define their ethnic heritage and that no gene exists exclusively within one race/ethnocultural group. Biomedical scientists remain divided on their opinion about "race" and how it may be used in treating human genetic conditions.

For a racial or subspecies classification scheme to be objective and biologically meaningful, researchers must decide carefully which heritable characteristics (passed to future generations genetically) will define the groups. Several principles are considered. First, the unique traits must be discrete and not continually changing by small degrees between populations. Second, everyone placed within a specific race must possess the selected trait's defining variant. All the selected characteristics are found consistently in each member of the group. For example, if blue eyes and brown hair are chosen as defining characteristics, everyone designated as belonging to that race must share both of those characteristics. Individuals placed in other races should not exhibit this particular combination. Third, individuals of the same race must have descended from a common ancestor, unique to those people. Many shared characteristics present in individuals of a race may be traced to that ancestor by heredity. Based on the preceding defining criteria (selection of discrete traits, agreement of traits, and common ancestry), pure representatives of each racial category should be detectable.

Most researchers maintain that traditional races do not conform to scientific principles of subspecies classification. For example, the traits used to define traditional human races are rarely discrete. Skin color, a prominent characteristic employed, is not a well-defined trait. Approximately eleven genes influence skin color significantly, but fifty or so are likely to contribute. Pigmentation in humans results from a complex series of biochemical pathways regulated by amounts of enzymes (molecules that control chemical reactions) and enzyme inhibitors, along with environmental factors. Moreover, the number of melanocytes (cells that produce melanin) do not differ from one person to another, while their level of melanin production does. Like most complex traits involving many genes, human skin color varies on a continuous gradation. From lightest to darkest, all intermediate pigmentations are represented.

On average, any two people of the same or a different race diverge genetically by a mere 0.2 percent, and only 0.012 percent contribute to traditional racial variations. Allelic variations account for most of the superficial differences perceived as race. (U.S. Department of Energy Human Genome Program, http://www.ornl.gov/hgmis)

Color may vary widely even within the same family. The boundary between black and white is an arbitrary, humanmade border, not one imposed by nature.

In addition, traditional defining racial characteristics, such as skin color and facial characteristics, are not found in all members of a race. For example, many Melanesians, indigenous to Pacific islands, have pigmentation as dark as any human but are not classified as "black." Another example is found in individuals of the Cherokee Nation that have Caucasoid facial features and very dark skin, yet have no European ancestry. When traditional racial characteristics are examined closely, many groups are left with no conventional racial group. No "pure" genetic representatives of any traditional race exist.

Common ancestry must also be considered. Genetic studies have shown that Africans do not belong to a single "black" heritage. In fact, several lineages are found in Africa. An even greater variance is found in African Americans. Besides a diverse African ancestry, on average 13 percent of African American ancestry is Northern European. Yet all black Americans are consolidated into one race.

The true diversity found in humans is not patterned according to accepted standards of a subspecies. Only at extreme geographical distances are notable differences found. Human populations in close proximity have more genetic similarities than distant populations. Well-defined genetic borders between human populations are not observed, and racial boundaries in classification schemes are most often formed arbitrarily.

History of Racial Classifications

Efforts to classify humans into a number of distinct types date back at least to the 19th Dynasty of Ancient Egypt. The sacred text Book of Gates described four distinct groups: "Egyptians," "Asiatics," "Libyans," and "Nubians" were defined using both physical and geographical characteristics. Applying scientific principles to divide people into distinct racial groups has been a goal for much of human history.

In 1758, the founder of biological classification, Swedish botanist Carolus Linnaeus, arranged humans into four principal races: *Americanus*, *Europeus*, *Asiaticus*, and *Afer.* Although geographic location was his primary organizing factor, Linnaeus also described the races according to subjective traits such as temperament. Despite his use of archaic criteria, Linnaeus did not give superior status to any of the races.

Johann Friedrich Blumenbach, a German naturalist and admirer of Linnaeus, developed a classification with lasting influence. Blumenbach maintained that the original forms, which he named "Caucasian," were those primarily of European ancestry. His final classification, published in 1795, consisted of five races: Caucasian, Malay, Ethiopian, American, and Mongolian. The fifth race, the Malay, was added to Linnaeus's classification to show a step-by-step change from the original body type.

After Linnaeus and Blumenbach, many variations of their categories were formulated, chiefly by biologists and anthropologists. Classification "lumpers" combined people into only a few races (for example, black, white, and Asian). "Splitters" separated the traditional groups into many different races. One classification scheme divided all Europeans into Alpine, Nordic, and Mediterranean races. Others split Europeans into ten different races. No one scheme of racial classification came to be accepted throughout the scientific community.

Genetics and Theories of Human Evolution

Advances in DNA technology have greatly aided researchers in their quest to reconstruct the history of *Homo sapiens* and its diversification. Analysis of human DNA has been performed on both nuclear and mitochondrial DNA. The nucleus is the organelle that contains the majority of the cell's genetic material. Mitochondria are organelles responsible for generating cellular energy. Each mitochondrion contains a single, circular DNA molecule. Research suggests that Africa was the root of all humankind and that humans first arose there 100,000 to 200,000 years ago. Several lines of research, including DNA analysis of humanoid fossils, provide further evidence for this theory.

Many scientists are using genetic markers to decipher the migrations that fashioned past and present human populations. For example, DNA comparisons revealed three Native American lineages. Some scientists believe one migration crossed the Bering Strait, most likely from Mongolia. Another theory states that three separate Asian migrations occurred, each bringing a different lineage.

Genetic Diversity Among Races

Three primary forces produce the genetic components of a population: natural selection, nonadaptive genetic change, and mating between neighboring populations. The first two factors may result in differences between populations, and reproductive isolation, either voluntary or because of geographic isolation, perpetuates the distinctions. Natural selection refers to the persistence of genetic traits favorable in a specific environment. For example, a widely held assumption concerns skin color, primarily a result of the pigment melanin. Melanin offers some shielding from ultraviolet solar rays. According to this theory, people living in regions with concentrated ultraviolet exposure have increased melanin synthesis and, therefore, dark skin color conferring protection against skin cancer. Individuals with genes for increased melanin have enhanced survival rates and reproductive opportunities. The reproductive opportunities produce offspring that inherit those same genes for increased melanin. This process results in a higher percentage of the population with elevated melanin production genes. Therefore, genes coding for melanin production are favorable and persist in these environments.

The second factor contributing to the genetic makeup of a population is nonadaptive genetic change. This process involves random genetic mutations (alterations). For example, certain genes are responsible for eye color. Individuals contain alternate forms of these genes, or alleles, which result in different eye color. Because these traits are impartial to environmental influences, they may endure from generation to generation. Different populations will spontaneously produce, sustain, and delete them.

The third factor, mating between individuals from neighboring groups, tends to merge traits from several populations. This genetic mixing often results in offspring with blended characteristics.

Several studies have compared the overall genetic complement of various human populations. On average, any two people of the same or a different race diverge genetically by a mere 0.1 percent. It is estimated that only 0.012 percent contributes to traditional racial variations. Hence, most of the genetic differences found between a person of African descent and a person of European descent are also different between two individuals with the same ancestry. The genes do not differ. It is the proportion of individuals expressing a specific allele that varies from population to population.

Upon closer examination, it was found that the continent of Africa is unequaled with respect to cumulative genetic diversity. Numerous races are found in Africa, Khoisan Africans of southern Africa being the most distinct. Therefore two people of different ethnicities who do not have recent African ancestry (for example Northern Europeans and South East Asians) have more similar genetics than any two distinct African ethnic groups. This finding supports theories of early human migration in which humans first evolved in Africa and a subset left the continent, experienced a population bottleneck, and then established the human populations around the world.

Human Genome Diversity Project and Advances in Research

Many scientists are attempting to reconcile the negativities associated with racial studies. The Human Genome Diversity Project (HGDP), was initiated by Stanford University in 1993 and functions independently from the Human Genome Project. The HGDP aims to collect and store DNA from ethnically diverse populations around the world, creating a library of samples to represent global human diversity. Results of future studies may aid in gene therapy treatments and greater success with organ transplantation. As a result, a more thorough understanding of the genetic diversity and unity in the species *Homo sapiens* will be possible.

At the population level, human diversity is greatest within racial/cultural groups rather than between them. Originally, geneticists who studied genetic diversity of human populations were limited to data from very few genetic loci (locations in the genome that are of interest); however, recent studies are able to simultaneously analyze hundreds to thousands of loci. It is currently estimated that 90 percent of genetic variation in human beings is found within each purported racial group, while differences between the groups only equate to the remaining 10 percent.

A second method of studying human genetic diversity is to compare ethnically diverse individuals and search for similarities and differences in their genomes. Early studies involved only a few dozen genetic loci and as a result did not find individuals to cluster (group together) based on their geographic origin. Recent studies, however, were able to analyze substantially more genetic loci and resulted in data with stronger statistical power. These studies focused on individuals from three distinct geographic areas: Europe, sub-Sahran Africa, and East Asia. Indeed, individuals clustered or shared more genetic similarities with others of the same geographic region. Participants from Africa were found to have the greatest diversity, which is in agreement with population studies. Another cluster consisted exclusively of Europeans, and a third comprised the Asian individuals. However, when individuals from neighboring regions were also analyzed, such as South Indians, the analysis showed similarities to both East Asians and Europeans. This finding may be explained by the numerous migrations between Europe and India during the past ten thousand years. Many individuals did not cluster with their geographic cohorts, demonstrating that individuals are not easily categorizeable into neat groups of races but tend to share more genetic similarities with people from their region.

Race or an individual's ancestry can sometimes provide useful information in medical decision making, as gender or age often do. Certain genetic conditions are more common among ethnocultural groups. For example, hemochromatosis is more prevalent within Northern Europeans and Caucasians, whereas sickle-cell disease is more often found in Africans and African Americans. Meanwhile other genetic diseases are equally prevalent across racial groups, as seen in spinal muscular atrophy (SMA). If a disease-causing gene is common, then it is likely to be relatively ancient and thus shared across ethnicities. Moreover, some genetic conditions remain prevalent in populations because they provide an adaptive advantage to the individual, as seen in sickle-cell disease carriers being protected against malarial infection. Likewise, an individual's response to drugs may be mediated by their genetic makeup. A gene called *CYP2D6* is involved in the metabolism or breakdown of many important drugs such as codeine and morphine. Some individuals have no working copy of this gene whereas others have one or two copies that function properly. The majority of individuals with no working copies are of European heritage (26 percent), whereas fewer Asian (6 percent) and African populations (7 percent) fall into this category. Thus it may be tempting to make medical decisions based on a patient's ethnic heritage; however, this may lead to inaccurate diagnoses (missing sickle-cell disease in an Asian individual) or inappropriate drug administration (prohibiting a Caucasian person from taking codeine). Ideally, each individual should have medical decisions made based on their genetic makeup in lieu of their ethnic heritage. Future patients may be able to first request an analysis of their genome, which would aid their physicians in making some genetically appropriate medical decisions.

Sociopolitical Implications

Race is often portrayed as a natural, biological division, the result of geographic isolation and adaptation to local environment. However, confusion between biological and cultural classification obscures perceptions of race. When individuals describe themselves as "black," "white," or "Hispanic," for example, they are usually describing cultural heredity as well as biological similarities. The relative importance of perceived cultural affiliations or genetics varies depending on the circumstances. Examples il-

Descendants of Sally Hemings, an African American slave of Thomas Jefferson who is known by DNA evidence to have had children by him, pose during a July, 2005, reunion. (AP/Wide World Photos)

lustrating the ambiguities are abundant. Nearly all people with African American ancestry are labeled black, even if they have a white parent. In addition, dark skin color designates one as belonging to the black race, including Africans and aboriginal Australians, who have no common genetic lineage. State laws, some on the books until the late 1960's, required a "Negro" designation for anyone with one-eighth black heritage (one black great-grandparent).

Unlike biological boundaries, cultural boundaries are sharp, repeatedly motivating discrimination, genocide, and war. In the early and mid-twentieth century, the eugenics movement, advocating the genetic improvement of the human species, translated into laws against interracial marriage, sterilization programs, and mass murder. Harmful effects include accusations of deficiencies in intelligence or moral character based on traditional racial classification.

The frequent use of biology to devalue certain races and excuse bigotry has profound implications for individuals and society. Blumenbach selected Caucasians (who inhabit regions near the Caucasus Mountains, a Russian and Georgian mountain range) as the original form of humans because in his opinion they were the most beautiful. All other races deviated from this ideal and were, therefore, less beautiful. Despite Blumenbach's efforts not to demean other groups based on intelligence or moral character, the act of ranking in any form left an ill-fated legacy.

In conclusion, race remains a contentious issue both in many fields of science and within the greater society. Recent genomic studies at both the individual and the population level have shown that the majority of human genetic composition is universal and shared across all ethnocultural groups. Shared genetics is most commonly found in individuals who originate from the same geographic region. However, there is no scientific support for the concept of distinct, "pure," and nonoverlapping races. Unfortunately throughout human history, the use and abuse of the term "race" has been pursued for sociopolitical gains or to justify bigotry toward and abuses of individuals. It is now known that human genetic diversity is a continuum, with natural selection, nonadaptive genetic change, and mating as the true driving forces for human genetic diversity.

Stacie R. Chismark, M.S.;
updated by Kayla Mandel Sheets, M.S.

Further Reading

Cavalli-Sforza, Luigi L. *The Great Human Diasporas: A History of Diversity and Evolution.* Translated by Serah Thorne. Reading, Mass.: Addison-Wesley, 1995. Argues that humans around the world are more similar than different.

_______, et al. *The History and Geography of Human Genes.* Princeton, N.J.: Princeton University Press, 1996. Often referred to as a "genetic atlas," this volume contains fifty years of research comparing heritable traits, such as blood groups, from more than one thousand human populations.

Fish, Jefferson M., ed. *Race and Intelligence: Separating Science from Myth.* Mahwah, N.J.: Lawrence Erlbaum, 2002. An interdisciplinary collection disputing race as a biological category and arguing that there is no general or single intelligence and that cognitive ability is shaped through education.

Garcia, Jorge J. E. *Race or Ethnicity? On Black or Latino Identity.* Ithaca, N.Y.: Cornell University Press, 2007. Essays discuss whether racial identity matters and consider issues associated with assimilation, racism, and public policy.

Gates, E. Nathaniel, ed. *The Concept of "Race" in Natural and Social Science.* New York: Garland, 1997. Argues that the concept of race, as a form of classification based on physical characteristics, was arbitrarily conceived during the Enlightenment and is without scientific merit.

Gibbons, A. "Africans' Deep Genetic Roots Reveal Their Evolutionary Story." *Science* 324 (2009): 575. Describes the largest study ever conducted of African genetic diversity, which reveals Africans are descendants from 14 distinct ancestral groups that often correlate with language and cultural groups.

Gould, Stephen Jay. *The Mismeasure of Man.* Rev. ed. New York: W. W. Norton, 1996. Presents a historical commentary on racial categorization and a refutation of theories espousing a single measure of genetically fixed intelligence.

Graves, Joseph L., Jr. *The Emperor's New Clothes: Biological Theories of Race at the Millennium.* New Brunswick, N.J.: Rutgers University Press, 2001. Argues for a more scientific approach to debates about race, one that takes human genetic diversity into account.

Herrnstein, Richard J., and Charles Murray. *The Bell Curve: Intelligence and Class Structure in America.* New York: Free Press, 1994. The authors maintain that IQ is a valid measure of intelligence, that intelligence is largely a product of genetic background, and that differences in intelligence among social classes play a major part in shaping American society.

Jorde, L. B., and S. P. Wooding. "Genetic Variation, Classification, and 'Race.'" *Nature Genetics* 36, no. 11 (2004): S28. A review article that provides an overview of human variation and discusses whether current data support historic ideas of race, and what these findings imply for biomedical research and medicine.

Kevles, Daniel J. *In the Name of Eugenics: Genetics and the Uses of Human Heredity.* Cambridge, Mass.: Harvard University Press, 1995. Discusses genetics both as a science and as a social and political perspective, and how the two often collide to muddy the boundaries of science and opinion.

Royal, C., and G. Dunston. "Changing the Paradigm from 'Race' to Human Genome Variation." *Nature Genetics* 36 (2004): S5-S7. Commentary suggests we begin to think outside the box and see ethnic groups as genomic diversity rather than distinct races.

Valencia, Richard R., and Lisa A. Suzuki. *Intelligence Testing and Minority Students: Foundations, Performance Factors, and Assessment Issues.* Thousand Oaks, Calif.: Sage Publications, 2000. Historical and multicultural perspective on intelligence and its often assumed relation with socioeconomic status, home environment, test bias, and heredity.

WEB SITES OF INTEREST

Genetics and Identity Project

http://www.bio ethics.umn.edu/genetics_and _identity

Project looks at the ways genetic research affects racial, ethnic, and familial identities.

Human Genome Project Information: Minorities, Race, and Genetics

http://www.ornl.gov/sci/techresources/Human _ Genome/elsi/minorities.shtml

This site provides up-to-date information on discoveries revealed by the Human Genome Project and how they impact what is known about race, early human history, and genetics.

International Hap Map Project

http://www.hapmap.org/abouthapmap.html

This project aims to expand knowledge of the human genome via scientists collaborating internationally.

National Academies Press: Evaluating Human Genetic Diversity

http://www.nap.edu

A free, downloadable book on human genetic diversity, which includes the chapter "Human Rights and Human Genetic-Variation Research."

See also: Biological determinism; Eugenics; Eugenics: Nazi Germany; Evolutionary biology; Genetic engineering: Social and ethical issues; Heredity and environment; Intelligence; Miscegenation and antimiscegenation laws; Sociobiology; Sterilization laws.

RB1 gene

CATEGORY: Molecular genetics

SIGNIFICANCE: *RB1*, the retinoblastoma susceptibility gene, was the first tumor-suppressor gene to be discovered as well as the first hereditary cancer gene to be cloned. Study of *RB1* has led to a greater understanding of cancer genes and the development of the "two-hit" model for anti-oncogene-associated cancers, as a result of retinoblastoma research.

KEY TERMS

allele: one of two or more different genes containing specific inheritable characteristics that occupy corresponding positions (loci) on paired chromosomes

anti-oncogene: tumor-suppressor gene

chromosome: a linear strand of DNA that carries genetic information

deletion: the loss of a portion of a chromosome

exon: one of the coding regions of the DNA of genes

expression: use of a gene to create the corresponding protein

hotspot: DNA positions where genetic recombination occurs with above-average frequency

locus: the position of a gene on a chromosome

mutation: a change in gene, potentially capable of being transmitted to offspring

retinoblastoma: most common intraocular malignancy (eye cancer) in children

GENETICS SUMMARY AND LOCATION

The *RB1* gene is an anti-oncogene (tumor suppressor) and was first recognized and indentified in patients with hereditary retinoblastoma by Thaddeus Dryja's laboratory at The Massachusetts Eye and Ear Infirmary at Harvard Medical School, originally isolated using the technique called chromosome walking. The cytogenic location of *RB1* is 13q14.2, which means it is on the long arm (q) of chromosome 13 at position 14.2, specifically located from base pair 47,775,911 to base pair 47,954,022. *RB1* is a very large gene, extending over 180 kilobase pairs, containing 27 exons. The promoter region of *RB1* gene is distinctive because it contains binding motifs for transcription factors Sp1 and ATF but no TATA or CAAT box.

The amino acid sequence for *RB1* product is more than 75 percent identical for humans, rats, mice, and frogs. *RB1* exists in all vertebrates, which may indicate that it has a critical biological role that has been maintained through evolution.

RB PROTEIN AND TUMOR SUPPRESSION

RB protein plays an integral role in cell cycle regulation. It controls auxiliary proteins, mediating their replication processes. During the G_1 (resting phase) phase of a cell cycle, RB protein binds to E2F factor and blocks the transcription of S-phase (phase in which DNA synthesis occurs) genes, mediating the G_1-to-S phase checkpoint. A loss of RB protein, due to *RB1* mutation, causes elimination of G_1-to-S phase checkpoint, increasing the odds of unregulated cellular production or cancer.

There are two distinct types of cancer genes, proto-oncogenes and anti-oncogenes. Proto-oncogenes are easier to identify and involve gene mutations that result in a gain-of-function, motivating cells to become cancerous. Anti-oncogenes are the most frequently mutated genes (except in leukemia and lymphoma) and cause a decreased expression, leading to cancer. *RB1* was the very first of only a few tumor-suppressing anti-oncogenes, that have been discovered and studied, and thus, *RB1* has served as an important model for study of anti-oncogene-associated cancers.

RB1 MUTATIONS

In 1971 Alfred Knudsen described the "two-hit" theory, a mutation model based on retinoblastoma observations and Mendelian genetics, which can be applied to inherited cancers. The theory states that in hereditary cancers, the predisposed individuals already carry one allele mutation and therefore, oncogenesis only requires one more mutation to occur. In nonhereditary cases, two or more mutational events are required. This hypothesis was validated with the first cloning of the *RB1* gene. Genotyping of retinoblastoma tumors concluded that mutations in both alleles of the *RB1* gene were required for tumor development to begin.

The majority of mutations of *RB1* occur between internal exons 2 and 25. There are more than 930 reported mutations, identified in more than 1000 families. The band of *RB1* mutations is very diverse, and does not demonstrate any distinct hotspots, although certain discrete hotspots of recurrent mutations have been characterized. The majority of these recurrent mutations are associated with C to T transitions in CGA codons. Normally in genetic mutations of large genes, the specific location of the stop codon will influence the phenotypic expression. This is not the case with *RB1*; the precise site of the mutation does not alter the phenotype.

The method of mutation analysis depends on the patient's family history and the specimens (blood or tumor) that are available. The mutations of *RB1* are distributed along almost the entire length of the gene, which suggests that no single analysis method will be fully sensitive. Approximately 90 percent of all *RB1* mutations can be detected by current molecular analysis procedures whereas 10 percent cannot.

RETINOBLASTOMA

Retinoblastoma, in its hereditary or nonhereditary form, is a malignant tumor, or tumors, which develop from the immature retina. In order for tumorgenesis to occur, mutations in both alleles of the *RB1* gene are required. Retinoblastoma occurs in early childhood and affects approximately 1 in 20,000 children. Only 6 to 10 percent of affected children have any previous family history.

Hereditary retinoblastoma is responsible for approximately 40 percent of all cases, is transmitted in an autosomal dominant fashion, and is characterized by multiple tumors in both eyes. Patients with the hereditary form are heterozygous for a mutation of *RB1* that has been inherited from one of their parents or has occurred in embryonic development (*RB1+/RB1-*). Tumors are formed when a single mutation in the functioning *RB1* allele occurs, creating a loss of heterozygosity at the *RB1* locus.

The nonhereditary form involves only one eye and a single tumor. These patients are wild-type homozygotes (*RB1+/RB1+*) and require two distinct mutational events to occur, causing a loss of function at both *RB1* loci in order for tumor formation to begin. The incidence of two mutational events occurring is rather low, in contrast to the single mutational event required for heterozygotes in order for tumor formation to be initiated.

If left untreated, retinoblastoma is consistently fatal, but with early diagnosis and current treatment methods, the present survival rate is more than 90 percent. Survivors of hereditary retinoblastoma have a vulnerability toward other cancers, notably osteosarcoma (37 percent) and soft-tissue sarcomas (16.8 percent), melanomas (7.4 percent), brain tumors (4.5 percent), and leukemia (2.4 percent). The incidence of retinoblastoma appears to be increasing as early diagnosis and improved treatment options have improved the survival of patients who have gone on to transmit this autosomal dominant trait to the next generation.

IMPACT

The discovery and characterization of the *RB1* gene, its mutations, and protein product have been fundamental in the understanding of the mechanisms of inherited cancers. Knudsen's "two-hit" theory established a critical model for tumorgenesis, uncovering new mechanisms for cancer origination. Research has concentrated on the structure and function of the *RB1* gene, as well as the regulation of the transcription and translational anti-oncogenic activity of the *RB1* protein. Numerous mutational categories have been established and aligned with phenotypic expression. These finding are significant as they expand the ability of physicians to offer improved diagnostic, prognostic, and genetic counseling to patients, as well as more specialized therapeutic interventions.

April D. Ingram

FURTHER READING

Klintworth, Gordon K., Alec Garner, and J. Godfrey Heathcote. *Garner and Klintworth's Pathobiology of Ocular Disease.* New York: Informa Healthcare, 2008. A scientific research tool and a quick reference guide to the causes and mechanisms of ocular disease.

McKusick, Victor, Marschall Stevens Runge, and Cam Patterson. *Principles of Molecular Medicine.* Totowa, N.J.: Humana Press, 2006. A comprehensive overview of molecular medicine including medical knowledge, diagnostic potentials, and therapeutic options.

Vogelstein, Bert, and Kenneth W. Kinzler, eds. *The Genetic Basis of Human Cancer.* 2d ed. New York: McGraw-Hill, 2002. This book provides an excellent account of recent progress in genetic cancer research and its impact on patient care.

WEB SITES OF INTEREST

Atlas of Genetics and Cytogenetics in Oncology and Haematology
http://atlasgeneticsoncology.org/Genes/RB1ID90.html

Cancer Genetics Web
http://www.cancerindex.org/geneweb/RB1.htm

U.S. National Library of Medicine, Genetics Home Reference: RB1
http://ghr.nlm.nih.gov/gene=rb1

See also: *BRAF* gene; *BRCA1* and *BRCA2* genes; *MLH1* gene; *PMS* genes; *RhoGD12* gene; *SCLC1* gene.

Refsum disease

CATEGORY: Diseases and syndromes

ALSO KNOWN AS: Phytanic acid oxidase deficiency; phytanic acid storage disease; heredopathia atactica polyneuritiformis; hereditary motor and sensory neuropathy IV (HMSN IV or HMSN4); classic Refsum disease (CRD); adult Refsum disease (ARD)

DEFINITION

Refsum disease is a severe, progressive inherited disorder that results in symptoms such as pain, difficulty walking, heart failure, skin issues, and/or loss of the ability to smell, see, and hear. The features of the disease are caused by the harmful buildup of fatty substances in the body and brain.

RISK FACTORS

Refsum disease is a genetic condition caused by the inheritance of the same nonworking gene that causes Refsum from both parents. Most Refsum disease cases have been reported in the United Kingdom and Norway. The incidence of the condition is very low. In the United Kingdom, the disease is found in 1 in 1,000,000 individuals. This condition is not caused by infections and cannot be transmitted by an affected individual.

Etiology and Genetics

Refsum disease is caused by the buildup of a substance called phytanic acid in the cells of the body. This abnormal buildup is caused by either a lack of an enzyme called phytanoyl-CoA hydroxylase or a decreased quantity of a critical cell part called the peroxisome-targeting signal type 2 receptor. When the enzyme or the receptor is missing, the accumulation of the phytanic acid in the cells, tissues, and body fluids results in the inability of the brain and body to work correctly.

Refsum disease is an autosomal recessive genetic condition, which occurs when a child receives two copies of a nonworking gene. Individuals with only one copy of a nonworking gene for a recessive condition are known as carriers and have no problems related to the condition. In fact, every individual carries between five and ten nonworking genes for harmful, recessive conditions. When two people with the same nonworking recessive gene meet, there is a chance, with each pregnancy, for the child to inherit two copies, one from each parent. That child then has no working copies of the gene and therefore has the signs and symptoms associated with the disease. There are two known genes that when nonworking cause Refsum disease: *PHYH* and *PEX7*.

Symptoms

The symptoms of Refsum disease involve progressive deterioration of motor and neurocognitive function including loss of smell, loss of physical milestones, a type of early-onset blindness called retinitis pigmentosa, nerve disorders (neuropathy), pain, deafness, stumbling gait, enlarged heart, heart arrhythmia, and cardiac failure. However, Refsum disease symptoms vary significantly in severity and time of onset from person to person. It is rare to see all the symptoms of Refsum disease in one individual.

The onset of Refsum disease can occur from seven months of age to mid-adulthood. The most common time of onset is late childhood. Symptoms often start with loss of the sense of smell (anosmia) and progressive loss of vision caused by retinal degeneration (retinitis pigmentosa). Over the next ten to fifteen years, as the disease progresses, affected individuals will often develop difficulty walking (ataxia), skin issues (ichthyosis), weakness, pain, and numbness from nerve malfunction (peripheral neuropathy), and possibly shortened finger or toe bones. Death most often occurs as a result of cardiac failure.

Screening and Diagnosis

In 2009, screening for Refsum disease is not part of routine testing in the prenatal or newborn periods of life. Diagnosis is most often made on the basis of disease signs and symptoms such as loss of smell, stumbling gait (ataxia), and early-onset blindness called retinitis pigmentosa. Biochemical testing is available to confirm the diagnosis through a combination of elevated phytanic acid and related chemicals and/or low or missing amounts of phytanoyl-CoA hydroxylase. However molecular genetic testing for nonworking *PHYH* and *PEX7* genes is the most definitive test available.

Treatment and Therapy

At this time, there is no cure for Refsum disease. Chronic treatment of the condition relies on a special diet that removes phytanic acid intake, avoidance of sudden weight loss, and careful monitoring for cardiac arrhythmias and other heart abnormalities by a cardiologist. Other symptoms would be treated individually, such as using hydrating creams for skin issues. When a fast decrease in symptom-causing phytanic acid levels is required, a treatment called plasmapheresis is sometimes used to quickly reduce the amount of phytanic acid in the blood. Use of ibuprofen must be avoided in individuals affected by Refsum disease as it may interfere with the breakdown of phytanic acid.

A few approaches for treating the underlying lack of enzyme that causes Refsum disease are under investigation, such as enzyme replacement therapy, but these are not yet approved by the Food and Drug Administration (FDA) for use in affected individuals.

Prevention and Outcomes

Carrier testing is available for individuals who are interested in learning if they carry a nonworking

PHYH or *PEX7* gene. Genetic counseling is available for parents who have an affected child or are concerned about being a carrier for a nonworking *PHYH* or *PEX7* gene. As the severity and symptoms of Refsum disease vary from individual to individual, life expectancy depends on the ability to follow the special low-phytanic-acid diet as well as the speed of progression of the disease.

Dawn A. Laney, M.S.

FURTHER READING

Gonick, Larry, and Mark Wheelis. *The Cartoon Guide to Genetics.* New York: HarperPerennial, 1991.

Parker, James, and Philip Parker, eds. *The Official Parent's Sourcebook on Refsum Disease: A Revised and Updated Directory for the Internet Age.* San Diego: Icon Health, 2002.

Willett, Edward. *Genetics Demystified.* New York: McGraw-Hill, 2005.

WEB SITES OF INTEREST

National Institute of Neurological Disorders and Stroke (NINDS): Refsum Disease Information Page
http://www.ninds.nih.gov/disorders/refsum/refsum.htm

United Leukodystrophy Foundation
http://www.ulf.org

See also: Adrenoleukodystrophy; Alexander disease; Alzheimer's disease; Amyotrophic lateral sclerosis; Arnold-Chiari syndrome; Ataxia telangiectasia; Canavan disease; Cerebrotendinous xanthomatosis; Charcot-Marie-Tooth syndrome; Chediak-Higashi syndrome; Dandy-Walker syndrome; Deafness; Epilepsy; Essential tremor; Friedreich ataxia; Huntington's disease; Jansky-Bielschowsky disease; Joubert syndrome; Kennedy disease; Krabbé disease; Leigh syndrome; Leukodystrophy; Limb girdle muscular dystrophy; Maple syrup urine disease; Metachromatic leukodystrophy; Myoclonic epilepsy associated with ragged red fibers (MERRF); Narcolepsy; Nemaline myopathy; Neural tube defects; Neurofibromatosis; Parkinson disease; Pelizaeus-Merzbacher disease; Pendred syndrome; Periodic paralysis syndrome; Prion diseases: Kuru and Creutzfeldt-Jakob syndrome; Spinal muscular atrophy; Vanishing white matter disease.

Reiter's syndrome

CATEGORY: Diseases and syndromes
ALSO KNOWN AS: Reactive arthritis

DEFINITION

Reiter's syndrome is an inflammatory reaction to an infection somewhere in the body. It usually follows a urogenital or intestinal infection. Symptoms of the disorder primarily involve three body systems: the joints, the eyes, and the urinary or genital tract.

RISK FACTORS

Risk factors include having family members with Reiter's syndrome and inheriting the genetic trait associated with Reiter's syndrome (*HLA-B27*). The disease is more common in males, homosexual or bisexual men, and individuals between the ages of twenty and forty. Other risk factors include having a sexually transmitted disease, recently getting a new sexual partner, and eating improperly handled food.

ETIOLOGY AND GENETICS

The causes of Reiter's syndrome are not well understood, but it seems clear that both genetic and environmental factors play contributing roles. Approximately 80 percent of affected individuals carry the *HLA-B27* gene, but not all individuals who express this gene will develop the disease. For example, while 50 percent of the children of an affected parent will inherit the *HLA-B27* gene, only about 25 percent of these will subsequently develop Reiter's syndrome. This gene is one of a family of genes located at the major histocompatability locus on the short arm of chromosome 6 (at position 6p21.3). It encodes a protein that is present on the surface of almost all cells and functions to display protein fragments (peptides) that have been exported from the cell to components of the immune system. If the antigens are recognized as foreign, an inflammatory response is triggered. The conditions under which the HLA-B27 protein initiates an inflammatory response resulting in reactive arthritis are not clear, and theories range from the improper presentation of peptides to the misfolding of the protein itself. This gene variant is also associated with a group of autoimmune disorders known as the seronegative spondarthritides, the most common of which are ankylosing spondylitis and psoriatic arthropathy.

The prevalence of the *HLA-B27* gene variant varies quite widely among different human populations. It is found in about 8 percent of Caucasians and only 3 to 5 percent of African Americans, but it is present in as many as 25 percent of Eskimo populations and 50 percent of Haida Indians. A simple blood test is available to determine if an individual carries this gene.

Symptoms

Symptoms occur in three main areas of the body: the joints, the eyes, and the urinary tract and genitals. Men and women may experience different symptoms. The disease may be milder in women. Symptoms may come and go. In rare cases, heart problems may develop later in the disease.

Specific symptoms in the joints include swelling, pain, and redness, especially in the knees, ankles, and feet; heel pain; shortening and thickening of fingers and toes; and back pain and stiffness. Symptoms in the eyes include redness, pain, irritation, blurred vision, tearing, discharge, and sometimes sun sensitivity or swollen eyelids. Symptoms in the male urinary tract and reproductive system include frequent urination, a burning sensation when passing urine, penile discharge, sores at the end of the penis, fever, and chills. Symptoms in the female urinary tract and reproductive system include a burning sensation when passing urine and an inflamed vagina and cervix. Other symptoms include rash, especially on the palms or soles; ulcers in the mouth or on the tongue; weight loss; poor appetite; fatigue; and fever.

Rare complications may include heart problems, such as heart conduction defects, including arrhythmia, a heart murmur (aortic insufficiency), and pericarditis (an inflammation of the outer lining of heart). Other complications may include lung problems, such as pneumonia, pulmonary fibrosis, and fluid on the lung (pleural effusion). There may also be nervous system problems, such as neuropathy, which may include a tingling or loss of sensation, and behavior changes.

Screening and Diagnosis

The doctor will ask about a patient's symptoms and medical history and will perform a physical exam. The doctor uses these findings to help make the diagnosis. There is no specific test to check for Reiter's syndrome.

Testing may include blood tests to check for signs of inflammation (sedimentation rate), signs of infection (complete blood count), and the genetic factor associated with Reiter's syndrome (*HLA-B27*). Additional tests may include culture, gram stain, or other tests to look for bacteria that commonly cause infections associated with Reiter's syndrome; the removal of synovial fluid from around the joints to check for infection; X rays, a test that uses radiation to take a picture of structures inside the body, such as the joints; ultrasound, a test that uses sound waves to examine the inside of the body; a magnetic resonance imaging (MRI) scan, a test that uses magnetic and radio waves to make pictures of the inside of the body; and a computed tomography (CT) scan, a type of X ray that uses a computer to make pictures of the inside of the body.

Treatment and Therapy

There is no cure for Reiter's syndrome. However, early treatment of the infection may slow or stop the course of the disease. Most patients recover from the initial episode within six months. However, some develop a mild, chronic arthritis. Some patients suffer from additional bouts of the disorder.

Treatment aims to relieve symptoms and may include short-term bed rest to take the strain off the joints, and exercise, including gentle range-of-motion exercises to improve flexibility, strengthening to build muscles that can better support the joints, and physical therapy with specific exercises to keep muscles strong and joints moving. Treatment also aims to protect the joints, which may include the use of assistive devices as recommended by the doctor and occupational therapy to learn how to take it easy on joints during daily activities.

The doctor may prescribe medications, including nonsteroidal anti-inflammatory drugs (NSAIDs), such as aspirin and ibuprofen (Motrin, Advil). Additional medications may include sulfasalazine (Azulfidine), steroid injections into the inflamed joint, topical steroid creams applied to skin lesions, and, in some cases, antibiotics to treat the triggering infection.

Immunosuppressive drugs (drugs that decrease the immune system's ability to function), such as azathioprine (Imuran) and methotrexate, and eyedrops may also be prescribed.

Prevention and Outcomes

The key to preventing Reiter's syndrome is to avoid the triggering infection. To do so, a patient should avoid sexually transmitted diseases (STDs), either by abstaining from sex or practicing safe sex. A patient can practice safe sex by always using a latex condom during sexual activity, asking sex partners about any history of sexual disease, having sex with only one partner who only has sex with the patient, not going back and forth between sexual partners, and having regular checkups for STDs.

Patients can also take steps to prevent chlamydia urogenital infections. If they are age twenty-five or younger, they should be tested for chlamydia annually. Pregnant women should also get tested for chlamydia.

Patients should also avoid intestinal infections by washing hands before eating or handling food and by only eating foods that have been stored and prepared properly.

Debra Wood, R.N.; reviewed by Jill D. Landis, M.D.
"Etiology and Genetics" by Jeffrey A. Knight, Ph.D.

Further Reading

EBSCO Publishing. *Health Library: Reiter's Syndrome.* Ipswich, Mass.: Author, 2009. Available through http://www.ebscohost.com.

Fauci, Anthony S., et al., eds. *Harrison's Principles of Internal Medicine.* 17th ed. New York: McGraw-Hill Medical, 2008.

Firestein, Gary S., ed. *Kelley's Textbook of Rheumatology.* 8th ed. Philadelphia: Saunders/Elsevier, 2009.

Goldman, Lee, and Dennis Ausiello, eds. *Cecil Medicine.* 23d ed. Philadelphia: Saunders Elsevier, 2008.

Goroll, Allan H., and Albert G. Mulley, Jr., eds. *Primary Care Medicine: Office Evaluation and Management of the Adult Patient.* 6th ed. Philadelphia: Wolters Kluwer Health/Lippincott Williams & Wilkins, 2009.

Noble, John, et al. *Textbook of Primary Care Medicine.* 3d ed. St. Louis: Mosby, 2001.

Web Sites of Interest

American College of Obstetrics and Gynecology: "Why Screen for Chlamydia?"
http://www.acog.org/departments/dept_notice.cfm?recno=7&bulletin=4805

Arthritis Foundation
http://www.arthritis.org

The Arthritis Society
http://www.arthritis.ca

National Institute of Arthritis and Musculoskeletal and Skin Diseases
http://www.niams.nih.gov

Public Health Agency of Canada
http://www.phac-aspc.gc.ca

Spondylitis Association of America
http://www.spondylitis.org

See also: Agammaglobulinemia; Allergies; Antibodies; Anthrax; Ataxia telangiectasia; Autoimmune disorders; Autoimmune polyglandular syndrome; Chediak-Higashi syndrome; Chronic granulomatous disease; Hybridomas and monoclonal antibodies; Immunodeficiency with hyper-IgM; Immunogenetics; Infantile agranulocytosis; Myeloperoxidase deficiency; Purine nucleoside phosphorylase deficiency.

Repetitive DNA

CATEGORY: Molecular genetics

SIGNIFICANCE: Eukaryotic nuclei contain repetitive DNA elements of different origin (polymorphisms) which constitute between 20 and 90 percent of the genome depending on the species. These repetitive DNA sequences are typically moved throughout the DNA by specific DNA interactions (transposons moved by transposition) or via RNA intermediates (retrotransposons moved by retrotransposition). The human genome is composed of about 42 percent retrotransposons and 2 to 3 percent transposons. The presence and type of repetitive DNA elements have provided insights into evolution, gene flow, gene mapping, forensic investigation, and biomedicine including disease diagnosis and detection.

Key terms

alleles: alternative forms of a gene located at a specific location or locus on a chromosome *nucleotides:* basic units of DNA, each consisting of a five-carbon sugar, a nitrogen-containing base, and a phosphate group

polymorphisms: different alleles for a particular gene locus among individuals of the same species

retrotransposition: movement of repetitive DNA sequences through RNA intermediates; the process of moving a subset of transposable DNA elements through RNA intermediates

tandem repetitive DNA (TR-DNA): nucleotide sequences including repeating sections oriented in "head-to-tail" arrays, such as (GGAAT)n

transposon: DNA segment able to replicate itself and insert the new DNA sequence into a new location either within the same chromosome or into another chromosome or plasmid

variable number tandem repeats (VNTRs): DNA sequences including short sequences repeated over and over; however, chromosomes from different individuals frequently have different numbers of the basic repeat, and if many of these variants are known, the sequence is termed a hypervariable

Types of Repetitive DNA

The eukaryotic nuclear genome is characterized by repetitive DNA elements including nucleotide sequences of varying length and base composition either localized to a particular region of the genome or dispersed throughout the genome (for example, on different chromosomes). Some repetitive DNA elements are found in the genome a few times, whereas others may be repeated millions or billions of times; thus, the percentage of the total genome represented by repetitive DNA varies widely among taxa. Tandem repeats are repetitive DNA sequences lying adjacent to each other in a block or array, whereas interspersed repeats are repetitive DNA sequences found throughout the genome surrounded by unique (nonrepetitive) DNA sequences.

The two major classes of tandem repetitive DNAs (TR-DNAs) are those that are localized to a particular region (or regions) of the genome and those that are dispersed throughout the genome. TR-DNAs include repeating units that are oriented in "head-to-tail" arrays. The repetitive units of an array may include genes, promoters, and intergenic spacers or repeats of simple nucleotide sequences. For example, in the kangaroo rat the simple sequence AAG is repeated 2.4 billion times.

Localized TR-DNA is often composed of members of multigene families. For example, in humans there are 350 copies of the ribosomal RNA (rRNA) genes on five different chromosomes that occur as tandemly repeated arrays. Transfer RNA (tRNA) and immunoglobin genes represent other examples of multigene families that are tandemly repeated. However, most localized TR-DNA consists of simple, noncoding repetitive DNA sequences that often, but not always, can be found in heterochromatic or centromeric regions.

Dispersed TR-DNA sequences are scattered throughout the genome and can be divided into two major groups: short interspersed elements (SINEs) and long interspersed repeats (LINEs). The most common examples of a SINE family are the *Alu* sequences with a copy occurring once in every 3 kilobase pairs (kb) or so or the human genome, making the *Alu* repeat the most abundant sequence in the human genome (copy number is about one million). TR-DNA sequences are sometimes referred to as "satellite DNAs" that are grouped and classified by repeat length size as satellites (up to 100 million bp), minisatellites (9-100 bp), and microsatellites (2-6bp).

Origin and Evolution of Dispersed Repetitive DNA Elements

Dispersed repetitive DNA is believed to be an evolutionary device that catalyzes the formation of new genes. Within a species, DNA sequences are thought to maintain similarity by gene conversion (a type of DNA repair mechanism during meiosis believed to maintain the DNA coding sequence of the organism being replicated) while repetitive sequences disrupt this process and allow new genes to evolve. SINEs are believed to disrupt gene conversions between chromosomes, while the longer LINE elements disrupt the gene conversions within the chromosome. SINEs are transposable elements capable of "jumping" from one locus to another via an RNA intermediate.

SINEs are called nonviral retropseudogenes because they are believed to have been derived from genes encoding small, untranslated RNAs (for example, tRNAs). SINES were created when the RNA transcript was reverse transcribed into DNA and then was inserted into the genome. In their current state, and although they resemble the genes from which they derived, they no longer function properly.

The best-characterized SINEs in humans are highly repetitive *Alu* sequences, so named because they are cleaved multiple times by the endonuclease AluI, derived from the bacterium *Arthrobacter luteus.* Between 500,000 and 1 million *Alu* copies are scattered across the human genome; each *Alu* sequence is approximately three hundred nucleotides in length. *Alu* sequences may constitute as much as 5 percent of the human genome. These *Alu* repeat sequences may have developed by retrotransposition of the 7SL RNA gene sequence. The *Alu* repeat sequence includes an RNA polymerase III promoter sequence that is sometimes transcribed.

LINEs are derived from a viral ancestor and are also capable of transposition. The most common LINE element in humans, constituting 5 percent of the human genome, is termed L1. There are about 200,000 copies of L1 in each diploid cell. Full-length, functional (that is, transpositionally competent) L1 elements are approximately 6 kb in length, but most copies of L1 are truncated at the 5′ end and incapable of moving. Full-length L1 copies contain two protein-coding regions, or open reading frames (ORFs): ORF-1 and ORF-2. ORF-1 encodes an RNA-binding protein, and ORF-2 codes for reverse transcriptase. The transposition of LINE elements into various parts of the genome is believed to contribute to evolution through the creation of new genes with altered ORFs.

Are Interspersed Repeated Elements "Junk" DNA?

Repeated DNA elements were once believed to be "selfish" or "junk" DNA, concerned only with their own proliferation within the host cell's genome. Recent studies, however, reveal that repetitive elements interact with the genome with profound evolutionary consequences. For example, satellite DNA found near the centromere may play a role in assembling and fusing chromosomal microtubules during cell division. In addition, transposable genetic elements such as SINEs, LINEs, and *Alu* sequences may have played a significant role in the evolution of particular proteins. For example, *Alu* elements flanking the primordial human growth hormone gene are believed to be responsible for the evolution of a relatively new member of the gene family, the chorionic somatomammotropin gene. Transposable repeated elements may have contributed substantially to the origin of new gene functions by initiating a copy of an existing gene (which, over time, may acquire a different function) or by creating a "composite" gene composed of domains from two or more previously unrelated genes.

Classification of Simple Tandem Repeats

Tandemly repeated simple DNA sequences are classified into four major groups based on three characteristics: the number of nucleotides in the repetitive unit, the number of times the unit is repeated, and whether the element is localized or scattered across the genome. The four groups include satellites, minisatellites, microsatellites, and dispersed *Alu* sequences. Satellite DNA is composed of tandemly repeated basic DNA sequences, ranging from two to hundreds of nucleotides in length and repeated more than one thousand times, locally, in the DNA. Satellite DNA represents an example of a localized simple repeat typically found in the centromeric region of a chromosome. Tandemly repeated basic DNA sequences, ranging from nine to one hundred nucleotides in length, repeated ten to one hundred times, and scattered throughout the genome are known as minisatellites. Microsatellites are also dispersed repetitive sequence elements; however, microsatellites are composed of short DNA sequence repeats of a basic unit one to six nucleotides in length that are tandemly repeated ten to one hundred times at each locus. The most common microsatellite loci in humans are dinucleotide arrays of $(CA)_N$. However, on average, at least one tri- or tetranucleotide microsatellite locus is found in each 10 kb of human genomic DNA. In a separate group, the basic unit of dispersed *Alu* sequences is one to five nucleotides in length, and this unit is repeated ten to forty times per locus.

Polymorphism at Loci Composed of Simple Tandem Repeats

For purposes of convenience, the four groups of simple tandem repeats discussed above (satellite DNA, minisatellites, microsatellites, and *Alu* sequences) are sometimes collectively referred to as variable number tandem repeats (VNTRs).

Separate VNTR loci are thought of as alleles; therefore, in humans each VNTR locus will be represented by two alleles, one paternal and the other maternally inherited. All VNTR loci exhibit high rates of mutation. For these reasons, VNTR loci are highly polymorphic, that is, a large number of alleles exist at any given locus. This polymorphism can be assayed using laboratory techniques such as polymerase chain reaction (PCR) or Southern blotting to examine the differences in the lengths of the alleles (repetitive elements) at a particular DNA locus. Length differences at VNTR loci arise as a result of mispairing of repeats during replication, mitosis, or meiosis, theoretically resulting in the loss or gain of one to many of the repeat units. Empirical studies and computer-based modeling experiments have demonstrated each mutation usually increases or decreases the number of repeated units of an allele in a "one-step" manner. In other words, most mutations result in the loss or gain of only one repeated unit.

The multiallelic variation arising through this variation in repeat copy number provides useful genetic markers for many different applications. For example, under conditions of random mating and because of high mutation rates at VNTR loci, most individuals within the human population are heterozygous at any selected VNTR locus. This observation directly led to the origin of DNA fingerprinting (or DNA profiling), which is now considered admissible forensic evidence in many judicial systems worldwide. Length variation of VNTRs creates a powerful tool for identity analysis (for example, paternity testing) and is routinely used by population geneticists to examine gene flow among populations. In the fields of genomics and biomedicine, VNTR loci are useful genetic landmarks for mapping the location of other genes of interest, including genes with a particular function or genes implicated in disease.

Transposable Elements and Human Disease

Retrotranspositions of LINEs and SINEs into coding or noncoding genomic DNAs represent major insertional mutations. The effects of such insertions vary but are usually deleterious, leading to debilitating human diseases. Among a growing list of diseases known in some cases to be caused by the insertion of LINEs or SINEs are Duchenne muscular dystrophy, Glanzmann thrombasthenia, hemophilia, hypercholesterolemia, neurofibromatosis, Sandhoff disease, and Tay-Sachs disease. Translocation of repeated sequences has also been demonstrated to "turn on" tumorogenic oncogenes (for example, one type of colon cancer is associated with the insertion of repetitive DNA).

Other studies suggest "unstable" VNTRs (including minisatellite, microsatellite, and *Alu* loci) can also cause disease. These studies suggest a threshold number of repeats of the basic nucleotide unit may exist that can be accommodated at a given locus. When this threshold number of repeats is exceeded by overamplification of the basic repeated unit, serious diseases may arise. Among the diseases attributed to overamplification of tandem repeats of simple DNA sequences are fragile X syndrome and Huntington's disease.

J. Craig Bailey, Ph.D.;
updated by Joy Frestedt, Ph.D., RAC, CCTI

Further Reading

Lander, E. S., L. M. Linton, B. Birren, et al. "Initial Sequencing and Analysis of the Human Genome." *Nature* 409, no. 6822 (2001): 860-921. The initial release of the human genome sequence.

Li, Wen-Hsiung. *Molecular Evolution.* Sunderland, Mass.: Sinauer Associates, 1997. Provides a basic introduction to the different types of variable tandem repeats, their uses in the biological sciences, and how they affect genome organization.

Maichele, A. J., N. J. Farwell, and J. S. Chamberlain. "A B2 Repeat Insertion Generates Alternate Structures of the Mouse Muscle Gamma-phosphorylase Kinase Gene." *Genomics* 16, no. 1 (1993): 139-149. An excellent example of how retrotransposition of repetitive DNA elements may alter the function of, or give rise to, new structural proteins.

Maraia, Richard J., ed. *The Impact of Short Interspersed Elements (SINEs) on the Host Genome.* Austin, Tex.: R. G. Landes, 1995. A comprehensive treatise on the origin, evolution, and functional roles that SINEs play in the biology of organisms and in biomedicine.

Shapiro, James A. "A 21st Century View of Evolution: Genome System Architecture, Repetitive DNA, and Natural Genetic Engineering." *Gene* 345 (2005): 91-100. A broad overview of genome organization focused on information storage and evolution. The author relates moveable and repetitive DNA sequences to system engineering processes.

Shapiro, James A., and R. von Sternberg. "Why Repetitive DNA Is Essential to Genome Function." *Biological Reviews* 80 (2005): 1–24. Describes the diversity, abundance, nuclear organization, and functional importance of repetitive DNA.

Strachan, Tom, and Andrew P. Read, eds. "7.1 General Organization of the Human Genome." In *Human Molecular Genetics 2.* New York: John Wiley and Sons, 1999. A comprehensive textbook defining the structure and function of the human genome.

WEB SITES OF INTEREST

DNA Repeat Sequences and Disease
http://neuromuscular.wustl.edu/mother/dnarep.htm

Function of Repetitive DNA
http://www.repetitive-dna.org

Junk DNA—Repetitive Sequences
http://biol.lf1.cuni.cz/ucebnice/en/repetitive_dna.htm

See also: Aging; Anthrax; Chromosome structure; Chromosome walking and jumping; DNA fingerprinting; Gene families; Genome size; Genomics; Human genetics; Model Organism: *Neurospora crassa*; Molecular clock hypothesis; Pseudogenes; RFLP analysis; Telomeres.

Restriction enzymes

CATEGORY: Genetic engineering and biotechnology; Molecular genetics

SIGNIFICANCE: Restriction enzymes are bacterial enzymes capable of cutting DNA molecules at specific nucleotide sequences. Discovery of these enzymes was a pivotal event in the development of genetic engineering technology, and they are routinely and widely used in molecular biology.

KEY TERMS

enzyme: a molecule, usually a protein, that is used by cells to facilitate and speed up a chemical reaction

methylation: the process of adding a methyl chemical group (one carbon atom and three hydrogen atoms) to a particular molecule, such as a DNA nucleotide

nuclease: a type of enzyme that breaks down the sugar-phosphate backbone of nucleic acids such as DNA and RNA

nucleotides: the building blocks of nucleic acids, composed of a sugar, a phosphate group, and nitrogen-containing bases

DISCOVERY AND ROLE OF RESTRICTION ENZYMES IN BACTERIA

Nucleases are a broad class of enzymes that destroy nucleic acids by breaking the sugar-phosphate backbone of the molecule. Until 1970, the only known nucleases were those that destroyed nucleic acids nonspecifically—that is, in a random fashion. For this reason, these enzymes were of limited usefulness for working with nucleic acids such as DNA and RNA. In 1970, molecular biologist Hamilton O. Smith discovered a type of nuclease that could fragment DNA molecules in a specific and therefore predictable pattern. This nuclease, *Hin*dII, was the first restriction endonuclease or restriction enzyme. Smith was working with the bacterium *Haemophilus influenzae* (*H. influenzae*) when he discovered this enzyme, which was capable of destroying DNA from other bacterial species but not the DNA of *H. influenzae* itself. The term "restriction" refers to the apparent role these enzymes play in destroying the DNA of invading bacteriophages (bacterial viruses), while leaving the bacterial cell's own DNA untouched. A bacterium with such an enzyme was said to "restrict" the host range of the bacteriophage.

As more restriction enzymes from a wide variety of bacterial species were discovered in the 1970's, it became increasingly clear that these enzymes could be useful for creating and manipulating DNA fragments in unique ways. What was not clear, however, was how these enzymes were able to distinguish between bacteriophage DNA and the bacterial cell's own DNA. A chemical comparison between DNA that could and could not be fragmented revealed that the DNA molecules differed slightly at the restriction sites (the locations the enzyme recognized and cut). Nucleotides at the restriction site were found to have methyl (CH_3) groups attached to them, giving this phenomenon the name DNA methylation.

The conclusion was that the methylation somehow protected the DNA from attack, and this could account for Smith's observation that *H. influenzae* DNA was not destroyed by its own restriction enzyme; presumably the enzyme recognized a specific methylation pattern on the DNA molecule and left it alone. Foreign DNA (from another species, for example) would not have the correct methylation pattern, or it might not be methylated at all, and could therefore be fragmented by the restriction enzyme. Hence, restriction enzymes are now regarded as part of a simple yet effective bacterial defense mechanism to guard against foreign DNA, which can enter bacterial cells with relative ease.

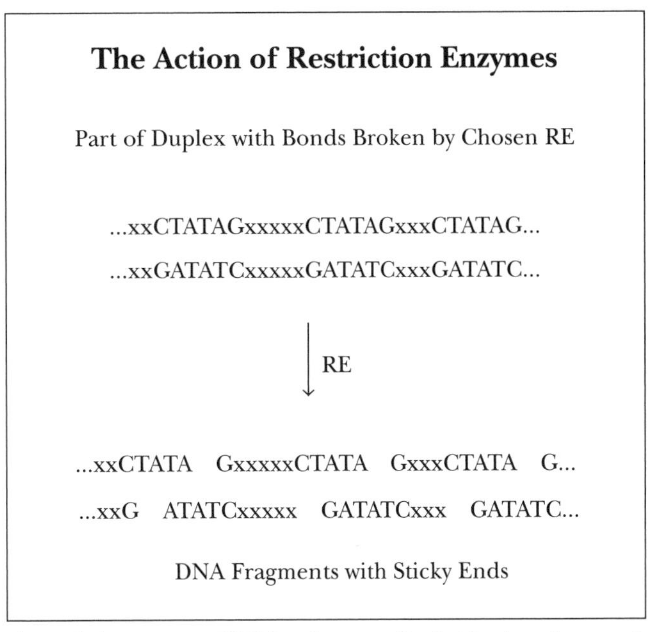

A restriction enzyme (RE) breaks part of a duplex into fragments with "sticky ends." Each x denotes an unspecified base in a nucleotide unit.

Mechanism of Action

To begin the process of cleaving a DNA molecule, a restriction enzyme must first recognize the appropriate place on the molecule. The recognition site for most restriction enzymes involves a short, usually four- to six-nucleotide, palindromic sequence. A palindrome is a word or phrase that reads the same backward and forward, such as "Otto" or "madam"; in terms of DNA, a palindromic sequence is one that reads the same on each strand of DNA but in opposite directions. *Eco*RI (derived from the bacterium *Escherichia coli*) is an example of an enzyme that has a recognition site composed of nucleotides arranged in a palindromic sequence:

——GAATTC——
——CTTAAG——

If the top sequence is read from left to right or the bottom sequence is read from right to left, it is always GAATTC.

An additional consideration in the mechanism of restriction enzyme activity is the type of cut that is made. When a restriction enzyme cuts DNA, it is actually breaking the "backbone" of the molecule, consisting of a chain of sugar and phosphate molecules. This breakage occurs at a precise spot on each strand of the double-stranded DNA molecule. The newly created ends of the DNA fragments are then informally referred to as "sticky ends" or "blunt ends." These terms refer to whether single-stranded regions of DNA are generated by the cutting activity of the restriction enzyme. For example, the enzyme *Eco*RI is a "sticky end" cutter; when the cuts are made at the recognition site, the result is:

—GAATTC— → —G AATTC—
—CTTAAG— → —CTTAA G—

The break in the DNA backbone is made just after the G in each strand; this helps weaken the connections between the nucleotides in the middle of the site, and the DNA molecule splits into two fragments. The single-stranded regions, where the bases TTAA are not paired with their complements (AATT) on the other strand, are called overhangs; however, the bases in one overhang are still capable of pairing with the bases in the other overhang as they did before the DNA strands were cut. The ends of these fragments will readily stick to each other if brought close together (hence the name "sticky ends").

Enzymes that create blunt ends make a flush cut and do not leave any overhangs, as demonstrated by the cutting site of the enzyme *Alu*I:

——AGCT—— → ——AG CT——
——TCGA—— → ——TC GA——

Because of the lack of overhanging single-strand regions, these two DNA fragments will not readily rejoin. In practice, either type of restriction enzyme may be used, but enzymes that produce sticky ends are generally favored over blunt-end-cutting enzymes because of the ease with which the resulting fragments can be rejoined.

IMPACT AND APPLICATIONS

It is no exaggeration to say that the entire field of genetic engineering would have been impossible without the discovery and widespread use of restriction enzymes. On the most basic level, restriction enzymes allow scientists to create recombinant DNA molecules (hybrid molecules containing DNA from different sources, such as humans and bacteria). No matter what the source, DNA molecules can be cut with restriction enzymes to produce fragments that can then be rejoined in new combinations with DNA fragments from other molecules. This technology has led to advances such as the production of human insulin by bacterial cells such as *Escherichia coli.*

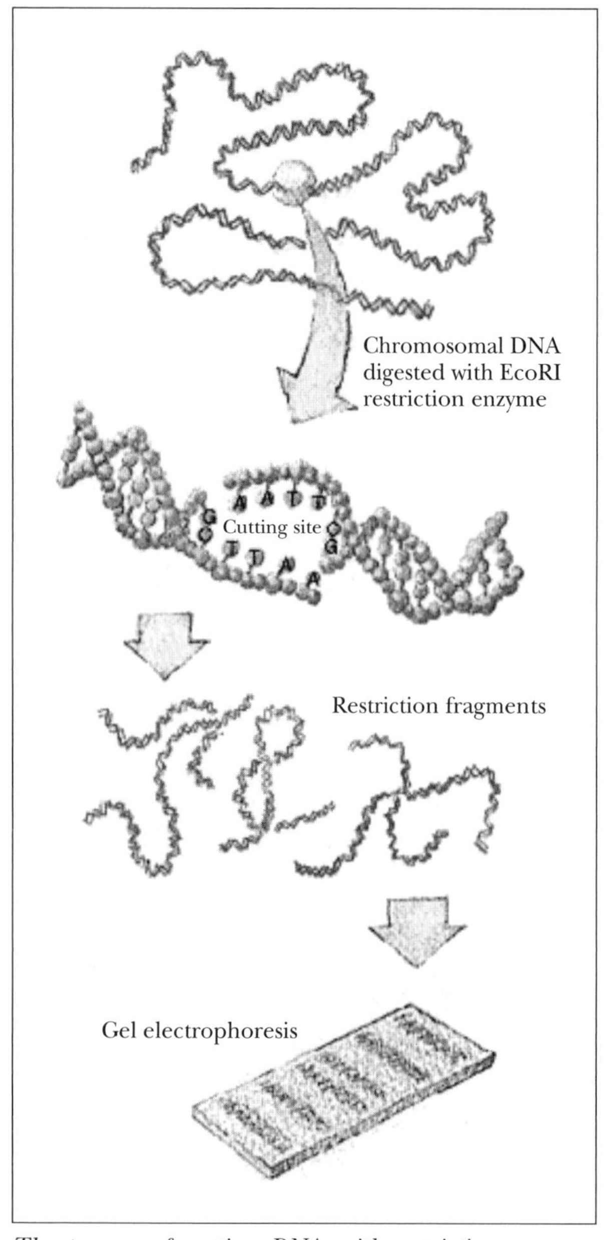

The process of cutting DNA with restriction enzymes. (U.S. Department of Energy Human Genome Program, http://www.ornl.gov/hgmis)

The DNA of most organisms is relatively large and complex; it is usually so large, in fact, that it becomes difficult to manipulate and study the DNA of some organisms, such as humans. Restriction enzymes provide a convenient way to cut large DNA molecules very specifically into smaller fragments that can then be used more easily in a variety of molecular genetics procedures.

Another area of genetic engineering that is possible because of restriction enzymes is the production of restriction maps. A restriction map is a diagram of a DNA molecule showing where particular restriction enzymes cut the molecule and the molecular sizes of fragments that are generated. The restriction sites can then be used as markers for further study of the DNA molecule and to help geneticists locate important genetic regions. Use of restriction enzymes has also revealed other interesting and useful markers of the human genome, called restriction fragment length polymorphisms (RFLP). RFLP refers to changes in the size of restriction fragments caused by mutations in the recognition site for a particular restriction enzyme. More specifically, the recognition site is mutated so that the restriction enzyme no longer cuts there; the result is one long fragment where, before the mutation, there would have been two shorter fragments. These changes in fragment length can then be used as markers for the region of DNA in question. Because they result from mutations in the DNA sequence, they are inherited from one generation to the next. Thus these mutations have been a valuable tool for molecular biologists in producing a map of human DNA and for those scientists involved in "fingerprinting" individuals by means of their DNA.

Randall K. Harris, Ph.D.; updated by Bryan Ness, Ph.D.

FURTHER READING

Allison, Lizabeth A. "Recombinant DNA Technology and Molecular Cloning." In *Fundamental Molecular Biology.* Malden, Mass.: Blackwell, 2007. Includes information about cutting and joining DNA and RFLP.

Drlica, Karl. *Understanding DNA and Gene Cloning: A Guide for the Curious.* 4th ed. New York: John Wiley and Sons, 2004. Provides basic information about restriction enzymes and their use in cloning. Illustrations, bibliography, index.

Karp, Gerald. "Techniques in Cell Molecular Biology." In *Cell and Molecular Biology: Concepts and Experiments.* 5th ed. Chichester, England: John Wiley and Sons, 2008. Includes information about restricted enzymes.

Lewin, Benjamin. *Genes IX.* Sudbury, Mass.: Jones and Bartlett, 2007. Provides a detailed yet highly readable explanation of restriction and methylation in bacteria. Illustrations, bibliography, index.

Watson, James D., et al. *Recombinant DNA—Genes and Genomes: A Short Course.* 3d ed. New York: W. H. Freeman, 2007. An excellent resource for the general reader wishing to understand the basics of genetic engineering. Illustrations, diagrams, bibliography, index.

Web Sites of Interest

Access Excellence, Biotech Chronicles
http://www.accessexcellence.org/AE/AEC/CC/restriction.php

Access Excellence, a national science education program, includes a background paper on restriction enzymes, as well as a chart listing examples of restriction enzymes, a classroom activity, and other information on restriction enzymes.

Dolan DNA Learning Center, DNA Restriction
http://www.dnalc.org/resources/animations/restriction.html

The site's biology animation library offers a brief definition of DNA restriction and enables users to download an animation that demonstrates the process.

Kimball's Biology Pages
http://users.rcn.com/jkimball.ma.ultranet/BiologyPages/R/RestrictionEnzymes.html

John Kimball, a retired Harvard University biology professor, includes a page about restriction enzymes, with links to information about recombinant DNA, in his online cell biology text.

See also: Bacterial genetics and cell structure; Bioinformatics; Biopharmaceuticals; Blotting: Southern, Northern, and Western; Cloning; Cloning vectors; DNA fingerprinting; Forensic genetics; Gender identity; Genetic engineering; Genetic engineering: Historical development; Genetic engineering: Social and ethical issues; Genomic libraries; Model organism: *Xenopus laevis*; Molecular genetics; Population genetics; RFLP analysis; Shotgun cloning; Synthetic genes.

Retinitis pigmentosa

Category: Diseases and syndromes
Also known as: RP

Definition

Retinitis pigmentosa (RP) is a group of inherited eye diseases that often leads to severe visual problems. The retina is a layer of light-sensitive tissue that lines the back of the eye. It converts visual images that people see into nerve impulses that it sends to the brain. Some types of RP are associated with other inherited conditions. This disorder is named for the irregular clumps of black pigment that usually occur in the retina with this disease.

Risk Factors

Having family members with RP and being male are risk factors for the disease.

Etiology and Genetics

Retinitis pigmentosa is an inherited disorder, yet the identification of the genes involved has proven to be an extraordinary challenge. Since retinal cells are so highly specialized, it is not surprising that they depend on a large number of specific genes and their protein products to create vision. Mutations in more than forty different genes have been identified that can cause retinitis pigmentosa, and one source estimates that the total number of genes in which mutations might occur may be as high as one hundred.

In the majority of cases, the disease is inherited in an autosomal recessive fashion, which means that both copies of a particular gene must be deficient in order for the individual to be afflicted. Typically, an affected child is born to two unaffected parents, both of whom are carriers of the recessive mutant allele. The probable outcomes for children whose parents are both carriers are 75 percent unaffected and 25 percent affected. Many other retinitis pigmentosa mutations are inherited in an autosomal dominant manner, meaning that a single copy of the mutation is sufficient to cause full expression of the disease. An affected individual has a 50 percent chance of transmitting the mutation to each of his or her children. Many cases of dominant retinitis pigmentosa, however, result from a spontaneous new mutation, so in these instances affected individuals will have unaffected parents.

Mutations in two genes on the X chromosome (*RP2* and *RP3*, at locations Xp11.3 and Xp21.1, respectively) are known to cause retinitis pigmentosa, and these show a sex-linked recessive pattern of inheritance. Mothers who carry the mutated gene on one of their two X chromosomes face a 50 percent chance of transmitting this disorder to each of their male children. Female children have a 50 percent chance of inheriting the gene and becoming carriers like their mothers. Affected males will pass the mutation on to all of their daughters but to none of their sons.

Finally, one rare form of retinitis pigmentosa results from mutations in the mitochondrial gene *MT-ATP6*. Each retinal cell contains anywhere from several to more than one hundred copies of mitochondrial DNA (deoxyribonucleic acid) eligible for testing, and each mitochondrial DNA molecule contains thirteen structural genes that encode protein components of respiratory chain complexes. Inheritance of mitochondrial DNA follows a pattern of strict maternal inheritance, since all of the mitochondria in a fertilized egg (zygote) come from the egg cell. Thus affected females will transmit the disease to all of their offspring, but affected males produce unaffected children.

Symptoms

Loss of vision is usually first noted in childhood or early adulthood. The disease gradually worsens. After a number of years, vision loss may become severe. Symptoms vary, depending on the type of retinal cell that is affected. Both eyes often experience similar vision loss.

It should be noted that RP is a slowly progressive disease over many years and that most patients never become completely blind. In fact, even though many people with RP are considered "legally blind," it is only because they have very constricted fields of vision (poor peripheral vision). Some still maintain excellent central visual acuity.

Overall, symptoms may include night blindness (the most common symptom); eyes taking longer to adjust to dim lighting; trouble seeing in foggy or rainy weather; eyes being slow to make the adjustment from bright sun to indoor lighting; and decreased peripheral vision and a narrowing field of vision, often called "tunnel vision." Additional symptoms include difficulty seeing colors, especially blue; visual loss, partial or complete, usually gradually progressive; and clumsiness from lack of sight, especially in narrow spaces, such as doorways. Blurry vision from cataracts may complicate RP later in the disease.

Screening and Diagnosis

The doctor will ask about a patient's symptoms and medical history and will perform an eye exam. The patient may be referred to an eye specialist, such as an ophthalmologist.

Vision tests may include visual field testing to check peripheral vision, which is how well a patient sees off to his or her side, rather than directly ahead, without moving his or her eyes.

Visual acuity tests check how well a patient can see progressively smaller objects, usually a row of letters or numbers. Additional tests may include dark adaptometry, which tests how a patient's vision adapts to darkness; color testing, which determines how well a patient can differentiate colors; and an electroretinogram (ERG), a test to measure electrical activity in the eye. An ERG identifies the loss of cell function in the retina and is used to track the progression of the disease.

Treatment and Therapy

There is no effective treatment or cure for retinitis pigmentosa. Treatment aims to help patients function with the vision that they have. The doctor can counsel patients about expected patterns of vision loss based on the type of RP they have.

Recommendations include the use of vitamin A. One study implied that large doses of vitamin A can slow the progression of RP by approximately 2 percent per year. However, the use of this finding is controversial. For one thing, it is a very mild effect given the large dose. Second, there may be side effects of such large doses of vitamin A. Patients should always talk to their doctors before taking any supplements.

Other recommendations include light avoidance. Although no direct link has been established, it is generally recommended that everyone, especially patients with disorders such as RP, wear dark ultraviolet (UV)-protected sunglasses and a wide-brimmed hat in bright, sunny conditions, such as while skiing or at the beach.

Aids for low vision may include magnifying glasses; electronic magnifiers, which project an enlarged image onto a screen; night vision scopes, which enlarge distant objects under conditions of low light; and lenses for distant vision (eyeglasses or contacts).

Some community organizations offer classes to help people with vision loss adjust and learn how to use vision aids. If a patient is considered legally blind, he or she is entitled to many low-vision services at no cost.

Prevention and Outcomes

Once RP has been inherited there are no known ways to prevent the disorder from occurring. Individuals who have RP or have a family history of the disorder can talk to a genetic counselor when deciding whether to have children.

Debra Wood, R.N.; reviewed by Christopher Cheyer, M.D.
"Etiology and Genetics" by Jeffrey A. Knight, Ph.D.

Further Reading

EBSCO Publishing. *Health Library: Retinitis Pigmentosa.* Ipswich, Mass.: Author, 2009. Available through http://www.ebscohost.com.

Fauci, Anthony S., et al., eds. *Harrison's Principles of Internal Medicine.* 17th ed. New York: McGraw-Hill Medical, 2008.

Foundation of the American Academy of Ophthalmology. *Retina and Vitreous, Basic and Clinical Science Course, 2008-2009.* San Francisco: American Academy of Ophthalmology, 2008.

Goldman, Lee, and Dennis Ausiello, eds. *Cecil Medicine.* 23d ed. Philadelphia: Saunders Elsevier, 2008.

Yanoff, Myron, et al. *Ophthalmology.* 3d ed. Edinburgh: Mosby Elsevier, 2009.

Web Sites of Interest

Canadian Association of Optometrists
http://www.opto.ca/en/public

Canadian National Institute for the Blind
http://www.cnib.ca/en/Default.aspx

Foundation Fighting Blindness
http://www.blindness.org

Genetics Home Reference
http://ghr.nlm.nih.gov

Prevent Blindness America
http://www.preventblindness.org

RP International
http://www.rpinternational.org

See also: Aniridia; Best disease; Choroideremia; Color blindness; Corneal dystrophies; Glaucoma; Gyrate atrophy of the choroid and retina; Macular degeneration; Norrie syndrome; Progressive external ophthalmoplegia; Retinoblastoma.

Retinoblastoma

Category: Diseases and syndromes

Definition

Retinoblastoma is a rare type of cancer found in the eye. In retinoblastoma, one or more tumors form in the retina. The retina is a layer of light-sensitive tissue that lines the back of the eye. It converts visual images into nerve impulses in the brain that allow people to see. If not treated, the tumors will continue growing. The cancer may grow along the optic nerve and reach the brain, or it may travel to other parts of the body.

Cancer occurs when cells in the body (in this case, retina cells) divide without control or order. Normally, cells divide in a regulated manner. If cells keep dividing uncontrollably when new cells are not needed, a mass of tissue forms, called a growth or tumor. Tumors can invade nearby tissue and spread to other parts of the body.

Risk Factors

Individuals who are less than five years old typically are at risk for retinoblastoma, as are family members with the disease.

Etiology and Genetics

The *RB1* gene, found on the long arm of chromosome 13 at position 13q14.1-q14.2, is the single gene that is responsible for almost all cases of retinoblastoma. It specifies a protein known as pRb that normally acts as a tumor-suppressor protein in retinal cells and in several other cell types. Tumor-suppressor proteins act in a variety of ways to regulate cell growth by preventing cells from dividing too fast or at inappropriate times. Specifically, the pRb protein represses the transcription of several other genes whose protein products are needed in order to initiate a new round of DNA replication in the cell. When functional pRb is unavailable as a result of mutations in the *RB1* genes, this control of cell growth is lost and tumor formation may develop.

Retinoblastomas are often categorized as being either germinal or nongerminal. In germinal retinoblastomas (40 percent of all cases), the mutated *RB1* genes are found in all body cells. These are the cases that can be inherited and in which there is often a family history of the disease. In nongerminal cases (60 percent), the mutations in *RB1* have been acquired over time in the retinal tissue only, and affected individuals are not at risk of passing them on to their offspring. In germinal retinoblastoma, the inheritance pattern is most often consistent with an autosomal dominant mode of transmission. A single inherited copy of the mutation is sufficient in most cases to cause expression of the disease. An affected individual has a 50 percent chance of transmitting the mutation to each of his or her children. Many cases of germinal retinoblastoma, however, result from a spontaneous new mutation, so in these instances affected individuals will have unaffected parents.

Symptoms

Symptoms of retinoblastoma are usually noticed by the parent or caregiver and may include, but are not limited to, eyes that do not look normal, often described as a glazed look or a "cat's eye."

Another symptom may include the pupil looking white rather than red when a light is directed at the eye; this is often noticed on a photograph. An additional symptom may be eyes that may appear to be crossed or looking in different directions. Less common symptoms may include eyes that may grow in size, eye pain, redness in the white part of the eye, tearing, a pupil that may not respond to light, and the iris (colored part of the eye) changing color.

Screening and Diagnosis

The doctor will ask about a child's symptoms and family medical history and will perform a physical exam. Many retinoblastomas are found during routine physical exams. If a tumor is suspected, the child will usually be referred to a specialist for a more complete eye exam. In children with a family history of the disease, eye exams often begin within a day or two of birth. Additional eye exams are scheduled at regular intervals thereafter.

Once retinoblastoma is found, staging tests are performed to find out if the cancer has spread and, if so, to what extent. Treatment depends on the stage of the cancer. The cancer may be localized to the eyes or it may have spread to tissues around the eye or to other parts of the body.

Tests may include, but are not limited to, an eye exam. In this test, the pupil is dilated with eyedrops; then the inside of the eye is examined with a lighted instrument that allows the examiner to view structures inside the eye. Other tests may include ultrasound, a test that uses sound waves to examine the inner part of the eye; and a magnetic resonance imaging (MRI) scan, which uses magnetic waves to make pictures of the inside of the body. An MRI can be used to check the spread of the cancer to the brain or other tissue. A computed tomography (CT) scan is a type of X ray that uses a computer to make pictures of the eye. CT scans of other areas of the body may be done to check if the cancer has spread. General anesthesia may be given to keep the child still during close examination and testing.

Treatment and Therapy

The child will likely be referred to a specialist for treatment. Without treatment, the cancer cells will continue to grow. Treatment aims to cure the cancer and preserve sight. Options vary, depending on whether the disease is limited to the eye or has spread and how large and where in the eye the tumor is located. Therapies may be used alone or in combination.

Treatments include surgery. This involves surgical removal of the entire eye and as much of the optic nerve as possible. The optic nerve is the nerve leading from the eye to the brain that is responsible for vision. Surgery may be used for a large tumor in one eye.

Radiation therapy is a form of treatment that involves the use of radiation to kill cancer cells and shrink tumors. Radiation may be external radiation therapy, in which radiation is directed at the tumor from a source outside the body, or internal radiation therapy, in which radioactive materials are placed into the body near the cancer cells.

Cryotherapy, used on small tumors, is the use of cold to freeze and destroy cancer cells. Thermotherapy uses heat to kill cancer cells. In photocoagulation, lasers are used to destroy a small tumor.

Chemotherapy is the use of drugs to kill cancer cells. This treatment may be given in many forms, including pill, injection, and through a catheter. The drugs enter the bloodstream and travel through the body killing mostly cancer cells, but also some healthy cells.

Prevention and Outcomes

Genetic counseling and close monitoring and screening for people at risk for retinoblastoma can help prevent the disease or detect it early if it occurs. Early diagnosis and treatment improve the chance of successful treatment.

Prevention and early detection techniques include screening. Children born into families with a history of retinoblastoma should have regular eye exams to screen for development of the tumor. All children should have regular eye screening by their doctor.

Genetic counseling may help determine a person's risk of developing retinoblastoma. If a patient has retinoblastoma or has a family history of the disorder, he or she can talk to a genetic counselor when deciding whether to have children.

Patients who have been treated for retinoblastoma require regular medical exams to assess the success of treatment and to check for recurrence or bilateral disease. Children with retinoblastoma are at increased risk for an associated brain tumor and other cancers in the body and should be monitored in order to check for other cancers.

Debra Wood, R.N.; reviewed by Christopher Cheyer, M.D.
"Etiology and Genetics" by Jeffrey A. Knight, Ph.D.

Further Reading

Abeloff, Martin D., et al., eds. *Abeloff's Clinical Oncology.* 4th ed. Philadelphia: Churchill Livingstone/ Elsevier, 2008.

EBSCO Publishing. *Health Library: Retinoblastoma.* Ipswich, Mass.: Author, 2009. Available through http://www.ebscohost.com.

Kleigman, Robert M., et al., eds. *Nelson Textbook of Pediatrics.* 18th ed. Philadelphia: Saunders Elsevier, 2007.

Pizzo, Philip A., and David G. Poplack. *Principles and Practice of Pediatric Oncology.* 5th ed. Philadelphia: Lippincott Williams & Wilkins, 2006.

Yanoff, Myron, et al. *Ophthalmology.* 3d ed. Edinburgh: Mosby Elsevier, 2009.

Web Sites of Interest

American Cancer Society
http://www.cancer.org

BC (British Columbia) Cancer Agency
http://www.bccancer.bc.ca/default.htm

Cancer Care Ontario
http://www.cancercare.on.ca

Genetics Home Reference
http://ghr.nlm.nih.gov

National Cancer Institute
http://www.cancer.gov

See also: Aniridia; Best disease; Choroideremia; Color blindness; Corneal dystrophies; Glaucoma; Gyrate atrophy of the choroid and retina; Macular degeneration; Norrie syndrome; Progressive external ophthalmoplegia; Retinitis pigmentosa.

Rett syndrome

Category: Diseases and syndromes
Also known as: Rett's disorder

Definition

Rett syndrome is a developmental nervous system disorder. It primarily affects girls. It is uncommon, but not rare. It occurs in 1 out of every 10,000-23,000 female births. Boys with the gene defect that causes this disorder are usually stillborn or die shortly after birth.

Rett syndrome can be classified into classic and atypical, depending on the symptoms.

Many people with Rett syndrome live into adulthood. Most have severe disabilities. While many cannot talk or walk, they usually have a full range of feelings and often communicate through their eyes. People with Rett syndrome usually need daily care throughout their lives.

RISK FACTORS

There are no known risk factors for Rett syndrome, except being female. The mutation that causes the syndrome appears to be sporadic.

ETIOLOGY AND GENETICS

Classic Rett syndrome is caused by mutations in the *MECP2* gene, which is located on the long arm of the X chromosome at position Xq28. The protein encoded by this gene is called the methyl CpG binding protein 2 (MeCP2), and it is normally present at high levels in mature nerve cells. While its exact molecular function remains unclear, it appears to be important for forming the proper synapses, or connections, between adjacent nerve cells so that essential chemical communication between cells can occur. MeCP2 also has a regulatory function, since it has been shown under some conditions to repress several other genes, thus preventing the accumulation of additional proteins when they are not necessary. It is not understood, however, how mutations that result in a missing or defective MeCP2 protein lead to the developmental problems and clinical symptoms of Rett syndrome.

An atypical form of Rett syndrome called the early-onset seizure variant is associated with mutations in another gene on the X chromosome called *CDKL5* (at position Xp22). This gene specifies the enzyme known as cyclin-dependent kinase-like 5. Like MeCP2, it is essential for normal brain development and appears to be involved in the regulation of gene activity of other genes in nerve cells.

The pattern of inheritance for mutations in both of these genes is the same: They are inherited in an X-linked dominant fashion. A single copy of the defective gene is sufficient to cause the severe disease symptoms noted in girls with Rett syndrome. One normal copy of the gene is required in order to have sufficient brain function to be consistent with life. That is why Rett syndrome is found only in girls, since an affected boy would not have a second X chromosome that carried the normal allele. There are only scattered reports in the literature of affected females bearing children and passing the defective gene on to daughters in the next generation. As a consequence, more than 99 percent of Rett syndrome cases are sporadic, in that they result from new mutations in females with no previous family history of the disorder.

SYMPTOMS

A girl with Rett syndrome will start developing normally. She will smile, move, and pick items up with her fingers. However, by eighteen months of age the developmental process seems to stop or reverse itself. The age of onset and the severity of symptoms can vary. There are four stages.

Stage I, the early-onset stage, occurs in children between the ages of six to eighteen months and lasts for months. Symptoms may include less eye contact with parents, less interest in toys and play, hand-wringing, slow head growth, and being a calm and quiet baby.

Stage II, the rapid destructive stage, occurs in children between the ages of one to four years old and lasts for weeks to months. Symptoms may include a small head; mental retardation; a loss of muscle tone; the inability to purposely use hands; the loss of (previous) ability to talk; repeatedly moving hands to mouth; and other hand movements, such as clapping, tapping, or random touching. Hand movements stop during sleep. Additional symptoms may include holding breath, gaps in breathing, or taking rapid breaths; irregular breathing stops during sleep. Teeth grinding, laughing or screaming spells, decreased social interactions, irritability, trouble sleeping, tremors, cold feet, and trouble crawling or walking may also be symptoms.

Stage III, the plateau stage, occurs in children who are in their preschool through school years and lasts for years. Symptoms may include difficulty controlling movement, seizures, and less irritability and crying. Communication may improve.

Stage IV, the late motor deterioration stage, occurs when stage III ceases, which can be anywhere from age five to twenty-five, and its duration lasts up to decades. Symptoms may include a decreased ability to walk, muscle weakness or wasting, stiffness of muscles, spastic movements, and scoliosis (curvature of the spine). Breathing trouble and seizures often decrease with age. Puberty usually begins at the expected age.

SCREENING AND DIAGNOSIS

The doctor will ask a parent about a child's symptoms and medical history, do a physical and neurological exam, and exclude other disorders, such as autism. The doctor can also do genetic testing, which can often confirm the diagnosis. The *MECP2* gene mutation is present in 95 percent of girls with Rett Syndrome and in 50 percent of those with the atypical form. However, not everyone with this mutation will have Rett syndrome. Some females may be normal or have only mild symptoms. However, these women can pass the gene to their daughters, and the daughters may then be more severely affected.

Some of the motor functions of Rett syndrome are similar to those of autism. Children with autism, who are more often boys, do not maintain person-to-person contact. Most girls with Rett syndrome, however, prefer human contact to focusing on inanimate objects. These differences may give the first clue in diagnosing Rett syndrome.

Aside from genetic testing, the diagnosis is confirmed by comparing the physical and developmental findings with those typically found in Rett syndrome. Tests may include a blood test to check for genetic mutation (*MECP2* gene); an electroencephalogram (EEG), a test that records the electrical activity of the brain; and a video EEG, a test that combines EEG with a video to see if some of the child's movements are caused by seizures.

Treatment and Therapy

There is no cure for Rett syndrome. Patients with this condition need to be monitored for skeletal problems and heart problems.

Treatment aims to control symptoms and includes medication, such as anticonvulsants to control seizure activity; stool softeners or laxatives, if the patient is constipated; drugs to help with breathing; and drugs to ease agitation. Scientists are investigating a group of medications called histone deacetylase inhibitors as a possible treatment for Rett syndrome.

Patients should eat small, frequent meals; take supplements; and may be tube fed if they are unable to consume enough food. Their diet should also include fluids and high-fiber foods to help control constipation.

Rehabilitation therapies may include occupational therapy to help patients learn to perform daily activities, such as dressing and eating; and physical therapy to help patients improve coordination and movement (can sometimes prolong the ability to walk). The use of braces and splints may be recommended. Speech therapy aides can help build communication skills, and social workers can help a family cope with caring for a child with Rett syndrome.

Parents can use various techniques to try to limit their children's problem behaviors. They can keep a diary of their children's behaviors and activities to help determine the cause of agitation. Other techniques that may help to prevent or control behavior problems include giving children warm baths and massages, playing soothing music, and creating a quiet environment.

Prevention and Outcomes

There is no way to prevent Rett syndrome. Individuals who have questions about the risk of Rett syndrome in their families can talk to a genetic counselor.

Debra Wood, R.N.; reviewed by Rimas Lukas, M.D.
"Etiology and Genetics" by Jeffrey A. Knight, Ph.D.

Further Reading

EBSCO Publishing. *Health Library: Rett Syndrome.* Ipswich, Mass.: Author, 2009. Available through http://www.ebscohost.com.

Kazantsev, A. G., and L. M. Thompson. "Therapeutic Application of Histone Deacetlyase Inhibitors for Central Nervous System Disorders." *Nature Reviews. Drug Discovery* 7, no. 10 (2008): 854-868.

Kleigman, Robert M., et al., eds. *Nelson Textbook of Pediatrics.* 18th ed. Philadelphia: Saunders Elsevier, 2007.

Singer, H. S., and S. Naidu. "Rett Syndrome: 'We'll Keep the Genes on for You.'" *Neurology* 56, no. 5 (March 13, 2001): 582-584.

Tasman, Allan, et al. *Psychiatry.* 3d ed. Hoboken, N.J.: Wiley-Blackwell, 2008.

Web Sites of Interest

Genetics Home Reference
http://ghr.nlm.nih.gov

Health Canada
http://www.hc-sc.gc.ca/index-eng.php

International Rett Syndrome Foundation
http://www.rettsyndrome.org

Ontario Rett Syndrome Association
http://www.rett.ca

See also: Adrenoleukodystrophy; Alexander disease; Alzheimer's disease; Amyotrophic lateral sclerosis; Arnold-Chiari syndrome; Ataxia telangiectasia; Autism; Canavan disease; Cerebrotendinous xanthomatosis; Charcot-Marie-Tooth syndrome; Chediak-Higashi syndrome; Dandy-Walker syndrome; Deafness; Epilepsy; Essential tremor; Friedreich ataxia; Huntington's disease; Jansky-Bielschowsky disease; Joubert syndrome; Kennedy disease; Krabbé disease; Leigh syndrome; Leukodystrophy; Limb girdle muscular dystrophy; Maple syrup urine disease; Metachromatic leukodystrophy; Myoclonic epilepsy associated with ragged red fibers (MERRF); Narcolepsy; Nemaline myopathy; Neural tube defects; Neurofibromatosis; Parkinson disease; Pelizaeus-Merzbacher disease; Pendred syndrome; Periodic paralysis syndrome; Prion diseases: Kuru and Creutzfeldt-Jakob syndrome; Refsum disease; Spinal muscular atrophy; Vanishing white matter disease.

Reverse transcriptase

CATEGORY: Genetic engineering and biotechnology; Molecular genetics

SIGNIFICANCE: Retroviruses infect eukaryotic cells, using reverse transcriptases (RTs) to turn their RNA genomes to DNA that enables their host to use the DNA to make new virus particles. Retroviral DNA, often dormant for years before new virus particles are released, can be oncogenic, giving infected cells high incidences of cancer. Purified RTs are used to make RNA into DNA for biotechnology.

KEY TERMS

deoxyribonucleoside triphosphate (dNTP): one of four monomers (dATP, dCTP, dGTP, dTTP) incorporated into DNA

DNA polymerase: an enzyme that catalyzes the formation of a DNA strand using a template DNA or RNA molecule as a guide

primer: A short piece of single-stranded DNA that can hybridize to denatured DNA and provide a start point for extension by a DNA polymerase

proofreading activity: enzyme activity in DNA polymerase that fixes errors made in copying templates

retroviruses: viruses that possess RNA genomes with genetic information that flows from RNA to host DNA via reverse transcriptases

GENETIC INFORMATION FLOW AND RETROVIRUSES

The central dogma of molecular genetics states that information flow is from DNA to RNA to proteins. RNA polymerase transcribes RNA using a DNA template. For structural genes, the transcribed RNA is a messenger RNA (mRNA), which is used by ribosomes to produce a protein. To maintain and reproduce its DNA, an organism uses RNA to make DNA, via DNA polymerase. It was long believed by geneticists that there were no exceptions to the central dogma.

Some viruses, retroviruses, possess RNA genomes with genetic information flow from RNA to DNA (via reverse transcriptases), and back, before translation. Retroviruses have been isolated from cancers and cancer tissue cultures from birds, rodents, primates, and humans, and some retroviruses cause a high incidence of certain cancers. Flow of retroviral genetic information from RNA to DNA was proposed in 1964 by Howard Temin (1934-1994) for Rous sarcoma virus. Temin, along with David Baltimore, jointly received the Nobel Prize in Physiology or Medicine in 1975 for independently discovering the enzyme reverse transcriptase (RT).

Rous sarcoma virus causes tumors in birds. Temin's hypothesis was based on effects of nucleic acid synthesis inhibitors on replication of the virus. First, the process was inhibited by actinomycin D, an inhibitor of DNA-dependent RNA synthesis. Furthermore, DNA synthesis inhibition by cytosine arabinoside, early after infection, stopped viral replication. Therefore, a DNA intermediate seemed involved in viral replication. The expected process was termed reverse transcription because RNA becomes DNA instead of DNA becoming RNA.

RT DISCOVERY AND PROPERTIES

Retrovirus infection begins with injection of RT and single-stranded RNA into host cells. RT (an RNA-dependent DNA polymerase) causes biosynthesis of viral DNA using an RNA template from the retrovirus HIV (human immunodeficiency virus), the causative agent of acquired immunodeficiency syndrome (AIDS). RTs have been purified from many retroviruses. Avian, murine, and human RTs have been studied most. All have ribonuclease H (RNase H) activity on the same protein as polymerase activ-

Howard M. Temin. (© The Nobel Foundation)

David Baltimore. (© The Nobel Foundation)

ity. Ribonuclease H degrades RNA strands of DNA-RNA hybrids. A nuclease that degrades DNA is later involved in retrovirus DNA integration into host cell DNA. Most biochemical properties of purified RTs are common to them and other DNA polymerases. For example, all require the following for DNA synthesis: a primer on which synthesis begins, a template that is copied, and a supply of the four dNTPs.

An RT converts a retroviral single-stranded RNA genome to "integrated double-stranded DNA" as follows: First a hybrid (DNA-RNA) duplex is made from viral RNA, as an antiparallel DNA strand is produced. The RNA-directed DNA polymerase activity of RT is primed by host cell transfer RNA, which binds to the viral RNA. Then, the viral RNA strand is destroyed by RNase H, and the first DNA strand now becomes the template for synthesis of a second antiparallel DNA strand. Resultant duplex DNA is next integrated into a host cell chromosome, where it is immediately used to make virus particles or, alternatively, it takes up residence in the host cell's genome, remaining unused—often for years—until it is activated and causes cancer or production of new viruses.

Importance of Reverse Transcriptases

RTs can use almost any RNA template for DNA synthesis. Low RT template specificity allows RT to be used to make DNA copies of a wide variety of RNAs in vitro. This has been very useful in molecular biology, especially in production of exact DNA copies of purified RNAs. Once the copies are made by RT, they can be cloned into bacterial expression vectors, where mass quantities of the gene product can be produced. It has also been shown that RT activity takes part in making telomeres (protective chromosome ends). Telomere formation and maintenance are essential cell processes, related to life span and deemed important to understanding cancer.

RTs are also important in treatment of acquired immunodeficiency syndrome (AIDS). The drugs most useful for AIDS treatment are RT inhibitors such as zidovudine, didanosine, zalcitabine, and stavudine. RTs are also associated with the difficulty in maintaining successful long-term AIDS treatment, due to rapid development of resistant HIV in individual AIDS patients. The resistance is postulated to be due to RT's lack of a proofreading component. Inadequate proofreading in sequential replication of HIV viral particles from generation to generation is believed to cause the rapid mutation of the viral genome.

Sanford S. Singer, Ph.D.

FURTHER READING

Goff, S. P. "Retroviral Reverse Transcriptase: Synthesis, Structure, and Function." *Journal of Acquired Immune Deficiency Diseases* 3, no. 8 (1990): 817-831. A solid, well-illustrated paper.

Joklik, Wolfgang K., ed. *Microbiology: A Centenary Perspective.* Washington, D.C.: ASM Press, 1999. Thoroughly reviews important microbiology issues, including reverse transcriptase papers by Howard Temin and David Baltimore.

Litvack, Simon. *Retroviral Reverse Transcriptases.* Austin, Tex.: R. G. Landes, 1996. Describes discovery, biosynthesis, structure, inhibitors, and action mechanistics of reverse transcriptase. Illustrations and bibliographic references.

Morel, Gérard, and Mireille Raccurt. "Reverse Transcription." In *PCR/RT-PCR In Situ Light and Electron Microscopy.* Boca Raton, Fla.: CRC Press, 2003. Describes how to use the techniques of polymerase chain reaction (PCR) and reverse transcription PCR (RT-PCR) in order to visualize DNA and RNA in tissues and cell cultures.

O'Connell, Joe, ed. *RT-PCR Protocols.* Totowa, N.J.: Humana Press, 2002. Collects several papers on the use of reverse transcription polymerase chain reaction in analysis of mRNA, quantitative methodologies, detection of RNA viruses, genetic analysis, and immunology. Tables, charts, index.

Shippen-Lentz, D., and E. H. Blackburn. "Functional Evidence for an RNA Template in Telomerase." *Science* 247, no. 4942 (February 2, 1990): 546-552. Points to the RT activity in telomerases. Illustrated.

Skowron, Gail, and Richard Ogden, eds. *Reverse Transcriptase Inhibitors in HIV/AIDS Therapy.* Totowa, N.J.: Humana Press, 2006. Examines how reverse transcriptase inhibitors are being used in drug discovery, pharmacology, development of drug resistance, toxicity, and prevention of mother-to-child transmission of HIV/AIDS. Reviews the role of reverse transcriptase in the life cycle of viruses.

Varmus, Harold. "Retroviruses." *Science* 240, no. 4858 (June 6, 1988): 1427-1435. The discoverer of oncogenes describes properties of different retroviruses, including the mechanism of reverse transcription.

WEB SITES OF INTEREST

Biotechnology Index, Reverse Transcriptases
http://www.vivo.colostate.edu/hbooks/genetics/biotech/enzymes/rt.html

The index, prepared by the department of biomedical sciences at Colorado State University, includes a page about reverse transcriptases.

Cells Alive!
http://www.cellsalive.com/hiv2.htm

Features an explanation of how reverse transcription converts viral RNA into DNA as part of its discussion of the human immunodeficiency virus (HIV).

RCSD Protein Data Bank
http://www.rcsb.org/pdb/static.do?p=education_discussion/molecule_of_the_month/pdb33_1.html

The Research Collaboratory for Structural Bioinformatics (RCSB) Protein Data Bank is an archive of information about proteins, nucleic acids, and "complex assemblies." The site provides an explanation of reverse transcriptases.

See also: cDNA libraries; Central dogma of molecular biology; Model organism: *Chlamydomonas reinhardtii*; Pseudogenes; Repetitive DNA; RNA isolation; RNA structure and function; RNA world; Shotgun cloning.

Reverse transcription polymerase chain reaction (RT-PCR)

CATEGORY: Molecular genetics

SIGNIFICANCE: This technique uses ribonucleic acid (RNA) and a heat-resistant enzyme to produce many copies of deoxyribonucleic acid (DNA). The development of RT-PCR has greatly advanced genetics research and is used in the detection of many inherited disorders and infections.

KEY TERMS

deoxyribonucleotide: the building block of DNA consisting of a nitrogen-containing molecule, deoxyribose containing sugar, and phosphate(s)

enzyme: a protein that allows a reaction to proceed more quickly or with less supplied energy, but which is not altered by the reaction

protein: a molecule composed of amino acids

primer: a fragment of nucleic acid that is used to initiate the replication of DNA

reverse transcriptase: enzyme that catalyzes the formation of DNA from RNA

RNA: ribonucleic acid; a form of genetic material composed of nucleotides

THE BASICS OF RT-PCR

DNA, the genetic material of most organisms, is used to make a form of nucleic acid called ribonucleic acid (RNA) that provides the blueprint to actually manufacture molecules such as proteins. This process is called transcription. Reverse transcription (RT) involves the use of RNA as the template to make DNA. This involves an enzyme called reverse transcriptase.

In RT-PCR, a strand of RNA is reverse transcribed to produce what is called complementary DNA (cDNA). It is the cDNA that is subsequently amplified in number in the PCR reaction.

RT-PCR used to be done by hand. Now, the process is carried out in an automated machine that can produce tens of thousands of copies of the selected sequence of DNA within hours. Organizations such as The Institute for Genomic Research (TIGR) that have dozens of PCR machines are capable of producing millions of copies of DNA in a day.

THE COMPONENTS OF PCR

PCR relies on the presence of a template, deoxyribonucleotides, a pair of primers, and DNA polymerase. The template is the RNA strand.

Deoxyribonucleotides are the building blocks of DNA. There are four deoxyribonucleotides: adenine, thymine, cytosine, and guanine. Their three-dimensional structures are such that adenine will bond with thymine, and cytosine will bond with guanine. In the double strand of DNA that forms the helical structure found in the cell, the two strands will be complementary to one another; thus, a sequence of adenine-thymine-guanine on one strand will be mirrored by thymine-adenine-cytosine on the other strand.

Primers are sequences of deoxyribonucleotides that have been deliberately constructed to be complementary with the sequences of a portion of each DNA strand. This causes the primers to associate with one or the other of the DNA strands. Binding of primer to DNA is important in the PCR process.

DNA polymerase is an enzyme that catalyzes the construction of DNA from the building blocks that are supplied. PCR is carried out at an elevated temperature that helps keep the DNA strands separated from each other, so the polymerase must be capable of functioning at the higher temperatures (many proteins including enzymes change their structure at higher temperature, causing lose of function). PCR became possible with the discovery of Taq polymerase, an enzyme made by the thermophilic ("heat-loving") bacterium *Thermus aquaticus,* which naturally lives in hot springs.

THE THREE STEPS OF PCR

Part of the sequence of the targeted DNA has to be known in order to design the primers. In the first step, the targeted double-stranded DNA is heated to more than 90 degrees Celsius (194 degrees Fahrenheit). During this process, the two strands of the targeted DNA separate from each other. Each strand is capable of being a template.

The second step involves a gradual decrease to a temperature at which the two DNA strands would normally reassociate, but primers are also present in the solution at a greater concentration than DNA. This causes reassociation to occur between the primers and the specific complementary regions on each DNA strand. Complete reassociation of the DNA strands cannot occur. The association between the primers and the DNA will produce a short region of two strands with a longer region having a single strand of the DNA.

The third step, amplification, focuses on these single strands of DNA. In the presence of the necessary deoxyribonucleotides and other compounds, DNA polymerase catalyzes the manufacture of a complementary strand of DNA. At the end of this process, the newly constructed strands can be separated at high temperature to begin the cycle again.

As the cycles are repeated again and again, the number of each target sequence doubles with each cycle (a logarithmic increase).

PCR reactions are carried out in an apparatus called a thermocycler, which is designed to change temperatures automatically. The various conditions of temperature and reaction times can be programmed and the procedure occurs without operator assistance.

Impact

American molecular biologist Kary Mullis developed PCR in the 1970's. The ability to quickly amplify DNA revolutionized molecular biology. In 1993, Mullis received the Nobel Prize in Physiology or Medicine. RT-PCR has also greatly advanced genetics research. For example, the technique was invaluable in the sequencing of the human genome. RT-PCR is also widely used in the detection of many inherited disorders.

Brian D. Hoyle, Ph.D.

Further Reading

Hodge, Russ. *The Future of Genetics: Beyond the Human Genome Project.* New York: Facts On File, 2009. A consideration of future advancements on human genetics research.

Read, Andrew. *New Clinical Genetics.* Oxfordshire, England: Scion, 2007. Concise summary of human genetics from a clinical standpoint.

Strachan, Tom. *Human Molecular Genetics.* London: Garland Science, 2003. Comprehensive consideration of human genetics.

Web Sites of Interest

HUMGEN International
http://www.humgen.org/int

National Institutes of Health, National Human Genome Research Institute
http://www.genome.gov

See also: Ancient DNA; Anthrax; Bioinformatics; Blotting: Southern, Northern, and Western; Central dogma of molecular biology; Cloning vectors; DNA fingerprinting; DNA sequencing technology; Forensic genetics; Genetic engineering: Historical development; Human Genome Project; In vitro fertilization and embryo transfer; Mitochondrial diseases; Molecular genetics; Paternity tests; Polymerase chain reaction (PCR); Repetitive DNA; RFLP analysis; RNA isolation.

RFLP analysis

Category: Techniques and methodologies

Significance: RFLP analysis was the first simple method available for distinguishing individuals based on DNA sequence differences. The conceptual basis for this technique is still widely used in genetics, although RFLP analysis has been largely supplanted by faster and more powerful techniques for the comparison of genetic differences.

Key terms

gel electrophoresis: a method for separating DNA molecules by size by applying electric current to force DNA through a matrix of agarose, which inhibits the migration of larger DNA fragments more than small DNA fragments

restriction enzymes: proteins that recognize specific DNA sequences and then cut the DNA, normally at the same sequence recognized by the enzymes

Southern blotting: a method for transferring DNA molecules from an agarose gel to a nylon membrane; once the DNA is on the membrane, it is incubated with a DNA containing an identifiable label and is then used to detect similar or identical DNA sequences on the membrane

The Procedure

Restriction fragment length polymorphism (RFLP) analysis is a method for distinguishing individuals and analyzing relatedness, based on genetic differences. RFLP analysis relies on small DNA sequence differences that lead to the loss or gain of restriction enzyme sites in a chromosome or to the change in size of a DNA fragment bracketed by restriction enzyme sites. These sequence differences lead to a different pattern of bands on a gel (reminiscent of a bar code) that varies from individual to individual.

RFLP analysis starts with the isolation of DNA. Typically, DNA isolation requires the use of detergents, protein denaturants, RNA degrading enzymes, and alcohol precipitation to separate the DNA from the other cellular components. This DNA could be isolated from a blood sample provided by an individual, from evidence left at the scene of a crime, or from other sources of cells or tissues.

The purified DNA is then digested with a molecular "scissors" called a restriction enzyme. Restriction enzymes recognize and cut precise sequences, typically six base pairs in length. If one base pair is changed in that recognition sequence, the enzyme will not cut the DNA at that point. However, if a sequence that is not recognized by a restriction enzyme is altered by mutation, so that it now is recognized, the DNA will be cleaved at that point. In other cases, the DNA sequences recognized by the restriction enzymes themselves are not changed, but the length of DNA between two restriction enzyme sites differs between individuals. These types of mutations occur with enough regularity that often even two closely related individuals will have some detectable differences in the sizes of DNA fragments produced from restriction enzyme digestion.

Once the DNA has been digested with a restriction enzyme, it is separated by size in an agarose gel. At this point, the DNA appears, to the eye, to be a smear of molecules of all sizes, and it is not generally possible to differentiate the DNAs from different individuals at this stage. The size-fractioned DNA is next transferred to a nylon membrane in a process called Southern blotting. The result of the transfer is that the location and arrangement of the DNA fragments in the gel is maintained on the membrane, but the DNA is now single-stranded (critical for the next step in the process) and much easier to handle.

The final step in the process is to detect specific DNA fragments on the membrane. This is done by using a DNA fragment that is labeled to act as a probe, to home in on and identify similar DNA sequences on the membrane. Before use, the probe is made single-stranded, so it can bind to the single-stranded DNA on the membrane. The probe DNA can be labeled with radioactivity, in which case it is detected using X-ray film. The probe DNA can also be labeled with molecules that are bound by proteins, and the proteins can then be detected either directly or indirectly.

In a case in which a restriction enzyme site has been added or removed, the probe is normally a DNA fragment that is found in only one location in the genome. In cases in which one is looking at the size of fragments bracketed by restriction enzyme sites, the probe DNA is normally a DNA molecule that is found in several sites in the genome, and the DNA fragments that are identified in this analysis are ones that tend to vary between individuals. In many cases, the probe DNA binds to regions of DNA that consist of variable number tandem repeats (VNTRs). The number of VNTRs tends to vary between different individuals and, consequently, these sequences are useful for identification.

Applications

One of the earliest uses of this technique in clinical medicine was in the prenatal diagnosis of sickle-cell disease. Previous work had shown that many individuals with the disease had a mutation in their DNA that eliminated a restriction enzyme site in a gene encoding a hemoglobin protein. This information was used to develop a diagnostic RFLP procedure. A section of the hemoglobin gene is used as a probe. The size of restriction enzyme fragments identified is different in individuals who have sickle-cell disease (and therefore have two mutant alleles) compared with individuals who carry either one mutant allele or have two unmutated hemoglobin alleles. This method allowed for the identification of affected fetuses using DNA from cells isolated from amniotic fluid (a much simpler and safer procedure than the previous method of diagnosis, which required isolating fetal red blood cells).

Another widely reported use of RFLP analysis has been in forensic science. RFLP methods have been critical in helping to identify criminals, and these methods have also helped exonerate innocent people. The first application of RFLP in forensic analysis was in the case of the murders of two young girls in England, in 1983 and 1986. Initially, a seventeen-year-old boy confessed to the murders. RFLP analysis, using DNA from the crime scene, indicated that he was not the murderer. After extensive investigation, including RFLP analysis of DNA from more than forty-five hundred men, a suspect was identified. Confronted with the evidence, the suspect pleaded guilty to both murders and was jailed for life. Since then, RFLP analysis has been used in thousands of criminal cases. Other forensic applications of RFLP include its use as evidence in court cases involving paternity determinations and its role in identifying the bodies of missing persons who otherwise could not be identified.

In addition to the clinical and forensic applications described above, RFLP analysis has been used in many subdisciplines of biology since the early 1980's. The applications of RFLP analysis range from the conservation of endangered species to the identification of strains of bacteria associated with disease outbreaks to basic research involving the classification of organisms.

Although RFLP analysis has been widely used since its inception, it is increasingly being displaced by polymerase chain reaction (PCR) methods, which typically are much faster and require much less DNA. RFLP analysis was, however, an important step in the introduction of modern DNA analysis into the biology laboratory and the courtroom. The guiding principle behind RFLP analysis—identifying individuals, strains, and species, based on DNA sequence differences—remains a part of newer techniques.

Patrick G. Guilfoile, Ph.D.

FURTHER READING

Allison, Lizabeth A. "Recombinant DNA Technology and Molecular Cloning." In *Fundamental Molecular Biology.* Malden, Mass.: Blackwell, 2007. Includes information about RFLP.

Chang, J. C., and Y. W. Kan. "A Sensitive New Prenatal Test for Sickle-Cell Anemia." *New England Journal of Medicine* 307, no. 1 (July 1, 1982): 30-32. A short and readable scientific paper describing one of the first applications of RFLP analysis to clinical medicine. The same issue of the journal also has another, somewhat more detailed, article on the same topic by S. H. Orkin et al.

Guilfoile, P. *A Photographic Atlas for the Molecular Biology Laboratory.* Englewood, Colo.: Morton, 2000. An illustrated guide to molecular biology techniques, including a substantial illustrated section on RFLP analysis.

Hartl, D. L., and Elizabeth W. Jones. "Types of DNA Markers Present in Genomic DNA." In *Genetics: Analysis of Genes and Genomes.* 7th ed. Sudbury, Mass.: Jones and Bartlett, 2009. This excellent introductory genetics textbook devotes a section of chapter 2 to a discussion of RFLP within the broader context of DNA structure and genetic variation.

Jeffreys, A., V. Wilson, and S. L. Thein. "Individual-Specific Fingerprints of Human DNA." *Nature* 316, no. 6023 (July 4-10, 1985): 76-79. A technical article that describes some of the background information that led to the use of RFLP analysis in forensic science.

Karp, Gerald. "Genetic Analysis in Molecular Biology." In *Cell and Molecular Biology: Concepts and Experiments.* 5th ed. Chichester, England: John Wiley and Sons, 2008. Includes information about RFLP.

Orkin, S. H., et al. "Improved Detection of the Sickle Mutation by DNA Analysis: Application to Prenatal Diagnosis." *New England Journal of Medicine* 307, no. 1 (July 1, 1982): 32-36. Appears in the same issue as another article on the topic by J. C. Chang and Y. W. Kan.

WEB SITES OF INTEREST

Kimball's Biology Pages
http://users.rcn.com/jkimball.ma.ultranet/BiologyPages/R/RFLPs.html

John Kimball, a retired Harvard University biology professor, includes a page about restriction fragment length polymorphisms in his online cell biology text.

NCBI, Restriction Fragment Length Polymorphism (RFLP)
http://www.ncbi.nlm.nih.gov/projects/genome/probe/doc/TechRFLP.shtml

The National Center for Biotechnology Information has prepared this explanation of RFLP.

RFLP Method: Restriction Fragment Length Polymorphism
http://www.bio.davidson.edu/COURSES/genomics/method/RFLP.html

An explanation of RFLP is featured in this site designed for a microbiology course at Davidson College.

See also: Blotting: Southern, Northern, and Western; Chromosome theory of heredity; DNA fingerprinting; Gender identity; Genetic engineering; Genetic testing; Model organism: *Arabidopsis thaliana*; Paternity tests; Polymerase chain reaction; Prenatal diagnosis; Restriction enzymes.

Rh incompatibility and isoimmunization

CATEGORY: Diseases and syndromes
ALSO KNOWN AS: RhD incompatibility

DEFINITION

One of the first tests performed at the beginning of a pregnancy is blood-type. This basic test determines a pregnant woman's blood type and Rh factor. People with different blood types have proteins specific to that blood type on the surface of their red blood cells. There are four blood types (A, B, AB, and O). Each of the four blood types is additionally classified according to the presence of another protein on the surface of the red blood cells that indicates an individual's Rh factor. If an individual carries this protein, he or she is Rh positive. If he or she does not carry the protein, he or she is Rh negative.

Most people, about 85 percent, are Rh positive. However, if a woman who is Rh negative and a man who is Rh positive conceive a baby, there is the potential for incompatibility. The baby growing inside the Rh-negative mother may have Rh-positive blood, inherited from the father. Statistically, at least 50 percent of the children born to an Rh-negative mother and an Rh-positive father will be Rh positive.

RISK FACTORS

Risk factors for Rh incompatibility and isoimmunization include being a pregnant woman with Rh-negative blood who had a prior pregnancy with a fetus that was Rh positive; being a pregnant woman who had a prior blood transfusion or amniocentesis; and being a pregnant woman with Rh-negative blood who did not receive Rh immunization prophylaxis during a prior pregnancy with an Rh-positive fetus.

ETIOLOGY AND GENETICS

The Rh locus on the short arm of chromosome 1 at position 1p36.2-p34 actually consists of two genes, *RHD* and *RHCE*, that are organized in tandem and are so similar in sequence that it is clear that they evolved from the duplication of a common ancestral gene. There are two phenotypes, Rh positive and Rh negative, that are distinguished by the presence or absence of the RhD antigen (the protein product of the *RHD* gene) on the surface of red blood cells. The *RHCE* gene encodes both the RhC and RhE proteins, but since these do not confer antigenic properties on the cell, they do not contribute to the Rh phenotype. Most individuals who are Rh negative have a deletion of the entire *RHD* gene and retain only the *RHCE* gene.

The Rh-positive allele is dominant to the Rh-negative allele, so Rh-positive individuals have either one or two copies of the *RHD* gene. Rh-negative individuals lack the *RHD* gene on both of their copies of chromosome 1. Two Rh-positive individuals who are both heterozygous (having only one copy of the *RHD* gene) have a 25 percent chance of having an Rh-negative child and a 75 percent chance of having an Rh-positive child. In all cases of Rh incompatability, the mother is Rh negative and both the father and the developing fetus are Rh positive.

SYMPTOMS

Symptoms and complications affect only the fetus and/or newborn. They occur when standard preventive measures are not taken and can vary from mild to very serious. The mother's health is not affected.

Symptoms of the fetus or newborn baby include anemia and swelling of the body (also called hydrops fetalis), which may be associated with heart failure and respiratory problems. Another symptom is kernicterus (a neurological syndrome), which can occur in stages.

Symptoms in the early stage include a high bilirubin level (greater than 18 milligrams/cubic centimeters), extreme jaundice, an absent startle reflex, a poor suck, and lethargy. Symptoms in the intermediate stage include a high-pitched cry, an arched back with neck hyperextended backward (opisthotonos), a bulging fontanel (soft spot), and seizures. Late-stage symptoms include high-pitched hearing loss, mental retardation, muscle rigidity, speech difficulties, seizures, and movement disorder.

Screening and Diagnosis

There are not any physical symptoms that would allow women to detect on their own if they are Rh incompatible with any given pregnancy. If women are pregnant, it is standard procedure for their health care providers to order a blood test that will determine whether they are Rh positive or Rh negative. If the blood test indicates that they have developed Rh antibodies, their blood will be monitored regularly to assess the level of antibodies it contains. If the levels are high, an amniocentesis would be recommended to determine the degree of impact on the fetus.

Treatment and Therapy

Since Rh incompatibility is almost completely preventable with the use of prophylactic immunization (immune globulin injection of RhoGAM), prevention remains the best treatment.

Women will be given an injection of Rho immune globulin at week twenty-eight of their pregnancy. This desensitizes their blood to Rh-positive blood. They will also have another injection of immune globulin within seventy-two hours after delivery (or miscarriage, induced abortion, or ectopic pregnancy). This injection further desensitizes their blood for future pregnancies.

Treatment of a pregnancy or newborn depends on the severity of the condition. If the condition is mild, treatment includes aggressive hydration and phototherapy using bilirubin lights.

Treatment for hydrops fetalis includes amniocentesis to determine severity, an intrauterine fetal transfusion, and early induction of labor. A direct transfusion of packed red blood cells (compatible with the infant's blood) and exchange transfusion of the newborn may be done to rid the infant's blood of the maternal antibodies that are destroying the red blood cells. Treatment may also include control of congestive failure and fluid retention.

Treatment for kernicterus includes exchange transfusion (which may require multiple exchanges) and phototherapy.

Full recovery is expected for mild Rh incompatibility. Both hydrops fetalis and kernicterus represent extreme conditions caused by the breakdown of red blood cells, called hemolysis. Both have guarded outcomes; hydrops fetalis has a high risk of mortality. Long-term problems can result from severe cases, including cognitive delays, movement disorders, hearing loss, and seizures.

Prevention and Outcomes

Rh incompatibility is almost completely preventable. Rh-negative mothers should be followed closely by their obstetricians during pregnancy. If the father of the infant is Rh-positive, the mother is given a midterm injection of RhoGAM and a second injection within a few days of delivery. These injections prevent the development of antibodies against Rh-positive blood. This effectively prevents the condition. Routine prenatal care should help identify, manage, and treat any complications of Rh incompatibility.

Rick Alan;
reviewed by Jeff Andrews, M.D., FRCSC, FACOG
"Etiology and Genetics" by Jeffrey A. Knight, Ph.D.

Further Reading

Beers, Mark H., ed. *The Merck Manual of Medical Information.* 2d home ed., new and rev. Whitehouse Station, N.J.: Merck Research Laboratories, 2003.

EBSCO Publishing. *Health Library: Rh Incompatibility and Isoimmunization.* Ipswich, Mass.: Author, 2009. Available through http://www.ebscohost.com.

Moise, K. J., Jr. "Management of Rhesus Alloimmunization in Pregnancy." *Obstetrics and Gynecology* 112, no. 1 (July, 2008): 164-176.

Smits-Wintjens, V. E., F. J. Walther, and E. Lopriore. "Rhesus Haemolytic Disease of the Newborn: Postnatal Management, Associated Morbidity, and Long-Term Outcome." *Seminars in Fetal and Neonatal Medicine* 13, no. 4 (August, 2008): 265-271.

Web Sites of Interest

American College of Obstetricians and Gynecologists
http://www.acog.org

Kids Health from Nemours: Rh Incompatability
http://kidshealth.org/parent/pregnancy_newborn/pregnancy/rh.html

March of Dimes Foundation: Rh Disease
http://www.marchofdimes.com/professionals/14332_1220.asp

The Society of Obstetricians and Gynaecologists of Canada
http://www.sogc.org/index_e.asp

Women's Health Matters
http://www.womenshealthmatters.ca/index.cfm

See also: ABO blood types; Chronic myeloid leukemia; Fanconi anemia; Hemophilia; Hereditary spherocytosis; Infantile agranulocytosis; Myelodysplastic syndromes; Paroxysmal nocturnal hemoglobinuria; Porphyria; Sickle-cell disease.

RhoGD12 gene

Category: Diseases and syndromes

Significance: The activity of the *RhoGD12* gene has been linked to stopping cancerous tumors from spreading to other areas of the body. People with cancer whose *RhoGD12* gene is active have much less chance of a cancer spreading from one part of the body to another. This gene's ability to act seems to be linked to an endothelin inhibitor and may be helped by the *SRC* gene.

Key terms

anti-oncogene: a gene that stops or suppresses uncontrolled cell growth that may lead to cancer

expression: the ability of a gene to influence the body to perform a certain way

metastasis: the spread of a cancer from its original site to other parts of the body

oncogene: a gene that is known to trigger some type of cancer

suppression: "turning off" a gene or suppressing the gene's normal actions

tumor suppression: stopping the spread of a cancerous tumor from one part of the body to another

Tumor Suppression

RhoGD12 is an anti-oncogene, or a gene that may help suppress the growth of cancer. In 2002, researchers discovered a link between *RhoGD12* and tumor suppression. Researchers put *RhoGD12* into some cells that were then injected into mice that had compromised immune systems. When compared to a control group of mice in which every mouse had many tumors that metastasized, only about half of the mice with *RhoGD12* got tumors, and those mice that had tumors had many fewer tumors than the control group. Current thinking is that this gene is able to affect the way cells move about the body, thus influencing the movement of cancerous cells between body systems.

ET-1 and *RhoGD12*

The impact of the *RhoGD12* gene has been studied extensively in the relationship between bladder cancer and lung cancer. According to studies, about half of patients with late-stage bladder cancer end up having a relapse of their cancer where it spreads to the lungs, which is often the cause of death. Researchers believed that this was caused by a suppression of the *RhoGD12* gene that allowed the cancer to spread. They believed that this suppression occurred because of the *RhoGD12* gene's complicated relationship with an endothelin inhibitor called ET-1. ET-1 is known to cause cells to grow and divide, particularly cells involved in the production of blood vessels. Due to the increased ineffectiveness of the *RhoGD12* gene to stop the spread of the cancer, ET-1 was able to allow tumors to grow in the lungs at an increased rate. Currently, researchers are unsure as to why or how ET-1 stops the *RhoGD12* gene from helping to suppress the spread of cancer throughout the body.

SRC and *RhoGD12SRC*

RhoGD12SRC is an oncogene, or a gene that has been known to trigger cancer. This gene was commonly thought to help all cancers grow and spread by allowing cells to grow and divide in an uncontrolled way, and many cancers were treated by using therapies that inhibited this gene from expression. Recent research has shown that this gene may work together with or somehow modify the *RhoGD12* gene to help cells use their natural ability to stop tumor growth in bladder cancer, and that the best treatment for bladder cancer may involve allowing SRC gene expression, even though this may not help in treatment for other types of cancer. This research is changing the way some scientists and doctors think about treating cancer—some genes that negatively influence some types of cancers could be a positive influence for other types of cancers, emphasizing the important of individually treating cancers in the way that is best for the patient and the type of cancer that he or she may have.

Impact

This gene seems to be able to stop metastasis or restrain the secondary growth and spread of any type of cancer. In some recent studies of cancer in humans, the presence of this gene seemed to track along with the stage and grade of the cancer. It is possible that this gene can be used as a marker to determine which cancers are likely to spread and which are likely to remain contained. Research currently focuses on discovering the mechanism by which this gene suppresses cancer growth and determining if that mechanism can be manipulated so this gene can be used in treating cancer or keeping it in check. Research is also ongoing on how this gene is influenced by other genes or substances in the body.

Marianne M. Madsen, M.S.

Further Reading

Gildea, J. J., et al. "*RhoGD12* Is an Invasion and Metastasis Suppressor Gene in Human Cancer." *Cancer Research* 62, no. 22 (November 15, 2002): 6418-6423. The description of the research that led to *RhoGD12* being identified as an anti-oncogene.

Stafford, L. J., K. S. Vaidya, and D. R. Welch. "Metastasis Suppressors Genes in Cancer." *The International Journal of Biochemistry & Cell Biology* 40, no. 5 (2008): 874-891. Provides a discussion of many types of anti-oncogenes.

Titus, B., et al. "Endothelin Axis Is a Target of the Lung Metastasis Suppressor Gene *RhoGD12*." *Clinical Cancer Research* 65, no. 16 (2005). This article is a generally cited primary research article describing the research linking *RhoGD12* to ET-1.

Wu, Y., et al. "Neuromedin U Is Regulated by the Metastasis Suppressor *RhoGDI2* and Is a Novel Promoter of Tumor Formation, Lung Metastasis, and Cancer Cachexia." *Oncogene* 26 (2007). This article describes further research linking this gene to tumor suppression in a variety of cancers.

Web Sites of Interest

American Cancer Society: Oncogenes and Tumor Suppressor Genes
http://www.cancer.org/docroot/ETO/content/ETO_1_4x_oncogenes_and_tumor_suppressor_genes.asp

National Center for Biotechnology Information, Molecular Cell Biology: Proto-Oncogenes and Tumor Suppressor Genes
http://www.ncbi.nlm.nih.gov/books/bv.fcgi?rid=mcb.section.7090

See also: *BRAF* gene; *BRCA1* and *BRCA2* genes; *MLH1* gene; *PMS* genes; *RB1* gene; *SCLC1* gene.

RNA interference

CATEGORY: Cell biology; Genetic engineering and biotechnology; Molecular genetics

SIGNIFICANCE: The term RNAi stands for "RNA interference," a newly discovered gene regulatory mechanism that downregulates gene expression in a sequence-specific manner by acting at the post-transcriptional level. This has undoubtedly been one of the biggest advances in genetics in decades, since it has added another perspective to the role of RNA in cells.

Key terms

Dicer: an RNase that cleaves dsRNA into short 21-23bp long siRNAs
dsRNA: double-stranded RNA
miRNA: micro RNA
RISC: RNA interference silencing complex
RNAi: RNA interference
siRNA: small interfering RNA

Background

In the late 1980's to early 1990's, plant biologist Richard Jorgensen was trying to make intense purple petunias by introducing extra copies of the purple pigment gene (Chalcone synthase *chs*A). He was quite surprised when the effort yielded instead white and variegated petunias. It seemed that somehow the introduced gene (the transgenes) had blocked (silenced) its own expression as well the expression of the endogenous purple pigment genes (termed "cosuppression"). Analysis of transcription rates in these cells using nuclear run-on assays did not indicate a decrease, implying that the phenomenon should be renamed as post-transcriptional gene silencing (PTGS). Following soon thereafter, were reports of "quelling" observed in the fungi *Neurospora crassa.* An effort to boost the orange phenotype *(al1+)* shown by wild-type (wt) *Neurospora* by transforming it with a plasmid containing a 1500 bp fragment of the coding sequence of the *al1+* gene, generated albino *Neurospora.* Once again in these *al1+* quelled strains, it was the level of the mature mRNA that was affected in a homology-dependent manner. These findings were better understood when in 1998 Andrew Fire and Craig Mello reported that introduction of dsRNA corresponding to the *unc-22* gene into *Caenorhabditis elegans* specifically disrupted the expression of the endogenous *unc-22* gene. *unc-22* in *C. elegans* codes for a nonessential myofilament protein, and a reduction in *unc-22* activity produced a twitching phenotype in these worms. If the dsRNA was introduced into the worm gonads, the phenotype was heritable since the progeny too demonstrated the classic twitching phenotype. This demonstration offered an easy method to generate loss-of-function mutants since the silencing mechanism in *C. elegans* could be initiated simply by soaking the worms in solutions containing dsRNA or by feeding the worms *E. coli* that expressed the dsRNA. The silencing of a functional gene by introduction of exogenous dsRNA, homologous to the target gene was renamed RNA interference (RNAi). Since then RNAi-related events have been identified in a multitude of eukaryotes ranging from protozoa, parasites, and fruit flies to mouse- and human-derived cell lines.

Molecular Mechanism

One of the earliest clues into the molecular basis of RNAi was provided by the discovery of short (20-25 nucleotides long) RNAs in the plant cells that matched the silencing target. Around the same time, in vitro RNAi assays using fruit-fly (*Drosophila melanogaster*) extracts revealed that the long dsRNA was being cleaved into 21-23 bp long RNA duplexes with 2 base overhangs at each end. These RNA species were called short interfering RNA (siRNA), and the enzyme responsible for producing these siRNAs from the dsRNA was called Dicer. Soon thereafter, the "sense" strand (sequence is identical to the target mRNA) is destroyed and the fate of the "antisense" strand thereon depends upon the organism. In mammals and fruit flies, the antisense strand is incorporated into the RISC (RNA-induced silencing complex), which is made up of RNA-binding proteins and RNA nucleases such as the Argonaute family of proteins found in *Arabidopsis thaliana*. The activated RISC (with the antisense strand of RNA) can now bind to the target mRNA and cleave it, thereby initiating sequence-specific mRNA degradation. This cleaving activity, whereby the RISC cleaves mRNA as directed by siRNA, is called the "Slicer" activity. In worms and plants, the next set of events following destruction of the siRNA sense strand are somewhat different even though the final outcome—RNA destruction—is similar. In these organisms the antisense strand pairs up with the complementary target mRNA, thus initiating synthesis of a long dsRNA by the enzyme RdRP (RNA dependent RNA polymerase). Then Dicer cleaves the long dsRNA and generates siRNAs which act via the RISC to target mRNAs for destruction, as described earlier.

Function

Over the years, substantial evidence has been collected that indicates a general role for PTGS (RNAi) in silencing parasitic genes such as viruses, transposons ("jumping genes") and repetitive genes (including transgenes). In plant viruses, the double-stranded RNA intermediate in RNA virus replication has been shown to initiate PTGS/RNAi. These findings were supported by the discovery of virus encoded genes that suppress PTGS by a variety of mechanisms such as reducing target mRNA degradation and counteracting the systemic spread of RNA silencing. Recently in 2005, Courtney Wilkins and her team working with a *C. elegans* infection model (utilizing mammalian VSV) have shown that VSV infection is more potent in RNAi defective worms (*rde-1* and *rde-4* mutants) and vice versa. RNase protection assays to investigate whether RNAi is indeed induced upon viral infection revealed the presence of virus-specific siRNAs only in the virus-infected wild-type cells. Transposon duplication and insertion has long been known to create "junk DNA" and thus destabilize the genome. Experiments wherein worms with mutations in the RNAi pathway genes were shown to be incapable of silencing transposons lead to the idea that RNAi has evolved as a cellular defense mechanism that silences transposable elements in order to maintain genome stability. RNAi has also been shown to limit gene expression in a dosage-dependent manner in plants as well as fruit flies. In both of these instances gene expression is enhanced by adding extra copies of the particular gene but only up to a point. Beyond that point, gene expression (especially transcription) is drastically reduced by modifying proteins that help to package DNA.

Impact

Over the years, experiments in *Arabidopsis*, *C. elegans*, and *Drosophila* have shown that besides their usual functions such as maintaining genome stability (by blocking transposition), the proteins involved in the RNAi machinery also play a role in the development of the organism. Research has shown that these genomically encoded short, 21- to 28-nucleotide RNAs that regulate temporal development are in fact members of the microRNA (miRNA) family, whose members have been identified across species as diverse as plants, flies, worms, and humans. Although miRNAs, like the siRNAs, are derived by Dicer activity on the dsRNA precursor and are found to be associated with RISC and their target mRNAs, unlike siRNAs, miRNAs are single-stranded and seem capable of affecting previously unknown, post-transcriptional steps such as RNA splicing, mRNA localization, and RNA turnover. An in-depth understanding of the siRNA-miRNA kinship could provide the key to several technological advancements from understanding gene functions by specific gene silencing (akin to genetic knockouts) to exploiting the therapeutic potential of this innate cellular defense mechanism to treat diseases like macular degeneration, caused by overexpression of certain genes.

Sibani Sengupta, Ph.D.

Further Reading

Agarwal, Neema, et al. "RNA Interference: Biology, Mechanism, and Applications." *Microbiology and Molecular Biology Reviews*, December, 2003, pp. 657-685.

Montgomery, Mary K., SiQun Xu, and Andrew Fire. "RNA as a Target of Double-Stranded RNA Mediated Genetics Interference in *C. elegans.*" *Proceedings of the National Academy of Sciences* 95 (December, 1998): 15502-15507.

Novina, Carl D., and Philip Sharp. "The RNAi Revolution." *Nature*, 430 (July 8, 2004): 161-164.

Wilkins, Courtney, et al. "RNA Interference Is an Antiviral Defence Mechanism in *Caenorhabditis elegans.*" *Nature* 436 (August 18, 2005): 1044-1047.

Web Sites of Interest

RNA Interference
http://www.hhmi.org/biointeractive/rna/rnai/index.html

RNAi Technique Animation Tutorial from Nature
http://www.nature.com/focus/rnai/animations/animation/animation.htm

See also: RNA isolation; RNA structure and function; RNA transcription and mRNA processing; RNA world.

RNA isolation

CATEGORY: Molecular genetics

SIGNIFICANCE: All cells in an organism or population of organisms of the same species contain the same (or nearly the same) set of genes. Therefore, understanding which genes are expressed under different conditions is critical to answering many questions in biology, including how cells differentiate into tissues, how cells respond to different environments, and which genes are expressed in tumor cells. The starting point for answering those questions is RNA isolation.

Key terms

cDNA library: a set of copies, or clones, of all or nearly all mRNA molecules produced by cells of an organism

complementary DNA (cDNA): also called copy DNA, DNA that copies RNA molecules, made using the enzyme reverse transcriptase

microarray analysis: a method, requiring isolated RNA, that allows simultaneous determination of which of thousands of genes are transcribed (expressed) in cells

reverse transcription polymerase chain reaction (RT-PCR): a technique, requiring isolated RNA, for quickly determining if a gene or a small set of genes are transcribed in a population of cells

RNA: ribonucleic acid, the macromolecule in the cell that acts as an intermediary between the genetic information stored as DNA and the manifestation of that genetic information as proteins

RNases: ribonucleases, or cellular enzymes that catalyze the breakdown of RNA

Cell Lysis

RNA isolation is a difficult proposition. RNA has a short life span in cells (as short as minutes in bacteria), and it is somewhat chemically unstable. In addition, enzymes that degrade RNA (RNases) are widespread in the environment, further complicating the task of separating intact RNA from other molecules in the cell.

The first step in RNA isolation is rapidly breaking open cells under conditions where RNA will not be degraded. One method involves freezing cells immediately in liquid nitrogen, then grinding the cells in liquid nitrogen in order to prevent any RNA degradation. Other methods involve lysing cells in the presence of strong protein denaturants so that any RNases present in the cell or the environment will be rapidly inactivated. The difficulty of the cell lysis step depends substantially on the type of cell involved. Bacterial and fungal cells are typically much more difficult to break open than cells from mammals. As a consequence, it is often more difficult to isolate intact RNA from bacteria and fungi.

Protein Denaturation and Further Purification

The next step in RNA isolation is to denature all proteins from the cell, to ensure that RNases will be inactive. In many cases, this is done at the same time as cell lysis. RNases are among the most resilient enzymes known, capable of being boiled or even autoclaved, yet retaining the ability to cleave RNA once they cool down. Consequently, the RNA next needs to be separated from RNases and other proteins to ensure that it will remain intact.

The separation of RNA from the rest of the macromolecules in the cell can be accomplished in a number of ways. One of the older methods for purifying RNA uses ultracentrifugation in very dense cesium chloride solutions. During high-speed centrifugation, these solutions create a gradient, with the greatest density at the bottom of the tube. RNA is the densest macromolecule in the cell, so it forms a pellet in the bottom of the ultracentrifuge tube. A newer technique for RNA purification involves the use of columns that bind RNA but not other macromolecules. The columns are washed to remove impurities, such as DNA and proteins, and then the RNA is eluted from the column matrix. A newer technique is based on the observation that, at an appropriate pH (level of acidity), RNA partitions into the water phase of a water-organic mixture. DNA and proteins either are retained at the boundary of the water-organic mixture or are dissolved in the organic phase.

Once the RNA is isolated, it needs to be handled carefully to ensure that it will not be degraded. Normally this involves resuspending the RNA in purified water, adding an alcohol solution, and storing it at −70 or −80 degrees Celsius (−94 or −112 degrees Fahrenheit). The purified RNA can then be used in a variety of techniques that help determine which genes are being transcribed in particular cells or tissues. These techniques include RT-PCR, Northern hybridization, microarray analysis, and the construction of cDNA libraries.

Special RNA Isolation Procedures

In some cases, a geneticist wants to isolate only RNA from the cytoplasm of the cell, since RNA from the nucleus may be more heterogeneous. In this case, cells are lysed using a gentle detergent that disrupts the cytoplasmic membrane, without disturbing the nuclear membrane. Centrifugation is used to separate the nuclei from the cytoplasm and then the cytoplasmic RNA is further purified as described above.

For some procedures, such as RT-PCR, the RNA sometimes needs to be further purified to ensure that no contaminating DNA is present. In this case, the RNA sample may be treated with the enzyme DNase I, which destroys DNA but leaves RNA intact.

For other procedures, like cDNA library construction, the RNA is often purified to remove ribosomal RNA (rRNA), transfer RNA (tRNA), and other stable RNAs, since the majority of RNA in the cell (typically more than 90 percent) is rRNA and tRNA. In this case, the RNA solution is treated by incubating it with single-stranded DNA containing a chain of eighteen to twenty thymine nucleotides, either on a column or in solution. Messenger RNA (mRNA) from eukaryotes contains runs of twenty to two hundred adenine nucleotides that bind to the single-stranded DNA and allow the mRNA to be purified away from the stable RNAs.

Like most techniques in genetics, RNA isolation methods have improved greatly over the years. With advances in methods for studying gene expression such as microarray analysis, isolating intact RNA is a technique that is more critical than ever in the genetics laboratory.

Patrick G. Guilfoile, Ph.D.

Further Reading

Ausubel, Fredrick, et al. *Current Protocols in Molecular Biology.* Hoboken, N.J.: John Wiley and Sons, 1998. Includes RNA isolation protocols from several different laboratories.

Avison, Matthew B. "Isolation and Analysis of RNA." In *Measuring Gene Expression.* New York: Taylor & Francis, 2007. Discusses the properties of different types of RNA and various methods of isolating, purifying, and stabilizing RNA. Describes six protocols for isolating RNA from animal cells and tissues, bacterial and yeast cells, plant and filamentous fungal cells, and tissue culture cells.

Clark, David P. "Nucleic Acids: Isolation, Purification, Detection, and Hybridization." In *Molecular Biology*. Boston: Elsevier Academic Press, 2005. Includes information about RNA isolation.

Farrell, Robert E., Jr. *RNA Methodologies: A Laboratory Guide for Isolation and Characterization*. 3d ed. Boston: Elsevier/Academic Press, 2005. Probably the definitive book on RNA techniques, including RNA isolation. Includes a substantial amount of background information, as well as detailed protocols.

Liu, Dongyou, ed. *Handbook of Nucleic Acid Purification*. Boca Raton, Fla.: CRC Press, 2009. Describes various nucleic acid isolation methods, with sections focusing on techniques to analyze viruses, bacteria, fungi, parasites, insects, mammals, and plants.

O'Connell, Joe, ed. *RT-PCR Protocols*. Totowa, N.J.: Humana Press, 2002. Collects several papers on the use of reverse transcription polymerase chain reaction in analysis of mRNA, quantitative methodologies, detection of RNA viruses, genetic analysis, and immunology. Tables, charts, index.

Sambrook, J., and D. W. Russell, eds. *Molecular Cloning: A Laboratory Manual*. 3d ed. Cold Spring Harbor, N.Y.: Cold Spring Harbor Laboratory Press, 2000. One of the most popular guides to molecular biology protocols. Includes several RNA isolation procedures.

WEB SITE OF INTEREST

Ambion/Applied Biosystems, The Basics: RNA Isolation
http://www.ambion.com/techlib/basics/rnaisol/index.html

Ambion, a division of Applied Biosystems, manufactures products for scientists conducting RNA research. Its Web site provides an illustrated description of the RNA isolation process with links to an accompanying article, "Cell Disruption: Getting the RNA Out."

See also: cDNA libraries; DNA isolation; DNA structure and function; Polymerase chain reaction; Reverse transcriptase; RNA structure and function; RNA transcription and mRNA processing.

RNA structure and function

CATEGORY: Molecular genetics

SIGNIFICANCE: Ribonucleic acid (RNA), a molecule that plays many roles in the storage and transmission of genetic information, exists in several forms, each with its own unique function. RNA acts as the messenger between genes in the DNA and their protein product, directing the assembly of proteins. RNA is also an integral part of ribosomes, the site of protein synthesis, and some RNAs have been shown to have catalytic properties. Understanding the structure and function of RNA is important to a fundamental knowledge of genetics; in addition, many developing medical therapies will undoubtedly utilize special RNAs to combat genetic diseases.

KEY TERMS

messenger RNA: a type of RNA that carries genetic instructions, copied from genes in DNA, to the ribosome to be decoded during translation

retrovirus: a special type of virus that carries its genetic information as RNA and converts it into DNA that integrates into the cells of the virus's host organism

ribosomal RNA: a type of RNA that forms a major part of the structure of the ribosome

ribosomes: organelles that function in protein synthesis and are made up of a large and a small subunit composed of proteins and ribosomal RNA (rRNA) molecules

ribozyme: an RNA molecule that can function catalytically as an enzyme

RNAi: abbreviation for "interference RNA," which hampers mRNA translation

small RNA: a class of RNA in which several subspecies are only twenty to twenty-five base pairs long and involved in the degradation or regulation of mRNA

transcription: the synthesis of an RNA molecule directed by RNA polymerase using a DNA template

transfer RNA: a form of RNA that acts to decode genetic information present in mRNA, carries a particular amino acid, and is vital to translation

translation: the synthesis of a protein molecule directed by the ribosome using information provided by an mRNA

The Chemical Nature of RNA

Ribonucleic acid (RNA) is a complex biological molecule that is classified along with DNA as a nucleic acid. Chemically, RNA is a polymer (long chain) consisting of subunits called ribonucleotides linked together by phosphodiester bonds. Each ribonucleotide consists of three parts: the sugar ribose (a five-carbon simple sugar), a negatively charged phosphate group, and a nitrogen-containing base. There are four types of ribonucleotides, and the differences among them lie solely in which of four possible bases each contains. The four bases are adenine (A), guanine (G), cytosine (C), and uracil (U).

The structures of DNA and RNA are very similar, with the following differences. The sugar found in the nucleotide subunits of DNA is deoxyribose, which differs slightly from the ribose found in the ribonucleotides of RNA. In addition, while DNA nucleotides also contain four possible bases, there is no uracil in DNA; instead, DNA nucleotides contain a different base called thymine (T). Finally, while DNA exists as a double-stranded helix in nature, RNA is almost always single-stranded. Like DNA, a single RNA strand has a 5′-to-3′ polarity. These numbers are based on which carbon atom is exposed at the end of the polymer, each of the carbon atoms being numbered around the sugar molecule.

The Folding of RNA Molecules

The function of an RNA molecule is determined by its nucleotide sequence, which represents information derived from DNA. This nucleotide sequence is called the primary structure of the molecule. Many RNAs also have an important secondary structure, a three-dimensional shape that is also important for the function of the molecule. The secondary structure is determined by hydrogen bonding between parts of the RNA molecule that are complementary. Complementary pairing is always between A and U ribonucleotides and C and G ribonucleotides. Hydrogen bonding results in double-stranded regions in the secondary structure.

Since RNA is single-stranded, it was recognized shortly after the discovery of some of its major roles that its capacity for folding is great and that this folding might play an important part in the functioning of the molecule. Base pairing often represents local interactions, and a common structural element is a "hairpin loop" or "stem loop." A hairpin loop is formed when two complementary regions are separated by a short stretch of bases so that when they fold back and pair, some bases are left unpaired, forming the loop. The net sum of these local interactions is referred to as the RNA's secondary structure and is usually important to an understanding of how the RNA works. All transfer RNAs (tRNAs), for example, are folded into a secondary structure that contains three stem loops and a fourth stem without a loop, a structure resembling a cloverleaf in two dimensions.

Finally, local structural elements may interact with other elements in long-range interactions, causing more complicated folding of the molecule. The full three-dimensional structure of a tRNA molecule from yeast was finally confirmed in 1978 by several groups independently, using X-ray diffraction. In this process, crystals of a molecule are bombarded with X rays, which causes them to scatter; an expert can tell by the pattern of scattering how the different atoms in the molecule are oriented with respect to one another. The cloverleaf arrangement of a tRNA undergoes further folding so that the entire molecule takes on a roughly *L*-shaped appearance in three dimensions. An understanding of the three-dimensional shape of an RNA molecule is crucial to understanding its function. By the late 1990's, the three-dimensional structures of many tRNAs had been worked out, but it had proven difficult to do X-ray diffraction analyses on most other RNAs because of technical problems. More advanced computer programs and alternate structure-determining techniques like mass spectroscopy, nuclear magnetic resonance and cryo electron microscopy are enabling research in this field to proceed, with the RNA structures catalogued online in the Nucleic Acid Database and Protein Data Bank.

Synthesis and Stability of RNA

RNA molecules of all types are continually being synthesized and degraded in a cell; even the longest-lasting ones exist for only a day or two. Shortly after the structure of DNA was established, it became clear that RNA was synthesized using a DNA molecule as a template, and the mechanism was worked out shortly thereafter. The entire process by which an RNA molecule is constructed using the information in DNA is called transcription. An enzyme called RNA polymerase is responsible for assembling the ribonucleotides of a new RNA complementary to a specific DNA segment (gene). Only one strand of the DNA is used as a template (the sense strand), and the ribonucleotides are initially arranged according to the base-pairing rules. A DNA sequence called the "promoter" is a site RNA polymerase can bind initially and allows the process of RNA synthesis to begin. At the appropriate starting site, RNA polymerase begins to assemble and connect the nucleotides according to the complementary pairing rules, such that for every A nucleotide in the DNA, RNA polymerase incorporates a U ribonucleotide into the RNA being assembled. The remaining pairing rules stipulate that a T in DNA denotes an A in RNA and that a C in DNA represents a G in RNA (and vice versa). This process continues until another sequence, called a "terminator," is reached. At this point, the RNA polymerase stops transcription, and a new RNA molecule is released.

Much attention is rightfully focused on transcription, since it controls the rate of synthesis of each RNA. It has become increasingly clear, however, that the amount of RNA in the cell at a given time is also strongly dependent on RNA stability (the rate at which it is degraded). Every cell contains several enzymes called ribonucleases (RNases) whose job it is to cut up RNA molecules into their ribonucleotides subunits. Some RNAs last only thirty seconds, while others may last up to a day or two. It is important to remember that both the rates of synthesis (transcription) and degradation ultimately determine the amount of functional RNA in a cell at any given time.

Three Classes of RNA

While all RNAs are produced by transcription, several classes of RNA are created, and each has a unique function. By the late 1960's, three major classes of RNAs had been identified, and their respective roles in the process of protein synthesis had been identified. In general, protein synthesis refers to the assembly of a protein using information encoded in DNA, with RNA acting as an intermediary to carry information and assist in protein building. In 1956, Francis Crick, one of the scientists who had discovered the double-helical structure of DNA, referred to this information flow as the "central dogma," a term that continues to be used, although exceptions to it are now known.

A messenger RNA (mRNA) carries a complementary copy of the DNA instructions for building a particular protein. Making up about 5 percent of the three RNA classes, in eukaryotes mRNA typically represents the information from a single gene and carries the information to a ribosome, the site of protein synthesis. The information must be decoded to make a protein. Nucleotides are read in groups of three (called codons). In addition, mRNAs contain signals that tell a ribosome where to start and stop translating.

Ribosomal RNA (rRNA) is part of the structure of the ribosome and makes up about 80 percent of the total RNA in a cell. Four different rRNAs interact with many proteins to form functional ribosomes that direct the events of protein synthesis. One of the rRNAs interacts with mRNA to orient it properly so translation can begin at the correct location. Another rRNA acts to facilitate the transfer of the growing polypeptide from one tRNA to another (peptidyl transferase activity).

Transfer RNA (tRNA) accounts for 15 percent of the three RNA categories and serves the vital role of decoding the genetic information. There are at least twenty and usually more than forty different tRNAs in a cell. On one side, tRNAs contain an "anticodon" loop, which can base-pair with mRNA codons according to their sequence and the base-pairing rules. On the other side, each contains an amino acid binding site, with the appropriate amino acid for its anticodon. In this way, tRNAs recognize the codons and supply the appropriate amino acids. The process continues until an entire new polypeptide has been constructed.

The attachment of the correct amino acids is facilitated by a group of enzymes called tRNA amino acyl synthetases. Each type of tRNA has a corresponding synthetase that facilitates the attachment of the correct amino acid to the amino acid binding site. The integrity of this process is crucial to translation; if only one tRNA is attached to an incorrect amino acid, the resulting proteins will likely be nonfunctional.

Split Genes and mRNA Processing in Eukaryotes

In bacterial genes, there is colinearity between the segment of a DNA molecule that is transcribed and the resulting mRNA. In other words, the mRNA sequence is complementary to its template and is the same length, as would be expected. In the late 1970's, several groups of scientists made a seemingly bizarre discovery regarding mRNAs in eukaryotes (organisms whose cells contain a nucleus, including all living things that are not bacteria): the sequences of mRNAs isolated from eukaryotes were not collinear with the DNA from which they were transcribed. The coding regions of the corresponding DNA were interrupted by seemingly random sequences that served no apparent function. These "introns," as they came to be known, were apparently transcribed along with the coding regions (exons) but were somehow removed before the mRNA was translated. This completely unexpected observation led to further investigations that revealed that mRNA is extensively processed, or modified, after its transcription in eukaryotes.

After a eukaryotic mRNA is transcribed, it contains several introns and is referred to as immature, or a "pre-mRNA." Before it can become mature and functional, three major processing events must occur: splicing, the addition of a 5′ cap, and a "tail." The process of splicing is complex and occurs in the nucleus with the aid of "spliceosomes," large complexes of RNAs and proteins that identify intervening sequences and cut them out of the pre-mRNA. In addition, spliceosomes rejoin the exons to produce a complete, functional mRNA. Splicing must be extremely specific, since a mistake causing the removal of even one extra nucleotide could change the final protein, making it nonfunctional. During splicing, capping and the addition of a poly-A tail take place. A so-called cap, which consists of a modified G nucleotide, is added to the beginning (5′ end) of the pre-mRNA by an unconventional linkage. The cap appears to function by interacting with the ribosome, helping to orient the mature mRNA so that translation begins at the proper end. A tail, which consists of many A nucleotides (often two hundred or more), is attached to the 3′ end of the pre-mRNA. This so-called poly-A tail, which virtually all eukaryotic mRNAs contain, seems to be one factor in determining the relative stability of an mRNA. These important steps must be performed after transcription in eukaryotes to produce a functional mRNA.

Ribozymes

The traditional roles of RNA in protein synthesis were originally considered its only roles. RNA in general, while considered an important molecule, was thought of as a "helper" in translation. This all began to change in 1982, when the molecular biologists Thomas Cech and Sidney Altman, working independently and with different systems, reported the existence of RNA molecules that had catalytic activity. This means that RNA molecules can function as enzymes; until this time, it was believed that all enzymes were protein molecules. The importance of these findings cannot be overstated, and Cech and Altman ultimately shared the 1989 Nobel Prize in Chemistry for the discovery of these RNA enzymes, or "ribozymes." Both of these initial ribozymes catalyzed reactions that involved the cleavage of other RNA molecules, which is to say they acted as nucleases. Subsequently, many ribozymes have been found in various organisms, from bacteria to humans. Some of them are able to catalyze different types of reactions, and there are new ones reported every year. Thus ribozymes are not a mere curiosity but play an integral role in the molecular machinery of many organisms. Their discovery also gave rise to the idea that at one point in evolutionary history, molecular systems composed solely of RNA, performing many roles, existed in an "RNA world."

RNAi Pathway and Small RNAs

At around the same time as these momentous discoveries, still other classes of RNAs were being discovered, each with its own specialized functions. In 1981, Jun-ichi Tomizawa discovered interference RNA whereby RNA has the ability to interfere with protein production. This hindering RNA is known as "antisense RNA" because it is the opposite complement of, and thus can bind to, the "sense" strand of protein-coding mRNA, resulting in the prevention or regulation of the mRNA's translation.

In 1990, the power of antisense RNA showed itself in the form of an unexpected white flower. Researcher Richard Jorgensen was hoping to intensify the color of petunias with the introduction of a transgene to overexpress the pigmentation enzyme when instead he produced a white flower. This tipped off scientists to the existence of the post-transcriptional gene silencing (PTGS) cellular mechanism and the idea that adding genes to an organism can affect its phenotype.

In 1998, Craig Mello and Andrew Fire uncovered another significant use of interference RNA, for which they were awarded the 2006 Nobel Prize in Physiology or Medicine. The researchers noticed that the addition of mRNA and antisense RNA did not turn off their targeted gene in *Caenorhabditis elegans* roundworms as effectively as an injection of double-stranded RNA (dsRNA). Putting the current name to this interference phenomenon, RNAi, the duo had uncovered a crucial piece of a cell's regulation capability and immune response known as the RNAi pathway. When dsRNA enters the cell, the Dicer enzyme chops it into small nucleotides of 20 to 25 bases. These small pieces, known as small interfering RNA (siRNA), are separated and fed into the RNA-induced silencing complex (RISC) where they are bound to their complementary mRNA for the pairs' dismantled demise.

Much excitement and attention toward these protein-regulating small RNAs has unveiled numerous different subspecies and the varying methods in which they are initiated within a cell. siRNA are generated in the cytoplasm from dsRNA and match almost completely to their doomed mRNA. Another small RNA, microRNA (miRNA), is also termed "noncoding RNA" because it initiates from the cell's own gene, but is not created for the traditional purpose of protein-coding, but instead with the intent to restrain protein production. Once made, miRNA gets enzymatically processed before being sent into the cytoplasm where it is transported into the RISC mechanism. In humans, miRNA tend to only moderately complement their targeted mRNA, resulting in a partial restriction of the mRNA's protein translation. This class of small RNA is practically absent in cancerous cells, and a recent study showed that its addition can profoundly reverse cancer cell growth.

This power to subdue out-of-control cells has revolutionized the search for therapeutic treatments of several human diseases and disorders. By delivering synthetic small RNA to trigger the RNAi pathway or fulfill other policing roles, researchers hope to silence genes that cause not only cancer but also neurodegenerative diseases, diabetes, asthma, and even infectious diseases such as hepatitis, influenza, and HIV. The act of delivering these nuggets of authority is easier said than done; nucleases and immune responses are always at the ready to pounce on these foreign nucleotides. Studies are being conducted to fuse them to delivery vehicles like lipids, antibodies, or polymers in order to transport small RNAs across the cellular membrane.

Other Important Classes of RNA and Specialized Functions

Another major class of RNAs, the small nuclear RNAs (snRNAs), was also discovered in the early 1980's. Molecular biologist Joan Steitz was working on the autoimmune disease systemic lupus when she began to characterize the snRNAs. There are six different snRNAs, now called U1-U6 RNAs. These RNAs exist in the nucleus of eukaryotic cells and play a vital role in mRNA splicing. They associate with proteins in the spliceosome, forming so-called ribonucleoprotein complexes (snRNPs, pronounced "snurps"), and play a prominent role in detecting proper splice sites and directing the protein enzymes to cut and paste at the proper locations.

It has been known since the late 1950's that many viruses contain RNA, and not DNA, as their genetic material. This is another fascinating role for RNA. The viruses that cause influenza, polio, and a host of other diseases are RNA viruses. Of particular note are a class of RNA viruses known as retroviruses. Retroviruses, which include human immunodeficiency virus (HIV), the virus that causes acquired immunodeficiency syndrome (AIDS) in humans, use a special enzyme called reverse transcriptase to make a DNA copy of their RNA when they enter a cell. The DNA copy is inserted into the DNA of the host cell, where it is referred to as a "provirus," and never leaves. This discovery represents one of the exceptions to the central dogma. In the central dogma, RNA is always made from DNA, and retroviruses have reversed this flow of information. Clearly, understanding the structures and functions of the RNAs associated with these viruses will be important in attempting to create effective treatments for the diseases associated with them.

An additional role of RNA was noted during the elucidation of the mechanism of DNA replication. It was found that a small piece of RNA, called a "primer," must be laid down by the enzyme primase, an RNA polymerase, before DNA polymerase can begin. RNA primers are later removed and replaced with DNA. Also, it is worth mentioning that the universal energy-storing molecule of all cells, adenosine triphosphate (ATP), is in fact a version of the RNA nucleotide containing adenine (A).

Impact and Applications

The discovery of the many functions of RNA, especially its catalytic ability, has radically changed the understanding of the functioning of genetic and biological systems and has revolutionized the views of the scientific community regarding the origin of life. The key to understanding how RNA can perform all of its diverse functions lies in elucidating its many structures, since structure and function are inseparable. Much progress has been made in establishing the three-dimensional structure of hundreds of RNA molecules.

Extensive research has been conducted on RNA folding, degradation, and regulation, as it has become clear that it plays a vital role in genetic disease, cancer, and retroviral infections. The revelation of the RNAi pathway has ignited hope for treatments of ailments ranging from AIDS to Parkinson disease. One of the first RNAi remedies, Macugen, to pass human trials involved the eye disease wet macular degeneration. This occurs when too many blood vessels are grown due to an overactive gene, and the vessels' leakiness damages vision. The eye has a low population of nucleases, so when naked siRNAs are directly injected into the vitreous cavity, they successfully stunt the overactive gene, and vision is restored.

Additionally, plants, bacteria, and animals have been genetically engineered to alter the expression of some of their genes, in many cases making use of the new RNA technology. An example is a genetically engineered tomato that does not ripen until it is treated at the point of sale. This tomato was created by inserting an antisense RNA gene; when it is expressed, it inactivates the mRNA that codes for the enzyme involved in production of the ripening hormone.

One thing is clear: RNA is one of the most structurally interesting and functionally diverse of all the biological molecules.

Matthew M. Schmidt, Ph.D., and Bryan Ness, Ph.D.; updated by Cherie Dewar

FURTHER READING

Eckstein, Fritz, and David M. J. Lilley, eds. *Catalytic RNA*. New York: Springer, 1996. Offers a comprehensive overview of ribozyme diversity and function. Illustrations (some color), bibliography, index.

Erickson, Robert P., and Jonathan G. Izant, eds. *Gene Regulation: Biology of Antisense RNA and DNA*. New York: Raven Press, 1992. Provides both a comprehensive overview of natural antisense RNA function and prospects for its uses in gene therapy. Illustrations, bibliography, index.

Murray, James A. H., ed. *Antisense RNA and DNA*. New York: Wiley-Liss, 1992. Presents experimental approaches. Illustrations, bibliography, index.

Nelson, David L., and Michael M. Cox. *Lehninger Principles of Biochemistry*. 5th ed. New York: W. H. Freeman, 2009. An updated college textbook that includes the current status of RNA research. A companion Web site, www.whfreeman.com/lehninger, has interactive and narrated graphics of biochemical mechanisms.

Shrivastava, Neeta, and Anshu Srivastava. "RNA Interference: An Emerging Generation of Biologicals." *Biotechnology Journal*, March, 2008, 339-353. Reviews the small RNA components of the RNAi pathway, and how this discovery can be used to treat human diseases and disorders.

Simons, Robert W., and Marianne Grunberg-Manago, eds. *RNA Structure and Function*. Cold Spring Harbor, N.Y.: Cold Spring Harbor Laboratory Press, 1997. An advanced text that takes a detailed look at the various structures of RNA, their relationships to function, and the techniques for determining RNA structure. Illustrations, bibliography, index.

Watson, James D., et al. *Molecular Biology of the Gene*. 5th ed. Menlo Park, Calif.: Benjamin Cummings, 2003. Discusses RNA structures and their relationship to function. Illustrations, bibliography, index.

WEB SITES OF INTEREST

Nucleic Acid Database: A Repository of Three-Dimensional Structural Information for Nucleic Acids
http://ndbserver.rutgers.edu/index.html

Protein Data Bank
http://www.rcsb.org/pdb/home/home.do

RNA World
http://www.imb-jena.de/RNA.html

Silencing Genome Laboratory Experiments
http://www.silencinggenomes.org

University of Nebraska's Antisense and Tomato Ripening Slideshow
http://agbiosafety.unl.edu/flash/antisense.swf

See also: cDNA libraries; DNA isolation; Polymerase chain reaction; Reverse transcriptase; RNA isolation; RNA structure and function; RNA transcription and mRNA processing.

RNA transcription and mRNA processing

CATEGORY: Molecular genetics

SIGNIFICANCE: Translation of messenger RNA molecules (mRNAs) occurs even while transcription is taking place in prokaryotes. In eukaryotes the process is much more complex, with transcription occurring in the nucleus, followed by multiple processing steps before a mature mRNA is ready to be translated. All of these extra steps are required for mRNAs to be transported out of the nucleus and for recognition by ribosomes in the cytoplasm.

KEY TERMS

messenger RNA (mRNA): the form of RNA that contains the coding instructions used to make a polypeptide by ribosomes

RNA polymerase: the enzyme that transcribes RNA using a strand of DNA as a template

transcription: the process that converts DNA code into a complementary strand of RNA (mRNA) containing code that can be interpreted by ribosomes

translation: the process, mediated by ribosomes, in which the genetic code in an mRNA is used to produce a polypeptide, the ultimate product of structural genes

RNA POLYMERASE

Transcription is the process whereby the directions for making a protein are converted from DNA-based instructions to RNA-based instructions. This step is required in the process of expressing a gene as a polypeptide, because ribosomes, which assemble polypeptides, can read only RNA-based messages. Although transcription is complicated and involves dozens of enzymes and proteins, it is much simpler in prokaryotes than in eukaryotes. Because prokaryotes lack a nucleus, transcription and translation are linked processes both occurring in the cytoplasm. In eukaryotes, transcription and translation occur as completely separate processes, transcription occurring in the nucleus and translation occurring in the cytoplasm. (It is now known that some translation also occurs in the nucleus, but apparently only a small amount, probably less than 10 percent of the translation occurring in a cell.)

In eukaryotes there are three different types of RNA polymerase that transcribe RNA using a strand of DNA as a template (there is a single type of RNA polymerase in prokaryotes). Two of them, called RNA polymerase I (pol I) and RNA polymerase III (pol III), specialize in transcribing types of RNA that are functional products themselves, such as ribosomal RNA (rRNA) and transfer RNA (tRNA). These RNAs are involved in translation. RNA polymerase II (pol II) transcribes RNA from structural genes, that is, genes that code for polypeptides. Pol II therefore is the primary RNA polymerase and the one that will be the focus of this article when discussing transcription in eukaryotes.

TRANSCRIPTION IN PROKARYOTES

The first step in transcription is for RNA polymerase to identify the location of a gene. In prokaryotes many genes are clustered together in functional groups called operons. For example, the lactose (*lac*) operon contains three genes, each coding for one of the enzymes needed to metabolize the sugar lactose. At the beginning of each operon are two control sequences, the operator and the promoter. The promoter is where RNA polymerase binds, in preparation for transcription. The operator is a control region that determines whether RNA polymerase will be able to bind to the promoter. The operator interacts with other proteins that determine when the associated operon should be expressed. They do this by either preventing RNA polymerase from binding to the promoter or by assisting it to bind.

RNA polymerase recognizes promoters by the specific base-pair sequences they contain. Assuming all conditions are correct, RNA polymerase binds to the promoter, along with another protein called the sigma factor (σ). The beginning of genes are detected with the aid of σ. Transcription begins at a leader sequence a little before the beginning of the first gene and continues until RNA polymerase reaches a termination signal. If the operon contains more than one gene, all of the genes are transcribed into a single long mRNA, each gene separated from its neighbors by a spacer region. The mRNA is put together by pairing ribonucleotides with their complementary nucleotides in the DNA template. In place of thymine (T), RNA uses uracil (U); otherwise the same bases are present in RNA and DNA, the others being adenine (A), guanine

(G), and cytosine (C). The pairing relationships are as follows, the DNA base listed first in each pair: A-U, T-A, G-C, and C-G.

RNA polymerase catalyzes the joining of ribonucleotides as they pair with the DNA template. Each mRNA is constructed beginning at the 5′ end (the phosphate end) and ending with the 3′ end (the hydroxyl end). Even while transcription is taking place, ribosomes begin binding to the mRNA to begin translation. As soon as RNA polymerase has completed transcribing the genes of an operon, it releases from the DNA and soon binds to another promoter to begin the process all over again.

Transcription in Eukaryotes

Transcription in eukaryotes differs from the process in prokaryotes in the following major ways: (1) Genes are transcribed individually instead of in groups; (2) DNA is complexed with many proteins and is highly compacted, and therefore must be "unwound" to expose its promoters; (3) transcription occurs in a separate compartment (the nucleus) from translation, most of which occurs in the cytoplasm; and (4) initially transcription results in a pre-messenger RNA (pre-mRNA) molecule that must be processed before it emerges as a mature mRNA ready for translation. Additionally, mRNAs are much longer-lived in eukaryotes.

The first step in transcription is for RNA polymerase to find a gene that needs to be transcribed. Only genes occurring in regions of the DNA that have been unwound are prepared for potential transcription. RNA polymerase binds to an available promoter, which is located just before a gene and has a region in it called the TATA box (all promoters have the consensus sequence TATAAAA in them). RNA polymerase is unable to bind to the promoter without assistance from more than a dozen other proteins, including a TATA-binding protein, several transcription factors, activators, and coactivators. There are other DNA sequences farther upstream than the promoter that control transcription too, thus accounting for the fact that some genes are transcribed more readily, and therefore more often, than others.

Once RNA polymerase has bound to the promoter, it begins assembling an RNA molecule complementary to the DNA code in the gene. It starts by making a short leader sequence, then transcribes the gene, and finishes after transcribing a short trailer sequence. Transcription ends when RNA polymerase reaches a termination signal in the DNA. The initial product is a pre-mRNA molecule that is much longer than the mature mRNA will be.

mRNA Processing in Eukaryotes

Pre-mRNAs must be processed before they can leave the nucleus and be translated at a ribosome. Three separate series of reactions play a part in producing a mature mRNA: (1) intron removal and exon splicing, (2) 5′ capping, and (3) addition of a poly-A tail. Not all transcripts require all three modifications, but most do.

The reason pre-mRNAs are much longer than their respective mature mRNAs has to do with the structure of genes in the DNA. The coding sequences of almost all eukaryotic genes are interrupted with noncoding regions. The noncoding regions are called introns, because they represent "intervening" sequences, and the coding regions are called exons. For an mRNA to be mature it must have all the introns removed and all the exons spliced together into one unbroken message. Special RNA/protein complexes called small nuclear ribonucleoprotein particles, or snRNPs (pronounced as "snurps" by geneticists), carry out this process. The RNAs in the snRNPs are called small nuclear RNAs or snRNAs. Several snRNPs grouped together form a functional splicing unit called a spliceosome. Spliceosomes are able to recognize short signal sequences in pre-mRNA molecules that identify the boundaries of introns and exons. When a spliceosome has found an intron, it binds correctly, and through formation of a lariat-shaped structure, it cuts the intron out and splices the exons that were on each side of the intron to each other. Genes may have just a few introns, or they may have a dozen or more. Why eukaryotes have introns at all is still an open question, as introns, in general, appear to have no function.

While intron removal and exon splicing are taking place, both ends of maturing mRNAs must also be modified. At the 5′ end (the end with an exposed phosphate) an enzyme adds a modified guanosine nucleotide called 7-methylguanosine. This special nucleotide is added so that ribosomes in the cytoplasm can recognize the correct end of mRNAs, and it probably also prevents the 5′ end of mRNAs from being degraded.

At the 3′ end of maturing mRNAs, another enzyme, called polyadenylase, adds a string of adenine nucleotides. Polyadenylase actually recognizes a special signal in the trailer sequence, at which it cuts and then adds the adenines. The result is what is called a poly-A tail. Initially geneticists did not understand the function of poly-A tails, but now it appears that they protect mRNAs from enzymes in the cytoplasm that could break them down. Essentially, poly-A tails are the main reason mRNAs in eukaryotes survive so much longer than mRNAs in prokaryotes.

Once the modifications have been completed, mRNAs are ready to be exported from the nucleus and will now travel through nuclear pores and enter the cytoplasm, where awaiting ribosomes will translate them, using the RNA code to build polypeptides.

Transcription and Disease

Ordinarily transcription works like a well-oiled machine, and only the right genes are transcribed at the right time so that just the right amount of protein product is produced. Unfortunately, due to the great complexity of the system, problems can occur that lead to disease. It has been estimated that about 15 percent of all genetic diseases may be due to improper intron removal and exon splicing in pre-mRNA molecules. Improper gene expression accounts for many other diseases, including many types of cancer.

Beta-thalassemia, a genetic disorder causing Cooley's anemia, is caused by a point mutation (a change in a single nucleotide) that changes a cutting and splicing signal. As a result, the mature mRNA has an extra piece of intron, making the

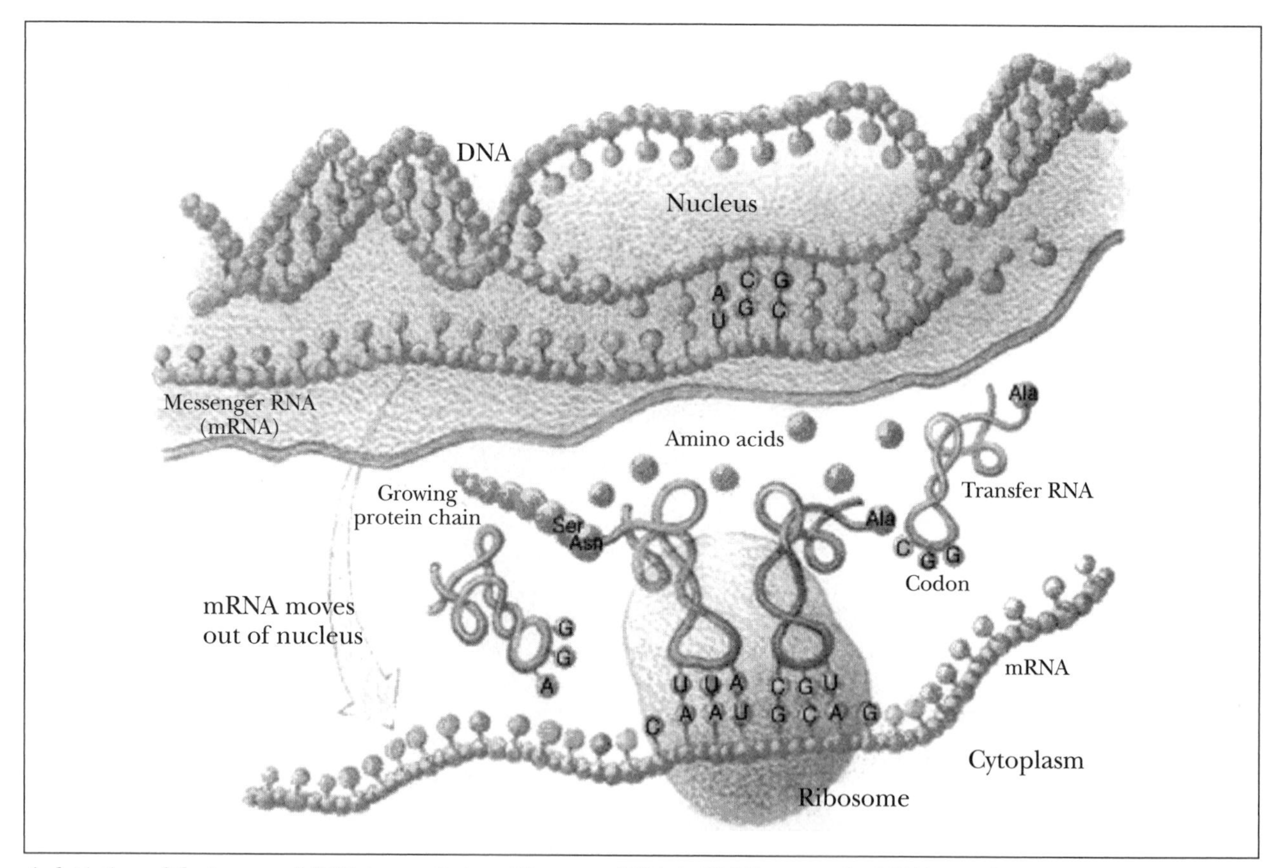

A depiction of the process of RNA transcription. Messenger RNA (mRNA) moves the DNA's template or instructions for protein synthesis (genetic code, or arrangement of bases) from the cell nucleus out into the cytoplasm, where it binds to a ribosome, the cell's "protein factory." Transfer RNA molecules then synthesize amino acids by linking to a codon on the mRNA and transferring the resulting amino acid to a growing chain of amino acids, the protein molecule. (U.S. Department of Energy Human Genome Program, http://www.ornl.gov/hgmis)

mRNA longer and causing a reading frame shift. A reading frame shift causes everything from the mutation forward to be skewed, so that the code no longer codes for the correct amino acids. Additionally, as in the case of Cooley's anemia, a reading frame shift often introduces a premature stop codon. The gene involved codes for the beta chain of hemoglobin, the protein that carries oxygen in the blood, and this mutation results in a shortened polypeptide that does not function properly.

A single point mutation in a splicing site can have even more far-reaching consequences. In 2000, researchers in Italy discovered an individual who was genetically male (having one X and one Y chromosome) but was phenotypically female. She had no uterus or ovaries and only superficial external female anatomy, making her a pseudohermaphrodite. This condition can be caused by defects either in androgen production or in the androgen receptor. In this case, the defect was a simple point mutation in the androgen receptor gene that led to one intron being retained in the mature mRNA. Within the intron was a stop codon, which meant when the mRNA was translated, a shorter, nonfunctional polypeptide was formed. The subject did show a very small response to androgen, so apparently some of the pre-mRNAs were being cut and spliced correctly, but not enough to produce the normal male phenotype.

The same kinds of mutations as those discussed above can lead to cancer, but mutations that change the level of transcription of proto-oncogenes can also lead to cancer. Proto-oncogenes are normal genes involved in regulating the cell cycle, and when these genes are overexpressed they become oncogenes (cancer-causing genes). Overexpression of proto-oncogenes leads to overexpression of other genes, because many proto-oncogenes are transcription factors, signal proteins that interact with molecules controlling intracellular growth and growth factors released by cells to stimulate other cells to divide.

Overexpression can occur when there is a mutation in one of the control regions upstream from a gene. For example, a mutation in the promoter sequence could cause a transcription factor, and thus RNA polymerase, to bind more easily, leading to higher transcription rates. Other control regions, such as enhancer sequences, often far removed from the gene itself, may also affect transcription rates.

Anything that causes the transcription process to go awry will typically have far-reaching consequences. Geneticists are just beginning to understand some of the underlying errors behind a host of genetic diseases, and it should be no surprise that some of them involve how genes are transcribed. Knowing what the problem is, unfortunately, does not usually point to workable solutions. When the primary problem is an excessive rate of transcription, specially designed antisense RNA molecules (RNA molecules that are complementary to mRNA molecules) might be designed that will bind to the overexpressed mRNAs and disable them. This approach is still being tested. In the case of point mutations that derail the cutting and splicing process, the only solution may be gene therapy, a technique still not considered technically possible and not expected to be feasible for some time to come.

Bryan Ness, Ph.D.

FURTHER READING

Courey, Albert J. "RNA Polymerases and the Transcription Cycle." In *Mechanisms in Transcriptional Regulation.* Malden, Mass.: Blackwell, 2008. Focuses on the fundamental concepts in transcription and its regulation, beginning with this chapter's discussion about RNA polymerases and the transcription cycle.

Hampsey, Michael. "Molecular Genetics of the RNA Polymerase II General Transcriptional Machinery." *Microbiology and Molecular Biology Reviews* 62, no. 2 (1998): 465-503. An overview of the role of RNA polymerase II.

Hartmann, Roland K., et al., eds. *Handbook of RNA Biochemistry.* Weinheim, Germany: Wiley-VCH, 2005. The numerous references to RNA transcription are listed in the index.

Latchman, David S. "RNA Polymerases and the Basal Transcription Complex." In *Eukaryotic Transcription Factors.* 5th ed. Boston: Elsevier/Academic Press, 2008. Provides an overview of the transcription process.

_______. "Transcription-Factor Mutations and Disease." *The New England Journal of Medicine* 334, no. 1 (January 4, 1996): 28-33. A general overview of the kinds of diseases caused by mutations in transcription factor genes. Includes an overview of potential treatments.

Macfarlane, W. M. "Transcription Demystified."

Journal of Clinical Pathology: Molecular Pathology 53 (February, 2000): 1-7. A general introduction with an emphasis on transcription factors and their role in transcription.

Ptashne, Mark, and Alexander Gann. *Genes and Signals.* Cold Spring Harbor, N.Y.: Cold Spring Harbor Laboratory Press, 2002. A nice overview of transcription and related topics, readable by advanced high school students and undergraduates.

Shatkin, Aaron J., and James L. Manley. "The Ends of the Affair: Capping and Polyadenylation." *Nature Structural Biology* 7, no. 10 (October, 2000): 838-842. Describes advances in understanding the enzymes and factors that participate in capping and polyadenylation, focusing on apparent evolutionary conservation.

White, Robert J. *Gene Transcription: Mechanisms and Control.* Malden, Mass.: Blackwell, 2001. An in-depth look at all aspects, including regulation, of transcription. Aimed mostly at upper undergraduate students, but begins with the basics.

WEB SITES OF INTEREST

Kimball's Biology Pages
http://users.rcn.com/jkimball.ma.ultranet/BiologyPages/T/Transcription.html#Gene_Transcription_DNA_to_RNA

John Kimball, a retired Harvard University biology professor, includes a page describing gene transcription and RNA processing in his online cell biology text.

Virtual Chembook, RNA Transcription
http://www.elmhurst.edu/~chm/vchembook/583rnatrans.html

The book, created by a professor at Elmhurst College, includes an illustrated explanation of RNA transcription and processing.

See also: Ancient DNA; Antisense RNA; Cancer; Chromosome structure; DNA isolation; DNA repair; DNA replication; DNA structure and function; Genetic code; Genetic code, cracking of; Molecular genetics; Noncoding RNA molecules; One gene-one enzyme hypothesis; Protein structure; Protein synthesis; Pseudohermaphrodites; Repetitive DNA; RNA isolation; RNA structure and function; RNA world.

RNA world

CATEGORY: Evolutionary biology; Molecular genetics

SIGNIFICANCE: The RNA world is a theoretical time in the early evolution of life, during which RNA molecules played important genetic and enzymatic roles that were later taken over by molecules of DNA and proteins. Ideas about RNA's ancient functions have led to new concepts of the origin of life and have important implications in the use of gene therapy to treat diseases.

KEY TERM

ribosomal RNA (rRNA): a type of RNA that forms a major part of the structure of the ribosome

ribosome: an organelle that functions in protein synthesis, composed of a large and a small subunit composed of proteins and ribosomal RNA molecules

ribozyme: an RNA molecule that can function catalytically as an enzyme

THE CENTRAL DOGMA AND THE MODERN GENETIC WORLD

Soon after the discovery of the double-helical structure of DNA in 1953 by James Watson and Francis Crick, Crick proposed an idea regarding information flow in cells that he called the "central dogma of molecular biology." Crick correctly predicted that in all cells, information flows from DNA to RNA to protein. DNA was known to be the genetic material, the "library" of genetic information, and it had been clear for some time that the enzymes that actually did the work of facilitating chemical reactions were invariably protein molecules. The discovery of three classes of RNA during the 1960's seemed to provide the link between the DNA instructions and the protein products.

In the modern genetic world, cells contain three classes of RNA that act as helpers in the synthesis of proteins from information stored in DNA, a process called translation. A messenger RNA (mRNA) is "transcribed" from a segment of DNA (a gene) that contains information about how to build a particular protein and carries that information to the cellular site of protein synthesis, the ribosome. Ribosomal RNAs (rRNAs) interacting with many proteins make up the ribosome, whose major job is to coordinate and facilitate the protein-building procedure. Transfer RNAs (tRNAs) act as decoding molecules, reading the mRNA information and correlating it with a specific amino acid. As the ribosome integrates the functions of all three types of RNA, polypeptides are built one amino acid at a time. These polypeptides, either singly or in aggregations, can then function as enzymes, ultimately determining the capabilities and properties of the cell in which they act.

While universally accepted, the central dogma led many scientists to question how this complex, integrated system came about. It seemed to be a classic "chicken and egg" dilemma: Proteins could not be built without instructions from DNA, but DNA could not replicate and maintain itself without help from protein enzymes. The two seemed mutually dependent upon each other in an inextricable way. An understanding of the origins of the modern genetic system seemed far away.

The Discovery of Ribozymes

In 1983, a discovery was made that seemed so radical it was initially rejected by most of the scientific community. Molecular biologists Thomas Cech and Sidney Altman, working independently and in different systems, announced the discovery of RNA molecules that possessed catalytic activity. This meant that RNA itself can function as an enzyme, obliterating the idea that only proteins could function catalytically.

Cech had been working with the protozoan *Tetrahymena*. In most organisms except bacteria, the coding portions of DNA genes (exons) are interrupted by noncoding sequences (introns), which are transcribed into mRNA but which must be removed before translation. Protein enzymes called nucleases are usually responsible for cutting out the introns and joining together the exons in a process called splicing. The molecule with which Cech was working was an rRNA that contained introns but could apparently remove them and rejoin the coding regions without any help. It was a self-splicing RNA molecule, which clearly indicated its enzymatic capability. Altman was working with the enzyme ribonuclease (RNase) P in bacteria, which is responsible for cutting mature tRNA molecules out of an immature RNA segment. RNase P thus also acts as a nuclease. It was known for some time that RNase P contains both a protein and an RNA constituent, but Altman was ultimately able to show that it was the RNA rather than the protein that actually catalyzed the reaction.

The importance of these findings cannot be overstated, and Cech and Altman ultimately shared the 1989 Nobel Prize in Chemistry for the discovery of these RNA enzymes, or ribozymes (joining the terms "ribonucleic acid" and "enzymes"). Subsequently, many ribozymes have been found in various organisms, from bacteria to humans. Some of them are able to catalyze different types of reactions, and new ones are periodically reported. Ribozymes have thus proven to be more than a mere curiosity, playing an integral role in the molecular machinery of many organisms.

At around the same time as these important discoveries, still other functions of RNA were being identified. While perhaps not as dramatic as the ribozymes, antisense RNAs, small nuclear RNAs, and a variety of others further proved the versatility of RNA. While understanding the roles of ribozymes and other unconventional RNAs is important to the understanding of genetic functioning in present-day organisms, these discoveries were more intriguing to many scientists interested in the origin and evolution of life. In a sense, the existence of ribozymes was a violation of the central dogma, which implied that information was ultimately utilized solely in the form of proteins. While the central dogma was not in danger of becoming obsolete, a clue had been found that might possibly allow a resolution, at least in theory, to questions about whether the DNA or the protein came first. The exciting answer: perhaps neither.

The RNA World Theory and the Origin of Life

Given that RNA is able to store genetic information (as it certainly does when it functions as mRNA) and the new discovery that it could function as an enzyme, there was no longer any need to invoke the presence of either DNA or protein as necessities in the first living system. The first living molecule would have to be able to replicate itself without any help, and just such an "RNA replicase" has been proposed as the molecule that eventually led to life as it is now known. Like the self-splicing intron of *Tetrahymena*, this theoretical ribozyme could have worked on itself, catalyzing its own replication. This RNA would therefore have functioned as both the genetic material and the replication enzyme, allowing it to make copies of itself without the need for DNA or proteins. Biologist Walter Gilbert coined the term "RNA world" for this interesting theoretical period dominated by RNA. Modern catalytic RNAs can be thought of as molecular fossils that remain from this period and provide clues about its nature.

How might this initial RNA have come into being in the first place? Biologist Aleksandr Oparin predicted in the late 1930's that if simple gases thought to be present in Earth's early atmosphere were subjected to the right conditions (energy in the form of lightning, for example), more complex organic molecules would be formed. His theory was first tested in 1953 and was resoundingly confirmed. A mixture of methane, ammonia, water vapor, and hydrogen gas was energized with high-voltage electricity, and the products were impressive: several amino acids and aldehydes, among other organic molecules. Subsequent experiments have been able to produce ribonucleotide bases. It seems reasonable, then, that nucleotides could have been present on the early Earth and that their random linkage could lead to the formation of an RNA chain.

After a while, RNA molecules would have found a way to synthesize proteins, which are able to act as more efficient and diverse enzymes than ribozymes by their very nature. Why are proteins better enzymes than ribozymes? Since RNA contains only four bases that are fundamentally similar in their chemical properties, the range of different configurations and functional capabilities is somewhat limited as opposed to proteins. Proteins are constructed of twenty different amino acids whose functional groups differ widely in terms of their chemical makeup and potential reactivity. It is logical to suppose, therefore, that proteins eventually took over most of the roles of RNA enzymes because they were simply better suited to doing so. Several of the original or efficient ribozymes would have been retained, and those are the ones that still can be observed.

How could a world composed strictly of RNAs, however, be able to begin protein synthesis? While it seems like a tall order, scientists have envisioned an early version of the ribosome that was composed exclusively of RNA. Biologist Harry Noller reported in the early 1990's that the activity of the modern ribosome that is responsible for catalyzing the formation of peptide bonds between amino acids is in fact carried out by rRNA. This so-called peptidyltransferase activity had always been attributed to one of the ribosomal proteins, and rRNA had been envisioned as playing a primarily structural role. Noller's discovery that the large ribosomal RNA is actually a ribozyme allows scientists to picture a ribosome working in roughly the same way that modern ones do, without containing any proteins. As proteins began to be synthesized from the information in the template RNAs, they slowly began to assume some of the RNA roles and probably incorporated themselves into the ribosome to allow it to function more efficiently.

The transition to the modern world would not be complete without the introduction of DNA as the major form of the genetic material. RNA, while well suited to diverse roles, is actually a much less suitable genetic material than DNA for a complex organism (even one only as complex as a bacterium). The reason for this is that the slight chemical differences between the sugars contained in the nucleotides of RNA and DNA cause the RNA to be more reactive and much less chemically stable; this is good for a ribozyme but clearly bad if the genetic material is to last for any reasonable amount of time. Once DNA initially came into existence, therefore, it is likely that the relatively complex organisms of the time quickly adopted it as their genetic material; shortly thereafter, it became double-stranded, which facilitated its replication immensely. This left RNA, the originator of it all, relegated to the status it now enjoys; molecular fossils exist that uncover its former glory, but it functions mainly as a helper in protein synthesis.

This still leaves the question of how DNA evolved from RNA. At least two protein enzymes were probably necessary to allow this process to begin. The first, ribonucleoside diphosphate reductase, converts RNA nucleotides to DNA nucleotides by reducing the hydroxyl group located on the 2′ carbon of ribose. Perhaps more important, the enzyme reverse transcriptase would have been necessary to transcribe RNA genomes into corresponding DNA versions. Examples of both of these enzymes exist in the modern world.

Some concluding observations are in order to summarize the evidence that RNA and not DNA was very likely the first living molecule. No enzymatic activity has ever been attributed to DNA; in fact, the 2′

hydroxyl group that RNA possesses and DNA lacks is vital to RNA's ability to function as a ribozyme. Furthermore, ribose is synthesized much more easily than deoxyribose under laboratory conditions. All modern cells synthesize DNA nucleotides from RNA precursors, and many other players in the cellular machinery are RNA-related. Important examples include adenosine triphosphate (ATP), the universal cellular energy carrier, and a host of coenzymes such as nicotinamide adenine dinucleotide (NAD), derived from B vitamins and vital in energy metabolism.

Impact and Applications

The discovery of ribozymes and the other interesting classes of RNAs has dramatically altered the understanding of genetic processes at the molecular level and has provided compelling evidence in support of exciting new theories regarding the origin of life and cellular evolution. The RNA world theory, first advanced as a radical and unsupported hypothesis in the early 1970's, has gained almost universal acceptance by scientists. It is the solution to the evolutionary paradox that has plagued scientists since the discovery and understanding of the central dogma: Which came first, DNA or proteins? Since they are inextricably dependent upon each other in the modern world, the idea of the RNA world proposes that, rather than one giving rise to the other, they are both descended from RNA, that most ancient of genetic and catalytic molecules. Unfortunately, the RNA world model is not without its problems.

In the mid- to late 1990's, several studies on the stability of ribose, the sugar portion of ribonucleotides, showed that it breaks down relatively easily, even in neutral solutions. A study of the decay rate of ribonucleotides at different temperatures also caused some concern for the RNA world theory. Most current scenarios see life arising in relatively hot conditions, at least near boiling, and the instability of ribonucleotides at these temperatures would not allow for the development of any significant RNA molecules. Ribonucleotides are much more stable at 0 degrees Celsius (32 degrees Fahrenheit), but evidence for a low-temperature environment for the origin of life is limited. Consequently, some evolutionists are suggesting that the first biological entities might have relied on something other than RNA, and that the RNA world was a later development. Therefore, although the RNA world seems like a plausible model, another model is now needed to establish the precursor to the RNA world.

Apart from origin-of-life concerns, the discoveries that led to the RNA world theory are beginning to have a more practical impact in the fields of industrial genetic engineering and medical gene therapy. The unique ability of ribozymes to find particular sequences and initiate cutting and pasting at desired locations makes them powerful tools. Impressive uses have already been found for these tools in theoretical molecular biology and in the genetic engineering of plants and bacteria. Most important to humans, however, are the implications for curing or treating genetically related disease using this powerful new RNA-based technology.

Gene therapy, in general, is based on the idea that any faulty, disease-causing gene can theoretically be replaced by a genetically engineered working replacement. While theoretically a somewhat simple idea, in practice it is technically very challenging. Retroviruses may be used to insert DNA into particular target cells, but the results are often not as expected; the new genes are difficult to control or may have adverse side effects. Molecular biologist Bruce Sullenger pioneered a new approach to gene therapy, which seeks to correct the genetic defect at the RNA level. A ribozyme can be engineered to seek out and replace damaged sequences before they are translated into defective proteins. Sullenger has shown that this so-called trans-splicing technique can work in nonhuman systems and, in 1996, began trials to test his procedure in humans.

Many human diseases could be corrected using gene therapy technology of this kind, from inherited defects such as sickle-cell disease to degenerative genetic problems such as cancer. Even pathogen-induced conditions such as acquired immunodeficiency syndrome (AIDS), caused by the human immunodeficiency virus (HIV), could be amenable to this approach. It is ironic and gratifying that an understanding of the ancient RNA world holds promise for helping scientists to solve some of the major problems in the modern world of DNA-based life.

Matthew M. Schmidt, Ph.D.;
updated by Bryan Ness, Ph.D.

Further Reading

De Duve, Christian. "The Beginnings of Life on Earth." *American Scientist* 83, no. 5 (September/

October, 1995): 428. Discusses several scenarios regarding RNA's possible involvement in life's origins.

Gesteland, Raymond F., Thomas R. Cech, and John F. Atkins, eds. *The RNA World: The Nature of Modern RNA Suggests a Prebiotic RNA.* 3d ed. Cold Spring Harbor, N.Y.: Cold Spring Harbor Laboratory Press, 2006. An advanced, detailed look at the theories behind the RNA world, the evidence for its existence, and the modern "fossils" that may be left from this historic biological period. Illustrations, including some in color.

Hart, Stephen. "RNA's Revising Machinery." *Bioscience* 46, no. 5 (May, 1996): 318. Discusses the use of ribozymes in gene therapy.

Hazen, Robert M. "The Emergence of Self-Replicating Systems." In *Genesis: The Scientific Quest for Life's Origin.* Washington, D.C.: Joseph Henry Press, 2005. Includes discussion of the pre-RNA and RNA world hypotheses.

Horgan, John. "The World According to RNA." *Scientific American* 274, no. 1 (January, 1996): 27. Summarizes the accumulated evidence that RNA molecules once served both as genetic and catalytic agents.

Luisi, Pier Luigi. "The 'Prebiotic' RNA World." In *The Emergence of Life: From Chemical Origins to Synthetic Biology.* Cambridge, England: Cambridge University Press, 2006. Traces the origins of life, beginning with prebiotic chemistry. Provides information about Oparin's theory.

Miller, Stanley L. *From the Primitive Atmosphere to the Prebiotic Soup to the Pre-RNA World.* NASA CR-2076334007116722. Washington, D.C.: National Aeronautics and Space Administration, 1996. Miller discusses his famous 1953 experiment, whereby he produced organic amino acids from inorganic materials in a laboratory, igniting a new understanding of how RNA and DNA work.

Rauchfuss, Horst. "The 'RNA World.'" In *Chemical Evolution and the Origin of Life.* Translated by Terence N. Mitchell. Berlin: Springer, 2008. This overview of chemical evolution, designed for the general reader, devotes a chapter to discussing the RNA world hypotheses.

Watson, James D., et al. *The Molecular Biology of the Gene.* 6th ed. San Francisco: Pearson/Benjamin Cummings, 2008. Includes a comprehensive discussion of all aspects of the RNA world.

WEB SITES OF INTEREST

Exploring Life's Origins, Understanding the RNA World
http://exploringorigins.org

This site, created by the Museum of Science in Boston, uses molecular illustration and animation to describe life's origins. It includes a section describing RNA and how the discovery of ribozymes led to the RNA world hypothesis.

Nobel Prize.org, The RNA World
http://nobelprize.org/nobel_prizes/chemistry/articles/altman

An article by Sidney Altman, who, with Thomas Cech, shared the 1989 Nobel Prize in Chemistry, discussing RNA world theory and his involvement in the discovery of ribozymes. The site also features an article by Cech, "The RNA World."

See also: Ancient DNA; Antisense RNA; Chromosome structure; DNA isolation; DNA repair; DNA replication; DNA structure and function; Genetic code; Genetic code, cracking of; Molecular genetics; Noncoding RNA molecules; One gene-one enzyme hypothesis; Protein structure; Protein synthesis; Repetitive DNA; RNA isolation; RNA structure and function; RNA transcription and mRNA processing.

Robert syndrome

CATEGORY: Diseases and syndromes

ALSO KNOWN AS: Roberts syndrome; Roberts-SC phocomelia syndrome; pseudothalidomide syndrome; SC-phocomelia syndrome; Appelt-Gerken-Lenz syndrome; hypomelia hypotrichosis facial hemangioma syndrome; RBS; SC syndrome; tetraphocomelia-cleft palate syndrome

DEFINITION

Robert syndrome is a developmental disorder that affects multiple organ systems, prenatal and postnatal growth, and mental development. It is an autosomal recessive condition caused by mutations in the gene *ESCO2.*

RISK FACTORS

Since Robert syndrome is a rare condition, carrier frequency and prevalence are unknown. Approximately 150 cases have been reported in individuals of a variety of ethnic backgrounds. Couples who are consanguineous or have a family history of Robert syndrome are at increased risk for having an affected child.

Etiology and Genetics

Robert syndrome is an autosomal recessive genetic condition. When both parents are carriers of a single mutation *ESCO2*, there is a 25 percent chance in each pregnancy that the child will inherit both mutation and will be affected with Robert syndrome.

In individuals with Robert syndrome, cytogenetic abnormalities have been noted in the process of cell division. Typically, during metaphase the two sister chromatids of each chromosome are bound closely together. The proteins that bind them together are broken down during anaphase and the sister chromatids separate into sister chromosomes. In individuals with Robert syndrome, the centromeres of the sister chromatids separate during metaphase rather than in anaphase, which is referred to as premature centromere separation. This gives the chromosomes in a metaphase spread a "railroad track" appearance due to the absence of the primary constriction at the centromere. In addition, there is a lack of cohesion in the heterochromatic (dense, genetically inactive) regions of the sister chromatids. This results in "heterochromatin repulsion" or "puffing" around the centromere and in heterochromatic regions, such as in the acrocentric chromosomes and the long arm of the Y chromosome.

Premature centromere separation and separation of the heterochromatic regions are thought to trigger activation of the mitotic spindle checkpoint in metaphase, which delays mitosis and impairs cell proliferation. This impaired proliferation is thought to cause the growth failure associated with Robert syndrome. The congenital anomalies characteristic of Robert syndrome may result from the loss of progenitor cells during embryogenesis.

The gene *ESCO2* (Establishment of Cohesion 1 homolog 2) is located at 8p21.1. It codes for the protein N-acetyltransferase ESCO2. This enzyme has been found to have autoacetylation activity in vitro. Most of the *ESCO2* mutations that have been described in families with Robert syndrome create a premature stop codon that would result in a truncated protein or mRNA instability. However, a point mutation that disrupts autoacetylation activity has also been described in an affected individual. Therefore, the inability of this protein to transfer acetyl groups from one compound to another in affected individuals is thought to be involved in the pathogenesis of Robert syndrome. It is thought that *ESCO2* is involved in regulating the establishment of sister chromatid cohesion during the S phase of the cell cycle.

Symptoms

Robert syndrome commonly affects the growth and development of the limbs and midface. Affected individuals are born with hypomelia, or incomplete development of the limbs, which is typically more severe in the upper limbs. They may also have cleft lip with or without cleft palate, prenatal onset growth deficiency, mental retardation, and other congenital anomalies.

Screening and Diagnosis

A diagnosis of Robert syndrome can be made with a routine chromosome analysis using Giemsa or C-banding. Chromosomes have a characteristic appearance due to premature centromere separation and separation of the heterochromatic regions.

In addition, molecular genetic testing for Robert syndrome is available. Sequence analysis of *ESCO2* can confirm a diagnosis in an affected individual. Since carrier status cannot be determined from cytogenetic testing, molecular genetic testing must be used for carrier testing. Prenatal diagnosis can be performed on an amniocentesis or chorionic villus sampling with either molecular genetic testing or cytogenetic testing.

Treatment and Therapy

Following an initial diagnosis, evaluations may include X rays and assessments of the limbs, hands, and face; ophthalmologic evaluation; cardiac evaluation; renal ultrasound; and developmental assessment. Treatment would be based on an individual's specific needs. Limb abnormalities may require reconstructive surgery or prostheses. Facial clefting may require multiple surgeries, speech therapy, and management of recurrent ear infections. Developmental delays may result in a need for special education. Growth and development should be closely monitored.

Prevention and Outcomes

The prognosis for an affected individual depends of the severity of the condition. Many severely affected pregnancies result in stillbirth or early infant death. Mildly affected individuals may survive to adulthood. Although prenatal diagnosis is available for pregnancies at risk based on ultrasound findings or family history, there is no effective means of preventing Robert syndrome. Genetic counseling should be made available to the families of affected individuals.

Laura Garasimowicz, M.S.

Further Reading

Canepa, G., Pierre Maroteaux, and V. Pietrogrande. *Dysmorphic Syndromes and Constitutional Disease of the Skeleton.* Padova, Italy: Piccin, 2001. Overview of clinical features of conditions involving the skeletal system.

Gorlin, Robert J., Michael M. Cohen, Jr., and Raoul C. M. Hennekam. *Syndromes of the Head and Neck.* 4th ed. New York: Oxford University Press, 2001. A classic text containing clinical information and illustrations of syndromes affecting the head and neck.

Jones, Kenneth Lyons. *Smith's Recognizable Patterns of Human Malformation.* 6th ed. Philadelphia: Elsevier, 2006. A clinical reference that includes descriptions and pictures of individuals with malformation syndromes.

Web Sites of Interest

Genetics Home Reference: Roberts Syndrome
http://ghr.nlm.nih.gov/condition=robertssyndrome

National Center for Biotechnology Information, GeneReviews: Roberts Syndrome
http://www.ncbi.nlm.nih.gov/bookshelf/br.fcgi?book=gene&part=rbs

National Center for Biotechnology Information, Online Mendelian Inheritance in Man: Roberts Syndrome; RBS
http://www.ncbi.nlm.nih.gov/entrez/dispomim.cgi?id=268300

See also: Apert syndrome; Brachydactyly; Carpenter syndrome; Cleft lip and palate; Congenital defects; Cornelia de Lange syndrome; Cri du chat syndrome; Crouzon syndrome; Down syndrome; Edwards syndrome; Ellis-van Creveld syndrome; Holt-Oram syndrome; Ivemark syndrome; Meacham syndrome; Opitz-Frias syndrome; Patau syndrome; Polydactyly; Rubinstein-Taybi syndrome.

Rubinstein-Taybi syndrome

Category: Diseases and syndromes

Also known as: Rubinstein syndrome; broad thumbs and great toes; broad thumb-hallux syndrome

Definition

Rubinstein-Taybi syndrome (RSTS) is a congenital anomaly and cognitive impairment syndrome with multisystem effects. It was first described as a syndrome by Jack Rubinstein and Hooshang Taybi in 1963 after examining multiple similar cases. RSTS is a dominant genetic condition displaying consistent physical and neurological findings. Some varying attributes are also reported in the literature.

Risk Factors

The occurrence of RSTS is approximately 1 in 100,000 newborns. The incidence increases to 50 percent of offspring of affected individuals. Various phenotypical penetrance is reported in families. Other than familial cases, there are no known risk factors.

Etiology and Genetics

RSTS is caused by a heterozygous germline (oocyte, spermatocyte) mutation. Approximately 55 percent of cases of RSTS are caused by mutations in one of two genes, the CREB binding protein (*CREBBP*) gene located on chromosome 16p13.3 and the *EP300* gene located on chromosome 22q12.2. Both genes encode histone acetyltransferases (HAT), which affect cellular signaling pathways and are important in cell growth and differentiation. HAT plays an important role in activation of transcription. HAT blocks the restricted access to deoxyribonucleic acid (DNA). In translation, DNA gives ribonucleic acid (RNA) orders to produce a certain needed protein. RNA then produces the requested protein in a process called transcription. These proteins are utilized for cellular processes and development. CREBBP also plays a part in long-term memory formation, and its mutation may result in the cognitive deficits seen in individuals with RSTS. Mutations of these genes are usually point mutations, although translocations, inversions, missense mutations, mosaic microdeletions, and large deletions have been reported. Most mutations are dominant de novo events. The expected recurrence rate in future pregnancies is 0.1 percent. The genetic etiology of the remaining 45 percent of cases has not yet been determined.

Symptoms

RSTS is characterized by neurological and physical features. Neurologically, failure to achieve developmental milestones and mental retardation are common. Later, children exhibit short attention span and self-stimulatory behaviors such as rocking or spinning. Physically, broad great toes and thumbs are hallmarks of the syndrome. In addition, distinct facial features include down-slanting palpebral fissures, hypertelorism, high-arched eyebrows, prominent beaked nose with the nasal septum extending below the nares, a grimacing mouth, and hypoplastic maxilla. On dental exam, 90 percent of affected individuals exhibit talon cusps. Failure to thrive may follow feeding difficulties in infancy and gastroesophageal reflux is common.

Screening and Diagnosis

Diagnosis is by two methods. Clinically, it includes family history, and patient history and physical examination findings. In the absence of positive genetic tests the diagnosis is based on clinical and radiological findings which include dysmorphological facial features, broad thumbs and first toes, and mental and physical growth delay. Genetic testing now allows confirmation of suspicion in approximately 55 percent of cases using fluorescence in situ hybridization (FISH), DNA sequence analysis, and polymerase chain reaction (PCR) testing. Currently there are no genetic tests that can detect the syndrome in the remaining 45 percent of affected individuals.

Treatment and Therapy

Early diagnosis permits treatment of medical issues as well as the provision of genetic counseling including information about RSTS, recurrence risks, and prenatal diagnosis. In addition to genetic evaluation, the individual should be screened for cardiac, renal, ophthalmologic, auditory, endocrine, and orthopedic issues. Because of the high incidence of dental problems, referral and follow-up by a pediatric dentist is needed. Because of feeding difficulties, growth and nutrition counseling is beneficial. Individuals with *CREBBP* and *EP300* gene mutations are at risk for future malignancies due to an imbalance of alleles implicated in malignancies such as hematologic, breast, ovarian, hepatic, and colorectal cancer and require close monitoring for development of these tumors.

Prevention and Outcomes

No identifiable prevention strategy is currently known to prevent the majority of cases as they are primarily due to a de novo mutation. Only one known case has been detected prior to birth by ultrasound technique. Thus, clinicians must perform thorough physical examinations on newborns and be alert to the dysmorphic and neurological features of this syndrome. The prognosis is dependent on the nature and severity of defects as well as early, consistent medical, dental, and development care for children with this syndrome.

Wanda Todd Bradshaw, R.N., M.S.N.

Further Reading

Foley, Patricia, et al. "Further Case of Rubinstein-Taybi Syndrome Due to a Deletion in *EP300*." *American Journal of Medical Genetics*, Part A 149A (2009): 997-1000.

Greco, Elena, Gabriella Sglavo, and Dario Paladini. "Prenatal Sonographic Diagnosis of Rubinstein-Taybi Syndrome." *Journal of Ultrasound Medicine* 28 (2009): 669-672.

Hennekam, Raoul C. M. "Practical Genetics: Rubinstein-Taybi Syndrome." *European Journal of Human Genetics* 14 (2006): 981-985.

Schorry, Elizabeth, et al. "Genotype-Phenotype Correlations in Rubinstein-Taybi Syndrome." *American Journal of Medical Genetics*, Part A 146A (2008): 2512-2519.

Wiley, Susan, et al. "Rubinstein-Taybi Syndrome Medical Guidelines." *American Journal of Medical Genetics* 119-A (2003): 101-110.

Web Sites of Interest

Dermatology Online Journal: Rubinstein-Taybi Syndrome (Broad Thumb-Hallux Syndrome) (Sherry H. Hsiung)
http://dermatology.cdlib.org/103/NYU/case_presentations/102103n2.html

Online Mendelian Inheritance in Man: Rubinstein-Taybi Syndrome; RSTS
http://www.ncbi.nlm.nih.gov/entrez/dispomim.cgi?id=180849

Proceedings for the 1998 International Conference on Rubinstein-Taybi Syndrome
http://www.rubinstein-taybi.org/bluebook/hisorical .html

Special Friends
http://www.specialfriends.org

U.S. Rubinstein-Taybi Support Group
http://www.rubinstein-taybi.org

See also: Apert syndrome; Brachydactyly; Carpenter syndrome; Cleft lip and palate; Congenital defects; Cornelia de Lange syndrome; Cri du chat syndrome; Crouzon syndrome; Down syndrome; Edwards syndrome; Ellis-van Creveld syndrome; Holt-Oram syndrome; Ivemark syndrome; Meacham syndrome; Opitz-Frias syndrome; Patau syndrome; Polydactyly; Robert syndrome.

S

Sandhoff disease

CATEGORY: Diseases and syndromes

ALSO KNOWN AS: Beta-hexosaminidase-beta-subunit deficiency; GM2 gangliosidosis, Type II; hexosaminidase A and B deficiency disease; Sandhoff-Jatzkewitz-Pilz disease; total hexosaminidase deficiency

Definition

Sandhoff disease is a rare, inherited disorder that causes abnormal accumulation of particular types of fats inside nerve cells, which eventually kills them. This disease culminates in a gradual and progressive deterioration of the nervous system.

Risk Factors

Anyone whose biological parents both carry mutations in the *HexB* gene is at risk for Sandhoff disease. This genetic disease equally affects males and females but more commonly occurs in the Creole population of northern Argentina; Metis Indians in Saskatchewan, Canada; and Christian Marionite communities from Cyprus.

Etiology and Genetics

Sandhoff disease is an autosomal recessive condition caused by mutations in the *HexB* gene. The *HexB* gene is located on the long arm of chromosome 5, in band region 13. People must inherit two copies of a mutant form of the *HexB* gene in order to have Sandhoff disease. Children whose parents both carry a mutant copy of the *HexB* gene have a 25 percent chance of being born with Sandhoff disease. The *HexB* gene encodes the information for the synthesis of a protein called hexosaminidase B.

Cells are bordered by the plasma membrane, which is composed of phosphate-containing lipids, proteins, and another groups of lipids called sphingolipids. Many sphingolipids have chains of sugars attached to them and are collectively called glycolipids. Glycolipids are very common in the membranes of nerve cells, but cells must constantly produce new glycolipids and degrade old ones in order to keep their plasma membranes in good working condition.

Once made, membrane lipids are loaded into spherical vesicles that fuse with the membrane. Membrane lipids are removed from the membrane by being loaded into vesicles that pinch off the membrane and fuse with a waiting vesicle called a lysosome. Lysosomes serve as the garbage disposal of cells, and they contain a cadre of enzymes called acid hydrolases that degrade the components brought to the lysosome. Hexosaminidase B is one of the lysosomal acid hydrolases that degrade glycolipids.

Hexosaminidase B is part of two different acid hydrolases. It forms a complex with another protein called hexosaminidase A to form an acid hydrolase called β-hexosaminidase A. Hexosaminidase A is encoded by the *HexA* gene, and mutations in *HexA* cause Tay-Sachs disease, which is clinically very similar to Sandhoff disease. β-hexosaminidase A degrades a glycolipid called the GM ganglioside. Hexosaminidase B also forms a complex with itself to make an acid hydrolase called β-hexosaminidase B, which degrades so called "neutral glycolipids."

Mutations in the *HexB* gene that inactivate hexosaminidase B abrogate the enzymatic activities of β-hexosaminidase A and B. Cells accumulate GM gangliosides and neutral glycolipids, which kills the cells. Because nerve cells make such extensive use of complex glycolipids, they are among the most sensitive to an inability to degrade them. Gradual death of nerve cells causes an inexorable and progressive deterioration of nerve functions.

Sandhoff disease belongs to a larger group of genetic diseases called lysosomal storage diseases, in which the activity of lysosomal acid hydrolases is compromised.

Symptoms

With the classical infantile form of Sandhoff disease, infants begin showing symptoms by six months. They lose muscle tone and movement (motor control) and show an exaggerated startle response to loud noises. As the nervous system deteriorates, babies suffer from seizures; loss of sight, hearing, and the ability to swallow; and paralysis.

In the juvenile form, symptoms start between three to ten years of age and include muscle weakness, loss of coordination (ataxia) and speech, and mental problems.

The adult-onset form occurs in older people, and the symptoms (muscle weakness) are milder and vary.

Screening and Diagnosis

Biochemical tests on blood or biopsied tissues are required to distinguish Sandhoff disease from Tay-Sachs disease. The absence of β-hexosaminidase A and B activity is diagnostic for Sandhoff disease.

Genetic tests for prenatal or preimplantation diagnoses are available for Sandhoff disease. Prenatal diagnoses use tissue derived from amniocentesis, chorionic villus sampling, placental biopsy, umbilical cord blood sampling, or fetal skin biopsy to screen unborn babies. Preimplantation genetic diagnosis screens individual blastomeres from embryos formed by in vitro fertilization.

Treatment and Therapy

No cure exists for Sandhoff disease. Antiseizure medication can mitigate seizures, and feeding tubes can prevent food aspiration into the lungs. Experimental treatments with stem cells and gene therapy are ongoing.

Prevention and Outcomes

Genetic screening of prospective parents can reduce the rates of Sandhoff disease.

The infantile form of Sandhoff disease usually causes death before the age of four, while the juvenile-onset form causes death by age fifteen. Adult-onset patients experience severe lifestyle restrictions.

Michael A. Buratovich, Ph.D.

Further Reading

Cobb, Bryan. "Sandhoff Disease." In *The Gale Encyclopedia of Neurological Disorders*, edited by Stacey L. Chamberlin and Brigham Narins. Farmington, Mich.: Gale Group, 2005. A brief but useful explanation of the clinical and genetic aspects of Sandhoff disease that is easily understood.

Nelson, David L., and Michael M. Cox. *Lehninger Principles of Biochemistry.* 5th ed. New York: W. H. Freeman, 2008. An excellent college-level biochemistry textbook that is clearly written and lavishly illustrated. Chapter 10 is the chapter on lipids that has a fine explanation of lysosomal storage diseases such as Sandhoff disease.

Parker, James N., and Philip M. Parker. *The Official Parent's Sourcebook on Sandhoff Disease: A Revised and Updated Directory for the Internet Age.* San Diego: Icon Health, 2002. A parent's guide that is accurate and nontechnical.

Web Sites of Interest

Disabled World: Sandhoff Disease Information
http://www.disabled-world.com/health/pediatric/sandhoff-disease.php

Genetics Home Reference: Sandhoff Disease
http://ghr.nlm.nih.gov/condition=sandhoffdisease

Sandhoff Disease
http://sandhoffdisease.webs.com/whatisit.htm

See also: Fabry disease; Gaucher disease; Gm1-gangliosidosis; Hereditary diseases; Hunter disease; Hurler syndrome; Inborn errors of metabolism; Jansky-Bielschowsky disease; Krabbé disease; Metachromatic leukodystrophy; Niemann-Pick disease; Pompe disease; Sanfilippo syndrome; Tay-Sachs disease.

Sanfilippo syndrome

CATEGORY: Diseases and syndromes
ALSO KNOWN AS: Mucopolysaccharidosis type III; MPS III

Definition

Sanfilippo syndrome is a progressive disorder that primarily affects the central nervous system. In individuals with Sanfilippo syndrome, gene mutations cause a deficiency of an enzyme needed for breakdown of heparan sulfate. There are four subtypes of Sanfilippo syndrome: A, B, C, and D.

Risk Factors

A family history of Sanfilippo syndrome is the most significant risk factor. The parents of a child with Sanfilippo syndrome have a 25 percent chance to have another child with the condition. Unaffected siblings of an individual with Sanfilippo syndrome have a 67 percent chance to be a carrier of the condition. There is a high incidence of Sanfilippo syndrome in individuals from the Cayman Islands.

Etiology and Genetics

Sanfilippo syndrome is an autosomal recessive condition. Humans typically have twenty-three pairs of chromosomes. The first twenty-two pairs are the autosomes, and are numbered 1 through 22. The twenty-third pair consists of the sex chromosomes, which are called X and Y. Each chromosome contains many genes. Humans have two copies of most genes. Individuals affected with an autosomal recessive condition have a mutation in both copies of a given gene. In autosomal recessive conditions, individuals with a mutation in one copy of a gene are considered carriers. Mutations in one of four genes (*SGSH, NAGLU, HGSNAT, GNS*) cause each type of Sanfilippo syndrome (A, B, C, D, respectively). *SGSH* and *NAGLU* are both located on chromosome 17. *HGSNAT* is located on chromosome 8, and the *GNS* gene is located on chromosome 12.

Each of the four genes for Sanfilippo syndrome provides instructions for a single enzyme, which is partially responsible for the breakdown of heparan sulfate. Heparan sulfate is a type of mucopolysaccharide, which are complex chains of sugar molecules. Because heparan sulfate is broken down in the lysosome of the cell, Sanfilippo syndrome is considered a lysosomal storage disease.

In individuals with Sanfilippo syndrome, heparan sulfate is stored in the body's tissues. However, the majority of heparan sulfate is stored in the brain. It is also believed that the storage of heparan sulfate interferes with the breakdown of a component of cell membranes, gangliosides. The combination of heparan sulfate and ganglioside accumulation causes the symptoms of Sanfilippo syndrome.

Symptoms

Type A, B, C, and D of Sanfilippo syndrome have similar symptoms, and the type cannot be determined based on symptoms. Sanfilippo syndrome is a progressive disorder that primarily affects the central nervous system. As the condition progresses, the neurological symptoms gradually change. The symptoms of Sanfilippo syndrome have been divided into three phases. The timing and severity of each phase vary from one individual to the next. Phase one is characterized by developmental delay, which is typically identified between twelve months and twenty-four months of age. The delay is typically most significant in speech. The second phase is usually from age three to age ten and is characterized by hyperactivity and behavioral problems. A need for very little sleep may develop during phase two. The third phase typically begins at the end of the first decade. Individuals slowly lose motor skills and eventually are no longer able to walk. Swallowing problems may also develop during the third phase.

Screening and Diagnosis

Mucopolysaccharides are also known as glycosaminoglycans (GAGs). Quantification of GAGs in urine is a screening test for all mucopolysaccharidoses. Thin-layer chromotography (TLC) separation is another screening test that may identify a specific mucopolysaccharidosis. However, if quantification of GAGs and TLC separation is abnormal, then confirmatory enzyme analysis is necessary.

Treatment and Therapy

Unfortunately, there is currently not a therapy or treatment for Sanfilippo syndrome. However, several potential treatments are currently being researched. Replacing the missing enzyme is a potential treatment; however, the enzyme is unable to get through the protective coating around the brain. Because the symptoms of Sanfilippo syndrome are primarily neurological, enzyme must be administered directly into the brain. Substrate deprivation therapy (SDT) is a pharmaceutical approach to reduce the amount of GAGS or gangliosides produced by the body. Two SDTs, miglustat and genistein, have been trialed in patients with Sanfilippo syndrome, with some stabilization of neurological symptoms. Gene therapy is another potential treatment and is currently being researched in animal models.

Prevention and Outcomes

Genetic testing is available for all four types of Sanfilippo syndrome. If gene mutations are identified in an individual with Sanfilippo syndrome, then preimplantation genetic diagnosis, chorionic villus sampling, and amniocentesis are available to the parents for prevention. Individuals with Sanfilippo syndrome do not typically survive past their late twenties or early thirties.

Beth M. Hannan, M.S.

FURTHER READING

Fernandes, John, Jean-Marie Saudubray, Georges van den Berghe, and John H. Walter. *Inborn Metabolic Diseases.* 4th ed. Germany: Springer, 2006.

Jakobkiewicz-Banecka, Joanna, Alicja Wegrzyn, and Grzegorz Wegrzyn. "Substrate Deprivation Therapy: A New Hope for Patients Suffering from Neuronopathic Forms of Inherited Lysosomal Storage Diseases." *Journal of Applied Genetics* 48, no. 4 (2007): 383-388.

Nussbaum, Robert L., Roderick R. McInnes, and Huntington F. Willard. *Thompson and Thompson Genetics in Medicine.* 7th ed. New York: Saunders, 2007.

Valstar, M. J., et al. "Sanfilippo Syndrome: A Mini Review." *Journal of Inherited Metabolic Disease* 31 (2008): 240-252.

WEB SITES OF INTEREST

Canadian Society for Mucopolysaccharide and Related Diseases
www.mpssociety.ca

National MPS Society
www.mpssociety.org

Society for Mucopolysaccharide Diseases
www.mpssociety.co.uk

See also: Fabry disease; Gaucher disease; Gm1-gangliosidosis; Hereditary diseases; Hunter disease; Hurler syndrome; Inborn errors of metabolism; Jansky-Bielschowsky disease; Krabbé disease; Metachromatic leukodystrophy; Niemann-Pick disease; Pompe disease; Sandhoff disease; Tay-Sachs disease.

Schizophrenia

CATEGORY: Diseases and syndromes

DEFINITION

Schizophrenia is a chronic, severe, disabling brain disorder that interferes with the way a person interprets reality. People with schizophrenia sometimes hear voices or see things that others do not, become paranoid that people are plotting against them, and experience cognitive deficits and social withdrawal. These and other symptoms make it difficult for them to have positive relationships with others. Schizophrenia is different from split or multiple personality disorders. It affects many different areas of the brain, causing a wide range of behavioral, emotional, and intellectual symptoms.

RISK FACTORS

Factors that increase an individual's risk of schizophrenia include having a parent or sibling with schizophrenia and having an abnormal brain structure. Risks for individuals in the Northern Hemisphere include being born during winter months and being born in the city. Additional risk factors include oxygen deprivation during pregnancy and issues at birth, such as a long labor, bleeding during pregnancy, prematurity, a low birth weight, maternal malnutrition, and infections during pregnancy. Loss of a parent during childhood is another risk factor.

Men typically develop symptoms in their late teens or early twenties, while onset for women tends to occur in their twenties or thirties. In rare cases, schizophrenia is seen in childhood.

ETIOLOGY AND GENETICS

Schizophrenia is an extraordinarily complex and variable clinical condition, and its etiology no doubt depends on a host of both genetic and environmental factors. It is safe to assume that there are many genes that contribute either to the susceptibility or the pathology of schizophrenia, but there is no single gene or mutation that exerts an overriding or self-sufficient effect in determining the development of the disease. Online Mendelian Inheritance in Man (http://www.ncbi.nlm.nih.gov/omim/), an authoritative database maintained by the National Center for Biotechnology Information, lists thirty-one genes scattered throughout the human genome that may be associated with susceptibility to schizophrenia.

In the summer of 2009, three research consortia reported results of their genomewide association studies on the genetics of schizophrenia, and all identified a region of particular importance on the short arm of chromosome 6 (at position 6p22.1). This region is known to contain a large cluster of genes (the major histocompatibility complex) that helps determine immunity and that also codes for proteins that are important for turning other genes on and off. Since mutations in this region are known to influence susceptibility to several autoimmune diseases, it may be likely that schizophrenia has an autoimmune component as well. Genetic variation in the control of such regulatory mechanisms might also account for the environmental component that has been repeatedly shown to contribute to disease susceptibility.

Evidence from additional molecular studies has suggested that large structural changes in DNA (deoxyribonucleic acid) called copy number variants (CNVs) may play a critical role in the development of schizophrenia. These CNVs are usually duplications or deletions of anywhere from hundreds to millions of base pairs in the DNA, and the largest deletions appear only in people with schizophrenia.

With so many genes possibly acting as contributing factors, it is not surprising that there is no predictable inheritance pattern regarding the development of schizophrenia. Many new cases appear in families with one or more other affected members, yet the majority of new cases appear in families in which there is no previous history of the disease.

SYMPTOMS

Symptoms usually start in adolescence or early adulthood. They often appear slowly and become more disturbing and bizarre over time, or they may occur in a matter of weeks or months.

Symptoms include hallucinations—seeing or hearing things/voices that are not there, and delusions—and strong but false personal beliefs that are not based in reality. Disorganized thinking; disorganized speech and a lack of ability to speak in a way that makes sense or to carry on a conversation; catatonic behavior, such as slow movement, repeating rhythmic gestures, pacing, walking in circles, negativism, and repetitive speech; emotional flatness, including flat speech, lack of facial expression, and general disinterest and withdrawal; paranoia, a psychosis characterized by systematized delusions of persecution or grandeur; inappropriate laughter; and poor hygiene and self-care are also symptoms. Associated conditions include obsessive-compulsive disorder, substance abuse (of drugs, alcohol, caffeine, or nicotine), and self-injury, including suicide.

SCREENING AND DIAGNOSIS

Early diagnosis is extremely important. Patients who are diagnosed early are able to stabilize their symptoms, decrease the risk of suicide, decrease alcohol and substance abuse, and reduce the chance of relapse and/or hospitalization.

A person must have active symptoms for at least two weeks and other symptoms for at least six months before a diagnosis can be made. The doctor will rule out other causes, such as drug use, medical illness, or a different mental condition.

TREATMENT AND THERAPY

Schizophrenia is not curable, but it is highly treatable. Hospitalization may be required during acute episodes. Symptoms are usually controlled with antipsychotic medications. These medications work by blocking certain chemicals in the brain, which helps control the abnormal thinking that occurs in people with schizophrenia. Determining a medication plan can be a complicated process. Often medications or dosages need to be changed until the right balance is found. This can take months or even years. Examples of medications include haloperidol (Haldol), thioridazine, or fluphenazine.

Relapse is common, even for patients taking medication. Treatment compliance can be a challenge, since people often stop taking their medication when they are feeling better. The side effects of traditional antipsychotics can also cause patients to discontinue treatment. The most common are physical side effects, such as slow and stiff movements, restlessness, facial tics, and protruding tongue.

New medications, called atypical antipsychotics, have fewer side effects and are better tolerated over long periods of time. However, they may cause weight gain, elevated blood sugar, and elevated serum cholesterol. Examples of these medications include aripiprazole (Abilify), clozapine (Clozaril), risperidone (Risperdal), olanzapine (Zyprexa), paliperidone (Invega), quetiapine (Seroquel), and ziprasidone (Geodon).

Medications may also be prescribed for coexisting conditions. Conditions often associated with schizophrenia include depression and anxiety. They may be treated with antidepressants, anxiolytic drugs, lithium, and anticonvulsants. Electroconvulsive therapy may be used to treat severe depression, suicidal ideation, or severe psychosis.

Schizophrenia is a lifelong condition. It can be confusing and frightening for the person with the disease and for family members. Individual and family therapy can address social skills, vocational guidance, community resources, coping with the family, living arrangements, general emotional support, and working with the family to help family members deal with the patient.

Prevention and Outcomes

There are no guidelines for preventing schizophrenia because the cause is unknown. However, studies show that early, aggressive treatment leads to better outcomes.

Julie Riley, M.S., RD; reviewed by Theodor B. Rais, M.D.
"Etiology and Genetics" by Jeffrey A. Knight, Ph.D.

Further Reading

EBSCO Publishing. *Health Library: Schizophrenia.* Ipswich, Mass.: Author, 2009. Available through http://www.ebscohost.com.

Frith, Christopher D., and Eve C. Johnstone. *Schizophrenia: A Very Short Introduction.* Oxford, England: Oxford University Press, 2003.

Moore, David P., and James W. Jefferson. *Handbook of Medical Psychiatry.* 2d ed. Philadelphia: Elsevier Mosby, 2004.

Temes, Roberta. *Getting Your Life Back Together When You Have Schizophrenia.* Oakland, Calif.: New Harbinger, 2002.

Torrey, E. Fuller. *Surviving Schizophrenia: A Manual for Families, Consumers, and Providers.* 4th ed. New York: Quill, 2001.

Web Sites of Interest

American Psychiatric Association
http://www.psych.org

Canadian Psychiatric Association
http://www.cpa-apc.org

Cochrane Reviews: "Electroconvulsive Therapy for Schizophrenia"
http://www.cochrane.org/reviews/en/ab000076.html

Mayo Clinic.com: Schizophrenia
http://www.mayoclinic.com/health/schizophrenia/DS00196

Mental Health Canada
http://www.mentalhealthcanada.com

National Institute of Mental Health
http://www.nimh.nih.gov

Schizophrenia.com
http://www.schizophrenia.com/index.html

World Fellowship for Schizophrenia and Allied Disorders
http://world-schizophrenia.org

See also: Attention deficit hyperactivity disorder (ADHD); Autism; Bipolar affective disorder; Dyslexia; Fragile X syndrome; Tourette syndrome.

SCLC1 gene

Category: Molecular genetics

Significance: The *SCLC1* gene is expressed in cells and tissues of individuals with lung cancer, particularly in those with small-cell cancer of the lung. Interactions between cigarette smoking and acquired chromosomal abnormalities facilitate expression of the *SCLC1* gene.

Key terms

allele: one member of a pair of different forms of a gene at a specific locus

autosomal dominant: the presence of one abnormal gene allows expression of that gene

chromosomal deletion: chromosomal abnormality in which part of the chromosome is missing or deleted

heterozygosity: different alleles of a given gene at a locus

locus: location on a chromosome occupied by one or more genes

loss of heterozygosity: loss of alleles at numerous chromosomal loci

oncogene: transforms a normal cell into a tumor cell when mutated or expressed

tumor-suppressor gene: a gene that inhibits tumor cell growth

Small-Cell Lung Carcinoma

In the United States, lung cancer is the most common cause of cancer deaths, with North Americans having the highest rates of lung cancer worldwide. There are two types of lung cancer: small-cell lung cancer (SCLC) and non-small-cell lung cancer (NSCLC), with the designation as SCLC or NSCLC based primarily on the appearance of the cancer cells under the microscope. Small-cell lung cancer accounts for approximately 110,000 cancer diagnoses annually. SCLC is highly aggressive and is associated with poor prognosis, as metastases are generally already present at the time of diagnosis. Only a small percentage of patients with limited-stage SCLC can be treated successfully, with the five-year survival rate only about 14 percent. Tobacco smoke is the most common carcinogen associated with SCLC, with approximately 98 percent of cases occurring in patients with a history of smoking.

Chromosome Location and Inheritance

The *SCLC1* gene, also known as *SCLC* or *SCCL*, is located on chromosome 3, between bands 21 and 23 (denoted 3p23-p21) of the short arm of this chromosome. The expression of this gene is associated with lung cancer development, with individuals carrying this gene predisposed to developing lung cancer. Inheritance of the *SCLC1* gene is autosomal dominant. Although expression of *SCLC1* may increase the risk for lung cancer, a combination of several genetic defects and environmental factors is required before cancer develops. SCLC, and other types of lung cancer, are examples of cancer induced by external factors, such as carcinogens in :igarette smoke, but associated with specific chronosomal changes.

olecular Abnormalities Associated with SCLC

Although SCLC is strongly tied to cigarette smok;, with an estimated 90 percent of lung cancers used by smoking, a number of genetic abnormalis are also observed in lung cancer cells. Some are arkers of disease progression whereas others play irect roles in lung cancer development. Lung cancers are characterized by multiple gene alterations, such as amplification of oncogenes and deletions or mutations in tumor suppressor genes during disease progression. Common chromosomal abnormalities associated with lung cancer include deletions on the short arms of chromosomes 3 [del(3p)] and 9 [del(9p)]. The most frequent genetic mutation associated with SCLC is deletion of a region on the short arm of chromosome 3 in an area that contains several genes potentially associated with lung cancer. Specifically, deletion of alleles located on bands 14 through 25—that is, 3p(14-25)—are associated with lung cancer development.

Loss of Heterozygosity and Tumor-Suppressor Genes

In hereditary cancer syndromes that give individuals a predisposition to certain forms of cancer, individuals are heterozygous, which means they have at least one gene pair with different alleles. The other allele of that pair is normal, so the disease phenotype is not immediately expressed. If the normal suppressive gene that inhibits expression of the disease phenotype is lost, such as through chromosomal deletion, a condition called loss of heterozygosity occurs. Loss of heterozygosity at multiple chromosomal loci has been identified in lung cancer, along with changes in tumor-suppressor genes and oncogenes. Most small-cell lung carcinomas are associated with large deletions in segments of 3p, leading to loss of heterozygosity at numerous loci, including many potential tumor-suppressor genes.

Impact

The *SCLC1* gene is associated with lung cancer development and is known as a cancer-predisposing gene. A specific acquired chromosomal abnormality, a deletion on the short arm of chromosome 3, is associated with SCLC. Many inherited cancer syndromes exhibit intricate interactions with environmental factors. Interactions between cigarette smoking and a predisposition to developing lung cancer are documented. Susceptibility to many inherited cancer syndromes is now detectable through genetic screening, and advances in molecular genetics associated with some forms of lung cancer can lead to new approaches to diagnosis, treatment, and prognosis.

C. J. Walsh, Ph.D.

FURTHER READING

Cooper, David N. *The Molecular Genetics of Lung Cancer.* Berlin: Springer-Verlag, 2005. Comprehensive review of the rapidly changing fields of genetics and lung cancer.

Cowell, John K. *Molecular Genetics of Cancer.* 2d ed. Oxford, England: BIOS Scientific, 2001. Includes a chapter on molecular genetics of lung cancer.

Katarjian, Hagop, Charles A. Koller, and Robert A. Wolff. *The MD Anderson Manual of Medical Oncology.* New York: McGraw-Hill Professional, 2006. A detailed guide for practitioners, this book includes information on many types of cancer.

Miller, Robert Hopkins, and Barbara Driscoll. *Molecular Pathology Methods and Reviews.* Vol. 1 in *Lung Cancer.* New York: Springer-Verlag, 2007. Leading scientists review genetic abnormalities associated with lung cancer.

Whang-Peng, C. S., et al. "Specific Chromosome Defect Associated with Human Small-Cell Lung Cancer: Deletion 3p(14-23)." *Science* 215 (1982): 181-182. The first study describing a link between chromosome 3 deletions and lung cancer.

WEB SITES OF INTEREST

National Cancer Institute
www.cancer.gov/cancertopics/types/lung

National Center for Biotechnology Information: Genes and Disease
http://www.ncbi.nlm.nih.gov/disease/Lung.html

See also: *BRAF* gene; *BRCA1* and *BRCA2* genes; Cancer; Chromosome mutation; *DPC4* gene testing; Harvey *ras* oncogene; *HRAS* gene testing; *MLH1* gene; Mutation and mutagenesis; Oncogenes; Tumor-suppressor genes.

Severe combined immunodeficiency syndrome

CATEGORY: Diseases and syndromes
ALSO KNOWN AS: SCID; boy in the bubble syndrome; XSCID; SCIDX1; Athabascan SCID; Swiss agammaglobulinemia

DEFINITION

Severe combined immunodeficiency syndrome (SCID) is a group of rare (1 in 100,000), congenital conditions defined by a deficiency of the immune response. A mutation in one of several possible genes causes a defect in the specialized white blood cells (lymphocytes) that defend against infection. Patients are prone to repeated and persistent infections, rarely surviving past their first year unless treated within the first three to four months of life with bone marrow transplantation.

RISK FACTORS

All forms of SCID are inherited. The most common form of SCID (approximately half of reported cases) occurs only in males as it is caused by mutations in the X chromosome (XSCID).

Though the *Artemis* form of SCID is very rare in the general population, it has a high incidence (1 in 2,000) in Navajo and Apache Native Americans. This form of SCID is also known as Athabascan SCID (SCIDA), after the Athabascan linguistic group of Native Americans.

ETIOLOGY AND GENETICS

The immune system is a network of organs, tissues, cells, and protein substances that work together, defending the body against attack by bacteria, viruses, fungi, parasites, or other foreign substances. When a major component of the immune system is not functioning correctly, the body cannot ward off infections. Lymphocytes (T, B, and NK cells) and the proteins they produce make up key components of the immune system. T cells are crucial in identifying invading agents and in activating and regulating other cells of the immune system. The defining characteristic for SCID is a severe defect in T-cell production and function. The different genetic forms of SCID vary as to the presence of defects in NK and B cells.

SCID results from mutations in any one of twelve known genes encoding components involved in lymphocyte development. Since these twelve genetic causes of SCID account for approximately 90 percent of cases, there are probably additional genetic forms of SCID. Note that reticular dysgenesis and Omenn syndrome are sometimes included as SCID disorders (both are characterized by T-cell deficiency); however, the World Health Organization considers these disorders distinct from SCID.

XSCID results from mutations in the interleukin-2 receptor gamma gene (*IL2RG*; cytogenetic location: Xq13.1). Defective IL receptors prevent the normal development of T cells. SCID also results from autosomal recessive gene defects. Mutations found on autosomal chromosomes have been identified in eleven other genetic forms of SCID with deficiencies in adenosine deaminase (*ADA*; cytogenetic location: 20q12-q13.11); Janus kinase 3 (*JAK3*; cytogenetic location: 19p13.1); IL-7 receptor alpha chain (*IL-7Rα*: cytogenetic location: 5p13); recombinase-activating gene (*RAG-1* or *RAG-2*; cytogenetic location: 11p13); Artemis (cytogenetic location: 10p13); ligase 4 (*LIG4*; cytogenetic location: 13q33-q34); CD3 delta, epsilon, and zeta chain (CD3δ, CD3ε, and CD3ζ; cytogenetic locations: 11q23, 11q23, and 1q22-q23); and CD45 (cytogenetic location: 1q31-q32). These various components of the immune system are critical in T, B, and/or NK cell development or function via their role in toxic byproduct disposal, DNA repair, signal transduction, or VDJ recombination (a form of genetic recombination which assembles pieces of genes encoding specific proteins that have key roles in the immune system).

ADA deficiency is the second most common form of SCID, accounting for about one-sixth of reported cases. Mutations in the *ADA* gene lead to a deficiency in the enzyme and accumulation of toxic byproducts. Though ADA is present in all cell types, immature lymphocytes in the thymus normally have the highest level of ADA in the body and are thus particularly sensitive to the toxic effects of these byproducts. Without ADA, immature lymphocytes fail to reach maturity and die.

Symptoms

Young children often catch colds. However, a baby with SCID develops more serious and life-threatening infections. Bacteria or viruses that typically cause little to no illness in normal children may cause serious illness in SCID children. For example, *Pneumocystis jiroveci* and Cytomegalovirus, both of which are found in healthy people, can cause a fatal pneumonia or hepatitis in SCID children. Similarly, adenovirus usually produces a sore throat and cold in people in the general population but can cause a fatal hepatitis in SCID children. SCID children may also develop severe fungal infections such as thrush (a Candida fungal infection of the mouth). Additionally, they are at risk from the live viruses used in certain vaccines. Other symptoms include persistent diarrhea (not necessarily infection-related), rash, and autoimmunity.

Screening and Diagnosis

Though SCID is considered rare, some experts believe it may be more common than thought. Children may be dying of SCID infections before being diagnosed. For example, some cases of sudden infant death syndrome (SIDS) may actually be due to SCID. Nonetheless, standard screening of

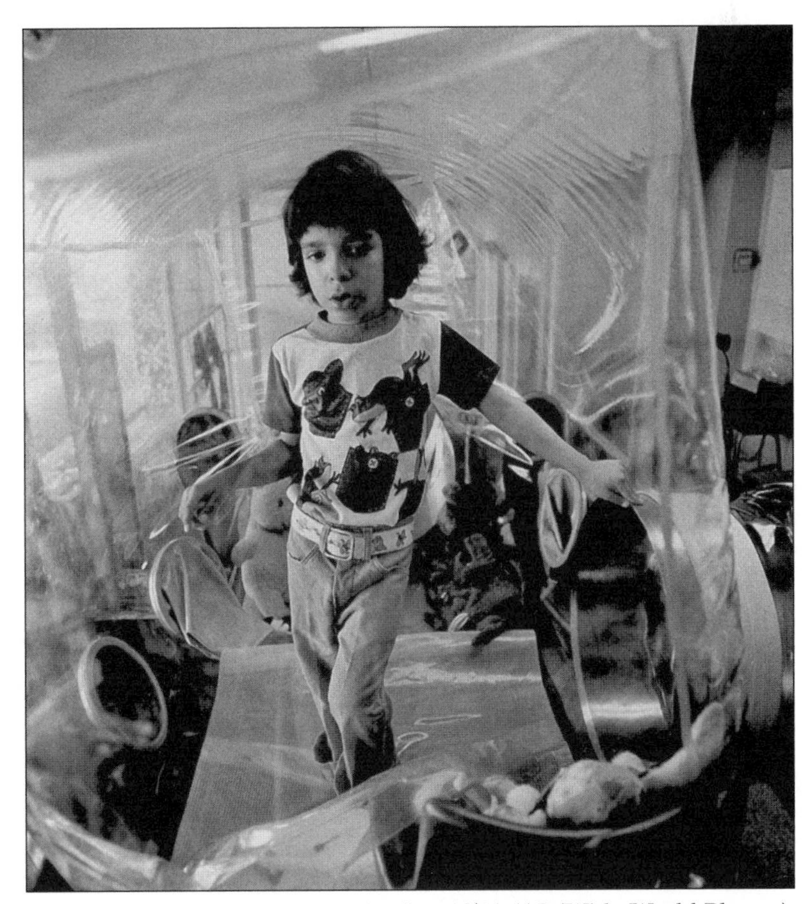

David Vetter, the original "bubble boy," in 1976. (AP/Wide World Photos)

newborns to diagnose SCID before infections occur is generally not done because the disease is rare and has a diverse genetic etiology and testing is expensive. Testing for SCID is performed if there is a known family history of the disease or the baby exhibits suggestive symptoms (such as persistent and recurrent infections). As of 2009, only two states screened newborns for SCID. Wisconsin used population-based screening with low TREC (T-cell receptor excision circles) as the marker for SCID. In Massachusetts, parents of newborns were offered the option of screening but were not required to have the test performed.

Current screening methods include measurement of lymphocyte counts (very low in SCID) and population-based screening using IL-7 or TREC. Ongoing studies are evaluating new methods for their value in screening newborns for the early diagnosis of SCID. Most promising is the DNA microarray gene chip that can detect the expression of thousands of genes simultaneously and may be able to detect both known and new mutations.

If SCID is suspected, the following tests may aid diagnosis: complete blood counts (low lymphocyte counts in SCID); T, B, and NK cell counts (T cells are absent or dysfunctional in all forms of SCID; NK and B cells may be absent depending on the form of SCID); and antibody levels (low in SCID; may be artificially high during the first three to four months of life due to maternal antibodies that cross the placenta during pregnancy). Specific genetic testing to identify underlying genetic defects is possible in about 90 percent of cases.

Treatment and Therapy

Initially, infections are treated with antibiotics, antifungal and antiviral drugs, and intravenous immunoglobulin (IVIg). However, the restoration of a functional immune system is necessary. The preferred treatment is bone marrow transplantation from a healthy donor. If performed within the first three to four months of life it is effective in the majority of patients. Blood-forming stem cells from the bone marrow of a healthy donor provide a functioning immune system that can protect the patient against infection. These stem cells can renew themselves as needed and produce a continuous supply of healthy immune cells.

In the ADA deficiency form of SCID, enzyme replacement therapy with weekly injections of polyethylene glycol-modified bovine ADA allows the immune cells to recover. This treatment is effective in the majority of cases.

Gene therapy is still a considered an experimental treatment option for SCID, and studies are ongoing. In this treatment, researchers correct the DNA mutation in a sample of the patient's T cells and then return these corrected T cells to the patient. Though this therapy has been successful in some patients, serious complications, such as a leukemia-like disorder, have occurred in others.

Prevention and Outcomes

Genetic counseling is recommended if there is a family history of the disease. Early diagnosis allows for early treatment and would significantly improve chances of a good outcome.

Anita P. Kuan, Ph.D.

Further Reading

Immune Deficiency Foundation. *IDF Patient and Family Handbook: For Primary Immunodeficiency Diseases.* 4th ed. Towson, Md.: Author, 2007.

"Newborn Screening for Severe Combined Immunodeficiency (SCID): A Review." *Frontiers in Bioscience* 10 (May 1, 2005): 1024-1039.

Web Sites of Interest

Biology News Net: "First SCID Gene Chip to Be Introduced at Academy Meeting on Immunodeficiencies"
http://www.biologynews.net/archives/2006/04/22/first_scid_gene_chip_to_be_introduced_at_academy_meeting_on_immunodeficiencies.html

Immune Deficiency Foundation
http://www.primaryimmune.org

National Institutes of Health: Office of Rare Disease Research
http://www.rarediseases.info.nih.gov

SCID Self-Help Support Group and Resource Guide
http://www.scid.net

See also: Agammaglobulinemia; Allergies; Antibodies; Anthrax; Ataxia telangiectasia; Autoimmune disorders; Autoimmune polyglandular syndrome; Chediak-Higashi syndrome; Chronic granulomatous disease; Hybridomas and monoclonal antibodies; Immunodeficiency with hyper-IgM; Immunogenetics; Infantile agranulocytosis; Myeloperoxidase deficiency; Purine nucleoside phosphorylase deficiency.

Shotgun cloning

Category: Genetic engineering and biotechnology

Significance: Shotgun cloning is the random insertion of a large number of different DNA fragments into cloning vectors. A large number of different recombinant DNA molecules are generated, which are then introduced into host cells, often bacteria, and amplified. Because a large number of different recombinant DNAs are generated, there is a high likelihood one of the clones contains a fragment of DNA of interest.

Key terms

cloning vector: a plasmid or virus into which foreign DNA can be inserted to amplify the number of copies of the foreign DNA

marker: a gene that encodes an easily detected product that is used to indicate that foreign DNA is in an organism

recombinant DNA: a novel DNA molecule formed by the joining of DNAs from different sources

restriction endonuclease: an enzyme that recognizes a specific nucleotide sequence in a piece of DNA and causes cleavage of the DNA; often simply called a restriction enzyme

Recombinant DNA Cloning and Shotgun Cloning

Before the development of recombinant DNA cloning, it was very difficult to study DNA sequences. Cloning a DNA fragment allows a researcher to obtain large amounts of that specific DNA sequence to analyze without interference from the presence of other DNA sequences. There are many uses for a cloned DNA fragment. For example, a DNA fragment can be sequenced to determine the order of its nucleotides. This information can be used to determine the location of a gene and the amino acid sequence of the gene's protein product. Cloned pieces of DNA are also useful as DNA probes. Because DNA is made of two strands that are complementary to each other, a cloned piece of DNA can be used to probe for copies of the same or similar DNA sequences in other samples. A cloned gene can also be inserted into an expression vector where it will produce the gene's protein product.

Shotgun cloning begins with the isolation of DNA from the organism of interest. In separate test tubes, the DNA to be cloned and the cloning vector DNA are digested (cut) with a restriction endonuclease that cuts the vector in just one location and the foreign DNA many times. Many restriction endonucleases create single-stranded ends that are complementary, so the end of any DNA molecule cut with that endonuclease can join to the end of any other DNA cut with the same endonuclease. When the digested vector and foreign DNA are mixed, they join randomly and are then sealed using DNA ligase, an enzyme that seals the small gap between two pieces of DNA. This creates recombinant DNA molecules composed of a copy of the vector and a random copy of foreign DNA. The recombinant DNA molecules are then introduced into host cells where the cloning vector can replicate each time the cell divides, which is approximately every twenty minutes in the case of *Escherichia coli.* The resulting collection of clones, each containing a potentially different fragment of foreign DNA, is called a genomic library. If a large collection of clones is produced, it is likely that every part of the genome from which the DNA came will be represented somewhere in the genomic library.

The presence of the cloning vector in host cells is determined by selecting for a marker gene in the cloning vector. Most vectors have two marker genes, and often both are different antibiotic resistance genes. A common example is the plasmid pBR322, which has a tetracycline and an ampicillin resistance gene. A restriction endonuclease cuts once somewhere in the tetracycline resistance gene, and if a foreign DNA fragment becomes incorporated, the resulting recombinant plasmid will have a nonfunctional tetracycline resistance gene. A bacterial cell transformed with a recombinant plasmid will therefore be resistant to ampicillin but will be sensitive to tetracycline. Many plasmids will not incorporate any foreign DNA and will be nonrecombinant. Cells that are transformed with a nonrecombinant plasmid will be resistant to both tetracycline and ampicillin. After the bacterial cells have been transformed, they are grown on a medium with ampicillin. The only cells that will survive will be those that have received a plasmid vector. To determine which cells have received a recombinant plasmid, the colonies are carefully transferred onto a new medium that has both ampicillin and tetracycline. On this medium, only cells with nonrecombinant plasmids will survive. Thus, colonies that grew on the first media, but not on the second, contain recombinant plasmids. Cells from these colonies are collected and grown, each in a separate tube, and these constitute a genomic library.

Once a genomic library has been produced, the DNA fragments contained in it can be screened and analyzed in various ways. Using the right techniques, specific genes can be found, which can then be used in future analyses and experiments.

Alternatives to Shotgun Cloning

In shotgun cloning, many different DNA fragments from an organism are cloned, and then the specific DNA clone of interest is identified. The number of clones can be reduced, making the search easier, if the DNA of interest is known to be in a restriction endonuclease fragment of a specific size. DNA can be size-selected before cloning using gel electrophoresis, in which an electric current carries DNA fragments through the pores or openings of an agarose gel. DNA migrates through the gel based on DNA fragment size, with the smaller fragments traveling more rapidly than the larger fragments. DNA of a specific size range can be isolated from the gel and then used for cloning. Finally, to clone a piece of DNA known to code for a protein, scientists can use an enzyme called reverse transcriptase to make DNA copies (called a complementary DNA or cDNA) of isolated messenger RNA (mRNA). The cDNA is then cloned, in a similar manner to that already discussed, to produce what is called a cDNA library. One of the advantages of this approach is that the number of clones is greatly reduced.

Susan J. Karcher, Ph.D.; updated by Bryan Ness, Ph.D.

Further Reading

Glick, Bernard R. *Molecular Biotechnology: Principles and Applications of Recombinant DNA.* 3d ed. Washington, D.C.: ASM Press, 2003. Covers the scientific principles of biotechnology and gives applications. Color illustrations.

Godiska, Ronald, et al. "Beyond pUC: Vectors for Cloning Unstable DNA." In *DNA Sequencing: Optimizing the Process and Analysis*, edited by Jan Kieleczawa. Sudbury, Mass.: Jones and Bartlett, 2005. Explains how to construct shotgun sequencing libraries.

Graham, Colin A., and Allison J. M. Hill, eds. *DNA Sequencing Protocols.* 2d ed. Totowa, N.J.: Humana Press, 2001. Includes two articles on shotgun sequencing: "The Universal Primers and the Shotgun DNA Sequencing Method" by Joachim Messing and "Shotgun DNA Sequencing" by Alan T. Bankier.

Kreuzer, Helen, and Adrianne Massey. *Recombinant DNA and Biotechnology: A Guide for Students.* 3d ed. Washington, D.C.: ASM Press, 2008. Designed for high school students. Provides introductory text and activities to learn the basics of molecular biology and biotechnology.

_______. *Recombinant DNA and Biotechnology: A Guide for Teachers.* 3d ed. Washington, D.C.: ASM Press, 2008. For high school and introductory college-level teachers. A guide to biotechnolgy with history, applications, simple protocols, and exercises.

Micklos, David, Greg A. Freyer, and David A. Crotty. *DNA Science: A First Course.* 2d ed. Cold Spring Harbor, N.Y.: Cold Spring Harbor Laboratory Press, 2003. Gives an introduction to molecular biology techniques for high school or beginning college students. Includes background text and laboratory procedures.

Sambrook, Joseph, and David W. Russell. "DNA Sequencing." In *Molecular Cloning: A Laboratory Manual.* 3d ed. Cold Spring Harbor, N.Y.: Cold Spring Harbor Laboratory Press, 2001. Discusses shotgun cloning techniques.

Watson, James D., et al. *The Molecular Biology of the Gene.* 6th ed. San Francisco: Pearson/Benjamin Cummings, 2008. Provides a description of shotgun cloning, as well as an overview of many other cloning methods. Illustrations, diagrams, bibliography, index.

Web Sites of Interest

Genome News Network, Sequencing the Genome
http://www.genomenewsnetwork.org/articles/06_00/sequence_primer.shtml

Compares the whole genome shotgun sequencing technique to other sequencing methods.

Scitable
http://www.nature.com/scitable/topicpage/Complex-Genomes-Shotgun-Sequencing-609

Scitable, a library of science-related articles compiled by the Nature Publishing Group, contains the article "Complex Genomes: Shotgun Sequencing," which discusses the sequencing technique in insects, rice, mice, and other organisms.

See also: cDNA libraries; Cloning; Cloning vectors; Genomic libraries; Genomics; Human Genome Project; Restriction enzymes.

Sickle-cell disease

CATEGORY: Diseases and syndromes
ALSO KNOWN AS: Sickle-cell anemia; hemoglobin S/S disease; sickle-hemoglobin C disease; sickle-beta thalassemia

DEFINITION

Sickle-cell disease, also known as sickle-cell anemia, is a hereditary blood disorder that affects red blood cells. Red blood cells undergo hemolysis (breakage) and change shape, which causes vascular occlusive crises (small blood vessel blockages). This results in acute and chronic pain, as well as multi-organ dysfunction.

RISK FACTORS

Individuals of African, Mediterranean, Middle Eastern, Indian, Central American, and Caribbean ancestry have the greatest chance of having a mutation in the beta globin gene (*HBB*). Approximately 1 in 12 individuals of African American ancestry have sickle-cell trait, which means approximately 1 in 600 babies born in the United States to African American parents have sickle-cell disease.

ETIOLOGY AND GENETICS

Sickle-cell disease was the very first example of a genetic disease being traced to its precise origin at the molecular level. The *HBB* gene, located on chromosome 11, encodes the main type of adult hemoglobin (hemoglobin A). Hemoglobin is the major protein in red blood cells whose properties allow it to bind with oxygen in the lungs and transport it to other parts of the body.

Sickle-cell anemia occurs when there is a single point mutation with one nucleotide (thymine instead of adenine) of the *HBB* gene. This causes a substitution of amino acids: Glutamic acid in the sixth codon is replaced by valine. The change of amino acids causes an absence in normal hemoglobin A production. Consequently, a structurally abnormal hemoglobin S is produced. For individuals with sickle-cell anemia, the mutation causes the hemoglobin molecules to stick to one another and changes the normally smooth and flexible, donut-shaped appearance of the red blood cell to a characteristic sickled crescent moon shape. This prevents red blood cells from traveling through tiny capillaries, especially under conditions of oxygen deprivation. On average, the life span of red blood cells of healthy individuals is 120 days before they are replaced by new cells. In individuals with sickle-cell disease, the red blood cells have a reduced life span of only 16 days.

Sickle-cell disease is inherited in an autosomal recessive manner. One mutation in the *HBB* gene means that a person is a carrier for sickle-cell disease, which is also known as having sickle-cell trait (hemoglobin A/S). If both parents have sickle-cell trait, then there is a 25 percent risk for each child to have sickle-cell disease because both parents may pass the trait on to a pregnancy (hemoglobin S/S). There is a 50 percent risk for each child to have sickle-cell trait because only one parent may pass on sickle-cell trait and the other parent may pass on the unaffected allele. Lastly, there is a 25 percent chance that the child will be unaffected and not have the trait because neither parent passes on sickle-cell trait (hemoglobin A/A). If only one parent has sickle-cell trait and the other parent is found to not be a carrier, then there is no chance to have an affected child but there does remain a 50 percent risk that the child may have sickle-cell trait.

Sickle-cell disease may occur in classical form, as described above for sickle-cell anemia, or may occur as other forms of the blood disorder. Hemoglobin S/S occurs in 60 to 70 percent of affected individuals. There are many mutations that have been reported in the beta globin gene which can interact with sickle-cell trait and cause abnormal hemoglobin to be produced, such as mutations for hemoglobin C trait or beta thalassemia trait. If an individual inherits sickle-cell trait from one parent and hemoglobin C trait from another, for example, then the

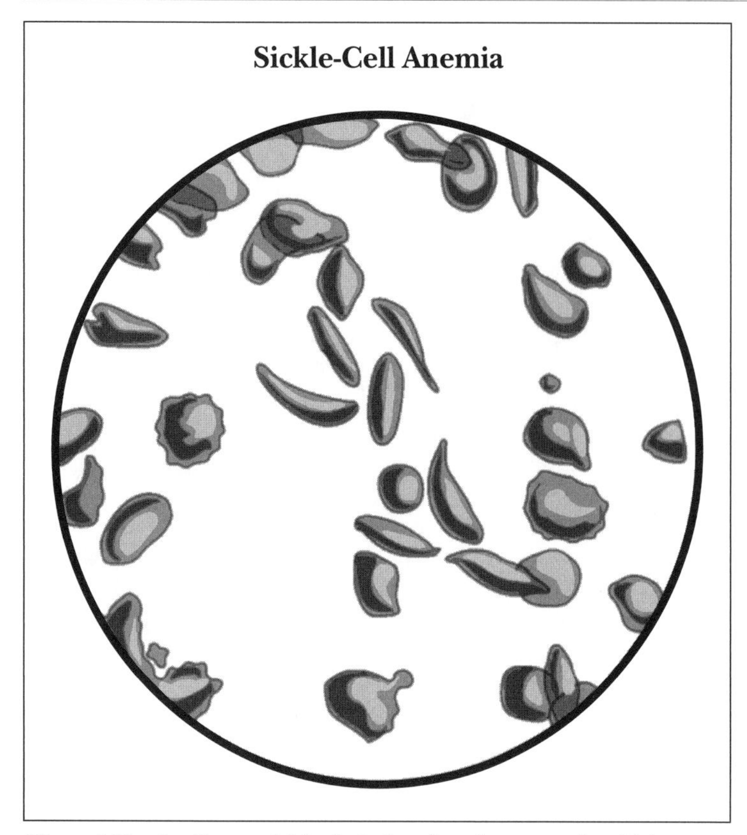

The red blood cells are sickle-shaped rather than round, which causes blockage of capillaries. (Hans & Cassidy, Inc.)

pregnancy is at 25 percent risk for sickle-hemoglobin C disease, which is often less severe than classical sickle-cell anemia but may show similar symptoms. HbVar, a database of hemoglobin variants, lists more than a thousand entries for globin gene mutations. These mutations may lead to hundreds of clinically significant blood disorders.

Sickle-cell disease is one of the best-documented examples of an evolutionary process known as "heterozygote advantage," in which carriers have a greater probability of surviving or reproducing than either an affected or unaffected, noncarrier individual. In most cases, hereditary diseases with such significant symptoms as those associated with sickle-cell disease are kept at low frequencies in populations by natural selection. However, having sickle-cell trait provides an advantage in environments such as Africa were malaria is prevalent. Malaria is a deadly, mosquito-borne disease which uses human red blood cells as hosts for part of its life cycle. Individuals who have hemoglobin A/A are vulnerable to the disease, and people who have sickle-cell disease have an increased risk for early death because of anemia and other complications. However, when the red blood cells of individuals with sickle-cell trait are invaded by the malarial parasite, the red blood cells adhere to blood vessel walls, become deoxygenated, and assume a sickle shape, prompting both their destruction and that of their parasitic invader for malaria. This provides a sickle-cell carrier with a natural resistance to malaria and explains the relatively high frequency of the sickle-cell gene in such environments. This is an important phenomenon from an evolutionary standpoint because it provides a mechanism by which genetic diversity in a population may be preserved.

SYMPTOMS

Individuals with sickle-cell trait typically display no symptoms, although some have been known to become ill under extreme circumstances, such as high altitudes with decreased oxygen supply. Symptoms of sickle-cell disease typically appear about six to twelve months after birth, when the last of fetal hemoglobin (hemoglobin F), a type of hemoglobin that increases the oxygen supply of blood in pregnancy, decreases and hemoglobin S increases. The severity of the illness varies widely among individuals.

Individuals have varying degrees of red cell breakdown, which may cause a decrease in the amount of red blood cells (hemolytic anemia), jaundice, and physical weakness. The change in the red blood cell shape leads to a vaso-occlusive crisis (sickle-cell crisis), which causes a lack of oxygen to be delivered to the body's tissues. Subsequently, this causes organ dysfunction and may cause significant pain in joints and bones. The organs that are most often involved include the spleen, lungs, brain, kidneys, and genitalia. Individuals typically have an enlarged spleen, and the compromised function predisposes mostly children but also adults to infections. The lungs may have significant life-threatening complications such as damage or changes to the lungs (acute chest syndrome), as well as a constriction of blood supply to the lungs (pulmonary hypertension). Other significant complications include the possibility of a stroke or blindness. Individuals may have priapism, a long and painful erection, as well as delayed growth and puberty. Despite the many treatments that exist, there remains an increased risk for medical problems and mortality. The mean age of death is forty-two years for males and forty-eight years for females.

SCREENING AND DIAGNOSIS

The diagnosis of sickle-cell disease may be made by several different laboratory techniques. High-performance liquid chromatography (HPLC), isoelectric focusing (IEF), hemoglobin electrophoresis, and peripheral blood smear are all available testing options. Almost all cases of sickle-cell disease are diagnosed at birth. All fifty states currently provide newborn screening for sickle-cell disease by one of these methods. Sickledex (sickleprep, solubility test) is an outdated method of testing for sickle-cell disease, as an abnormal test result cannot distinguish sickle-cell trait from an affected individual, nor can it tell if an individual has a different form of sickle-cell disease from hemoglobin S/S. Molecular genetic testing is available to detect mutations in the HBB gene associated with sickle-cell disease. Mutation analysis is typically reserved for when sickle-cell trait is suspected to be inherited with another hemoglobin variant or for purposes of diagnosing a fetus in pregnancy.

Any individual planning a pregnancy or who is currently pregnant that is from the aforementioned ethnic backgrounds should be offered screening for sickle-cell disease. If a pregnancy is found to be at 25 percent risk for sickle-cell disease, then prenatal diagnosis is available. Prenatal diagnosis includes testing options of a chorionic villus sampling (CVS) at ten to twelve weeks of gestation or amniocentesis at approximately fifteen to twenty weeks of gestation. In CVS, a sample from the placenta is obtained, and in amniocentesis a sample of fetal cells from amniotic fluid is withdrawn. Both diagnostic testing options provide a more than 99 percent detection rate of sickle-cell disease.

Genetic counseling is available for individuals at risk for having a pregnancy with sickle-cell disease or for individuals who are diagnosed by newborn screening with sickle-cell disease. Genetic counselors may review testing options and participate in the multidisciplinary management and education of children or adults with the diagnosis.

PREVENTION AND OUTCOMES

Despite advancing treatments, there still exists no cure for sickle-cell disease. Stem cell transplantation, whereby an individual receives stem cells from an unaffected individual, has the potential to cure symptoms of sickle-cell disease. Clinical trials with animal models and human cell lines are ongoing. Bone marrow transplantation is similar in theory to stem cell transplantation, but an individual donating bone marrow must be related to the affected individual and be an immunological match. There are significant risks with this procedure that put the individual at risk for developing complications, such as bone marrow rejection, infection, bleeding, and possible death. Therefore, this is currently reserved for a select population of individuals.

Much early research on sickle-cell disease was performed by Linus Pauling (right), with George Beadle, c. 1952. (California Institute of Technology)

It may be possible to prevent the onset of symptoms by proper fluid hydration, avoiding high altitudes with poor oxygenation and climates with extreme temperatures, and taking preventive medications. Penicillin is administered to all children to reduce infections, and vaccinations are also recommended to reduce the chance of other illnesses. A folic acid supplement may also be prescribed.

Current treatments include oral medication and fluid hydration for vaso-occlusive pain and crises. Typical medications that are prescribed for patients to manage their pain crisis at home include acetaminophen or ibuprofen. However, more severe pain crises require hospitalization for stronger medications, including morphine, and intravenous fluids. Blood transfusions may be performed during severe pain episodes or in an attempt to prevent a further complication, such as a stroke. Hydroxyurea is a chemotherapy agent used to increase the amount of fetal hemoglobin and break down sickled cells. It has proven successful at reducing the amount of pain episodes and need for blood transfusions and hospitalizations. The drawbacks of this treatment are that it does not prevent the brain complications of sickle-cell disease and that the treated individual is at a higher risk for infection and possible other complications such as leukemia. Much research is ongoing regarding additional potential medications and treatment for sickle-cell disease.

Prevention of sickle-cell disease itself may be possible in pregnancy via the technology of preimplantation genetic diagnosis (PGD). PGD is a laboratory technique of testing embryos or egg cells (oocytes) prior to fertilization. Only unaffected embryos or oocytes, with or without sickle-cell trait, would be implanted in the uterus for pregnancy by in vitro fertilization. The main advantage of PGD technology is that parents who would not end a pregnancy with sickle-cell disease diagnosed by prenatal diagnosis have a high probability of not having an affected child.

Janet Ober Berman, M.S., C.G.C.

FURTHER READING

ACOG Committee on Obstetrics "Hemoglobinopathies in Pregnancy." *Obstetrics and Gynecology* 109 (2007): 229-237. Practice recommendations for carrier screening for sickle-cell disease and other blood disorders in or prior to pregnancy. Illustrations for testing algorithm.

MacMillin, M., ed. *Ancestry Based Carrier Screening.* Chicago: National Society of Genetic Counselors, 2005. A guide for which populations should be offered screening for sickle-cell disease and other genetic disorders. Tables for carrier frequency.

Pack-Mabien, A. "A Primary Care Provider's Guide to Preventative and Acute Care Management of Adults and Children with Sickle-Cell Disease." *Journal of the American Academy of Nurse Practitioners* 21 (2009): 250-257. In-depth coverage for symptoms and management of sickle-cell disease as well as new treatment options.

Serjeant, Graham R., and Beryl E. Serjeant. *Sickle Cell Disease.* 3d ed. New York: Oxford University Press, 2001. Discusses the biology of sickle-cell disease and the disease's management. Illustrations, maps, bibliography, index.

Steinberg, Martin H., et al., eds. *Disorders of Hemoglobin: Genetics, Pathophysiology, and Clinical Management.* Foreword by H. Franklin Bunn. New York: Cambridge University Press, 2001. Covers the diseases' molecular and genetic bases, epidemiology and genetic selection, and diagnoses and treatments. Illustrations (some color), bibliography, index.

WEB SITES OF INTEREST

American Sickle Cell Anemia Association
www.ascaa.org

Dolan DNA Learning Center, Your Genes Your Health
http://www.ygyh.org

Gene Reviews
http://www.ncbi.nlm.nih.gov/bookshelf/br.fcgi?book=gene&part=sickle

HbVar
http://globin.cse.psu.edu/hbvar/menu.html

Sickle Cell Disease Association of America
www.sicklecelldisease.org

Sickle Cell Information Center
http://www.scinfo.org

See also: ABO blood types; Amniocentesis; Biopharmaceuticals; Chorionic villus sampling; Chronic myeloid leukemia; Fanconi anemia; Genetic engineering; Genetic screening; Genetic testing; Hardy-Weinberg law; Hemophilia; Hereditary spherocytosis; Incomplete dominance; Infantile agranulocytosis; Mutation and mutagenesis; Myelodysplastic syndromes; Paroxysmal nocturnal hemoglobinuria; Porphyria; RFLP analysis; Rh incompatibility and isoimmunization; Shotgun cloning; Von Willebrand disease.

Signal transduction

CATEGORY: Molecular genetics

SIGNIFICANCE: Signal transduction consists of all of the molecular events that occur between the arrival of a signaling molecule at a target cell and its response. A significant proportion of the genome in animals consists of genes involved in cell signaling. The protein products of these genes allow cells to communicate with each other in order to coordinate their metabolism, movements, and reproduction. Failure of cells to communicate properly can lead to cancer, defects in embryological development, and many other disorders.

KEY TERMS

cell cycle: the orderly sequence of events by which a cell grows, duplicates its chromosomal DNA, and partitions the DNA into two new cells

cell signaling: communication between cells that occurs most commonly when one cell releases a specific "signaling" molecule that is received by another cell

receptors: molecules in target cells that bind specifically to a particular signaling molecule

target cell: the cell that receives and responds to a signaling molecule

SIGNAL TRANSDUCTION PATHWAYS

Signal transduction can occur by a number of different, often complex, sequences of molecular events called signal transduction pathways, which result in several kinds of target cell response, including the turning on of genes, the activation of metabolic pathways, and effects on the cell cycle. Among the signaling molecules found in higher organisms are hormones, local mediators that produce local physiological effects, growth factors that act locally to promote growth, and survival factors that act locally to repress cell suicide (apoptosis). Growth factors and survival factors are particularly important during embryological development, when they orchestrate the changes in cell types, positions, and numbers that give rise to the new organism.

TYPES OF RECEPTORS

Most signal transduction pathways begin with the binding of signaling molecules to specific receptors in target cells. Signaling molecules are often referred to as receptor ligands. The binding of the ligand to its receptor initiates a signal transduction pathway. A cell can respond to a particular signaling molecule only if it possesses a receptor for it.

Receptors are protein molecules. There are two categories of them, based on location in the cell: receptors that are intracellular and receptors that are anchored in the cell's surface membrane. The membrane-anchored receptors can be further divided based on the steps of the signal transduction pathway that they initiate: receptors that bind to and activate GTP-binding proteins (G proteins), receptors that are enzymes, and receptors that are ion channels. Receptors that are channels bind neurotransmitters or hormones and increase or decrease the flow of specific ions into the cell, leading to a physiological response by the cell. These receptors generally do not have a direct effect on gene expression (although changes in a cell's calcium ion concentrations can influence gene expression). Each of the other receptor types stands at the head of a signal transduction pathway that is characteristic for each receptor type and can lead to gene expression. In what follows, some of the more common transduction pathways that can lead to gene expression are described.

INTRACELLULAR RECEPTORS

Intracellular receptors include the receptors for lipid-soluble hormones such as steroid hormones. Some of these receptors are in the cell's cytoplasm, and some are in the nucleus. Hormone molecules enter the cell by first diffusing across the membrane and then binding to the receptor. Before the hormones enter the cell, the receptors are attached to "chaperone" proteins, which hold the receptor in a configuration that allows hormone binding but prevents it from binding to DNA. Hormone molecules displace these chaperone molecules, enabling the receptor to bind to DNA. If the receptor is a cytoplasmic receptor, the hormone-receptor complex is first transported into the nucleus, where it binds to a specific DNA nucleotide sequence called a hormone response element (HRE) that is part of the promoter of certain genes. In most cases the receptors bind as dimers; that is, two hormone-receptor complexes bind to the same HRE. The receptor-hormone complex functions as a transcription factor, promoting transcription of the gene and production of a protein that the cell was not previously producing. The hormone hydrocortisone, for example, triggers the synthesis of the enzymes aminotransferase and tryptophan oxygenase. A single hormone such as hydrocortisone can turn on synthesis of two or more proteins if each of the genes for the proteins contains an HRE. In some cases, when hormone-receptor complexes bind to an HRE, they suppress transcription rather than promote it.

G Protein-Binding Receptors

Many hormones, growth factors, and other signaling molecules bind to membrane receptors that can associate with and activate heterotrimeric G proteins when a signaling molecule is bound to the receptor. Heterotrimeric G proteins are a family of proteins that are present on the cytoplasmic surface of the cell membrane. Many cell types in the body contain one or more of these family members, and different cell types contain different ones. All heterotrimeric G proteins are made up of three subunits: the alpha, beta, and gamma subunits. The alpha subunit has a binding site for GTP or GDP (hence the name G proteins) and is the principal part of the protein that differs from one heterotrimeric G protein family member to another. When the receptor is empty (no signal molecule attached), these G proteins have GDP bound to the alpha subunit and the G protein is not bound to the receptor.

However, when a signaling molecule binds to the receptor, the cytoplasmic domain of the receptors changes shape so that it now binds to the G protein. In binding to the receptor, the G protein also changes shape, causing GDP to leave and GTP to bind instead. Simultaneously, the alpha subunit detaches from the beta-gamma subunit, and both the alpha subunit and the beta-gamma subunit detach from the receptor. The alpha subunit or the beta-gamma subunit (depending upon the particular G protein family member involved and the cell type) then activates (or with some G protein family members, inhibits) one of several enzymes, most commonly adenylate cyclase or phospholipase C. Alternatively, they can open or close a membrane ion channel, altering the electrical properties of the cell; for example, potassium ion channels in heart muscle cells can be opened by G proteins in response to the neurotransmitter acetylcholine.

In cases where adenylate cyclase or phospholipase C is activated, these enzymes catalyze reactions that produce molecules called second messengers, which, through a series of steps, activate proteins that lead to a physiological response (such as contraction of smooth muscle), a biochemical response (such as glycogen synthesis), or a genetic response (such as activating a gene).

Activation of adenylate cyclase causes it to catalyze the conversion of adenosine triphosphate (ATP) to the second messenger cyclic adenosine monophosphate (cAMP), which in turn activates a protein called protein kinase A, which, in some cells, moves into the nucleus and phosphorylates and activates transcription factors such as CREB (CRE-binding protein). CREB binds to a specific DNA sequence in the promoter of certain genes called the CRE (cAMP-response element), as well as to other transcription factors, to activate transcription of the gene. In other cells, protein kinase A activates enzymes or other proteins involved in physiological or metabolic responses.

Activation of phospholipase C catalyzes the breakdown of a glycolipid component of the cell membrane called phosphatidylinositol bisphosphate (PIP2) into two second messengers, inositol triphosphate (IP3) and diacylglycerol (DAG). DAG activates a protein called protein kinase C (PK-C), which in turn activates other proteins, leading to various cell responses, including, in certain cells of the immune system, activation of transcription factors which turn on genes involved in the body's immune response to infection. IP3 causes the release of calcium ions stored in the endoplasmic reticulum. These ions bind and activate the protein calmodulin, which activates a variety of proteins, leading in most cases to a physiological response in the cell.

Catalytic Receptors

Catalytic receptors are receptors that function as enzymes, catalyzing specific reactions in the cell.

The part of the receptor that is in the cytoplasm (the cytoplasmic domain) has catalytic capability. Binding of a signaling molecule to the external domain of the receptor activates the catalytic activity of the cytoplasmic domain. There are several kinds of catalytic receptors based on the type of reaction they catalyze; these include receptor tyrosine phosphatases, receptor guanylate cyclases, receptor serine/threonine kinases, and receptor tyrosine kinases. Receptor tyrosine kinases (RTKs) are the most common of these.

RTKs are the receptors for many growth factors and at least one hormone. For example, they are the receptors for fibroblast growth factor (FGF), epidermal growth factor (EGF), platelet-derived growth factor (PDGF), nerve growth factor (NGF), and insulin. RTKs play a role in regulating many fundamental processes, such as cell metabolism, the cell cycle, cell proliferation, cell migration, and embryonic development. In most cases, when a ligand binds to this type of receptor, a conformational (shape) change occurs in the receptor so that it binds to another identical receptor-ligand complex to produce a double, or dimeric, receptor. The dimeric receptor then catalyzes a cytoplasmic reaction in which several tyrosine amino acids in the cytoplasmic domain of the receptor itself are phosphorylated. The phosphorylated tyrosines then function as docking sites for several other proteins, each of which can initiate one of the many branches of the RTK signal transduction pathway, leading to the various cell responses. One of the major branches of the RTK pathway that in many cases results in gene expression begins with the binding of the G protein ras (ras is not one of the trimeric G proteins discussed above) to the activated RTK receptor via adapter proteins. Binding of ras to the adapter proteins activates it by allowing it to bind GTP instead of GDP. Activated ras then phosphorylates the enzyme MEK, which phosphorylates and activates an enzyme of the MAP kinase family. In cases where this enzyme is MAP kinase itself, the enzyme dimerizes, moves into the nucleus, and activates genes, usually many genes, by phosphorylating and activating their transcription factors.

Signal Transduction and the Cell Cycle

The biochemical machinery that produces the cell cycle consists of several cyclins whose concentrations rise and fall throughout the cell cycle. Cyclins activate cyclin-dependent kinases (cdk), which activate the proteins that carry out the events of each stage of the cell cycle. In higher organisms, control of the cell cycle is carried out primarily by growth factors. In the absence of growth factors, many cells will stop at a point in the cell cycle known as the G_1 checkpoint and cease dividing. The cell cycle is started when the cells are exposed to a growth factor. For example, some growth factors start cell division by binding to a membrane receptor and initiating the RTK/MAP kinase signal transduction pathway. The activated transcription factor that results from this pathway activates a gene called *myc.* The protein that is produced from this gene is itself a transcription factor, which activates the cyclin D gene, which produces cyclin D, an important component of the cell cycle biochemical machinery. Cyclin D activates cyclin-dependent kinase 4 (cdk4), which drives the cell into the G_1 phase of the cell cycle. cdk4 also causes an inhibiting molecule called pRB to be removed from a transcription factor for the cyclin E gene. Cyclin E is then produced and activates cyclin-dependent kinase 2 (cdk2), which drives the cell into the S phase of the cell cycle, during which chromosomal DNA is replicated, leading to cell division by mitosis.

Signal Transduction and Cancer

Cancer is caused primarily by uncontrolled cell proliferation. Since many signal transduction pathways lead to cell proliferation, it is not surprising that defects in these pathways can lead to cancer. For example, as described above, many growth factors promote cell proliferation by activating the RTK/MAP kinase signal transduction pathway. In that pathway, a series of proteins is activated (ras, MAP kinase, and so on). If a mutation occurred in the gene for one of these, ras for example, such that the mutant ras protein is always activated rather than being activated only when it binds to the receptor, then the cell would always be dividing and cancerous growth could result. Another example would be if the gene for pRB that binds to and inhibits the cyclin E transcription factor were mutated such that the pRB could never bind to the transcription factor; then the cell would divide continuously. Mutations in both ras and pRB are in fact known to cause cancer in humans.

Robert Chandler, Ph.D.

FURTHER READING

Alberts, Bruce, et al. "Mechanics of Cell Communication." In *Molecular Biology of the Cell.* 5th ed. New York: Garland Science, 2008. One of the standard textbooks in the field of cell biology. The chapter on cell communication focuses on cell signaling.

Cell 103, no. 2 (October 13, 2000): 181-320. A special issue of the journal *Cell* devoted entirely to the topic of cell signaling. The cited pages contain three "minireviews" and eleven reviews of the primary literature. A good entry into the primary literature.

Gomperts, Bastien D., Peter E. R. Tatham, and Ijsbrand M. Kramer. *Signal Transduction.* 2d ed. San Diego: Academic Press, 2009. Provides comprehensive yet readable coverage of signal transduction. Contains excellent illustrations and citations of other literature.

Hoch, James A., and Thomas J. Silhavy, eds. *Two-Component Signal Transduction.* Washington, D.C.: ASM Press, 1995. Written for microbiologists working in the areas of gene expression, pathogenesis, and bacterial metabolism. Covers the molecular and cellular biology of a wide variety of two-component signal transduction systems in bacteria. Illustrated.

Krauss, Gerhard. *Biochemistry of Signal Transduction and Regulation.* 4th, enl. and improved ed. Weinheim, Germany: Wiley-VCH Verlag, 2008. Textbook explains the molecular basis of signal transduction, as well as regulated gene expression, the cell cycle, and apoptosis.

Lodish, Harvey, et al. *Molecular Cell Biology.* 6th ed. New York: W. H. Freeman, 2008. One of the standard textbooks in the field of cell biology. Chapter 20 provides detailed information on signal transduction.

Marks, Friedrich, Ursula Klingmüller, and Karin Müller-Decker. *Cellular Signal Processing: An Introduction to the Molecular Mechanisms of Signal Transduction.* New York: Garland Science, 2009. Designed for undergraduate students, this textbook examines the cell as a "data processing unit in which the proteins form an extremely complex and ever-changing network of interactions."

Nelson, John. *Structure and Function in Cell Signalling.* Hoboken, N.J.: John Wiley and Sons, 2008. An accessible introduction to signal transduction for students.

WEB SITES OF INTEREST

BioChemWeb.org: Signal Transduction
http://www.biochemweb.org/signaling.shtml
An annotated list of online resources about cell signaling and related topics.

Kimball's Biology Pages
http://users.rcn.com/jkimball.ma.ultranet/BiologyPages/C/CellSignaling.html
John Kimball, a retired Harvard University biology professor, includes a page describing cell signaling in his online cell biology text.

UCSD-Nature Cell Signaling Gateway
http://www.signaling-gateway.org
The gateway is sponsored by the University of California, San Diego, and the Nature Publishing Group and provides information about cell signaling research, covers news developments, and offers other resources regarding cell signaling.

See also: Burkitt's lymphoma; Cancer; Cell cycle; Cell division; DNA replication; Gene regulation: Bacteria; Model organism: *Saccharomyces cerevisiae*; Oncogenes; One gene-one enzyme hypothesis; Steroid hormones; Tumor-suppressor genes.

Small-cell lung cancer

CATEGORY: Diseases and syndromes

ALSO KNOWN AS: Oat cell carcinoma; small-cell anaplastic carcinoma; undifferentiated small-cell carcinoma; intermediate cell type; and mixed small-cell/large-cell carcinoma.

DEFINITION

Small-cell lung cancer (SCLC) is one of the two major types of lung cancer, the other being non-small-cell lung cancer (NSCLC). SCLC originates from nerve cells or hormone-producing cells in the lungs, whereas NSCLC originates from epithelial cells that make up the lining of the lungs.

RISK FACTORS

Smoking is the greatest risk factor for SCLC, with an estimated 87 percent of lung cancers caused by smoking. Other risk factors are exposure to radon, asbestos, beryllium, cadmium, silica, vinyl chloride, nickel- or chromium-containing compounds, coal products, mustard gas, chloromethyl esters, diesel exhaust, radiation to the chest, arsenic, air pollution, radioactive substances like uranium, and genetic factors.

Etiology and Genetics

The National Cancer Institute estimates that 219,440 new lung cancer cases and 159,390 deaths will occur in 2009, with SCLC accounting for about 15 percent of these. Cancer in general is considered to be a genetic disease because cancer is associated with genetic changes in somatic cells that lead to uncontrolled cell growth. Lung cancer is not caused by a single gene. The human genome is made up of pairs of chromosomes. When tissue from the lungs of smokers is genetically analyzed, several areas of the DNA that makes up chromosomes and contain the genes are found to be damaged, duplicated, or missing. However, not all smokers with genetic damage in their lung tissue have lung cancer, suggesting that multiple sites of DNA damage must accumulate before lung cancer develops. Not every SCLC case has the same set of mutated genes present, but damage to specific regions on chromosomes 1, 3, 4, 5, 8, 9, 10, 11, 13, 15, 16, 17, 18, and 20 occurs frequently in SCLCs. In SCLC and other cancers, one of the most commonly mutated genes is called *p53*, a gene that usually causes damaged cells to undergo programmed cell death, known as apoptosis. When the *p53* gene is mutated, the damaged cells do not die and can go on to become malignant cells.

While 85 percent of all lung cancers occur in former or current smokers, fewer than 20 percent of long-term smokers develop lung cancer, suggesting that genetic factors make some people more susceptible to lung cancer. In 2009, researchers identified two sites of genetic variation on chromosome 15 that people can inherit from their parents that are associated with an increased risk of lung cancer in smokers and former smokers. These genetic variations, called single nucleotide polymorphisms (SNPs), are places in the human genome that vary by a single DNA nucleotide. People who have one or two copies of either of these SNPs and who have ever smoked have increased risks of 28 percent to 81 percent of developing lung cancer. The two SNPs identified are near five genes on chromosome 15. Interestingly, three of these five genes encode proteins that serve as docking sites for nicotine, the addictive agent in tobacco. These findings suggest that nicotine, in addition to several known carcinogens present in tobacco smoke, may be involved in lung cancer development. Another of the genes near the SNPs encodes a component of the proteasome, an important organelle in cells that degrades other proteins. The function of the fifth gene is unknown.

Symptoms

In early-stage SCLC, patients often do not have symptoms. Symptoms of lung cancer include a cough that does not go away; chest pain that often gets worse with deep breathing, coughing, or laughing; hoarseness; unexplained weight loss or lack of appetite; bloody or rust-colored spit or phlegm (also called sputum); shortness of breath; recurrent infections like bronchitis and pneumonia; and new onset of wheezing. Small-cell lung cancer commonly spreads or metastasizes to other organs. Symptoms of cancer that has spread include bone pain; weakness or numbness in the limbs; headache, dizziness, or seizure; jaundice; and lumps near the surface of the body such as lymph nodes in the neck. Those who experience these symptoms should see a physician right away.

Screening and Diagnosis

Chest X rays and checking sputum under a microscope to look for cancer cells was studied as a screening tool for lung cancer for several years. Unfortunately, this kind of screening is not very effective, with very few lung cancers found early enough to improve a person's chances for a cure. Because this kind of screening is not very effective, it is not done routinely, even in smokers who are at high risk. A new screening method, spiral computed tomography (CT) scanning, is being tested in the National Lung Screening Trial to find early lung cancer in smokers and former smokers. However, until the research is completed, it is unknown whether this test will detect lung cancers at an early stage and improve a patient's chances of surviving lung cancer.

Treatment and Therapy

Small-cell lung cancer has usually spread or metastasized to other organs before it is diagnosed; therefore, it is usually not treated with surgery. Patients with advanced-stage disease are usually treated with platinum-based chemotherapy (cisplatin or carboplatin), which may be combined with the drugs etoposide or camptothecin-11. Patients whose tumors respond very well to chemotherapy may also be offered radiation therapy.

Prevention and Outcomes

The most important prevention for SCLC is to not smoke, or to quit smoking. After quitting smoking, it takes about ten years to reduce the risk of lung cancer in former smokers, but the risk never drops to the level of those who never smoked.

Nancy E. Price, Ph.D.

Further Reading

Broderick, Peter, et al. "Deciphering the Impact of Common Genetic Variation on Lung Cancer Risk: A Genome-Wide Association Study." *Cancer Research* 69 (2009): 6633-6641. Discusses the research that identified the two inherited SNPs on chromosome 15 that make some people more susceptible to lung cancer.

Girard, L., et al. "Genome-Wide Allelotyping of Lung Cancer Identifies New Regions of Allelic Loss, Differences Between Small Cell Lung Cancer and Non-small Cell Lung Cancer, and Loci Clustering." *Cancer Research* 60 (2000): 4894-4906. Discusses differences in the locations of chromosomal damage in SCLC and NSCLC, and lists the most frequent sites of damage in both lung cancer types.

Hansen, Heine H., and Paul Bunn. *Lung Cancer Therapy Annual 4.* New York: Taylor and Francis, 2005. Discusses recent developments in the diagnosis and treatment of lung cancer.

Travis, William D., Elizabeth Brambilla, H. Konrad Müller-Hermelink, and Curtis C. Harris, eds. *World Health Organization Classification of Tumours: Pathology and Genetics Tumours of the Lung, Pleura, Thymus, and Heart.* Lyon, France: IARC Press, 2004. A reference on the histological and genetic typing of tumors of the lung.

Web Sites of Interest

American Cancer Society: Learn About Lung Cancer—Small Cell
http://www.cancer.org/docroot/lrn/lrn_0.asp

Management of Small Cell Lung Cancer: ACCP Evidence-Based Clinical Practice Guidelines (2d ed.)
http://www.guideline.gov/summary/summary.aspx?doc_id=11420

National Cancer Institute: General Information About Small Cell Lung Cancer
http://www.cancer.gov/cancertopics/pdq/treatment/small-cell-lung/patient

See also: Bloom syndrome; *BRAF* gene; *BRCA1* and *BRCA2* genes; Breast cancer; Burkitt's lymphoma; Chemical mutagens; Chromosome mutation; Chronic myeloid leukemia; Colon cancer; Cowden syndrome; *DPC4* gene testing; Familial adenomatous polyposis; Gene therapy; Harvey *ras* oncogene; Hereditary diffuse gastric cancer; Hereditary diseases; Hereditary leiomyomatosis and renal cell cancer; Hereditary mixed polyposis syndrome; Hereditary non-VHL clear cell renal cell carcinomas; Hereditary papillary renal cancer; Homeotic genes; *HRAS* gene testing; Hybridomas and monoclonal antibodies; Li-Fraumeni syndrome; Lynch syndrome; Mutagenesis and cancer; Multiple endocrine neoplasias; Mutation and mutagenesis; Nondisjunction and aneuploidy; Oncogenes; Ovarian cancer; Pancreatic cancer; Prostate cancer; Tumor-suppressor genes; Wilms' tumor aniridia-genitourinary anomalies-mental retardation (WAGR) syndrome.

Smallpox

CATEGORY: Diseases and syndromes; Viral genetics

Definition

Smallpox is a poxvirus, any of the family of viruses that produce pustules on the surface of the skin. It is a disease of humans existing in two forms. The more virulent and frequently lethal form is *Variola major,* and a milder form is *Variola minor.* Smallpox is very contagious, requiring strict quarantine measures and aggressive vaccination programs to contain and eradicate outbreaks.

Smallpox is a member of the *Poxviridae* family of viruses, which are the largest and most complex of all known viruses. Poxviruses are named for the characteristic rash or pox lesions that occur during most infections. The poxviruses include a number of familiar diseases, such as smallpox, cowpox, rabbitpox, sheeppox, and fowlpox. Two subfamilies of poxviruses are recognized based on their hosts. The orthopoxvirus subfamily comprises viruses that affect vertebrates and includes smallpox; the poxviruses of the subfamily parapoxviruses infect invertebrates, primarily insects. There are two types of variola, the poxvirus that causes smallpox: *Variola major* causes the more virulent and lethal form of smallpox in humans, and *V. minor* causes a milder form of smallpox. Both varieties infect only humans and monkeys. Other names or synonyms for smallpox include alastrim, amaas, Kaffir mil pox, West Indian modified smallpox, and para-smallpox.

Humans have had a long and unfortunate history of association with smallpox. The disease apparently originated in India and spread westward into the Middle East and northern Africa several thousand years ago. An Egyptian mummy of the Twentieth Dynasty shows the characteristic scarring associated with smallpox. Warriors returning from the Crusades brought the disease back with them. In the following centuries, smallpox became endemic throughout much of Europe and became a rite of passage for much of the population—those who contracted smallpox and survived were marked by its scars throughout life. In time, the population built up a partial immunity to the disease. Smallpox was carried by Europeans to the New World and to Australia during the Age of Exploration. It was spread to the immunologically defenseless Amerindians of North America and Aboriginals of Australia with devastating effect and may have contributed to the ease of European settlement following the decimation of tribal peoples in both areas, as it caused widespread death and devastation among the indigenous populations and was at least partly responsible for the depopulation of natives in the newly discovered lands. Before its eradication, smallpox was endemic throughout the world, with major centers of the disease in Africa, Asia, and the Middle East.

Risk Factors

Smallpox is a highly contagious disease, and therefore anyone who comes in contact with it is susceptible to it. Patients who have weak immune systems—such as those with human immunodeficiency virus (HIV) or cancer and organ transplant recipients—are at risk for severe cases of smallpox and would have higher mortality rates. People born after 1971 who did not receive primary immunization would also be highly susceptible.

Etiology and Genetics

The poxviruses are the largest and most complex of all the viruses that have so far been identified in animals. The variola virus that causes smallpox has a brick-shaped outer envelope and a dumbbell-shaped core that contains the smallpox genome. The smallpox genome is composed of linear, double-stranded DNA containing more than two hundred genes. Chemically, the smallpox virion consists of 90 percent protein, 3 percent DNA, and 5 percent lipid. The DNA genome codes for several hundred polypeptides, including several transcriptases responsible for replication of the virus within the cells of the host.

Replication of smallpox begins when the virus attaches to the surface of a host cell. After binding to receptors on the plasma membrane of the host cell, the host cell passes the virus into the cytoplasm by endocytosis. Once inside the cell, the virus becomes trapped in a lysosome vesicle in the cytoplasm. The first step in removing its viral coat probably occurs at this stage, as host cell enzymes dissolve the viral envelope. The viral core, containing the DNA, then exits the lysosome and enters the cytoplasm, where the viral genome can be expressed. One of the first steps involves the production of enzymes that degrade the proteins of the viral core, which releases the naked viral DNA into the cytoplasm. Additional transcription takes place, initially producing structural proteins and enzymes, including DNA polymerase, which promotes the replication of the viral DNA. Finally, the late messenger RNA (mRNA) is transcribed, producing additional structural proteins and assembly enzymes that complete virion construction. During viral replication, most host-cell protein synthesis is blocked, because transport of host-cell mRNA molecules though the nuclear envelope into the cytoplasm is prevented.

Newly completed virons exit the host cell through microvilli on the cell surface or fuse with the cell membrane, after which they exit the cell by the process of exocytosis. Once in the tissue, fluids, and bloodstream, the newly released and highly infectious viral particles can invade and replicate in other host cells.

Smallpox is transmitted from one human to another, either by direct contact or via droplets released into the air during sneezing and coughing fits. The virus does not live long outside the human body and does not reproduce outside the human body. No natural animal carriers of variola other than monkeys, which are also susceptible, are known for the smallpox disease. In extremely rare cases smallpox is transmitted by carriers that are themselves immune to the disease but can transmit the disease to others. Still, only a few droplets settling on another person are sufficient to transmit smallpox. Because of the virulence and mode of transmission, public health regulations specify decontamination procedures. Living quarters, bedding, clothes, and other articles of infected persons must be thoroughly cleansed by heat or with formaldehyde, or destroyed altogether.

SYMPTOMS

Historically one of the most devastating and lethal of all human diseases, smallpox is named for the small pustules that occur as a rash over the skin of the victim. Smallpox symptoms include a rash that spreads over the entire body, high fever, chills, aches and pains, and vomiting. The most lethal form, black or hemorrhagic smallpox, results in death within two to six days. The fatality rate varies with health and previous exposure of the local population but ranges from 30 to 90 percent.

Infection occurs when the variola virus enters the respiratory mucosa of the nasal or pharyngeal region of the upper respiratory tract of humans. Apparently, only a few viral particles are needed to produce an infection. After a few hours or a few days, the virus migrates to and invades cells in the lymph

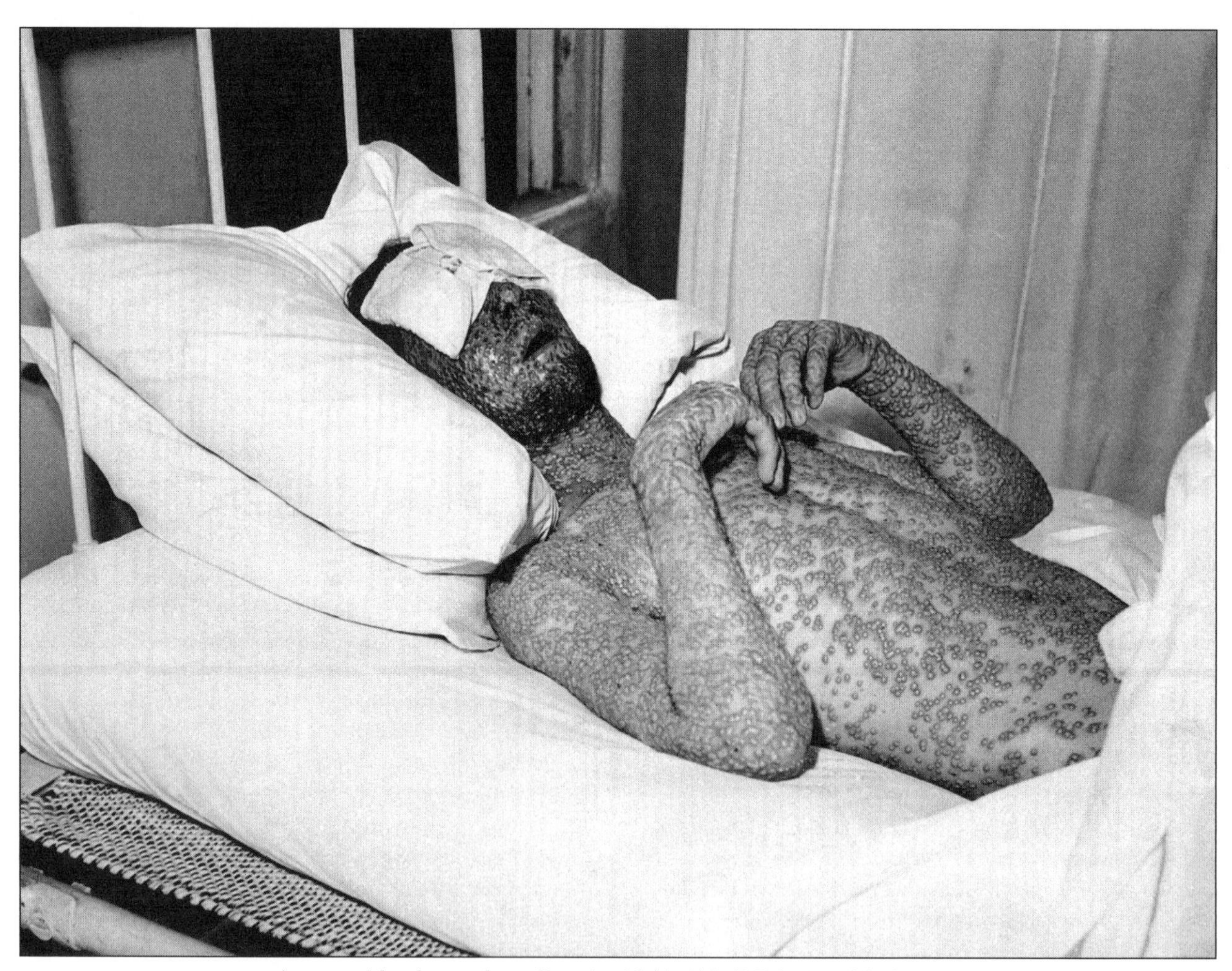

A man with advanced smallpox in 1941. (AP/Wide World Photos)

nodes of the nasopharyngeal region, where it enters the cells, following which rapid reproduction occurs. After a few days, it enters the bloodstream, a condition called viremia. At this time, symptoms of smallpox appear. The virus spreads into lymph nodes, spleen, and bone marrow, where reproduction continues rapidly. By the eighth day of infection, the virus is contained in white blood cells or leukocytes, which transmit it to the small blood vessels in the dermis of the skin as well as in the mucosa that lines the mouth and pharynx.

Following an incubation period of about two weeks (the range is between seven and seventeen days), symptoms appear, including high fever, headache, nausea, malaise, and often backache. Accompanying these symptoms is a rash that begins in the mouth and spreads across the face, forearms, trunk, and legs. The rash is first confined to a reddish or purplish swelling of the blood vessels but soon becomes pustular as little round nodules appear on the surface of the body. If the patient recovers, the pustules crust over and the resultant scabs eventually split, which causes scarring of the face.

Death occurs within a few days following the appearance of the rash, most commonly from toxemia caused by variola antigens and various immune complexes circulating in the blood. In some cases the disease is followed by encephalitis. Smallpox fatalities typically occur because of complications such as pneumonia, septicemia, and nephritis (kidney failure). Survivors often suffer from general scarring, ulcers, scarring of the cornea leading to blindness, and skin abscesses. Treatment of survivors with chemotherapy has reduced the severity of many of these complications.

The considerably less virulent form of smallpox, *Variola minor*, produces a much less severe illness characterized by fever, chills, and a milder rash. The same conditions are sometimes seen in patients who have previously been vaccinated or even as a response to vaccinations.

Screening and Diagnosis

The Centers for Disease Control and Prevention (CDC) or a CDC Laboratory Response Network-designated variola testing facility can conduct a definitive examination for smallpox by testing a tissue sample from a lesion on the skin of an infected person. Even a single confirmed case of smallpox would be deemed an international emergency because of the devastating nature of the disease and concern that it may be used as a biological weapon.

Treatment and Therapy

Despite decades of research, there is no specific treatment for smallpox other than bed rest and application of antibiotics to prevent secondary infections. Therefore, only prevention of spread by quarantine of infected persons prevents epidemics. Immediate recognition of the disease remains the strongest control measure, followed by vaccination of all health care personnel and others who may come in contact with infected persons.

Widespread and aggressive inoculation programs conducted during the first half of the twentieth century eradicated smallpox from most regions of the world, including North America, Eurasia, and Oceania, largely as a result of the success of the vaccination process originally developed by Edward Jenner. By 1967 smallpox was found only in thirty-three countries and had an annual infection rate of 10 million to 15 million cases. In that year the World Health Organization (WHO) initiated a campaign to eliminate smallpox completely as a human disease, concentrating in Africa, India, and Indonesia. The last case of smallpox in Asia was reported in Bangladesh in 1975, and the last known smallpox victim was recorded in Somalia in October, 1977. Eradication was considered accomplished by 1979. The cost of the eradication campaign was $150 million.

Most researchers conclude that the effective eradication of smallpox was made possible for several reasons: Smallpox cases could be quickly and positively identified, there are no natural carriers that serve as disease reservoirs, humans were the only carriers, individuals who survived did not continue to harbor the virus, and the smallpox vaccine proved highly effective.

Vaccinia viruses can absorb comparatively large amounts of foreign DNA without losing their ability to replicate, giving rise to the idea that they may provide a vehicle for providing immunity for other viral diseases of humans. One of several ongoing investigations involves insertion of 22-25 kilobase pairs into vaccinia. Experiments using this technique have produced vaccinia strains that encode surface proteins (antigens) of a number of important viruses, including influenza, hepatitis B, and herpesvirus. One possible outcome of these recombinant DNA experiments is the production of vaccinia strains that can serve as vaccines for several viral diseases simultaneously.

Prevention and Outcomes

Although smallpox was eradicated globally in 1979, at least two research stocks exist, and there is concern that clandestinely held stocks of the virus may be used as weapons of bioterrorism, the use of living organisms as instruments of terror, such as the deliberate introduction of diseases into civilian populations.

Since its official eradication in 1979, only two stocks of smallpox officially remain: One stock is held at the Centers for Disease Control in Atlanta, Georgia, and the other is kept at VECTOR, Novosibirsk, in central Russia. However, there remains the possibility that clandestine stocks still exist, and these stocks may serve as potential bioterrorism weapons, either to be used against military or civilian populations or to be mounted as international threats. The use of smallpox as a bioterrorism weapon would be classed as an international crime, but prevention of its use is difficult unless all existing stocks can be identified and destroyed.

Smallpox is a potential bioterrorism weapon because of its transmissibility, its known lethality, and the general lack of immunity of much of the global population. Because of its bioterrorism potential, research is now centered on rapid identification methods that enable the early detection of smallpox, as well as aggressive vaccination programs for individuals most at risk, who have been identified as health care workers. In addition, smallpox vaccinations were reinstated in 2002 for some U.S. military personnel and some health care workers, essentially those considered at highest risk. The vaccine is made from live but weakened vaccinia virus that is pricked into the skin. The characteristic blisters scab over within three weeks. During this time it is possible to transmit the virus to other parts of the body and to other people. Reactions to the vaccine range from a mild soreness around the vaccination site to more severe effects that may include brain inflammation and a rare and progressive bacterial inflammation called vaccinia that is sometimes fatal. For these reasons, mass vaccinations of the general public have been discouraged.

Dwight G. Smith, Ph.D.

Further Reading

Anderson, R. M., and R. M. May. *Infectious Diseases of Humans: Dynamics and Control.* Oxford, England: Oxford University Press, 1992. Smallpox and other major diseases of humans are described and discussed.

Brooks, George F., et al. *Jawetz, Melnick, and Adelberg's Medical Microbiology.* 24th ed. New York: McGraw-Hill Medical, 2007. Includes a summary of biological and medical properties of the virus that causes smallpox.

Fenner, F., et al. *Smallpox and Its Eradication.* Geneva, Switzerland: World Health Organization Report, 1988. The detailed story of the global eradication of smallpox as a human disease. Some of this report is technical, but the effort to eradicate smallpox is thoroughly described.

Henderson, D. A. *Smallpox: The Death of a Disease—the Inside Story of Eradicating a Worldwide Killer.* Amherst, N.Y.: Prometheus Books, 2009. Henderson, a physician, headed the successful campaign by the World Health Organization (WHO) to eradicate smallpox throughout the world. He provides a brief historical overview of the disease and efforts to cure it and then focuses on a detailed account of WHO's eradication efforts.

Miller, Judith, Stephen Engelberg, and William Broad. *Germs: Biological Weapons and America's Secret War.* New York: Simon & Schuster, 2001. This book, written by three *New York Times* reporters, explores the ideas and actions of scientists and politicians involved in the past, present, and future of germ warfare. Includes forty-two pages of notes and a select bibliography.

Preston, R. *The Demon in the Freezer: A True Story.* New York: Random House. 2002. This book, available in both print and audio, explores the use of smallpox stocks for research and evaluates the potential of genetically engineered smallpox as a weapon of mass destruction.

Tucker, Jonathan B. *Scourge: The Once and Future Threat of Smallpox.* New York: Grove Press, 2001. Examines the continuing debate about the destruction of smallpox. Describes the process used to eradicate the naturally occurring form of the disease and measures the pros and cons of destroying laboratory stockpiles of the smallpox virus.

U.S. Department of Defense. *Twenty-first Century Bioterrorism and Germ Weapons: U.S. Army Field Man-*

ual for the Treatment of Biological Warfare Agent Casualties (Anthrax, Smallpox, Plague, Viral Fevers, Toxins, Delivery Methods, Detection, Symptoms, Treatment, Equipment). Washington, D.C.: U.S. Department of Defense Manual, 2002. Available to the public, this is the standard reference manual for members of the Armed Forces Medical Services.

World Health Organization. *Future Research on Smallpox Virus Recommended.* Geneva, Switzerland: World Health Organization Press, 1999. This press release emphasizes the need for smallpox research in the light of its potential use as a weapon in the bioterrorism arsenal.

WEB SITES OF INTEREST

Centers for Disease Control
http://www.bt.cdc.gov/agent/smallpox
The CDC's Web page on smallpox includes information on the disease and on smallpox vaccines.

Medline Plus, Smallpox
http://www.nlm.nih.gov/medlineplus/smallpox.html
Offers a brief description of the disease and numerous links to other online resources.

National Organization for Rare Disorders
http://www.rarediseases.org
Site searchable by type of disorder. Includes background information on smallpox and a list of related resources.

World Health Organization
http://www.who.int/mediacentre/factsheets/smallpox/en
Provides a comprehensive overview of smallpox, including information about its historical significance, forms and clinical features of the disease, and vaccines and other treatments.

See also: Anthrax; Bacterial resistance and super bacteria; Biological weapons; Emerging and reemerging infectious diseases; Gene regulation: Viruses; Hereditary diseases; Viral genetics.

Smith-Lemli-Opitz syndrome

CATEGORY: Diseases and syndromes
ALSO KNOWN AS: 7-Dehydrocholesterol reductase deficiency; RSH syndrome; SLOS; SLO syndrome

DEFINITION

Smith-Lemli-Opitz syndrome (SLOS) is a developmental disorder that affects multiple organ systems, prenatal and postnatal growth, and mental development. It is caused by mutations in the *DHCR7* gene, that result in deficiency of the enzyme 7-dehydrocholesterol reductase.

RISK FACTORS

A pregnancy is at risk for SLOS when both parents are carriers of mutations in the *DHCR7* gene. SLOS is most common in individuals of northern or central European ancestry and is less common among individuals of African or Asian ancestry.

ETIOLOGY AND GENETICS

SLOS is an autosomal recessive genetic condition. When both parents are carriers of a single mutation in the *DHCR7* gene, there is a 25 percent chance in each pregnancy that the child will inherit both mutations and will be affected with SLOS.

The *DHCR7* gene is located at 11q12-13. It codes for the enzyme 7-dehydrocholesterol (7-DHC) reductase, which normally catalyzes the conversion of 7-DHC to cholesterol. Mutations in the *DHCR7* gene result in a loss of function of 7-DHC reductase. This blocks the final step in the cholesterol biosynthesis pathway.

Individuals who have a mutation in one copy of the *DHCR7* gene may have slightly increased 7-DHC levels but do not experience any clinical consequences. Individuals who have mutations in both copies of the *DHCR7* gene are affected with SLOS. Affected individuals have abnormally low cholesterol levels and elevated levels of the cholesterol precursors 7-DHC and 8-DHC.

Cholesterol deficiency is thought to be the cause of the developmental defects seen in SLOS. During fetal life, very little cholesterol is transported across the placenta from the mother. The fetus relies primarily on biosynthesis of cholesterol. Cholesterol is a major structural component of cell membranes and mitochondrial membranes. It is needed for bile acid, steroid hormone, and vitamin D metabolism. Cholesterol has also been found to bind to hedgehog proteins, a family of signaling proteins that direct embryonic development.

Clinical severity of SLOS can be variable and is inversely associated with cholesterol levels. Mortality is highest in individuals with the lowest cholesterol levels. The variability in severity is partially due to the more than 120 disease-causing mutations that have been described in the *DHCR7* gene. The twelve most common mutations account for approximately 69 percent of the alleles found in affected individuals.

Historically, SLOS was classified as type II, which is more severe and often lethal in the neonatal period, and classic type I. However, it is now clear that the difference in severity is related to the genotype. The most severely affected individuals typically have two mutations that result in very little residual 7-DHC reductase activity. Individuals with classic SLOS often have one of these severe mutations and a second missense mutation that results in higher levels of residual 7-DHC reductase activity. Individuals who have two copies of less severe missense mutations may be mildly affected.

Symptoms

Individuals with SLOS have moderate to severe mental retardation, growth retardation, multiple congenital anomalies, and behaviors characteristic of autism. Birth defects are often seen involving the heart, lungs, kidneys, gastrointestinal tract, genitalia, palate, fingers, and toes.

Screening and Diagnosis

A diagnosis of SLOS can be made by checking blood or tissues for an elevated concentration of 7-DHC. Most affected individuals have low cholesterol levels as well, but some affected individuals have cholesterol levels that are within the normal range. Sequencing of the *DHCR7* gene is able to identify disease-causing mutations in approximately 96 percent of affected individuals.

Prenatal testing for SLOS should be considered when there is a family history of SLOS, ultrasound findings are suggestive of SLOS, or maternal serum unconjugated estriol levels are low. Prenatally, a diagnosis of SLOS can be made by detecting an elevated concentration of 7-DHC in amniotic fluid or in tissue obtained from chorionic villus sampling (CVS). Enzyme testing can also be performed on cultured cells from an amniocentesis or CVS to confirm the diagnosis.

Treatment and Therapy

Following an initial diagnosis, evaluations may include a developmental assessment, ophthalmologic evaluation, cardiac evaluation, genitourinary examination, evaluation for functional gastrointestinal problems, nutritional assessment, MRI or other cranial imaging, renal ultrasound, and hearing evaluation. Treatment is directed to the particular problems identified in each child with SLOS. Early intervention and physical, occupational, and speech therapies are often needed.

Cholesterol supplementation has been shown to improve growth and increase nerve conduction velocities. It may improve behavioral problems as well. However, since many of the findings associated with SLOS are due to embryologic developmental defects, cholesterol supplementation cannot cure this condition.

Prevention and Outcomes

Stillbirth is common, and approximately 20 percent of affected children die within the first year of life. Although prenatal screening and diagnosis is available, there is no effective means of preventing SLOS. Genetic counseling should be made available to the families of affected individuals.

Laura Garasimowicz, M.S.

Further Reading

Cassidy, Suzanne B., and Judith E. Allanson, eds. *Management of Genetic Syndromes.* 2d ed. Hoboken, N.J.: John Wiley & Sons, 2005. A practical guide to the diagnosis and management of genetic conditions.

Jones, Kenneth Lyons. *Smith's Recognizable Patterns of Human Malformation.* 6th ed. Philadelphia: Elsevier, 2006. A clinical reference that includes descriptions and pictures of individuals with malformation syndromes.

Parker, Philip M. *Smith-Lemli-Opitz Syndrome: A Bibliography and Dictionary for Physicians, Patients, and Genome Researchers.* San Diego: Icon Group International, 2007. A research aid that defines medical terms and expressions relating to SLOS.

Web Sites of Interest

Genetics Home Reference: Smith-Lemli-Opitz Syndrome
http://ghr.nlm.nih.gov/condition=smithlemliopitzsyndrome

National Center for Biotechnology Information: GeneReviews
http://www.ncbi.nlm.nih.gov/bookshelf/br.fcgi?book=gene&part=slo

Smith-Lemli-Opitz/RSH Foundation
http://www.smithlemliopitz.org

See also: Apert syndrome; Brachydactyly; Carpenter syndrome; Cleft lip and palate; Congenital defects; Cornelia de Lange syndrome; Cri du chat syndrome; Crouzon syndrome; Down syndrome; Edwards syndrome; Ellis-van Creveld syndrome; Holt-Oram syndrome; Ivemark syndrome; Meacham syndrome; Opitz-Frias syndrome; Patau syndrome; Polydactyly; Robert syndrome; Rubinstein-Taybi syndrome.

Sociobiology

Category: History of genetics; Human genetics and social issues; Population genetics

Significance: Sociobiology attempts to explain social interactions among members of animal species from an evolutionary perspective. The application of the principles of sociobiology to human social behavior initiated severe criticism and accusations of racism and sexism.

Key terms

altruism: the capacity of one individual to behave in a way that benefits another individual of the same species at some cost to the actor

eusociality: an extreme form of altruism and kin selection in which most members of the society do not reproduce but rather feed and protect their relatives

kin selection: a special type of altruistic behavior in which the benefactor is related to the actor

reciprocal altruism: a type of altruism in which the benefactor may be expected to return the favor of the actor

society: a group of individuals of the same species in which members interact in relatively complex ways

History

Sociobiology is best known from the works of Edward O. Wilson, especially his 1975 book *Sociobiology: The New Synthesis.* This work both synthesized the concepts of the field and initiated the controversy over the application of sociobiological ideas to humans. However, the concepts and methods of sociobiology did not start with Wilson; they can be traced to Charles Darwin and others who studied the influence of genetics and evolution on behavior. Sociobiologists attempt to explain the genetics and evolution of social activity of all types, ranging from flocking in birds and herd formation in mammals to more complex social systems such as eusociality. "The new synthesis" attempted to apply genetics, population biology, and evolutionary theory to the study of social systems.

When sociobiological concepts were applied to human sociality, many scientists, especially social scientists, feared a return to scientific theories of racial and gender superiority. They rebelled vigorously against such ideas. Wilson was vilified by many of these scientists, and some observers assert that the term "sociobiology" generated such negative responses that scientists who studied in the field began using other names for it. At least one scientific journal dropped the word "sociobiology" from its title, perhaps in response to its negative connotations. However, the study of sociobiological phenomena existed in the social branches of animal behavior and ethology long before the term was coined. Despite the criticism, research has continued under the name sociobiology as well as other names, such as "behavioral ecology."

Sociobiology and the Understanding of Altruism

Sociobiologists have contributed to the understanding of a number of aspects of social behavior, such as altruism. Illogical in the face of evolutionary theory, apparently altruistic acts can be observed in humans and other animal groups. Darwinian evolution holds that the organism that leaves the largest number of mature offspring will have the greatest influence on the characteristics of the next generation. Under this assumption, altruism should disappear from the population as each individual seeks to maximize its own offspring production. If an individual assists another, it uses energy, time, and material it might have used for its own survival and reproduction and simultaneously contributes energy, time, and material to the survival and reproductive effort of the recipient. As a result, more members of the next generation should be like the assisted organism than like the altruistic one. Should this continue generation after generation, altruism would decrease in the population and selfishness would increase. Yet biologists have cataloged a number of altruistic behaviors.

When a prairie dog "barks," thus warning others of the presence of a hawk, the prairie dog draws the hawk's attention. Should it not just slip into its burrow, out of the hawk's reach? When a reproductively mature acorn woodpecker stays with its parents to help raise the next generation, the woodpecker is bypassing its own reproduction for one or more years. Should it not leave home and attempt to set up its own nest and hatch its own young? Eusocial species, such as honeybees and naked mole rats, actually have many members who never reproduce; they work their entire lives to support and protect a single queen, several reproductive males, and their offspring. It would seem that all these altruistic situations should produce a decrease in the number of members of the next generation carrying altruistic genes in favor of more members with "selfish" genes.

Sociobiologists have reinterpreted some of these apparently altruistic acts as camouflaged selfishness. The barking prairie dog, for example, may be notifying the hawk that it sees the predator and that it is close to its burrow and cannot be caught; therefore, the hawk would be better off hunting someone else. Perhaps the young acorn woodpecker learns enough from the years of helping to make its fewer reproductive years more successful than its total reproductive success without the training period.

It is difficult, however, to explain the worker honeybee this way. The worker bee never gets an opportunity to reproduce. Sociobiologists explain this and other phenomena by invoking kin selection. Since the worker bees are closely related to the queen (as sisters or daughters), to reproductive males, and to other workers they help feed and protect, they share a large number of genes with them. If they help raise enough brothers and sisters (especially males and queens) to more than make up for the offspring they do not produce themselves, they will actually increase the proportion of individuals similar to themselves more than if they "selfishly" reproduced.

The prairie dog's behavior might be explained this way as well. The organisms the prairie dog is warning are primarily relatives. By warning them, the prairie dog helps preserve copies of its own genes in its relatives. If the cost of the behavior (an occasional barking prairie dog being captured by a hawk because the warning call drew the hawk's attention) is more than compensated for by the number of relatives saved from the hawk by the warning, kin selection will preserve the behavior. The helper acorn woodpecker's behavior may be explained in similar ways, not as an altruistic act but as a selfish act to favor copies of the helper's genes in its relatives. Another explanation of altruism set forth by sociobiologists is reciprocity or reciprocal altruism: If the prairie dog is sometimes warned by others and returns the favor by calling out a warning when it sees a predator, the prairie dog town will be safer for all prairie dogs.

Opposition to the Application of Sociobiology to Humans

Wilson's new synthesis attempted to incorporate biology, genetics, population biology, and evolution into the study and explanation of social behavior. When the analyses turned to human sociality, critics feared that they would lead back to the sexist, racist, and determinist viewpoints of the early twentieth century. The argument over the relative importance of heredity or environment (nature or nurture) in determining individual success had been more or less decided in favor of the environment, at least by social scientists. Poor people were not poor because they were inherently inferior but because the environment they lived in did not give them an equal chance. Black, Hispanic, and other minority people were not inordinately represented among the poor because they were genetically inferior but because their environment kept them from using their genetic capabilities.

Sociobiologists entered the fray squarely on the side of an appreciable contribution from genetic and evolutionary factors. Few, if any, said that the environment was unimportant in the molding of racial, gender, and individual characteristics; rather, sociobiologists claimed that the genetic and evolutionary history of human individuals and groups played an important role in determining their capabilities, just as they do in other animals. Few, if any, claimed that this meant that one race, gender, or

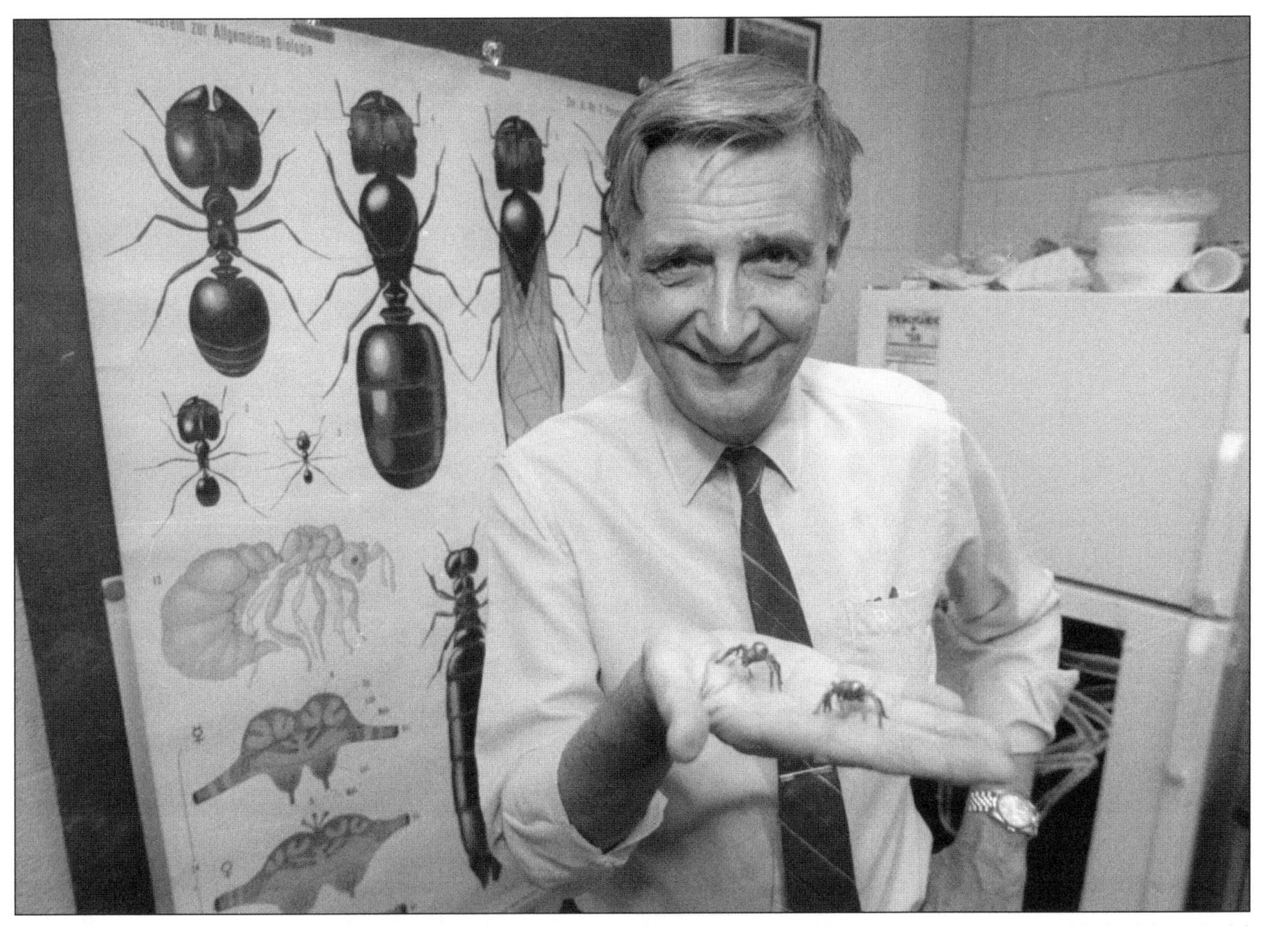

Edward O. Wilson's studies of insect behavior influenced his controversial theories of sociobiology. (AP/Wide World Photos)

group was superior to another. However, many (if not all) sociobiologists were accused of promoting racist, sexist, and determinist ideas with their application of sociobiological concepts to humans.

Extremists on both sides of the question have confused the issues. Such extremists range from opponents of sociobiological ideas who minimize genetic or evolutionary influence on the human cultural condition to sociobiologists who minimize the role of environmental influences. In at least some minds, extremists in the sociobiological camp have done as much damage to sociobiology as its most ardent opponents. Sociobiology (by that or another name) will continue to contribute to the understanding of the social systems of animals and humans. The biological, genetic, and evolutionary bases of human social systems must be studied. The knowledge obtained may prove to be as enlightening as has sociobiology's contribution to the understanding of social systems in other animals.

Carl W. Hoagstrom, Ph.D.

FURTHER READING

Alcock, John. *The Triumph of Sociobiology.* Reprint. New York: Oxford University Press, 2003. Reviews the history of the controversies and debates surrounding Edward O. Wilson's ideas on sociobiology. Illustrations, bibliography, index.

Baxter, Brian. *A Darwinian Worldview: Sociobiology, Environmental Ethics, and the Work of Edward O. Wilson.* Burlington, Vt.: Ashgate, 2007. Considers the various sociobiological theories that view the human brain as a product of evolution. Argues that Edward O. Wilson's sociobiological ideas exemplify a Darwinian worldview and thoroughly examines the views of Wilson and his major critics.

Blackmore, Susan J. *The Meme Machine.* Foreword by Richard Dawkins. New York: Oxford University Press, 1999. Argues that human behavior and cultural production, such as habits and the making of songs, ideas, and objects, are memetic; that is, they replicate in the form of memes, as do genes, within and between populations. Maintains that memes serve as the foundation of culture. Bibliography, index.

Cartwright, John. *Evolution and Human Behavior: Darwinian Perspectives on Human Nature.* 2d ed., updated and expanded ed. Cambridge, Mass.: MIT Press, 2008. Offers an overview of the key theoretical principles of human sociobiology and evolutionary psychology and shows how they illuminate the ways human beings think and behave. Argues that humans think, feel, and act in ways that once enhanced the reproductive success of their ancestors.

Cronk, Lee. *That Complex Whole: Culture and the Evolution of Human Behavior.* Boulder, Colo.: Westview Press, 1999. Discusses the links between behavioral and social scientists, who may not have a basic understanding of the import of culture on human behavior, and anthropologists, who in turn may lack a complete understanding of evolutionary biology. Bibliography, index.

Cziko, Gary. *The Things We Do: Using the Lessons of Bernard and Darwin to Understand the What, How, and Why of Our Behavior.* Cambridge, Mass.: MIT Press, 2000. Contrary to the Isaac Newton-inspired idea that humans react to the environment, Cziko argues that humans are less passive and reactive and more active beings, acting on their environments in order to shape their perceptions of the world. Illustrations, bibliography, index.

Richerson, Peter J., and Robert Boyd. *Not by Genes Alone: How Culture Transformed Human Evolution.* Chicago: University of Chicago Press, 2005. Draws upon new ideas about natural selection and evolutionary psychology to argue that culture plays a special role in the evolution of the human mind and human behavior.

Segerstråle, Ullica. *Defenders of the Truth: The Battle for Science in the Sociobiology Debate and Beyond.* New York: Oxford University Press, 2000. Addresses Wilson's *Sociobiology* and the ensuing debates on determinism versus free will, nature versus nurture, and adaptationism versus environmentalism. Bibliography, index.

Van der Dennen, Johan M. G., David Smillie, and Daniel R. Wilson, eds. *The Darwinian Heritage and Sociobiology.* Westport, Conn.: Praeger, 1999. Interdisciplinary approach to Darwin's influence on sociobiology, and discussions of sociobiological perspectives on war and other forms of conflict, marital relations, and utopia. Illustrations, bibliography, index.

Wilson, Edward O. "The Biological Basis of Morality." *Atlantic Monthly* 281, no. 4 (April, 1998): 53-70. Wilson argues that ethical and moral reasoning comes not from outside human nature, as if God-given, but from human nature itself in an ever-changing world.

_______. *On Human Nature.* 1978. Reprint. Cambridge, Mass.: Harvard University Press, 2004. A look at the significance of biology and genetics on how people understand human behaviors, including aggression, sex, and altruism and the institution of religion.

_______. *Sociobiology: The New Synthesis.* 1975. Reprint. Cambridge, Mass.: Belknap Press of Harvard University Press, 2000. The text that brings together Wilson's theories on the genetic, biological, and evolutionary basis of social systems.

WEB SITES OF INTEREST

Center for Evolutionary Psychology
http://www.psych.ucsb.edu/research/cep/index.html

Located at the University of California, Santa Barbara, the center conducts research about evolutionary psychology and related disciplines. The center's Web site provides a primer on evolutionary psychology, information on research in the field, and access to journal articles about the subject.

Social Psychology Basics
http://webspace.ship.edu/cgboer/socpsy.html

George Boeree, a professor of psychology at Shippenburg University, has written this primer that explains the various theories of social psychology. The primer includes a page on sociobiology.

Stanford Encyclopedia of Philosophy, Sociobiology
http://plato.stanford.edu/entries/sociobiology

Describes the key assumptions of sociobiology, sociobiological research on selfishness and altruism, and the philosophical implications of sociobiology.

See also: Aggression; Alcoholism; Altruism; Behavior; Biological clocks; Biological determinism; Criminality; Developmental genetics; Eugenics; Gender identity; Genetic engineering: Medical applications; Genetic engineering: Social and ethical issues; Genetic screening; Genetic testing; Genetic testing: Ethical and economic issues; Heredity and environment; Homosexuality; Human genetics; Inbreeding and assortative mating; Intelligence; Klinefelter syndrome; Knockout genetics and knockout mice; Miscegenation and antimiscegenation laws; Natural selection; Twin studies; XYY syndrome.

Speciation

Category: Population genetics

Significance: Speciation, the biological formation of new species, has produced the wide variety of living things on Earth. Although speciation can be caused by other forces or events, natural selection is considered the primary mechanism promoting speciation.

Key terms

allopatric speciation: the genetic divergence of populations caused by separation from each other by a geographic barrier such as a mountain range or an ocean

population: a group of organisms of the same species in the same place at the same time and thus potentially able to mate; populations are the basic unit of speciation

reproductive isolating mechanism: a characteristic that prevents an individual of one species from interbreeding (hybridizing) with a member of another species

species: a class of organisms with common attributes; individuals are usually able to produce fertile offspring only when mating with members of their own species

sympatric speciation: the genetic divergence of populations that are not separated geographically

Species Concepts

Before the time of Charles Darwin, physical appearance was the only criterion for classifying an organism. This "typological species concept" was associated with the idea that species never change (fixity of species). This way of defining a species causes problems when males and females of the same species look different (as with peacocks and peahens) or when there are several different color patterns among members of a species (as with many insects). Variability within species, whether it is a visible part of their anatomy, an invisible component of their biochemistry, or another characteristic such as behavior, is an important element in understanding how species evolve.

The "biological species concept" uses reproduction to define a species. It states that a species is composed of individuals that can mate and produce fertile offspring in nature. This concept cannot be used to classify organisms such as bacteria, which do not reproduce sexually. It also cannot be used to classify dead specimens or fossils. This definition emphasizes the uniqueness of each individual (variability) in sexually reproducing species. For example, in the human species (*Homo sapiens*), there are variations in body build, hair color and texture, ability to digest milk sugar (lactose), and many other anatomical, biochemical, and behavioral characteristics. All of these variations are the result of genetic mutations, or changes in genes.

According to evolutionary scientist Ernst Mayr, to a "population thinker," variation is reality and type is an abstraction or average; to a "typological thinker," variation is an illusion and type is the reality. Typological thinking is similar to typecasting or stereotyping, and it cannot explain the actual variability seen in species, just as stereotyping does not recognize the variability seen in people. Additional definitions, such as the "evolutionary species concept," include the continuity of a species' genes through time or other factors not addressed by the biological species concept.

Isolation and Divergence of Populations

Species are composed of unique individuals that are nevertheless similar enough to be able to mate and produce fertile offspring. However, individuals of a species are infrequently in close enough proximity to be able to choose a mate from all opposite-sex members of the same species. Groups of individuals of the same species that are at least potential mates because of proximity are called populations.

The basic type of speciation in most sexually reproducing organisms is believed to be "allopatric," in which geographic isolation (separation) of the species into two or more populations is followed by accumulation of differences (divergence) between the populations that eventually prevent them from interbreeding. These differences are caused primarily by natural selection of characteristics advantageous to populations in different environments. If both populations were in identical environments after geographic isolation, they would be much less likely to diverge or evolve into new species.

Another type of speciation is "sympatric," in which populations are not separated geographically, but reproduction between them cannot occur (reproductive isolation) for some other reason. For example, one population may evolve a mutation that makes the fertilized egg (zygote) resulting from interbreeding with the other population incapable of surviving. Another possibility is a mutation that changes where or when individuals are active so that members of the different populations never encounter one another.

Darwin thought that divergence, and thus speciation, occurred gradually by the slow accumulation of many small adaptations "selected" by the environment. Geneticists now recognize that a very small population, or even a "founder" individual, may be the genetic basis of a new species that evolves more rapidly. This process, called genetic drift, is essentially random. For example, which member of an insect species is blown to an island by a storm is not determined by genetic differences from other members of the species but by a random event (in this case, the weather). This individual (or small number of individuals) is highly unlikely to contain all of the genetic diversity of the entire species. Thus the new population begins with genetic differences that may be enhanced by its new environment. Speciation proceeds according to the allopatric model, but faster. However, extinction of the new population may also occur.

Plants are able to form new species by hybridization (crossbreeding) more often than are animals. When plants hybridize, postmating incompatibility between the chromosomes of the parents and the offspring may immediately create a new, fertile species rather than a sterile hybrid, as in animals such as the mule. A frequent method of speciation in plants is polyploidy, in which two or more complete sets of chromosomes end up in the offspring. (Usually, one complete set is made up of half of each parent's chromosomes.)

Many species reproduce asexually (without the exchange of genes between individuals that defines sexual reproduction). These include bacteria and some plants, fish, salamanders, insects, rotifers, worms, and other animals. In spite of the fact that reproductive isolation has no meaning in these organisms, they are species whose chromosomes and genes differ from those of their close relatives.

Impact and Applications

Environmentalists and scientists recognize that the biodiversity created by speciation is essential to the functioning of Earth's life-support systems for humans as well as other species. Some practical benefits of biodiversity include medicines, natural air and water purification, air conditioning, and food.

The impact of understanding the genetic basis of evolving species cannot be underestimated. Artificial selection (in which humans decide which individuals of a species survive and reproduce) of plants has produced better food crops (for example, modern corn from teosinte) and alleviated hunger in developing nations by creating new varieties of existing species (for example, rice). Hybridization of animals has resulted in mules and beefaloes for the farm (both of which are sterile hybrids rather than species). Artificial selection of domesticated animals has produced the many breeds of horses, dogs, and cats (each of which is still technically one species). Genetic engineering promises to create crops that resist pests, withstand frost or drought, and contain more nutrients. Finally, understanding the genetics of the evolving human species has broad implications for curing disease and avoiding birth defects.

Barbara J. Abraham, Ph.D.

Further Reading

Coyne, Jerry A., and H. Allen Orr. *Speciation*. Sunderland, Mass.: Sinauer Associates, 2004. Textbook providing an overview of speciation in both plants and animals. Includes explanations of the reality and concepts of speciation, the role of hybridization in speciation, the search for genes causing reproductive isolation, and speciation versus macroevolution.

Crow, Tim J., ed. *The Speciation of Modern Homo Sapiens*. Oxford, England: Oxford University Press, 2002. Chapters cover sexual selection, the question of whether or not *Homo sapiens* speciates on the Y chromosome, and what the Y chromosome might reveal about the origin of humans. Illustrations, bibliography.

Giddings, L. V., Kenneth Y. Kaneshiro, and Wyatt W. Anderson, eds. *Speciation, and the Founder Principle.* New York: Oxford University Press, 1989. Seventeen internationally known geneticists cover general principles of speciation among both plants and animals, with emphasis on the founder principle.

Hey, Jody. *Genes, Categories, and Species: The Evolutionary and Cognitive Causes of the Species Problem.* New York: Oxford University Press, 2001. Argues that the answer to the species problem lies not with the processes and patterns of biological diversity but in the way the human mind perceives and categorizes this diversity.

Hey, Jody, Walter M. Fitch, and Francisco Ayala, eds. *Systematics and the Origin of Species: On Ernst Mayr's 100th Anniversary.* Washington, D.C.: National Academies Press, 2005. In 2004, the National Academy of Sciences hosted a colloquium to honor Mayr, a scientist who argued that population variation played a significant role in the evolutionary process and the origin of species. This book collects sixteen papers on speciation that were presented at the conference.

Margulis, Lynn, and Dorion Sagan. *Acquiring Genomes: A Theory of the Origins of Species.* New York: Basic Books, 2003. Margulis, a microbiologist, takes issue with the neo-Darwinists, who argue that random gene mutation is the source of new species. She maintains that new species are created by a process of "genome ingestion."

Mayr, Ernst. *One Long Argument: Charles Darwin and the Genesis of Modern Evolutionary Thought.* Cambridge, Mass.: Harvard University Press, 1991. Includes a chapter, "How Species Originate," that points out that Darwin's explanation of speciation was limited by his lack of understanding of the origin of genetic variation (mutation and recombination). Illustrations, bibliography, index.

Web Sites of Interest

Evolution 101, Speciation
http://evolution.berkeley.edu/evosite/evo101/VSpeciation.shtml

This site, created by the University of California Museum of Paleontology, provides information about evolution for teachers and their students. It includes a section about speciation, which defines what a species is and explains the biological species concept and other theories of speciation.

Kimball's Biology Pages
http://users.rcn.com/jkimball.ma.ultranet/BiologyPages/S/Speciation.html

John Kimball, a retired Harvard University biology professor, includes a page about speciation in his online cell biology text.

See also: Artificial selection; Evolutionary biology; Hardy-Weinberg law; Hybridization and introgression; Lateral gene transfer; Natural selection; Polyploidy; Population genetics; Punctuated equilibrium.

Spinal muscular atrophy

Category: Diseases and syndromes

Also known as: Infantile Werdnig-Hoffmann disease; Kugelberg-Welander disease; spinobulbar muscular atrophy (SBMA); Kennedy disease

Definition

Spinal muscular atrophy (SMA) is a progressive neurodegenerative disorder caused by the loss of function of motor neurons, which leads to muscular atrophy. There are five clinical subtypes that range in age of onset and the severity of disease. SMA Type I is the most common subtype, which combined with other childhood forms affects an estimated 1 in 10,000 live births.

Risk Factors

SMA has a purely genetic etiology and can occur with or without a family history of the disorder. SMA affects all ethnicities at an equal rate with an estimated 1 in 45 being asymptomatic carriers. There are no environmental risk factors reported.

Etiology and Genetics

SMA is transmitted in an autosomal recessive pattern whereby a person simultaneously inherits two copies of the disease-causing gene (called *SMN1*, for survival motor neuron) at conception. Carriers remain symptom-free because one working copy of *SMN1* produces sufficient SMN protein for motor neuron cells to function.

The second gene implicated with SMA is called *SMN2* (survival motor neuron 2). This gene can range in copy number from zero to five copies in each cell. *SMN1* and *SMN2* are very similar in their composition, differing by only one nucleotide. This nucleotide difference does not change the amino acid generated but does alter the gene splicing (paring down mRNA by editing out noncoding regions of DNA). Usually the protein produced by *SMN1* is longer than *SMN2*, but splicing is not a perfect process, and at times an equally long protein is produced from *SMN2* and functions much like SMN1 protein. Thus, a person with SMA that has a high copy number of *SMN2* genes is predicted to have a less severe case than someone who does not have any *SMN2* genes.

Symptoms

With the exception of Type 0, most infants with SMA are born seemingly healthy and have no immediate symptoms. There is a tremendous range of severity. Generally earlier onset is predictive of a worse prognosis; thus, Type 0 is most severe, while Type IV is comparatively mild. Type 0 symptoms include a lack of fetal movements (at thirty to thirty-six weeks gestation), as well as joint contractures plus difficulty breathing and swallowing as a newborn. Most infants with Type 0 do not survive longer than two months of age. Type I symptoms begin between birth and six months of age and include hypotonia (low muscle tone and strength), paralysis, and mild joint contractures which result in difficulty sitting without support, breathing, and swallowing. Death occurs most commonly before two years of age, typically caused by complications such as respiratory infections. Type II (onset from six to twelve months) symptoms include an inability to sit, but standing or walking unaided is possible. Type III occurs in childhood (over one year of age) and is considerably more mild. Type IV has adult onset of symptoms that include muscle weakness, tremor, and twitching. All forms of SMA have normal cognition and a high rate of mortality and morbidity.

Screening and Diagnosis

Screening and diagnostic testing for SMA are readily available through commercial laboratories. Carrier screening involves evaluating the *SMN1* gene for mutations and determining the number of *SMN2* genes via sequencing and quantitative polymerase chain reaction (PCR), respectively. Diagnostic testing is recommended if an individual is symptomatic and can help confirm a clinical diagnosis. Confirmation of a mutation may also aid in determining if extended family members should pursue carrier screening. Prenatal diagnosis requires testing fetal cells and thus performing an amniocentesis to sample the amniotic fluid or a chorionic villus sampling (CVS) to biopsy a small portion of the placenta.

Treatment and Therapy

There is no cure for SMA, but treatment is available to ease symptoms. Treatment of recurrent pulmonary infections, as well as tracheotomies, may be necessary. Alternative therapies include long-term ventilation, tube feeding, and physical therapy, which can help in prolonging survival. SMA Type I is a fatal condition; thus, palliative care is also provided.

Prevention and Outcomes

SMA is best prevented by carrier screening and other testing services. With two carrier parents, the risk is 25 percent for future pregnancies also being affected with SMA. There is also a 50 percent risk of having a future child that is an asymptomatic carrier. Lastly, there is a 25 percent chance of having a future child that is completely unaffected.

Prenatal diagnosis is available with the option of ending the pregnancy should it be affected. Preimplantation genetic diagnosis (PGD) is available to diagnose embryos prior to implantation if pursuing in vitro fertilization; however, the procedure is rarely pursued because insurance coverage is extremely limited.

Kayla Mandel Sheets, M.S.

Further Reading

Cummings, Michael. *Human Heredity: Principles and Issues.* 8th ed. Pacific Grove, Calif.: Brooks/Cole, 2008. A comprehensive text for referencing basic concepts of human genetics.

Firth, Helen V., and Jane A. Hurst. *Oxford Desk Reference: Clinical Genetics.* Oxford, England: Oxford University Press, 2005. An overview of human genetic conditions, testing, and treatment.

Harper, Peter S. *Practical Genetic Counseling.* 6th ed. London: Hodder Arnold, 2004. A human genetics reference written by a professor of medical genetics.

Web Sites of Interest

Families of Spinal Muscular Atrophy
www.fsma.org
International support group and resource center for this disease. Includes current research.

SMA Foundation
www.smafoundation.org
Mission of the spinal muscular atrophy (SMA) foundation is to accelerate the development of a treatment or cure for spinal muscular atrophy.

NINDS: Spinal Muscular Atrophy Information Page
www.ninds.nih.gov/disorders/sma/sma.htm
Spinal muscular atrophy (SMA) information page compiled by the National Institute of Neurological Disorders and Stroke (NINDS).

See also: Adrenoleukodystrophy; Alexander disease; Alzheimer's disease; Amyotrophic lateral sclerosis; Arnold-Chiari syndrome; Ataxia telangiectasia; Canavan disease; Cerebrotendinous xanthomatosis; Charcot-Marie-Tooth syndrome; Chediak-Higashi syndrome; Dandy-Walker syndrome; Deafness; Epilepsy; Essential tremor; Friedreich ataxia; Huntington's disease; Jansky-Bielschowsky disease; Joubert syndrome; Kennedy disease; Krabbé disease; Leigh syndrome; Leukodystrophy; Limb girdle muscular dystrophy; Maple syrup urine disease; Metachromatic leukodystrophy; Myoclonic epilepsy associated with ragged red fibers (MERRF); Narcolepsy; Nemaline myopathy; Neural tube defects; Neurofibromatosis; Parkinson disease; Pelizaeus-Merzbacher disease; Pendred syndrome; Periodic paralysis syndrome; Prion diseases: Kuru and Creutzfeldt-Jakob syndrome; Refsum disease; Vanishing white matter disease.

Spinocerebellar ataxia

Category: Diseases and syndromes
Also known as: SCA; SCA1 (SCA2; etc.); spinocerebellar degeneration; olivopontocerebellar atrophy (OPCA); Machado-Joseph disease (SCA3)

Definition

The spinocerebellar ataxias are hereditary movement disorders characterized by ataxia, or loss of coordination, with variable ages of onset and possible development of other neurological features. There are currently about thirty distinct types of spinocerebellar ataxias, typically labeled as SCA1, SCA2, and so on. Each has a unique genetic cause leading to cell loss in the cerebellum and possibly the spine and brain stem. Dentatorubralpallidalluysian atrophy (DRPLA) may also be categorized as a subtype of SCA, as the symptoms and genetics of this condition are similar.

Risk Factors

The only risk factors are a known gene mutation or a family history of SCA. The SCAs can occur in all ethnicities, although some occur at a higher frequency among certain ethnic backgrounds, possibly due to a founder effect. Some SCAs are incredibly rare and have been identified only in a single family of one particular ethnic background. Men and women are equally likely to be affected. There is an average age of onset for each type of SCA, with onset in childhood or young adulthood, while others may not occur until late in life.

Etiology and Genetics

Each of the thirty SCAs is inherited in an autosomal dominant manner and has a distinct genetic cause, although the exact gene has not been identified for all. First-degree relatives of an affected individual are at a 50 percent risk to develop the same type of SCA. Some SCAs result from single point mutations, but many are trinucleotide repeat expansion disorders causing variation in age of onset, severity, and disease course exhibited among family members with the same disorder. Certain genes have a set number of repeating trinucleotides, but in some SCAs, the number of these repeated trinucleotides has expanded, causing an unstable gene. For those individuals who have a number of trinucleotide repeats within the reduced penetrance range, it is not possible to predict whether or not SCA will develop. Anticipation is common with some of the SCAs, meaning the number of trinucleotide repeats in an already unstable gene will continue to expand in future generations into the full mutation range. The condition will likely develop if the individual with a full mutation lives long enough. Signs and symptoms may occur at earlier ages of onset and with greater severity in succeeding generations due to anticipation. Family history may appear negative because of reduced penetrance, late onset, or early death of relatives. New gene mutations that are not inherited are also possible but rare.

Machado-Joseph disease (SCA3) is the most common SCA, and combined with SCA1, 2, and 6 is responsible for at least 50 percent of all autosomal dominant ataxias. It is estimated that 1 to 5 in 100,000 individuals are affected with an SCA.

While the exact function of each gene and its associated protein is unknown, it appears that the abnormal protein causes nerve cell damage in the brain and spinal cord. Current research focuses on understanding the effect of these gene mutations on the brain and further characterization of each SCA.

Symptoms

All types of SCAs include ataxia. Loss of coordination leads to difficulty with balance and walking, and to clumsiness of the hands. Speech may become slurred, and swallowing may become increasingly difficult. Visual disturbances and eye movement abnormalities are possible. Some SCAs have characteristic features that may aid in diagnosis, such as cognitive impairment, neuropathy, hearing loss, seizures, parkinsonism, other abnormal movements, or specific neurological signs detected on exam. It is essential to see a physician when symptoms interfere with daily activities, injury occurs after a fall, or swallowing becomes difficult.

Screening and Diagnosis

Family history, neurological exam, brain MRI, and genetic testing are all useful in diagnosis. The MRI may reveal cerebellar atrophy, or loss of tissue, common to all SCAs, in addition to spinal cord or brain stem atrophy. Only genetic testing can establish a diagnosis of a specific SCA, although testing is not available for every known type. Other clinical studies may be necessary to detect neurological findings characteristic of some SCAs.

Treatment and Therapy

There are no cures or specific treatments, although assistive devices and therapy may be beneficial. Medications may help alleviate some features, but individual response varies.

Prevention and Outcomes

Diagnostic genetic testing can help to provide an accurate diagnosis in affected individuals. With a known familial mutation, predictive genetic testing is available to determine whether an unaffected individual has inherited the gene mutation and will develop SCA.

There is no known prevention. Prenatal diagnosis and preimplantation genetic diagnosis may be possible with known familial mutations.

Some SCAs have a slowly progressive course and normal life span, but many progress more rapidly, with ten to twenty years passing from onset until extensive assistance with activities of daily living is required.

Katherine L. Howard, M.S.

Further Reading

Nussbaum, Robert L., Roderick R. McInnes, and Huntington F. Willard. *Thompson and Thompson Genetics in Medicine.* 7th ed. New York: Saunders, 2007. Thorough review of genetics for professionals and the layperson.

Pulst, Stefan. *Genetics of Movement Disorders.* San Diego: Elsevier Science, 2003. Detailed review of the hereditary aspects of movement disorders in particular.

Watts, Ray, and William C. Koller. *Movement Disorders: Neurologic Principles and Practices.* 2d ed. New York: McGraw-Hill, 2004. Thorough review of movement disorders aimed at professionals.

Web Sites of Interest

National Ataxia Foundation
http://www.ataxia.org

Neuromuscular Disease Center, Washington University, St. Louis
http://www.neuro.wustl.edu/neuromuscular/ataxia/aindex.html

Online Mendelian Inheritance in Man
http://www.ncbi.nlm.nih.gov/sites/entrez

Spinocerebellar Ataxia: Making an Informed Choice About Genetic Testing
http://depts.washington.edu/neurogen/SpinoAtaxia.pdf

University of Washington, Seattle: GeneTests
http://www.genereviews.org

See also: Adrenoleukodystrophy; Alexander disease; Alzheimer's disease; Amyotrophic lateral sclerosis; Arnold-Chiari syndrome; Ataxia telangiectasia;

Canavan disease; Cerebrotendinous xanthomatosis; Charcot-Marie-Tooth syndrome; Chediak-Higashi syndrome; Dandy-Walker syndrome; Deafness; Epilepsy; Essential tremor; Friedreich ataxia; Huntington's disease; Jansky-Bielschowsky disease; Joubert syndrome; Kennedy disease; Krabbé disease; Leigh syndrome; Leukodystrophy; Limb girdle muscular dystrophy; Maple syrup urine disease; Metachromatic leukodystrophy; Myoclonic epilepsy associated with ragged red fibers (MERRF); Narcolepsy; Nemaline myopathy; Neural tube defects; Neurofibromatosis; Parkinson disease; Pelizaeus-Merzbacher disease; Pendred syndrome; Periodic paralysis syndrome; Prion diseases: Kuru and Creutzfeldt-Jakob syndrome; Refsum disease; Sandhoff disease; Sanfilippo syndrome; Spinal muscular atrophy; Vanishing white matter disease.

SRY gene

Category: Classical transmission genetics; Molecular genetics

Significance: The *SRY* gene is on the Y chromosome and encodes a transcription factor that triggers development of the male phenotype. The absence of this protein results in development of the female phenotype. Most mutations in *SRY* that disable the protein's DNA-binding domain result in XY females.

Key terms

sex determination: the biological mechanism that determines the sex of an individual

sex reversal: an XY female or XX male

XO chromosome constitution: an individual with one X chromosome and no corresponding X or Y chromosome

XXY chromosome constitution: an individual with two X chromosomes and one Y chromosome

SRY and Sex Determination

The biological mechanisms of sex determination vary widely throughout the animal and plant kingdoms, ranging from purely nongenetic to fully genetic mechanisms. The XX-XY genetic system, in which females have a pair of homomorphic X chromosomes and males a pair of heteromorphic X and Y chromosomes, was investigated most extensively in *Drosophila melanogaster* in the early part of the twentieth century. In 1925, Calvin Bridges published abundant evidence that the ratio of X chromosomes to autosomes determines sex in *Drosophila.* Of particular note was the observation that XXY flies are fertile females, and XO flies are infertile males. The *Drosophila* Y chromosome plays no role in sex determination although it is essential for male fertility.

In 1921, Theophilus Painter identified the Y chromosome in human and opossum males, and demonstrated that it pairs with and segregates from the X chromosome during meiosis. These discoveries suggested an evolutionarily ancestral XX-XY system for both eutherian (placental) and metatherian (marsupial) mammals, which has since been confirmed.

The observation that both *Drosophila* and humans have XX-XY systems of sex determination led most geneticists to assume that the mechanism of sex determination in humans was the same as in *Drosophila.* This assumption persisted until 1959, when C. E. Ford and K. W. Jones found that a woman with Turner syndrome had an XO chromosome constitution, while P. A. Jacobs and J. A. Strong showed that a man with Klinefelter syndrome had an XXY chromosome constitution. That same year, W. J. Welshons and L. B. Russell demonstrated that XO mice are female. The evidence from these and subsequent studies pointed to the presence or absence of the Y chromosome as the principal sex-determining factor in mammals.

People with sex-reversal phenotypes (XY females and XX males) offered clues regarding the molecular mechanism of sex determination in humans. Most XX males had a portion of the Y chromosome translocated onto one of their X chromosomes. The translocated region of the Y chromosome apparently triggered development of the male phenotype and was named the testis determining factor, or *TDF.*

Analysis by Peter Goodfellow and coworkers of these Y-derived translocated sequences in XX males narrowed the region on the Y chromosome containing *TDF* to a 35-kilobase (kb) segment. Within this segment is an open-reading frame, which they named *SRY,* for sex-determining region Y. Evidence that *SRY* is *TDF* in humans accumulated as numerous XY females were found to have mutations in the region of *SRY* encoding the DNA-binding domain. Moreover, Goodfellow and coworkers provided powerful evidence that *SRY* is the testis determining factor when they recovered a phenotypically male XX mouse that was transgenic for the region containing the mouse orthologue *Sry*. The accumulated evidence points to *SRY* as the initial trigger that determines male phenotypic development, although other genes on the Y chromosome, the X chromosome, and autosomes are essential for development of male and female phenotypes and for fertility.

Molecular Genetics and Evolution

SRY is a relatively small (3.8-kb) intronless gene that encodes a transcription factor with an HMG (high mobility group) DNA-binding domain. This region encoding this domain is highly conserved in all therian mammals. The sequences within the gene on either side of this conserved region, however, are highly variable among species, indicative of very rapid evolution. The platypus, by contrast, has a different sex-determination mechanism, governed by five X chromosomes and five Y chromosomes. These chromosomes are most closely related to the bird Z chromosome. The therian X and Y chromosomes probably evolved from an ancient autosomal pair of chromosomes ancestral to platypus chromosome 6 after the divergence of therian mammals and monotremes, approximately 166 million years ago.

Impact

The most significant clinical effect of mutations in *SRY* is sex reversal. Nearly all nonsynonymous mutations within the region that encodes the HMG domain result in complete sex reversal, observed as an XY individual with an unambiguous female phenotype. However, a particular mutation in this region has been associated with true hermaphroditism and has also been found in fully fertile male relatives, indicating that it is not fully penetrant.

Daniel J. Fairbanks, Ph.D.

Further Reading

Goodfellow, Peter N., and Robin Lovell-Badge. "*SRY* and Sex Determination in Mammals." *Annual Review of Genetics* 27 (1993): 71-92. An excellent review of the discovery and function of *SRY* by two of the leading researchers.

Sinclair, A. H., et al. "A Gene from the Human Sex-Determining Region Encodes a Protein with Homology to a Conserved DNA-Binding Motif." *Nature* 346 (1990): 240-245. Publication reporting the discovery of *SRY*.

Veyrunes, F., et al. "Bird-Like Sex Chromosomes of Platypus Imply Recent Origin of Mammal Sex Chromosomes." *Genome Research* 18 (2008): 965-973. Evidence of the evolutionary origin of the mammalian X and Y chromosomes and the *SRY* gene.

Web Site of Interest

Online Mendelian Inheritance in Man: SRY gene
http://www.ncbi.nlm.nih.gov/entrez/dispomim.cgi?id=480000

See also: also: Androgen insensitivity syndrome; Gender identity; Hermaphrodism; Homosexuality; Klinefelter syndrome; Metafemales; Pseudohermaphrodites; Steroid hormones; Turner syndrome; X inactivation; XYY syndrome.

Stargardt's disease

CATEGORY: Diseases and syndromes
ALSO KNOWN AS: Fundus flavimaculatus

Definition

Stargardt's disease is the most common form of inherited juvenile macular degeneration. It is almost always inherited through an autosomal recessive mode of inheritance, but up to 10 percent of cases may be inherited in an autosomal dominant mode.

Risk Factors

Stargardt's disease is an inherited disease; there are no steps that can be taken to avoid developing the disease.

Etiology and Genetics

Stargardt's disease is usually diagnosed after the age of six but before the age of twenty; however, in some cases, the vision loss may not be noticed until age thirty or forty. Men and women are affected with Stargardt's disease equally. The gene for the autosomal recessive form of Stargardt's disease was found on the short arm of chromosome 1 and is called the *ABCR* or *ABCA4* gene. The specific gene has been localized to 1p21-22. This gene produces an ATP-binding transport protein; with the mutated gene, the defective protein is unable to perform its transport function. This protein is expressed in the inner rod segments, and the defect leads to degeneration of these cells in the retina and vision loss.

However, only 60 percent of people with Stargardt's, or fundus flavimaculatus, have a mutation in the *ABCA4* gene. Further research has shown that this locus is involved in the development of age-related macular degeneration, autosomal recessive retinitis pigmentosa, and autosomal recessive rod-cone dystrophy. It is thought that these diseases are part of a continuum.

In autosomal dominant Stargardt's disease, the defect has been mapped to chromosome 6q16.6 and is associated with the mutation located on the *ELOVL4* gene. This gene codes for a membrane-bound protein that is involved in long-chain fatty acid biosynthesis. Other retinal diseases have been traced back to chromosome 6, such as recessive retinitis pigmentosa, Leber congenital amaurosis, and North Carolina macular dystrophy.

Symptoms

The main symptom is bilateral decreased central vision starting in childhood and young adulthood. This vision loss manifests itself as bilateral central scotomas. Color vision is also impacted, but it usually is not noticeable until the later stages of the disease. People with Stargardt's disease generally do not have any decrease in peripheral vision or any problems with night vision.

Screening and Diagnosis

For people whose parents or genetically related siblings have a diagnosed inherited macular dystrophy, there should be a genetic screening test done to rule out Stargardt's disease or one of the other related conditions. For many patients, however, this may not be an option, and the disease may be misdiagnosed in the early stages because of lack of changes in the retina. As the disease progresses, there may be yellowish "pisciform" flecks at the level of the retinal pigment epithelium at the posterior pole; a bull's-eye type pattern of retinal degeneration may also become evident later in the disease. Areas of regional geographic atrophy also give the macula what is known as a "beaten bronze" appearance.

Diagnostic tests that may be used to diagnose Stargardt's disease include fluoroscein angiography (FA) and electrooculography. In Stargardt's disease, the FA shows what is known as a dark or silent choroid, irregular hyperfluorescent spots, and the bull's-eye pattern of retinal degeneration. The electrooculography test tends to be subnormal. Occasionally, visual field tests will be obtained to create a map of the visual field.

Treatment and Therapy

At this time, there is no known treatment for Stargardt's disease. Because of the role of *ABCA4* in the processing of vitamin A, some researchers feel that excess vitamin A might make the disease worse.

Prevention and Outcomes

Genetic screening in couples where one person is known to have the disease or where family members are known to have the disease may help determine if any future children are at risk of developing the disease. It is thought that bright lights and ultraviolet rays may accelerate the progression of Stargardt's disease, so it is suggested that people with Stargardt's disease should avoid bright lights and wear UV-blocking sunglasses as much as possible.

An important factor in improving outcomes in people with Stargardt's disease is to learn about available low vision services and aids that can help a person live a full and functional life.

Dominique Walton Brooks, M.D., M.B.A.

Further Reading

Hartnett, Mary Elizabeth, et al. *Pediatric Retina: Medical and Surgical Approaches.* Illustrated ed. Philadelphia: Lippincott Williams & Wilkins, 2004. This book looks at pediatric retinal diseases.

Ho, Allen C., et al. *Retina: Color Atlas and Synopsis of Clinical Ophthalmology.* Illustrated ed. New York: McGraw-Hill Professional, 2003. A combination of illustrated atlas and reference book that is easy to understand.

Regillo, Carl D., Gary C. Brown, and Harry W. Flynn. *Vitreoretinal Disease: The Essentials.* Illustrated ed. New York: Thieme Medical, 1999. This book makes it easier to understand recent ophthalmologic advances.

Yanoff, Myron, Jay S. Duker, and James J. Augsburger. *Ophthalmology.* 3d ed. New York: Elsevier Health Sciences, 2009. Updated edition of reference book that offers information about almost every ophthalmic condition.

WEB SITES OF INTEREST
Lighthouse International
http://www.lighthouse.org/medical/eye-disorders/stargardts-disease/

Macular Degeneration Support
http://www.mdsupport.org/library/stargrdt.html

See also: Aniridia; Best disease; Choroideremia; Color blindness; Corneal dystrophies; Glaucoma; Gyrate atrophy of the choroid and retina; Macular degeneration; Norrie syndrome; Progressive external ophthalmoplegia; Retinitis pigmentosa; Retinoblastoma.

Stem cells

CATEGORY: Cellular biology; Human genetics and social issues

SIGNIFICANCE: Stem cells, which are self-maintaining cell populations that can give rise to multiple types of more specialized cells, have therapeutic potential for a variety of diseases and injuries that have destroyed a patient's cells, tissues, or organs. Stem cells have also been used to gain a better understanding of how genetic factors work in the early stages of cell development and homeostasis and may play a role in the testing and development of drugs and cell therapies.

KEY TERMS

adult stem cell: a multipotent undifferentiated cell found among differentiated cells in a tissue or organ of an adult organism

blastocyst: a preimplantation embryo consisting of a hollow ball of two layers of cells

cell differentiation: the process whereby a more primitive precursor cell produces progeny that have different gene expression patterns and more specialized functions

embryonic stem cell: an undifferentiated cell derived from the inner cell mass of a blastocyst

multipotency: the ability of a cell to give rise to multiple types (but not all) of more specialized tissue in the organism; compare to pluripotency

pluripotency: the ability of a cell to give rise to all of the differentiated cell types in the organism

totipotency: the ability of a single cell to give rise to all of the tissues of both the organism and the extraembryonic tissues (e.g., placenta) necessary for development

TYPES OF STEM CELLS

Stem cells are defined by their ability to renew themselves (that is, their ability to produce more stem cells) and their ability to diversify into other cell types. There are three major classes of stem cells: totipotent, pluripotent, and multipotent, with each of these types having progressively less developmental potential.

After fertilization, the fertilized egg (zygote) and subsequent early embryo are composed of totipotent cells, which can differentiate to become all the cells that make up the embryo, all the extraembryonic support tissues (for example, the placenta and umbilical cord), and all the cells of the adult organism. After about four days, the embryo reaches a stage of development consisting of a hollow sphere. The approximately fifty to one hundred cells on the inner side of the sphere are called the inner cell mass; these cells have lost the ability to make extraembryonic tissues and are now said to be pluripotent, and they will continue developing to form the embryo. The cells on the outer surface will give rise to the extraembryonic tissues. Importantly, the pluripotent cells of the inner cell mass are the cells scientists use to create immortalized embryonic stem cell lines. As the embryo develops, multipotent stem cells arise; these cells can self-renew but are capable of giving rise to a smaller range of differentiated cell types. For example, hematopoietic stem cells can give rise to all the types of blood cells (lymphocytes, granulocytes, red blood cells, platelets, macrophages), but not to bone or skin cells.

In practice, the two main types of stem cells that researchers and physicians work with are multipotent ("adult") stem cells and pluripotent ("embryonic") stem cells. An additional type of stem cell is the theoretical cancer stem cells, based on the idea that tumors might contain or arise from stem-cell-like cancer cells.

ADULT STEM CELLS

Multipotent stem cells have been identified in a wide variety of tissues in adult mammals and are specialized stem cells committed to replenishing cells of a particular function. Hematopoietic stem cells constantly replenish the blood over the life of an animal, and mesenchymal stem cells in the bone marrow give rise to bone, cartilage, fat, and connective tissue cells. Neural stem cells have been shown to give rise to neurons as well as two types of non-neuronal support cells (astrocytes and oligodendrocytes), and skin stem cells occur at the base of hair follicles and replenish the protective keratinocyte layer of the skin. Other types of adult stem cells that have been identified include liver, kidney, intestinal, retinal, muscle, dental, and pancreatic stem cells.

Stem cells in adult tissues represent only a tiny fraction of the adult cells, and thus, sophisticated techniques have been developed to identify and isolate these cells from the differentiated, non-stem cells. Adult stem cells are relatively quiescent cells, usually dividing at a slow rate to balance the loss of tissue to physiologic turnover or wear and tear. Under conditions of severe stress, adult stem cells can be induced to divide very rapidly. In the organism, adult stem cells have a practically unlimited life span; however, once isolated and cultured in a laboratory, adult stem cells are much harder to maintain in a primitive state and will ultimately lose their stem cell properties in the dish. Despite their limited ability to be grown outside the body, adult stem cells nonetheless have enormous therapeutic potential, particularly in the areas of gene therapy and transplantation medicine.

Embryonic Stem Cells

In contrast to adult stem cells, embryonic stem cells are said to be immortalized; that is, using the right culture conditions, embryonic stem cells can be propagated indefinitely in a laboratory setting. The ability to culture and expand embryonic stem cells, coupled with their pluripotent phenotype, makes embryonic stem cell technology a powerful tool that has captured imaginations and sparked wide-ranging controversy.

Embryonic stem cells were first derived from mice in 1981 and enabled the development of technologies, such as the generation of mice defective for one particular gene, that are now fundamental and indispensable to the study of gene function and disease. In 1998, stem cell derivation techniques were successfully applied to human embryos, and the generation of human embryonic stem cells (hESC) ushered in a new era of research and controversy.

Potential Therapeutic Uses

Although stem cells have significant use as models for early embryonic development, another major research thrust has been for therapeutic uses. Stem cell therapy has been limited almost exclusively to multipotent stem cells obtained from umbilical cord blood, bone marrow, or peripheral blood. These stem cells are most commonly used to assist in hematopoietic (blood) and immune system recovery following high-dose chemotherapy or radiation therapy for malignant and nonmalignant diseases such as leukemia and certain immune and genetic disorders. For stem cell transplants to succeed, the donated stem cells must repopulate or engraft the recipient's bone marrow, where they will provide a new source of essential blood and immune system cells.

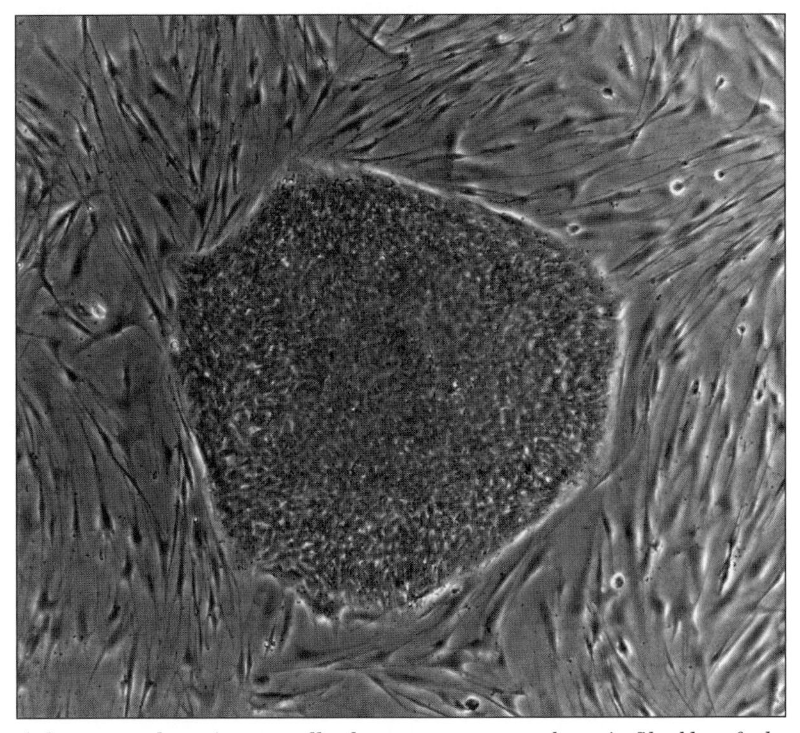

A human embryonic stem cell colony on a mouse embryonic fibroblast feeder layer. (NIH)

In addition to the uses of stem cells in cancer treatment, the isolation and characterization of stem cells and in-depth study of their molecular and cellular biology may help scientists understand why cancer cells, which have certain properties of stem cells, survive despite very aggressive treatments. Once the cancer cell's ability to renew itself is understood, scientists can develop strategies for circumventing this property.

Research efforts are under way to improve and expand the use of stem cells in treating and potentially curing human diseases. Possible therapeutic uses of stem cells include treatment of autoimmune diseases such as muscular dystrophy, multiple sclerosis, and rheumatoid arthritis; repair of tissues damaged during stroke, spinal cord injury, or myocardial infarction; treatment of neurodegenerative diseases such as amyotrophic lateral sclerosis (ALS, commonly called Lou Gehrig's disease) and numerous neurological conditions such as Parkinson's, Huntington's, and Alzheimer's diseases; and replacement of insulin-secreting cells in diabetics.

Stem cells may also find use in the field of gene therapy, where a gene that provides a missing or necessary protein is introduced into an organ for a therapeutic effect. One of the most difficult problems in gene therapy studies has been the loss of expression (or insufficient expression) following introduction of the gene into more differentiated cells. Introduction of the gene into stem cells to achieve sufficient long-term expression would be a major advance. In addition, the stem cell is clearly a more versatile target cell for gene therapy, since it can be manipulated to become theoretically any tissue. A single gene transfer into a pluripotent stem cell could enable scientists to generate stem cells for blood, skin, liver, or even brain targets.

Ethical Issues Concerning Use

Stem cell research, particularly embryonic stem cell research, has unleashed a storm of controversy. One primary controversy surrounding the use of embryonic stem cells is the belief by its opponents that a fertilized egg is fundamentally a human being with rights to be protected. Those who oppose stem cell research object to the destruction of embryos for research purposes. Others accept the special status of an embryo as a potential human being yet argue that the respect due to the embryo increases as it develops and that this respect, in the very early stages in particular, may be weighed against the potential benefits arising from the proposed research.

In addition to the concerns over the use of embryos for research purposes, another ethical controversy arises out of the potential use of cloning techniques to produce human embryonic cell lines. Currently, hESC cells are produced from surplus embryos that have been produced for the assisted reproduction technique of in vitro fertilization. However, a potential future source of embryonic material might be embryos produced by somatic cell nuclear transfer (SCNT), also called therapeutic cloning. In SCNT, genetic material from an adult cell is fused with an enucleated egg cell. With the right conditions, this new cell can develop into an embryo from which stem cells could be harvested. Opponents argue that therapeutic cloning is the first step on the slippery slope to reproductive cloning, the use of SCNT to create a new adult organism. Proponents maintain that producing stem cells by SCNT using genetic material from the patient will eliminate the possibility of rejection when the resulting stem cells are returned to the patient.

Infamously, in 2004-2005, the South Korean scientist Hwang Woo-Suk published a series of papers claiming to have derived the first hESC lines using SCNT. Even more incredibly, Dr. Hwang claimed to have generated multiple stable lines from patients with specific diseases, appearing to usher in the long-anticipated era of medically useful therapeutic cloning. Unfortunately, this work proved to be a spectacular example of scientific misconduct—all the reported cell lines were discovered to be fabricated, with the subsequent retraction of the papers and disgrace of Dr. Hwang.

In the first decade of the twenty-first century, researchers tried to develop several methods for generating stem cells that would assuage or circumvent the complex ethical issues surrounding embryonic stem cell research. A seminal breakthrough in this regard came in 2006, when Dr. Shinya Yamanaka and his team were able to start with fibroblasts (skin cells) and go on to create cell lines that had embryonic stem cell properties (pluripotency, culture immortalization). The technique involved using viral vectors to force skin cells to express four specific genes which reprogrammed the epigenetic tags that had shut down the cell's ability to express genes necessary for pluripotency. This technique represents a way to create induced pluripotent stem cells

President Barack Obama signs an Executive Order reversing a ban on funding stem cell research in March, 2009. (AP/Wide World Photos)

(iPS), which have the essential properties of hESC, without creating or destroying any embryos and has since been demonstrated to work in humans.

An important caveat to this work is that the current techniques for creating these cells lead to an increased risk of cancer arising from the transformed cells. This and other hurdles must be overcome before human iPS cells can be used therapeutically, yet this technique has enormous promise. Not only does it obviate the concerns of those who oppose embryo destruction, but its use of adult skin cells (potentially from a sick patient) as starting material also means that the technically and ethically challenging method of SCNT may also fall by the wayside as well.

LEGAL STATUS

In 2001, amid much controversy, President George W. Bush issued an executive order allowing federal funds to support research on hESC lines that were already in existence but prohibiting the use of such funds to develop or work with any new hESC lines. This attempt to balance ethical considerations with potential research benefits was met with heavy criticism by both sides of the debate. In ensuing years, many states (notably California and Michigan) passed funding initiatives to support stem cell research at the state level.

In March, 2009, newly inaugurated President Barack Obama issued an executive order which largely reversed prior policy. Under the new order, federal funds can be used to support research on hESC lines if the cells in question were obtained from extra embryos created for reproductive purpose. This decision was beset by controversy as well, both by those who continue to harbor ethical objections to the use of hESC and by those who feel that the new executive order remains too restrictive. Specifically, the requirement of the new executive order for extensive documentation of the consent process for embryo donation may have the practical effect of excluding current hESC lines that were created with state or private funding that did not require the same rigorous level of documentation.

Lisa M. Sardinia, Ph.D.;
updated by David C. Weksberg, M.D., Ph.D.

FURTHER READING

Cyranoski, David. "Verdict: Hwang's Human Stem Cells Were All Fakes." *Nature* 122/123 (January 12, 2006). An accounting of the controversy surrounding the fraudulent claims of Korean scientists to have created human embryonic stem cell lines using therapeutic cloning.

Holland, Suzanne, Karen Lebacqz, and Laurie Zoloth, eds. *The Human Embryonic Stem Cell Debate: Science, Ethics, and Public Policy (Basic Bioethics).* Cambridge, Mass.: MIT Press, 2001. A collection of twenty essays organized into four sections: basic science and history of stem cell research, ethics, religious perspectives, and public policy.

Kaji, Eugene H., and Jeffrey M. Leiden. "Gene and Stem Cell Therapies." *Journal of the American Medical Association* 285, no. 5 (2001): 545-550. An overview of stem cells from a clinical viewpoint. Includes discussion of the feasibility of stem cell therapy, future research, and ethical issues.

Kiessling, Ann, and Scott C. Anderson. *Human Embryonic Stem Cells: An Introduction to the Science and Therapeutic Potential.* Boston: Jones and Bartlett, 2003. In the context of the social debate and public policy of the George W. Bush administration, addresses the various forms of stem cell research from the perspectives of many disciplines, from cell biology, embryology, and endocrinology to transplantation medicine.

Marshak, Daniel R., Richard L. Gardner, and David Gottlieb, eds. *Stem Cell Biology.* Cold Spring Harbor, N.Y.: Cold Spring Harbor Laboratory Press, 2002. Contains papers on early embryonic development, cell cycle controls, embryonal carcinoma cells as embryonic stem cells, stem cells of human adult bone marrow, intestinal epithelial stem cells, and much more, designed for researchers new to the field of stem cell biology.

Solter, Davor. "From Teratocarcinomas to Embryonic Stem Cells and Beyond: A History of Embryonic Stem Cell Research." *Nature Reviews Genetics* 7, no. 4 (April, 2006): 319-327. An excellent review of the history and development of embryonic stem cell technologies. Somewhat aimed at scientists but very readable for the interested layperson.

Yamanaka, Shinya. "A Fresh Look at iPS cells." *Cell Stem Cell* 137, no. 1 (April 3, 2009): 13-17. A review of the developments in the revolutionary and fast-changing field of induced pluripotency stem cells, written by the author of the original paper describing the technology, designed for scientists.

WEB SITES OF INTEREST

International Society for Stem Cell Research: Stem Cell Information for the Public
http://www.isscr.org/public/index.htm

Informational site from one of the leading organizations of scientists conducting stem cell research, with links to many resources.

National Institutes of Health: Stem Cell Information
http://stemcells.nih.gov/info/basics

Government site covering stem cell basics, the science of stem cell research, and links to related resources.

University of Michigan: Stem Cell FAQ
http://www.umich.edu/stemcell/faq

Site of the University of Michigan's stem cell center, addressing stem cell basics as well as ethical issues.

See also: Aging; Alzheimer's disease; Autoimmune disorders; Biochemical mutations; Bioethics; Cancer; Cell culture: Animal cells; Cell culture: Plant cells; Cell cycle; Cell division; Cloning; Cloning: Ethical issues; Cloning vectors; Cystic fibrosis; Developmental genetics; Eugenics; Eugenics: Nazi Germany; Gene therapy; Gene therapy: Ethical and economic issues; Genetic engineering: Medical applications; Huntington's disease; In vitro fertilization and embryo transfer; Infertility; Knockout genetics and knockout mice; Model organism: *Mus musculus*; Organ transplants and HLA genes; Totipotency; Transgenic organisms.

Sterilization laws

Category: Human genetics and social issues

Significance: Forced sterilization for eugenic reasons became legal throughout much of the United States and many parts of the world during the first half of the twentieth century. Though sterilization is an ineffective mechanism for changing the genetic makeup of a population, sterilization laws remain in effect in many states in the United States and other countries throughout the world.

Key terms

negative eugenics: the effort to improve the human species by discouraging or eliminating reproduction among those deemed to be socially or physically unfit

positive eugenics: the effort to encourage more prolific breeding among "gifted" individuals

sterilization: an operation to make reproduction impossible; in tubal ligation, doctors sever the Fallopian tubes so that a woman cannot conceive a child

The Eugenics Movement and Sterilization Laws

The founder of the eugenics movement is considered to be Sir Francis Galton (1822-1911), who carried out extensive genetic studies of human traits. He thought that the human race would be improved by encouraging humans with desirable traits (such as intelligence, good character, and musical ability) to have more children than those people with less desirable traits (positive eugenics). With the development of Mendelian genetics shortly after the beginning of the twentieth century, research on improving the genetic quality of plants and animals was in full swing. Success with plants and domestic animals made it inevitable that interest would develop in applying those principles to the improvement of human beings. As some human traits became known to be under the control of single genes, some geneticists began to claim that all sorts of traits (including many behavioral traits and even social characteristics and preferences) were under the control of a single gene, with little regard for the possible impact of environmental factors.

The Eugenics Record Office at Cold Spring Harbor, New York, was set up by Charles Davenport to gather and collate information on human traits. The eugenics movement became a powerful political force that led to the creation and implementation of laws restricting immigration and regulating reproduction. Some geneticists and politicians reasoned that since mental retardation and other "undesirable" behavioral and physical traits were affected by genes, society had an obligation and a moral right to restrict the reproduction of individuals with "bad genes" (negative genetics).

The state of Indiana passed the first sterilization law in 1907, which permitted the involuntary sterilization of inmates in state institutions. Inmates included not only "imbeciles," "idiots," and others with varying degrees of mental retardation (described as "feeble-minded") but also people who were committed for behavioral problems such as criminality, swearing, and slovenliness. By 1911, similar laws had been passed in six states, and, by the end of the 1920's, twenty-four states had similar sterilization laws. Although not necessarily strictly enforced, twenty-two states currently have sterilization laws on the books.

The U.S. Supreme Court, in its 1927 *Buck v. Bell* decision, supported the eugenic principle that states could use involuntary sterilization to eliminate genetic defects from the population. The vote of the Court was eight to one. The court's reasoning went as follows:

> We have seen more than once that the public welfare may call upon the best citizens for their lives. It would be strange if it could not call upon those who already sap the strength of the state for these lesser sacrifices, often not felt to be such by those concerned, in order to prevent our being swamped with incompetence. It is better for all the world, if instead of waiting to execute degenerate offspring for crime, or to let them starve for their imbecility, society can prevent those who are manifestly unfit from continuing their kind. The principle that sustains compulsory vaccination is broad enough to cover cutting the Fallopian tubes.

Ironically, the sterilization laws of the United States and Canada served as models for the eugenics movement in Nazi Germany in its program to ensure so-called racial purity and superiority.

Impact and Applications

Two problems associated with eugenics are the subjective nature of deciding which traits are desir-

In 1927, the U.S. Supreme Court, in its Buck v. Bill *decision, supported the eugenic principle that states could use involuntary sterilization to eliminate genetic defects from the population. The result was the sterilization of more than sixty thousand mainly young people deemed to be weak, "feebleminded," or otherwise genetically inferior. Two sterilized residents of Lynchburg, Virginia, where many such sterilizations occurred, unveil a historical marker that commemorates the tragic decision.* (AP/Wide World Photos)

able and determining who should decide. These concerns aside, the question of whether there is a sound scientific basis for the desire to manipulate the human gene pool remains. Does the sterilization of individuals who are mentally retarded or who have some other mental or physical defect improve the human genetic composition? Involuntary sterilization of affected individuals would quickly reduce the incidence of dominant genetic traits. Individuals who were homozygous for recessive traits would also be eliminated. However, most harmful recessive genes are carried by individuals who appear normal and, therefore, would not be "obvious" for sterilization purposes. These "normal" people would continue to pass on the "bad" gene to the next generation, and a certain number of affected people would again be born. It would take an extraordinary number of generations to significantly reduce the frequency of harmful genes.

Although the number of involuntary sterilizations in the United States is now minimal, the impact sterilization laws had on the population through 1960 was far-reaching, as nearly sixty thousand people were sterilized. Other countries also had laws that allowed forced sterilizations, with many programs continuing into the 1970's. The province of Alberta, Canada, sterilized three thousand people before its law was repealed. Another sixty thousand were sterilized in Sweden. The story of sterilization and "euthanasia" in Germany needs no retelling. With the ability to decipher the human genome and implement improved genetic testing procedures, a danger exists that new programs of eugenics and involuntary sterilization might once again emerge.

Donald J. Nash, Ph.D.

Further Reading

Campbell, Annily. *Childfree and Sterilized: Women's Decisions and Medical Responses.* New York: Cassell, 1999. Explores the lives of twenty-three women who chose sterilization over bearing children and the prejudices and stereotypes they faced from the medical profession, which often deemed sterilization pathological. Bibliography, index.

Gallagher, Nancy L. *Breeding Better Vermonters: The Eugenics Program in the Green Mountain State.* Hanover, N.H.: University Press of New England, 2000. A biologist looks at the science of eugenics and the social, ethnic, and religious tensions brought about by the Eugenics Survey of Vermont, an organization in existence from 1925 to 1936.

Kevles, Daniel J. *In the Name of Eugenics: Genetics and the Uses of Human Heredity.* Cambridge, Mass.: Harvard University Press, 1995. A comprehensive introduction to the history of the eugenics movement and the development of sterilization laws. Discusses genetics both as a science and as a social and political perspective, and how the two often collide to muddy the boundaries of science and opinion.

Lombardo, Paul A. *Three Generations, No Imbeciles: Eugenics, the Supreme Court, and Buck v. Bell.* Baltimore: Johns Hopkins University Press, 2008. Chronicles the history of the *Buck v. Bell* decision, turning up new information. Lombardo, a law professor and historian, finds evidence that the Buck family was not feebleminded and that Carrie Buck's attorney did not sufficiently defend his client.

Mason, J. K. "Unsuccessful Sterilization." *The Troubled Pregnancy: Legal Wrongs and Rights in Reproduction.* New York: Cambridge University Press, 2007. Examines litigation involving the unexpected birth of a healthy child due to failed sterilization.

Meyers, David W. *The Human Body and the Law: A Medico-Legal Study.* New Brunswick, N.J.: Aldine Transaction, 2006. Compares the law and medical practices regarding voluntary sterilization and compulsory sterilization and castration in various countries, particularly the United States and the United Kingdom.

Web Sites of Interest

Center for Individual Freedom
http://www.cfif.org/htdocs/freedomline/current/in_our_opinion/un_sterile_past.html

The center, which works to protect individual freedoms guaranteed by the U.S. Constitution, includes a page entitled "The Sterilization of America: A Cautionary Tale" on its Web site. The page provides a history of the eugenics movement in the United States, including *Buck v. Bell*, and explores the aftermath of that decision.

Cold Spring Harbor Laboratory, Image Archive on the American Eugenics Movemement
http://www.eugenicsarchive.org

Comprehensive and extensively illustrated site that covers the eugenics movement in the United States, including sterilization laws.

Harry Laughlin's "Model Eugenical Sterilization Law"
http://www.people.fas.harvard.edu/~wellerst/laughlin

Alex Wellerstein, a professor in the history of science department at Harvard University, includes a copy of the law on his Web site. Laughlin was the director of the eugenics records office at the Cold Spring Harbor Laboratory from 1910 to1939 and an influential advocate of eugenics. He crafted a model sterilization law that was the template for the law adopted by the state of Virginia and subsequently upheld in the *Buck v. Bell* decision.

Race and Membership: The Eugenics Movement
http://www.facinghistorycampus.org/campus/rm.nsf

Facing History and Ourselves, an organization offering support to teachers and students in the areas of history and social studies, created this site that traces the history of the eugenics movement in the United States and Germany. It includes a page about sterilization, with information about sterilization laws and the *Buck v. Bell* opinion.

University of Vermont, Vermont Eugenics: A Documentary History
http://www.uvm.edu/~eugenics/sterilizationdl.html
A listing of original documents related to sterilization and eugenics in the United States, including a statement from the American Eugenics Society (1926) and related newspaper articles.

See also: Criminality; Eugenics; Eugenics: Nazi Germany; Hardy-Weinberg law; Miscegenation and antimiscegenation laws; Race.

Steroid hormones

CATEGORY: Developmental genetics; Molecular genetics

SIGNIFICANCE: Steroid hormones—hormones containing a steroid ring derived from cholesterol—are important for many processes that control sex determination, reproduction, behavior, and metabolism. Mutations in the genes that produce or regulate the action of specific steroid hormones may lead to infertility, sterility, sex determination, osteoporosis, autoimmune diseases, heart abnormalities, and breast, uterine, and prostate cancer.

KEY TERMS

anabolic steroids: drugs derived from androgens and used to enhance performance in sports

androgens: steroid hormones that cause masculinization

estrogens: steroid hormones that produce female characteristics

glucocorticoids: steroid hormones that respond to stress and maintain sugar, salt, and body fluid levels

hormones: chemical messengers produced by endocrine glands and secreted into the blood

mineralocorticoids: a group of steroid hormones important for maintenance of salt and water balance

progestins: steroid hormones important for pregnancy and breast development

testosterone: the principal androgen, produced by the testes and responsible for male secondary sexual characteristics

STEROID HORMONE CHARACTERISTICS AND FUNCTION

Steroid hormones represent a group of hormones that all contain a characteristic "steroid" ring structure. This steroid ring is derived from cholesterol, and cholesterol is the starting material for the production of different steroid hormones. Steroid hormones, like other types of hormones, are secreted by endocrine glands into the bloodstream and travel throughout the body before having an effect. All steroid hormones, although specific for the regulation of certain genes, function in a similar manner. Because steroid hormones are derived from cholesterol, they have the unique ability to diffuse through a cell's outer plasma membrane. Inside the cell, the steroid hormone binds to its specific receptor in the cytoplasm. Upon binding, the newly formed hormone-receptor complex relocates to the nucleus. In the nucleus, the hormone-receptor complex binds to the DNA in the promoter region of certain genes at specific nucleotide sequences termed hormone-responsive elements. The binding of the hormone-receptor complex to hormone-responsive elements causes the increased production of transcription and protein production in most cases. In some instances, binding to a specific hormone-responsive element will stop the production of proteins that are usually made in the absence of the hormone.

There are two types (sex steroid and adrenal steroid) and five classes of steroid hormones. The sex steroid hormones include the androgens, estrogens, and progestins and are produced by the male testes (androgens) and female ovaries. Adrenal steroid hormones include glucocorticoids and mineralocorticoids and are produced by the adrenal glands.

SEX STEROID HORMONES

Sex steroid hormone genes are responsible for determining the sex and development of males and females. Androgens are a group of steroid hormones that cause masculinization. The principal androgen is testosterone, which is produced by the testes and is responsible for male secondary sexual characteristics (growth of facial and pubic hair, deepening of voice, sperm production). Estrogens are sex steroid hormones produced in the ovaries and cause femininization. In addition, estrogens control calcium content in the bones, modulate other hormones produced in the ovary, modify sexual behavior, regulate secondary sex characteristics (menstrual periods, breast development, growth of pubic hair), and are essential for pregnancy to occur. The most potent estrogen is 17-beta estradiol. Progestins, including progesterone, are also sex steroid hormones. Progesterone is important for proper breast development and normal and healthy pregnancies; it functions in the mother to alter endometrial cells so the embryo can implant. The loss of progesterone at the end of a pregnancy aids in the beginning of uterine contractions.

Anabolic steroids are drugs derived from the male steroid hormone testosterone and were developed in the late 1930's to treat hypogonadism in men, a condition that results in insufficient testosterone production by the testes. During this same period, scientists discovered that anabolic steroids also increased the muscle mass in animals. These findings led to the use of anabolic steroids by bodybuilders, weightlifters, and other athletes to increase muscle mass and enhance performance. Anabolic steroid use can seriously affect the long-term health of an individual and in women results in masculinization.

Adrenal Steroid Hormones

Adrenal steroid hormones are secreted from the adrenal cortex and are important for many bodily functions, including response to stress and maintenance of blood sugar levels, fluid balance, and electrolytes. The glucocorticoids represent one class of adrenal steroid hormone. The most important, cortisol, performs critically important functions; it helps to maintain blood pressure and can decrease the response of the body's immune system. Cortisol can also elevate blood sugar levels and helps to control the amount of water in the body. Elevated cortisol helps the body respond to stress. The glucocorticoids cortisone and hydrocortisone are used as anti-inflammatory drugs to control itching, swelling, pain, and other inflammatory reactions. Prednisone and prednisolone, also members of the glucocorticoid class of hormones, are the broadest anti-inflammatory and immunosuppressive medications available.

The second class of adrenal steroid hormones is the mineralocorticoids, including aldosterone, which helps maintain salt and water balance and increases blood pressure. Aldosterone is crucial for retaining sodium in the kidney, salivary glands, sweat glands, and colon.

Genetic Defects Affecting Sex Steroid Hormones

Defects in the genes involved in the production of sex steroid hormones can have serious consequences. Mutations in the androgen receptor, the receptor for testosterone, result in androgen insensitivity syndrome. In this syndrome, the individual has the genes of a male (XY) but develops, behaves, and appears female. Other gene defects in androgen biosynthesis often result in sterility. Genetic defects in estrogen receptors or estradiol biosynthesis lead to infertility. Reduced levels of estradiol have also been linked to bone loss (osteoporosis) and infertility, whereas excessive levels are associated with an increased risk of breast and uterine cancer. Similarly, genetic mutations in the progesterone production pathway or the progesterone receptor are associated with infertility. In addition, bone loss is one of the most serious results of progesterone deficiency, made worse by inappropriate diet and lack of exercise.

Genetic Defects Affecting Adrenal Steroid Hormones

Genetic abnormalities in adrenal steroid hormone biosynthesis are known to cause hypertension in some cases of congenital adrenal hyperplasia (CAH). In people with this condition, hypertension usually accompanies a characteristic phenotype with abnormal sexual differentiation. CAH is a family of autosomal recessive disorders of adrenal steroidogenesis. Each disorder has a specific pattern of hormonal abnormalities resulting from a deficiency of one of the enzymes necessary for cortisol synthesis. The most common form of CAH is 21-hydroxylase deficiency; however, in all forms, cortisol production is impaired, which results in an increase in adrenocorticotropin and the overproduction of androgen steroids.

There are two major forms of 21-hydroxylase deficiency. Classic CAH deficiency results in masculinized girls who are born with genital ambiguity and may possess both female and male genitalia. Nonclassic 21-hydroxylase deficiency does not produce ambiguous genitalia in female infants but may result in premature puberty, short stature, menstrual irregularities or lack of a menstrual cycle, and infertility. Familial glucocorticoid deficiency (FGD) is an extremely rare, genetic autosomal recessive condition in which a part of the adrenal glands is destroyed. These changes result in very low levels of cortisol. Although this disease is easily treatable if recognized, when left untreated it is often fatal or can lead to severe mental disability.

The genetic basis of four forms of severe hypertension transmitted on an autosomal basis has also been determined. All of these conditions are characterized by salt-sensitive increases in blood pressure, indicating an increased mineralocorticoid effect. The four disorders—aldosteronism, mineralocorticoid excess syndrome, activating mutation of the mineralocorticoid receptor, and Liddle syndrome—are a consequence of either abnormal biosynthesis, abnormal metabolism, or abnormal action of steroid hormones and the development of hypertension. Adrenal insufficiency is known as Addison's disease and causes death within two weeks unless treated. Classical Addison's disease results from a loss of both cortisol and aldosterone secretion as a result of the near total or total destruction of both adrenal glands.

Thomas L. Brown, Ph.D.

FURTHER READING

Burnstein, Kerry L., ed. *Steroid Hormones and Cell Cycle Regulation.* Boston: Kluwer Academic, 2002. Collection of technical articles examining the role of steroid hormones in regulating cell proliferation and differentiation.

Ethier, Stephen P., ed. *Endocrine Oncology.* Humana Press, 2000. Experts provide chapters on cancers of the breast, prostate, endometrium, and ovary from cellular and molecular perspectives, including the way that steroid hormones function in both normal processes and pathogenesis.

Freedman, Leonard P., and M. Karin, eds. *Molecular Biology of Steroid and Nuclear Hormone Receptors.* Boston: Birkhauser, 1999. A molecular perspective on steroid functions in both normal and cancerous cells.

Jameson, J. Larry, ed. *Harrison's Endocrinology.* New York: McGraw-Hill Medical, 2006. Discusses steroid hormones in the chapters about disorders of the adrenal cortex and disorders of the ovary and female reproductive tract. The index lists other references to steroid hormones.

Khan, Sohaib A., and George M. Stancel, eds. *Protooncogenes and Growth Factors in Steroid Hormone Induced Growth and Differentiation.* Boca Raton, Fla.: CRC Press, 1994. Experts from cancer centers discuss the roles of steroid hormones in cancer from the perspectives of biochemistry, physiology, development, genetics, endocrinology, and other disciplines.

Moudgil, V. K., ed. *Steroid Hormone Receptors: Basic and Clinical Aspects.* Boston: Birkhauser, 1994. A scientific researcher examines the structural and functional alterations in steroid hormone receptors induced by phosphorylation, and hormonal and antihormonal ligands.

National Institutes of Health. *Steroid Abuse and Addiction.* NIH 00-3721. Bethesda, Md.: Author, 2000. This pamphlet outines the dangers of steroid use for unapproved purposes, such as bodybuilding.

Salway, Jack G. "Steroid Hormones: Aldosterone, Cortisol, Androgens, and Oestrogens." In *Medical Biochemistry at a Glance.* 2d ed. Malden, Mass.: Blackwell, 2006. Contains an illustration and a concise description of the hormones' functions and medical conditions related to them.

Strauss, Jerome F., III, and Robert L. Barbieri, eds. *Yen and Jaffe's Reproductive Endocrinology: Physiology, Pathophysiology, and Clinical Management.* 6th ed. Philadelphia: Saunders/Elsevier, 2009. One chapter discusses the synthesis and metabolism of steroid hormones, and another chapter describes steroid hormone action.

Tilly, J., J. F. Strauss III, and M. Tenniswood, eds. *Cell Death in Reproductive Physiology.* New York: Springer, 1997. Describes the selective death of steroid-producing tissues.

Wynn, Ralph M., and W. Jollie, eds. *Biology of the Uterus.* 2d rev. ed. Boston: Kluwer Academic, 1989. Reviews the basic biology of pregnancy and the role of sex steroid hormones in pregancy.

WEB SITES OF INTEREST

Kimball's Biology Pages
http://users.rcn.com/jkimball.ma.ultranet/BiologyPages/S/SteroidREs.html

John Kimball, a retired Harvard University biology professor, includes a page about steroid hormone receptors and their response elements in his online cell biology text.

Medical Biochemistry Page, Steroid Hormones and Receptors
http://themedicalbiochemistrypage.org/steroid-hormones.html

Medical Biochemistry Page was created by Michael W. King, a professor of medical and developmental biology at the Indiana University School of Medicine. The site contains a page with information about steroid hormone synthesis, steroid hormones of the adrenal cortex, gonadal steroid hormones, and steroid hormone receptors.

St. Edward's University, Department of Chemistry
http://www.cs.stedwards.edu/chem/Chemistry/CHEM43/CHEM43/Steroids/Steroids.HTML
Offers an overview of the structure, function, and regulation of steroid hormones.

See also: Aggression; Allergies; Androgen insensitivity syndrome; Autoimmune disorders; Behavior; Cancer; Gender identity; Heart disease; Hermaphrodites; Human genetics; Metafemales; Pseudohermaphrodites; X chromosome inactivation; XYY syndrome.

Sulfite oxidase deficiency

CATEGORY: Diseases and syndromes
ALSO KNOWN AS: Isolated sulfite oxidase deficiency; sulphite oxidase deficiency

DEFINITION

Sulfite oxidase deficiency is a rare deficiency disease in which there is a defect in the enzyme sulfite oxidase, which metabolizes sulfur amino acids. Isolated sulfite oxidase deficiency is characterized by severe neurological abnormalities, seizures, mental retardation, physical deformities, and dislocation of ocular lenses and is often fatal in infancy.

RISK FACTORS

Inheritance of sulfite oxidase deficiency is autosomal recessive. Individuals who are heterozygous for this condition have no symptoms. Individuals with this disorder belong to both genders and to various ethnic groups. There is a 25 percent risk of sulfite oxidase deficiency occurring in future children of parents who already have a child with this disorder.

ETIOLOGY AND GENETICS

Sulfite oxidase catalyzes the final reaction in oxidative degradation of sulfur amino acids. Defects in enzyme activity result in sulfite oxidase deficiency. Two conditions which affect enzyme activity include mutations in the enzyme itself and a genetic defect that results in absence of a required molybdenum cofactor.

Normally, the amino acids methionine and cysteine are metabolized to sulfite followed by oxidation to sulfite through the activity of the enzyme sulfite oxidase in a step that requires a molybdenum cofactor. Deficiency of either the enzyme or the cofactor results in similar disease symptoms. A defect in the pathway results in buildup of sulfite, which increases activity of alternate metabolic pathways for sulfite degradation. This results in formation of metabolites, S-sulfocysteine and thiosulfate, which can be used as diagnostic indicators of sulfite oxidase deficiency. The metabolite S-sulfocysteine probably substitutes for cysteine in connective tissues. This substitution weakens the zonule, a tissue in the lens that is normally rich in cysteine, and results in characteristic dislocated lenses associated with this disorder. The pathogenesis of brain damage associated with sulfite oxidase deficiency is not known but is likely related to toxic sulfite accumulation in the brain. Many cases of sulfite oxidase deficiency are actually associated with deficiency of the molybdenum cofactor. Individuals with isolated sulfite oxidase deficiency lack sulfite oxidase activity, but are normal with respect to molybdenum cofactor.

Inheritance of isolated sulfite oxidase deficiency is autosomal recessive. Mutations in the gene that codes for sulfite oxidase, *SUOX*, and in genes that code for molybdenum cofactor components (*MOCS1*, *MOCS2*, or *GEPH*) have been described, although no mutation is clearly predominant. Some mutations in the sulfite oxidase gene have been characterized at the molecular level, with deletions, insertions, and both nonsense and missense mutations identified.

SYMPTOMS

In isolated sulfite oxidase deficiency, severe convulsions begin soon after birth. Individuals with this disorder generally show symptoms such as seizures, neurological disorders, mental retardation, and physical deformities as newborns. Other symptoms include hypotonia and myclonus. Choreiform movements and symptoms similar to cerebral palsy may be present in individuals with milder forms. Brain pathology in affected individuals shows brain degradation, severe encephalopathy, marked neuronal loss, and white matter demyelination.

Screening and Diagnosis

Although the frequency of sulfite oxidase deficiency is unknown, this disorder is likely underdiagnosed. The condition is rare, with only about fifty cases reported worldwide. A predominance of reported cases in Europe and the United States most likely reflects increased recognition in these areas. Sulfite oxidase deficiency has traditionally been identified in patients during the neonatal period, although an increasing number of diagnoses are being reported with later onset. Elevated urinary sulfite is one indicator used to diagnose sulfite oxidase deficiency, and it can be measured using commercially available dipsticks. Enzyme activity in fibroblasts and cofactor levels in liver biopsies are used for confirmation. Sophisticated laboratory techniques can identify characteristic metabolites, S-sulfocysteine and thiosulfate, present in urine or plasma of patients with this disorder. Cranial CT scans or MRI may show characteristic brain abnormalities such as cerebral atrophy, decreased white matter density, neuronal loss, or cerebral edema. Sulfite oxidase activity in chorionic villi or DNA analysis in families known to carry a mutation in the sulfite oxidase gene have been used in prenatal diagnosis.

Treatment and Therapy

There are no effective medical treatments known for sulfite oxidase deficiency, particularly for patients diagnosed in infancy. Treatment for this disorder is supportive. Although diets restricted in cysteine and methionine have been tried and resulted in biochemical improvements, clinical gains were not substantial. Various drug treatments have also been tried, with limited success. High-dose thiamine has been used to replace thiamine destroyed by sulfite. Compounds such as cysteamine and penicillamine have been used to chelate excess sulfite, with some biochemical effects but few or no clinical effects. Some success in controlling seizures has been achieved with the antiepileptic drug vigabatrin.

Prevention and Outcomes

Sulfite oxidase deficiency is almost always associated with severe brain damage and death in early childhood, and is generally fatal in infancy. Survival beyond two years is often associated with development of dislocation of the ocular lens and profound mental retardation. Some cases of individuals with later presentation of sulfite oxidase deficiency symptoms have more favorable outcomes.

C. J. Walsh, Ph.D.

Further Reading

Clarke, Joe T. R. *A Clinical Guide to Inherited Metabolic Diseases.* Cambridge, England: Cambridge University Press, 2002. A clinical handbook for health care providers involved in diagnosis of metabolic disorders.

Fernandes, John, Jean-Marie Saudubray, Geroges Van Den Berghe, and John H. Walter. *Inborn Metabolic Diseases.* New York: Springer-Verlag, 2006. A classical textbook geared toward those involved in study and treatment of this disorder.

Milunksy, Aubrey. *Genetic Disorders and the Fetus.* Baltimore: Johns Hopkins University Press, 2004. A reference for physicians and scientists interested in human fetal development.

Web Sites of Interest

eMedicine from WebMD
http://emedicine.medscape.com/article/949303-overview

The Merck Manuals Online Medical Library
http://www.merck.com/mmpe

The Online Metabolic and Molecular Bases of Inherited Disease
http://www.ommbid.com

See also: Alkaptonuria; Andersen's disease; Diabetes; Diabetes insipidus; Fabry disease; Forbes disease; Galactokinase deficiency; Galactosemia; Gaucher disease; Glucose galactose malabsorption; Glucose-6-phosphate dehydrogenase deficiency; Glycogen storage diseases; Gm1-gangliosidosis; Hemochromatosis; Hereditary diseases; Hereditary xanthinuria; Hers disease; Homocystinuria; Hunter disease; Hurler syndrome; Inborn errors of metabolism; Jansky-Bielschowsky disease; Kearns-Sayre syndrome; Krabbé disease; Lactose intolerance; Leigh syndrome; Lesch-Nyhan syndrome; McArdle's disease; Maple syrup urine disease; Menkes syndrome; Metachromatic leukodystrophy; Niemann-Pick disease; Ornithine transcarbamylase deficiency; Orotic aciduria; Phenylketonuria (PKU); Pompe disease; Tarui's disease; Tay-Sachs disease.

Synthetic antibodies

CATEGORY: Immunogenetics

SIGNIFICANCE: Synthetic antibodies are artificially produced replacements for natural human antibodies. They are used to treat a variety of illnesses and promise to be an important part of medical technology in the future.

KEY TERMS

antibody: a protein molecule that binds to a substance in order to remove, destroy, or deactivate it

antigen: the substance to which an antibody binds

B cells: white blood cells that produce antibodies

monoclonal antibodies: identical antibodies produced by identical B cells

THE DEVELOPMENT OF ANTIBODY THERAPY

Natural antibodies are protein molecules produced by white blood cells known as B cells in response to the presence of foreign substances. A specific antibody binds to a specific substance, known as an antigen, in a way that renders it harmless or allows it to be removed from the body or destroyed. A person will produce antibodies naturally upon exposure to harmless versions of an antigen, a process known as active immunization. Active immunization was the first form of antibody therapy to be developed and is used to prevent diseases such as measles and polio.

The oldest method of producing therapeutic antibodies outside the human body is known as passive immunization. This process involves exposing an animal to an antigen so that it develops antibodies to it. The antibodies are separated from the animal's blood and administered to a patient. Passive immunization is used to treat diseases such as rabies and diphtheria. A disadvantage of antibodies derived from animal blood is the possibility that the patient may develop an allergic reaction. Because the animal's antibodies are foreign substances, the patient's own antibodies may treat them as antigens, leading to fever, rash, itching, joint pain, swollen tissues, and other symptoms. Antibodies derived from human blood are much less likely to cause allergic reactions than antibodies from the blood of other animals. This led researchers to seek a way to develop synthetic human antibodies.

A major breakthrough in the search for synthetic antibodies was made in 1975 by Cesar Milstein and Georges Köhler. They developed a technique that allowed them to produce a specific antibody outside the body of a living animal. This method involved exposing an animal to an antigen, causing it to produce antibodies. Instead of obtaining the antibodies from the animal's blood, they obtained B cells from the animal's spleen. These cells are then combined with abnormal B cells known as myeloma cells. Unlike normal B cells, myeloma cells can reproduce identical copies of themselves an unlimited number of times. The normal B cells and the myeloma cells fuse to form cells known as hybridoma cells. Hybridoma cells are able to reproduce an unlimited number of times and are able to produce the same antibodies as the B cells. Those hybridoma cells that produce the desired antibody are separated from the others and allowed to reproduce. The antibodies produced this way are known as monoclonal antibodies.

Because human B cells do not normally form stable hybridoma cells with myeloma cells, B cells from mice are usually used. Because mouse antibodies are not identical to human antibodies, they may be treated as antigens by the patient's own antibodies, leading to allergic reactions. During the 1980's and 1990's, researchers began to develop methods of producing synthetic antibodies that were similar or identical to human antibodies. An antibody consists of a variable region, which binds to the antigen, and a constant region. The risk of allergic reactions can be reduced by combining variable regions derived from mouse hybridoma cells with constant regions from human cells. The risk can be further reduced by identifying the exact sites on the mouse variable region that are necessary for binding and integrating these sites into human variable regions. This method produces synthetic antibodies that are very similar to human antibodies.

Other methods exist to produce synthetic antibodies that are identical to human antibodies. A species of virus known as the Epstein-Barr virus can be used to change human B cells in such a way that they will fuse with myeloma cells to form stable hybridoma cells that produce human antibodies. Another method involves using genetic engineering to produce mice with B cells that produce human antibodies rather than mouse antibodies. One of the most promising techniques involves creating a "library" of synthetic human antibodies. This is done

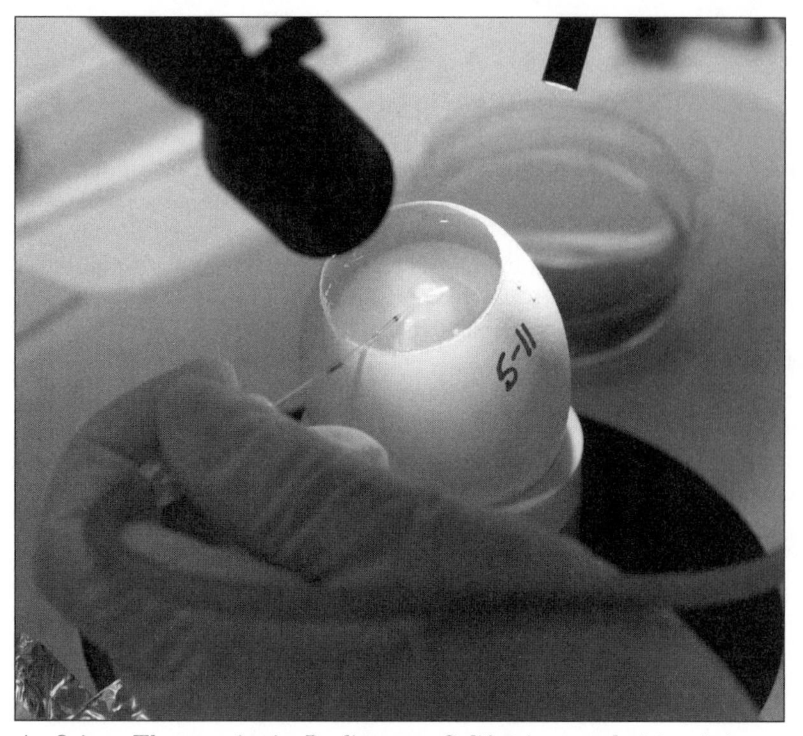

At Origen Therapeutics in Burlingame, California, a technician injects a chicken embryo with stem cells of another chicken embryo to which human antibodies have been added in order to make additional antibodies for pharmaceutical use. (AP/Wide World Photos)

by using the polymerase chain reaction (PCR) to produce multiple copies of the genetic material within B cells. This genetic material contains the information that results in the production of proteins that come together to form antibodies. By causing these proteins to be produced and allowing them to combine at random, researchers are able to produce millions of different antibodies. The antibodies are then tested to detect those that bind to selected antigens.

Impact and Applications

Some synthetic antibodies are used to help prevent the rejection of transplanted organs. An antibody that binds to the heart drug digoxin can be used to treat overdoses of that drug. Antibodies attached to radioactive isotopes are used in certain diagnostic procedures. Synthetic antibodies have also been used in patients undergoing a heart procedure known as a percutaneous transluminal coronary angioplasty (PTCA). The use of a particular synthetic antibody has been shown to reduce the risk of having one of the blood vessels that supply blood to the heart shut down during or after a PTCA. Researchers also hope to develop synthetic antibodies to treat acquired immunodeficiency syndrome (AIDS) and septic shock, a syndrome caused by toxic substances released by certain bacteria.

The most active area of research involving synthetic antibodies in the 1990's was in the treatment of cancer. On November 26, 1997, the U.S. Food and Drug Administration approved a synthetic antibody for use in non-Hodgkin's lymphoma, a cancer of the white blood cells. It was the first synthetic antibody approved for use in cancer therapy.

Antibodies are important for immune system protection against pathogens. They are also important tools in biological research, as with the use of antibodies to help determine the structure of cellular proteins. There may be a role for synthetic antibodies in determining the structure of RNA as well. RNA is genetic material that works with DNA. DNA is the master genetic code, but the role of RNA is a crucial one. More information on RNA structure may unveil more information on RNA function.

New approaches to therapeutic synthetic antibodies are in development. One approach involves the production of symphobodies. Symphobodies are made up of several different synthetic antibodies attacking a variety of antigens on the same target. These targets can be tumor or cancer cells or whole organisms such as viruses. Multiple synthetic antibody attacks on the same pathogen could result in more effective treatment.

Rose Secrest;
updated by Richard P. Capriccioso, M.D.

Further Reading

Coghlan, Andy. "A Second Chance for Antibodies." *New Scientist* 129 (February 9, 1991). An early discussion of the history and future of antibody therapy.

Doerr, Allison. "RNA Antibodies: Upping the Ante." *Nature Methods* 5, no. 3 (March, 2008): 220. A discussion on the use of RNA antibodies to find the structure of RNA.

Kontermann, Roland, and Stefan Dübel, eds. *Antibody Engineering.* New York: Springer, 2001. A detailed look at basic methods, protocols for analysis, and recent and developing technologies. Illustrations, bibliography, index.

Mayforth, Ruth D. *Designing Antibodies.* San Diego: Academic Press, 1993. Methods of synthetic antibody production are described in detail. Illustrations, bibliography, index.

"The Next Generation of Antibody Therapeutics?" *Pharmaceutical & Diagnostic Innovation* 4, no. 11 (2006): 13-14. A discussion on the use of several different synthetic antibodies in one therapeutic agent to better combat pathogens.

Web Sites of Interest

Patent Abstract: "Preparation of Synthetic Enzymes and Synthetic Antibodies and Use"
www.patentstorm.us/patents/5110833.html

A specific U.S. patent for the preparation of synthetic antibodies using the molecular imprinting method. A variety of U.S. patents can be searched at this site, with many different filters, such as date or inventor.

"Synthetic Antibodies from a Four-Amino-Acid Code: A Dominant Role for Tyrosine in Antigen Recognition" (Frederic A. Fellouse, Christian Wiesmann, and Sachdev S. Sidhu)
www.pubmedcentral.nih.gov/articlerender.fcgi?artid=515084

An article related to synthetic antibodies from the *Proceedings of the National Academy of Sciences of the USA.* PubMed Central is a searchable database with many research tools and categories that enable a better understanding of scientific research.

See also: Allergies; Anthrax; Antibodies; Autoimmune disorders; Biopharmaceuticals; Blotting: Southern, Northern, and Western; Burkitt's lymphoma; Cancer; Central dogma of molecular biology; Cloning; Diabetes; Genetic engineering: Historical development; Genetic engineering: Industrial applications; Hybridomas and monoclonal antibodies; Immunogenetics; Molecular genetics; Multiple alleles; Oncogenes; Organ transplants and HLA genes; Prion diseases: Kuru and Creutzfeldt-Jakob syndrome; Transgenic organisms.

Synthetic genes

Category: Genetic engineering and biotechnology

Significance: Synthetic genes have been shown to function in biological organisms. Scientists hope that it will prove possible to restore normal function in diseased humans, animals, and plants by replacing defective natural genes with appropriately modified synthetic genes.

Key terms

restriction enzyme: an enzyme that cleaves, or cuts, DNA at specific sites with sequences recognized by the enzyme; also called restriction endonucleases

reverse transcription: the synthesis of DNA from RNA

A Brief History

In 1871, Swiss physician Johann Friedrich Miescher reported that the chief constituent of the cell nucleus was nucleoprotein, or nuclein. Later it was established that the nuclei of bacteria contained little or no protein, so the hereditary material was named nucleic acid. At the end of the nineteenth century, German biochemist Albrecht Kossel identified the four nitrogenous bases: the purines adenine (A) and guanine (G) and the pyrimidines cytosine (C) and uracil (U). In the 1920's, Phoebus A. Levene and others indicated the existence of two kinds of nucleic acid: ribonucleic acid (RNA) and deoxyribonucleic acid (DNA); the latter contains thymine (T) instead of uracil.

The chemical identity of genes began to unfold in 1928, when Frederick Griffith discovered the phenomenon of genetic transformation. Oswald Avery, Colin MacLeod, and Maclyn McCarty (in 1944) and Alfred Hershey and Martha Chase (in 1952) demonstrated that DNA was the hereditary material. Following the elucidation of the structure of DNA in 1953 by James Watson and Francis Crick, pioneering efforts by several scientists led to the eventual synthesis of a gene. The successful enzymatic synthesis of DNA in vitro (in the test tube) in 1956, by Arthur Kornberg and colleagues, and that of RNA by Marianne Grunberg-Manago and Severo Ochoa, also contributed to the development of synthetic genes. In 1961, Marshall Nirenberg and Heinrich Matthaei synthesized polyphenylalanine chains using a synthetic messenger RNA (mRNA). In 1965,

Robert W. Holley and colleagues determined the complete sequence of alanine transfer RNA (tRNA) isolated from yeast. The interpretation of the genetic code by several groups of scientists throughout the 1960's was also clearly important.

In 1970, Har Gobind Khorana, along with twelve associates, synthesized the first gene: the gene for an alanine tRNA in yeast. There were no automatic DNA synthesizers available then. In 1976, Khorana's group synthesized the tyrosine suppressor tRNA gene of *Escherichia coli* (*E. coli*). The *lac* operator gene (twenty-one nucleotides long) was also synthesized, introduced into *E. coli*, and demonstrated to be functional. It took ten years to synthesize the first gene; by the mid-1990's, gene machines could synthesize a gene in hours.

Gene Synthesis

Protein engineering is possible by making targeted changes in a DNA sequence to produce a different product (protein) polypeptide with different properties, such as stress tolerance. The process of targeting a specific change in the nucleotide sequence (site-directed mutagenesis) allows the correlation of gene structure with protein function. Rapid sequencing with modern capillary DNA sequencers facilitates determination of the order of nucleotides that make up a gene in a matter of hours.

Once the sequence of a gene is known, it can be synthesized from nucleotides using gene machines. A gene machine is simply a chemical synthesizer made up of tubes, valves, and pumps that bonds nucleotides together in the right order under the direction of a computer. An intelligent person with a minimum of training can produce synthetic genes. A gene may be isolated from an organism using restriction enzymes (any of the several enzymes found in bacteria that serve to chop up the DNA of invading viruses), or it may be made on a gene machine. For example, the chymosin gene (an enzyme used in cheese making) in calves can be synthesized from its known nucleotide sequence instead of isolating it from calf DNA using restriction enzymes. Alternatively, chymosin mRNA can be obtained from calf stomach cells, which can be transformed into DNA through reverse transcription.

New or modified genes may be manufactured to obtain a desired product. Gene synthesis, coupled with automated rapid sequencing and protein analysis, has yielded remarkable dividends in medicine and agriculture. Genetic engineers are designing new proteins from scratch to learn more about protein function and architecture. With synthetic genes, the process of mutagenesis can be explored in greater depth. It is possible to produce various alterations at will in the nucleotide sequence of a gene and observe their effects on protein function. Such studies carry the potential to unravel many biochemical and genetic pathways that could be the key to a better understanding of health and disease.

Manjit S. Kang, Ph.D.

Further Reading

Aldridge, Susan. *The Thread of Life: The Story of Genes and Genetic Engineering.* New York: Cambridge University Press, 1996. Provides a guide to DNA and genetic engineering.

Carlson, Robert. *Biology Is Technology: The Promise, Peril, and New Business of Engineering Life.* Cambridge, Mass.: Harvard University Press, 2009. Describes how new mathematical, computational, and laboratory tools can facilitate the engineering of biological artifacts, including organisms and ecosystems. Questions to what end these synthetic biology techniques should be used and who should be allowed to use them.

Gibson, Daniel G., et al. "Complete Chemical Synthesis, Assembly, and Cloning of a *Mycoplasma genitalium* Genome." *Science* 319, no. 5867 (February 29, 2008): 1215-1220. Scientists at the Craig Venter Institute describe how they synthesized the complete genome for a small parasitic bacterium. They maintain their techniques can be used to construct large DNA molecules from chemically synthesized pieces or from combinations of natural and synthetic DNA.

Greber, David, and Martin Fussenegger. "Synthetic Gene Networks." In *Systems Biology and Synthetic Biology*, edited by Pengcheng Fu and Sven Panke. Hoboken, N.J.: John Wiley and Sons, 2009. Discusses the characteristics of synthetic gene networks, the network building blocks, and the genetic transcriptional components of these networks.

Henry, Robert J. *Practical Applications of Plant Molecular Biology.* New York: Chapman and Hall, 1997. Gives protocols for important plant molecular biology techniques. Illustrations, bibliography, index.

Pollack, Andrew. "How Do You Like Your Genes? Biofabs Take Orders." *The New York Times*, September 12, 2007, p. 4. Describes how foundries are producing made-to-order genes that can be used to genetically create bacteria or other cells in order to make proteins or to conduct various types of research.

Richardson, Sarah M., et al. "GeneDesign: Rapid, Automated Design of Multikilobase Synthetic Genes." *Genome Research* 16, no. 4 (April, 2006): 13. Researchers at Johns Hopkins University describe how they developed a technique "for rapid synthetic gene design for multikilobase sequences."

Specter, Michael. "A Life of Its Own: Where Will Synthetic Biology Lead Us?" *The New Yorker* 85, no. 30 (September 28, 2009): 56-65. An overview of the emerging field of synthetic biology, chronicling its origins and development and weighing the benefits and perils of its applications.

WEB SITES OF INTEREST

Bio-Bricks Foundation
http://bbf.openwetware.org
The foundation registers and develops BioBricks—standard biological parts for assembling DNA. Its Web site catalogs, explains, and discusses these biological parts.

Biobuilder.org
http://www.biobuilder.org
Provides animations and activities for teaching and learning about synthetic biology.

Synthetic Biology
http://syntheticbiology.org/FAQ.html
Frequently asked questions and answers that explain what synthetic biology is, why this field is useful, and the ethical and moral questions that it generates.

Synthetic Biology Project
http://www.synbioproject.org
A clearinghouse for information on synthetic biology, providing news and information about research, publications, and other aspects of the field.

See also: Biopharmaceuticals; Cell culture: Plant cells; Cloning; Cloning vectors; DNA sequencing technology; Gene therapy; Genetic engineering; Protein synthesis; Restriction enzymes; Reverse transcriptase; Synthetic antibodies.

T

T-cell immunodeficiency syndrome

CATEGORY: Diseases and syndromes

DEFINITION

T-cell immunodeficiency syndrome refers to the group of immunodeficiencies that increase susceptibility to infection as a result on an immune system with deficient, absent, or defective T cells. This immunodeficiency can be expressed as a primary (congenital) or a secondary (acquired) disorder. Primary T-cell immunodeficiency results from autosomal or X-linked genetic defects.

RISK FACTORS

Secondary T-cell immunodeficiency is the most common and develops as a result of chronic infection, malnutrition, systemic disease, malignancy, or drug therapy. Although primary T-cell immunodeficiency is less common than secondary forms, they are rare genetic disorders and generally develop in infancy or early childhood accompanied by recurrent unusual infections.

ETIOLOGY AND GENETICS

Considering that T cells (a type of white blood cells) are the dominant type of lymphocytes in circulating blood, it is very likely that a significant T-cell deficiency usually may cause a decrease in the number of blood lymphocytes. Several genetic abnormalities may prevent T cells from identifying and destroying foreign or abnormal cells circulating in the body. An impaired T-cell immunity damages the immune system's ability to protect the body against bacterial, viral, fungal, or cancer cells attacks. The two general reasons contributing to the development of T-cell immunodeficiency syndrome are either autosomal or X-linked gene disorders resulting from mutations or deletions. These hereditary genetic disorders give rise to partial or absolute defects in T-cell function resulting in a defective immune system at birth or early in life. The most common forms of T-cell immunodeficiency disorders are severe combined immune deficiency (SCID), X-linked lymphoproliferative syndrome, X-linked hyper-IgM immunodeficiency, hyper-IgE syndrome, DiGeorge syndrome, ataxia telangectasia, Nijmegen breakage syndrome, and Wiskott-Aldrich syndrome.

Severe combined immune deficiency (SCID or bubble boy syndrome) is the most serious immunodeficiency disorder. A mutation in the *IL-2R* gene of the X chromosome causes SCID. This results in low levels of antibodies (immunoglobulins) and no T cells. The severity of this deficiency results in the development of more serious infections with infants not growing or developing normally. Untreated children frequently die before the age of one.

In DiGeorge syndrome, a deletion of chromosome 22 results in a thymus gland absent or underdeveloped at birth. The fetus has heart, face, thymus, and parathyroid gland abnormalities. The X-linked lymphoproliferative syndrome (Duncan disease) is caused by mutations in the *SHD2D1A* gene of the X chromosome. It results in defective T cells and natural killer (NK) cells and is characterized by extreme sensitivity to Epstein-Barr (mononucleosis) virus infection, resulting in liver failure, immunodeficiency, and malignant lymphoma.

Hyper-IgM syndrome is an antibody deficiency known by normal or elevated IgM levels and decreased levels or absence of other serum antibodies. It is caused by a mutation in the *CD40LG* gene of the X chromosome and leads into susceptibility to bacterial infections.

Hyper-IgE syndrome (Buckley or Job syndrome) is caused by a mutation in the *STAT3* gene of chromosome 17, which results in high levels of IgE and normal levels of the other antibody classes. These patients suffer from chronic eczema, recurrent lung infections, weak bones, coarse facial features, and pus abscesses.

The mutation in the *ATM* gene of chromosome 11, results in ataxia telangiectasia (AT). AT is characterized by dilated capillaries, higher susceptibility to infections, body movement incoordination, and malignancy predisposition.

A mutation in the *NBS1* gene located on chromosome 8 is responsible for the Nijmegen (Berlin) breakage syndrome, which is characterized by microcephaly, sinopulmonary infections, immunodeficiency, and high risk for lymphoma cancer. Because Wiskott-Aldrich syndrome is caused by mutation in the *WAS* gene on the X chromosome, immunodeficiencies are characterized by thrombocytopenia, eczema, and recurrent respiratory infections. There is abnormal antibody production and defective T-cell function, affecting boys only. There is a higher incidence of developing cancers such as lymphoma and leukemia.

Symptoms

A history of recurrent unusual infections might suggest T-cell immunodeficiency syndrome. Frequently, respiratory infections arise first and reappear, becoming severe and persistent and leading to complications. Generally, the earliest the symptoms in life arise in children, the more severe will be the T-cell immunodeficiency syndrome. Other different ailments may contrast or change based on the infections' severity and length. Chronic diarrhea may cause infants or young children not to grow or develop (failure to thrive) due to weight loss.

Screening and Diagnosis

Doctors suspecting a primary T-cell immunodeficiency will run a battery of tests to identify the specific genetic abnormality, immunoglobulin levels, white blood microscopic irregularities, skin tests, number of circulating B and T cells, and proper B and T cell function. Upon physical examination, rashes, weight loss, chronic cough, hair loss, and/or enlarged spleen or liver may suggest a particular disorder to doctors based on disease-specific clinical manifestations. An early onset of recurring or unusual infections is a key determinant in identifying the type of immunodeficiency disorder. Adverse prognosis follows a delayed diagnosis. The particular form of infection will also help healthcare professionals pinpoint the T-cell immunodeficiency variant.

Treatment and Therapy

Symptoms suggest the type of genetic disorder, and treatment strategies are tailored to meet specific immunodeficiencies needs. Many patients will require prompt aggressive treatment. General guidelines for these patients include: periodic IV immunoglobulin replacement therapy, practicing excellent personal hygiene, avoiding undercooked food, drinking boiled water, and avoiding contact with infected people. SCID patients are kept in protected environments, preventing exposure to pathogens and are treated with antibiotics, antivirals, and antibodies. Bone marrow stem cell transplantation from an unaffected sibling matching the same tissue type is the only effective treatment for SCID. Thymus transplantation for DiGeorge syndrome patients can cure the immunodeficiency, and corrective heart surgery is done for severe heart conditions.

Prevention and Outcomes

Prevention and outcome strategies depend on the type of T-cell immunodeficiency disorder diagnosed and time of diagnosis. Some T-cell immunodeficiency syndromes shorten life span while others can be managed throughout life. When the disorder does not impair antibody production, vaccination is recommended only with killed viral and bacterial vaccines since live vaccines might cause disease in immunodeficient patients. There is no effective form of prevention; therefore, genetic testing and counseling should be provided to people with family history previously identified for these types of hereditary immunodeficiencies. Early diagnosis is critical in improving patient outcome since many children die young if untreated. The role of alert healthcare professionals to recognize T-cell immunodeficiency clinical findings can play an essential part confirming rapid accurate diagnosis, ultimately resulting in improved patient outcome.

Ana Maria Rodriguez-Rojas, M.S.

Further Reading

Bonilla, F. A., et al. "Practice Parameter for the Diagnosis and Management of Primary Immunodeficiency." *Annals of Allergy, Asthma & Immunology* 94, no. 5 Suppl 1 (May, 2005): S1-63.

Boztug, K., et al. "Multiple Independent Second-Site Mutations in Two Siblings with Somatic Mosaicism for Wiskott-Aldrich Syndrome." *Clinical Genetics* 74 (2008): 68-74.

Conley, Mary Ellen. "Antibody Deficiencies." In *The Metabolic and Molecular Bases of Inherited Disease,* edited by Charles Scriver et al. 8th ed. New York: McGraw-Hill, 2001.

Cooper, Megan A., Thomas L. Pommering, and Katalin Koranyi. "Primary Immunodeficiencies." *American Family Physician* 68 (2003): 2001-2008, 2011.

Edgar, J. D. "T Cell Immunodeficiency." *Journal of Clinical Pathology* 61 (2008): 988-993.

Kingsmore, S. F., et al. "Identification of Diagnostic Biomarkers for Infection in Premature Neonates." *Molecular & Cellular Proteomics* 7, no. 10 (October, 2008): 1863-1875.

WEB SITES OF INTEREST

Immune System Poster
http://www.info4pi.org/aboutPI/pdf/ImmuneSystemPoster.pdf

Immunodeficiency Disorders
http://www.nlm.nih.gov/medlineplus/ency/article/000818.htm

Jeffrey Modell Foundation: Primary Immunodeficiency FAQs
http://www.info4pi.org/aboutPI/index.cfm?section=aboutPI&content=faq&CFID=32813331&CFTOKEN=3862055

National Institute of Child Health and Human Development (NICHD): Primary Immunodeficiency
http://www.nichd.nih.gov/publications/pubs/primary_immuno.cfm#WhatisPrimaryImmunodeficiency

Nemours Foundation: Immune System
http://kidshealth.org/parent/general/body_basics/immune.html

Warning Signs of Primary Immunodeficiency
http://www.aafp.org/afp/20031115/2011ph.html

See also: Ataxia telangectasia; DiGeorge syndrome; Immunodeficiency with hyper-IgM; Severe combined immune deficiency; Wiskott-Aldrich syndrome.

Tangier disease

CATEGORY: Diseases and syndromes

ALSO KNOWN AS: A-alphalipoprotein neuropathy; alpha high-density lipoprotein deficiency disease; analphalipoproteinemia; cholesterol thesaurismosis; familial high-density lipoprotein deficiency disease; familial hypoalphalipoproteinemia; HDL lipoprotein deficiency disease; lipoprotein deficiency disease, HDL, familial; Tangier disease neuropathy; Tangier hereditary neuropathy

DEFINITION

Tangier disease is a rare, inherited disease that severely reduces blood concentrations of high-density lipoprotein. Cells lack the ability to export fats and cholesterol, and cholesterol accumulates inside cells, causing premature cardiovascular disease and other systemic pathologies.

RISK FACTORS

Because Tangier disease is inherited, it tends to run in families. A person cannot contract Tangier disease unless both biological parents have at least one mutant copy of the *ABCA1* gene. This disease is quite rare, with less than one hundred cases known worldwide, but is it found in many different countries and occurs equally in men and women.

ETIOLOGY AND GENETICS

Mutations in the *ABCA1* gene cause Tangier disease, and this disorder is inherited as an autosomal recessive trait. *ABCA1* is located on the long arm of chromosome 9 in band 31.

To inherit Tangier disease, both parents must carry at least one mutant copy of the *ABCA1* gene. Each sibling born to parents who both carry one mutant copy of the *ABCA1* gene has a 25 percent chance of contracting Tangier disease.

The *ABCA1* gene provides the instructions for the synthesis of a protein called the ATP-binding cassette transporter A1. This protein is inserted into cell membranes. The cell membrane delimits the interior of the cell from its exterior, and is composed of phosphate-containing lipids. To function properly, biological membranes must maintain a particular level of fluidity. Animal cells maintain the fluidity of their membranes by interspersing cholesterol molecules between the membrane lipid molecules.

Cholesterol, however, is too bulky a molecule to move across cell membranes without a transport protein. The ATP-binding cassette transporter A1 protein serves as a cholesterol transporter, and it exports lipid and cholesterol molecules from cells.

Since human cells make their own cholesterol and use dietary cholesterol, they must dispose of excess cholesterol. A process called reverse cholesterol transport moves excess cholesterol from the peripheral tissues to the liver where it is converted into bile salts, secreted into the gastrointestinal tract, disposed of in feces. The ATP-binding cassette transporter A1 mediates reverse cholesterol transport.

Reverse cholesterol transport begins when the liver secretes a protein called apoA1 into the blood. ApoA1 binds to blood lipids and forms a nascent discoidal high-density lipoprotein (ndHDL). Cells use the ATP-binding cassette transporter A1 to export lipids and cholesterol into the extracellular spaces and ndHDLs absorb these molecules and use an enzyme called lecithin cholesterol acyltransferase (LCAT) to convert cholesterol into cholesterol esters. This enlarges the ndHDLs and transforms them into spherical, mature HDL particles, which are called HDL-C particles. HDL-C particles are taken up by the liver, and their cholesterol esters are converted into bile salts that are excreted into the small intestine for disposal.

Mutations in the *ABCA1* gene that inactivate the ATP-binding cassette transporter A1 protein prevent cells from ridding themselves of excess cholesterol. Without the increased levels of cholesterol near the surfaces of cells, ApoA1 protein molecules never make HDL-C particles and are, instead, rapidly degraded. HDL levels in the blood are extremely low, and cholesterol builds up in cells throughout the body, eventually killing them.

Symptoms

The tonsils are greatly enlarged and have an orange or yellow color. Cholesterol buildup in nerves causes nervous system abnormalities (neuropathy). The large blood vessels show premature buildup of lipids and cholesterol within the inner lining of the vessel wall (atherosclerosis). Complications include an enlarged spleen (splenomegaly) and liver (hepatomegaly), clouding of the cornea of the eyes, and early-onset cardiovascular disease.

Screening and Diagnosis

Cholesterol deposits in the cornea of the eyes, tonsils, or rectal tissues are diagnostic, especially if combined with unexplained enlargement of the spleen or liver and neurological disturbances. Biochemically, patients show hypoalphalipoproteinemia, or unusually low blood HDL-C levels (less than 5 mg/dL). No commercially available DNA test exists to date, but particular research laboratories can detect mutations in the *ABCA1* gene.

Treatment and Therapy

Presently, there is no cure for Tangier disease. Treatment regimes vary, but organs with excessive cholesterol deposits, like the tonsils and spleen, are usually removed. Treatment of arteriosclerosis requires angioplasty, and coronary artery disease is treated with bypass surgery. Medications typically show no efficacy.

Prevention and Outcomes

The prognosis for Tangier disease is, in most cases, rather good. If patients develop heart disease, then they might have a decreased life span, but this depends on the severity of their disease and the quality of medical care they that receive.

Michael A. Buratovich, Ph.D.

Further Reading

Andres, Lisa Marie. "Tangier Disease." In *Gale Encyclopedia of Genetic Disorders*, edited by Stacey L. Blachford. Farmington Hills, Mich.: Thomson Gale, 2005. A focused and easily understood summary of Tangier disease from a genetic and clinical perspective.

Maxfield, Frederick R., and Ira Tabas. "Role of Cholesterol and Lipid Organization in Disease." *Nature* 438, no. 7068 (2005): 612-621. A technical but very well organized review of cholesterol trafficking and the role of genetic variation in the onset of cardiovascular diseases.

Oram, John. "Tangier Disease and ABCA1." *Biochimica et Biophysica Acta* 1529 (2000): 321-330. A somewhat technical, biochemical discussion of the link between Tangier disease and the transport protein encoded by the *ABCA1* gene. It nicely reviews the research on Tangier disease and other cholesterol-induced pathologies.

WEB SITES OF INTEREST

Genetics Home Reference: Tangier Disease
http://ghr.nlm.nih.gov/condition=tangierdisease

Medline Plus: HDL Test
http://www.nlm.nih.gov/medlineplus/ency/article/003496.htm

National Organization for Rare Disorders: Tangier Disease
http://www.rarediseases.org/search/rdbdetail_abstract.html?disname=Tangier%20Disease

See also: Atherosclerosis; Barlow's syndrome; Cardiomyopathy; Heart disease; Holt-Oram syndrome; Long QT syndrome.

Tarui's disease

CATEGORY: Diseases and syndromes
ALSO KNOWN AS: Glycogenosis type VII; muscle phosphofructokinase deficiency

DEFINITION

Tarui's disease is both a glycogen storage disease and a glycolytic disorder. In glycogen storage diseases, ATP production in muscle tissue is impaired and excessive glycogen amounts are stored in muscles and other tissues. Glycolytic disorders feature a defect in one or more steps in glycogen metabolism. Tarui's disease is caused by an inherited defect in the gene coding for phosphofructokinase (PFK), an enzyme that catalyzes the rate-limiting step in glycolysis.

RISK FACTORS

The main risk factor for Tarui's disease is having a family member who has the disorder. The disease is especially prevalent among Ashkenazi Jewish individuals, with disease onset usually occurring in childhood.

ETIOLOGY AND GENETICS

Tarui's disease is a glycogen storage disease in which the conversion of glycogen to glucose is impaired, resulting in the accumulation of glycogen in tissues. The defective glycogen metabolism is caused by a deficiency in PFK, which catalyzes the rate-limiting conversion of fructose-6-phosphate and fructose-1-phosphate to fructose-1,6-diphosphate. PFK occurs in three forms in humans, the muscle (PFK-M), liver (PFK-L), and platelet (PFK-P) isozymes. PFK-M is the most common form in skeletal muscle, heart, and brain and the only PFK isozyme that metabolizes glucose in muscle. Red blood cells, however, use both PFK-M and PFK-L.

Tarui's disease is inherited as an autosomal recessive trait, meaning that an affected individual will receive one copy of the mutated gene from each parent. J. B. Sherman and colleagues found that the mutation occurs in two major forms in Ashkenazi Tarui's disease patients, as a splicing mutation and as a deletion mutation. The splicing mutation, delta5, is a G to A point mutation at the 5′-splice donor site of intron 5 of the PFK muscle subunit (PFK-M) gene, resulting in the complete deletion of exon 5 from the PFK-M mRNA transcript. The delta5 mutation was detected in eleven out of eighteen (60 percent) mutated genes in nine Ashkenazi families affected by the disease. The deletion mutation, which involves deletion of a single base in exon 22, was observed in six out of seven (86 percent) mutated genes in the nine Ashkenazi families. Together, these two mutations accounted for 94 percent of the total mutated genes detected in the members of these families. In contrast, the delta5 gene was detected in only one out of 250 normal Ashkenazi individuals. Mutations in other exons and in an intron between exon 10 and exon 11 have also been associated with the Tarui's disease phenotype.

These pathogenic PFK-M mutations cause the muscle symptoms of Tarui's disease, including muscle weakness and exercise intolerance. Because PFK-M is used by red blood cells to metabolize glucose, these mutations can also cause red blood cell impairment in the form of reticulocytosis and hemolysis. Currently, patients with symptoms of Tarui's disease may be genetically tested by a sequencing polymerase chain reaction (PCR) assay on DNA extracted from muscle tissue. Healthcare practitioners will usually offer patients the option of genetic counseling if they are considering genetic testing, especially because there is no specific treatment for Tarui's disease.

SYMPTOMS

Tarui's disease symptoms include muscle weakness, cramps, and exercise intolerance, referring to the excessive fatigue experienced in response to exercise. Some patients experience hemolysis with jaundice or severe renal impairment. This results in muscle necrosis and myoglobinuria, in which myoglobin is released from muscle cells and excreted in the urine, which appears rust-colored. In some cases, myoglobin accumulates in the kidney tubules, causing severe renal insufficiency. The phosphofructokinase defect also affects red blood cells, thus some patients have hemolysis with hyperbilirubinemia or jaundice.

SCREENING AND DIAGNOSIS

Physical examination often does not reveal abnormalities. Taking a patient's history, including response to exercise and family medical history, and lab testing are the most useful tools for diagnosing this disease. If risk factors are identified, laboratory testing is performed. Frequently, a creatine kinase assay is performed on a blood specimen and a myoglobinuria assay is conducted on a urine specimen. Elevated creatine kinase and positive myoglobinuria, especially after exercise, are warning signs of Tarui's disease. Another common assay determines the presence of the PFK enzyme by immunohistochemistry, which involves the binding of PFK-specific antibodies to PFK, if present in the specimen. The absence of PFK implies Tarui's disease. The ischemic forearm test is also frequently performed as a Tarui's disease screening test. In this assay, the patient grasps an object once or twice per second for two to three minutes while a blood pressure cuff is inflated on his or her upper arm. After the patient relaxes his or her grip, creatinine kinase, ammonia, and lactate levels are assayed at five-, ten-, and twenty-minute intervals. A positive ischemic forearm test, suggesting Tarui's disease, is the lack of lactate elevation as ammonia levels rise. Healthy subjects are expected to demonstrate increases in both lactate and ammonia levels. If one or more of these laboratory tests are positive, a muscle biopsy is required for a definitive diagnosis. When the muscle sample is examined under a microscope, the presence of subcarcolemmal vacuoles, abnormal polysaccharide deposits in muscle fibers, or abnormal red blood cells are positive findings for Tarui's disease.

TREATMENT AND THERAPY

Currently, there is no treatment for Tarui's disease, although diet modifications, rest, and exercise avoidance have been shown to be effective for decreasing clinical symptoms. Oral intake of glucose or fructose may also improve exercise tolerance.

PREVENTION AND OUTCOMES

Tarui's disease symptoms can be avoided by minimizing strenuous exercise and by observing a low-carbohydrate, high-protein diet, especially for less severe cases where disease onset occurs later in childhood or in adulthood.

Ing-Wei Khor, Ph.D.

FURTHER READING

Gray, R. G. F., et al. "Inborn Errors of Metabolism as a Cause of Neurological Disease in Adults: An Approach to Investigation." *Journal of Neurology, Neurosurgery & Psychiatry* 69 (2000): 5-12.

Raben, N., et al. "Various Classes of Mutations in Patients with Phosphofructokinase Deficiency (Tarui's Disease)." *Muscle & Nerve* 3 (1995): S35-S38.

Sherman, J. B., et al. "Common Mutations in the Phosphofructokinase-M Gene in Ashkenazi Jewish Patients with Glycogenesis vII—and Their Population Frequency." *The American Journal of Human Genetics* 55 (1994): 305-313.

Vasconcelos, O., et al. "Nonsense Mutation in the Phosphofructokinase Muscle Subunit Gene Associated with Retention of Intron 10 in One of the Isolated Transcripts in Ashkenazi Jewish Patients with Tarui Disease." *The Proceedings of the National Academy of Sciences* 92 (1995): 10322-10326.

WEB SITES OF INTEREST

Association for Glycogen Storage Disease
http://www.agsdus.org

eMedicine: Glycogen Storage Disease, Type VII
http://emedicine.medscape.com/article/119947-overview

See also: Andersen's disease; Forbes disease; Galactokinase deficiency; Galactosemia; Gaucher disease; Glucose galactose malabsorption; Glucose-6-phosphate dehydrogenase deficiency; Glycogen storage diseases; Hereditary diseases; Hers disease; Inborn errors of metabolism; McArdle's disease; Pompe disease.

Tay-Sachs disease

CATEGORY: Diseases and syndromes
ALSO KNOWN AS: Hexosaminidase A deficiency; GM2-Gangliosidosis; TSD

DEFINITION

Tay-Sachs disease (TSD) is a lethal disease inherited as an autosomal recessive disorder. Affected children are normal at birth, and symptoms are usually noticed by six months of age, after which they progressively worsen; the child usually dies at or before four years of age. There is no cure for this severe disorder of the nervous system, but an understanding of the genetic nature of the disorder has led to effective population screening, prenatal diagnosis, and genetic counseling.

RISK FACTORS

One of the interesting features of TSD, as is true of some other genetic disorders, is its variation across ethnic groups. While people of any ethnic descent can be TSD carriers, there is a greatly increased risk among people of Ashkenazi Jewish or Eastern European descent. Approximately 1 in 27 Ashkenazi Jews is a TSD carrier, compared to perhaps 1 in 250 for the rest of the world's population (including Sephardic Jews). Other groups with a higher incidence of TSD include French Canadians who live near the St. Lawrence River and members of the Cajun community of Louisiana.

ETIOLOGY AND GENETICS

Tay-Sachs disease (TSD) is named after Warren Tay, an English ophthalmologist, and Bernard Sachs, an American neurologist, who first described the disorder. TSD is one of the lysosomal storage disorders, as are Hurler syndrome, Hunter syndrome, Gaucher disease, and Fabry disease. Lysosomes are organelles found in the cytoplasm of cells and contain many enzymes that digest the cell's food and waste. TSD is caused by the lack of the enzyme hexosaminidase A (Hex A), which facilitates the breakdown of fatty substances and gangliosides in the brain and nerve cells. When Hex A is sufficiently lacking, as in TSD, gangliosides accumulate in the body and eventually lead to the destruction of the nervous system.

All forms of TSD are inherited as autosomal recessive disorders. Individuals with only one copy of the defective gene will not manifest the disease, while individuals with two copies of the gene will be afflicted with the disease. Couples in which only one partner carries the defective gene have no risk of having a baby with TSD, while couples in which both partners have the gene have a 25 percent risk of having a baby with TSD.

SYMPTOMS

Children with TSD appear normal at birth and up to six months of age. During this time, they may show an exaggerated startle response to sound. Shortly after six months, more obvious symptoms appear. The child may show poor head control and an involuntary back-and-forth movement of the eyes. Also distinctive of TSD is a "cherry red spot" on the retina of the eye, first described by Tay, that usually appears after one year of age as atrophy of the optic nerve head occurs. The symptoms are progressive, and the child loses all the motor and mental skills developed to that point. Convulsions, increased motor tone, and blindness develop as the disease progresses. The buildup of storage material in the brain causes the head to enlarge, and brain weight may be 50 percent greater than normal at the time of death. There is no cure for TSD, and death usually occurs between two and four years of age, with the most common cause of death being pneumonia.

There are several forms of Tay-Sachs disease in addition to the classical, or infant, form already described. There is a juvenile form in which similar symptoms appear between two and five years of age, with death occurring around age fifteen. A chronic form of TSD has symptoms beginning at age five that are far milder than those of the infant and juvenile forms. Late-onset Tay-Sachs disease (LOTS) is a rare form in which there is some residual Hex A activity so that symptoms appear later in life and the disease progresses much more slowly.

SCREENING AND DIAGNOSIS

Screening and educational programs among at-risk groups (in particular, the Ashkenazi Jews) have been very effective. Couples with advance knowledge of their carrier status can receive genetic counseling so that they understand their risk of having a child affected with TSD.

Diagnosis of TSD often occurs when a "cherry-red" spot is noted during an eye exam. Blood testing will reveal decreased hexosaminidase A activity. The parents' blood may be tested to determine their carrier status.

TREATMENT AND THERAPY

Unfortunately, there is no treatment available for TSD. Researchers continue to work towards some form of gene therapy or enzyme replacement therapy that could slow or halt the relentless progression of TSD in an affected baby.

Therapy is aimed at providing comfort for the affected child and support for the family. The child may be made more comfortable by helping to clear accumulated mucus from the lungs via chest physiotherapy, thus improving breathing and decreasing the risk of infection. A nasogastric tube or percutaneous esophago-gastrostomy tube may be required for feeding, in order to avoid aspiration of liquids or foods into the lungs. Antiseizure medications may be necessary, and physical therapy may be provided in order to prevent joint stiffness and/or muscle contractures.

PREVENTION AND OUTCOMES

Although much has been learned about the genetics of the Tay-Sachs gene and the protein deficiency that causes the disease, there is still no cure. Nevertheless, TSD provides an excellent example of how the medical community can assist a susceptible population in confronting an incurable genetic disease. The effective screening of populations at risk for TSD and prenatal detection of fetuses with TSD have served to dramatically reduce the overall incidence of this terrible disease.

Individuals who want to determine whether they carry the defective gene can undergo a blood plasma assay that identifies differences in Hex A activity. If one or both parents are not TSD carriers, then they will not be at risk of having an affected child. However, if both couples are carriers, then they have a 25 percent risk of bearing a child with TSD. Once a woman is pregnant, prenatal diagnosis (chorionic villus sampling at ten to twelve weeks of pregnancy, or amniocentesis at approximately fifteen to twenty weeks of pregnancy) can be used to ascertain whether the fetus has Tay-Sachs disease, allowing the couple to receive counseling regarding therapeutic pregnancy termination.

Donald J. Nash, Ph.D.;
updated by Rosalyn Carson-DeWitt, M.D.

FURTHER READING

Bach G., J. Tomczak, N. Risch, and J. Ekstein. "Tay-Sachs Screening in the Jewish Ashkenazi Population: DNA Testing Is the Preferred Procedure." *American Journal of Medical Genetics* 99 (February 15, 2001). Argues for using DNA testing as the most cost-effective and efficient way to screen for TSD.

Bembi, B. "Substrate Reduction Therapy in the Infantile Form of Tay-Sachs." *Neurology* 66 (January 24, 2006). The researchers explored the use of miglustat to treat TSD; neurological deterioration was not arrested, although the drug concentrated in the cerebrospinal fluid (CSF) and did decrease the risk of macrocephaly.

Desnick, Robert J., and Michael M. Kaback, eds. *Tay-Sachs Disease.* San Diego: Academic, 2001. Detailed analysis of Tay-Sachs. Illustrations, bibliography, index.

Leavitt, J. A. "The 'Cherry Red' Spot." *Pediatric Neurology* 37 (July 1, 2007). Discusses the clinical appearance and etiology of this pathognomonic finding.

National Tay-Sachs and Allied Diseases Association. *A Genetics Primer for Understanding Tay-Sachs and the Allied Diseases.* Brookline, Mass.: Author, 1995. An introductory overview of Tay-Sachs disease.

Parker, James N., and Philip M. Parker, eds. *The Official Parent's Sourcebook on Tay-Sachs Disease: A Revised and Updated Directory for the Internet Age.* San Diego: Icon Press, 2002. Topics include the essentials on Tay-Sachs disease, parents' rights, and insurance.

Shapiro, B. E. "Late-Onset Tay-Sachs Disease: Adverse Effects of Medications and Implications for Treatment." *Neurology* 67 (September 12, 2006). The authors studied the use of about 350 different medications in individuals with late-onset Tay-Sachs disease, to identify which medications are relatively safe, and which may actually worsen the clinical picture.

Zallen, Doris Teichler. *Does It Run in the Family? A Consumer's Guide to DNA Testing for Genetic Disorders.* New Brunswick, N.J.: Rutgers University Press, 1997. Covers the applications and social implications of testing for genetic disorders. Bibliography, index.

WEB SITES OF INTEREST

Dolan DNA Learning Center: Your Genes Your Health
http://www.ygyh.org

Sponsored by the Cold Spring Harbor Laboratory, this site, a component of the DNA Interactive

Web site, offers information on more than a dozen inherited diseases and syndromes, including Tay-Sachs disease.

National Human Genome Research Institute
http://www.genome.gov/10001220

Provides a vast amount of information regarding advances made in the study of genetics as related to human disease.

National Institute of Neurologic Disorders and Stroke
http://www.ninds.nih.gov/disorders/taysachs/taysachs.htm

A division of the National Institutes of Health, this Web site provides information on a variety of neurological disorders, including Tay-Sachs disease.

See also: Fabry disease; Gaucher disease; Genetic counseling; Genetic screening; Genetic testing; Genetic testing: Ethical and economic issues; Hereditary diseases; Hunter disease; Hurler syndrome; In vitro fertilization and embryo transfer; Inborn errors of metabolism; Prenatal diagnosis.

Telomeres

CATEGORY: Cellular biology

SIGNIFICANCE: Telomeres, the ends of the arms of chromosomes of eukaryotes, become shorter as organisms age. They are thought to act biologically to slow chromosome shortening, which can lead to cell death caused by the loss of genes and may be related to aging and diseases such as cancer. Telomeric DNA may also be involved in chromosome movement and localization in the nucleus, as well as transcriptional regulation of genes near the telomeres.

KEY TERMS

eukaryote: a unicellular or multicellular organism with cells that contain a membrane-bound nucleus, multiple chromosomes, and membrane-bound organelles

prokaryote: a unicellular organism with a single chromosome and lacking a nucleus or any other membrane-bound organelles

EUKARYOTIC CHROMOSOMES AND TELOMERES

The DNA of bacteria and other related simple organisms (prokaryotes) consists of one double-stranded DNA molecule. Structurally and functionally, the prokaryotic chromosome contains one copy of most genes as well as DNA regions that control expression of these genes. Prokaryotic gene expression depends primarily upon a cell's moment-to-moment needs. An entire prokaryotic chromosome, its genome, usually encodes about one thousand genes.

The genomes of eukaryotes are much more complex and may include 100,000 or more genes. The number of chromosomes in different types of eukaryotes can range from just a few to several hundred. Each of these huge DNA molecules is linear rather than the circular molecule seen in prokaryotes. In addition, many individual segments of eukaryotic DNA exist in multiple copies. For example, about 10 percent of the DNA of a eukaryote consists of "very highly repetitive segments" (VRS's), units that are less than ten deoxyribonucleotides long that are repeated up to several million times per cell. DNA segments that are several hundred deoxyribonucleotide units long represent about 20 to 25 percent of the DNA. They are repeated one thousand times or more per cell. The rest of the eukaryote DNA (from 65 to 70 percent of the total) consists of larger segments repeated once or a few times, the genes, and the DNA regions that control the expression of the genes.

Much of the repetitive DNA, called satellite DNA, does not seem to be involved in coding for proteins or RNAs involved in making proteins. Telomeres are part of this DNA and consist of pieces of DNA that are several thousand deoxyribonucleotide units long, found at both chromosome ends. They are believed to act to stabilize the ends of chromosomes and protect them from exonuclease enzymes that degrade DNA from the ends. Researchers have concluded this for two reasons. First, the enzymes that make two chromosomes every time a cell reproduces are unable to operate at the chromosome ends. Hence, the repeated reproduction of a eukaryote cell and its DNA will lead to the creation of shorter and shorter chromosomes, a process that can cause cell death when essential genes are lost. Second, as organisms age, the telomeres of their cells become shorter and shorter.

Telomerase Enzymes

When chromosomes are replicated in preparation for cell division, the internal segments are replicated by a complex process involving the enzymes primase and DNA polymerase. Primase lays down a small segment of RNA on the template strand of DNA, and DNA polymerase uses the primer to start replication. Making the end of a linear chromosome is a problem, however, because primers cannot consistently be produced at the very ends of the chromosomes. Consequently, with each cell division a small portion of the ends of newly replicated chromosomes is single-stranded and is trimmed off by exonucleases. This problem is solved by enzymes known as telomerases, which add telomeres to eukaryote chromosomes. Each telomerase contains a nucleic acid subunit, known as human telomerase RNA component (hTERC), which is about 150 ribonucleotides long. This length is equivalent to 1.5 copies of the appropriate repeat in the DNA telomere to be made.

The catalytic subunit of telomerase is called the human telomerase reverse transcriptase enzyme (hTERT). hTERT uses this piece of RNA as a template to make the desired DNA strand of the telomere. At the telomere, these two telomerase subunits are present along with other telomere-associated proteins and telomere-binding proteins, which affect the localization and activity of telomerase. The exact mechanism by which the DNA strand is made is not yet confirmed; however, it is thought that telomerase uses its RNA sequence (AACCCC) to bind to the target complementary DNA sequence (TTGGGG) at the end of the parent DNA strand. The polymerase activity and RNA template allows for the addition of nucleotides to the telomere. After six new nucleotides have been added (TTGGGG), the telomerase unit moves down the DNA parent strand and continues adding nucleotides. How the telomerase in any given species identifies the correct length of telomere repeat for a specific chromosome is not clearly understood, although it may be regulated by various telomere-associated proteins. After the addition of telomeric DNA sequences, the parent DNA strand will be longer than the complementary daughter DNA strand. This so-called "end-replication problem" is hypothesized to be solved by the enzyme primase, which uses the extended telomere to create a primer on the daughter DNA strand that DNA polymerase can then extend to fill in the gap.

Telomerase activity can be lost in certain strains of simple eukaryotes, such as protozoa. When this happens to a given cell line, each cell division leads to the additional shortening of its telomeres. This procedure continues for a fixed number of cell divisions; it then ends with the death of the telomerase-deficient cell line in a process known as replicative senescence.

A related observation has been made in humans. It has been shown that when human fibroblasts are grown in tissue culture, telomere length is longest when cells are obtained from young individuals. They are shorter in cells taken from the middle-aged, and very short in cells taken from the aged. Similar observations have been made with the fibroblasts from other higher eukaryotes as well as with other human cell types. In contrast, the process of telomere shortening does not happen when

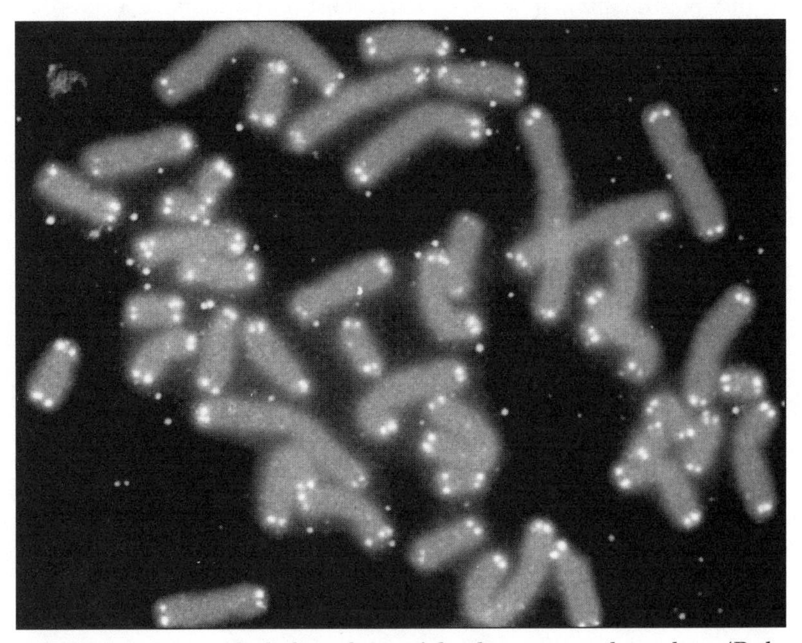

Telomeres appear as the lightened tips of the chromosomes shown here. (Robert Moyzis, University of California, Irvine, CA; U.S. Department of Energy Human Genome Program, http://www.ornl.gov/hgmis)

Telomere Length in Clones

Dolly the sheep, the first mammal to be cloned from adult cells, was born on July 5, 1996. While Dolly ushered in a new era of mammalian cloning, her tenure as the cloning community's lovable mascot was, quite literally, short-lived. Dolly was euthanized on February 14, 2003, after being diagnosed with a progressive lung disease; she had already been suffering from debilitating arthritis. While Dolly's health problems could have resulted from "natural causes," both ailments are more characteristic of much older sheep. Sheep normally live to an age of about twelve, Dolly was only half that.

Dolly's early demise was actually foreshadowed in 1999, when the group which cloned her reported that Dolly's telomeres were shorter than expected for a sheep of her age. Dolly's telomeres were about the length one would expect if her cells had been six years old on the day she was born (Dolly was cloned from a six-year-old ewe). Since telomere length acts as a "molecular clock" that determines the age of a cell, researchers had hoped that this clock would somehow be "reset" upon transfer of an adult nucleus to a host ovum.

While clearly not the case for Dolly, this resetting of telomere length has been demonstrated in cloned cows. In 2000, Robert Lanza and colleagues reported that cloned cows had longer-than-normal telomeres. Will these "super cows" be able to live appreciably longer than normal cows? Only time will tell, since cows have a normal life span of about twenty years. What accounts for the difference between Dolly and these cloned cows? Subsequent research has shown that the type of cell used in the cloning process may be an important factor. In 2002, Norikazu Miyashita and colleagues reported that cows cloned from mammary gland cells (like Dolly) had shorter-than-normal telomeres, clones obtained from skin fibroblasts (connective tissue precursors, those used by Lanza) had longer-than-normal telomeres, and clones obtained from muscle cells showed no significant differences in telomere length.

The telomere length of clones may also be species-specific. Teruhiko Wakayama and colleagues cloned mice sequentially for six generations but saw no difference in telomere length in any of the clones produced. Mice, however, are known to have extremely long telomeres to begin with; also, unlike the somatic cells of cows or sheep, many of the somatic cells of mice are known to express the telomerase enzyme.

More research is necessary to understand exactly why certain animal clones are produced with shortened telomeres and others are not. Currently, our lack of knowledge on the subject remains one of the more compelling reasons not to attempt to clone a human at this time. A human clone produced with unusually short telomeres may, like Dolly, meet with an untimely death. In fact, patients with a human genetic disease called Hutchinson-Gilford progeria have skin fibroblasts with greatly reduced telomere lengths; persons affected with this disease live to an average age of about thirteen years.

James S. Godde, Ph.D.

germ-cell lines—which in the whole organism produce sperm and ova—are grown in tissue culture. This suggests a basis for differences in longevity of the germ cells and the somatic cells that make up other human tissues.

Impact and Applications

The discovery and study of telomeres and telomerases produced new insights into DNA synthesis, the number of times a cell can reproduce, and the aging process. The circular DNA of bacteria (which are prokaryotes) allows them to undergo many more cycles of reproduction than the somatic cells of the eukaryotes. The linear eukaryote chromosome may have evolved because such DNA molecules were too large to survive as circular molecules given their rigidity and fragility. In addition, the observation of telomere shortening in simple and complex eukaryotes raises the fascinating possibility that the life spans of organisms may be related to the conservation of telomeres associated with the replication of these structures by telomerases.

The role of telomere length in longevity is uncertain, but apparently significant. Cells grown in cell culture typically divide only a predictable number of times, and once this limit is reached they can no longer divide. At the same time, telomere length shortens with each division. Sometimes cells in culture will go through what is called a "crisis," after which they become "immortalized" and are able to divide an indefinite number of times. Immortal cells also actively express telomerases and maintain constant telomere lengths. Cancer cells typically exhibit these same characteristics. A better understanding of telomeres and telomerase expression might provide insights into aging and cancer, leading to a potential cure for cancer and age-related diseases.

Multiple strategies to inhibit the telomere/telomerase complex are under investigation. For example, antisense oligonucleotides and gene-directed enzyme pro-drug therapy have been tested in vitro and in animals for inhibitory effects on hTERC. The first clinically tested hTERC inhibitor is GRN163L, which is being studied in patients with solid tumors and lymphoproliferative diseases. Small molecule inhibitors of hTERT have also been identified; however, issues with specificity and lengthy time to produce cell death have hampered clinical development. Vaccines are another strategy to target hTERT. In this case, either short protein fragments (known as peptides or epitopes) or whole cells engineered to overexpress hTERT are given to patients with adjuvants (molecules that help stimulate immune responses) so that tumor cells expressing hTERT are killed by cytotoxic T lymphocytes (also known as killer-T cells).

Sanford S. Singer, Ph.D.;
updated by Elizabeth A. Manning, Ph.D.

FURTHER READING

Abstracts of Papers Presented at the 2001 Meeting on Telomeres and Telomerase. Arranged by Elizabeth H. Blackburn, Titia De Lange, and Carol Grieder. Cold Spring Harbor, N.Y.: Cold Spring Harbor Laboratory Press, 2001. Synopses of research and studies on telomeres and telomerase. Bibliography, index.

Blackburn, Elizabeth H. "Telomeres, Telomerase, and Cancer." *Scientific American*, February, 1996. Provides background on telomeres, telomerases, and their potential importance in carcinogenesis.

Blackburn, Elizabeth H., and Carol W. Greider, eds. *Telomeres.* Cold Spring Harbor, N.Y.: Cold Spring Harbor Laboratory Press, 1995. Covers the discovery, synthesis, and potential effects of telomeres on normal life, aging, neoplasms, and other pathologies. Illustrations, bibliography, index.

Chan, S. R. W. L., and E. H. Blackburn. "Telomeres and Telomerase." *Philosophical Transactions of the Royal Society B: Biological Sciences* 359 (2004): 109-121. Discusses "the molecular consequences of telomere failure, and the molecular contributors to telomere function, with an emphasis on telomerase."

Double, John A., and Michael J. Thompson, eds. *Telomeres and Telomerase: Methods and Protocols.* Totowa, N.J.: Humana Press, 2002. A laboratory guide for exploring the world of the telomerase. Illustrations, bibliography, index.

Kipling, David, ed. *The Telomere.* New York: Oxford University Press, 1995. Describes telomeres, telomerases, relationships to cancer, and other aspects of potential telomere action.

Krupp, Guido, and Reza Parwaresch, eds. *Telomerases, Telomeres, and Cancer.* New York: Kluwer Academic/Plenum, 2003. Considers the way telomeres function in cancer. Illustrations, bibliography, index.

Lewis, Ricki. "Telomere Tales." *Bioscience* 48, no. 12 (December, 1998). An overview of some of the molecular research that supports the telomere shortening model of cellular aging.

Phatak, P., and A. M. Burger. "Telomerase and Its Potential for Therapeutic Intervention." *British Journal of Pharmacology* 152, no. 7 (December, 2007): 1003-1011. Reviews "telomere and telomerase biology and the various approaches which have been developed to inhibit the telomere/telomerase complex."

WEB SITES OF INTEREST

Telomeres
http://mcb.berkeley.edu/courses/mcb135k/telomeres.html

Telomeres and Telomerase—Solving the End-Replication Problem: A Tutorial for BMB 210 and BMB 330
http://web.centre.edu/bmb/movies/Telomeres.html

See also: Aging; Animal cloning; Chromosome mutation; Chromosome structure; Cloning vectors; DNA replication; Molecular genetics; Reverse transcriptase.

Thalassemia

CATEGORY: Diseases and syndromes
ALSO KNOWN AS: Mediterranean anemia; Cooley's anemia; thalassemia major; thalassemia minor

DEFINITION

Thalassemia is an inherited disorder. It leads to the decreased production and increased destruction of red blood cells. Hemoglobin in the red blood cells carry oxygen for all organs in the body. The loss of red blood cells results in low hemoglobin. This leads to anemia. The decreased oxygen will impair the ability to maintain normal functions.

Hemoglobin is made of two separate amino acid chains. They are alpha and beta. Thalassemias are categorized by the specific chain and number of genes affected.

In alpha thalassemia, the alpha chain is affected. Silent carrier thalassemia affects one gene; thalassemia trait affects two genes; Hemoglobin H disease affects three genes; alpha hydrops fetalis, the most severe form, affects four genes and results in fetal or newborn death.

In beta thalassemia, the beta chain is affected. Thalassemia minor involves one abnormal gene; thalassemia major (Cooley's anemia) involves two abnormal genes.

Risk Factors

Risk factors that increase an individual's chance of thalessemia include the geographic location of his or her ancestors. Individuals with ancestors from southeast Asia, Malaysia, and southern China are at risk for alpha thalassemias; individuals of southeast Asian, Chinese, and Filipino ancestry are at risk for alpha hydrops fetalis; those with ancestors from Africa, areas surrounding the Mediterranean Sea, and southeast Asia are at risk for beta thalassemias. A family history of the disorder is another risk factor.

Etiology and Genetics

Mutations in the *HBB* gene, located on the short arm of chromosome 11 at position 11p15.5, are the cause of beta thalassemia. This gene encodes the beta-globin polypeptide, which is one of the major components of hemoglobin. Some mutations completely inactivate the gene and stop the production of beta-globin (β^0 thalassemia), while others result in only a decreased amount of beta-globin in red blood cells (β^+ thalassemia). In either case, insufficient amounts of hemoglobin are formed, and anemia and other growth problems associated with beta thalassemia result. The inheritance pattern for this disease is usually autosomal recessive, meaning that both copies of the gene must be deficient in order for the individual to be afflicted. Typically, an affected child is born to two unaffected parents, both of whom are carriers of the recessive mutant allele. The probable outcomes for children whose parents are both carriers are 75 percent unaffected and 25 percent affected.

The inheritance of alpha thalassemia is more complex, since two genes are involved. The *HBA1* and *HBA2* genes are virtually identical copies of each other, and they are found in adjacent positions on the short arm of chromosome 16 at position 16p13.3. Each encodes an alpha-globin polypeptide, one of the four subunits of a mature hemoglobin molecule. Alpha thalassemia results when one or more copies of these genes are deleted from the genome. Since each individual has two copies of each gene, there are four functional genes that encode alpha-globin in healthy individuals. Different degrees or severities of alpha thalassemia are recognized when one, two, three, or all four copies of the gene are missing. If both parents are missing at least one copy of either *HBA1* or *HBA2*, their children could inherit anywhere from zero to four alpha-globin genes and present clinically with anywhere from the most severe form of alpha thalassemia to no thalassemia at all. The precise risk in each particular case depends entirely on the number and orientation of the missing genes in each parent.

Symptoms

Symptoms most often begin within three to six months of birth. Symptoms may include anemia, which may be mild, moderate, or severe; jaundice; an enlarged spleen; fatigue (tiredness); listlessness; and reduced appetite. Another symptom may be enlarged and fragile bones, including thickening and roughening of facial bones, bones that break easily, and teeth that do not line up properly. Growth problems, increased susceptibility to infection, and skin that is paler than usual may also be symptoms. Additional symptoms may be hormone problems, such as delayed or absent puberty, diabetes, and thyroid problems. Heart failure, shortness of breath, liver problems, and gallstones may be other symptoms.

Alpha thalassemia usually causes milder forms of the disease with varying degrees of anemia.

Beta thalassemia can be asymptomatic or be a mild form of disease. The mild form is known as thalassemia intermedia. This form rarely needs extensive medical care.

Beta thalassemia major (Cooley's anemia) has symptoms within the first two years of a child's life. Children are pale and listless. They often have poor appetites. They grow slowly and often develop jaundice (yellowing of skin). It is a serious disease. It requires regular blood transfusions and extensive medical care.

Without treatment, the spleen, liver, and heart soon become very enlarged. Bones become thin and brittle. Abnormal deposits of iron in body organs can lead to organ failure. This is called secondary hemochromatosis. It most often affects the heart, liver, and pancreas. Heart failure and infection are the leading causes of death among children with untreated thalassemia major.

Screening and Diagnosis

The doctor will ask about a patient's symptoms and medical history. A physical exam will be done. Blood tests may include a complete blood count, which is a count of the different types of blood cells; a blood smear; hemoglobin electrophoresis; quantitative hemoglobin analysis; and iron levels.

Treatment and Therapy

Treatment may include blood transfusions. These are done to replace abnormal red blood cells with healthy new ones.

Iron chelation therapy is another form of treatment. Excess iron can accumulate in the body after repeated blood transfusions. Too much iron can damage the heart, liver, and other vital organs. A drug called deferoxamine (Desferal) can be given to bind to excess iron in the body. It is then carried out through the urine. This drug is given through the skin or by vein using a small infusion pump. There is a new drug for the therapy called deferasirox (Exjade). It can be given as a drink or mixed with water or juice.

Splenectomy is a surgical procedure done to remove the spleen. It may help reduce the number of blood transfusions that are needed.

In a bone marrow transplant, healthy stem cells from a donor's bone marrow are injected into a patient's vein. The new cells travel through the bloodstream to the bone cavities; there they can produce new blood cells, including red blood cells. This is usually only done in severe cases. A compatible sibling donor is required.

Prevention and Outcomes

This disease is inherited. Blood tests and family genetic studies will show if an individual is a carrier. A genetic counselor can discuss the risks of passing on the disease. The counselor can also provide information on testing.

Jenna Hollenstein, M.S., RD;
reviewed by Igor Puzanov, M.D.
"Etiology and Genetics" by Jeffrey A. Knight, Ph.D.

Further Reading

EBSCO Publishing. *Health Library: Thalassemia.* Ipswich, Mass.: Author, 2009. Available through http://www.ebscohost.com.

Steinberg, Martin H., et al., eds. *Disorders of Hemoglobin: Genetics, Pathophysiology, and Clinical Management.* New York: Cambridge University Press, 2001.

Weatherall, D. J., and J. B. Clegg. *The Thalassaemia Syndromes.* 4th ed. Malden, Mass.: Blackwell Science, 2001.

Web Sites of Interest

Canadian Hemophiliac Society
http://www.hemophilia.ca/en

Centers for Disease Control and Prevention
http://www.cdc.gov

Genetics Home Reference
http://ghr.nlm.nih.gov

National Heart, Lung, and Blood Institute
http://www.nhlbi.nih.gov

Northern California Comprehensive Thalassemia Center
http://www.thalassemia.com

The Thalassemia Foundation of Canada
http://www.thalassemia.ca

See also: ABO blood group system; Fanconi anemia; Hereditary diseases; Sickle-cell disease.

Thalidomide and other teratogens

Category: Diseases and syndromes

Significance: Teratogenesis is the development of defects in the embryo or fetus caused by exposure to chemicals, radiation, or other environmental conditions. Thalidomide, a sedative whose ingestion by pregnant women led to the birth of abnormal babies in the late 1950's and early 1960's, is one of the more publicized examples of a chemical teratogen.

KEY TERMS

congenital defect: a defect or disorder that occurs during prenatal development

peromelia: the congenital absence or malformation of the extremities caused by abnormal development of the limb bud from about the fourth to the eighth week after conception; the ingestion of thalidomide by pregnant women can cause this disorder in fetuses

TERATOGENESIS AND ITS CAUSES

Teratogenesis is the development of structural or functional abnormalities in an embryo or fetus due to the presence of a toxic chemical or other environmental factor. The term is derived from the Greek words *teras* (monster) and *genesis* (birth). The phenomenon is usually attributed to exposure of the mother to some causative agent during the early stages of pregnancy. These may include chemicals, excessive radiation exposure, viral infections, or drugs. Approximately 3 percent of the developmental abnormalities are attributed to drugs. Drugs that are taken by the father may be teratogenic only if they damage the chromosomes of a spermatozoan that then joins with the egg to form a zygote.

For many centuries, the impression that malformed babies were conceived as a result of the intercourse between humans and devils or animals dominated society. Seventeenth century English physiologist William Harvey attributed teratogenesis to embryonic development. In the nineteenth century, the French brothers Étienne and Isidore Geoffroy Saint-Hilaire outlined a systematic study on the science of teratology. In the United States, the importance of teratogens was first widely covered during the 1940's, when scientists discovered that pregnant women who were affected by German measles (rubella) often gave birth to babies that had one or more birth defects. In the 1940's and 1950's, the consumption of diethylstilbestrol (DES) before the ninth week of gestation to prevent miscarriage was found to produce cancer in the developing fetus. Animal studies have also shown that defective offspring result from the use of hallucinogens such as lysergic acid diethylamide (LSD).

A broader definition of teratogenesis may include other minor birth defects that are more likely to be genetically linked, such as clubfoot, cleft lip, and cleft palate. These defects can often be treated in a much more effective way than those caused by toxic substances. Clubfoot, for example, which can be detected by the unusual twisted position of one or both feet, may be treated with surgery and physical therapy within the first month after birth. Brachydactyly (short digits) in rabbits has been linked to a recessive gene that causes a local breakdown of the circulation in the developing bud of the embryo, which is followed by necrosis (tissue death) and healing. In more extreme cases of agenesis, such as limb absence, a fold of amnion (embryonic membrane) was found to cause strangulation of the limb. Agenesis has been observed with organs such as kidneys, bladders, testicles, ovaries, thyroids, and lungs. Other genetic teratogenic malformations include anencephaly (absence of brain at birth), microcephaly (small-size head), hydrocephaly (large-size head caused by accumulation of large amounts of fluids), spina bifida (failure of the spine to close over the spinal cord), cleft palate (lack of fusion in the ventral laminae), and hermaphrodism (presence of both male and female sexual organs).

THALIDOMIDE AND ITS IMPACT

Thalidomide resembles glutethimide in its sedative action. Laboratory studies of the late 1950's and early 1960's had shown thalidomide to be a safe sedative for pregnant women. As early as 1958, the West German government made the medicine available without prescription. Other Western European countries followed, with the medicine available only upon physician's prescription. It took several years for the human population to provide the evidence that laboratory animals could not. German physician Widukind Lenz established the role of thalidomide in a series of congenital defects. He proved that administration of the drug during the first twelve weeks of the mother's pregnancy led to the development of phocomelia, a condition characterized by peromelia (the congenital absence or malformation of the extremities caused by the abnormal formation and development of the limb bud from about the fourth to the eighth week after conception), absence or malformation of the external ear, fusion defects of the eye, and absence of the normal openings of the gastrointestinal system of the body.

The United States escaped the thalidomide tragedy to a great extent because of the efforts of Frances O. Kelsey, M.D., of the U.S. Food and Drug Administration (FDA). She had serious doubts about the drug's safety and was instrumental in banning the approval of thalidomide for marketing in the United States. Other scientists such as Helen Brooke Taussig, a pioneer of pediatric cardiology and one of the physicians who outlined the surgery on babies with the Fallot (blue baby) syndrome, played a key role in preventing the approval of thalidomide by the FDA. It is estimated that about seven thousand births were affected by the ingestion of thalidomide.

The thalidomide incident made all scientists more skeptical about the final approval of any type of medicine, especially those likely to be used during pregnancy. The trend intensified the fight against any chemicals that might affect the fetus during the first trimester, when it is particularly vulnerable to teratogens. Alcohol and tobacco drew many headlines in the media in the 1990's. Both have been shown to create congenital problems in mental development and learning abilities. At the same time, regulation of new FDA-approved medicine became much stricter, and efforts to study the long-term effects of various pharmaceuticals increased. Surprisingly, thalidomide itself has been used successfully in leprosy cases and, in conjunction with cyclosporine, to treat cases of the immune reaction that appears in many bone-marrow transplant patients. There is also a movement to use thalidomide in the treatment of acquired immunodeficiency syndrome (AIDS).

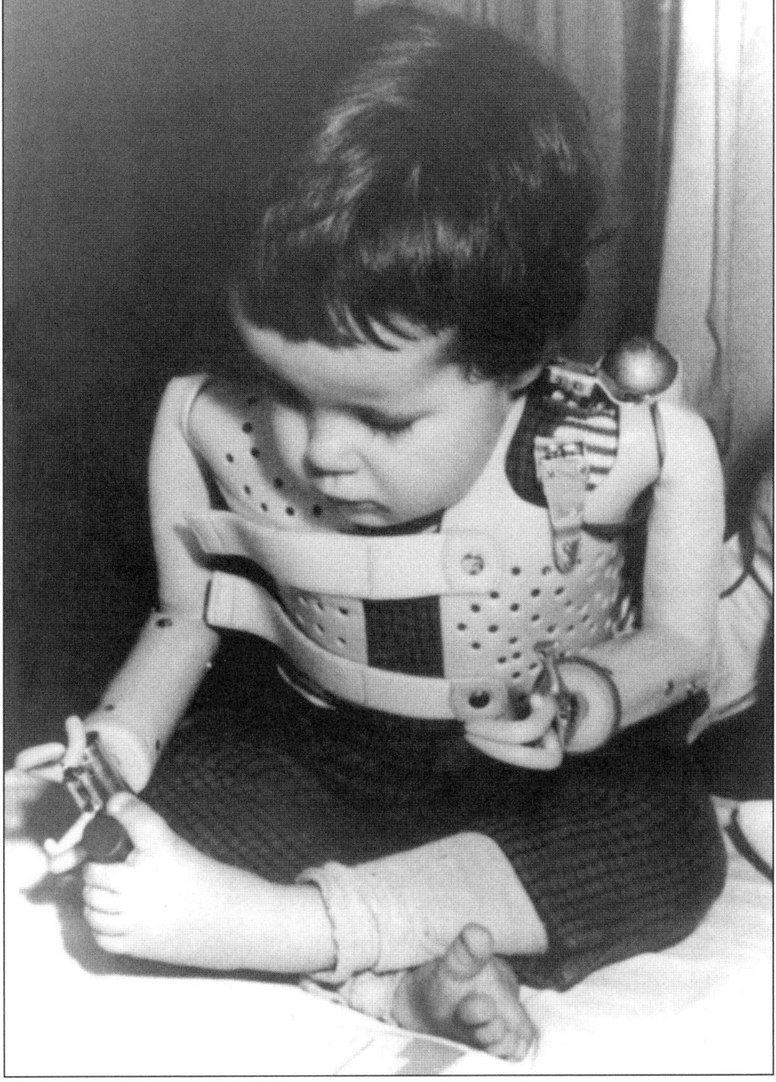

This three-year-old girl was born without arms in 1962 to a German mother who had taken thalidomide during her pregnancy. (AP/Wide World Photos)

In addition to drugs, many other agents can affect fetal development. Essentially, any factor with the potential to cause DNA mutations has a high probability of being teratogenic. Consequently, early in pregnancy, women are advised to limit their exposure to a variety of potential teratogens, such as excess radiation, toxic chemicals, tobacco, alcohol, and other drugs. Prevention might even include work reassignment to limit or eliminate the woman's normal exposure to teratogens. Unfortunately, teratogenesis can occur early in the pregnancy, before the woman is even aware that she is pregnant. Prevention by avoidance is therefore essential.

Soraya Ghayourmanesh, Ph.D.;
updated by Bryan Ness, Ph.D.

FURTHER READING

Ferretti, Patrizia, et al., eds. *Embryos, Genes, and Birth Defects.* 2d ed. Hoboken, N.J.: Wiley, 2006. Collection of articles that aim to bridge the gap between embryology and cellular, molecular, and developmental biology. Includes discussions of the identification and analysis of genes involved in congenital malformation syndromes, chemical teratogens, and birth defects in specific organs of the body.

Holmes, L. B. "Teratogen-Induced Limb Defects." *American Journal of Medical Genetics* 112, no. 3 (October 15, 2002): 297-303. Discusses limb defects, a common effect of human teratogens.

McCredie, Janet. *Beyond Thalidomide: Birth Defects Explained.* London: Royal Society of Medicine Press, 2007. A thorough examination of the history of the thalidomide scandal and the causes and pathology of the birth defects related to the drug. McCredie, who has spent a lifetime researching this subject, concludes that these birth defects are related to damage in the embryo's sensory nerves.

Schardein, James L. *Chemically Induced Birth Defects.* 3d ed., rev. and expanded. New York: Marcel Dekker, 2000. Comprehensive catalog of available information on drugs and chemicals that have the potential to cause birth defects in humans and animals. The majority of the book describes the impact of various drugs on the developing fetus, while the remainder of the book focuses on the effects of metal, industrial solvents, plastics, and other chemicals.

Stephens, Trent D., and Rock Brynner. *Dark Remedy: The Impact of Thalidomide and Its Revival as a Vital Medicine.* Cambridge, Mass.: Perseus, 2001. Surveys the history of the birth defects epidemic from the 1960's through the end of the twentieth century. Discusses the search for an alternative to thalidomide that retains its curative effects. Bibliography, index.

WEB SITES OF INTEREST

Center for the Evaluation of Risks to Human Reproduction
http://cerhr.niehs.nih.gov/index.html
The center, administered by the National Toxicology Program of the Department of Health and Human Services, provides information about the potential for adverse effects on reproduction and development caused by exposure to teratogenic agents.

March of Dimes
http://www.marchofdimes.com
This organization, which works to prevent birth defects, includes information about birth defects and genetic conditions on its Web site.

Organization of Teratology Specialists
http://www.otispregnancy.org/hm
The organization provides information to patients and medical professionals about exposures to teratogenic agents during pregnancy and lactation. Its Web site contains fact sheets about medications, maternal medical conditions, infections and viruses, and other common exposures, as well as other educational materials.

Teratology Society
http://www.teratology.org
The Teratology Society is a multidisciplinary scientific organization that studies the causes and biological processes leading to abnormal development and birth defects and appropriate measures for prevention.

See also: Congenital defects; Prenatal diagnosis.

Totipotency

CATEGORY: Cellular biology

SIGNIFICANCE: Totipotency is the ability of a living cell to express all of its genes to regenerate a whole new individual. Totipotent cells from plants have been used in tissue-culture techniques to produce improved plant materials that are pathogen-free and disease-resistant. Totipotent cells from animals are now being used to clone mammals, although ethical questions remain over whether cloning a human should be done.

KEY TERMS

multipotent cell: a stem cell capable of forming multiple differentiated tissues

parthenogenesis: asexual reproduction from a single egg without fertilization by sperm *pluripotent cell:* a stem cell that forms all types of differentiated tissues

unipotent cell: a stem cell that forms only one differentiated tissue

Egg and Sperm Cells

In plants and animals, a whole organism is sexually reproduced from a zygote, a product of fusion between egg and sperm. Zygotes are totipotent. A zygote in the seed of a plant or in the uterus of a mammal divides by mitosis and has the potential to produce more cells, called embryonic cells, before developing into an adult individual. During embryonic cell division, the cells begin to differentiate. Once differentiated, these specialized cells still possess all the genetic materials inherited from the zygote. Differentiated cells express or use some of their genes (not all) to produce their own specific proteins. For example, epidermal cells in human beings produce fibrous proteins called keratin to protect the skin, and red blood cells produce hemoglobin to help transport oxygen. Due to the differences in gene expression, differentiated cells have their own distinct structures and functions, and some differentiated cells are totipotent.

A whole organism can be asexually reproduced from a single egg without the sperm by a process called parthenogenesis. This occurs naturally in some insects, snakes, lizards, and amphibians, as well as in some plants. In this type of reproduction, the haploid chromosomes within an unfertilized egg duplicate, and the embryo develops as if the egg had been fertilized. The pseudo-fertilized eggs are totipotent and generate all female individuals. The females can reproduce under favorable environmental conditions without waiting for a mate. Like in vitro fertilization, parthenogenesis is used as a technique to create an embryo in the laboratory. Chromosomal duplication is induced in the egg cell to reproduce female individuals. However, no parthenogenic mammals had been developed.

It may become possible to produce males through a process called androgenesis. In the laboratory, the haploid chromosomes from one sperm may be induced to duplicate. As in animal cloning, the duplicated chromosomes, which are diploid, can be implanted into an enucleated egg cell (a cell from which the nucleus has been removed). Although androgenesis holds some promise, so far it has not produced normal embryos.

Cell Differentiation

Cell differentiation is a process whereby genetically identical cells become different or specialized for their specific functions. During differentiation, enzymes and other polypeptides, including other large molecules, are synthesized. Ribosomes and other cell structures are assembled. Differentiated cells express only some of their genes to make enzymes and other proteins.

Tissue differentiation is usually triggered by mitosis, followed by cytokinesis. Then differentiation occurs in the daughter cells. Often the two daughter cells have different structures and functions, but both retain the same genes. For example, the epidermal cell mitotically divides to produce one large and one small cell on the root surface; the large one maintains the role of epidermal cell as a root covering, whereas the small one becomes the root hair.

Totipotent Cells in Plant-Tissue Culture

Cuttings of plants and tissue-culture techniques have proven that many plant cells are totipotent. Tissue culture, however, helps to identify what specific type of cell is totipotent, because the technique uses a very small piece of known tissue. For example, if pith tissues from tobacco (*Nicotiana tobaccum*), soybean (*Glycine max*), and other dicot stems are cut off and cultured aseptically on an agar medium with proper nutrients and hormones, a clump of unspecialized and loosely arranged cells, called a callus, is formed. Each cell from the callus begins to divide and differentiate, forming a multicellular embryoid. One test tube can accommodate thousands of cells, and each embryoid has the potential to become a complete plantlet. Plantlets can be transplanted into the soil to develop into adult plants.

The phloem tissues from the roots of carrots (*Daucus* species) also exhibit totipotency. Cells in pollen grains of tobacco are totipotent, and they produce haploid plants. Using meristem tissues of shoot and root tips, the cells regenerate new plants that are free of viruses, bacteria, and fungi. Pathogen elimination is possible because vascular tissues (xylem and phloem), in which viruses move, do not reach the root or shoot apex. The protoplasts (cells without cell walls) from mesophyll cells of the leaf regenerate new plants.

Plant Hormones

Totipotency of plant cells is enhanced by the presence of hormones, such as auxins and cytokinins, in the culture media. Addition of auxins influences the expression of genes and causes physiological and morphological changes in plants. Addition of cytokinins promotes cell division, cytokinesis, and organ formation. If these are present in the proper ratio, callus from many plant species can be made to develop into an entire new plant. If the cytokinin-to-auxin ratio is high, cells in the callus divide and give rise to the development of buds, stems, and leaves. If the cytokinin-to-auxin ratio is low, root formation is favored. Totipotency of some plant cells is promoted by the addition of coconut water to the culture media—an indication that coconut water has the right proportion of cytokinin and auxin to regenerate an entire plant.

Importance of Totipotency in Plants

Clonal propagation of plants using tissue culturing is used commercially to mass-produce numerous ornamentals, vegetables, and forest trees. A major use of pathogen-free plants is for the storage of germ plasm and for transport of plant materials into different countries. It is also posssible to generate plants with desirable traits, such as resistance to herbicides and environmental stressors or tolerance of soil salinity, soil acidity, and heavy-metal toxicity. It is easier to select resistant or tolerant plants from a thousand cells than from a thousand plants.

Somatic Cells in Animal Cloning

Animals are more difficult to reproduce asexually than plants are. Somatic cells of animals become totipotent when used as donor cells in cloning. The first animal successfully cloned was a frog, *Xenopus laevis.* This cloning involved the use of a nucleus from the intestinal epithelial cells of a tadpole and an egg cell from a mature frog. In the laboratory, the nucleus from the egg cell was removed (enucleated) by micropipette. The tadpole's nucleus (the donor cell) was inserted into the enucleated frog's egg cell. The nuclei-injected egg cell underwent a series of embryonic developmental stages, including the blastula stage, developing into tadpoles that later died before becoming adults.

Cloning of Dolly the sheep (*Ovis* species) used the mammary cell of a six-year-old ewe as the donor cell. It was injected into the enucleated sheep's egg cell. Cloning a mammal requires a surrogate mother. The blastula stage of the embryo was developed in vitro and was implanted into a surrogate mother. After five months, a lamb was born. The lamb was genetically identical to the sheep from which the mammary cell was taken. Cloning is now done by scientists to produce other animals, including cattle, pigs, monkeys, cats, and dogs. Cloning a human seems possible, but there are many ethical and moral questions about whether it should or should not be done.

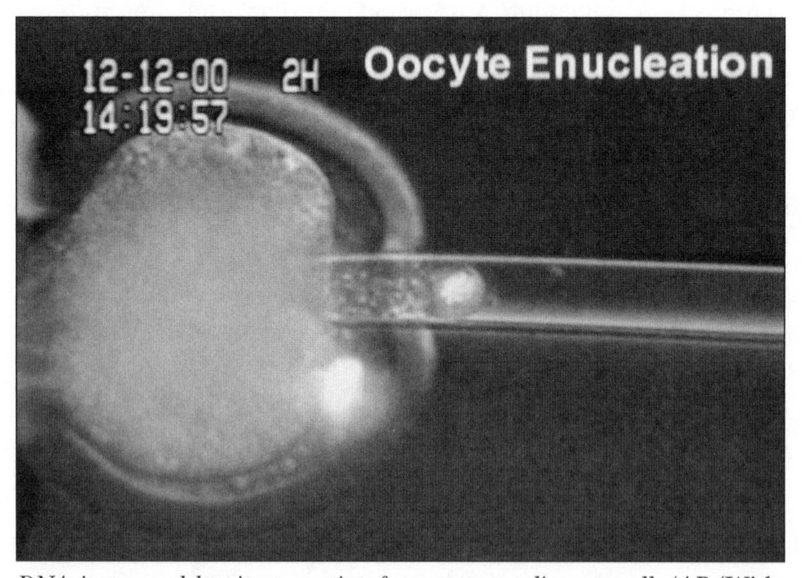

DNA is removed by pipette suction from a mammalian egg cell. (AP/Wide World Photos)

Stem Cells in Animal Cloning

Stem cells exhibit totipotency because they can generate new types of tissues. Some sources of stem cells are the blastocyst (the immature embryo), the fetus, the placenta, bone marrow, blood, skeletal muscle, and brain. Because there is no proof yet whether a single embryonic stem cell has the ability to regenerate into a complete individual, stem cells are generally only partially totipotent. A unipotent stem cell can form only one differentiated tissue. A multipotent stem cell can form multiple differentiated tissues. For example, stem cells from blood can form platelets, white blood cells, or red blood cells. The stem cells from skeletal muscle can form smooth muscle, cardiac muscle, bone, or cartilage. A pluripotent stem cell from embryo, brain, or bone marrow has the ability to develop all types of differentiated tissues of the body. For example, brain stem cells can be turned into all tissue types, including brain, muscles, blood cells, and nerves.

Domingo M. Jariel, Ph.D.

Further Reading

Control and Regulation of Stem Cells. Meeting organized by Bruce Stillman, David Stewart, and Terri Grodzicker. Woodbury, N.Y.: Cold Spring Harbor Laboratory Press, 2008. The information in this book is based on presentations by investigators who attended the seventy-third annual Cold Spring Harbor Symposium on Quantitative Biology. One of the chapters discusses germ cells and totipotency.

Gibbs, Melissa. "Plant Cell Totipotency." In *A Practical Guide to Developmental Biology.* New York: Oxford University Press, 2003. This textbook discusses plant and animal development and includes a chapter on plant cell totipotency.

Lanza, Robert, et al., eds. *Essentials of Stem Cell Biology.* 2d ed. San Diego: Academic Press, 2009. Includes information on pluripotent and multipotent stem cells.

Prentice, David A. *Stem Cells and Cloning.* New York: Benjamin Cummings, 2003. Discusses partial totipotency and differentiation of stem cells, with illustrations showing the sources of stem cells and how embryos can be developed by cloning techniques.

Russell, Peter J. *Genetics.* San Francisco: Benjamin Cummings, 2002. Discusses totipotency of some plants and animals. Includes illustrations showing how cells from plants are used in tissue-culture techniques and how cells from animals are used in animal cloning.

Smith, Roberta H. *Plant Tissue Culture Techniques and Experiments.* New York: Academic Press, 2000. Discusses the totipotency of different types of cells in regenerating new plants in vitro using culture growth media. Includes an illustration of explant preparation and discusses media preparation and transplantation of regenerated plants.

Web Sites of Interest

Biology, Sixth Edition
http://www.dwm.ks.edu.tw/bio/activelearner/16/ch16intro.html

The chapter "How Does a Single-Celled Zygote Give Rise to a Complex Organism with Many Specialized Parts?" discusses totipotency and determination and the relationship between totipotency and differentiation.

Explore Stem Cells, Totipotent Stem Cells
http://www.explorestemcells.co.uk/TotipotentStemCells.html

One of the articles in this site describes the changes in and special qualities and benefits of totipotent stem cells.

See also: Cell culture: Plant cells; Stem cells.

Tourette syndrome

Category: Diseases and syndromes
Also known as: Tourette's syndrome

Definition

Tourette syndrome is a chronic, neurological disorder that is a member of a larger group of primary tic disorders. It is characterized by motor and vocal tics. Both motor and vocal tics must be present, though not necessarily at the same time. Tics must be present for more than one year, and its onset must be prior to age eighteen. Tics are rapid, involuntary movements or sounds that occur repeatedly.

Risk Factors

Patients should tell their doctors if they have any of the risk factors, which include a family history of Tourette syndrome, other tic disorders, or obsessive-compulsive disorder. Males are three to four times more likely to be affected. Another risk factor is strep infection (may be a risk in some children).

There are many secondary causes of tics, including hereditary disorders, carbon monoxide poisoning, traumatic brain injury, cerebral infections, medications, and illegal drugs.

Etiology and Genetics

The causes of Tourette syndrome are complex and poorly understood, but it is likely that both genetic and environmental factors are involved. The characteristic tics expressed by individuals with Tourette syndrome probably result from changes in levels or activities of any of several different neurotransmitters, chemicals produced in the brain that help produce and control voluntary motion.

The only specific gene reported to be associated with cases of Tourette syndrome is the *SLITRK1* gene, found on the long arm of chromosome 13 at position 13q31.1. The normal protein product of this gene is a member of a family of proteins that orchestrate the growth and development of neurons. Specifically, the SLITRK1 protein may be important for the development of the long extensions of nerve cells, called dendrites and axons, which facilitate chemical communication between cells. It is unclear, however, how mutations in this gene might lead to the behavioral features characteristic of the syndrome. The majority of people with Tourette syndrome do not have mutations in the *SLITRK1* gene, so it is clear that a variety of genetic and environmental factors must be involved as opposed to changes in this one single gene. While some cases of the syndrome do appear to cluster in families, there are also many examples of isolated diagnoses in which there is no previous family history. As a result, it is not possible to reliably predict inheritance patterns.

Symptoms

Symptoms range from mild to severe, but most cases are mild. They can occur suddenly, and the length of time they last can vary. Tics may temporarily decrease with concentration or distraction. During times of stress, they may occur more often.

Tics are divided into motor and vocal, and then subdivided into simple and complex. Common examples of motor tics include simple tics, such as eye blinking, facial grimacing, head jerking, and arm or leg thrusting; and complex tics, such as jumping, smelling, touching things or other people, and twirling around. Common examples of vocal tics include simple tics, such as throat clearing, coughing, sniffing, grunting, yelping, and barking; and complex tics, such as saying words or phrases that do not make sense in a given situation and saying obscene or socially unacceptable words (called coprolalia).

Many people with Tourette syndrome also have one or more of the following problems: obsessions, compulsions, and ritualistic behaviors; attention deficit disorder, with or without hyperactivity (ADD or ADHD); learning disabilities; difficulties with impulse control; and sleep disorders.

While tics may occur throughout life, older teens may find that symptoms improve. In less than 10 to 40 percent of cases, people have remission from symptoms.

Screening and Diagnosis

The doctor will ask about a patient's symptoms and medical history and will perform a physical exam. Tourette syndrome is diagnosed by observing the symptoms and reviewing when they began and how they progressed. There are no blood or neurological tests to diagnose Tourette syndrome. Some doctors may order a magnetic resonance imaging (MRI) scan, a computed tomography (CT) scan, an electroencephalogram (EEG), or blood tests to rule out other disorders.

Treatment and Therapy

Most people with Tourette syndrome do not need medical treatment. If patients have a tic that disrupts daily activities, there are treatments. The most common is medication. No single treatment is helpful for all people with Tourette syndrome. Treatment may not completely eliminate symptoms.

Several medications can help control tics, including clonidine (Catapres), pimozide (Orap), risperidone (Risperdal), and haloperidol (Haldol), as well as newer antipsychotic medications, such as aripiperazole (Abilify) and ziprasidone (Geodon). Obsessive-compulsive symptoms may be treated with fluoxetine (Prozac), clomipramine (Anafranil), sertraline (Zoloft), or other similar medications. Symptoms of ADHD may be treated with stimulants, such as methylphenidate (Ritalin), pemoline (Cylert), dextroamphetamine sulfate (Dexedrine), or tricyclic antidepressants.

Other forms of treatment include behavior therapy, which can help people with Tourette syndrome learn to substitute their tics with alternative movements or sounds that are more acceptable. Cognitive behavioral therapy can help reduce obsessive-compulsive symptoms. Relaxation, biofeedback, and exercise can reduce the stress that often makes symptoms worse.

Psychotherapy can help people with Tourette syndrome and their families cope with the disorder. In tic disorders associated with pediatric autoimmune neuropsychiatric disorders associated with streptococcal (PANDAS), intravenous immunoglobulin therapy has been used with some success in a small number of patients, but this therapy is still considered experimental.

PREVENTION AND OUTCOMES

There is currently no known way to prevent Tourette syndrome.

Laurie Rosenblum, M.P.H.; reviewed by Rimas Lukas, M.D. "Etiology and Genetics" by Jeffrey A. Knight, Ph.D.

FURTHER READING

Bruun, Ruth Dowling, and Bertel Bruun. *A Mind of Its Own: Tourette's Syndrome—A Story and a Guide.* New York: Oxford University Press, 1994.

EBSCO Publishing. *DynaMed: Tourette Syndrome.* Ipswich, Mass.: Author, 2009. Available through http://www.ebscohost.com/dynamed.

_______. *Health Library: Tourette Syndrome.* Ipswich, Mass.: Author, 2009. Available through http://www.ebscohost.com.

Mell, L. K., R. L. Davis, and D. Owens. "Association Between Streptococcal Infection and Obsessive-Compulsive Disorder, Tourette's Syndrome, and Tic Disorder." *Pediatrics* 116, no. 1 (July, 2005): 56-60.

Menkes, John H., and Harvey B. Sarnat, eds. *Child Neurology.* 6th ed. Philadelphia: Lippincott Williams & Wilkins, 2000.

Samuels, Martin A., and Steven K. Feske, eds. *Office Practice of Neurology.* 2d ed. Philadelphia: Churchill Livingstone, 2003.

WEB SITES OF INTEREST

About Kids Health
http://www.aboutkidshealth.ca

Genetics Home Reference
http://ghr.nlm.nih.gov

National Institute of Neurological Disorders and Stroke: NINDS Tourette Syndrome Information Page
http://www.ninds.nih.gov/disorders/tourette/tourette.htm

Tourette Syndrome Association
http://www.tsa-usa.org

Tourette Syndrome Foundation of Canada
http://www.tourette.ca

See also: Attention deficit hyperactivity disorder (ADHD); Autism; Bipolar affective disorder; Dyslexia; Fragile X syndrome; Schizophrenia.

Transgenic organisms

CATEGORY: Genetic engineering and biotechnology
SIGNIFICANCE: Implanting genes from one organism into the genome of another enables scientists to study basic genetic mechanisms and inherited diseases and to create plants and animals with traits that are beneficial to humans.

KEY TERMS

genome: the complete genetic material carried by an individual
plasmid: a circular piece of bacterial DNA that is often used as a vector
transformation: integration of foreign DNA into a cell
transgene: the foreign gene incorporated into a cell's DNA during transformation
vector: a carrier molecule that introduces foreign genetic materials into a cell

ENGINEERING ORGANISMS

Domestication and selective breeding of animals and plants began before recorded history. In fact, historians propose, the shaping of organisms to fit human needs contributed to the rise of settled complex cultures. Until late in the twentieth century, farmers and scientists could breed novel strains only from closely related species or subspecies because the DNA had to be compatible in order to produce offspring that in turn were fertile.

In the late 1970's and early 1980's, molecular biologists learned how to surpass the limitations of selective breeding. They invented procedures for combining the DNA of species as distantly related as microbes, plants, and animals. These organisms carrying the novel heritable genetic material derived from different species using recombinant molecu-

These two rhesus monkeys were born from cloned embryos in 1996. (AP/Wide World Photos)

lar biology techniques are termed as transgenic or genetically modified organisms (GMO). Genetic engineering makes it possible to design novel organisms for genetic and biochemical research and for medical, agricultural, and ecological innovations. However commercial use of transgenic organisms created worldwide controversy because of their potential threat to human health and the environment.

Transgenesis seeks to produce an entirely modified organism by incorporating the transgene into all the cells of the mature organism and changing the genome. This is done by transforming not only the somatic (body) cells of the host organism but also the germ cells, so that when the organism reproduces, the transgene will pass to the next generation. Transgenes can block the function of a host gene, replace the host gene with one that codes for a variant protein, or introduce additional proteins.

Transgenic Animals

In recent years the use of transgenic organisms has become widespread and a large variety of animals and plants has been successfully engineered. Some examples of transgenic mammals include mice, rabbits, pigs, sheep, and cows. Birds such as chickens as well as fishes like salmon and trout, and even amphibians like frogs and toads have been made transgenic. Among invertebrates, transgenic fruit flies (*Drosophila melanogaster*) and nematodes (*Caenorhabditis elegans*) have become indispensable for research.

The first transgenic mice were created by Rudolf Jaenisch in 1974, showing that SV40 viral DNA had integrated into mouse genome. In 1982, Richard Palmiter and Ralph Brinster created "supermice" that grew much larger than ordinary mice because these had received the rat growth hormone gene. Most of these transformations were generated by microinjection of DNA directly into cells. Later, scientists were able to deliver foreign genes into hosts by several other methods: incorporating them into retroviruses and then infecting target cells; electroinfusion, whereby an electric current passed the foreign DNA through the relatively flimsy animal cell wall; and biolistics, a means of mechanically shooting a DNA bullet into cells. Two methods, developed at first for mice, are particularly successful in creating genetically modified animals. The first entails injecting transformed embryonic stem cells into a blastocyst (an early spherical form of an embryo). In the second, the DNA is inserted into the pronucleus of a freshly fertilized egg. The blastocyst or egg is then implanted into a foster mother for gestation.

Initially, the modified transgenic animals were intended for scientific research studies. After disabling a specific gene, scientists could study its effect on the structure, metabolic processes, and health of the mature animal. By 2003, thousands of genes had been tested. Studies on mice transformed with human DNA enabled scientists to identify genes associated with breast and prostate cancers, cystic fibrosis, Alzheimer's disease, and severe combined immunodeficiency disorder (SCID). In 2001, the first transgenic primate, a rhesus monkey, expressing the green fluorescent protein was created. It took quite a few years before the first transgenic primate model for the human disorder, Huntington's disease, was generated in rhesus macaque in 2008. However, both these animals were unable to give birth to offspring carrying the transgene. In 2009, scientists in the laboratory of Tatsuji Nomura created transgenic marmosets that are the first transgenic primates that were able to pass the foreign gene to their offspring, potentially supplying a research model genetically much more similar to humans than mice.

Beginning in the late 1990's, transgenic animals were developed for production of proteins that can be used in pharmaceutical drugs to treat human disease. Accordingly, they have become known as "pharm animals." This new technology is known as biopharming and has a potential for producing large quantities of pharmaceutical products and vaccines at much lower costs. A number of companies are developing bioproducts using transgenic domestic lifestock. They include blood clotting factors, malarial antigens for vaccine against malaria, and spider silk protein for fiber development. One example of a transgenic animal is lactating transgenic mice producing tissue plasminogen activator in their milk. Similarly, transgenic sheep supply blood coagulation factor IX and alpha^1-antitrypsin, transgenic pigs produce human hemoglobin, and transgenic cows make human lactoferrin. Scientists have also developed transgenic pigs that may supply tissue and organs for transplantation into humans without tissue rejection.

TRANSGENIC PLANTS

Plant cells present greater difficulties for transformation because their cell walls are sturdier than animal cell walls. Microinjection and biolistics are possible, but tricky and slow. A breakthrough for plant transgenesis came in 1983, when three separate teams of scientists used plasmids as vectors (carrier molecules) to infect plants with foreign DNA. The achievement came about because of research into plant tumors caused by crown gall disease. The pathogen, the soil bacterium *Agrobacterium tumefaciens,* caused the disease by ferrying bits of its own DNA into the genome of plants via plasmids, circular bits of extranuclear DNA. Scientists found that they could take the same plasmid, cut out bits of its DNA with enzymes and insert transgenes, and then use the altered plasmids as vectors to transform plants. Subsequently, scientists discovered that liposomes can be vectors. A liposome is a tiny ball of lipids that binds readily to a cell wall, opens a passage, and delivers any DNA that has been put inside it.

A great variety of transgenic plants have been designed for agriculture to produce genetically modified (GM) foods. The first to be marketed was a strain of tomato that ripened slowly so that it gained flavor by staying longer on the vine and remained ripe longer on supermarket shelves. However this Flavr Savr tomato was not a commercial success. Other crops have been made resistant to herbicides so that weeds can be easily killed without harming the food plants. Corn, cotton, soybeans, potatoes, and papayas received a gene from the bacterium *Bacillus thuringiensis* (*Bt*) that enables them to make a caterpillar-killing toxin; these are frequently referred to as *Bt* crops. Herbicide resistant (HT) soybeans were cultivated in 87 percent of the total U.S. soybean acreage in 2006. Other transgenic crops such as pest-resistant (*Bt*) cotton was grown in 52 percent of cotton acreage while *Bt* corn was grown in 35 percent of the total corn acreage. Despite the expanding acreage, the diversity of crop types and traits in commercial production is tightly regulated by government agencies.

Like transgenic animals, some transgenic crops promise to deliver pharmaceuticals at lower costs and more conveniently than factory-made drugs. In 2000, scientists reported invention of rice and wheat strains that produce anticancer antibodies. Golden rice, a transgenic strain that contains vitamin A, was developed to ward off blindness from vitamin A deficiency, which is a problem in countries that subsist largely on rice. Another strain has elevated iron levels to combat anemia. Also in a bid to reduce the health risk from smoking, a tobacco company developed a strain free of nicotine. In recent years, maize, potato, soybean, and tomato have been used to produce vaccines for both humans and animals against pneumonic and bubonic plague. A vaccine for hepatitis B has been developed in transgenic potato that raises an immunological response in humans. Also edible rice-based vaccine targeted to allergic diseases such as asthma, seasonal allergies, and atopic dermatitis has been developed.

THE DEBATE OVER TRANSGENESIS

Transgenic organisms offer great benefits to humankind: deeper understanding of the genetic component in disease and aids in diagnosis; new, cheaper, more easily produced drugs; and crops that could help alleviate the growing hunger in the world. Yet during the 1990's protests against transgenesis began that are as contentious as any since the controversy over the pesticide DDT during the 1960's.

Some opponents object to the very fact that organisms are modified strictly for human benefit. They find such manipulations of life's essential code blasphemous and arrogant, or at the very least unethical and reckless. Furthermore, animal rights groups regard the production of transgenic pharm and research animals as cruel and in violation of the natural rights of other species.

The greater portion of opponents, however, are concerned with specific dangers that transgenic organism may pose. Many consumers, most noticeably those in Europe, worry that GM foods contain hidden health risks. After transgenes were found to escape from crops and become part of wild plants, environmentalists proposed that there could be unforeseen and harmful ecological consequences, especially in the destruction of natural species and reduction of biodiversity.

Even those who welcome the creation of transgenic animals and plants are concerned about the legal and social effects. Principally, because biotechnology corporations can patent transgenic organisms, they potentially have great influence on agribusiness, perhaps to the detriment of small farmers and consumers.

Therefore, as with the development of any new technology, it is critical to conduct extensive studies and peer reviewed research for harnessing the power of genetic engineering. The close scrutiny and appropriate policies would allow the human society to reap its immense benefits without compromising biodiversity or creating environmental imbalance.

Roger Smith, Ph.D.;
updated by Poonam Bhandari, Ph.D.

FURTHER READING

Brown, Kathryn, Karen Hopkin, and Sasha Nemecek. "GM Foods: Are They Safe?" *Scientific American* 284, no. 4 (2001): 52-57. Describes the risks and benefits in growing genetically modified foods and human consumption of them. Accompanied by graphics and tables that summarize and clarify technical matters.

Houdebine, Louis-Marie. "Production of Pharmaceutical Proteins by Transgenic Animals." *Comparative Immunology, Microbiology, and Infectious Diseases* 32 (2009): 107–121. Explains the diverse range of trangenics available and the methods for production of biopharmaceutical products.

Lurquin, Paul F. *The Green Phoenix: A History of Genetically Modified Plants.* New York: Columbia University Press, 2001. Written by a pioneer in the field of transgenic plants, this technically detailed but readable book requires a basic familiarity with microbiology and genetics. The author discusses the ecological and ethical controversies with insight and balance.

Nicholl, Desmond S. T. *An Introduction to Genetic Engineering.* 2d ed. New York: Cambridge University Press, 2002. A thorough, lucidly structured survey of the techniques and applications of genetic engineering. One chapter is devoted to transgenic plants and animals.

Rehbinder, E., et al. *Pharming: Promises and Risks of Biopharmaceuticals Derived from Genetically Modified Plants and Animals.* Berlin: Springer-Verlag, 2009. This book provides an extensive account of the applications of GMOs in research and industry. In addition the ethical issues relating to the environment and human society are discussed in detail.

Velander, William, et al. "Transgenic Livestock as Drug Factories." *Scientific American* 276 (January, 1997). Explains how genetic engineering methods have resulted in the production of "pharm" animals whose milk contains large amounts of medicinal proteins.

Winston, Mark L. *Travels in the Genetically Modified Zone.* Cambridge, Mass.: Harvard University Press, 2002. A popular account of agribusiness, government oversight, and the science of genetically modified plants and animals. The science is explained cursorily for general readers.

WEB SITES OF INTEREST

Council for Biotechnology Information
http://www.whybiotech.com

Presents information about the benefits and safety of agricultural biotechnology for the general public.

Human Genome Project Information: Genetically Modified Food and Organisms
http://www.ornl.gov/sci/techresources/Human_Genome/elsi/gmfood.shtml
A searchable professional database about lines of genetically modified animals, methods used to create them and descriptions of the modified DNA, the expression of transgenes, and how transgenes are named.

Information Systems for Biotechnology
http://www.isb.vt.edu/indexmain.cfm
Searchable database pertaining to the development, testing, and regulatory review of genetically modified plants, animals, and microorganisms within the United States and abroad.

International Service for Acquisition of Agri-Biotech Applications (ISAAA)
http://www.isaaa.org/kc
A comprehensive searchable database addressing the use of biotechnology to benefit resource-poor farmers in developing countries. With links and a photo gallery.

University of Michigan: Transgenic Animal Model Core
http://www.med.umich.edu/tamc
A professional Web site for researchers seeking a host animal to test transgenes. However, it contains much useful general information about transgenics (especially transgenic rats), vectors, and laboratory procedures.

University of Nebraska, Lincoln: AgBiosafety
http://agbiosafety.unl.edu
This site provides information regarding the procedures for evaluation of GMO and the development of regulations.

See also: Antibodies; Biopesticides; Biopharmaceuticals; Genetic engineering; Genetic engineering: Agricultural applications; Genetic engineering: Medical applications; Genetic engineering: Risks; Genetic engineering: Social and ethical issues; Genetically modified foods; Genomics; Human growth hormone; Hybridization and introgression; Knockout genetics and knockout mice; Lateral gene transfer; Model organism: *Drosophila melanogaster*; Model organism: *Mus musculus*; Model organism: *Xenopus laevis*; Molecular genetics; Viroids and virusoids.

Transposable elements

Category: Bacterial genetics; Molecular genetics

Significance: Transposable elements are discrete DNA sequences that have evolved the means to move (transpose) within the chromosomes. Transposition results in mutation and potentially large-scale genome rearrangements. Transposable elements contribute to the problem of multiple antibiotic resistance by mobilizing the genes of pathogenic bacteria for antibiotic resistance.

Key terms

composite transposon: a transposable element that contains genes other than those required for transposition

resistance plasmid (R plasmid): a small, circular DNA molecule that replicates independently of the bacterial host chromosome and encodes a gene for antibiotic resistance

selfish DNA: a DNA sequence that has no apparent purpose for the host and that spreads by forming additional copies of itself within the genome

transposase: an enzyme encoded by a transposable element that initiates transposition by cutting specifically at the ends of the element and randomly at the site of insertion

Jumping Genes

Transposable elements are DNA sequences that are capable of moving from one chromosomal location to another in the same cell. In some senses, transposable elements have been likened to intracellular viruses. The first genetic evidence for transposable elements was described by Barbara McClintock in the 1940's. She was studying the genetics of the pigmentation of maize (corn) kernels and realized that the patterns of inheritance were not following Mendelian laws. Furthermore, she surmised that insertion and excision of genetic material were responsible for the genetic patterns she observed. McClintock was recognized for this pioneering work with a Nobel Prize in Physiology or Medicine in 1983. It was not until the 1960's that the jumping genes that McClintock postulated were isolated and characterized. The first transposable elements to be well characterized were found in the bacteria *Escherichia coli* but have subsequently been found in the cells of many bacteria, plants, and animals.

Barbara McClintock

Though best known for her research on mobile genetic elements (for which she won the first unshared Nobel Prize in Physiology or Medicine awarded to a woman), Barbara McClintock's contributions to the field of genetics were many. McClintock's career in genetics spanned the development of the field itself. While an undergraduate at Cornell University (from which she earned her bachelor of science degree in 1923), she was invited to participate in a graduate course in genetics, then a fledgling discipline. She continued at Cornell as a graduate student (earning her Ph.D. in 1927), combining her interests in the microscopic internal structure of the cell (cytology) with the transmission of heritable traits (genetics).

McClintock's keen observational skills and her holistic approach to science allowed her to make significant advancements. It had only recently been established that the chromosomes (visible under the microscope) were the carriers of Gregor Mendel's "factors," or genes. McClintock used information gleaned from characteristics of the corn plant, *Zea mays* (maize), in conjunction with changes in its chromosomes to elucidate many aspects of genetic control. Her first major contribution was the identification and naming of the maize chromosomes. Shortly after, using cytological markers on the chromosomes, McClintock and graduate student Harriet Creighton demonstrated the correlation between patterns of inheritance and chromosomal crossover—the exchange of material between chromosomes. Breeding experiments focusing on "linkage groups" allowed McClintock to associate each of corn's ten chromosomes with the genes they carry.

Barbara McClintock (© The Nobel Foundation)

At that point, Lewis Stadler, who was studying the mutagenic effects of X rays, sent some irradiated corn to McClintock. McClintock demonstrated that the resultant broken chromosomes can fuse into a ring, and she then hypothesized the existence of the telomere, a protective stabilizing structure at the end of the chromosome. At the University of Missouri, McClintock observed the ability of such broken chromosomes to go through a series of breakages and fusions (the breakage-fusion-bridge cycle) and identified concomitant chromosomal inversions and deletions. She ultimately discovered that certain genes could transfer from cell to cell and between chromosomes, thereby influencing the color patterns in the leaves and kernels of corn.

In 1941, McClintock moved to Cold Spring Harbor, New York, where she would remain. Doing much of her work before the discovery of the double helical structure of DNA, she rejected the simplistic one-way flow from DNA to RNA to protein outlined in the central dogma of molecular biology, seeking instead an explanation for the spatial and temporal variation in gene expression needed to link genetics to developmental and evolutionary change. Her own work showed that both the location and direction of genetic material, as well as the presence of other "controlling elements," had important effects on the expression of the gene.

In addition to this better-known work, McClintock identified the chromosomes of the bread mold *Neurospora* and described its meiotic cycle. She also headed a study aimed at conserving indigenous corn varieties in the Americas. In recognition of her place as one of the most distinguished scientists of the twentieth century, *The Barbara McClintock Papers* are available through the National Library of Medicine through its "Profiles in Science" Web site.

Lee Anne Martínez, Ph.D.

Transposable elements are discrete DNA sequences that encode a transposase, an enzyme that catalyzes transposition. Transposition refers to the movement within a genome. The borders of the transposable element are defined by specific DNA sequences; often the sequences at either end of the transposable element are inverted repeats of one another. The transposase enzyme cuts the DNA sequences at the ends of the transposable element to initiate transposition and cuts the DNA at the insertion site. The site for insertion of the transposable element is not specific. Therefore, transposition results in random insertion into chromosomes and often results in mutation and genome rearrangement. In many organisms, transposition accounts for a significant fraction of all mutation. Although the details of the mechanism may vary, there are two basic mechanisms of transposition: conservative and replicative. In conservative transposition, the transposable element is excised from its original site and inserted at another. In replicative transposition, a copy of the transposable element is made and is inserted in a new location. The original transposable element remains at its initial site.

A subset of the replicative transposable elements includes the retrotransposons. These elements transpose through an RNA intermediate. Interestingly, their DNA sequence and organization are similar to those of retroviruses. It is likely that either retroviruses evolved from retrotransposons by gaining the genes to produce the proteins for a viral coat or retrotransposons evolved from retroviruses that lost the genes for a viral coat. This is one of the reasons that transposons are likened to viruses. Viruses can be thought of as transposons that gained the genes for a protein coat and thus the ability to leave one cell and infect others; conversely, transposons can be thought of as intracellular viruses.

Genetic Change and Selfish DNA

Transposition is a significant cause of mutation for many organisms. When McClintock studied the genetic patterns of maize kernel pigmentation, she saw the results of insertion and excision of transposable elements into and out of the pigment genes. Subsequently, it has been well established that mutations in many organisms are the result of insertion of transposable elements into and around genes. Transposition sometimes results in deletion mutations as well. Occasionally the transposase will cut at one end of the transposable element but skip the other end, cutting the DNA further downstream. This can result in a deletion of the DNA between the end of the transposable element and the cut site.

In addition to these direct results, it is believed that transposable elements may be responsible for large-scale rearrangements of chromosomes. Genetic recombination, the exchange of genetic information resulting in new combinations of DNA sequences, depends upon DNA sequence homology. Normally, recombination does not occur between nonhomologous chromosomes or between two parts of the same chromosome. However, transposition can create small regions of homology (the transposable element itself) spread throughout the chromosomes. Recombination occurring between homologous transposable elements can create deletions, inversions, and other large-scale rearrangements of chromosomes.

Scientists often take advantage of transposable elements to construct mutant organisms for study. The random nature of insertion ensures that many different genes can be mutated, the relatively large insertion makes it likely that there will be a complete loss of gene function, and the site of insertion is easy to locate to identify the mutated region.

Biologists often think of natural selection as working at the level of the organism. DNA sequences that confer a selective advantage to the organism are increased in number as a result of the increased reproductive success of the organisms that possess those sequences. It has been said that organisms are simply DNA's means of producing more DNA. In 1980, however, W. Ford Doolittle, Carmen Sapienza, Leslie Orgel, and Francis Crick elaborated on another kind of selection that occurs among DNA sequences within a cell. In this selection, DNA sequences are competing with each other to be replicated. DNA sequences that spread by forming additional copies of themselves will increase relative to other DNA sequences. There is selection for discrete DNA sequences to evolve the means to propagate themselves. One of the key points is that this selection does not work at the level of the organism's phenotype. There may be no advantage for the organism to have these DNA sequences. In fact, it may be that there is a slight disadvantage to having many of these DNA sequences. For this reason, DNA sequences that are selected because of their tendency to make additional copies of themselves are referred to as "selfish" DNA. Transposable elements are often cited as examples of selfish DNA.

Composite Transposons and Antibiotic Resistance

Some transposable elements have genes unrelated to the transposition process located between the inverted, repeat DNA sequences that define the ends of the element. These are referred to as composite transposons. Very frequently, bacterial composite transposons contain a gene that encodes resistance to antibiotics. The consequence is that the antibiotic resistance gene is mobilized: It will jump along with the rest of the transposable element to new DNA sites. Composite transposons may be generated when two of the same type of transposable elements end up near each other and flanking an antibiotic resistance gene. If mutations occurred to change the sequences at the "inside ends" of the transposable elements, the transposase would then only recognize and cut at the two "outside end" sequences to cause everything in between to be part of a new composite transposon.

Resistance to antibiotics is a growing public health problem that threatens to undo much of the progress that the antibiotic revolution made against infectious disease. Transposition of composite transposons is part of the problem. Transposition can occur between any two sites within the same cell, including between the chromosome and plasmid DNA. Plasmids are small, circular DNA molecules that replicate independently of the bacterial host chromosome. Resistance plasmids (R plasmids) are created when composite transposons carrying an antibiotic resistance gene insert into a plasmid. What makes this particularly serious is that some plasmids encode fertility factors (genes that promote the transfer of the plasmid from one bacterium to another). This provides a mechanism for rapid and widespread antibiotic resistance whenever antibiotics are used. The great selective pressure exerted by antibiotic use results in the spread of R plasmids throughout the bacterial population. This, in turn, increases the opportunities for composite transposon insertion into R plasmids to create multiple drug-resistant R plasmids. The first report of multiple antibiotic resistance caused by R plasmids was in Japan in 1957 when strains of *Shigella dysenteriae*, which causes dysentery, became resistant to four common antibiotics all at once. Some R plasmids encode resistance for up to eight different antibiotics, which often makes treatment of bacterial infection difficult. Furthermore, some plasmids are able to cause genetic transfer between bacterial species, limiting the usefulness of many antibiotics.

Craig S. Laufer, Ph.D.

Further Reading

Bushman, Frederic. *Lateral DNA Transfer: Mechanisms and Consequences.* Cold Spring Harbor, N.Y.: Cold Spring Harbor Laboratory Press, 2002. Discusses mobile genes and how they transfer DNA between unrelated cells. Examines the biological and health consequences of this DNA transfer. Illustrations.

Capy, Pierre, et al. *Dynamics and Evolution of Transposable Elements.* New York: Chapman and Hall, 1998. Addresses the structure of transposable elements, heterochromatin, host phylogenies, the origin and coevolution of retroviruses, evolutionary links between telomeres and transposable elements, and population genetics models of transposable elements. Illustrations, maps, bibliography.

Galun, Esra. *Transposable Elements: A Guide to the Perplexed and the Novice, with Appendices on RNAi, Chromatin Remodeling, and Gene Tagging.* Boston: Kluwer Academic, 2003. Provides historical background on transposable elements. Describes all the known transposable elements, how they are transpositioned, and their use in medicine, genetics, and agriculture.

Keller, Evelyn Fox. *A Feeling for the Organism: The Life and Work of Barbara McClintock.* 10th anniversary ed. New York: W. H. Freeman, 1993. A now-classic look at McClintock and her pathbreaking work.

McClintock, Barbara. *The Discovery and Characterization of Transposable Elements: The Collected Papers of Barbara McClintock.* New York: Garland, 1987. Presents the papers on transposable elements that McClintock wrote between 1938 and 1984. Illustrations, bibliography.

McDonald, John F., ed. *Transposable Elements and Genome Evolution.* London: Kluwer Academic, 2000. Includes the seminal papers presented in 1992 at the University of Georgia during a meeting of molecular, population, and evolutionary geneticists to discuss the relevance of their research to the role played by transposable elements in evolution. Illustrations, bibliography, index.

Web Sites of Interest

Kimball's Biology Pages
http://users.rcn.com/jkimball.ma.ultranet/BiologyPages/T/Transposons.html

John Kimball, a retired Harvard University biology professor, includes a page about transposons and mobile DNA in his online cell biology text.

Microbial Genetics, Transposable Elements
http://www.sci.sdsu.edu/~smaloy/MicrobialGenetics/topics/transposons

This site evolved from a course at San Diego State University and includes a section on transposable elements, with information on insertion sequences and transposons, transposon derivatives, and the uses of transposons. These pages also feature bibliographies and links to other online resources.

Transposable Genetic Elements
http://www.ndsu.nodak.edu/instruct/mcclean/plsc431/transelem/trans1.htm

Philip McClean, a professor in the department of plant science at North Dakota State University, includes a section on transposable genetic elements in his course-related Web site. Some of the pages discuss Barbara McClintock and the transposable elements of corn and transposable elements in yeast, fruit flies, and bacteria.

See also: Antisense RNA; Archaea; Bacterial resistance and super bacteria; Immunogenetics; Lateral gene transfer; Model organism: *Escherichia coli*; Molecular genetics; Mutation and mutagenesis; Plasmids; Repetitive DNA.

Tuberous sclerosis

Category: Diseases and syndromes
Also known as: Tuberous sclerosis complex (TSC)

Definition

Tuberous sclerosis (TS) is a rare autosomal dominant genetic disease whose name comes from *tuber* (Latin for "swelling") and *skleros* (Greek for "hard"). The condition results from mutations in one of two genes, *TSC1* (coding for hamartin) or *TSC2* (coding for tuberin). TS is a multiorgan disease and is characterized by tumorlike lesions called hamartomas in a variety of organs such as the brain, skin, eyes, heart, lungs, and kidneys.

Risk Factors

The mutation occurs spontaneously in people with no family history for unknown reasons (in two-thirds of cases) or may be inherited as an autosomal dominant trait (familial; one-third of cases). The estimated prevalence of TS is 1 in 10,000 people. It affects males and females equally, and there is 50 percent chance that the defective gene will be passed on to their offspring. There are no identified risk factors.

Etiology and Genetics

Mutations that render the *TSC1* and *TSC2* genes inactive may be missense, nonsense, deletions, or insertions. Studies have demonstrated that mutations in *TSC2* are about five times more common than mutations in *TSC1* in the sporadic TS population. There is no clearly identified "hotspot" (preferred site) for mutations; more than two hundred different mutations are known in the *TSC1* gene and approximately seven hundred different mutations in the *TSC2* gene. Any particular mutation accounts for around 1 percent of the disease in the affected population. The *TSC* genes are located on two different chromosomes with *TSC1* on chromosome 9 (9q34) and *TSC2* on chromosome 16 (16p13.3).

The harmartin and tuberin proteins have been shown to form heterodimers that play a role in cell division and the production of tumor-suppressor proteins. One particular downstream protein cascade is affected by the pathogenesis of the disease; the pathway of the mammalian target of rapamycin (mTOR). mTOR detects signals of nutrient availability, hypoxia, or growth factor stimulation and is part of many cell processes, such as cell-cycle progression, transcription and translation control, and nutrient uptake. Tuberin and hamartin form an intracellular complex that reduces mTOR stimulation.

In families with more than one child with TS complex, no extended family history, and no clinical features of the TS complex, germline mosaicism is the likely explanation. Germ-line mosaicism is when the individual's eggs or sperm may carry the mutation, even though it is absent from the somatic cells. The carrier of germ-line mosaicism may be asymptomatic or may present with various symptoms of the disease.

Symptoms

The symptoms of tuberous sclerosis vary from person to person. The most common presentation is seizures in infancy or early childhood, particularly infantile spasms. Delayed development and skin lesions usually appear shortly after birth but may remain very discrete in children. Other manifestations include kidney and lung disease, skin abnormalities, and vision problems. Some affected individuals have normal intelligence and no seizures. Others have severe retardation, serious tumors, or difficult-to-control seizures.

Screening and Diagnosis

Diagnosis of TS is usually based on clinical and radiological findings; however, no single feature of TS is diagnostic. Therefore, an evaluation including all clinical features is necessary. The clinical manifestations of TS appear at distinct developmental points. For example, cortical tubers and cardiac rhabdomyomas form during embryogenesis and are typical in infancy. Skin lesions are detected at all ages in more than 90 percent of patients. Hypopigmented macules (formerly known as ash-leaf spots) are generally detected in infancy or early childhood, whereas the so-called shagreen patch is identified with increasing frequency after the age of five. Ungual fibromas typically appear after puberty and may develop in adulthood.

DNA testing for either of the two genes that cause this disease (*TSC1* or *TSC2*) is available. Molecular diagnosis using DNA-based testing is not yet routinely available, however, but it could be developed in the future. It would help to identify patients at increased or decreased risk for particular complications.

Treatment and Therapy

There is no specific treatment for TS. Because the disease can differ from person to person, treatment is based on the symptoms. Finding the right medications to control seizures is often difficult. Depending on the severity of the mental retardation, the child may need special education.

Small growths on the face may be removed by laser treatment, but tend to come back, making repeat treatments necessary. Rhabdomyomas commonly disappear after puberty, so surgery is usually not necessary.

A multidisciplinary team approach is useful to address the many organ systems that may be affected.

Prevention and Outcomes

Molecular genetic testing of the *TSC1* and *TSC2* genes is complicated by the large size of the two genes and the large number of disease-causing mutations. Genetic counseling for families with one affected child should include a small (1 to 2 percent) possibility of recurrence, even for parents without evidence of TS after a thorough diagnostic evaluation.

Prognosis for people with TS can range depending on the severity of symptoms. Individuals with mild forms do not have a shortened life expectancy, while individuals with more severe forms may have serious disabilities. Currently, there is no cure for TS. However, with appropriate medical care (such as lung transplant or surgical removal of tumors) most individuals with the disorder can look forward to a normal life expectancy.

Maria Mavris, Ph.D.

Further Reading

Crino, P. B., K. L. Nathanson, and E. P. Henske. "The Tuberous Sclerosis Complex." *New England Journal of Medicine* 355 (2006): 1345-1356.

Curatolo, P., R. Bombardieri, and S. Jozwiak. "Tuberous Sclerosis." *Lancet* 372 (2008): 657–668.

Web Sites of Interest

European Organisation for Rare Diseases (EURORDIS)
http://www.eurordis.org

National Organization for Rare Diseases (NORD)
http://www.rarediseases.org

Orphanet: The Portal for Rare Diseases and Orphan Drugs
http://www.orphanet.org

See also: Albinism; Chediak-Higashi syndrome; Epidermolytic hyperkeratosis; Hermansky-Pudlak syndrome; Ichthyosis; Melanoma; Palmoplantar keratoderma.

Tumor-suppressor genes

CATEGORY: Molecular genetics

SIGNIFICANCE: Molecular analysis of tumor-suppressor genes has provided important information on mechanisms of cell cycle regulation and patterns of growth control in normal dividing cells and cancer cells. Tumor-suppressor genes represent cell cycle control genes that inhibit cell division and initiate cell death (apoptosis) processes in abnormal cells. Tumor-suppressor genes cause cancer when they are inactivated (turned off), unlike oncogenes which cause cancer when they are activated (turned on). Tumor-suppressor genes can be inherited as well as acquired, unlike oncogenes, which are typically acquired by mutation or improper expression of normal genes (proto-oncogenes). Mutations in tumor-suppressor genes have been identified in many types of human cancer and play a critical role in the genetic destabilization (e.g., through DNA rearrangement) and loss of growth control characteristic of malignancy.

KEY TERMS

cell cycle: a highly regulated series of events critical to the initiation and control of cell division processes

oncogenes: mutated or improperly expressed genes known to cause cancer; the normal form of an oncogene is called a proto-oncogene, which is usually involved in regulating the cell cycle; however, activating or "turning on" the proto-oncogene (through mutation or improper expression) results in cancer

p53 gene: a tumor-suppressor gene implicated in many types of cancer

Rb gene: a tumor-suppressor gene, when inactivated, is associated with several cancers including the formation of a rare and lethal eye tumor (retinoblastoma) in infants

DISCOVERY OF TUMOR-SUPPRESSOR GENES

The existence of genes playing critical roles in cell cycle regulation by inhibiting cell division was predicted by several lines of evidence. In vitro studies involving the fusion of normal and cancer cell lines were observed to result in suppression of the malignant phenotype, suggesting normal cells contained inhibitors with the ability to reprogram the abnormal growth behavior of the cancer cell lines. In addition, back in 1971, studies by Alfred Knudsen on inherited and noninherited forms of retinoblastoma, a childhood cancer associated with tumor formation in the eye, suggested the inactivation of recessive genes as a consequence of mutation, resulted in the loss of function of inhibitory gene products critical to cell division control. With the advent of molecular methods of genetic analysis, the gene whose inactivation is responsible for retinoblastoma was identified and designated *Rb*.

Additional tumor-suppressor genes were identified by studies of DNA tumor viruses whose cancer-causing properties were found to result, in part, from the ability of specific viral gene products to inactivate host cell inhibitory gene products involved in cell cycle regulation. By inactivating these host cell proteins, the tumor virus removes the constraints on viral and cellular proliferation. The most important cellular gene product to be identified in this way is the p53 protein, named after its molecular weight. Genetic studies of human malignancies have implicated mutations in the *p53* gene in up to 75 percent of tumors of diverse tissue origin, including an inherited disorder called Li-Fraumeni syndrome, which is associated with a higher risk for developing sarcomas, brain tumors, adrenal gland cancers, and leukemias. In addition, studies of other rare inherited malignancies have led to the identification of about thirty recessive tumor-suppressor genes whose inactivation contributes to oncogenic or cancer-causing cell proliferation. The accompanying table provides examples of the chromosomal locations

Chromosomal Locations for Select Tumor-Suppressor Genes

Gene	Chromosomal location
Rb1	13q14
p53	17p13
BRCA1	17q21
BRCA2	13q12
NF1	17q11
p16	9p21
APC	5q21

for several tumor-suppressor genes. Included in the list are the *BRCA1* and *BRCA2* genes in breast cancer, the *NF1* gene in neurofibromatosis, the *p16* gene in melanoma, and the *APC* gene in colorectal carcinoma. Each of these genes has also been implicated in nonhereditary cancers. Three types of tumor-suppressor genes are generally recognized: genes controlling cell division, DNA repair, and cell death.

The Properties of Tumor-Suppressor Genes

Molecular analyses of the genetic and biochemical properties of tumor-suppressor genes have suggested some of these gene products play critical but distinct roles in regulating processes involved in cellular division and proliferation. The *Rb* gene product represents a prototype tumor-suppressor gene of this type. The *Rb* gene product normally blocks progression of the cell cycle and cell division by binding to transcription factors in its active form. In order for cell division to occur in response to growth factor stimulation, elements of the signal cascade inactivate *Rb*-mediated inhibition by a mechanism involving the addition of phosphate to the molecule, a reaction called phosphorylation. In cancer, loss of *Rb* function results as a consequence of mutation which removes the brakes on this form of normal inhibitory control; the cell division machinery proceeds without appropriate initiation by growth factors or other stimuli.

Like the *Rb* tumor-suppressor gene, most tumor-suppressor genes become active due to a loss of heterozygosity (LOH). Since chromosomes (and the genes residing within the chromosomes) are paired, a mutation in one gene of the pair is called heterozygosity. For *Rb*, a single gene mutation (the heterozygous condition for the trait encoded by the gene pair) is inherited as a recessive condition where no cancer results (since the infant has one normal gene and one mutated *Rb* gene). During development however, a random mutation can occur in the normal *Rb* gene. As soon as the second *Rb* gene mutates in a single cell, cancer can begin.

Tumor-suppressor genes are also found among the genes specifically involved in DNA repair. Whenever a cell divides the DNA must be duplicated, and errors in the duplication process must be repaired in order for the new cell to function properly. The protein products of DNA repair gene repair errors in duplicated DNA by proofreading the DNA sequence; however, if the DNA repair genes are mutated, errors will slip by and these mutations can create proteins which may activate oncogenes, or the mutations may create abnormal, mutated tumor-suppressor genes. For example, when DNA repair genes fail to repair DNA errors, some endometrial cancers and hereditary nonpolyposis colon cancer (HNPCC) can result (about 5 percent of all colon cancers).

If the cell's DNA is damaged beyond repair by the DNA repair proteins, then the *p53* tumor-suppressor gene is among the family of genes responsible for destroying the cell by causing cell death (also called "cell suicide" or "programmed cell death" or "apoptosis"). The *p53* tumor-suppressor gene product is a DNA-binding protein with the ability to regulate the expression of other genes in response to genetic damage or other abnormal events occurring during cell cycle progression. In response to p53 activation, the cell will normally arrest the process of cell division (by indirectly blocking Rb inactivation). This cell cycle arrest may allow time for the cell machinery to repair genetic damage before proceeding further along the cell cycle; alternatively, if the damage is too great, the *p53* gene product may initiate a process of cell death called apoptosis. The loss of *p53* activity in the cell as a consequence of mutation results in genetic destabilization and the failure of cell death mechanisms to eliminate damaged cells from the body; both events appear to be critical to late-stage oncogenic mechanisms. Mutations of the *p53* gene may be inherited (as in the Li-Fraumeni syndrome, LFS); however, mutations in *p53* are also routinely found in many sporadic (noninherited) cancers (including lung, colon, breast, and other cancer types).

Impact and Applications

The discovery of tumor-suppressor genes has revealed the existence of inhibitory mechanisms critical to the regulation of cellular proliferation. Mutations that destroy the functional activities of these gene products cause the loss of growth control characteristic of cancer cells. Taken together, research on the patterns of oncogene activation and the loss of tumor-suppressor gene function in many types of human malignancy suggest a general model of oncogenesis. Molecular analyses of many tumors show multiple genetic alterations involving both oncogenes and tumor-suppressor genes, suggesting that oncogenesis (development of cancer) requires unregulated stimulation of cellular proliferation pathways along with a loss of inhibitory activities that operate at cell cycle checkpoints.

Some tumor-suppressor gene mutations are inherited and, in rare cases, testing for these gene mutations (which have been found often enough in the general population, like *BRCA 1* and *BRCA 2*) may be helpful to determine which people are at higher risk for certain cancers. Genetic testing for families at risk requires careful screening and counseling and might allow surgical intervention, lifestyle changes or more frequent cancer screening steps to minimize the individual's familial cancer risk.

Genetic tests for oncogenes and tumor-suppressor genes have become commonplace in cancer diagnosis and treatment. Some genetic tests have been shown to help diagnose the type of cancer, guide cancer treatment, predict survival for patients with cancer, and sometimes to produce novel therapies for cancer patients.

With respect to clinical applications, restoration of *p53* tumor-suppressor gene function by gene therapy appears to result in tumor regression in some experimental systems; however, much more work needs to be done in this area to achieve clinical relevance. More important, research on the mechanism of action of standard chemotherapeutic drugs suggests that cytotoxicity may be caused by p53-induced cell death; the absence of functional p53 in many tumors may account for their resistance to chemotherapy. Promising research is investigating new ways to elicit cell death in tumor cells lacking functional *p53* gene product in response to chemotherapy. The clinical significance of activating these p53-independent cell death mechanisms may be extraordinary.

Sarah Crawford Martinelli, Ph.D.;
updated by Joy Frestedt, Ph.D., RAC, CCTI

Further Reading

DeCaprio, J. A. "How the Rb Tumor-Suppressor Structure and Function Was Revealed by the Study of Adenovirus and SV40." *Virology* 384, no. 2 (February 20, 2009): 274-284. This work reviews how the study of DNA tumor viruses led to discoveries about the Rb protein.

Ehrlich, Melanie, ed. *DNA Alterations in Cancer: Genetic and Epigenetic Changes.* Natick, Mass.: Eaton, 2000. A comprehensive overview of the numerous and varied genetic alterations leading to the development and progression of cancer. Topics include oncogenes, tumor-suppressor genes, cancer predisposition, DNA repair, and epigenetic alteration such as methylation.

Fearon, Eric, and Bert Vogelstein. "Tumor Suppressor Gene Defects in Human Cancer." In *Cancer Medicine,* edited by Robert C. Bast, Jr., et al. 5th ed. Atlanta: American Cancer Society, 2000. Topics include genetic basis for tumor formation, HNPCC, retinoblastoma, *p53, APC, BRCA1* and *BRCA2,* and additional examples including the INK4A locus and the $P16_{INK4a}$ and $P19_{ARF}$ genes, WT1, NF1 and 2, VHL, and others.

Fisher, David E., ed. *Tumor Suppressor Genes in Human Cancer.* Totowa, N.J.: Humana Press, 2001. Covers models used in suppressor gene studies, cancer drugs, and gene descriptions. Illustrated (some color).

Habib, Nagy A. *Cancer Gene Therapy: Past Achievements and Future Challenges.* New York: Kluwer Academic/Plenum, 2000. Reviews forty-one preclinical and clinical studies in cancer gene therapy, organized into sections on the vectors available to carry genes into tumors, cell cycle control, apoptosis, tumor-suppressor genes, antisense and ribozymes, immunomodulation, suicidal genes, angiogenesis control, and matrix metalloproteinase.

Iglehart, J. Dirk, and Daniel P. Silver. "Synthetic Lethality: A New Direction in Cancer-Drug Development." *New England Journal of Medicine* 361, no. 2 (July 9, 2009): 189-191. This editorial describes a small study of *BRCA1* or *BRCA2* tumors pointing to a new direction in cancer-drug treatment. Using "synthetic lethality" (when a mutation in either of two genes alone is not lethal but mutations in both cause the death of the cell) *BRCA1* or *BRCA2* cells were stabilized by an oral *PARP1* enzyme inhibitor.

Lattime, Edmund C., and Stanton L. Gerson, eds. *Gene Therapy of Cancer: Translational Approaches from Preclinical Studies to Clinical Implementation.* 2d ed. San Diego: Academic Press, 2002. Provides a comprehensive review of the basis and approaches involved in the gene therapy of cancer.

Maruta, Hiroshi, ed. *Tumor-Suppressing Viruses, Genes, and Drugs: Innovative Cancer Therapy Approaches.* San Diego: Academic, 2002. An international field of experts address potential alternative cancer

treatments, such as viral and drug therapy and gene therapy with tumor-suppressor genes.
Oliff, Alan, et al. "New Molecular Targets for Cancer Therapy." *Scientific American* 275 (September, 1996). Summarizes novel genetic approaches to cancer treatment.
Ruddon, Raymond. *Cancer Biology.* 3d ed. New York: Oxford University Press, 1995. A good general text.

WEB SITES OF INTEREST

American Cancer Society
http://www.cancer.org
Site has searchable information on tumor-suppressor genes. For example, http://www.cancer.org/docroot/ETO/content/ETO_1_4x_oncogenes_and_tumor_suppressor_genes.asp This site describes recent activity in genetic and molecular biological research including detailed descriptions of mutations, oncogenes and tumor-suppressor genes.

National Human Genome Research Institute
http://www.genome.gov/27530882
Site has links to search for additional information on tumor-suppressor genes.

Nature Education
http://www.nature.com/scitable/topicpage/Tumor-Suppressor-TS-Genes-and-the-Two-887
Site has links to numerous articles about tumor suppression, the two-hit hypothesis and loss of heterozygosity to explain mechanisms of tumor-suppression gene inactivation including links to further discussion of commonly inherited tumor-suppression genes.

A New Gene Map of the Human Genome, Gene Map '99
http://www.ncbi.nlm.nih.gov/projects/genome/genemap99
Site has searchable information on chromosomal locations for genes at the Online Mendelian Inheritance in Man (OMIM) database.

See also: Aging; Breast cancer; Cancer; Cell cycle; Cell division; DNA repair; Human genetics; Human Genome Project; Model organism: *Mus musculus*; Oncogenes; *RB1* gene; *RhoGD12* gene.

Turner syndrome

CATEGORY: Diseases and syndromes

DEFINITION

Turner syndrome is caused by a chromosomal abnormality and is one of the most common genetic problems in women, affecting 1 out of every 2,000 to 2,500 women born. Henry H. Turner, an eminent clinical endocrinologist, is credited with first describing Turner syndrome. In 1938, he published an article describing seven patients, ranging in age from fifteen to twenty-three years, who exhibited short stature, a lack of sexual development, arms that turned out slightly at the elbows, webbing of the neck, and low posterior hairline. He did not know what caused this condition.

In 1959, C. E. Ford discovered that a chromosomal abnormality involving the sex chromosomes caused Turner syndrome. He found that most girls with Turner syndrome did not have all or part of one of their X chromosomes and argued that this missing genetic material accounted for the physical findings associated with the condition.

RISK FACTORS

Only females are at risk for Turner syndrome. The loss or alteration of the X chromosome that characterizes the disorder is not inherited but takes place spontaneously. For this reason, parents who have one child with Turner syndrome will most likely not have another child with this disorder.

ETIOLOGY AND GENETICS

Turner syndrome begins at conception. The disorder results from an error during meiosis in the production of one of the parents' sex cells, although the exact cause remains unknown.

SYMPTOMS

Shortness is the most common characteristic of Turner syndrome. The incidence of short stature among women with Turner syndrome is virtually 100 percent. Women who have this condition are, on average, 4 feet 8 inches (1.4 meters) tall. The cause of the failure to grow is unclear. However, growth-promoting therapy with growth hormones has become standard. Most women with the syndrome also experience ovarian failure. Since the ovaries normally produce estrogen, women with Turner syndrome lack this essential hormone. This deficit results in infertility and incomplete sexual development. Cardiovascular disorders are the single source of increased mortality in women with this condition. High blood pressure is common.

Other physical features often associated with Turner syndrome include puffy hands and feet at birth, a webbed neck, prominent ears, a small jaw, short fingers, a low hairline at the back of the neck, and soft fingernails that turn up at the ends. Some women with Turner syndrome have a tendency to become overweight. Many women will exhibit only a few of these distinctive features, and some may not show any of them. This condition does not affect general intelligence. Girls with Turner syndrome follow a typical female developmental pattern with unambiguous female gender identification. However, another possible symptom is poor spatial perception abilities. For example, women with this condition may have difficulty driving, recognizing subtle social clues, and solving nonverbal mathematics problems; they may also suffer from clumsiness and attention deficit disorder.

Screening and Diagnosis

Girls suspected of having Turner syndrome, usually because of their short stature, often undergo chromosomal analysis. A simple blood test and laboratory analysis called a karyotype are done to document the existence of an abnormality.

Additional tests to diagnose Turner syndrome include measuring blood hormone levels, an echocardiogram (the use of ultrasound to measure heart defects), ultrasound examinations of reproductive organs and kidneys, and a pelvic exam. Pregnant women can receive chromosome analysis to determine whether their children will have the disorder.

Treatments and Therapy

No treatment is available to correct the chromosome abnormality that causes this condition. However, injections of human growth hormone can restore most of the growth deficit. Unless they undergo hormone replacement therapy, girls with Turner syndrome will not menstruate or develop breasts and pubic hair. In addition to estrogen replacement therapy, women with Turner syndrome are often advised to take calcium and exercise regularly. Although infertility cannot be altered, pregnancy may be made possible through in vitro fertilization (fertilizing a woman's egg with sperm outside the body) and embryo transfer (moving the fertilized egg into a woman's uterus). Individuals with Turner syndrome can be healthy, happy, and productive members of society.

Nevertheless, because of its relative rarity, a woman with Turner syndrome may never meet another individual with this condition and may suffer from self-consciousness, embarrassment, and poor self-esteem. The attitudes of parents, siblings, and relatives are important in helping develop a strong sense of identity and self-worth. The Turner Syndrome Society of the United States is a key source of information and support groups.

Prevention and Outcomes

Although there is no treatment to cure the chromosomal abnormality, advances in chromosomal analysis have proved helpful in the diagnosis and management of Turner syndrome. In addition, new developments in hormonal therapy for short stature and ovarian failure, combined with advances in in vitro fertilization, have significantly improved the potential for growth, sexual development, and fertility for afflicted individuals.

Fred Buchstein, M.A.

Further Reading

Albertsson-Wikland, Kerstin, and Michael B. Ranke, eds. *Turner Syndrome in a Life Span Perspective: Research and Clinical Aspects.* New York: Elsevier, 1995. Focuses on molecular genetic evaluation, prenatal diagnosis, growth, medical and psychosocial management, pediatric and adult care, estrogen substitution therapy, bone mineralization, and hearing. Illustrations, bibliography, index.

Broman, Sarah H., and Jordan Grafman, eds. *Atypical Cognitive Deficits in Developmental Disorders: Implications for Brain Function.* Hillsdale, N.J.: Lawrence Erlbaum, 1994. Includes multidisciplinary research about Turner syndrome. Illustrations, bibliography, index.

Powell, M. Paige, and Kimberly R. Snapp. "Turner Syndrome." In *Handbook of Neurodevelopmental and Genetic Disorders in Adults,* edited by Sam Goldstein and Cecil R. Reynolds. New York: Guilford Press, 2005. Traces the developmental course of Turner syndrome in adults, discusses the neurobiological basis and clinical characteristics of the syndrome, and describes ways to help adults with this disorder.

Rieser, Patricia A., and Marsha Davenport. *Turner Syndrome: A Guide for Families.* Houston: Turner Syndrome Society, 1992. Describes the syndrome's causes and features. Advises how families can help girls with the condition cope with the physical, social, and emotional concerns.

Rosenfeld, Ron G. *Turner Syndrome: A Guide for Physicians.* 2d ed. Houston: Turner Syndrome Society, 1992. Definition, etiology, and management of Turner syndrome.

Saenger, Paul. "Turner Syndrome." In *Pediatric Endocrinology,* edited by Mark A. Sperling. Rev. 3d ed. Philadelphia: Saunders/Elsevier, 2008. Provides guidance to help medical practitioners provide basic and clinical care for the syndrome.

Saenger, Paul, and Anna-Maria Pasquino, eds. *Optimizing Health Care for Turner Patients in the Twenty-first Century.* New York: Elsevier, 2000. Covers the management of health care from infancy through the adult years. Discusses molecular genetics, prenatal diagnosis, cognitive function, and reproduction. Illustrations, index.

Sybert, Virginia P. "Turner Syndrome." In *Management of Genetic Syndromes,* edited by Suzanne B. Cassidy and Judith E. Allanson. 2d ed. Hoboken, N.J.: Wiley-Liss, 2005. Offers information for medical practitioners about the nature, evaluation, treatment, incidence, diagnostic testing, etiology, and other aspects of the syndrome.

WEB SITES OF INTEREST

Eunice Kennedy Shriver National Institute of Child Health and Human Development
http://turners.nichd.nih.gov
Site provides information on the genetic and clinical features of the syndrome.

Genetics Home Reference
http://ghr.nlm.nih.gov/condition=turnersyndrome
Describes the genes related to and the inheritance patterns of Turner syndrome and offers links to additional online resources.

Intersex Society of North America
http://www.isna.org
The society is "devoted to systemic change to end shame, secrecy, and unwanted genital surgeries for people born with an anatomy that someone decided is not standard for male or female." Its Web site includes links to information on Turner syndrome and other medical conditions.

Johns Hopkins University, Division of Pediatric Endocrinology, Syndromes of Abnormal Sex Differentiation
http://www.hopkinschildrens.org/intersex
Site provides a guide to the science and genetics of sex differentiation, with information about Turner syndrome and other syndromes of sex differentiation.

Medline Plus
http://www.nlm.nih.gov/medlineplus/turnersyndrome.html
Offers a brief description of Turner syndrome and numerous links to additional information.

The Turner Syndrome Society of the United States
http://www.turner-syndrome-us.org
The main national support organization, offering resources and information.

See also: Amniocentesis; Chorionic villus sampling; Dwarfism; Hereditary diseases; Infertility; Klinefelter syndrome; Mutation and mutagenesis; Nondisjunction and aneuploidy; X chromosome inactivation; XYY syndrome.

Twin studies

CATEGORY: Techniques and methodologies

SIGNIFICANCE: Studies of twins are widely considered to be the best way to determine the relative contributions of genetic and environmental factors to the development of human physical and psychological characteristics.

KEY TERMS

dizygotic: developed from two separate zygotes; fraternal twins are dizygotic because they develop from two separate fertilized ova (eggs)

monozygotic: developed from a single zygote; identical twins are monozygotic because they develop from a single fertilized ovum that splits in two

zygosity: the degree to which two individuals are genetically similar
zygote: a cell formed from the union of a sperm and an ovum

The Origin of Twin Studies

Sir Francis Galton, an early pioneer in the science of genetics and a founder of the theory of eugenics, conducted some of the earliest systematic studies of human twins in the 1870's. Galton recognized the difficulty of identifying the extent to which human traits are biologically inherited and the extent to which traits are produced by diet, upbringing, education, and other environmental influences. Borrowing a phrase from William Shakespeare, Galton called this the "nature vs. nurture" problem. Galton reasoned that he could attempt to find an answer to this problem by comparing similarities among people who obviously shared a great deal of biological inheritance, with similarities among people sharing less biological inheritance. Twins offered the clearest example of people who shared common biological backgrounds.

Galton contacted all of the twins he knew and asked them to supply him with the names of other twins. He obtained information on ninety-four sets of twins. Of these, thirty-five sets were very similar, people who would now be called identical twins. These thirty-five pairs reported that people often had difficulty telling them apart. Using questionnaires and interviews, Galton compared the thirty-five identical pairs with the other twins. He found that the identical twins were much more similar to one another in habits, interests, and personalities, as well as in appearance. They were even much more alike in physical health and susceptibility to illness. The one area in which all individuals seemed to differ markedly was in handwriting.

Modern Twin Studies

Since Galton's time, researchers have discovered how biological inheritance occurs, and this has made possible an understanding of why twins are similar. It has also enabled researchers to make more sophisticated use of twins in studies that address various aspects of the nature vs. nurture problem. Parents pass their physical traits to their children by means of genes in chromosomes. Each chromosome carries two genes (called alleles) for every hereditary trait. One allele comes from the father and one comes from the mother. Any set of full brothers and sisters will share many of the same alleles, since all of their genes come from the same parents. However, brothers and sisters usually also differ substantially; each zygote (ovum, or egg, fertilized by a sperm cell) will combine alleles from the father and the mother in a unique manner, so different zygotes will develop into unique individuals. Even when two fertilized eggs are present at the same time, as in the case of dizygotic or fraternal twins, the two will have different combinations of genes from the mother and the father.

Identical twins are an exception to the rule of unique combinations of genes. Identical twins develop from a single zygote, a cell created by one union of egg and sperm. Therefore, monozygotic twins (from one zygote) will normally have the same genetic makeup. Differences between genetic twins, researchers argue, must therefore be produced by environmental factors following birth.

The ideal way to conduct twin studies is to compare monozygotic twins who have been reared apart from each other in vastly different types of families or environments. This is rarely possible, however, because the number of twins separated at birth and adopted is relatively small. For this reason, researchers in most twin studies use fraternal twins as a comparison group, since the major difference between monozygotic and dizygotic twins is that the former are genetically identical. Statistical similarities among monozygotic twins that are not found among dizygotic twins are therefore believed to be caused by genetic inheritance.

Researchers use several types of data on twins to estimate the extent to which human characteristics are the consequence of genetics. One of the main sources for twin studies is the Minnesota Twin Registry. In the 1990's, this registry consisted of about 10,500 twins in Minnesota. They were found in Minnesota birth records from the years 1936 through 1955, and they were located and recruited by mail between 1985 and 1990. A second major source of twin studies is the Virginia Twin Registry. This is a register of twins constructed from a systematic review of public birth records in the Commonwealth of Virginia. A few other states also maintain records of twins. Some other organizations, such as the American Association of Retired Persons (AARP), keep records of twins who volunteer to participate and make these records available to researchers.

In the background, the co-director of the Twins Reared Apart project, Nancy Segal, with twins Mark Newman and Gerald Levey, separated at birth. (AP/Wide World Photos)

Zygosity, or degree of genetic similarity between twins, is usually measured by survey questions about physical similarity and by how often other people mistake one twin for the other. In some cases, zygosity may be determined more rigorously through analysis of DNA samples.

PROBLEMS WITH TWIN STUDIES

Although twin studies are one of the best available means for studying genetic influences in human beings, there are a number of problems with this approach. Although twin studies assume that monozygotic twins are biologically identical, some critics have claimed that there are reasons to question this assumption. Even though these twins tend to show greater uniformity than other people, developmental differences may emerge even in the womb after the splitting of the zygote.

Twins who show a great physical similarity may also be subject to environmental similarities so that traits believed to be caused by genetics may, in fact, be a result of upbringing. Some parents, for example, dress twins in matching clothing. Even when twins grow up in separate homes without being in contact with each other, their appearances and mannerisms may evoke the same kinds of responses from others. Physical attractiveness, height, and other characteristics often affect how individuals are treated by others so that the biologically based resemblances of twins can lead to common experiences.

Finally, critics of twin studies point out that twins constitute a special group of people and that it may be difficult to apply findings from twin studies to the population at large. Some studies have indicated that intelligence quotient (IQ) scores of twins, on average, are about five points below IQ scores in the general population, and twins may differ from the general population in other respects. It is conceivable that genetics plays a more prominent role in twins than in most other people.

Impact and Applications

Twin studies have provided evidence that a substantial amount of human character and behavior may be genetically determined. In 1976, psychologists John C. Loehlin and Robert C. Nichols published their analyses of the backgrounds and performances of 850 sets of twins who took the 1962 National Merit Scholarship test. Results showed that identical twins showed greater similarities than fraternal twins in abilities, personalities, opinions, and ambitions. A careful examination of backgrounds indicated that these similarities could not be explained by the similar treatment received by identical twins during upbringing.

Later twin studies continued to provide evidence that genes shape many areas of human life. Monozygotic twins tend to resemble each other in probabilities of developing mental illnesses, such as schizophrenia and depression, suggesting that these psychological problems are partly genetic in origin. A 1996 study published in the *Journal of Personality and Social Psychology* used a sample from the Minnesota Twin Registry to establish that identical twins are similar in probabilities of divorce. A 1997 study in the *American Journal of Psychiatry* indicated that there is even a great resemblance between twins in intensity of religious faith. Twin studies have offered evidence that homosexual or heterosexual orientation may be partly a genetic matter, although researcher Scott L. Hershberger has found that the genetic inheritance of sexual orientation may be greater among women than among men.

Carl L. Bankston III, Ph.D.

Further Reading

Hershberger, Scott L. "A Twin Registry Study of Male and Female Sexual Orientation." *Journal of Sex Research* 34, no. 2 (1997). Discusses the results of a Minnesota study of twins, homosexual orientation, sibling environment, and the genetic influence on sexuality.

Joseph, Jay. *The Gene Illusion: Genetic Research in Psychiatry and Psychology Under the Microscope.* New York: Algora, 2004. Joseph critiques prevailing genetic theories, arguing that twin studies and other research have been tainted by researcher bias, unsound methodology, and unsupported assumptions. Chapter 2, "Twin Research: Misunderstanding Twins, from Galton to the Twenty-first Century," provides a historical review of twin research. Chapters 3 and 4 critically review the methodology of twin studies and genetic studies of twins reared apart.

Kendler, Kenneth S., and Carol A. Prescott. *Genes, Environment, and Psychopathology: Understanding the Causes of Psychiatric and Substance Use Disorders.* New York: Guilford Press, 2006. Summarizes the results of the Virginia Adult Twin Study of Psychiatric and Substance Use Disorder, which obtained data on nine thousand people. The authors describe what the survey's findings revealed about the risk for depression, anxiety, substance abuse, and other psychiatric disorders.

Loehlin, John C., and Robert C. Nichols. *Heredity, Environment, and Personality: A Study of 850 Sets of Twins.* Austin: University of Texas Press, 1976. One of the most influential books on twin studies. Graphs, bibliography, index.

Piontelli, Alessandra. *Twins: From Fetus to Child.* New York: Routledge, 2002. A longitudinal study of the everyday lives of thirty pairs of twins, from life in the womb to age three. Illustrations, bibliography, index.

Spector, Tim D., Harold Snieder, and Alex J. MacGregor, eds. *Advances in Twin and Sib-Pair Analysis.* New York: Oxford University Press, 2000. Discusses background and context of twin and sibling-pair analysis and epidemiological and biostatistical perspectives on the study of complex diseases. Illustrations, bibliography, index.

Steen, R. Grant. *DNA and Destiny: Nurture and Nature in Human Behavior.* New York: Plenum, 1996. Steen, a medical researcher and popular science writer, summarizes evidence regarding the relative contributions of genetics and environment in shaping human personalities. Bibliography, index.

Wright, Lawrence. *Twins: And What They Tell Us About Who We Are.* New York: John Wiley and Sons, 1997. An overview of the use of twin studies in behavioral genetics, written for general readers. Bibliography, index.

Web Sites of Interest

Center for the Study of Multiple Birth
http://www.multiplebirth.com/index.html
Features articles that describe the center's medi-

cal and social research and provides links to other resources.

Gene Watch UK, Twin Studies
http://www.genewatch.org/sub-564682

GeneWatch UK investigates the impact of genetic science on society, government, agriculture, and health. The organization's Web site includes a page on twin studies that describes missing heritability, how twin studies are conducted, and the debate over the usefulness of these studies. The page also expresses the organization's support of twin studies for some diseases, but not for others.

Minnesota Twin Family Study
http://www.psych.umn.edu/psylabs/mtfs

Site of an ongoing research study into the genetic and environmental factors of psychological development. Includes the discussion, "What's Special About Twins to Science?"

See also: Aging; Animal cloning; Behavior; Cloning: Ethical issues; Diabetes; DNA fingerprinting; Gender identity; Genetic testing; Genetics in television and films; Heredity and environment; Homosexuality; Intelligence; Prenatal diagnosis; Quantitative inheritance.

Tyrosinemia type I

CATEGORY: Diseases and syndromes
ALSO KNOWN AS: Hepatorenal tyrosinemia; fumarylacetoacetase deficiency

DEFINITION

Tyrosinemia type I is a hereditary disorder affecting primarily the liver and kidneys. It is caused by mutations in the *FAH* gene that lead to an inability to fully break down the amino acid tyrosine. It can lead to liver failure, liver cancer, rickets, pain crises, and death if untreated.

RISK FACTORS

Tyrosinemia type I is an autosomal recessive disorder and affects males and females equally. Parents who are carriers of a mutation in the *FAH* gene have a 25 percent risk of having an affected child. Overall, the disorder has a prevalence of 1 in 100,000 live births but is more common in the Saguenay Lac St-Jean region of Quebec, Canada, where the prevalence is 1 in 1,846 and 1 in 25 are carriers.

ETIOLOGY AND GENETICS

Amino acids are the building blocks of protein. The amino acid tyrosine is broken down via several steps to form fumarate and acetoacetate, which are then sent through other metabolic pathways to create energy. Fumarylacetoacetate hydrolase (FAH) is the last enzyme in this pathway and is defective in tyrosinemia type I. FAH deficiency leads to the accumulation of fumarylacetoacetate (FAA), which is converted to succinylacetoacetate and succinylacetone. The accumulated FAA is predicted to cause cellular damage and premature cell death in the liver and kidneys. Succinylacetone inhibits two enzymes: parahydroxyphenylpyruvic acid dioxygenase (*p*-HPPD) which increases plasma tyrosine concentration, and PBG synthase, which leads to reduced heme synthesis and increased delta-aminolevulinic acid which can induce neurologic pain episodes.

The FAH enzyme is encoded by the *FAH* gene, located on the long arm of chromosome 15. More specifically, *FAH* is located at 15q23 – q25. The *FAH* gene consists of fourteen exons and is primarily expressed in the liver and kidney, but it can be found at low levels in all tissues. There are several mutations found more frequently in specific populations due to founder effects or genetic drift. The p.P261L mutation accounts for virtually 100 percent of alleles in affected individuals of Ashkenazi Jewish descent. The common French Canadian mutation is c.1062+5GA (IVS 12+5 GA). There are common mutations in Finland, Turkey, Pakistan, Northern Europe, Southern Europe, and individuals of Scandinavian descent. Reported mutations are missense, nonsense, and splice-site mutations, and small deletions have been reported but no large multi-exon deletions that cannot be picked up by standard DNA sequencing methodology.

In the childhood-onset chronic form, the liver nodules show a high incidence of allele reversion—that is, the mutant allele self-corrects to the normal allele, yielding a normal genotype and normal expression of the FAH enzyme in those cells. Milder expression of symptoms can be correlated to the amount of allelic reversion; the corrected cells modify the phenotype. Gene reversion may explain the variability of symptoms within the same family.

Symptoms

Infants presenting earlier than six months of age have severe liver dysfunction. Symptoms include markedly reduced ability to clot, ascites, jaundice, and gastrointestinal bleeding. Untreated affected infants can die from liver failure within weeks of first symptoms. Children presenting at older than six months can have chronic liver dysfunction with significant risk of hepatocellular carcinoma (HCC) and renal dysfunction causing rickets. Episodes of neurologic pain crises are not uncommon and can be severe enough to require mechanical breathing assistance. If untreated, death occurs typically before the age of ten from liver failure, neurologic crisis, or hepatocellular carcinoma.

Screening and Diagnosis

The presence of succinylacetone in the urine and blood in the setting of the patient with liver disease is the hallmark of tyrosinemia type I. Succinylacetone can be detected on urine organic acid analysis.

Final confirmation of diagnosis can be performed by DNA sequencing of the *FAH* gene. Measuring FAH enzyme activity in cultured skin fibroblasts will also provide confirmation but it is not readily available in the United States.

More recently, tyrosinemia type I has been added to state newborn screening programs. Measuring succinylacetone from the newborn blood spot by tandem mass spectroscopy is a more sensitive indicator of tyrosinemia type I than measuring tyrosine.

Treatment and Therapy

Treatment consists of medication and dietary restriction. Orfadin (nitisinone) formerly known as NTBC, prevents FAA and succinylacetone formation but elevates blood tyrosine levels. A low tyrosine diet is required to keep levels in a safe range. Liver transplantation was once the only treatment, but it is used much less frequently now, given the effectiveness of nitisinone and the significant risks associated with liver transplant. Patients should be monitored with regular lab tests for liver and kidney function, plasma amino acids, succinylacetone, blood nitisinone levels, and yearly liver imaging for risk of HCC.

Prevention and Outcomes

Response to nitisinone has markedly increased life spans, reduced the incidence of HCC, and dramatically lowered the need for liver transplant. Patients who have had transplants may still require nitisinone to prevent ongoing renal damage from FAA and succinylacetone that is still being produced in kidney cells.

Lisa Sniderman King, M.Sc., CGC, CCGC

Further Reading

Nussbaum, Robert L., Roderick R. McInnes, and Huntington F. Willard. *Thompson and Thompson Genetics in Medicine.* 7th ed. New York: Saunders, 2007. A standard genetics text for undergraduates or medical students.

Web Sites of Interest

GeneTests, Medical Genetics Information Resource, University of Washington, Seattle. GeneReviews: Tyrosinemia Type 1 (L. Sniderman King, C. Trahms, and C. R. Scott)
http://www.genetests.org

National Society of Genetic Counselors
http://www.nsgc.org

Tyrosinemia Parent Information
http://www.newbornscreening.info/Parents/aminoaciddisorders/Tyrosinemia.html

University of Washington Tyrosinemia Homepage
http://depts.washington.edu/tyros

See also: Alkaptonuria; Andersen's disease; Diabetes; Diabetes insipidus; Fabry disease; Forbes disease; Galactokinase deficiency; Galactosemia; Gaucher disease; Glucose galactose malabsorption; Glucose-6-phosphate dehydrogenase deficiency; Glycogen storage diseases; Gm1-gangliosidosis; Hemochromatosis; Hereditary diseases; Hereditary xanthinuria; Hers disease; Homocystinuria; Hunter disease; Hurler syndrome; Inborn errors of metabolism; Jansky-Bielschowsky disease; Kearns-Sayre syndrome; Krabbé disease; Lactose intolerance; Leigh syndrome; Lesch-Nyhan syndrome; McArdle's disease; Maple syrup urine disease; Menkes syndrome; Metachromatic leukodystrophy; Niemann-Pick disease; Ornithine transcarbamylase deficiency; Orotic aciduria; Phenylketonuria (PKU); Pompe disease; Sulfite oxidase deficiency; Tarui's disease; Tay-Sachs disease.

U

Usher syndrome

Category: Diseases and syndromes

Definition

Usher syndrome is a rare inherited disorder that involves loss of both hearing and sight. Hearing loss is usually present at birth or soon thereafter. It is due to an impaired ability of the auditory nerves to transmit sensory input to the brain. It is called sensorineural hearing loss.

The vision loss, called retinitis pigmentosa (RP), begins later in childhood, usually after age ten. It slowly gets worse over time. During the teen years, loss of vision is characterized by night blindness and loss of peripheral vision. RP is a deterioration of the retina. The retina is a layer of light-sensitive tissue that lines the back of the eye. It converts visual images into nerve impulses in the brain that allow people to see.

Three types of Usher syndrome have been identified: types I, II, and III. The age of onset and severity of symptoms distinguish the different types.

Approximately 4 out of 100,000 infants born in the United States have Usher syndrome. Usher syndrome accounts for 3 to 6 percent of all deaf children and 3 to 6 percent of all hard-of-hearing children.

Risk Factors

The only known risk factor for Usher syndrome is having parents with the disorder and/or parents who carry the genes for the disorder. If both parents carry the abnormal gene, their child has a 25 percent chance of inheriting both of these abnormal genes and therefore developing Usher syndrome.

Etiology and Genetics

Ten genes have now been identified that are associated with Usher syndrome, and at least three additional genes are known to be involved, although they have not yet been characterized. All these genes specify proteins that function in the development or maintenance of specialized sensory cells in the inner ear and light-sensing cells in the retina (rods and cones). While the exact functions of some of these proteins in the processes of hearing and vision may be poorly understood, the effect of mutations in the associated genes is most often observed as the gradual loss of rods and cones in the retina and of hair cells in the inner ear.

Type I Usher syndrome can result from mutations in any of the following genes: *MYO7A* (located on chromosome 11 at position 11q13.5), *USH1C* (at position 11p15.1), *CDH23* (at position 10q21-q22), *USH1H* (at position 15q22-q23), *USH1G* (at position 17q24-q25), or *PCHD15* (at position 10q21-q22). Mutations resulting in Usher syndrome type II occur in the following genes: *USH2A* (at position 1q41), *WHRN* (at position 9q32-q34), and *GPR98* (at position 5q14). The gene known as *USH3A* (at position 3q21-q25) is the only one so far identified in which mutations leading to type III Usher syndrome occur.

All forms of Usher syndrome are inherited in an autosomal recessive fashion, which means that both copies of the particular gene must be deficient in order for the individual to be afflicted. Typically, an affected child is born to two unaffected parents, both of whom are carriers of the recessive mutant allele. The probable outcomes for children whose parents are both carriers are 75 percent unaffected and 25 percent affected. If one parent has Usher syndrome and the other is a carrier of the same type, there is a 50 percent probability that each child will be affected.

Symptoms

The main symptoms of Usher syndrome are hearing and vision loss. Some people also have balance problems due to inner ear problems.

RP begins in the early teenage years as night blindness and leads to blindness by mid-adulthood. The progression of RP limits a person's ability to see in dim light or in the dark (night blindness). It also causes a person to lose peripheral (side) vision slowly over time. Eventually, any vision left is only in a small tunnel-shaped area. Almost everyone with RP becomes legally blind. There is no known way of predicting when or how quickly a person will lose vision.

Symptoms of type I Usher syndrome include being deaf at birth and receiving little or no benefit from hearing aids, having severe balance problems, being slow to sit without support, and rarely learning to walk before the age of eighteen months. RP begins by age ten with difficulty seeing at night, and it quickly progresses to blindness.

Symptoms of type II Usher syndrome include being born with moderate to severe hearing loss, benefitting from hearing aids, and not having balance problems. RP begins in the teenage years.

Symptoms of type III include being born with normal hearing that gets worse in the teenage years and leads to deafness by mid- to late adulthood and being born with near-normal balance.

Screening and Diagnosis

The doctor will ask about symptoms and medical history and will perform a physical exam. Hearing loss is determined with standard hearing tests. Balance problems can be detected with a test called electronystagmography (ENG). In this test, the doctor flushes the ears with warm and then cool water. This causes nystagmus, which is rapid eye movements that can help the doctor detect a balance disorder.

An eye doctor will perform an eye exam, which will likely include a visual field test to check side vision, a test to check for ability to adapt to seeing in the dark, and a test to check sensitivity to color and contrast.

If any problems are found on these tests, an electroretinography (ERG) is done. This test confirms a diagnosis of RP. It measures the electricity given off by the nerves in the retina. The test is done while wearing special contact lenses and looking at a flashing light.

Treatment and Therapy

There is no cure for Usher syndrome. The best treatment is to identify the disorder as early as possible and begin educational programs and other services right away. This helps reduce the communication and learning problems that can result from hearing and vision loss.

The specific programs and services depend on the severity of the hearing, vision, and balance problems and the person's age and abilities. Options include hearing aids; assistive listening devices; cochlear implant, in which a small device is surgically put under the skin behind the ear to give deaf people some ability to hear; adjustment and career counseling; training to help with balance and movement; low-vision services; communications training; and skills in living independently.

Prevention and Outcomes

Currently, there is no known way to prevent Usher syndrome.

Laurie Rosenblum, M.P.H.;
reviewed by Kari Kassir, M.D.
"Etiology and Genetics" by Jeffrey A. Knight, Ph.D.

Further Reading

EBSCO Publishing. *Health Library: Usher Syndrome.* Ipswich, Mass.: Author, 2009. Available through http://www.ebscohost.com.

Sutton, Amy L. *Eye Care Sourcebook: Basic Consumer Health Information About Eye Care and Eye Disorders.* Detroit: Omnigraphics, 2008.

Williams, David. "Retinal Degeneration in Usher Syndrome." In *Retinal Degenerations: Biology, Diagnostics, and Therapeutics,* edited by Joyce Tombran-Tink and Colin J. Barnstable. Totowa, N.J.: Humana Press, 2007.

Web Sites of Interest

Boys Town National Research Hospital
http://www.boystownhospital.org/home.asp

Canadian National Institute for the Blind
http://www.cnib.ca/en/Default.aspx

Genetics Home Reference
http://ghr.nlm.nih.gov

National Eye Institute
http://www.nei.nih.gov

National Institute on Deafness and Other Communication Disorders: Usher Syndrome
http://www.nidcd.nih.gov/health/hearing/usher.asp

See also: Aniridia; Best disease; Choroideremia; Color blindness; Corneal dystrophies; Deafness; Glaucoma; Gyrate atrophy of the choroid and retina; Macular degeneration; Norrie syndrome; Pendred syndrome; Progressive external ophthalmoplegia; Retinitis pigmentosa; Retinoblastoma; Stargardt's disease.

V

Vanishing white matter disease

CATEGORY: Diseases and syndromes

ALSO KNOWN AS: Myelinosis centralis difusa; childhood ataxia with central nervous system hypomyelination (CACH); Cree leukoencephalopathy; leukoencephalopathy with vanishing white matter; VWM

DEFINITION

Vanishing white matter (VWM) disease is a rare, progressive, panethnic autosomal recessive disorder that affects the central nervous system predominantly in children, but also in young adults. Approximately 150 cases have been reported thus far. The disease leads to a deterioration of the white matter (leukodystrophy). The fatty myelin sheath that insulates nerve fibers, giving white matter its color, is affected.

RISK FACTORS

Family history is an important risk factor.

ETIOLOGY AND GENETICS

The disease is caused by mutations in one of the five genes coding for the eukaryotic initiation factor 2B (eIF2B): *EIF2B1*, *EIF2B2*, *EIF2B3*, *EIF2B4*, and *EIF2B5*, located on chromosomes 12q24.3, 14q24, 1p34.20, 2p23.3, and 3q27, respectively. The genes encode the five subunits of EIF2B complex, which regulates translation initiation and ultimately protein synthesis in the cell. A defective EIF2B leads to an impaired cellular ability to regulate protein synthesis under normal conditions, as well as in response to physical, chemical, and oxidative stress. This might explain the exacerbation of clinical symptoms under stress conditions.

To date, more than eighty mutations have been identified, including punctiform mutations (the majority), deletions, and insertions. The process of identification of the first two genes involved in VWM, *EIF2B5* and *EIF2B2*, was facilitated by two founder effects in the Dutch population. At present, it appears that gene *EIF2B5* contains most mutations (57 percent), frequently R113H. Interestingly, brain glial cells (astrocytes and oligodendrocytes) are particularly susceptible to *EIF2B5* mutations. Neuropathologic examination reveals a unique and selective disruption of the glial cells, with spared neurons. White matter rarefaction, cavitating lesions, and increased numbers of "foamy" oligodendrocytes are observed. Studies show that, in astrocytes, *EIF2B5* alterations lead to cellular maturation defects, but the basis of their selective vulnerability is still poorly understood. Other frequent changes are present on genes *EIF2B4* (17 percent) and *EIF2B2* (15 percent), followed by *EIF2B3* (7 percent) and *EIF2B1* (4 percent). A recent report identified several novel mutations, including missense, and estimated the frequency of *EIF2B3* alterations to be 20 percent. Of note, many mutations, regardless of type, involve highly conserved amino acids. Half of the amino acids involved are conserved in human, rat, and mouse and a quarter are conserved starting with insects. A correlation between genotype and the severity of the phenotype is difficult to establish. However, an association between the mildest phenotype, with onset after five years of age, and mutations in nonconserved amino acids (R113H and E213G) was suggested. In addition, a severe phenotype with mutations in highly conserved amino acids (R195H and V309L) was reported.

The pattern of inheritance is autosomal recessive, signifying that both copies of the gene in each cell display the mutation. Patients are homozygotes or compound heterozygotes for mutations that are on the same gene. The parents of a patient each carry a copy of the affected gene without showing any signs of disease.

SYMPTOMS

Frequently reported clinical characteristics include spasticity, cerebellar ataxia, optic atrophy (inconstant), ovarian insufficiency, and seizures. The first signs may occur antenatally, in infancy, early childhood, late childhood or adulthood. The severity of clinical presentation is highly variable. Mental abilities are relatively preserved. The progression is slow, with exacerbations (triggered by injuries, febrile illness, even fright) leading to partial recovery or coma and death. Lengthy periods of relative stability and even mild improvements are possible.

Screening and Diagnosis

The initial diagnosis relies on clinical symptoms and magnetic resonance imaging (MRI) findings. Imaging reveals diffuse abnormal signal of the cerebral white matter and cystic degeneration (cavitation). The diagnosis also involves detection of *EIF2B* mutations via sequence analysis or mutation scanning. Carrier testing is available for family members once mutation has been identified in a proband. Routine studies of cerebrospinal fluid (CSF), blood, and urine are unremarkable. The lymphoblasts of these patients show a decreased intrinsic activity of the eIF2B factor. A marked reduction in the amount of asialo-transferrin in CSF is a recently described biochemical abnormality that can become a clinical diagnostic biomarker.

Treatment and Therapy

At present, there is no cure for this disease. Treatment is symptomatic. Antibiotics and vaccines should be used rigorously. In children with frequent upper respiratory infections, daily low-dose antibiotics can be considered. Corticosteroids are sometimes useful in the acute phase. Minimizing "cellular stress" is important. Affected individuals should avoid psychological stressors, trauma (contact sports), and high body temperature.

Prevention and Outcomes

Genetic counseling, prenatal testing, and testing of family members can be pursued by individuals with relevant family history. Prognosis appears to correlate with the time of onset, with early forms being more severe (for example, Cree leukoencephalopathy, a fatal infantile form). The phenotype varies widely, from early demise in antenatal onset to slowly progressive disease in the adult form. Most patients survive for a few years after diagnosis.

Mihaela Avramut, M.D., Ph.D.

Further Reading

Nussbaum, Robert L., Roderick R. McInnes, and Huntington F. Willard. *Thompson and Thompson Genetics in Medicine.* 7th ed. New York: Saunders, 2007. A classic medical school textbook.

Rosenberg, Roger N., et al., eds. *The Molecular and Genetic Basis of Neurologic and Psychiatric Disease.* 4th ed. Philadelphia: Lippincott Williams and Wilkins, 2007. Comprehensive review of molecular, genetic, and genomic features, with fresh insights into pathogenesis.

ten Donkelaar, Hans J., Martin Lammens, and Akira Hori. *Clinical Neuroembriology: Development and Developmental Disorders of the Human Central Nervous System.* New York: Springer, 2006. Introduction to developmental disorders, in the context of neurogenetics and neuropathology advances.

Web Sites of Interest

National Library of Medicine, Genetics Home Reference: Leukoencephalopathy with Vanishing White Matter
http://ghr.nlm.nih.gov/condition=leukoencephalopathywithvanishingwhitematter

NINDS Leukodystrophy Information Page
http://www.ninds.nih.gov/disorders/leukodystrophy/leukodystrophy.htm

United Leukodystrophy Foundation
http://www.ulf.org/index.html

See also: Adrenoleukodystrophy; Alexander disease; Alzheimer's disease; Amyotrophic lateral sclerosis; Arnold-Chiari syndrome; Ataxia telangiectasia; Canavan disease; Cerebrotendinous xanthomatosis; Charcot-Marie-Tooth syndrome; Chediak-Higashi syndrome; Dandy-Walker syndrome; Deafness; Epilepsy; Essential tremor; Friedreich ataxia; Huntington's disease; Jansky-Bielschowsky disease; Joubert syndrome; Kennedy disease; Krabbé disease; Leigh syndrome; Leukodystrophy; Limb girdle muscular dystrophy; Maple syrup urine disease; Metachromatic leukodystrophy; Myoclonic epilepsy associated with ragged red fibers (MERRF); Narcolepsy; Nemaline myopathy; Neural tube defects; Neurofibromatosis; Parkinson disease; Pelizaeus-Merzbacher disease; Pendred syndrome; Periodic paralysis syndrome; Prion diseases: Kuru and Creutzfeldt-Jakob syndrome; Refsum disease; Sandhoff disease; Sanfilippo syndrome; Spinal muscular atrophy; Spinocerebellar ataxia.

Viral genetics

CATEGORY: Viral genetics

SIGNIFICANCE: The composition and structures of virus genomes are more varied than any identified in the entire bacterial, botanical, or animal kingdoms. Unlike the genomes of all other cells, which are composed of DNA, virus genomes may contain their genetic information encoded in either DNA or RNA. Viruses cannot replicate on their own but must instead use the reproductive machinery of host cells to reproduce themselves.

KEY TERMS

capsid: the protective protein coating of a virus particle

ribosome: a cytoplasmic organelle that serves as the site for amino acid incorporation during the synthesis of protein

virions: mature infectious virus particles

WHAT IS A VIRUS?

Viruses are submicroscopic, obligate intracellular parasites. This definition differentiates viruses from all other groups of living organisms. There exists more biological diversity within viruses than in all other known life-forms combined. This is the result of viruses successfully parasitizing all known groups of living organisms. Viruses have evolved in parallel with other species by capturing and using genes from infected host cells for functions that they require to produce their progeny, to enhance their escape from their host's cells and immune system, and to survive the intracellular and extracellular environment. At the molecular level, the composition and structures of virus genomes are more varied than any others identified in the entire bacterial, botanical, or animal kingdoms. Unlike the genomes of all other cells composed of DNA, virus genomes may contain their genetic information encoded in either DNA or RNA. The nucleic acid comprising a virus genome may be single-stranded or double-stranded and may occur in a linear, circular, or segmented configuration.

THE NEED FOR A HOST

It must be understood that virus particles themselves do not grow or undergo division. Virus particles are produced from the assembling of preformed components, whereas other agents actually grow from an increase in the integrated sum of their components and reproduce by division. The reason is that viruses lack the genetic information that encodes the apparatus necessary for the generation of metabolic energy or for protein synthesis (ribosomes). The most critical interaction between a virus and a host cell is the need of the virus for the host's cellular apparatus for nucleic acid and for the synthesis of proteins. No known virus has the biochemical or genetic potential to generate the energy necessary for producing all biological processes. Viruses depend totally on a host cell for this function.

Viruses are therefore not living in the traditional sense, but they nevertheless function as living things; they do replicate their own genes. Inside a host cell, viruses are "alive," whereas outside the host they are merely a complex assemblage of metabolically inert chemicals—basically a protein shell. Therefore, while viruses have no inner metabolism and cannot reproduce on their own, they carry with them the means necessary to get into other cells and then use those cells' own reproductive machinery to make copies of themselves. Viruses thrive at the host cells' expense.

REPLICATION

The sole goal of a virus is to replicate its genetic information. The type of host cell infected by a virus has a direct effect on the process of replication. For viruses of prokaryotes (bacteria, primarily), reproduction reflects the physical simplicity of the host cell. For viruses with eukaryotic host cells (plants and animals), reproduction is more complex. The coding capacity of the genome forces the virus to choose a reproductive strategy. The strategy might involve near-total reliance on the host cell, resulting in a compact genome encoded for only a few essential proteins (+), or could involve a large, complex virus genome encoded with nearly all the information necessary for replication, relying on the host cell only for energy and ribosomes. Those viruses with an RNA genome plus messenger RNAs (mRNAs) have no need to enter the nucleus of their host cell, although during replication many often do. DNA genome viruses mostly replicate in the host cell's nucleus, where host DNA is replicated and the biochemical apparatus required for this process is located. Some DNA viruses (poxviruses) have evolved to contain the biochemical capacity to replicate in their host's cytoplasm, with a minimal need for the host cell's other functions.

Virus replication involves several stages carried out by all types of viruses, including the onset of infection, replication, and release of mature virions from an infected host cell. The stages can be defined in eight basic steps: attachment, penetration, uncoating, replication, gene expression, assembly, maturation, and release.

The first stage, attachment, occurs when a virus interacts with a host cell and attaches itself—binds with a virus-attachment protein (antireceptor)—to a cellular receptor molecule in the cell membrane. The receptor may be a protein or a carbohydrate residue. Some complex viruses, such as herpesviruses, use more than one receptor and therefore have alternate routes of cellular invasion.

Shortly after attachment the target cell is penetrated. Cell penetration is usually an energy-dependent process, and the cell must be metabolically active for penetration to occur. The virus bound to the cellular receptor molecule is translocated across the cell membrane by the receptor and is engulfed by the cell's cytoplasm.

Uncoating occurs after penetration and results in the complete or partial removal of the virus capsid and the exposure of the virus genome as a nucleoprotein complex. This protein complex can be a simple RNA genome or can be highly complex, as in the case of a retrovirus containing a diploid RNA genome responsible for converting a virus RNA genome into a DNA provirus.

How a virus replicates and the resulting expression of its genes depends on the nature of its genetic materials. Control of gene expression is a vital element of virus replication. Viruses use the biochemical apparatus of their infected host cells to express their genetic information as proteins and do this by using the appropriate biochemical language recognized by the host cell. Viruses include double-stranded DNA viruses such as papoviruses, poxviruses, and herpesviruses; single-stranded sense DNA viruses such as parvoviruses; double-stranded RNA reoviruses; single-stranded sense RNA viruses such as flaviviruses, togaviruses, and claiciviruses; single-stranded antisense RNA such as filoviruses and bunyaviruses; single-stranded sense RNA with DNA intermediate retroviruses; and double-stranded DNA with RNA intermediate-like hepadnaviruses.

During assembly, the basic structure of the virus particle is formed. Virus proteins anchor themselves to the cellular membrane, and, as virus proteins and genome molecules reach a critical concentration,

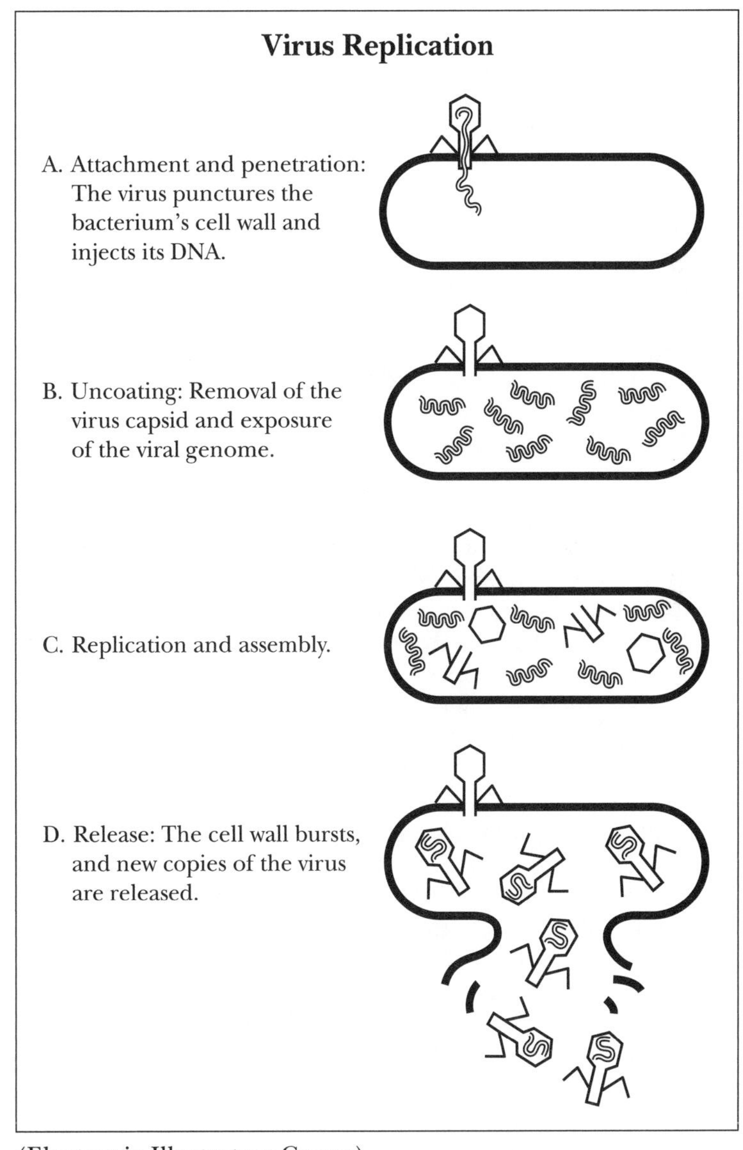

(Electronic Illustrators Group)

assembly begins. The result is that a genome is stuffed into a completed protein shell. The process of maturation prepares the virus particle for infecting subsequent cells and usually involves the cleavage of proteins to form matured products or conformational structural changes.

For most viruses, release is a simple matter of breaking open the infected cell and exiting. The breakage normally occurs through a physical interaction of proteins against the inner surface of the host cell membrane. A virus may also exit a cell by budding. Budding involves the creation of a lipoprotein envelope around the virion prior to the virion's being extruded out through the cell membrane.

Randall L. Milstein, Ph.D.

FURTHER READING

Becker, Yechiel, and Gholamreza Darai, eds. *Molecular Evolution of Viruses: Past and Present.* Boston: Kluwer Academic, 2000. Detailed research of the evolution of viruses by acquisition of cellular RNA and DNA, and how virus genes evade the host immune responses.

Dimmock, N. J., A. J. Easton, and K. N. Leppard. *Introduction to Modern Virology.* 6th ed. Malden, Mass.: Blackwell, 2007. Part 2 of this textbook focuses on virus growth in cells, with information about the replication of viral DNA, genome replication in RNA viruses, and the replication of RNA viruses with a DNA intermediate and vice versa.

Domingo, Esteban, Colin R. Parrish, and John J. Holland, eds. *Origin and Evolution of Viruses.* 2d ed. Boston: Elsevier/Academic Press, 2008. An interdisciplinary reference book consisting of chapters authored by leading researchers in the fields of RNA and DNA viruses. Deals with the simplest, as well as the most complex, viral genomes known.

Holland, J. J., ed. *Current Topics in Microbiology and Immunology: Genetic Diversity of RNA Viruses.* New York: Springer-Verlag, 1992. Detailed collection of papers concerning the genetic and biological variabilities of RNA viruses, replicase error frequencies, the role of environmental selection pressures in the evolution of RNA populations, and the emergence of drug-resistant virus genomes.

Shors, Teri. *Understanding Viruses.* Sudbury, Mass.: Jones and Bartlett, 2009. Chapter 3 of this undergraduate textbook focuses on viral replication cycles.

Yang, Decheng, ed. *RNA Viruses: Host Gene Responses to Infections.* Hackensack, N.J.: World Scientific, 2009. Examines human and animal gene responses to RNA viruses, including West Nile virus, influenza virus, severe acute respiratory syndrome (SARS), and human immunodeficiency virus (HIV).

Zimmer, Carl. *Parasite Rex: Inside the Bizarre World of Nature's Most Dangerous Creatures.* New York: Free Press, 2000. A fascinating, general publication on parasites and their effects on their host organisms, with many references to the historical ramifications of viruses on biological evolution and the development of human culture.

WEB SITES OF INTEREST

Microbial Genetics, Genetics of Viruses
http://student.ccbcmd.edu/courses/bio141/lecguide/unit6/genetics/virus/virus.html

Gary Kaiser, a teacher at Community College of Baltimore County, includes a page on viral genetics in the site he created for his microbiology course.

Microbiology and Immunology On-Line, Virology
http://pathmicro.med.sc.edu/book/virol-sta.htm

The virology section of this site, which was created by the University of South Carolina School of Medicine, contains a page on viral genetics, pages on DNA and RNA virus replication strategies, and other information about viruses.

See also: DNA structure and function; Gene regulation: Viruses; Genetic engineering; Hybridomas and monoclonal antibodies; Oncogenes; Organ transplants and HLA genes; RNA structure and function; Viroids and virusoids.

Viroids and virusoids

CATEGORY: Viral genetics

SIGNIFICANCE: Viroids are naked strands of RNA, 270 to 380 nucleotides long, that are circular and do not code for any proteins. However, some viroids are catalytic RNAs (ribozymes), able to cleave and ligate themselves. In spite of their simplicity, they are able to cause disease in susceptible plants, many of them economically important. Virusoids are similar to viroids, except that they require a helper virus to infect a plant and reproduce.

Key terms

RNA polymerase: an enzyme that catalyzes the joining of ribonucleotides to make RNA using DNA or another RNA strand as a template

RNase: an enzyme that catalyzes the cutting of an RNA molecule

General Characteristics of Viroids and Virusoids

Viroids, and some virusoids, are circular, single-stranded RNA molecules, which normally appear as rods but when denatured by heating appear as closed circles. The rod-shaped structure is formed by extensive base pairing within the RNA molecule, and the secondary structure is divided into five structural domains. One domain is called the pathogenicity (P) domain, because differences among variant strains of the same species of viroid seem to correlate with differences in pathogenicity. Virusoids may also comprise linear RNA or, rarely, double-stranded circular RNA.

The difference between viroids and virusoids is in their mode of transmission. Viroids have no protective covering of any kind and are no more than the RNA that makes up their genetic material. They depend on breaks in a plant's epidermis or can travel with pollen or ovules to gain entry. Virusoids, also known as satellite RNAs, are packaged in the protein coat of other plant viruses, referred to as helpers, and are therefore dependent on the other virus.

Viroids are typically divided into two groups based on the nature of their RNA molecule. Group A is the smallest group, and their RNA has the ability to self-cleave. These include the avocado sunblotch and peach latent mosaic viroids. Group B contains all the other viroids, and their RNA is not capable of self-cleavage. Species in group B include the potato spindle tuber, coconut cadang, tomato plant macho, and citrus bent leaf viroids.

Virusoids are less well studied than viroids and, although more diverse, are most similar to group B viroids in that they cannot self-cleave. Examples include the tomato black ring virus viroid, the peanut stunt virus viroid, and the tobacco ringspot virus viroid. Because so little is known about virusoids, the remainder of this article will focus on viroids.

Viroid Pathogenesis

If infected leaves are homogenized in a blender and passed through an "ultrafilter" fine enough to exclude bacteria, the infection is easily transmitted to another plant by painting some of the filtrate on a leaf. Even billionfold dilutions of the filtrate retain the ability to cause infection, suggesting that it is being replicated. RNase destroys infectivity, suggesting that the genetic material (RNA) is exposed to the medium, unlike viruses, which have a protective protein coat. When isolated from other cell components, an absorbance spectrum shows that viroids are pure nucleic acid, lacking a protein coat.

Although viroids are structurally simple and do not code for any proteins, they still cause disease. Although the molecular mechanisms of viroid pathogenesis are unknown, it is clear that the pathogenesis domain (P domain) is primarily responsible.

Changes in the sequence of nucleotides in the P domain have been correlated with pathogenicity. Some research suggests that the pathogenicity of a viroid strain is related to the resistance of the P domain to heat denaturation, with stability of this region being inversely related to severity. However, some evidence suggests that this may not be entirely true. In a series of nucleotide substitutions introduced by researchers into the P region of an intermediate strain (that is, intermediate in pathogenicity) of potato spindle tuber viroid (PSTVd), four showed viroid infectivity and pathogenicity that were the same as those of a previously reported severe strain of PSTVd. Altogether, eight different mutant strains were analyzed, and resistance to denaturation and PSTVd pathogenicity were not correlated in all cases.

Research is under way to understand how viroids move from cell to cell and traverse the cytoplasm to the nucleus, where many viroids replicate. There is evidence that a possible interaction might involve viroid RNA activating an RNA-activated protein kinase in response to a nucleotide sequence similar to that of the normal RNA activator. Protein kinases are integral to intracellular signaling pathways that control many aspects of cell metabolism. Once researchers understand the signals that viroids use to get around, it may be possible to devise treatments against them. A better understanding of the process may also shed light on normal biochemical communication pathways in plant cells.

VIROID REPLICATION

Viroids replicate by a rolling circle mechanism, a method also used by some viruses. The original strand is referred to as the "+ strand," and complementary copies of it are called "– strands." Type A and B viroids replicate slightly differently. In type A viroids, the circular + strand is replicated by RNA-dependent RNA polymerase to form several linear copies of the RNA – strand connected end to end. Site-specific self-cleavage produces individual – strands later circularized by a host RNA ligase. Each – strand is finally copied by the RNA polymerase to make several linear copies of + strand RNA. Cleavage of this last strand makes individual RNA + strands, which are then circularized. Self-cleavage in viroids represents one of the cases in which RNA acts as an enzyme. The RNA forms a "hammerhead" structure that enzymatically cleaves the longer RNAs at just the right sites.

Replication of type B viroids is apparently mediated by normal host DNA-dependent RNA polymerase, which mistakes the viroid RNA for DNA. The overall process is similar to what happens with type A viroids, except that the – strand is not cleaved but instead is copied directly, yielding a + strand that is cleaved by host RNase to form individual copies that are ligated to become circular.

ECONOMIC IMPACT OF VIROIDS

Genetically engineered plants in the future might make proteins that would essentially confer immunity by preventing viroids from entering the nucleus. With no access to the nucleus, a viroid would be incapable of replicating, effectively preventing the damage normally associated with viroid infection. Currently, no such transgenic plants exist, and viroids can reduce agricultural productivity if outbreaks are not checked quickly. The typical treatment is simply to destroy the affected plants, as there is no cure.

Although predominantly negative, viroids may have some potentially positive benefits. They have already been used in unique ways to study plant genetics, and they may provide insights into how plant proteins and nucleic acids move in and out of cell nuclei. It may also be possible to harness the benefits of viroid infection for certain agricultural applications, such as dwarfing citrus trees. Considerably more will need to be learned about viroids before they can be adequately controlled or used for human benefit.

Bryan Ness, Ph.D.

FURTHER READING

Diener, T. O., R. A. Owens, and R. W. Hammond. "Viroids, the Smallest and Simplest Agents of Infectious Disease: How Do They Make Plants Sick?" *Intervirology* 35, nos. 1-4 (1993): 186-195. A review of the process whereby viroids cause pathology in plants.

Duran-Vila, Núria, et al. "Structure and Evolution of Viroids." In *Origin and Evolution of Viruses*, edited by Esteban Domingo, Colin R. Parrish, and John J. Holland. 2d ed. Boston: Elsevier/Academic Press, 2008. Includes information about the origins, structure, and biological properties of viroids and about viroids in various plant species.

Hadidi, Ahmed, et al., eds. *Viroids.* Enfield, N.H.: Science, 2003. Examines the properties and detection of viroids, diseases and viroids associated with plants, and the control and economic impact of viroid diseases.

Hammond, R. W. "Analysis of the Virulence Modulating Region of Potato Spindle Tuber Viroid (PSTVd) by Site-Directed Mutagenesis." *Virology* 187, no. 2 (April, 1992): 654-662. Report of experiemental results on the potato spindle tuber viroid.

Owens, R. A., W. Chen, Y. Hu, and Y-H. Hsu. "Suppression of Potato Spindle Tuber Viroid Replication and Symptom Expression by Mutations, Which Stabilize the Pathogenicity Domain." *Virology* 208, no. 2 (April 20, 1995): 554-564. Aimed at researchers.

Shors, Teri. "What About Prions and Viroids?" and "Plant Viruses." In *Understanding Viruses.* Sudbury, Mass.: Jones and Bartlett, 2009. These two chapters contain a definition of viroids, a discussion of their pathogenesis and structure, and information about viroid diseases.

Wassenegger, M., et al. "RNA-Directed *De Novo* Methylation of Genomic Sequences in Plants." *Cell* 76, no. 3 (February 11, 1994): 567-576. Report on an experiment involving viroids that demonstrated the possibility that a mechanism of de novo methylation of genes might exist that can be targeted in a sequence-specific manner by their own mRNA.

Web Sites of Interest

Microbiology Bytes, Viroids and Virusoids
http://www.microbiologybytes.com/virology/Viroids.html

Alan Cann, a professor at the University of Leicester, includes a page about viroids and virusoids in his informative microbiology Web site.

University of Hamburg, Department of Biology
http://www.biologie.uni-hamburg.de/bzf/mppg/agviroid.htm

Written in both English and German, this site provides information about plant diseases, including a page explaining viroids, with an electron microscopic photograph of a potato spindle tuber viroid.

See also: DNA structure and function; Gene regulation: Viruses; Genetic engineering; Hybridomas and monoclonal antibodies; Oncogenes; Organ transplants and HLA genes; RNA structure and function; Viral genetics.

Von Gierke disease

Category: Developmental genetics
Also known as: Glycogen storage disease type I; von Gierke's disease

Definition

Von Gierke disease is a rare inherited glycogen storage disease. It is characterized by the inability to break down glycogen into glucose because of the absence of the enzyme glucose-6-phosphatase (G6P). There are two types of von Gierke disease, Ia and Ib. Type Ia is caused by lack of G6P, and type Ib is caused by the lack of G6P translocase, the main transporter substance for G6P.

Risk Factors

Von Gierke disease is an autosomal recessive trait that requires that both parents pass on the mutations. Eighty percent of cases are type Ia, and 20 percent are type Ib. The incidence is about one case per 100,000 births. Type Ia is more common in Ashkenazi Jewish populations, where the incidence is 1 case per 20,000 births.

Etiology and Genetics

Normally, the body stores excess sugar as glycogen in the liver and kidneys. As the glucose is needed for energy and to support body functions, it is freed from the glycogen by a process that requires the enzyme G6P. A similar process converts proteins and fats to glucose. Without G6P, blood glucose levels are unstable, and the body develops large stores of glycogen. The drop in blood glucose levels several hours after meals or at night causes severe hypoglycemia, which can be fatal. Chronic hypoglycemia causes other problems of metabolism. Levels of lactic acid, triglycerides, and uric acid are elevated.

Von Gierke type 1a is caused by mutations on chromosome 17 at 17q21. More than sixty mutations have been identified. Mutations tend to vary according to country of origin. The more common mutations are *R83C, Q347X, 459 insTA*, and *R83H.* These mutations stop production of G6P. There have been no genotype-phenotype correlations established.

Von Gierke type 1b is caused by the absence of translocator1 gene (*G6PT1*) located on chromosome 11 at 11q23. Some of the mutations are due to splicing bases from one exon to another. More than fifty mutations have been identified. The more common mutations are *1211delCT, G339C*, and *W188Rv.*

Symptoms

Symptoms of von Gierke disease develop right after birth and can be fatal if not treated. The infant can demonstrate symptoms of irritability, tremors, cyanosis, seizures, apnea, and coma. Usually, the disease is diagnosed at this time. If not, the child exhibits lethargy, difficult arousal from overnight sleep, overwhelming hunger, poor growth, increase in abdominal girth, easy bruising, and puffy cheeks as a result of fat deposits.

Older patients have poor tolerance of fasting, severe hepatomegaly, growth retardation, osteoporosis, gout, and enlarged kidneys. With type Ib, there are frequent infections.

Screening and Diagnosis

Due to its rarity, von Gierke disease is not screened for unless there is a family history. The diagnosis includes the presenting symptoms. Blood tests of glucose, lactic acid, triglycerides, and uric acid are performed four hours after eating. Typically, the blood glucose is quite low, while lactic acid, triglycerides, and uric acid are elevated. For type Ib, a complete blood count with white cell differential is performed to evaluate for decreased neutrophils. Ultrasounds of the liver and kidneys demonstrate organ enlargement.

The liver is biopsied and examined for G6P activity and deposits of glycogen. Growth and development is behind schedule. Genetic testing is performed for the chromosomal changes that occur with von Gierke disease.

TREATMENT AND THERAPY

Von Gierke disease is treated by a diet that is high in starches and glucose. This diet should be 65 to 70 percent carbohydrate, 10 to 15 percent protein, and 20 to 25 percent fat. Galactose, fructose, sucrose, and lactose should be avoided, since these sugars require G6P to be converted to glucose. Carbohydrates, such as glucose or corn starch, are provided during the night. This is done using a gastric tube with continuous infusion of glucose or starch, or by night feedings with uncooked corn starch.

Elevated uric acid levels are treated with the drug allopurinal, which interferes with its production. During childhood, episodes of metabolic acidosis can occur, during minor illnesses or vomiting. They are treated with intravenous fluids in order to restore glucose levels. If all else fails, a liver transplant can be performed to treat von Gierke disease.

PREVENTION AND OUTCOMES

There is no way to prevent von Gierke disease unless there is a family history of the condition. In this case, genetic testing is performed.

There is no cure for the disease. Patients should be taught to observe for hypoglycemia, and how to treat it. Long-term complications of von Gierke disease include seizures, kidney failure, hepatic tumors, kidney stones, and brain damage. Type Ib can also lead to inflammatory bowel disease, frequent lung and skin infections, secondary diabetes mellitus, and acute myelogenous leukemia.

Christine M. Carroll, R.N., B.S.N., M.B.A.

FURTHER READING

Nussbaum, Robert, Roderick R. McInnes, and Huntington F. Willard. *Thompson and Thompson Genetics in Medicine.* 7th ed. Philadelphia: Saunders, 2007. This text provides a clear understanding of the principles of genetics.

Pritchard, Dorian J., and Bruce R. Korf. *Medical Genetics at a Glance.* 2d ed. Hoboken, N.J.: Wiley-Blackwell, 2007. This book includes all the information needed to understand genetics and genetic disease.

WEB SITES OF INTEREST

Association for Glycogen Storage Disease
http://www.agsdus.org

Children's Fund for Glycogen Storage Disease Research
http://www.cureGSD.org

NIH/National Institute of Diabetes, Digestive, and Kidney Disease
http://www.niddk.nih.gov

See also: Andersen's disease; Forbes disease; Galactokinase deficiency; Galactosemia; Gaucher disease; Glucose galactose malabsorption; Glucose-6-phosphate dehydrogenase deficiency; Glycogen storage diseases; Hereditary diseases; Hers disease; Inborn errors of metabolism; McArdle's disease; Pompe disease; Tarui's disease.

Von Hippel-Lindau syndrome

CATEGORY: Diseases and syndromes
ALSO KNOWN AS: Von Hippel-Lindau disease; VHL

DEFINITION

Von Hippel-Lindau (VHL) syndrome is a rare genetic disorder (affecting approximately 1 in 36,000 people) in which some blood vessels grow in an abnormal way and cause tumors in parts of the body that are rich in blood vessels.

RISK FACTORS

The only known risk factor for VHL is having family members with this syndrome.

ETIOLOGY AND GENETICS

A single gene, *VHL,* located on the short arm of chromosome 3 at position 3p26-p25, is known to be associated with Von Hippel-Lindau syndrome. There are actually two different proteins specified by this gene, depending on how the coding portions of the messenger ribonucleic acid (RNA) are spliced together. These proteins normally move back and forth in cells between the nucleus and the cytoplasm, and they form complexes with several other proteins in a pathway that targets yet other proteins to be degraded when they are no longer needed. One such protein that is often targeted for degradation is the hypoxia-inducible factor (HIF), which acts to control cell division and the formation of new blood vessels. When mutations in the *VHL* gene result in missing or altered VHL proteins, an excess of HIF may develop. This can lead to uncontrolled cell division with superfluous blood vessels, resulting in the characteristic tumors and cysts seen in patients with VHL.

The inheritance of VHL follows a classic autosomal dominant pattern, meaning that a single copy of the mutation is sufficient to cause full expression of the syndrome. An affected individual has a 50 percent chance of transmitting the mutation to each of his or her children. In about 20 percent of cases, however, the disease results from a spontaneous new mutation, so in these instances affected individuals will have unaffected parents. Curiously, it has been shown that at the molecular level within cells, both copies of the *VHL* gene must be mutated in order to initiate the formation of tumors and cysts. One of these mutations is the inherited one that is present in all cells, while the second occurs randomly during normal cell growth in tissues, such as the kidney, brain, or retina. It is this acquired second mutation that triggers tumor or cyst formation in the affected tissue.

Symptoms

There is wide variation in the age at which VHL begins, the organs where problems occur, and the types and severity of symptoms. These differences occur even among members of the same family.

Although there is no consistent set of symptoms, the most common ones are vision problems (retinal angiomatosis), headaches, signs of elevated intracranial pressure, and trouble walking (cerebellar hemangioblastoma). Less common findings include pheochromocytoma, a tumor of the adrenal gland that leads to many problems, including very high or spiking blood pressure; and tumors and/or cysts in the spinal cord, lungs, liver, pancreas, and epididymis (part of the scrotum). Renal cell carcinoma is the most common cause of death.

Few people with VHL have all these problems. Full-blown symptoms usually occur in adulthood, but they can begin in childhood.

Screening and Diagnosis

The doctor will ask about a patient's symptoms and medical history and will perform a physical exam. A blood test that analyzes DNA may be done to determine if the patient has the VHL gene. Not all families with VHL have an identifiable VHL mutation. If members of the patient's family are positive for the gene and the patient is not, the patient does not need any further testing.

However, if other family members have been diagnosed with VHL despite a negative genetic test, or if the patient tests positive for the VHL gene, the patient needs to have regular medical exams and tests to uncover early signs. Even in the absence of symptoms, screening should begin in childhood and continue periodically throughout life.

Screening for VHL complications includes a physical exam with special attention to the patient's eyes and nervous system. Tests may include a magnetic resonance imaging (MRI) scan, a test that uses powerful magnets and radio waves to generate images of the inside of the body; a computed tomography (CT) scan, a type of X ray that uses a computer to generate images of the inside of the body; ultrasound, a test that uses high-pitched sound waves to examine the inside of the body; angiography, X rays taken after a dye is injected into the arteries allowing their interior to be seen; and a twenty-four-hour urine test for elevated levels of hormones. Depending on the test results, the doctor will tell the patient what symptoms to watch for and if the patient needs treatment.

If the patient has any VHL symptoms, he or she should consider being tested for the gene. This is advised even if the patient has no known family history of the disease. The patient could be the first person in his or her family to have VHL, or he or she could be the first one to have it properly diagnosed since many people are not aware they have it.

Treatment and Therapy

There is no known cure for VHL. Treatment depends on a patient's specific symptoms, test results, and general health. Retinal angiomas may be treated with photocoagulation or cryocoagulation. When treatment is needed (for example, for cerebellar lesions), it usually involves surgery to remove tumors.

However, tumors are usually removed only if they are cancerous or causing other problems, such as preventing an organ from working properly. If tumors are not removed, they must be watched carefully for further growth.

Prevention and Outcomes

There is no known way to prevent the VHL gene from causing its many manifestations. Therefore, genetic counseling is advised for families with known VHL or who test positive for the gene.

Individuals who have a family history of the disease or know they have the gene can reduce their risks of serious health problems by having regular screening tests to detect VHL complications early, watching carefully for any suspicious symptoms, and getting treatment as soon as symptoms occur. Individuals can also take steps to reduce their risk of the cancers associated with VHL, such as eating a diet high in fruits and vegetables, not smoking, and limiting their consumption of alcohol.

Laurie Rosenblum, M.P.H.;
reviewed by Rosalyn Carson-DeWitt, M.D.
"Etiology and Genetics" by Jeffrey A. Knight, Ph.D.

Further Reading

Choyke, P. L., et al. "Von Hippel-Lindau Disease: Genetic, Clinical, and Imaging Features." *Radiology* 194, no. 3 (March, 1995): 629-642.

EBSCO Publishing. *Health Library: Von Hippel-Lindau Syndrome.* Ipswich, Mass.: Author, 2009. Available through http://www.ebscohost.com.

Kleigman, Robert M., et al., eds. *Nelson Textbook of Pediatrics.* 18th ed. Philadelphia: Saunders Elsevier, 2007.

Schmike, R. Neil, and Debra L. Collins. "Von Hippel-Lindau Syndrome." In *Management of Genetic Syndromes,* edited by Suzanne B. Cassidy and Judith E. Allanson. Hoboken, N.J.: Wiley-Liss, 2005.

Web Sites of Interest

Canadian Cardiovascular Society
http://www.ccs.ca/home/index_e.aspx

Genetics Home Reference
http://ghr.nlm.nih.gov

Heart and Stroke Foundation
http://ww2.heartandstroke.ca

National Institute of Neurological Disorders and Stroke: "NINDS Von Hippel-Lindau Disease (VHL) Information Page"
http://www.ninds.nih.gov/disorders/von_hippel_lindau/von_hippel_lindau.htm

VHL Family Alliance
http://www.vhl.org

See also: Atherosclerosis; Barlow's syndrome; Cardiomyopathy; Heart disease; Holt-Oram syndrome; Long QT syndrome; Wolff-Parkinson-White syndrome.

Von Willebrand disease

CATEGORY: Diseases and syndromes
ALSO KNOWN AS: vWD

Definition

Von Willebrand disease (vWD) is an inherited blood disorder. It decreases the blood's ability to clot. As a result, bleeding lasts longer than usual.

Von Willebrand disease is the most common hereditary bleeding disorder, affecting about 1 percent of the population. It affects both sexes approximately equally. There are no racial or ethnic associations with this disorder.

Risk Factors

The only risk factor for vWD is having family members with this disease.

Etiology and Genetics

Von Willebrand disease is caused by mutations in a gene called *VWF,* which is found on the short arm of chromosome 12 at position 12p13.3. This gene encodes a protein known as the von Willebrand factor (vWF), which plays an important role in the formation of blood clots. It is the von Willebrand factor that causes platelets to aggregate together and to help plug the holes in damaged blood vessel walls, a critical first step in clot formation. Some mutations in the gene cause a reduced amount of von Willebrand factor to be produced, and these result in the mildest (type 1) form of the disease. Type 2 disease usually results from different mutations that do not reduce the amount of protein but instead compromise the ability of the protein to perform its normal function. In the severest form (type 3), mutations typically result in an abnormally short protein that is completely nonfunctional.

All cases of type 1 vWD and a minority of type 2 cases are inherited in an autosomal dominant fashion, meaning that a single copy of the mutation is sufficient to cause full expression. An affected individual has a 50 percent chance of transmitting the mutation to each of his or her children. Some of these cases, however, result from a spontaneous new mutation, so in these instances affected individuals will have unaffected parents. All cases of type 3 disease and most instances of type 2 disease are inherited in an autosomal recessive manner, which means that both copies of the gene must be deficient in order for the individual to be afflicted. Typically, an affected child is born to two unaffected parents, both of whom are carriers of the recessive mutant allele. The probable outcomes for children whose parents are both carriers are 75 percent unaffected and 25 percent affected

Symptoms

Many people with the vWD gene have very mild symptoms or none at all. When symptoms do occur, the severity varies from person to person. Many people only notice symptoms after taking aspirin or similar medications that interfere with clotting. Symptoms usually begin in childhood and fluctuate throughout life.

Common symptoms include easy bruising, frequent or prolonged nosebleeds, prolonged bleeding from the gums and minor cuts, heavy or prolonged bleeding during menstrual periods, and bleeding in the digestive system. Another symptom is prolonged bleeding after injury, childbirth, surgery, or invasive dental procedures.

Screening and Diagnosis

The doctor will ask about a patient's symptoms and medical history and will perform a physical exam. Blood tests may be done to check for bleeding time, which in vWD will be prolonged, particularly after the administration of aspirin. Blood tests can also check for the factor VIII antigen. This indirectly measures levels of vWF in the blood; in vWD, it will be reduced. Ristocetin cofactor activity can also be checked by a blood test. This shows how well a patient's vWF works; in vWD, it will be decreased. In addition, a blood test can check the von Willebrand factor multimer. This examines the different structural types of vWF in the blood; in vWD, it will be reduced.

Treatment and Therapy

Many people with vWD do not need treatment. For patients who do, their treatment will depend on the type and severity of their vWD. In many cases, treatment is only necessary for patients who are having a surgical or dental procedure that is likely to cause bleeding.

Treatment may include desmopressin. This medicine usually controls bleeding in mild cases of type I vWD by raising the level of vWF in the blood. It can be taken as a nasal spray (Stimate) or an injection (DDAVP). Intravenous infusions control a patient's bleeding. These infusions are concentrates that contain factor VIII and the von Willebrand factor. Birth control pills may be used to control heavy menstrual periods in women with type 1 vWD. Antifibrinolytic medicine (often Amicar) can be used for bleeding in the nose or mouth. It helps keep a clot that has already formed from being dissolved.

Prevention and Outcomes

There are no guidelines for preventing vWD. Genetic counseling can be helpful to review a detailed family history and discuss risks and tests available for von Willebrand disease.

Laurie Rosenblum, M.P.H.;
reviewed by Rosalyn Carson-DeWitt, M.D.
"Etiology and Genetics" by Jeffrey A. Knight, Ph.D.

Further Reading

EBSCO Publishing. *Health Library: Von Willebrand Disease.* Ipswich, Mass.: Author, 2009. Available through http://www.ebscohost.com.

Kouides, Peter A. "Von Willebrand Disease." In *Inherited Bleeding Disorders in Women*, edited by Christine A. Lee, Rezan A. Kadir, and Kouides. Hoboken, N.J.: Wiley-Blackwell, 2009.

Longe, Jacqueline L., and the Gale Group. *The Gale Encyclopedia of Medicine.* 3d ed. Detroit: Thomson Gale, 2006.

Strozewski, Susan M. "Von Willebrand's Disease." *American Journal of Nursing* 100, no. 2 (February, 2000): 24AA-24BB, 24DD.

WEB SITES OF INTEREST

Canadian Hemophilia Society
http://www.hemophilia.ca/en

Genetics Home Reference
http://ghr.nlm.nih.gov

Health Canada
http://www.hc-sc.gc.ca/index-eng.php

MedLinePlus: Bleeding Disorders
http://www.nlm.nih.gov/medlineplus/bleedingdisorders.html

National Heart, Lung, and Blood Institute
http://www.nhlbi.nih.gov

National Hemophilia Foundation: Von Willebrand Disease
http://www.hemophilia.org/bdi/bdi_types3.htm

See also: ABO blood types; Chronic myeloid leukemia; Fanconi anemia; Hemophilia; Hereditary spherocytosis; Infantile agranulocytosis; Myelodysplastic syndromes; Paroxysmal nocturnal hemoglobinuria; Porphyria; Rh incompatibility and isoimmunization; Sickle-cell disease.

W

Waardenburg syndrome

Category: Diseases and syndromes

Also known as: Waardenburg type I (WSI); type II (WSII); type III (WSIII, Klein-Waardenburg syndrome); type IV (WSIV, Shah-Waardenburg syndrome, Waardenburg-Hirschsprung disease)

Definition

Waardenburg syndrome (WS) is a genetic disorder first described in 1951 by Petrus Johannes Waardenburg. Since the original publication, at least four types (I-IV) have been described. All forms are classified as hereditary auditory-pigmentary syndromes characterized by congenital sensorineural hearing loss, patchy depigmentation, and defects of neural-crest-derived tissues. WS is caused by mutations in many genes, including *EDN3*, *EDNRB*, *MITF*, *PAX3*, *SNAI2*, and *SOX10*.

Risk Factors

Individuals without a family history presumably have de novo (sporadic) mutations. Advanced paternal age is a risk factor for sporadic mutations including those that cause WS. The new mutation rate is estimated at 0.4 per 100,000. Both sexes and people of all races are affected equally with an overall prevalence of 1 in 32,400-42,000.

Etiology and Genetics

WS is a genetically heterogeneous group of conditions. Approximately 96 percent of individuals who meet the diagnostic criteria for WSI have a detectable mutation (90 percent) or a whole or partial gene deletion (6 percent) of the *PAX3* gene located on chromosome 2q35. *PAX3* mutations have also been found in individuals with WSIII. Mutations in the microphthalmia-associated transcription factor (*MITF*) gene, located on chromosome 3p14.1-p12.3, are found in approximately 15 percent of individuals with WSII. Additional subtypes of WSII have been linked to other loci including 1p, 8p23, 8q11 (*SNAI2*), and 22q13 (*SOX10*). WSIV is associated with mutations in the endothelin-B-receptor gene (*EDNRP*) on chromosome 13q22, the endothelin-3 gene (*EDN3*) on 20q13.2-q13.3, and the *SOX 10* gene on 22q13.

WS is inherited in both an autosomal dominant (AD) and autosomal recessive (AR) manner. AD refers to a type of condition that can be inherited by a single parent. A person who is heterozygous (one working gene and one nonworking gene) is affected with the condition. The recurrence risk for the offspring of individuals who are affected is 50 percent. AR refers to a type of condition that has to be inherited by both parents. A person who is heterozygous is referred to as a carrier and is unaffected. When two carriers of the same nonworking gene have children, they have a 25 percent chance of having a child homozygous (with two nonworking genes) and thus, of having a child affected with the condition. WS types I, II, and in some cases IV (when *SOX10*) are AD. When WSIII occurs in families it is AD; however, it is usually sporadic. There have also been cases reported of individuals with WSIII who are homozygous for *PAX3* mutations. WS type IV (when *EDNRP* and *EDN3*) is AR.

The genes that cause WS are important in the regulation of melanocyte development. Melanocytes produce melanin, which contributes to the pigmentation of skin, hair, and eye color and is important for the proper function of the cochlea. Mutations in any of these genes may result in hearing loss and changes in pigmentation. Genes responsible for WSIV (*SOX10*, *EDN3*, or *EDNRB*) are involved in the development of nerve cells in the large intestine, leading to problems related to Hirschsprung's disease. All forms of WS show marked interfamilial and intrafamilial variability, indicating that modifier genes probably play a role in the expression of the disease.

SYMPTOMS

Major features of WS include sensorineural hearing loss, a white forelock or premature graying of the hair, different colored eyes (one blue and one brown), and patchy skin dipigmentation. WSI is associated with dystopia canthorum, a condition in which the inner corners of the eyes are spaced farther apart then normal. WSII appears similar to WSI without dystopia canthorum. The principal feature of WSIII is musculoskeletal anomalies of the upper limbs, including hypoplasia (underdevelopment), contractures, and syndactyly (fused fingers). WSIV includes features of type II combined with Hirschsprung's disease.

SCREENING AND DIAGNOSIS

The Waardenburg Consortium has proposed clinical diagnostic criteria. Molecular genetic testing may be available to confirm the diagnosis, clarify risks for family members, and provide prenatal diagnosis. Testing is offered clinically for all types. Population screening for WS is not available.

TREATMENT AND THERAPY

There is no cure for WS. Treatment depends on manifestations as appropriate. Often the most significant complication is hearing loss; benefits are gained from early detection and treatment with hearing aids or cochlear implants. Individuals may also require orthopedic interventions or treatment for Hirschsprung's disease. For optimal care, patients should see a variety of specialists, including a clinical geneticist, genetic counselor, otolaryngologist, audiologist, speech-language pathologist, and possibly an ophthalmologist, dermatologist, craniofacial surgeon, and gastrointestinal specialist.

PREVENTION AND OUTCOMES

Prenatal or preimplantation genetic diagnoses are available if the cause is known. Women at risk of having a child with WSI are recommended to take folic acid supplementation. Most individuals with WS have normal intelligence and live long, productive lives.

Amber M. Mathiesen, M.S.

FURTHER READING

Farrer, L. A., et al. "Waardenburg Syndrome (WS) Type I Is Caused by Defects at Multiple Loci, One of Which Is Near ALPP on Chromosome 2: First Report of the WS Consortium." *The American Journal of Human Genetics* 50 (1992): 902-913.

Newton, V. E. "Clinical Features of the Waardenburg Syndromes." *Advances in Oto-Rhino-Laryngology* 61 (2002): 201-208

WEB SITES OF INTEREST

American Society for Deaf Children
www.deafchildren.org

My Baby's Hearing: Newborn Screening
www.babyhearing.org

See also: Albinism; Chediak-Higashi syndrome; Epidermolytic hyperkeratosis; Hermansky-Pudlak syndrome; Hirschsprung's disease; Ichthyosis; Melanoma; Palmoplantar keratoderma.

Waldenström macroglobulinemia (WM)

CATEGORY: Diseases and syndromes
ALSO KNOWN AS: Lymphoplasmacytic lymphoma

DEFINITION

Waldenström macroglobulinemia (WM) is a rare and indolent form of non-Hodgkin's lymphoma and is one of the malignant monoclonal gammopathies. In WM, lymphoplasmacytic cells that are in the process of maturing from B cells into plasma cells multiply abnormally and produce large amounts of monoclonal immunoglobulin M antibody (IgM). High levels of IgM create hyperviscosity of the serum, and lymphoplasmacytic cells infiltrate into the bone marrow, spleen, and lymph nodes.

RISK FACTORS

The most significant risk factor for WM is age. Median age at diagnosis is sixty-three. Men are at greater risk than women, and white men are at higher risk than African Americans. People who have monoclonal gammopathy of undetermined significance (MGUS) have a higher risk of developing non-Hodgkin's lymphoma and WM within twenty years. Having a first-degree relative with WM may also increase the risk.

ETIOLOGY AND GENETICS

The cause of WM remains unknown. There is some evidence that IgM in some patients with WM may have specificity for the same myelin-associated glycoprotein that has been associated with demyelinating diseases such as multiple sclerosis. This may explain why some patients with WM develop peripheral neuropathy before the onset of neoplastic disease. Speculation exists that this immunologic response may be triggered by viral infection. Hepatitis C, hepatitis G, and human herpes virus 8 have been suspected in the etiology of WM, but the data are not strong enough for these viruses to be considered causative at this time.

Evidence of a genetic etiology for WM comes from reports of family history. In one study approximately 20 percent of people with WM were found tohave a first-degree relative with WM or a similar B cell-associated disease. Various other studies have documented immunoglobulin abnormalities in first-degree relatives of people diagnosed with WM, including monoclonal gammopathy, polyclonal gammopathy, and decreased levels of immunoglobulins. In a recent report published in *Clinical Cancer Research*, a follow-up study of twenty-two first-degree relatives of patients with WM found that IgM macroglobulinemia seems to be a phenotypic marker of WM in some families and carries a high risk of progression to WM. The frequency of IgM macroglobulinemia was found to be greater than thirty times what would normally be expected in a general population. Of five relatives who initially had IgM monoclonal gammopathy, after a median time period of seventeen years, all had persisted or progressed and three had gone on to develop WM.

Several types of chromosomal abnormalities have been observed in patients with WM. Deletions of chromosome 6q, including 6q21-22, have been reported in about 75 percent of patients with WM. Prognostic indications of these findings is uncertain, and there is not enough evidence to link WM with any consistent chromosomal or genetic change.

SYMPTOMS

Signs and symptoms of WM are caused by secretion of IgM, which causes hyperviscosity and vascular complications because of the physical, chemical, and immunologic properties of the IgM paraprotein, and by direct infiltration of neoplastic lymphoplasmacytic cells into bone marrow, spleen, and lymph nodes.

The onset of WM may be gradual and nonspecific, and many asymptomatic patients are diagnosed by routine blood work abnormalities. The most common presenting symptoms are weakness, severe fatigue, anorexia, peripheral neuropathy, weight loss, fever, and Raynaud's phenomenon. Common symptoms due to hyperviscosity may include abnormal bleeding from the nose, gums, or gastrointestinal tract, headache, blurred vision, dizziness, or hearing loss.

Common signs of WM found on physical examination include enlargement of the spleen and liver as well as generalized lymphadenopathy. Examination of the skin may reveal signs of increased bleeding tendency such as bruising or purpura. Ophthalmoscopic evaluation may reveal dilation of retinal veins, which is characteristic of hyperviscosity.

Infiltration of the central nervous system can cause Bing-Neel syndrome, which is characterized by lethargy, stupor, confusion, memory loss, and motor abnormalities. The neuropathy of WM is typically distal, sensorimotor, symmetrical, and slowly progressive. The constellation of polyneuropathy, organomegaly, endocrinopathy, M protein, and skin changes is known as POEMS syndrome and can be seen in patients with WM.

The signs and symptoms of WM are most similar to multiple myeloma (MM). WM can be differentiated from MM by enlargement of the spleen and liver, which is common in WM but rare in MM. Lytic bone lesions and renal disease that are commonly seen in MM are rarely seem in WM.

SCREENING AND DIAGNOSIS

WM is a rare condition with about 1,500 cases diagnosed each year in the United States accounting for about 2 percent of hematologic malignancies. Diagnosis is based on finding evidence of a monoclonal spike on serum electrophoresis. Immunoelectrophoresis and immunofixation studies can help identify the spike as M component. About 80 percent of patients will present with a normocytic normochromic anemia. Thrombocytopenia and leukopenia are also common.

Malignant lymphocytes are usually present in peripheral blood. About 10 percent of macroglobulins in WM are cryoglobulins, but unlike the mixed cryoglobulins seen in rheumatoid arthritis or other autoimmune diseases, the cryoglobulins seen in WM are purely composed of IgM. Like the mixed cryoglobulins, the cryoglobulins of WM may cause Raynaud's phenomenon or other vascular symptoms precipitated by cold.

Bone marrow aspiration and biopsy are needed to confirm the diagnosis of WM. Bone marrow examination will show lymphoplasmacytic and small lymphocyte infiltration. Infiltration may be described on histology as diffuse, nodular, or packed. Nodular infiltration has the best prognosis and packed infiltration has the poorest prognosis.

Urinary immunoglobulins are usually light chain kappa type. Proteinuria is found in 40 percent of patients but exceeds 1 gram per day in only 3 percent of patients. Imaging studies including chest radiographs, computed tomography (CT), and magnetic resonance imaging (MRI) may be done to rule out infiltrates and organomegaly. Mental status symptoms should be evaluated by obtaining cerebrospinal fluid to rule out the presence of IgM.

Treatment and Therapy

Although there is no cure for WM, treatment can be effective and there are many treatment options available. Treatment for symptomatic hyperviscosity is considered to be a medical emergency and should be instituted promptly. The treatment of choice is plasmapheresis since 80 percent of IgM is confined in the intravascular space. Treatment of macroglobulinemia is by alkylating agents, nucleoside analogues, monoclonal antibodies, or combination therapy. The alkylating agent chlorambucil has been used for more than forty years. The response rate is about 60 percent when chlorambucil is used with or without prednisone. In patients who are resistant to alkylating agents, the purine nucleosides fludarabine or cladribine are often effective. The monoclonal antibody rituximab produces response rates of 20 to 50 percent. Combination therapies have reached response rates of 75 percent in some studies.

Other therapies that have been effective for WM include thalidomide, autologous blood cell transplantation, and interferon alpha. Autologous transplantation is reserved for younger patients who have not responded to other treatment. Interferon alpha given for six months has shown a response rate of 50 percent for a duration of more than two years. In cases where other treatments are not effective, splenectomy is occasionally performed because it removes a major reservoir of IgM.

Prevention and Outcomes

There is no way to prevent WM. Some patients meet the diagnostic criteria for WM but remain asymptomatic. These patients are considered to have smoldering or indolent disease and can be followed carefully without treatment for many years. Complications of WM include renal disease, amyloidosis, cardiac failure, infections, and increased risk for lymphomas, myelodysplasia, and leukemia.

Median survival for all patients with WM is 78 months. About 80 percent of patients respond to some type of chemotherapy, and their median survival rate exceeds three years. In patients whose macroglobulin level can be reduced by 75 percent, median survival increases to 7.7 years. Incidence of WM rises sharply with age. Half of all patients with WM are diagnosed after the age of sixty-three. Older age, male sex, general symptoms, and cytopenia are associated with a worse prognosis.

Chris Iliades, M.D.

Further Reading

Cashen, Amanda F., and Tanya M. Wildes, eds. *The Washington Manual Hematology and Oncology Subspecialty Consult.* 2d ed. Philadelphia: Wolters Kluwer Health/Lippincott Williams & Wilkins, 2008. A user-friendly resource that covers inpatient and outpatient management of hematology and oncology.

Fauci, Anthony S. *Harrison's Principles of Internal Medicine.* 17th ed. New York: McGraw-Hill, 2008. The definitive textbook for internal medicine.

Web Sites of Interest

International Waldenstrom Macroglobulinemia Foundation
http://www.iwmf.com

National Organization of Rare Disorders
http://www.rarediseases.org

National Cancer Institute
http://www.nci.nih.gov

See also: Burkitt's lymphoma; Cancer; Chronic myeloid leukemia; Multiple endocrine neoplasias; Mutagenesis and cancer; Mutation and mutagenesis; Oncogenes; Tumor-suppressor genes.

Weill-Marchesani syndrome

Category: Diseases and syndromes
Also known as: WMS; WM syndrome; spherophakia-brachymorphia syndrome; congenital mesodermal dysmorphology

Definition

Weill-Marchesani syndrome (WMS) is a genetic disorder first reported by George Weill in 1932 and subsequentially delineated by Oswald Marchesan in 1939. WMS is classified as a disorder of connective tissue, which binds together, supports, and strengthens muscles, joints, organs, and skin. Short stature, brachydactyly, joint stiffness, and characteristic eye anomalies including microspherophakia, ectopia of the lens, severe myopia, and glaucoma are all characteristic of WMS. Both autosomal recessive and dominant modes of inheritance have been described. Although not all causes are known, mutations in both *ADAMTS10* and *FBN1* genes have been found.

Risk Factors

The offspring of consanguineous relationships are at a higher risk for autosomal recessive conditions. Many people reported with autosomal recessive WMS were members of consanguineous families. The majority of individuals with the autosomal dominant form have a family history. If no family history is found, it is likely a result of a de novo (sporadic) mutation. There are no other reported factors associated with an increased risk for having a child with WMS. The prevalence has been estimated at 1 in 100,000 people.

Etiology and Genetics

The etiology of WMS syndrome depends on the mode of inheritance. WMS is inherited in both an autosomal dominant (AD) and autosomal recessive (AR) manner. AD refers to a type of condition that can be inherited by a single parent. A person who is heterozygous (one working gene and one nonworking gene) is affected with the condition. The recurrence risk for the offspring of individuals who are affected is 50 percent. AR refers to a type of condition that has to be inherited by both parents. In an AR condition, a person who is heterozygous is referred to as a carrier and is unaffected. When two carriers of the same nonworking gene have children, they have a 25 percent chance of having a child who is homozygous (with two nonworking genes) and thus affected with the condition. The AR mode of inheritance seems to be more frequent then the AD form.

Homozygous mutations within the *ADAMTS10* gene located on chromosome locus 19p13.3-p13.2 are responsible for the AR form. The *ADAMTS10* gene codes for a protein product belonging to the ADAMTS family of proteins, which are involved in the extracellular matrix. The protein is expressed in the skin, fetal chondrocytes (a cartilage cell), and in the fetal and adult heart. Mutations in the *ADAMTS10* gene are associated with impairment of support, adhesion, movement, and regulation of cells.

Heterozygous mutations within the *FBN1* gene located on chromosome locus 15q21.1 have been found in a large Lebanese family where mode of inheritance is AD. The significance of this finding remains unclear in the etiology of WMS. *FBN1* codes for the protein fibrillin-1, involved in making fibers that help strengthen and allow for flexibility of the connective tissues. Mutations lead to a nonfunctional protein product, leaving the connective tissue compromised.

Symptoms

Despite the genetically heterogeneous nature of WMS there is clinical homogeneity between all forms. The major features of WMS involve ocular (eye) problems including microspherophakia (small lens) in 84 percent, ectopia (abnormal placement) of the lens in 73 percent, severe myopia (nearsightedness) in 94 percent, glaucoma in 80 percent, and cataracts in 23 percent. Usually myopia is the first finding, while glaucoma is the most serious and may lead to blindness. Other features include brachydactyly (short fingers) in 98 percent, joint stiffness in 62 percent, and short stature in 98 percent with an average adult height of 130 to 157 centimeters (cm) for females and 142 to 169 cm for males. Cardiac anomalies are seen in 24 percent of patients, including pulmonary stenosis, mitral valve insufficiency, aortic valve stenosis, ductus ateriosus, and ventricular septal defects. Intelligence is usually normal.

Screening and Diagnosis

The diagnosis of WMS is made clinically; however, no diagnostic criteria have been proposed. Molecular genetic testing may be available to confirm the diagnosis, clarify risks for family members, and allow for prenatal diagnosis. Currently testing is offered clinically for *ADAMTS10.* The detection rate is unknown, however, and many people with a clinical diagnosis have no mutation found. Carrier and prenatal testing require prior identification of the disease-causing mutation. Population screening for WMS is not available.

Treatment and Therapy

There is no cure for WMS. Treatment depends on manifestations as appropriate. Often the most significant complication is the optical manifestations of this condition and cardiac abnormalities when seen. For optimal care, patients should see a variety of specialists, including a clinical geneticist, genetic counselor, ophthalmologist, physical therapist, and cardiologist.

Prevention and Outcomes

Prenatal or preimplantation genetic diagnosis is available if the cause is known. Benefits are gained from early detection and treatment of ocular manifestations. The use of ophthalmic miotics and mydriatics should be avoided, as they can induce pupillary block. Although mild intellectual deficits have been reported, most individuals have normal intelligence and live long and productive lives.

Amber M. Mathiesen, M.S.

Further Reading

Dagoneau, N., C. Benoist-Lasselin, C. Huber, et al. "*ADAMTS10* Mutations in Autosomal Recessive Weill-Marchesani Syndrome." *American Journal of Human Genetics* 75, no. 5 (2005): 801-806. A article discussing more specifically the mutations found in the *ADAMTS10* gene.

Faivre, L., et al. "Clinical Homogeneity and Genetic Heterogeneity in Weill-Marchesani Syndrome." *American Journal of Medical Genetics Part C* 123 (2003): 204-207. Provides information regarding manifestations of WMS and how they differ depending on mode of inheritance.

Web Sites of Interest

GeneReviews
http://www.ncbi.nlm.nih.gov/bookshelf/br.fcgi?book=gene&part=weill-ms

NIH/National Eye Institute
http://www.nei.nih.gov

See also: Aniridia; Best disease; Brachydactyly; Choroideremia; Color blindness; Corneal dystrophies; Glaucoma; Gyrate atrophy of the choroid and retina; Macular degeneration; Norrie syndrome; Progressive external ophthalmoplegia; Retinitis pigmentosa; Retinoblastoma; Stargardt's disease.

Werner syndrome

CATEGORY: Diseases and syndromes
ALSO KNOWN AS: Progeroid syndrome; adult progeria; adult premature aging syndrome; Werner's syndrome; WS

Definition

Werner syndrome (WS), named after the German scientist who first described it, is a rare, hereditary disorder characterized by premature aging. Patients typically have thin arms and legs, a thick torso, and a characteristic facial appearance described as "bird-like."

Risk Factors

The Werner syndrome (*WRN*) gene is transmitted as an autosomal recessive trait. The risk for Werner syndrome is related to the level of consanguinity (close relation) in a population.

In the United States, Werner syndrome occurs in approximately 1 in 200,000 births. However, it is more common in the Japanese population (approximately 1 in 20,000 to 1 in 40,000; based on the frequencies of detectable heterozygous mutations). This is likely the result of a founder mutation in the Japanese population.

Etiology and Genetics

The *WRN* gene is located on chromosome 8 (cytogenetic location: 8p12-p11.2) and codes for a DNA helicase belonging to the RecQ family. This DNA helicase normally plays an important role in cell division, maintenance, and repair of DNA, and the regulation of telomeres. DNA helicases are critical for preserving genome integrity. The molecular role of the *WRN* gene in Werner syndrome is not yet known.

Mutations in the *WRN* gene (also called *RECQL2* or *REQ3*) lead to the production of an abnormally short, nonfunctional Werner protein. This mutated form of *WRN* cannot interact with DNA, probably because of defective transport of the WRN protein to the nucleus. It is thought that cells with the mutated form of *WRN* may divide more slowly or stop dividing earlier than normal, thus causing growth problems. Additionally, without a functional WRN protein, DNA damage may accumulate that could prevent normal cell functions and cause the health problems seen in Werner syndrome. The genome of Werner syndrome cells is highly unstable, with large, spontaneous DNA deletions. Telomere length dynamics are abnormal—that is, as cells divide, telomeres shorten more rapidly than normal. The unstable genome of Werner syndrome cells may play a key role in the pathogenesis of Werner syndrome.

So far, fifty distinct mutations have been discovered since the *WRN* gene was first cloned in 1996. All these mutations result in the elimination of the nuclear localization signal, thus impeding any functional interactions in the nucleus.

Symptoms

The age of onset of Werner syndrome is variable, but patients develop normally until they reach puberty. One of the first signs of Werner syndrome is the lack of a growth spurt in the early teen years, which results in short stature. After puberty, patients begin to age rapidly, and by age forty, they appear several decades older than they are.

The cardinal signs and symptoms (onset after age ten years) of Werner syndrome include cataracts in both eyes; characteristic skin patterns (tight skin, atrophic skin, pigmentary alterations, ulceration, hyperkeratosis, regional subcutaneous atrophy); characteristic birdlike faces (nasal bridge appears pinched and subcutaneous tissue is diminished); short stature; premature graying and/or thinning of scalp hair; and parental consanguinity (third cousin or closer) or affected sibling.

Additional signs and symptoms include Type II diabetes mellitus; hypogonadism (secondary sexual underdevelopment, diminished fertility, testicular or ovarian atrophy); osteoporosis; osteosclerosis of distal phalanges of fingers and/or toes; soft-tissue calcification; premature atherosclerosis (history of myocardial infarction); neoplasms; abnormal voice (high-pitched, squeaky, or hoarse); and flat feet.

Screening and Diagnosis

In families in which the disease-causing mutations have been identified in an affected family member, prenatal screening (genetic testing) is available and family members may be tested. Diagnosis is based on the presence of the cardinal signs and symptoms of Werner syndrome.

Treatment and Therapy

There is no cure for Werner syndrome. Therapy is based on the treatment of symptoms: aggressive treatment of skin ulcers with standard therapies; control of Type II diabetes mellitus (promising results have been reported with the use of pioglitazone); cholesterol-lowering drugs if the lipid profile is abnormal; surgical treatment of ocular cataracts; treatment of malignancies in a standard fashion (except that radiation therapy is not used, as it may do more harm than good to those with Werner syndrome). Current clinical trials are studying the treatment of symptoms. Preliminary laboratory findings suggest restoring normal WRN function in cells from a patient with Werner syndrome may correct the molecular defects.

Prevention and Outcomes

Early diagnosis is useful so that screening for cancer and associated diseases such as diabetes can then be performed on a regular basis. Genetic counseling for patients and their family is advised.

Patients usually live to their late forties or early fifties, the most common cause of death being cancer or atherosclerosis.

Anita P. Kuan, Ph.D.

Further Reading

Ariyoshi, Kentaro, et al. "Introduction of a Normal Human Chromosome 8 Corrects Abnormal Phenotypes of Werner Syndrome Cells Immortalized by Expressing an hTERT Gene." *Journal of Radiation Research* 50, no. 3 (May, 2009): 253-259.

Chen, Lishan, and Junko Oshima. "Werner Syndrome." *Journal of Biomedicine and Biotechnology* 2, no. 2 (2002): 46-54.

Goto, Makoto, Robert W. Miller, and Nihon Gan Gakkai. *From Premature Gray Hair to Helicase-Werner Syndrome: Implications for Aging and Cancer.* Tokyo: Japan Scientific Societies Press, 2001.

Muftuoglu, Meltem, et al. "The Clinical Characteristics of Werner Syndrome: Molecular and Biochemical Diagnosis." *Human Genetics* 124, no. 4 (November, 2008): 369-377.

Parker, Philip M. *Werner Syndrome: A Bibliography and Dictionary for Physicians, Patients, and Genome Researchers.* San Diego: Icon Group International, 2007.

WEB SITES OF INTEREST

Genetic and Rare Diseases (GARD) Information Center
http://www.genome.gov/10000409

International Registry of Werner Syndrome
http://www.wernersyndrome.org

National Library of Medicine, Genetics Home Reference
http://ghr.nlm.nih.gov

NIH/National Institute on Aging
http://www.nih.gov/nia

See also: Aging; Congenital defects; Diabetes; Hereditary diseases.

Williams syndrome

CATEGORY: Diseases and syndromes
ALSO KNOWN AS: Williams-Beuren syndrome; WBS; WS; Elfin Facies syndrome

DEFINITION

Williams syndrome (WS) is a rare developmental disorder that affects many parts of the body. The condition is characterized by a distinctive cheerful facial appearance, unique behavioral and cognitive traits, mild to moderate intellectual disabilities, and cardiovascular problems. Williams syndrome is due to a deletion of several genes on chromosome 7.

RISK FACTORS

There are no known risk factors. The incidence of Williams syndrome is estimated to be less than 1 in 10,000 people. The disease almost always occurs sporadically. In the few reported examples of familial transmission, the deletion acts as an autosomal dominant mutation. Williams syndrome affects males and females equally.

ETIOLOGY AND GENETICS

Williams syndrome is caused by a loss of DNA from band 7q11.23 of chromosome 7. Therefore while the normal number of chromosomes is maintained, approximately twenty-five genes are lost because of the deletion. This region is referred to as the Williams-Bueren syndrome critical region (WBSCR). The deletion of the WBSCR in chromosome 7 is believed to be the result of unequal crossover events during meiosis. This region is flanked by low copy repeats that increase the likelihood of nonallelic homologous recombination. The deletion arises with equal frequency on the maternally or the paternally inherited chromosome 7. The size of the deletion can vary from 1.5 to 1.8 Mb pairs of DNA. Amongst the deleted genes is the elastin (*ELN*) gene. Loss of this gene is associated with the connective tissue abnormalities, cardiovascular disease (specifically supravalvular aortic stenosis), and facial dysmorphology found in people with this disease.

Other genes contained in this region include *LIMK1* (lim kinase 1), *GTF1IRD1* (part of the TFII-1 transcription family), and *GTF2I* (general transcription factor II, I). Their deletion may be responsible for the characteristic difficulties with visual-spatial tasks. Evidence also exists that the loss of the *CLIP2* gene, among others, may contribute to the learning disabilities and other cognitive difficulties seen in Williams syndrome.

Most cases of Williams syndrome occur as random events during the formation of reproductive cells in a parent of an affected individual. These cases occur in people with no history of the disorder in their family. Williams syndrome is an autosomal dominant mutation. Williams syndrome should be distinguished from other syndromes that include developmental delay, short stature, distinctive facies, and congenital heart disease, such as Noonan syndrome (deletion in chromosome 22q11) and Smith-Magenis syndrome (deletion in chromosome 17p11.2).

SYMPTOMS

The onset of symptoms usually occurs just after birth or during infancy and begins with physical characteristics, irritability, colic, and feeding problems. Almost all cases of Williams syndrome have typical facial features that can be recognized even at birth. Williams patients usually have problems with spatial-visual tasks but have strong abilities in the area of music and spoken language. They often have outgoing and engaging personalities. Children often have delays in walking and speaking; however, the latter improves with age. Motor difficulties persist at all ages, Symptoms progress to abdominal pain in adolescents, diabetes, high blood pressure, heart failure (specifically supravalvular aortic stenosis, SVAS, and supravalvular pulmonary stenosis, SVPS), and hearing loss in adults.

Elastin arteriopathy is present in about 75 percent of Williams individuals, with the most common being SVAS. In most cases, morbidity is due to this aortic narrowing, which can lead to elevated left heart pressure, cardiac hypertrophy, and eventually cardiac failure.

Two-year-old Aubrie Dixon was born with Williams syndrome, which causes physical and mental disabilities. (AP/Wide World Photos)

Screening and Diagnosis

The clinical manifestations include a distinct facial appearance, which can be observed early after birth. Other characteristics such as cardiovascular anomalies and hypocalcemia, can be detected via electroencephalogram and routine blood examinations. The deletion in chromosome 7 is not detected through standard karyotyping but rather through fluorescence in situ hybridization (FISH). Testing is routinely performed on peripheral blood leukocytes. FISH testing and karyotype are performed in cytogenetics laboratories.

Prenatal testing is clinically available but is rarely used because most cases occur in a single family member only and no prenatal indicators exist for low-risk pregnancies.

Treatment and Therapy

There is currently no cure for Williams syndrome. Williams syndrome is a complex multisystem medical condition that requires the attention of multiple health care professionals. Initial care often centers on failure to thrive, hypocalcemia, or repair of the cardiac lesion. Physical therapy is helpful to patients with joint stiffness. Developmental and speech therapy can also help affected children; for example, verbal strengths can help make up for other weaknesses. Other treatments are based on a patient's symptoms.

Prevention and Outcomes

There is no known way to prevent the genetic problem that causes Williams syndrome. Prenatal testing is available for couples with a family history of Williams syndrome who wish to conceive. About 75 percent of those with Williams syndrome have some mental retardation. Most patients will not live as long as normal, due to complications. Most patients require full-time caregivers and often live in supervised group homes.

Maria Mavris, Ph.D.

Further Reading

Bayes, M., et al. "Mutational Mechanisms of Williams-Beuren Syndrome Deletions." *American Journal of Human Genetics* 73 (2003): 131-151.

Morris, C., H. Lenhoff, and P. Wang. *Williams-Beuren Syndrome: Research, Evaluation, and Treatment.* Baltimore: Johns Hopkins University Press, 2006.

Morris, C. A., et al. "Natural History of Williams Syndrome: Physical Characteristics." *Journal of Pediatrics* 113 (1988): 318-326.

Web Sites of Interest

European Organisation for Rare Diseases (EURORDIS)
www.eurordis.org

National Organization for Rare Diseases (NORD)
www.rarediseases.org

Orphanet: The Portal for Rare Diseases and Orphan Drugs
www.orphanet.org

See also: Apert syndrome; Brachydactyly; Carpenter syndrome; Cleft lip and palate; Congenital defects; Cornelia de Lange syndrome; Cri du chat syndrome; Crouzon syndrome; Down syndrome; Edwards syndrome; Ellis-van Creveld syndrome; Holt-Oram syndrome; Ivemark syndrome; Meacham syndrome; Opitz-Frias syndrome; Patau syndrome; Polydactyly; Robert syndrome; Rubinstein-Taybi syndrome; Werner syndrome.

Wilms' tumor

CATEGORY: Diseases and syndromes
ALSO KNOWN AS: Nephroblastoma; kidney tumor

DEFINITION

Wilms' tumor is a type of kidney cancer that predominantly affects children. It usually occurs in the first five years of life, especially around ages three and four. In most cases, Wilms' tumor affects only one of the two kidneys. The frequency of this tumor is 1 in 200,000-250,000 children.

Wilms' tumor is generally divided into two types based on how it looks under the microscope. Favorable types of Wilms' tumor generally have a better outcome and require less aggressive treatment. The unfavorable or anaplastic histology of Wilms' tumor is still curable but requires more aggressive chemotherapy and higher doses of radiation therapy. Despite the more aggressive therapies, survival is generally more limited.

RISK FACTORS

One risk factor for Wilms' tumor is having a family member with a Wilms' tumor. Certain birth defects also increase the risk of a Wilms' tumor. These include WAGR, a syndrome that includes Wilms' tumor; aniridia—a completely or partially missing iris (colored part) of the eyes; genitourinary abnormalities, such as defects of the kidneys, urinary tract, penis, scrotum, testicles, clitoris, or ovaries; and mental retardation.

Additional risk factors include having Beckwith-Wiedemann syndrome, in which patients have a larger than normal tongue, larger than normal internal organs, and one arm or leg that may be larger than the other; having Denys-Drash syndrome, which is the absence of the penis, scrotum, and testicles; and hemihypertrophy, in which one side of the body is larger than the other. Other genetic defects, including defects on chromosome 11, also increase risk. Patients with these abnormalities should be regularly screened for Wilms' tumor during childhood.

ETIOLOGY AND GENETICS

The great majority of cases of Wilms' tumor are sporadic (nonfamilial), and they are believed to arise as the result of genetic mutations in the kidney tissue either before or shortly after birth. Only about 5 percent of all cases are inherited, and even among these it appears that mutations in any of several different genes may be responsible. About one-third of patients with familial Wilms' tumor have mutations in the *WT1* gene, located on the short arm of chromosome 11 at position 11p13. The protein encoded by this gene is a regulatory molecule that acts to turn other genes on at appropriate times during the development of kidney and reproductive tissue. Failure to do so as a result of mutations in the *WT1* gene may result in the uncontrolled growth of kidney tumors characteristic of the disease.

Other genes reported to be associated with genetic predisposition to Wilms' tumor include *WT2* (at position 11p15.5), *WT3* (at position 16q), *WT4* (at position 17q12-q21), *WT5* (at position 7p14-p13), *WTX* (at position Xq11.1), *GPC3* (at position Xq26), and *BRCA2* (at position 13q12.3). A small percentage of patients with Wilms' tumor have identifiable genetic syndromes in which this malignancy is one manifestation of a larger panoply of clinical presentations. These include Denys-Drash syndrome, Beckwith-Wiedemann syndrome, and WAGR syndrome.

The inheritance of familial Wilms' tumor follows a classic autosomal dominant pattern, meaning that a single copy of the mutation is sufficient to cause full expression of the disease. An affected individual has a 50 percent chance of transmitting the mutation to each of his or her children. In some cases, however, the disease results from a spontaneous new mutation, so in these instances affected individuals will have unaffected parents. Curiously, it has been shown that at the molecular level within cells, both copies of the *WT1* gene must be mutated in order to initiate the formation of kidney tumors. One of these mutations is the inherited one that is present in all cells, while the second occurs randomly during normal cell growth in kidney tissue. It is this acquired second mutation that triggers tumor formation in the affected tissue.

Symptoms

The first noticeable symptom is usually a large lump or hard mass in the abdomen. Other symptoms may include stomach pain; fever; blood in the urine; high blood pressure; and loss of appetite, nausea, vomiting, and constipation. Erythrocytosis (a condition where there are too many red cells in the blood) may also be a presenting sign that the doctor finds when taking a blood test. This is because Wilms' tumor makes a protein that causes increased production of red cells. Wilms' tumors may grow larger without causing any pain or other symptoms.

Screening and Diagnosis

The doctor will ask about a child's symptoms and medical history and will perform a physical exam. Blood and urine tests may be done. The child will need one or more tests in order to look for tumors. These tests provide pictures of the kidney, surrounding blood vessels, and other organs to which the cancer may have spread.

These tests include ultrasound, a test that uses sound waves to examine the kidneys. A computed tomography (CT) scan, a type of X ray that uses a computer to make pictures of the inside of the body, may also be done. In some CT scans, a dye is first injected into a vein. This dye makes structures in the body more visible on an X ray. Additional tests include a magnetic resonance imaging (MRI) scan, a test that uses magnetic waves to make pictures of the inside of the body; a chest X ray, a test that uses radiation to look for the spread of cancer to the lungs; a bone scan, in which a small amount of a radioactive material is injected into a vein to highlight any cancer that may have spread to the bones; and a biopsy, the removal of a sample of tissue to test for cancer cells. In Wilms' tumor, the biopsy may actually be a major surgical procedure to remove the kidney.

Except for removal of the kidney, these tests are not invasive but require the child to remain still. Sedation may be needed.

Children who have risk factors for Wilms' tumor should have a physical exam with a specialist and an ultrasound every three months until age six or seven. This screening should be done even if they do not have symptoms. It can help find tumors while they are small and have not yet spread to other parts of the body.

Treatment and Therapy

Wilms' tumor can be cured in most children. The specific treatment depends on if the cancer has spread beyond the kidney to other parts of the body and, if so, how far. The process for determining this, called staging, uses the results of the diagnostic tests. Tumor size, cell type, whether the tumor is favorable or unfavorable, and the child's age and health are also considered in choosing treatment. In general, tumors with a favorable histology are treated only with combinations of chemotherapy, and those with unfavorable patterns or that are recurrent often require the addition of radiation therapy.

Treatment may include surgery. The main treatment for Wilms' tumor is a type of surgery called nephrectomy. This is the removal of the kidney with the tumor. The tissue around the kidney may also be removed, as well as some nearby lymph nodes. The remaining kidney will take care of all of the needed functions for the body.

Chemotherapy is the use of drugs to kill cancer cells. Chemotherapy can be given in many forms, including pill, injection, and via a catheter placed in a blood vessel. The drugs enter the bloodstream and travel through the body killing mostly cancer cells, but also some healthy cells.

Radiation therapy is the use of radiation to kill cancer cells and shrink tumors. Radiation may be external radiation therapy, in which radiation is directed at the tumor from a source outside the body.

Prevention and Outcomes

Currently, there is no known way to prevent Wilms' tumor.

Laurie Rosenblum, M.P.H.;
reviewed by Mohei Abouzied, M.D.
"Etiology and Genetics" by Jeffrey A. Knight, Ph.D.

Further Reading

Driskoll, K. M. Isakoff, and F. Ferrer. "Update on Pediatric Genitourinary Oncology." *Current Opinion in Urology* 17, no. 4 (July, 2007): 281-286.

EBSCO Publishing. *Health Library: Wilms' Tumor.* Ipswich, Mass.: Author, 2009. Available through http://www.ebscohost.com.

Kalapurakol, John A., and Patrick R. M. Thomas. "Wilms' Tumor." In *Perez and Brady's Principles and Practice of Radiation Oncology*, edited by Edward C. Halperin, Carlos A. Perez, and Luther W. Brady. Philadelphia: Wolters Kluwer Health/Lippincott Williams & Wilkins, 2008.

Web Sites of Interest

American Cancer Society
http://www.cancer.org

Childhood Cancer Foundation
http://www.candlelighters.ca/index.html

Genetics Home Reference
http://ghr.nlm.nih.gov

MedLine Plus: Wilms' Tumor
http://www.nlm.nih.gov/medlineplus/wilmstumor.html

National Cancer Institute
http://www.nci.nih.gov

Pediatric Oncology Resource Center
http://www.acor.org/ped-onc

Sick Kids
http://www.sickkids.ca

See also: Alport syndrome; Bartter syndrome; Cancer; Chemical mutagens; Chromosome mutation; Hereditary leiomyomatosis and renal cell cancer; Hereditary non-VHL clear cell renal cell carcinomas; Hereditary papillary renal cancer; Mutagenesis and cancer; Mutation and mutagenesis; Oncogenes; Polycystic kidney disease; Tumor-suppressor genes.

Wilms' tumor aniridia-genitourinary anomalies-mental retardation (WAGR) syndrome

Category: Diseases and syndromes
Also known as: 11p deletion syndrome; chromosome 11p deletion syndrome; WAGR complex; Wilms' tumor-aniridia-gonadoblastoma-mental retardation syndrome

Definition

WAGR syndrome is a rare genetic condition. Children born with this syndrome often have an unusual group of developmental abnormalities. Generally, they have problems with theirs eyes, are at risk for developing certain types of cancer, and may be at risk for mental retardation.

Risk Factors

WAGR syndrome usually occurs spontaneously during fetal development. Several genes on chromosome 11 have been linked to this condition. Therefore, individuals who have a deleted portion of chromosome 11 are at risk of developing WAGR syndrome. Wilms' tumor occurs in approximately 8 per 1 million white children in the United States, and the incidence is somewhat higher in black children.

Etiology and Genetics

Several different disorders are associated with WAGR syndrome, each of which make up the acronym WAGR. *W*ilms' tumor is the most common type of pediatric kidney cancer, and children with WAGR syndrome have a 50 percent chance of developing it. *A*niridia is characterized by a missing iris, the colored part of the eye. *G*enitourinary anomalies include undesecended testes, abnormal locations of the opening for urination, and urinary problems in girls. Mental *R*etardation may also occur and can vary in severity between individuals. The majority of children with WAGR syndrome experience at least two of these conditions, but they do not always have all the described conditions.

WAGR syndrome occurs when part of chromosome 11 is missing, and therefore it sometimes is referred to as a contiguous gene deletion syndrome.

This designation means that WAGR syndrome develops when an individual does not have certain genes located on a specific part of chromosome 11 (11p13). In the majority of cases, this mutation is de novo, meaning that it occurs spontaneously during the early stages of fetal development. In other cases, this dysfunctional chromosome is passed from parent to offspring. This occurs when the chromosomes within a parent undergo a translocation event. Translocation of two chromosomes (one being chromosome 11) can lead to a loss of some of the genes on the chromosome. When such a translocation event occurs and is passed on to a new generation, the offspring may contain normal cells and ones that have the 11p13 deletion, a condition referred to as mosaic WAGR syndrome.

It is not surprising that the severity of WAGR syndrome will depend on how large the chromosomal deletion is (how many and which genes are missing). Individuals with smaller deletions usually have aniridia and Wilms' tumor. Larger deletions are responsible for genitourinary defects and mental retardation. The deleted region of chromosome 11 contains an ocular development gene called *PAX6*. When absent, this gene causes aniridia. Some studies suggest that the *PAX6* gene may also be responsible for pancreatic and brain deficiencies. The Wilms' tumor-suppressor gene, *WT1*, also is deleted in individuals with WAGR syndrome. Although it is best known for causing Wilms' tumor when it is missing, the *WT1* gene also may be associated with some of the genitourinary manifestations of WAGR syndrome.

Symptoms

Symptoms of WAGR syndrome manifest soon after birth. Aniridia is commonly apparent first in both males and females. In boys, undescended testes also are noticeable soon after the baby is born. The specific symptoms experienced by each person who has WAGR syndrome depend on the combination of disorders that are present, since each individual may have varying physical and mental disorders.

Screening and Diagnosis

Aniridia is a rare symptom that manifests almost immediately and may indicate that a patient has WAGR syndrome. To diagnose WAGR syndrome definitively, genetic tests such as karyotyping are performed to analyze the chromosomes and determine whether the chromosome 11p13 deletion is present. Another test that may be employed during the genetic testing is a fluorescence in situ hybridization (FISH) test, which also will evaluate whether the deletion is present.

Treatment and Therapy

Treatment of children with WAGR syndrome depends on the appearance of Wilms' tumor. The features and the stage of the tumor determine the appropriate management for each individual. Close to 50 percent of Wilms' tumors are diagnosed before reaching stage II, meaning that they are confined to the kidney and can be surgically excised. Approximately 90 percent of children with Wilms' tumor are cured. Depending on how far the cancer has spread, chemotherapy and radiation may also be used.

Prevention and Outcomes

WAGR syndrome patients can survive long-term, and early detection of the disorder improves their prognosis. However, lifelong disabilities such as vision loss and mental retardation may afflict these patients.

Kelly L. McCoy

Further Reading

Fischbach, Bernard V., et al. "WAGR Syndrome: A Clinical Review of 54 Cases." *Pediatrics* 116 (2005): 984-988. A review of various cases of WAGR syndrome including a description of different presentations of the disease.

Kliegman, Robert M., Richard E. Behrman, Hal B. Jenson, and Bonita F. Stanton. *Kliegman: Nelson Textbook of Pediatrics.* 18th ed. Philadelphia: Saunders, 2007. A comprehensive textbook of common pediatric disorders, including those that involve sexual development.

Kumar, Vinay, Abul K. Abbas, and Nelson Fausto. *Robbins and Cotran Pathologic Basis of Disease.* 7th ed. Philadelphia: Saunders Elsevier, 2005. Provides a detailed overview of childhood cancers including Wilms' tumor.

Web Sites of Interest

International WAGR Syndrome Association
http://www.wagr.org

National Human Genome Research Institute: Learning About WAGR Syndrome
http://www.genome.gov/26023527

National Institutes of Health, Office of Rare Diseases Research
http://rarediseases.info.nih.gov/GARD/Disease.aspx?PageID=4&diseaseID=5528

See also: Alagille syndrome; Attention deficit hyperactivity disorder (ADHD); Autism; Cockayne syndrome; Congenital defects; Dwarfism; Dyslexia; Fragile X syndrome; Hermaphrodism; Smith-Lemli-Opitz syndrome; Werner syndrome; Williams syndrome.

Wilson disease

CATEGORY: Diseases and syndromes
ALSO KNOWN AS: Wilson's disease

DEFINITION

Wilson disease is a rare, inherited, genetic disorder of copper metabolism. It occurs in one out of every thirty thousand people.

Copper is a trace mineral that the human body needs in small amounts. Most people get a lot more copper from food than they need. However, most people are also able to excrete the excess copper. People with Wilson disease cannot excrete the copper they do not need because of a deficiency in ceruloplasmin, a copper-carrying protein.

As a result, copper begins to build up in the liver right after birth and eventually damages the organ. When the liver can no longer hold the excess copper, the mineral goes into the bloodstream. It travels to other organs and may damage the brain, central nervous system, kidneys, and eyes. This disease is fatal unless it is treated before serious illness develops.

RISK FACTORS

The only known risk factor for Wilson disease is a family history of the disease. It tends to be most common in eastern Europeans, Sicilians, and southern Italians.

ETIOLOGY AND GENETICS

Wilson disease results from mutations in a gene called *ATP7B*, which is located on the long arm of chromosome 13 at position 13q14.2-q21. This gene belongs to a family of genes called the ATPase superfamily, and it specifies a protein known as copper-transporting ATPase 2. Found mainly in the liver, this protein normally acts to transport copper from the liver to other parts of the body where it is needed in minute quantities as a cofactor for some enzymatic reactions. Additionally, if copper levels in the liver get too high, this protein can facilitate the transport of excess copper to bile sacs in the liver where it can be eliminated. If concentrations of this protein are drastically reduced or eliminated altogether because of mutations in the *ATP7B* gene, removal of the excess copper is compromised. Copper may then accumulate in tissues to toxic levels, causing the clinical symptoms characteristic of Wilson disease.

Mutations in a second gene, *PRNP* (found on chromosome 20 at 20pter-p12), have been found that delay the age of onset of Wilson disease and ameliorate the symptoms somewhat. This gene encodes the prion protein, which is most active in the brain. In addition to its other functions, prion protein may also be involved in the intercellular transport of copper.

Wilson disease is inherited in a classic autosomal recessive manner, which means that both copies of the *ATP7B* gene must be deficient in order for the individual to be afflicted. Typically, an affected child is born to two unaffected parents, both of whom are carriers of the recessive mutant allele. The probable outcomes for children whose parents are both carriers are 75 percent unaffected and 25 percent affected.

SYMPTOMS

Symptoms most commonly appear in people less than forty years of age. In children, the symptoms usually begin to be expressed around four years of age.

Symptoms of excess copper in the liver include jaundice, a swelled abdomen, pain in the abdomen, nausea, and vomiting blood. Symptoms of excess copper in the brain include depression, anxiety, mood swings, aggressive or other inappropriate behaviors, difficulty speaking and swallowing, tremors, rigid muscles, and problems with balance and walking. Symptoms of excess copper in the eyes include Kayser-Fleischer rings (rusty or brown-colored ring around the iris).

Screening and Diagnosis

Wilson disease is easy to diagnose when it is suspected. However, because it is relatively rare, common signs, such as psychiatric symptoms or hepatitis, may initially be attributed to other causes. A patient may appear healthy even while his or her liver is getting damaged. Sometimes the liver symptoms are mistaken for infectious hepatitis or mononucleosis. Doctors may not recognize psychiatric symptoms caused by Wilson disease. However, it is very important to get diagnosed and treated early to avoid organ damage and early death.

The doctor will ask about a patient's symptoms and medical history and will perform physical and mental exams. Tests may include blood and urine tests to measure levels of copper and ceruloplasmin (a copper-carrying protein); an eye exam to look for brown, ring-shaped color in the cornea (Kayser-Fleischer rings); and a liver biopsy, in which a small sample of liver tissue is removed and tested for excess copper.

When there is a known family history of Wilson disease, early testing may prevent symptoms and organ damage. Genetic testing may be possible if a family member with the diagnosis of Wilson disease has identifiable changes in the gene. Because there are so many gene mutations that result in Wilson disease, there is no single, simple test for everyone.

Nonetheless, recent advances in the understanding of the *ATP7B* gene have increasingly allowed a diagnosis to be made by direct genetic analysis. This technique is particularly useful when other tests are negative or equivocal. Experts now think that when a condition known as "fatty liver" occurs in people who do not consume large amounts of alcohol (excessive alcohol intake is the most common cause of fatty liver), specific testing for Wilson disease is recommended.

Since diagnosis of this disease can be very difficult, experts have devised a scoring system that combines many of the available tests into a single score that is positive in over 90 percent of people who truly have the disease and negative in nearly 97 percent of people whose symptoms are caused by some other condition. Genetic counseling may be helpful to review risks and discuss appropriate testing and management.

Treatment and Therapy

The goals of treatment are to remove the excess copper, prevent copper from building up again, and improve all associated symptoms of copper overload. Treatment cannot cure the underlying problem of copper accumulation; therefore, patients must continue treatment throughout their lives.

Medications used to treat Wilson disease include zinc acetate, which blocks the absorption of copper in the intestinal tract; penicillamine (chelates, or binds, with copper, causing its increased urinary excretion); tetrathiomolybdolate (may be better than a similar drug called trientine); and dimercaprol. Penicillamine is probably the best-studied treatment and is commonly used, especially in severely symptomatic persons. Zinc has gained increasing importance in recent years because it is often effective in long-term maintenance and has fewer side effects than penicillamine. The role of tetrathiomolybdolate has not yet been clearly established.

If a patient has severe liver damage, he or she may need a liver transplant. Liver transplantation allows the body to correct its copper metabolism and can at least prevent the disease from worsening. Transplantation also affords an effective treatment for patients who cannot tolerate the sometimes serious side effects of penicillamine.

Prevention and Outcomes

Currently, there are no guidelines to prevent Wilson disease. However, when identified early, treatment can prevent the development of symptoms.

Laurie Rosenblum, M.P.H.;
reviewed by Rosalyn Carson-DeWitt, M.D.
"Etiology and Genetics" by Jeffrey A. Knight, Ph.D.

Further Reading

Brewer, G. J., et al. "Treatment of Wilson Disease with Ammonium Tetrathiomolybdate, IV: Comparison of Tetrathiomolybdate and Trientine in a Double-Blind Study of Treatment of the Neurologic Presentation of Wilson Disease." *Archives of Neurology* 63, no. 4 (April, 2006): 521-527.

EBSCO Publishing. *Health Library: Wilson Disease.* Ipswich, Mass.: Author, 2009. Available through http://www.ebscohost.com.

Ferenci, P. "Wilson's Disease." *Clinical Gastroenterology and Hepatology* 3, no. 8 (August, 2005): 726-733.

Parker, James N. *The Official Patient's Sourcebook on Wilson's Disease.* San Diego: Icon Health, 2002.

Wakim-Fleming, Jamile, and Kevin D. Mullen. "Wilson's Disease." In *Practical Management of Liver*

Diseases, edited by Zobair M. Younossi. New York: Cambridge University Press, 2008.

WEB SITES OF INTEREST

American Association for the Study of Liver Diseases
http://www.aasld.org/Pages/Default.aspx

Canadian Liver Foundation
http://www.liver.ca/Home.aspx

Genetics Home Reference
http://ghr.nlm.nih.gov

Health Canada
http://www.hc-sc.gc.ca/index-eng.php

Mayo Clinic.com: Wilson's Disease
http://www.mayoclinic.com/health/wilsons-disease/DS00411

Wilson's Disease Association
http://www.wilsonsdisease.org

See also: Alkaptonuria; Andersen's disease; Diabetes; Diabetes insipidus; Fabry disease; Forbes disease; Galactokinase deficiency; Galactosemia; Gaucher disease; Glucose galactose malabsorption; Glucose-6-phosphate dehydrogenase deficiency; Glycogen storage diseases; Gm1-gangliosidosis; Hemochromatosis; Hereditary diseases; Hereditary xanthinuria; Hers disease; Homocystinuria; Hunter disease; Hurler syndrome; Inborn errors of metabolism; Jansky-Bielschowsky disease; Kearns-Sayre syndrome; Krabbé disease; Lactose intolerance; Leigh syndrome; Lesch-Nyhan syndrome; McArdle's disease; Maple syrup urine disease; Menkes syndrome; Metachromatic leukodystrophy; Niemann-Pick disease; Ornithine transcarbamylase deficiency; Orotic aciduria; Phenylketonuria (PKU); Pompe disease; Sulfite oxidase deficiency; Tarui's disease; Tay-Sachs disease; Tyrosinemia, type I; Von Gierke disease.

Wiskott-Aldrich syndrome

CATEGORY: Diseases and syndromes
ALSO KNOWN AS: Eczema-thrombocytopenia-immunodeficiency syndrome; Aldrich syndrome

DEFINITION

Wiskott-Aldrich syndrome is an inherited disorder in which cells of the blood function improperly. The disease is caused by mutation of the gene encoding for Wiskott-Aldrich syndrome protein (WASP), which controls the shape, attachment, division, and movement of blood cells. Patients with Wiskott-Aldrich syndrome have problems with blood clotting and increased rates of infection, autoimmune disease, and cancers of the immune system.

RISK FACTORS

Wiskott-Aldrich syndrome is caused by a rare, recessive, sex-linked allele. The disease is seen almost exclusively in males. Cases in females are exceedingly rare. The global incidence of Wiskott-Aldrich syndrome is estimated at 1 to 10 in 1,000,000 males. There are no environmental risk factors.

ETIOLOGY AND GENETICS

The *WAS* gene encoding for WASP is found on the X chromosome (Xp11.22-23). Wiskott-Aldrich syndrome is caused by a rare recessive allele that causes disease predominantly in males because they have only one copy of the X chromosome. Although most cases of Wiskott-Aldrich syndrome are inherited from the mother, as many as one-third arise from spontaneous mutation. WASP is expressed predominantly in blood cells where its cellular function is to transmit signals to the cytoskeleton, thereby controlling the shape, attachment, division, and movement of these cells. Expression of nonfunctional WASP results in defects in platelets and cells of the immune system.

Many mutations in WASP have been found to cause Wiskott-Aldrich syndrome, and the severity of disease corresponds to the specific mutation that is present. Some patients with mild mutations may exhibit only low platelet counts, small platelet size, and bleeding, while others with more severe phenotypes may exhibit eczema, infections, autoimmune disease, and/or cancers of the immune system.

A specific activating mutation in the *WAS* gene is associated with a distinct disease, X-linked neutropenia.

SYMPTOMS

All patients with Wiskott-Aldrich syndrome exhibit platelets that are reduced in number and in size. Platelets play important roles in blood clotting, and bruising and bleeding may be present from birth. The spleen, which is responsible for removing damaged platelets, is sometimes enlarged. Eczema is also common in Wiskott-Aldrich patients. Many patients are immunodeficient, and at increased risk for bacterial, viral, and fungal infections due to defects in the activity of T cells and B cells. Patients may also exhibit autoimmune disease that can affect a variety of tissues. Patients are also at a higher risk for leukemia and lymphoma.

Screening and Diagnosis

Because patients with Wiskott-Aldrich syndrome invariably have platelets that are small and reduced in number, careful analysis of platelet number and size can be useful in determining diagnosis. Antibody levels are also abnormal in patients due to improper immune cell function and can be analyzed to aid diagnosis. Levels of IgG are low and levels of IgA and sometimes IgE are elevated. Genetic testing that confirms the presence of a mutation in the *WAS* gene provides the most definitive diagnosis.

Treatment and Therapy

Treatments vary depending on the severity of the disease. Patients with severe bleeding may require transfusions of red blood cells or platelets. Splenectomy results in increased platelet numbers and reduced bleeding but may make patients more susceptible to infection. Young boys may wear helmets to protect from head injuries that might result in bleeding in the brain. Susceptibility to various infectious agents may be a major problem in Wiskott-Aldrich syndrome patients, and proper antimicrobial therapy is essential. Patients may also benefit from immunoglobulin replacement therapy, in which antibodies from many donors are administered intravenously to provide some protection against infection, raise platelet levels, and reduce autoimmune disease. Patients should not receive live virus vaccines because they may cause infection.

Bone marrow transplantation or cord blood stem cell transplantation is the only cure for Wiskott-Aldrich syndrome. Because this technology has the potential to eliminate disease, a search for a matched donor usually occurs immediately following confirmation of diagnosis. In transplantation, the blood-producing cells of the patient are destroyed and replaced with those from a nondiseased individual. The new cells make normal platelets and immune cells, thereby eliminating symptoms. Although success rates of transplantation are high and improving, not all patients are good candidates for transplantation and significant risks are associated with transplantation.

Prevention and Outcomes

There is no means to prevent Wiskott-Aldrich syndrome. One-half of the brothers of patients with Wiskott-Aldrich syndrome also have the disease, and one-half of sisters are unaffected carriers of the disease-causing allele. Genetic counseling should be made available for parents of an affected child. Genetic testing can detect the presence of the disease in the fetus and determine if unaffected females are carriers. Outcome depends on the severity of disease and the response to treatment. Before modern treatment, the life expectancy of Wiskott-Aldrich patients was between two and three years. Today, nearly 90 percent of patients that receive tissue-matched bone marrow from unaffected siblings can be cured. The overall cure rate is lower because patients with infection or cancer may not be good candidates for transplantation.

Kyle J. McQuade, Ph.D.

Further Reading

Lewis, Ricki. *Human Genetics.* 8th ed. New York: McGraw-Hill, 2007. An introductory human genetics reference text written for nonscientists.

Rezaie, Nima, Asghar Aghamohammadi, and Luigi D. Notarangelo. *Primary Immunodeficiency Diseases.* Berlin: Springer Verlag, 2008. A comprehensive discussion of primary immunodeficiency diseases that focuses on practical diagnosis and disease management but that may provide useful information for patients' families.

Web Sites of Interest

Immune Deficiency Foundation
http://www.primaryimmune.org

Jeffrey Modell Foundation: National Primary Immunodeficiency Resource Center
http://www.info4pi.org

Johns Hopkins University: Online Mendelian Inheritance in Man
http://www.ncbi.nlm.nih.gov/entrez/dispomim.cgi?id=301000

See also: ABO blood types; Agammaglobulinemia; Allergies; Anthrax; Antibodies; Ataxia telangiectasia; Autoimmune disorders; Autoimmune polyglandular syndrome; Chediak-Higashi syndrome; Chronic granulomatous disease; Chronic myeloid leukemia; Fanconi anemia; Hemophilia; Hereditary spherocytosis; Hybridomas and monoclonal antibodies; Immunodeficiency with hyper-IgM; Immunogenetics; Infantile agranulocytosis; Myelodysplastic syndromes; Myeloperoxidase deficiency; Paroxysmal nocturnal hemoglobinuria; Porphyria; Purine nucleoside phosphorylase deficiency; Rh incompatibility and isoimmunization; Severe combined immunodeficiency syndrome (SCID); Sickle-cell disease; Wolf-Hirschhorn syndrome.

Wolf-Hirschhorn syndrome

CATEGORY: Diseases and syndromes
ALSO KNOWN AS: 4p-syndrome; monosomy 4p; del-(4p); includes Pitt-Rogers-Danks syndrome

DEFINITION

Wolf-Hirschhorn syndrome (WHS) is caused by a partial deletion of the short arm of human chromosome 4. The syndrome is characterized by multiple congenital malformations accompanied by mental retardation and is estimated to occur in at least 1 out of 50,000 births.

RISK FACTORS

Two-thirds of existing cases occur in females. About 75 percent of WHS cases can be linked to a paternal origin, which is likely the result of the preponderance of subtelomeric genetic recombination that is known to occur during meiosis in the male germ line.

ETIOLOGY AND GENETICS

A case report of a child exhibiting symptoms now associated with WHS was first published in 1961 by the laboratory of Kurt Hirschhorn, an Austrian-born American geneticist and pediatrician who also linked this disorder with a deletion on either human chromosome 4 or 5. Three years later, a partial deletion of the short arm of chromosome 5 was demonstrated to be the causal factor of cri du chat syndrome, and the following year WHS was described in detail in simultaneous reports by Hirschhorn's group as well as by German geneticist Ulrich Wolf and colleagues.

As techniques were later developed that allowed for the physical mapping of human chromosomes, scientists sought to delineate the minimal deletion which would define WHS, as well as to characterize the genes that contribute to the disorder. Noting that about 20 percent of WHS cases have deletions restricted to 4p16.3, the terminal 2 percent of the chromosome, a WHS "critical region" was defined that encompassed this area. The subsequent description of rare patients who had submicroscopic deletions was able to reduce this critical region by more than half, an area in which the human genome sequencing project revealed there were at least a dozen genes, including *WHSC1* (Wolf-Hirschhorn syndrome candidate 1), *LETM1* (Leucine zipper-EF-hand containing transmembrane protein 1), and *FGFR3* (Fibroblast growth factor receptor 3).

The most promising research to determine how the WHS critical region contributes to the disorder comes from mouse models in which the corresponding homologous genes have been mutated experimentally. Mice with mutations in *WHSC1* are seen to have growth retardation accompanied by craniofacial deformations and congenital heart defects, while those with mutations in *FGFR3* demonstrate specific skeletal defects. Although no mice-bearing mutations in the *LETM1* homolog have been reported, this gene has been linked with the appearance of seizures by examining patients with atypical patterns of deletion. It is clear that experiments involving mice will play a central role in elucidating the etiology of this syndrome in the future.

SYMPTOMS

WHS can display a wide range of symptoms, but just four are considered to define the core set of minimal diagnostic criteria. The first of these is a facial appearance which has been described as a "Greek warrior helmet." This includes a broad bridge of the nose which continues to the forehead, a prominent space between the eyebrows, an increased distance between the eyes, and high-arched eyebrows. Other facial features include a small head with an undersized jaw, a "carp-shaped" mouth with a short infranasal depression, and low-set ears. The other core symptoms are growth delay, mental retardation, and seizures (or at least detectable EEG anomalies). In addition to these core characteristics, some patients display a set of symptoms which have been characterized as "midline defects" and which include a cleft lip/palate; specific defects in such organs as the heart, esophagus, and diaphragm; and skeletal malformations such as scoliosis and clubfoot.

Screening and Diagnosis

Initial screening techniques used to detect WHS included basic cytogenetic procedures such as human karyotyping accompanied by staining to detect chromosome banding patterns, but these tended to miss about 40 percent of WHS cases. High-resolution banding techniques or FISH (fluorescence in situ hybridization), using probes which bind to the WHS critical region, are required to detect the smaller deletions which can lead to WHS.

Treatment and Therapy

The two most consistent treatable medical problems in children with WHS are major feeding difficulties and seizures. The former is usually treated by gastrostomy, creating a surgical opening into the stomach for feeding via a tube, while the latter has often been controlled by the use of antiepileptic drugs. Surgical intervention is also available to treat clubfoot and scoliosis, as well as to repair congenital heart defects.

Prevention and Outcomes

There is presently no cure for WHS. Over time, all cases of WHS will be expected to make improvements in communication skills, with about half of the cases learning how to walk, a fifth helping with simple household tasks, and a tenth becoming self-feeders. A tenth of WHS cases will also go on to achieve fecal continence during the day.

James S. Godde, Ph.D.

Further Reading

Battaglia, Agatino, John C. Carey, and Tracy J. Wright, "Wolf-Hirschhorn (4p-) Syndrome." *Advances in Pediatrics* 48 (2001): 75-113. A comprehensive review of WHS.

Nimura, Keisuke, et al. "A Histone H3 Lysine 36 Trimethyltransferase Links Nkx2-5 to Wolf-Hirschhorn Syndrome." *Nature*, 460 (July 9, 2009): 287-291. The first description of a *WHSC1*-deficient mouse.

Simon, Ruth, and Andrew D. Bergemann. "Mouse Models of Wolf-Hirschhorn Syndrome." *American Journal of Medical Genetics Part C (Seminars in Medical Genetics)* 148C (2008): 275-280. A review of the other mouse models which have been used to study WHS.

Web Sites of Interest

eMedicine: Wolf-Hirschhorn Syndrome (H. Chen)
http://www.emedicine.com/ped/toppic2446.htm

GeneReviews: "Wolf-Hirschhorn Syndrome" (A. Battaglia, J. C. Carey, and T. J. Wright).
http://www.ncbi.nlm.nih.gov/bookshelf/br.fcgi?book=gene&part=whs

OMIM (Online Mendelian Inheritance in Man): Wolf-Hirschhorn Syndrome; WHS (V. A. McKusick et al.)
http://www.ncbi.nlm.nih.gov/entrez/dispomim.cgi?id=194190

See also: ABO blood types; Agammaglobulinemia; Allergies; Anthrax; Antibodies; Ataxia telangiectasia; Autoimmune disorders; Autoimmune polyglandular syndrome; Chediak-Higashi syndrome; Chronic granulomatous disease; Chronic myeloid leukemia; Fanconi anemia; Hemophilia; Hereditary spherocytosis; Hybridomas and monoclonal antibodies; Immunodeficiency with hyper-IgM; Immunogenetics; Infantile agranulocytosis; Myelodysplastic syndromes; Myeloperoxidase deficiency; Paroxysmal nocturnal hemoglobinuria; Porphyria; Purine nucleoside phosphorylase deficiency; Rh incompatibility and isoimmunization; Severe combined immunodeficiency syndrome (SCID); Sickle-cell disease; Wiskott-Aldrich syndrome.

Wolff-Parkinson-White syndrome

Category: Diseases and syndromes
Also known as: WPW syndrome

Definition

Wolff-Parkinson-White (WPW) syndrome is a disorder of the heart's electrical activity. It causes the heart to beat with an irregular rhythm and faster than normal. This is called tachyarrhythmia.

Risk Factors

There are no known risk factors for WPW syndrome.

Etiology and Genetics

The etiology of Wolff-Parkinson-White syndrome is complex, involving both genetic and environmental determinants, most of which are not at all well understood. The great majority of cases are sporadic, and these individuals have no apparent family history of the disease. In a small minority of cases, mutations in the *PRKAG2* gene, located on the long arm of chromosome 7 at position 7q36.1, have been shown to cause the condition. This gene encodes a protein known as protein kinase, AMP-activated, gamma 2 non-catalytic subunit (AMPK). An important cellular regulatory molecule, AMPK senses and responds to energy demands within the cell by helping to regulate the concentrations of the cell's main energy source, adenosine triphosphate (ATP). AMPK is known to be active during heart development in the fetus, although its exact role in the process is not well understood. Mutations in the *PRKAG2* gene may result in either an overactive or underactive AMPK protein, but in either case heart function is compromised.

The familial form of WPW syndrome, caused by mutations in the *PRKAG2* gene, is inherited in an autosomal dominant fashion. This means that a single copy of the mutation is sufficient to cause full expression of the syndrome. An affected individual has a 50 percent chance of transmitting the mutation to each of his or her children. Many cases of familial WPW syndrome, however, result from a spontaneous new mutation, so in these instances affected individuals will have unaffected parents.

Symptoms

Some people with WPW syndrome never have tachyarrhythmia and its associated symptoms. In those who do, symptoms usually begin between ages eleven and fifty. The frequency and severity of the tachyarrhythmia varies from one person to another and may be associated with any or all of these symptoms: palpitations (sensation of a pounding heartbeat), chest pain or tightness, dizziness, fainting, and shortness of breath. In rare cases, a person will go into cardiac arrest (the heart stops pumping) and lose consciousness.

Screening and Diagnosis

The doctor will ask about a patient's symptoms and medical history and will perform a physical exam. If the patient has a tachyarrhythmia due to WPW syndrome, he or she will have normal or low blood pressure and a heart rate of 150-250 beats per minute. (A normal heart rate is 60-100 beats per minute.)

If a patient is not having irregular heart rhythms during the exam, the results of the exam may be normal. In either case, an electrocardiogram (a test that records the heart's activity by measuring electrical currents through the skin) will usually show a "delta wave" that signals an extra electrical pathway.

Other tests may include monitoring with a Holter monitor for twenty-four to forty-eight hours to check for any episodes of irregular heartbeat and an electrophysiology study, in which a catheter (a thin tube designed to be inserted into a blood vessel) is passed to the interior of the heart, where it takes detailed measurements of its electrical activity. This study will detect the extra pathway.

Treatment and Therapy

The goal of treatment is to reduce or eliminate episodes of tachyarrhythmia and associated symptoms. If a patient does not have symptoms, treatment is usually not necessary.

If a patient does need treatment, it may include medication. Antiarrhythmics may be given to coordinate the heart's electrical signals. This can control or prevent episodes of rapid heartbeat. However, the patient must take the medicine carefully because it can sometimes make an abnormal heart rhythm worse.

Radiofrequency ablation is a procedure in which a catheter delivers energy at a particular radio frequency to the heart. This destroys (ablates) the abnormal electrical pathway. In most cases, ablation is successful and ends the need to take medicine.

Open-heart surgery is done to destroy the abnormal pathway. However, this procedure is rarely done.

Defibrillation is done in the case of cardiac arrest, which is rare. Defibrillation gives the heart a brief electric shock. This procedure converts a rapid, irregular heartbeat back into a normal heartbeat.

Prevention and Outcomes

There is no known way of preventing WPW syndrome. However, symptoms can be prevented with proper treatment.

Laurie Rosenblum, M.P.H.;
reviewed by Michael J. Fucci, D.O.
"Etiology and Genetics" by Jeffrey A. Knight, Ph.D.

Further Reading

Beers, Mark H., et al. *The Merck Manual of Diagnosis and Therapy*. 18th ed. Whitehouse Station, N.J.: Merck Research Laboratories, 2006.

EBSCO Publishing. *Health Library: Wolff-Parkinson-White Syndrome*. Ipswich, Mass.: Author, 2009. Available through http://www.ebscohost.com.

Web Sites of Interest

American Heart Association
http://www.americanheart.org

Cleveland Clinic: Wolff-Parkinson-White Syndrome (WPW)
http://my.clevelandclinic.org/heart/disorders/electric/wpw.aspx

Genetic Home Reference: Wolff-Parkinson-White Syndrome
http://ghr.nlm.nih.gov/condition=wolffparkinsonwhitesyndrome

MedLine Plus: Wolff-Parkinson-White Syndrome
http://www.nlm.nih.gov/medlineplus/ency/article/000151.htm

See also: Atherosclerosis; Barlow's syndrome; Cardiomyopathy; Heart disease; Holt-Oram syndrome; Long QT syndrome.

Wolman disease

Category: Diseases and syndromes

Also known as: Lysosomal acid lipase deficiency; acid lipase deficiency; cholesteryl ester storage disease; familial xanthomatosis; LAL deficiency; LIPA deficiency

Definition

Wolman disease is a rare, autosomal recessive disorder caused by deficiency of the enzyme lysosomal acid lipase (LAL). Insufficiency of this enzyme causes accumulation of cholesteryl esters and triglycerides throughout the body, resulting in multisystem organ damage and death.

Risk Factors

Individuals with mutations in both copies of the *LIPA* gene are at risk for either Wolman disease or a milder condition known as cholesteryl ester storage disease (CESD). Full siblings of affected individuals have a 25 percent risk of inheriting the disorder. Males and females are affected with equal frequency. Cases have been reported in a variety of different ethnic groups; given the rarity of the condition, it is difficult to ascertain whether a higher prevalence truly exists among any particular group.

Etiology and Genetics

Located on chromosome 10q23.2q23.3, the *LIPA* gene produces the LAL enzyme, which is involved in the breakdown of cholesteryl esters and triglycerides within the lysosomes of the cell. This process is part of a tightly regulated system allowing the body to use these and other lipids (fats) appropriately, then recycle or dispose of them as needed.

Deficient LAL activity prevents the breakdown of cholesteryl esters and triglycerides; the effects of their subsequent accumulation within the lysosomes disrupt lipid processing throughout the body. The lysosomes become engorged, ultimately destroying the cell and eventually impairing the function of involved organs, such as the liver.

Mutations in the *LIPA* gene and subsequent deficiency of LAL can also result in the comparatively milder CESD phenotype. Whereas the Wolman phenotype is characterized by the tissue storage of both cholesteryl esters and triglycerides, CESD is characterized by cholesteryl ester storage. While Wolman disease is typically diagnosed in infancy and characterized by rapid progression and death, CESD (while quite variable) can remain undiagnosed into adulthood and be compatible with a normal life span.

The factors determining the expression of the Wolman disease phenotype versus the CESD phenotype have not been completely elucidated. Studies of *LIPA* gene mutations indicate the presence of one common mutation involved with CESD not typically seen in individuals with Wolman disease. This mutation (called Δ254-277) affects how the *LIPA* gene is put together (spliced) and read by the body and is believed to result in the production of both normal and unstable LAL enzyme. The production of at least some normal LAL enzyme is thought to play a role in the milder phenotype of CESD.

Mutation type, however, may not be the only factor that determines the clinical expression of this condition, as evidenced by the clinical variability among individuals with the common mutation. It is postulated that environmental risk factors and genetic background may also play a role, though the specific factors remain to be determined.

Symptoms

Wolman disease initially manifests in infancy with excessive vomiting and diarrhea. Affected infants typically have markedly distended abdomens due to enlargement of the liver and spleen (hepatosplenomegaly), in addition to progressive anemia and developmental regression. Calcification of the adrenal glands, a characteristic feature, can be observed on certain imaging studies, such as computed tomography (CT) scan. In contrast, hepatosplenomegaly is often the only physical sign of CESD, though affected individuals are at increased risk for premature atherosclerosis. Adrenal calcifications are rarely noted.

Screening and Diagnosis

Aside from looking for adrenal calcifications on abdominal imaging, there is no screening test that suggests a diagnosis of Wolman disease. Though tissue lipid levels are remarkably high, plasma lipids are typically within normal range in Wolman disease. If suspected based on clinical presentation, both Wolman disease and CESD can be diagnosed by assaying LAL enzyme activity in white blood cells, liver biopsy, or fibroblasts. Further confirmation of the diagnosis can be achieved by sequencing of the *LIPA* gene.

Treatment and Therapy

The only treatment currently available for infants diagnosed with Wolman disease is hematopoietic stem cell transplantation (HSCT). Donor stem cells with normal LAL activity are transplanted into an affected individual; if successful, the LAL expressed by the donor cells is enough to constitute normal activity within the host. Few reports are available regarding outcomes of HSCT therapy in Wolman disease. These reports indicate that, among those that survive the HSCT process and for whom the procedure is successful (complete donor cell engraftment), the features of Wolman syndrome are largely ameliorated. Widely used for standard hypercholesterolemia therapy, cholesterol-lowering drugs have been used in the treatment of CESD.

Prevention and Outcomes

Without treatment with HSCT, affected infants typically die within the first year of life. The life span after successful HSCT treatment is unknown, though ongoing survival into early adolescence has been documented. Though widely variable, individuals with CESD can have a normal life span with symptomatic treatment. There is no effective way to prevent Wolman disease or CESD, though prenatal diagnostic options and genetic counseling should be made available to those individuals with a family history of these conditions.

Erin Rooney Riggs, M.S.

Further Reading

Anderson, R., et al. "Lysosomal Acid Lipase Mutations That Determine Phenotype in Wolman Disease and Cholesterol Ester Storage Disease." *Molecular Genetics and Metabolism* 68 (1999): 333-345.

Assman, G., and U. Seedorf. "Acid Lipase Deficiency: Wolman Disease and Cholesteryl Ester Storage Disease." In *Metabolic and Molecular Bases of Inherited Disease*, edited by Charles Scriver et al. 8th ed. New York: McGraw-Hill, 2001.

Tolar, J., et al. "Long-Term Metabolic, Endocrine, and Neuropsychological Outcome of Hematopoietic Cell Transplantation for Wolman Disease." *Bone Marrow Transplantation* 43 (2009): 21-27.

Web Sites of Interest

Genetics Home Reference: Wolman Disease
http://ghr.nlm.nih.gov/condition=wolmandisease

Wolman Disease Support Group
http://www.mdjunction.com/wolman-disease

See also: Fabry disease; Gaucher disease; Gm1-gangliosidosis; Hereditary diseases; Hunter disease; Hurler syndrome; Inborn errors of metabolism; Jansky-Bielschowsky disease; Krabbé disease; Metachromatic leukodystrophy; Niemann-Pick disease; Pompe disease; Sandhoff disease; Tay-Sachs disease.

X

X chromosome inactivation

CATEGORY: Developmental genetics

SIGNIFICANCE: Normal females have two X chromosomes, and normal males have one X chromosome. In order to compensate for the potential problem of doubling of gene products in females, one X chromosome is randomly inactivated in each cell.

KEY TERMS

Barr body: a highly condensed and inactivated X chromosome visible in female cells as a darkly staining spot in a prepared microscope slide

dosage compensation: an equalization of gene products that can occur whenever there are more or fewer genes for specific traits than normal

mosaic: an individual possessing cells with more than one type of genetic constitution

sex chromosomes: the X and Y chromosomes; females possess two X chromosomes, while males possess one X and one Y chromosome

THE HISTORY OF X CHROMOSOME INACTIVATION

In 1961, Mary Lyon hypothesized that gene products were found in equal amounts in males and females because one of the X chromosomes in females became inactivated early in development. This hypothesis became known as the Lyon hypothesis, and the process became known as Lyonization, or X chromosome inactivation. Prior to this explanation, it was recognized that females had two X chromosomes and males had only one X chromosome, yet the proteins encoded by genes on the X chromosomes were found in equal amounts in females and males because of dosage compensation.

The principles of inheritance dictate that individuals receive half of their chromosomes from their fathers and the other half from their mothers at conception. Therefore, a female possesses two different X chromosomes (one from each parent). In addition to hypothesizing the inactivation of one X chromosome in each cell, the Lyon hypothesis also implies that the event occurs randomly. In any individual, approximately one-half of the paternal X chromosomes and one-half of the maternal X chromosomes are inactivated. Thus, females display a mosaic condition since half of their cells express the X chromosome genes inherited from the father and half of their cells express the X chromosome genes inherited from the mother. In fact, this situation can be seen in individuals who inherit an allele for a different form of a protein from each parent: Some cells express one parent's protein form, while other cells express the other parent's protein form.

Prior to Lyon's hypothesis, it was known that a densely staining material could be seen in cells from females that was absent in cells from males. This material was termed a "Barr body," after Murray Barr. Later, it was shown that Barr bodies were synonymous with the inactivated X chromosome. Other observations led scientists to understand that the number of Barr bodies in a cell was always one less than the number of X chromosomes in the cell. For example, one Barr body indicated the presence of two X chromosomes, and two Barr bodies indicated the presence of three X chromosomes.

CLINICAL SIGNIFICANCE

The significance of Barr bodies became apparent with the observation that females lacking one Barr body or possessing more than one Barr body developed an abnormal appearance. Particularly intriguing were females with Turner syndrome. These females possess only one X chromosome per cell, a condition that is not analogous to normal females, who possess only one functional X chromosome per cell as a result of inactivation. The difference in the development of a Turner syndrome female and a normal female lies in the fact that both X chromosomes are active in normal females during the first few days of development. After this period, inactivation occurs randomly in each cell, as hypothesized by Lyon. In cases in which inactivation is not random, individuals may have a variety of developmental problems. Therefore, there is apparently a critical need for both X chromosomes to be active in females in early development for normal development to occur.

It is equally important that there not be more than two X chromosomes present during this early development. Females possessing three X chromosomes, and therefore two Barr bodies, are sometimes called superfemales or metafemales because of a tendency to be taller than average. These females are also two to ten times more likely to suffer from mild to moderate mental retardation.

The same phenomenon has been observed in males who possess Barr bodies. Barr bodies are not normally present in males because they have only one X chromosome. The presence of Barr bodies indicates the existence of an extra X chromosome that has become inactive. Just as in females, extra X chromosomes are also expressed in early development, and abnormal amounts of gene products result in abnormal physical characteristics and mental retardation. Males with Klinefelter syndrome have two X chromosomes and a Y chromosome. In cases in which males have more than two X chromosomes, the effects are even more remarkable.

Mechanism of X Inactivation

While it has been apparent since the 1960's that X inactivation is required for normal female development, the mechanism has been elusive. Only with the development of techniques to study the molecular events of the cell and its chromosomes has progress been made in understanding the process of inactivation. One process involved in turning off a gene (thus "shutting down" the process of transcription) is the alteration of one of the molecules of DNA known as cytosine. When a methyl group is added to the cytosine, the gene cannot produce the RNA necessary to make a protein. It is thought that this methyl group blocks the proteins that normally bind to the DNA, so that transcription cannot occur. When methyl groups are removed from cytosines, the block is removed and transcription begins. This is a common means of regulating transcription of genes. Methylation is significantly higher in the inactivated X chromosome than in the activated X chromosome. As the genes on the chromosome become inactive, the chromosome condenses into the tightly packed mass known as the Barr body. However, the process of methylation alone cannot entirely account for inactivation.

A region on the X chromosome called the X inactivation center (XIC) is considered the control center for X inactivation. In this region is a gene called the X inactivation specific transcripts (*XIST*) gene. At the time of its discovery, this gene was the only gene known to be functional in an inactivated chromosome. It produces an RNA that remains inside the nucleus.

Evidence in humans supports the hypothesis that the *XIST* gene is turned on and begins to make its RNA when the egg is fertilized. Studies with mice have shown that RNA is produced, at first, in low levels and from both X chromosomes. It has been shown in mice, but not humans, that prior to inactivation, Xist (lowercased when referring to mouse genes) RNA is localized at the XIC site only, thus suggesting a potential role prior to actual inactivation of the chromosome. At this point, one X chromosome will begin to increase its production of XIST RNA; shortly thereafter, XIST RNA transcription from the other X chromosome ceases. It is not clear how XIST RNA initiates the process of inactivation and condensing of the inactive chromosome, but XIST RNA binds along the entire length of the inactive X chromosome in females. These results suggest that inactivation spreads from the XIC region toward the end of the chromosome and that XIST RNA is required to maintain an inactive state. If a mouse's *Xist* gene is mutated and cannot produce its RNA, inactivation of that X chromosome is blocked. Other studies have suggested that a product from a nonsex chromosome may interact with the XIC region, causing it to remain active. As expected, but not explained, the *XIST* gene is repressed, or expresses XIST RNA at only very low levels, in males with only one X chromosome.

No difference has been detected between maternally and paternally expressed *XIST* genes in humans. This has led scientists to suspect that *XIST* gene RNA may not be responsible for determining which X chromosome becomes inactivated. It is also not clear how the cell knows how many X chromosomes are present. The search for other candidates for these roles is under way. Finally, there are a few genes besides the *XIST* gene that are also active on the inactive X chromosome. How they escape the inactivation process and why this is necessary are also questions that must be resolved.

Linda R. Adkison, Ph.D.

Further Reading

Bainbridge, David. "The Double Life of Women." In *The X in Sex: How the X Chromosome Controls Our Lives.* Cambridge, Mass.: Harvard University Press, 2003. Provides information about X chromosome inactivation, including descriptions of Barr bodies, mosaicism, and the *XIST* gene.

Erbe, Richard W. "Single-Active-X Principle." *Scientific American Medicine* 2, section 9:IV (1995). Reviews the significance of gene dosage compensation in humans.

Latham, Keith E. "X Chromosome Imprinting and Inactivation in the Early Mammalian Embryo." *Trends in Genetics* 12, no. 4 (April, 1996): 134-138. Discussion of observations on embryos with sex chromosomes from only one parent.

Migeon, Barbara R. *Females Are Mosaics: X Inactivation and Sex Differences in Disease.* New York: Oxford University Press, 2007. Describes the molecular mechanism and the genetic and medical consequences of X chromosome inactivation. Includes a brief chapter on the Lyon hypothesis.

"X in a Cage." *Discover* 15, no. 3 (March, 1994): 14. Summarizes a mechanism for X chromosome inactivation.

Web Sites of Interest

Genetics Home Reference, X Chromosome
http://ghr.nlm.nih.gov/chromosome=X

This user-friendly source of genetics information includes a page on the X chromosome that provides information about X chromosome inactivation, diseases related to the X chromosome, and links to additional resources.

Intersex Society of North America
http://www.isna.org

The society is "devoted to systemic change to end shame, secrecy, and unwanted genital surgeries for people born with an anatomy that someone decided is not standard for male or female." Its Web site includes links to information on such conditions as clitoromegaly, micropenis, hypospadias, ambiguous genitals, early genital surgery, adrenal hyperplasia, Klinefelter syndrome, and androgen insensitivity syndrome.

Johns Hopkins University, Division of Pediatric Endocrinology: Syndromes of Abnormal Sex Differentiation
http://www.hopkinschildrens.org/intersex

Site provides a guide to the science and genetics of sex differentiation, with information about syndromes of sex differentiation.

Kimball's Biology Pages
http://users.rcn.com/jkimball.ma.ultranet/BiologyPages/S/SexChromosomes.html

John Kimball, a retired Harvard University biology professor, includes a page about sex chromosomes, with information about X chromosome inactivation, in his online cell biology text.

Scitable
http://www.nature.com/scitable/topicpage/X-Chromosome-X-Inactivation-323

Scitable, a library of science-related articles compiled by the Nature Publishing Group, contains the article "Chromosome X: X Inactivation," which contains text and illustrations about the *XIST* gene, random X inactivation, and future perspectives for X inactivation research.

See also: Androgen insensitivity syndrome; Fragile X syndrome; Gender identity; Hermaphrodites; Infertility; Klinefelter syndrome; Metafemales; Pseudohermaphrodites; Turner syndrome; XYY syndrome.

Xenotransplants

Category: Genetic engineering and biotechnology

Significance: Xenotransplants are transplants of organs or cellular tissue between different species of animals, such as between pigs and humans.

Key terms

hyperacute rejection: when a transplanted organ or tissue is immediately (within minutes) rejected by the body of the transplant recipient

neural crest cells: embryonic cells that eventually form part of the nervous system and other tissues
stem cells: multipotent cells that can differentiate into a diverse range of specialized cells
transgenic: involving the deliberate incorporation of foreign genes into an animal, such as human genes incorporated into the pig genome
xenosis: the transmission of infectious agents between species

History

The idea of xenotransplants is actually quite old. During the eighteenth century, for example, transfusions of sheep's blood were believed to be therapeutic for certain human illnesses. As the science of organ transplants between humans progressed, researchers became increasingly interested in experimenting with using animals as donors. Organ transplantation became an accepted medical treatment in humans, but there are not enough donor organs from humans available to treat every patient who could benefit from the procedure. In 2006, more than 98,000 people in the United States were waiting for an organ. As the demand for transplant surgery grows, the pool of available donor organs shrinks in relation. One of the ethical dilemmas inherent in human organ transplant is that, in most cases, for one person to receive a transplant, another person must die. Bone marrow, partial liver, and kidney transplants are among the few exceptions; donors can usually donate bone marrow, a lobe of their liver, or one kidney and still survive. Nevertheless, for most organs the dilemma remains.

Researchers are interested in using organs harvested from compatible animals, eliminating both the need to wait for a compatible human donor and the shortage of usable organs. This is one form of xenotransplant, and it could eliminate the shortage of donor organs. However, researchers have encountered two major barriers to xenotransplantation; organ rejection by the human body's immune system and the risk of xenosis, or the transmission of infectious agents, like viruses, between species.

Early research focused on potential donor animals that were similar to humans—that is, primates such as baboons and chimpanzees. Perhaps the most publicized example of such a xenotransplant was the 1984 Baby Fay case, in which doctors in California transplanted a baboon heart into a newborn human infant who otherwise had no chance for survival. The infant survived for several days before succumbing to complications. In 1993, a baboon liver was transplanted to a patient with liver failure. The patient lived for 70 days before dying of an infection. Researchers quickly learned that primates were not ideal candidates for organ donation; not only might the human body reject these organs, but primate organs are small in comparison to those of humans. Chimpanzee kidneys, for example, are too small to perform adequately in an adult human.

A human ear grows on the body of a mouse, engineered in Shanghai, China, in 1997. Human cells were used to grow the ear and then were inserted into the mouse body. (AP/Wide World Photos)

Researchers then turned their attention to pigs as possible donors. Swine make ideal donor candidates because they are physically large enough to have organs that can sustain humans, have a short gestation cycle, produce large litters of offspring, and, because they are routinely raised for meat production, are viewed as expendable by the general public. Pig heart valves are already routinely used in humans, with thousands implanted annually, and pig skin is often grafted onto burns. In a clinical trial, pig pancreas cells were injected into patients with Type I (juvenile) diabetes mellitus and several patients required less insulin afterward. People with liver failure have also been hooked up to pigs' livers to keep them alive until a human liver was available for transplant.

In preliminary trials in humans conducted in the 1990's, neural crest cells (embryonic cells that eventually form part of the nervous system) from fetal pigs were injected into the brains of patients who had experienced a stroke or who had Parkinson's disease. While a few patients had side effects that could have been the result of this treatment, some patients also seemed to get better. Later, another clinical trial was conducted in patients with Parkinson's disease. In this trial, all patients underwent a surgical procedure, but only half of the eighteen patients in the study had pig cells injected into their brains, while the other half did not. Neither the patients nor the neurologists who evaluated the patients for improvement knew who had gotten the cell treatment. When the trial was complete, the results showed that there was no difference between patients who got the cells and those who did not, indicating that the pig cells did not improve Parkinson's disease.

Despite the possible benefit of using pig cells to treat some human diseases, the use of larger organs from pigs has been less successful. Transplant recipients experienced hyperacute rejection; that is, their bodies immediately reacted to the foreign tissue by shutting off the flow of blood to it.

TRANSGENIC PIGS

In 1992, researchers in the United Kingdom announced the creation of Astrid, the first transgenic pig. A transgenic animal is one who has been genetically modified with genes from another species. Since then, several groups have worked to create pigs with organs that have been genetically modified to be human compatible so that they are not rejected by the human immune system. Another goal of this kind of research is to create pigs with organs that are not susceptible to pig viruses.

Animal rights advocates and other activists protest the use of animals as "spare parts" factories for human organs in this demonstration in Munich, Germany, in November, 2000. (AP/Wide World Photos)

ETHICAL AND MEDICAL CONCERNS

Xenotransplantation presents a number of ethical and medical dilemmas. The possibility that a virus, harmless to the donor animal, might be transmitted to the human host and then prove fatal is a major concern. Scientists worry that a potentially deadly disease epidemic could result from using organs or other tissue from either swine or primates. While many researchers are confident that careful screening of donor animals would eliminate or minimize such risks, critics remain convinced that it is possible a virus could lie dormant and undetected in animals, causing problems only after the transplants occurred. Baboons, for example, carry a virus that has the potential to cause cancer in humans.

In addition to the medical issues raised, many bioethicists question the morality of using animals as a source of "spare parts" for humans. They are particularly troubled by the idea of genetically altering a species such as swine in order to make their organs more compatible with human hosts. Proponents of xenotransplants counter these arguments by noting that humans have selectively bred animals for various purposes for thousands of years to eliminate certain characteristics while enhancing others. In addition, animals such as swine are already routinely slaughtered for human consumption.

Finally, there is the problem of human perceptions. While many people support the idea of xenotransplants on the genetic or cellular level, they are less enthusiastic about possible organ transplants. That is, while a majority of people surveyed said they would have no problem accepting a xenotransplant if it were part of gene therapy, far fewer were interested in possibly receiving a pig's heart if the need arose. If researchers do achieve successful xenotransplants using such organs, however, public perceptions could change. It is easy to question a medical procedure when it is still theoretical; it becomes much more difficult to do so after it becomes a reality.

Impact

An immediate and plentiful source of organs from swine could improve and save many human lives. Overcoming the immunological problems and infectious disease risks are serious barriers to xenotransplantation, and remain active areas of research. Creating transgenic pigs with human-compatible organs is a main thrust of research in xenotransplantation. However, human stem cells from unused embryos or adult stem cells is a competing line of research that may offer an alternative source of organs for transplantation, or a means to repair diseased ones.

Nancy Farm Männikkö, Ph.D.; updated by Nancy E. Price, Ph.D.

Further Reading

Bassett, Pamela. *Emerging Markets in Tissue Engineering: Angiogenesis, Soft and Hard Tissue Regeneration, Xenotransplant, Wound Healing, Biomaterials, and Cell Therapy.* Southborough, Mass.: D & MD Reports, 1999. Looks at xenotransplants from an economic perspective and discusses the potential for growth for biomedical firms entering the field.

Bloom, E. T., et al. "Xenotransplantation: The Potential and the Challenges." *Critical Care Nurse* 19 (April, 1999): 76-83. Looks at the potential impact of xenotransplantation on patient care as well as the effects it might have on nursing responsibilities.

Boneva, R. S., and T. M. Folks. "Xenotransplantation and Risks of Zoonotic Infections." *Annals of Medicine* 36 (2004): 504-517. Reviews infectious disease risks associated with xenotransplantation.

Cruz, J., et al. "Ethical Challenges of Xenotransplantation." *Transplant Proceedings* 32 (December, 2000): 2687. Members of a medical transplant team discuss questions about potential moral contradictions in using animal organs in human patients.

Daar, A. S. "Xenotransplants: Proceed with Caution." *Nature* 392 (March 5, 1998): 11. Sounds a warning about some of the potential risks involved in xenotransplants.

Ghebremariam, Y. T., et al. "Intervention Strategies and Agents Mediating the Prevention of Xenorejection." *Annals of the New York Academy of Science* 1056 (2005): 123-143. Discusses different approaches to overcoming rejection of xenotransplanted organs.

Hagelin, J. "Public Opinion Surveys About Xenotransplantation." *Xenotransplantation* 11 (2004): 551-558. Reviews the general public's opinion of xenotransplantation.

Inui, Akio. *Epigenetic Risks of Cloning.* Boca Raton, Fla.: CRC Press, 2006. Leading researchers who are intimately involved with various aspects of cloning present a detailed accounting of cloning methods, an objective review of current findings, and a discussion of potential concerns.

Persson, M. D., et al. "Xenotransplantation Public Perceptions: Rather Cells than Organs." *Xenotransplantation* 10 (2003): 72-79. Describes a public opinion survey done to gauge the public's willingness to support xenotransplant research and implementation.

"U.S. Decides Close Tabs Must Be Kept on Xenotransplants." *Nature* 405 (June 8, 2000): 606-607. Discussion of federal regulations regarding xenotransplant research.

WEB SITES OF INTEREST
eMedicine: Xenotransplantation
http://emedicine.medscape.com/article/432418-overview

PBS Frontline: Organ Farm
http://www.pbs.org/wgbh/pages/frontline/shows/organfarm

See also: Animal cloning; Cancer; Cloning; Cloning: Ethical issues; Gene therapy: Ethical and economic issues; Genetic engineering: Historical development; Heart disease; Immunogenetics; In vitro fertilization and embryo transfer; Model organism: *Mus musculus*; Model organism: *Xenopus laevis*; Organ transplants and HLA genes; Stem cells; Synthetic antibodies; Totipotency; Transgenic organisms.

Xeroderma pigmentosum

CATEGORY: Diseases and syndromes
ALSO KNOWN AS: Melanosis lenticularis progressiva; Kaposi disease; atrophoderma pigmentosum; XP

DEFINITION

Xeroderma pigmentosum is a group of rare skin disorders characterized by extreme sensitivity to sunlight (photosensitivity), with possible ocular and neurologic involvement. An inherited disease, it results from a defect in the genes involved in the repair of DNA damage caused by ultraviolet light.

RISK FACTORS

A child born to parents who each carry a copy of the defective DNA repair gene responsible for this disorder has a 25 percent chance of being affected. The disorder usually manifests at age one to two years and affects both sexes across ethnic groups, although a higher incidence is reported in Japan and the Middle East. Typically, it results from mutations in genes *XPA, ERCC3, XPC, ERCC2, DDB2, ERCC4, ERCC5*, and *ERCC1*—involved in DNA repair—or in the gene *POLH* that encodes an enzyme (a specialized polymerase) that replicates ultraviolet-damaged DNA.

ETIOLOGY AND GENETICS

Xeroderma pigmentosum is an autosomal recessive disease that manifests as hypersensitivity to the sun's ultraviolet rays. Exposure to ultraviolet light induces the formation of pyrimidine dimers by favoring the cross-linking of adjacent pyrimidine bases present in DNA. These structures distort the DNA double helix, thus preventing replication and gene expression. Normally, several distinct enzymes, involved in a process termed "nucleotide excision repair," work in concert to remove the lesion. The resulting gap is then filled in by a DNA polymerase that restores the original base sequence.

Persons with xeroderma pigmentosum carry a mutation in any of the known genes that specify the enzymes that participate in nucleotide excision repair. Several genetic subtypes of the disorder exist. They differ in disease severity and frequency of occurrence and are classified according to the defective repair gene involved. Thus, the classic form of xeroderma pigmentosum, known as XPA, results from a mutation in gene *XPA*. The other subtypes, designated XPB through XPH, are defective in other DNA repair genes and a variant form, known as XPV, is defective in gene *POLH*. Subgroups XPA and XPC encompass about half of xeroderma pigmentosum cases, with very rare cases reported of XPB, XPE, and XPH.

Accumulation of damaged DNA causes progressive degenerative alterations of the skin, thus greatly increasing the likelihood of mutagenesis and skin cancers, such as basal cell carcinoma, squamous cell carcinoma, or melanoma. Classic XPA, which shows the lowest level of DNA repair, is more likely to be associated with progressive neurologic complications. Individuals with subtypes XPB, XPD, XPF, XPG, and XPH show varying degrees of neurologic abnormalities, whereas individuals with subtypes XPC, XPE, and XPV generally do not. Although form XPV is clinically indistinguishable from classic XPA, it is genetically distinct from the other subtypes. In fact, XPV cells have the ability to repair DNA damage induced by ultraviolet light but lack the specialized polymerase that replicates ultraviolet-damaged DNA.

SYMPTOMS

Persons with xeroderma pigmentosum typically experience unusually severe sunburn with skin redness and blistering after only brief exposure to sunlight. The condition is characterized by excessive skin dryness, pigmentary changes, skin atrophy, lesions termed solar keratoses, and a 2,000-fold risk of developing sunlight-induced skin cancer before age twenty. Eye damage from ultraviolet radiation causes photophobia and irritation in about 80 percent of cases and may lead to ocular tumors. Neurologic complications are present in 20 to 30 percent of cases and include poor coordination, deafness, developmental delay, and mental retardation.

Screening and Diagnosis

Clinical diagnosis usually is based on a visual examination of the skin and eyes and on a history of photosensitivity, freckling, and skin tumors in young children. A blood test or skin biopsy can reveal a DNA repair defect. For persons with a family history of xeroderma pigmentosum, prenatal tests, including amniocentesis, are available, as are genetic counseling and genetic testing to determine carrier status for individuals with subtypes XPA and XPC.

Treatment and Therapy

Currently, no treatment for xeroderma pigmentosum exists. Disease management consists of preventing sun-induced damage and ensuring the prompt removal of precancerous and cancerous lesions. Premalignant solar keratoses may be removed by cryotherapy or dermabrasion. More aggressive tumors, such as melanoma, may require radiation and chemotherapy in addition to surgical excision. Retinoid therapy, moisturizers, and vitamin D supplements are recommended. A new topical lotion containing a DNA repair enzyme is reported to reduce the development of solar keratoses and basal cell carcinoma and is currently in clinical trials.

Prevention and Outcomes

Avoidance of all sources of ultraviolet radiation, including fluorescent lights, can improve outcome. Protective clothing, a wide-brimmed hat, ultraviolet-absorbing sunglasses, and a sunscreen with high sun-protection factor are mandatory when outdoors. Medications that increase sensitivity to sunlight should be used with caution and tobacco smoking avoided. Regular checkups by a dermatologist (every three months) and an ophthalmologist (annually) are important for disease surveillance. By taking measures that reduce deadly skin cancer, affected persons with no neurologic complications can achieve a normal life span.

Anna Binda, Ph.D.

Further Reading

Karp, Gerald. "DNA Replication and Repair." In *Cell and Molecular Biology: Concepts and Experiments.* 5th ed. Hoboken, N.J.: John Wiley & Sons, 2008. An illustrated text in cell biology with a relevant perspective on human disorders.

Snustad, D. Peter, and Michael J. Simmons. "Mutation, DNA Repair, and Recombination." In *Principles of Genetics.* 5th ed. Hoboken, N.J.: John Wiley & Sons, 2009. An introductory genetics text particularly suited to undergraduate students.

Webb, Sandra. "A Patient's Journey: Xeroderma Pigmentosum." *British Medical Journal* 336 (2008): 444-446. An article by the founder of the XP Support Group about the search to find reliable care for her son's skin condition.

Web Sites of Interest

Understanding Xeroderma Pigmentosum
http://www.xpfamilysupport.com/docs/UnderstandingXP.pdf

Xeroderma Pigmentosum Society
http://www.xps.org

XP Family Support Group
http://www.xpfamilysupport.org

See also: Albinism; Chediak-Higashi syndrome; Epidermolytic hyperkeratosis; Hermansky-Pudlak syndrome; Ichthyosis; Melanoma; Palmoplantar keratoderma.

XYY syndrome

CATEGORY: Diseases and syndromes

Definition

Males with XYY syndrome bear an extra Y chromosome. The syndrome occurs at fertilization and represents one of several human sex chromosome abnormalities.

Risk Factors

Only males are at risk for XYY syndrome. Most cases of the disorder are not inherited but result from a random event during the formation of sperm cells.

Etiology and Genetics

All normal human cells contain forty-six chromosomes consisting of twenty-three pairs; one member of each pair is contributed by the female parent and one by the male. Of these forty-six chromosomes, two chromosomes, designated X and Y, are known as the sex chromosome pair. Individuals with an XX pair are female, while those with an XY pair are male. Unlike the other twenty-two chromosome pairs, the X and Y chromosomes are strikingly different from each other in both size and function. While the Y chromosome is primarily concerned with maleness, the X chromosome contains information important to both genders.

During formation of sperm and eggs in the testes and ovaries, respectively, a unique form of nuclear division, known as meiosis (or reductional division), occurs during cell division that halves the chromosome number from forty-six to twenty-three. Sperm and eggs thus carry only one member of each pair of chromosomes, and the original number will be restored during fertilization. Because females have an XX pair, their eggs can have only an X chromosome, while males, having an XY pair, produce sperm bearing an X or a Y chromosome.

A common genetic error during sperm or egg production is known as nondisjunction, which is the improper division of chromosomes between the daughter cells. Nondisjunction in the production of either gamete can result, at fertilization, in embryos without the normal forty-six chromosomes. XYY syndrome is one of several of these aneuploid conditions (possession of one or a few more or less than the normal number of chromosomes) that involve the sex chromosome pair. While Klinefelter syndrome (an XXY male) and Turner syndrome (an X female) are more widely studied and recognized genetic diseases, the XYY male occurs with a frequency of 1 in 1,000 male births in the United States. Caused by a YY-bearing sperm fertilizing a normal X-bearing egg, the XYY embryo develops along a seemingly normal route and, unlike most other sex chromosome diseases, is not apparent at birth. The only physical clue is unusually tall stature; otherwise, an affected male will be normal in appearance. The XYY male is also fertile, unlike those with aneuploidies involving other combinations of sex chromosomes, which usually result in sterility.

Symptoms

Boys with XYY syndrome are usually tall, have difficulties with language, and may have slightly lower intelligence quotients (IQs) than other members of their families. They may also develop learning disabilities, hyperactivity, attention deficit disorder, and minor behavioral disorders.

Interest in the association between aggression and the Y chromosome began in the years following World War II. Both psychologists and geneticists began intensive scrutiny of the genes that were located on the male sex chromosome. Men with multiple copies of the Y chromosome thus became the subjects of much of this research. Genetic links to violent, aggressive, and even criminal behavior were found, although many argued that below-average intelligence played a greater role. Many males with XYY syndrome do perform lower than average on standard intelligence tests and have a greater incidence of behavioral problems. The majority, however, lead normal lives and are indistinguishable from XY males.

The controversy surrounding this research began with a study at Harvard University that began in the early 1960's and ended in 1973 because of pressure from both public and scientific communities. The researchers screened all boys born at a Boston hospital, identifying those with sex chromosomal abnormalities. Because the parents of XYY boys were told of their children's genetic makeup and the possibility of lower intelligence and bad behavior, critics claimed that the researchers had biased the parents against their sons, causing the parents to treat the children differently. The environment would thus play a greater role than genetics in their behavior. Subsequent research has shown that the original hypothesis is at least partially accurate. There is a disproportionately large number of XYY males in prison populations, and they are usually of subaverage intelligence compared to other prisoners. It must be emphasized, however, that the majority of XYY males show neither low intelligence nor criminal behavior.

Scientists, doctors, geneticists, and psychologists now agree that the extra Y chromosome does cause above-normal height, reading and math difficulties, and, in some cases, severe acne, but the explanation of the high prevalence of XYY men in prison populations has changed its focus from genes to environment. Large body size during childhood, adolescence, and early adulthood will no doubt cause people to treat these individuals differently, and they may in turn have learned to use their size defensively. Aggressive behavior, coupled with academic difficulties, may lead to further problems. Clearly, however, the majority of XYY males do well. The issue would be much easier to resolve if a YY or Y male existed, but because lack of an X chromosome results in spontaneous miscarriage, no YY or Y male embryo could ever survive.

Screening and Diagnosis

Identification of this disorder after a child is born requires that the child receive genetic testing or screening and is often discovered accidentally as a consequence of results from another genetic test. A pregnant woman's cells can be examined by either amniocentesis or chorionic villus sampling to determine if her child will have XYY syndrome. In amniocentesis, a sample of cells is taken from the amniotic fluid; chorionic villus sampling examines cells in the placenta.

Treatment and Therapy

Men with XYY syndrome may require help in school to deal with learning disabilities and behavioral difficulties. Speech, physical, and occupational therapies can also prove beneficial. The support of family members is another important element in helping men cope with this disorder.

Prevention and Outcomes

There is no cure for XYY syndrome. However, help in school and a supportive family can reduce the educational and behavioral problems associated with the disorder. Most males with XYY syndrome lead normal lives.

Connie Rizzo, M.D., Ph.D.

Further Reading

Chen Harold. "XYY Syndrome." In *Atlas of Genetic Diagnosis and Counseling*. Totowa, N.J.: Humana Press, 2006. Provides an overview of the syndrome, including its genetics, clinical features, diagnostic tests, and genetic counseling issues.

Mader, Sylvia S. *Human Reproductive Biology*. 3d ed. Dubuque, Iowa: McGraw-Hill Higher Education, 2005. Provides an excellent introduction to cell division, genetics, and sex from fertilization through birth. Illustrations, bibliography, index.

Simpson, Joe Leigh, and Sherman Elias. "Sex Chromosomal Polysomies (47,XXY; 47,XYY; 47,XXX), Sex Reversed (46,XX) Males, and Disorders of the Male Reproductive Ducts." In *Genetics in Obstetrics and Gynecology*. 3d ed. Philadelphia: Saunders, 2003. Describes how XYY syndrome and other genetic anomalies arise, how they are identified, and the implications of these genetic disorders for patients.

Tamarin, Robert H. *Principles of Genetics*. Boston: McGraw-Hill, 2002. A well-written reference text on genetics with complete discussions on aneuploidy, the sex chromosomes, genes, and abnormalities. Includes a thorough reading list, illustrations, maps (some color), index.

Vernice, Mirta, and Anna Cremante. "Life Span Development in XYY." In *Life Span Development in Genetic Disorders: Behavioral and Neurobiological Aspects*. New York: Nova Biomedical Books, 2008. Charts how persons with genetic disorders progress through their life span, from infancy through old age. Discusses the diagnosis and treatment of XYY and other genetic disorders.

Web Sites of Interest

Genetics Home Reference: 47,XYY Syndrome
http://ghr.nlm.nih.gov/condition=47xyysyndrome

A fact sheet on the disorder, describing what it is and the genes related to it. Offers links to additional information.

Johns Hopkins University, Division of Pediatric Endocrinology: Syndromes of Abnormal Sex Differentiation
http://www.hopkinschildrens.org/intersex

Site provides a guide to the science and genetics of sex differentiation, with information about syndromes of sex differentiation.

Klinefelter Syndrome and Associates (KS&A), Inc.: XYY Syndrome
http://www.genetic.org/knowledge/support/action/201

KS&A provides information and support to persons who have one or more extra X or Y chromosomes. Its Web site offers access to a range of information about XYY syndrome.

Texas Department of State Health Services: Birth Defects Risk Factor Series
http://www.dshs.state.tx.us/birthdefects/risk/risk26-xyy.shtm
Describes 47,XYY syndrome and provides information about its prenatal diagnosis, prevalence, pregnancy outcome, and demographic and reproductive factors.

See also: Aggression; Androgen insensitivity syndrome; Behavior; Criminality; Intelligence; Klinefelter syndrome; Metafemales; Nondisjunction and aneuploidy; Steroid hormones; X chromosome inactivation.

Z

Zellweger syndrome

CATEGORY: Diseases and syndromes

ALSO KNOWN AS: Zellweger syndrome spectrum (including infantile Refsum disease and neonatal adrenoleukodystrophy); cerebrohepatorenal syndrome; peroxisomal biogenesis disorder

DEFINITION

The Zellweger syndrome spectrum (ZSS) includes a group of disorders caused by a deficiency in peroxisomal biogenesis. Insufficient peroxisomal function results in multisystem organ dysfunction and subsequent death, typically in childhood.

RISK FACTORS

The ZSS disorders are inherited in an autosomal recessive manner. Individuals with mutations in both copies of any one of the twelve different *PEX* genes known to be associated with ZSS disorders are affected. Full siblings of these individuals have a 25 percent risk of being affected. Males and females are affected with equal frequency. Cases of ZSS disorders have been observed worldwide; no particular ethnic preponderance has been noted.

ETIOLOGY AND GENETICS

Peroxisomes are small, membrane-bound organelles found in cells throughout the body. They are involved in a variety of metabolic processes, including the breakdown of toxic substances, such as hydrogen peroxide, and the production of bile acids, necessary for digestion of dietary fats. Peroxisomes are also involved in degradation of very long chain fatty acids (VLCFAs) in a process known as beta-oxidation, essential for energy production. Peroxisomes play a role in the synthesis of certain types of lipids, such as plasmalogens, which are important components of cell membranes in certain tissues (such as the brain).

Proteins necessary for peroxisome function and membrane formation are made within the cell, then transported into existing peroxisomes. Importing these essential proteins prompts growth and division, resulting in the formation of new peroxisomes. In the ZSS disorders, these proteins are unable to be transported into peroxisomes, resulting in decreased biogenesis and function.

ZSS disorders are caused by mutations in the *PEX* genes, which encode these essential proteins, known as peroxins. Though approximately 29 *PEX* genes have been identified, ZSS disorders have been associated with mutations in only 12 (located at the following chromosomal positions): *PEX1* (7q21q22) *PEX2* (8q21.1), *PEX3* (6q23q24), *PEX5* (12p13.3), *PEX6* (6p21.1), *PEX10* (1p36.32), *PEX12* (17q12), *PEX13* (2p15), *PEX14* (1p36.2), *PEX16* (11p12p11.2), *PEX19* (1q22), and *PEX26* (22q11.2). In some cases, the particular gene involved and the type of mutations present are predictive of the ultimate effect on biogenesis and can be associated with clinical severity.

SYMPTOMS

Though once considered separate entities, it is now known that Zellweger syndrome (ZS), neonatal adrenoleukodystrophy (NALD), and infantile Refsum disease (IRD) share a common etiology and represent differing degrees of severity along the Zellweger syndrome spectrum. There is little distinction between these conditions, though historically severely affected infants were characterized as having ZS and those surviving more than one year and making some type of developmental progress were thought to have NALD or IRD.

The ZSS disorder typically appears during childhood; though extremely rare, presentation in adulthood has been reported. Affected neonates often present with severe hypotonia, failure to thrive, seizures, and liver dysfunction. Some are noted to have distinctive facial features and neuronal migration defects. Bony stippling may also be seen on radiographs. Older children may have additional issues such as hearing loss and/or retinal dystrophy. Affected individuals are noted to have decreased levels of docosahexaenoic acid (DHA), thought to be important in brain development, as well as decreased bile acids, important in normal digestion. Liver dysfunction can result in coagulation problems, and renal failure, thought to be due to renal cysts, can also occur. Severely affected individuals usually make no developmental progress; individuals with milder disease have a wide range of intellectual abilities, with most exhibiting developmental delays. Loss of previously acquired skills occurs in some due to progressive leukodystrophy.

Screening and Diagnosis

Analysis of levels of VLCFAs in the plasma is a useful initial screening test. Affected individuals typically have elevated concentrations of C26:0 and C26:1 and elevated C24/C22 and C26/C22 ratios. Measurements of plasmalogens, phytanic and pristanic acids, pipecolic acid, and plasma bile acids are also helpful in diagnosing ZSS and distinguishing this from other peroxisomal disorders. Confirmation of abnormalities noted on blood samples should be performed on cultured fibroblasts. Documentation of causative mutations in one of the *PEX* genes associated with ZSS can also further confirm the diagnosis.

Treatment and Therapy

Treatment for ZSS disorders remains largely symptomatic. Supplementation with bile acids, deficient in ZSS, is currently under investigation. Supplemental DHA has been proposed as a treatment for ZSS, but its clinical effects have not been proven. Liver transplantation has been attempted in few cases, though long-term clinical outcomes are currently unavailable.

Prevention and Outcomes

Most severely affected individuals die before one year of age. Evidence suggests that, of those that survive beyond this point with an apparently stable course, most will survive through early childhood, and survival into young adulthood has been documented. Death in older individuals is typically the result of respiratory complications and/or liver/renal failure.

There is no effective means of prevention for the ZSS disorders. Prenatal testing is typically available for those in which the causative mutations in the index case have been identified or the biochemical abnormalities have been demonstrated on cultured fibroblasts. Individuals with a family history of these disorders should be offered genetic counseling.

Erin Rooney Riggs, M.S.

Further Reading

Poll-The, B., et al. "Peroxisome Biogenesis Disorders with Prolonged Survival: Phenotypic Expression in a Cohort of 31 Patients." *American Journal of Medical Genetics* 126A, no. 4 (2004): 333-338.

Steinberg, S., et al. "Peroxisome Biogenesis Disorders." *Biochimica et Biophysica Acta* 1763 (2006): 1733-1748.

Wanders, R., et al. "Peroxisomal Disorders I: Biochemistry and Genetics of Peroxisome Biogenesis Disorders." *Clinical Genetics* 67, no. 2 (2005): 107-133.

Web Sites of Interest

GeneReviews: Peroxisome Biogenesis Disorders, Zellweger Syndrome Spectrum
http://www.ncbi.nlm.nih.gov/bookshelf/br.fcgi?book=gene&part=pbd

National Institute of Neurological Disorders and Stroke: Zellweger Syndrome
http://www.ninds.nih.gov/disorders/zellweger/zellweger.htm

United Leukodystrophy Foundation
http://www.ulf.org/types/Zellweger.html

See also: Adrenoleukodystrophy; Alexander disease; Alzheimer's disease; Amyotrophic lateral sclerosis; Arnold-Chiari syndrome; Ataxia telangiectasia; Canavan disease; Cerebrotendinous xanthomatosis; Charcot-Marie-Tooth syndrome; Chediak-Higashi syndrome; Dandy-Walker syndrome; Deafness; Epilepsy; Essential tremor; Friedreich ataxia; Huntington's disease; Jansky-Bielschowsky disease; Joubert syndrome; Kennedy disease; Krabbé disease; Leigh syndrome; Leukodystrophy; Limb girdle muscular dystrophy; Maple syrup urine disease; Metachromatic leukodystrophy; Myoclonic epilepsy associated with ragged red fibers (MERRF); Narcolepsy; Nemaline myopathy; Neural tube defects; Neurofibromatosis; Parkinson disease; Pelizaeus-Merzbacher disease; Pendred syndrome; Periodic paralysis syndrome; Prion diseases: Kuru and Creutzfeldt-Jakob syndrome; Refsum disease; Sandhoff disease; Sanfilippo syndrome; Spinal muscular atrophy; Spinocerebellar ataxia; Vanishing white matter disease.

Appendixes

BIOGRAPHICAL DICTIONARY OF IMPORTANT GENETICISTS

Adams, Jerry McKee (1940-): His research focus is on the genetics of haemopoietic cell differentiation and malignancy. He was a pioneer developer, with Professor Suzanne Cory, his wife, of the gene cloning techniques used to clone mammalian genes successfully. He also codiscovered the mechanism by which antibody genes combine to help the body fight infections.

Allis, Charles David (1951-): His research on covalent modifications of histone tails including acetylation, phosphorylation, ubiquitination, and methylation is the basis for the histone code that may be responsible for the epigenetic regulation of genetic information or gene expression (transcription) from the DNA code.

Alport, Cecil A. (1880-1959): English-born geneticist and physician who first identified a genetic disorder named after him, Alport syndrome, in a British family in the 1920's. This condition, the second most common inherited cause of kidney failure, usually affects young men but can also affect women and older people.

Alström, Carl-Henry (1907-1993): Swedish psychiatrist who first described Alström syndrome, an inherited condition characterized by a progressive loss of vision and hearing, dilated cardiomyopathy, obesity, Type II diabetes mellitus, and short stature.

Altman, Sidney (1939-): Won the 1989 Nobel Prize in Chemistry, with Thomas R. Cech. Working independently, Altman and Cech discovered that RNA, like proteins, can act as a catalyst; moreover, they found that when ribosomal RNA participates in translation of messenger RNA (mRNA) and the synthesis of polypeptides, it acts as a catalyst in some steps.

Ames, Bruce (1928-): He is a recipient of the 1998 National Medal of Science award and inventor of the test for determining if a chemical or compound is a mutagen. The test is named after him, the Ames Test. Ames is also known for his research on cancer and aging.

Anfinsen, Christian B. (1916-1995): Won the 1972 Nobel Prize in Chemistry. Anfinsen, studying the three-dimensional structure of the enzyme ribonuclease, proved that its conformation was determined by the sequence of its amino acids and that to construct a complete enzyme molecule no separate structural information was passed on from the DNA in the cell's nucleus.

Arber, Werner (1929-): First to isolate enzymes that modify DNA and enzymes that cut DNA at specific sites. Such restriction enzymes were critical in the developing field of molecular biology. Arber was awarded the 1978 Nobel Prize in Physiology or Medicine.

Aristotle (384-322 B.C.E.): Greek philosopher and scientist. Aristotle's *De Generatione* was devoted in part to his theories on heredity. Aristotle believed that the semen of the male contributes a form-giving principle (*eidos*), while the menstrual blood of the female is shaped by the *eidos*. The philosophy implied it was the father only who supplied form to the offspring.

Attardi, Giuseppe (1923-2008): Italian American molecular biologist who was awarded the 1998 Gairdner Foundation International Award for his pioneering contribution to the understanding of the structure of the human mitochondrial genome and its role in human disease.

Auerbach, Charlotte (1899-1994): German-born geneticist who fled to England following the rise of the Nazi Party. Demonstrated that the mutations produced by mustard gases and other chemicals in *Drosophila* (fruit flies) were similar to those induced by X rays, suggesting a common mechanism.

Avery, Oswald Theodore (1877-1955): Immunologist and biologist who determined DNA to be the genetic material of cells. Avery's early work involved classification of the pneumococci, the common cause of pneumonia in the elderly. In 1944, he reported that the genetic information in these bacteria is DNA.

Bailey, Catherine Hayes (1921-): By applying methods of selective breeding, she developed new varieties of fruits.

Baltimore, David (1938-): Along with Howard Temin, Baltimore isolated the enzyme RNA-directed DNA polymerase (reverse transcriptase), demonstrating the mechanism by which RNA tumor viruses can integrate their genetic material into the cell chromosome. Baltimore was awarded the 1975 Nobel Prize in Physiology or Medicine.

Barr, Murray Llewellyn (1908-1995): Canadian geneticist who discovered the existence of the Barr (Barr's) body, an inactive X chromosome found in cells from a female. The existence or absence of the body has been used in determining the sex of the individual from whom the cell originated.

Bateson, William (1861-1926): Plant and animal geneticist who popularized the earlier work of Gregor Mendel. In his classic *Mendel's Principles of Heredity* (1909), Bateson introduced much of the modern terminology used in the field of genetics. Bateson suggested the term "genetics" (from the Greek word meaning "descent") to apply to the field of the study of heredity.

Beadle, George Wells (1903-1989): Beadle's studies of the bread mold *Neurospora* demonstrated that the function of a gene is to encode an enzyme. Beadle and Edward Tatum were awarded the 1958 Nobel Prize in Physiology or Medicine for their one gene-one enzyme hypothesis.

Beckwith, Jonathan R. (1935-): Determined role of specific genes in regulating bacterial cell division. During the 1960's, he was among the first to isolate a specific gene. Beckwith is also known as a social activist in his arguments for the use of science for improvement of society.

Bell, Julia (1879-1979): British geneticist who applied statistical analysis in understanding hereditary medical disorders of the nervous system and limbs.

Benacerraf, Baruj (1920-): Won the 1980 Nobel Prize in Physiology or Medicine, with Jean Dausset and George D. Snell. Benacerraf, Dausset, and Snell each explained the genetic components of the major histocompatibility complex (MHC), the key to a person's immune system, and how the system produces antibodies to such a wide variety of foreign molecules and pathogens, such as viruses, fungi, and bacteria.

Berg, Paul (1926-): Developed DNA recombination techniques for insertion of genes in chromosomes. The techniques became an important procedure in understanding gene function and for the field of genetic engineering. Berg was awarded the 1980 Nobel Prize in Chemistry.

Bishop, John Michael (1936-): Determined that oncogenes, genetic information initially isolated from RNA tumor viruses, actually originate in normal host cells. Bishop was awarded the 1989 Nobel Prize in Physiology or Medicine for his discovery.

Bluhm, Agnes (1862-1943): German physician whose controversial theories on improvement of the "Aryan race" through eugenics and fertility selection provided a basis for Nazi race theories. Among other aspects of her theory was the use of enforced sterilization.

Boring, Alice Middleton (1883-1955): Confirmed existing theories of the chromosomal basis of heredity. Her professional career consisted primarily of serving as a biology teacher to students in China between the world wars.

Borlaug, Norman (1914-2009): Winner of the 1970 Nobel Peace Prize. Borlaug was a key figure in the Green Revolution in agriculture. Working as a geneticist and plant physiologist in a joint Mexican-American program, he developed strains of high-yield, short-strawed, disease-resistant wheat. His goal was to increase crop production and alleviate world hunger.

Botstein, David (1942-): Developed methods of localized mutagenesis for understanding the relationship between the structure and function of proteins. His development of a linkage map involving human genes contributed to the progress of the Human Genome Project.

Boyer, Herbert W. (1936-): His isolation of restriction enzymes that produced a staggered cut on the DNA allowed for creation of so-called "sticky ends," which allowed DNA from different sources or species to be spliced together.

Brenner, Sydney (1927-): Molecular geneticist whose observations of mutations in nematodes (long, unsegmented worms) helped in understanding the design of the nervous system. Brenner was among the first to clone specific genes. He was awarded the Nobel Prize in Physiology or Medicine in 2002.

Bridges, Calvin Blackman (1889-1938): American geneticist whose research on the sex-linked traits in the fruit fly led to the hypothesis, which was eventually proved, that the chromosomes contain genes.

Brink, Royal Alexander (1897-1984): Plant geneticist and breeder at the University of Wisconsin-Madison. His effort led to the development of the Wisconsin maize breeding program, and he created new varieties of clover and alfalfa. Brink was elected to the National Academy of Sciences.

Brown, Michael S. (1941-): By studying the role of cell receptors in uptake of lipids from the blood, Brown discovered the genetic defect in humans associated with abnormally high levels of cholesterol. He was awarded the 1985 Nobel Prize in Physiology or Medicine.

Buchwald, Manuel (1940-): Canadian human molecular geneticist who is a part of the team that owns the U.S. patent for "methods of detecting cystic fibrosis gene by nucleic acid hybridization."

Burnet, Frank Macfarlane (1899-1985): Proposed a theory of clonal selection to explain regulation of the immune response. Burnet was awarded the 1960 Nobel Prize in Physiology or Medicine.

Cairns, John (1922-): British virologist whose investigations of rates and mechanisms of DNA replication helped to lay the groundwork in studying the replication process.

Capecchi, Mario Renato (1937-): Molecular geneticist who studied the homeobox genes in the mouse. He was a cowinner of the 2007 Nobel Prize in Physiology or Medicine for the discoveries of principles for introducing specific gene modifications in mice by the use of embryonic stem cells.

Carroll, Christiane Mendrez (1937-1978): French geneticist and paleontologist most noted for her taxonomic interpretations of early reptiles.

Cech, Thomas R. (1947-): Won the 1989 Nobel Prize in Chemistry, with Sidney Altman. Working independently, Cech and Altman discovered that RNA, like proteins, can act as a catalyst; moreover, Cech found that when ribosomal RNA participates in translation of mRNA and the synthesis of polypeptides, it acts as a catalyst in some steps.

Chargaff, Erwin (1905-2002): Determined that the DNA composition in a cell is characteristic of that particular organism. His discovery of base ratios, in which the concentration of adenine is equal to that of thymine, and guanine to that of cytosine, provided an important clue to the structure of DNA.

Church, George (1954-): American molecular geneticist who developed the first commercial genome sequence. He is also the architect of the Human Genome Project and the Personal Genome Project.

Clarke, Sir Cyril Astley (1907-2000): British physician and geneticist and a Fellow of the Royal College of Physicians of London. His pioneering research on Rh disease of the newborn led to the development of prevention techniques.

Cohen, Stanley Norman (1935-): Developed the techniques for transfer of DNA between species, a major factor in the process of genetic engineering.

Collins, Francis Sellers (1950-): In 1989, Collins identified the gene that, when mutated, results in the genetic disease cystic fibrosis. Collins was instrumental in the identification of a number of genes associated with genetic diseases.

Correns, Carl Erich (1864-1935): German botanist who confirmed Gregor Mendel's laws through his own work on the garden pea. Correns was one of several geneticists who rediscovered Mendel's work in the early 1900's.

Crick, Francis Harry Compton (1916-2004): Along with James Watson, Crick determined the double-helix structure of DNA. Crick was awarded the Nobel Prize in Physiology or Medicine in 1962.

Darlington, Cyril (1903-1981): British geneticist who demonstrated changes in chromosomal patterns which occur during meiosis, leading to an understanding of chromosomal distribution during the process. He also described a role played by crossing over, or genetic exchange, in changes of patterns.

Darwin, Charles Robert (1809-1882): Naturalist whose theory of evolution established natural selection as the basis for descent with modification, more commonly referred to as evolution. His classic work on the subject, *On the Origin of Species by Means of Natural Selection* (1859), based on his five-year voyage during the 1830's on the British ship HMS *Beagle*, summarized the studies and observations that initially led to the theory. Darwin's pangenesis theory, first noted in *The Variation of Animals and Plants Under Domestication* (1868), later became the basis for the concept of the gene.

Darwin, Erasmus (1731-1802): British physician, inventor, and writer. In his classic *Zoonomia*, he advanced a theory of the role of the environment on genetic changes in organisms. A similar theory was later developed by Jean-Baptiste Lamarck. Darwin was the grandfather of both Charles Darwin and Francis Galton.

Dausset, Jean (1916-2009): Won the 1980 Nobel Prize in Physiology or Medicine, with Baruj Benacerraf and George D. Snell. Dausset, Benacerraf, and Snell each explained the genetic components of the major histocompatibility complex (MHC), the key to a person's immune system, and how the system produces antibodies to such a wide variety of foreign molecules and pathogens, such as viruses, fungi, and bacteria.

Delbrück, Max (1906-1981): Leading figure in the application of genetics to bacteriophage research, and later, with *Phycomyces*, a fungal organism. His bacteriophage course, taught for decades at Cold Spring Harbor, New York, provided training for a generation of biologists. He was awarded the 1969 Nobel Prize in Physiology or Medicine.

Demerec, Milislav (1895-1966): Croatian-born geneticist who was among the scientists who brought the United States to the forefront of genetics research. Demerec's experiments, based on the genetics of corn, addressed the question of what a gene represents. His work with bacteria included the determination of mechanisms of antibiotic resistance, as well as the existence of operons, closely linked genes which are coordinately regulated. Demerec was director of the biological laboratories in Cold Spring Harbor, New York, for many years among the most important sites of genetic research.

De Vries, Hugo (1848-1935): Dutch botanist whose hypothesis of intracellular pangenesis postulated the existence of pangenes, factors which determined characteristics of a species. De Vries established the concept of mutation as a basis for variation in plants. In 1900, de Vries was one of several scientists who rediscovered Mendel's work.

Dobzhansky, Theodosius (1900-1975): Russian-born American geneticist who established evolutionary genetics as a viable discipline. His book *Genetics and the Origin of Species* (1937) represented the first application of Mendelian theory to Darwinian evolution.

Doherty, Peter C. (1940-   ): Australian geneticist who, with Rolf Zinkernagel, described the mechanism of the cell-mediated immune defense system of the body. They were awarded the 1996 Nobel Prize in Physiology or Medicine for their discoveries.

Dulbecco, Renato (1914-   ): Among the first to study the genetics of tumor viruses. Dulbecco was awarded the 1975 Nobel Prize in Physiology or Medicine.

Ferguson, Margaret Clay (1863-1951): Plant geneticist whose use of *Petunia* as a model helped explain life cycles of various plants. Also noted for her description of the life cycle of pine trees.

Fink, Gerald R. (1940-   ): Isolation of specific mutants in yeast allowed the use of genetics in understanding biochemical mechanisms in that organism.

Fisher, Ronald Aylmer (1890-1962): British biologist whose application of statistics provided a means by which use of small sampling size could be applied to larger interpretations. Fisher's breeding of small animals led to an understanding of genetic dominance. He later applied his work to the study of inheritance of blood types in humans.

Franklin, Rosalind Elsie (1920-1958): British crystallographer whose X-ray diffraction studies helped confirm the double-helix nature of DNA. Franklin's work, along with that of Maurice Wilkins, was instrumental in confirming the structure of DNA as proposed by James Watson and Francis Crick. Franklin's early death precluded her receiving a Nobel Prize for her research.

Galton, Francis (1822-1911): British scientist who was an advocate of eugenics, the belief that human populations could be improved through "breeding" of desired traits. Galton was also the first to observe that fingerprints were unique to the individual.

Garnjobst, Laura Flora (1895-1977): Following her training under Nobel laureate Edward Tatum at both Stanford and Yale, Garnjobst spent her career in the study of genetics of the mold *Neurospora.*

Garrod, Archibald Edward (1857-1936): Applying his work on alkaptonuria, Garrod proposed that some human diseases result from a lack of specific enzymes. His theory of inborn errors of metabolism, published in 1908, established the genetic basis for certain hereditary diseases.

Gartler, Stanley Michael (1923-   ): He described the polymorphic nature of some types of protein and, with Walter Nelson-Rees, discovered the contamination of cell lines by HeLa cells.

Gartner, Carl Friedrich von (1772-1850): German plant biologist and geneticist. Though Gartner did not generalize as to the significance of his work, his results provided the experimental basis for questions later developed by Gregor Mendel and Charles Darwin.

Giblett, Eloise R. (1921-2009): Discoverer of numerous genetic markers useful in defining blood groups and serum proteins. In the 1970's, Giblett discovered that certain immunodeficiency diseases result from the absence of certain enzymes necessary for immune cell development.

Gilbert, Walter (1932-): Developed method of sequencing DNA. With Paul Berg and Frederick Sanger, awarded the 1980 Nobel Prize in Chemistry.

Gilman, Alfred G. (1941-): Discovered the role of "G" proteins in regulating signal transduction in eukaryotic cells. With Martin Rodbell, won the Nobel Prize in Physiology or Medicine for 1994.

Goldschmidt, Richard B. (1878-1958): German-born geneticist who proposed that the chemical makeup of the chromosome determines heredity rather than the quantity of genes. He theorized that large mutations, or "genetic monsters," were important in generation of new species.

Goldstein, Joseph L. (1940-): Won the 1985 Nobel Prize in Physiology or Medicine, with Michael S. Brown. Brown and Goldstein conducted extensive research in the regulation of cholesterol in humans. They showed that in families with a history of high cholesterol, individuals who carry two copies of a mutant gene (homozygotes) have cholesterol levels several times higher than normal and those who have one mutant gene (heterozygotes) have levels about double normal. Their discoveries proved invaluable in managing heart disease and other cholesterol-related ailments.

Griffith, Frederick (1877-1941): British microbiologist who in 1928 reported the existence of a "transforming principle," an unknown substance that could change the genetic properties of bacteria. In 1944, Oswald Avery determined the substance to be DNA, three years after Griffith was killed during the German bombing of London.

Gruhn, Ruth (1907-1988): German geneticist who applied mathematical principles in the breeding of poultry and pigs.

Haeckel, Ernst Heinrich (1834-1919): German zoologist whose writings were instrumental in the dissemination of Charles Darwin's theories. Haeckel's "biogenetic law," since discarded, stated that "ontogeny repeats phylogeny," suggesting that embryonic development mirrors the evolutionary relationship of organisms.

Haldane, John Burdon Sanderson (1892-1964): British physiologist and geneticist who proposed that natural selection, and not mutation per se, was the driving force of evolution. Haldane was the first to determine an accurate rate of mutation for human genes, and he later demonstrated the genetic linkage of hemophilia and color blindness.

Hanafusa, Hidesaburo (1929-): Japanese-born scientist who played a key role in elucidating the role of oncogenes found among the RNA tumor viruses in transforming mammalian cells.

Hanawalt, Philip C. (1931-): In 1963, discovered the existence of a repair mechanism associated with DNA replication in bacteria. He later found a similar mechanism in eukaryotic cells. Hanawalt's later work included development of the technique of site mutagenesis in gene mapping.

Hardy, Godfrey Harold (1877-1947): British mathematician who, along with Wilhelm Weinberg, developed the Hardy-Weinberg law of population genetics. In a 1908 letter to the journal *Science*, Hardy used algebraic principles to confirm Mendel's theories as applied to populations, an issue then currently in dispute.

Hartwell, Leland H. (1939-): Discovered genes that regulate the movement of eukaryotic cells through the cell cycle. With Tim Hunt and Sir Paul Nurse, won the Nobel Prize in Physiology or Medicine in 2001.

Haynes, Robert Hall (1931-1998): Canadian molecular biologist who carried out much of the early work in the understanding of DNA repair mechanisms.

Hershey, Alfred Day (1908-1997): Molecular biologist who played a key role in understanding the replication and genetic structure of viruses. His experiments with Martha Chase confirmed that DNA carried the genetic information in some viruses. Hershey was awarded the 1969 Nobel Prize in Physiology or Medicine.

Herskowitz, Ira (1946-2003): His studies of gene conversion pathways in yeast led to an understanding of gene switching in control of mating types.

Hertwig, Paula (1889-1983): German embryologist who studied the effects of radiation on embryonic development of fish and animals.

Hippocrates (c. 460-377 B.C.E.): Greek physician who proposed the earliest theory of inheritance. Hippocrates believed that "seed material" was carried by body humors to the reproductive organs.

Hogness, David S. (1925-): One of the first to clone a gene from *Drosophila* (fruit flies). His technique of chromosomal "walking" allowed for the isolation of any known gene based on its ability to mutate. Also involved in identification of homeotic genes, genes which regulate development of body parts.

Holley, Robert William (1922-1993): Determined the sequence of nucleotide bases in transfer RNA (tRNA), the molecule that carries amino acids to ribosomes for protein synthesis. Holley's work provided a means for demonstrating the reading of the genetic code. He was awarded the Nobel Prize in Physiology or Medicine in 1968.

Horvitz, H. Robert (1947-): Harvard neurobiologist whose study of cell regulation in the nematode *Caenorhabditis* led to the discovery of genes that regulate cell death during embryonic development. With Sydney Brenner and John Sulston, he was awarded the Nobel Prize in Physiology or Medicine in 2002.

Hunt, R. Timothy (1943-): Discovered the existence and role of proteins called cyclins, which regulate the cell cycle in eukaryotic cells. With Leland Hartwell and Sir Paul Nurse, won the Nobel Prize in Physiology or Medicine in 2001.

Jacob, François (1920-): French geneticist and molecular biologist who, along with Jacques Monod, elucidated a mechanism of gene and enzyme regulation in bacteria. The Jacob-Monod theory of gene regulation became the basis for understanding a wide range of genetic processes; they were awarded the 1965 Nobel Prize in Physiology or Medicine.

Jaenisch, Rudolf (1942-): German geneticist and pioneer of transgenic research including the creation of transgenic mice in the study of cancer and neurological disorders. Rudolf is a leader in the field of therapeutic cloning, and he created the first transgenic mice.

Jeffreys, Sir Alec (1950-): British biochemist who discovered the existence of introns in mammalian genes. His study of the pattern of repeat sequences in DNA was shown to be characteristic of individuals, and became the theoretical basis for DNA fingerprinting and DNA profiles.

Johannsen, Wilhelm L. (1857-1927): Danish botanist who introduced the term "genes," derived from "pangenes," factors suggested by Hugo de Vries to determine hereditary characteristics in plants. Johannsen also introduced the concepts of phenotype and genotype to distinguish between physical and hereditary traits.

Kabat, Elvin Abraham (1914-2000): Pioneer of modern quantitative immunochemistry who described the structure of antibodies as gamma globulins and the structural variation of their combining sites. He was also involved in the characterization of the structures of blood group antigens.

Kenyon, Cynthia (1955-): Discovered the role of specific genes in regulation of cell migration and the aging process in the nematode *Caenorhabditis*, helping to clarify similar processes in more highly evolved eukaryotic organisms.

Kerr, Warwick Estevam (1922-): Brazilian entomologist and geneticist who applied genetics to the sex determination of bees. Through his research, he created Africanized bees by the hybridization of the African bee and the Italian bee and was responsible for the 1957 release of his experimental, Africanized bee queens, often called "killer bees."

Khorana, Har Gobind (1922-): Developed methods for investigating the structure of DNA and deciphering the genetic code. Khorana synthesized the first artificial gene in the 1960's. He was awarded the Nobel Prize in Physiology or Medicine in 1968.

King, Helen Dean (1869-1955): By selective breeding of rodents, developed a method for production of inbred strains of animals for laboratory studies. The methodology was later applied to development of more desirable breeds of horses.

King, Mary-Claire (1946-): American human geneticist who identified breast cancer genes and demonstrated the genetic similarity of the chimpanzees to humans.

Klug, Aaron (1759-1853): Won the 1982 Nobel Prize in Chemistry. Klug used X-ray crystallography to investigate biochemical structures, especially that of viruses. He was able to link the assembly of viral protein subunits with specific sites on viral RNA, which helped in fighting viruses that cause disease in plants and, more basically, in understanding the mechanism of RNA transfer of genetic information. He also determined the structure of transfer RNA (tRNA), which has a shape similar to that of a bent hairpin.

Knight, Thomas Andrew (1759-1853): Plant biologist who first recognized the usefulness of the garden pea for genetic studies because of its distinctive traits. Knight was the first to characterize dominant and recessive traits in the pea, though, unlike Gregor Mendel, he never determined the mathematical relationships among his crosses.

Kölreuter, Josef Gottlieb (1733-1806): Forerunner of Gregor Mendel, Kölreuter demonstrated the sexual nature of plant fertilization, in which characteristics were derived from each member of the parental generation in equivalent amounts.

Kornberg, Arthur (1918-2007): Carried out the first purification of DNA polymerase, the enzyme that replicates DNA. His work on the synthesis of biologically active DNA in a test tube culminated with his being awarded the 1959 Nobel Prize in Physiology or Medicine.

Kossel, Albrecht (1853-1927): Won the 1910 Nobel Prize in Physiology or Medicine. Isolated and described molecular constituents of the cell's nucleus, notably cytosine, thymine, and uracil. These molecules later proved to be constituents of the codons in DNA and RNA. Thus, Kossel's research prepared the way for understanding the biochemistry of genetics.

Lamarck, Jean-Baptiste (1744-1829): French botanist and evolutionist who introduced many of the earliest concepts of inheritance. Lamarck proposed that hereditary changes occur as a result of an organism's needs; his theory of inherited characteristics, since discredited, postulated that organisms transmit acquired characteristics to their offspring.

Leder, Philip (1934-   ): Along with Marshall Nirenberg, identified the genetic code words for amino acids. His later work has involved the transplantation of human oncogenes into mice, for the purpose of studying the effects of such genes in development of cancer.

Lederberg, Joshua (1925-2008): Established the occurrence of sexual reproduction in bacteria. Lederberg demonstrated that genetic manipulation of the DNA during bacterial conjugation could be used to map bacterial genes. He was awarded the 1958 Nobel Prize in Physiology or Medicine.

Lejeune, Jérôme Jean Louis Marie (1926-1994): French pediatrician and geneticist who discovered the link of diseases to chromosome abnormalities and the link between inadequate intake of folic acid in pregnancy and neural tube defects. He also described trisomy, an extra copy of a chromosome (as with chromosome 21, or Down syndrome).

Levene, Phoebus Aaron (1869-1940): American biochemist who determined the components found in DNA and RNA. Levene described the presence of ribose sugar in RNA and of 2′-deoxyribose in DNA, thereby differentiating the two molecules. He also identified the nitrogen bases found in nucleic acid, though he was never able to determine the acid's molecular structure.

Lewis, Edward B. (1918-2004): Through the use of X-ray-induced mutations in *Drosophila* (fruit flies), Lewis was able to discover and map genes that regulate embryonic development. Among Lewis's discoveries was the existence of homeotic genes, genes that regulate development of body parts. Along with Christiane Nüsslein-Volhard and Eric Wieschaus, awarded the Nobel Prize in Physiology or Medicine in 1995.

Linnaeus, Carolus (1707-1778): Swedish naturalist and botanist most noted for establishing the modern method for classification of plants and animals. In his *Philosophia Botanica* (1751; *The Elements of Botany*, 1775), Linnaeus proposed that variations in plants or animals are induced by environments such as soil.

Luria, Salvador E. (1912-1991): Pioneer in understanding replication and genetic structure in viruses. The Luria-Delbrück fluctuation test, developed by Luria and Max Delbrück, demonstrated that genetic mutations precede environmental selection. Luria was awarded the 1969 Nobel Prize in Physiology or Medicine.

Lwoff, André (1902-1994): French biochemist and protozoologist. Lwoff's early work demonstrated that vitamins function as components of living organisms. He is best known for demonstrating that the genetic material of bacteriophage can become part of the host bacterium's DNA, a process known as lysogeny. Lwoff was awarded the 1965 Nobel Prize in Physiology or Medicine.

Lykken, David Thoreson (1928-2006): He was a behavioral geneticist and a principal investigator on the Minnesota Twin Family Study, which seeks to identify the genetic and environmental influences on the development of psychological traits in identical and fraternal twins through a birth record-based registry.

Lyon, Mary Frances (1925-   ): British cytogeneticist who proposed what became known as the Lyon hypothesis, that only a single X chromosome is active in a cell. Any other X chromosomes are observed as Barr bodies.

McClintock, Barbara (1902-1992): Demonstrated the existence in plants of transposable elements, or transposons, genes that "jump" from one place on a chromosome to another. The process was discovered to be widespread in nature. McClintock was awarded the 1983 Nobel Prize in Physiology or Medicine.

Macklin, Madge Thurlow (1893-1962): Developed a method to apply statistical analysis to understanding congenital diseases in human families. Her arguments were used to introduce genetics as a component of the curriculum in medical schools. Her support of eugenics for improvement of humans later made her views controversial.

McKusick, Victor A. (1921-2008): Cataloged and indexed many of the genes responsible for disorders that are passed in Mendelian fashion.

Margulis, Lynn (1938-   ): Developed the endosymbiont theory, which suggests that internal eukaryotic organelles, such as mitochondria and chloroplasts, originated as free-living prokaryotic ancestors. She proposed that free-living bacteria became incorporated in a larger, membrane-bound structure and developed a symbiotic relationship within the larger cell.

Mendel, Johann Gregor (1822-1884): The "father of genetics," Mendel was an Austrian monk whose studies on the transmission of traits in the garden pea established the mathematical basis of inheritance. Mendel's pioneering theories, including such fundamental genetic principles as the law of segregation and the law of independent assortment, were published in 1866 but received scant attention until the beginning of the twentieth century.

Meselson, Matthew Stanley (1930-   ): Demonstrated the nature of DNA replication, in which the two parental DNA strands are separated, each passing into one of the two daughter molecules. Also noted as a social activist.

Meyerowitz, Elliot M. (1951-   ): Discovered roles played by specific genes in differentiation of the plant *Arabidopsis*, as well as genes which regulate flowering in plants.

Miescher, Johann Friedrich (1844-1895): In 1869, Miescher discovered and purified DNA from cell-free nuclei obtained from white blood cells and gave the name "nuclein" to the extract. The substance was later known as nucleic acid.

Mintz, Beatrice (1921-   ): Noted for studies of the role of gene control in differentiation of cells and disease in humans. Developed a mouse model for the understanding of melanoma development in humans.

Monod, Jacques Lucien (1910-1976): French geneticist and molecular biologist who, with François Jacob, demonstrated a method of gene regulation in bacteria that came to be known as the Jacob-Monod model. Jacob and Monod were jointly awarded the 1965 Nobel Prize in Physiology or Medicine.

Moore, Stanford (1913-1982): Won the 1972 Nobel Prize in Chemistry, with William H. Stein. Moore and Stein supplemented Christian Anfinsen's research by identifying the sequence of amino acids in ribonuclease, a clue to the structure of the gene responsible for it.

Morgan, Lilian Vaughan (1870-1952): Discovered the attached X and ring X chromosomes in *Drosophila* (fruit flies). Later contributed to studying the effects of polio vaccines in primates. Married to Thomas Hunt Morgan.

Morgan, Thomas Hunt (1866-1945): Considered the father of modern genetics, an embryologist whose studies of fruit flies (*Drosophila melanogaster*) established the existence of genes on chromosomes. Through his selective breeding of flies, Morgan also established concepts such as gene linkage, sex-linked characteristics, and genetic recombination. Won the 1933 Nobel Prize in Physiology or Medicine.

Müller, Hermann Joseph (1890-1967): Geneticist and colleague of Thomas Hunt Morgan. Muller's experimental work with fruit flies established the gene as the site of mutation. His work with X rays demonstrated a means of artificially introducing mutations into an organism. Won the 1946 Nobel Prize in Physiology or Medicine.

Mullis, Kary Banks (1944-   ): Devised the polymerase chain reaction (PCR), a method for duplicating small quantities of DNA. The PCR procedure became a major tool in research in the fields of genetics and molecular biology. Mullis was awarded the 1993 Nobel Prize in Chemistry.

Nathans, Daniel (1928-1999): Applied the use of restriction enzymes to the study of genetics. Nathans developed the first genetic map of SV40, among the first DNA viruses shown to transform normal cells into cancer. Nathans was awarded the 1978 Nobel Prize in Physiology or Medicine.

Neel, James Van Gundia (1915-2000): Considered to be the father of human genetics. Among his discoveries was the recognition of the genetic basis of sickle-cell disease. He was also noted for his study of the aftereffects of radiation on survivors of the atomic attack on Hiroshima and Nagasaki in World War II. He was the first to propose what was referred to as the thrifty-gene hypothesis, the idea that potentially lethal genes may have been beneficial to the human population earlier in evolution.

Neitz, Jay (1953-): Ophthalmologist and color vision researcher in the field of functional genomics. With his wife Maureen, codiscovered the genes responsible for color blindness and defined examples of nervous system defect for which the DNA can be used to predict the occurrence and the severity of color blindness.

Neitz, Maureen (1957-): Ophthalmologist and color vision researcher who codiscovered the genes responsible for color blindness and defined examples of nervous system defect for which the DNA can be used to predict the occurrence and the severity of color blindness.

Nelson, Oliver Evans, Jr. (1920-2001): During the 1950's, carried out the first structural analysis of a gene in higher plants (corn), at the same time confirming the existence of transposable elements. His later work demonstrated the genetic significance of enzymatic defects in maize.

Neufeld, Elizabeth F. (1928-): French-born biochemist who found that many mucopolysaccharide storage diseases resulted from the absence of certain metabolic enzymes. Her work opened the way for prenatal diagnosis of such diseases.

Nirenberg, Marshall Warren (1927-): Molecular biologist who was among the first to decipher the genetic code. He later demonstrated the process of ribosome binding in protein synthesis and carried out the first cell-free synthesis of protein. Nirenberg was awarded the 1968 Nobel Prize in Physiology or Medicine.

Nurse, Sir Paul M. (1949-): British scientist who discovered the role of chemical modification (phosphorylation) in regulation of the cell cycle. With Tim Hunt and Leland Hartwell, he was awarded the Nobel Prize in Physiology or Medicine in 2001.

Nüsslein-Volhard, Christiane (1942-): German biologist whose genetic studies in *Drosophila* (fruit flies) led to the discovery of genes that regulate body segmentation in the embryo. Along with Edward Lewis and Eric Wieschaus, won the Nobel Prize in Physiology or Medicine in 1995.

Ochoa, Severo (1905-1993): Won the 1959 Nobel Prize in Physiology or Medicine, with Arthur Kornberg. Ochoa and Kornberg isolated enzymes involved in the synthesis of DNA and RNA, representing the first steps in decoding the biochemical instructions preserved in the structure of genes.

Ohno, Susumu (1928-2000): Asian American geneticist and evolutionary biologist. In 1970, he postulated that gene duplication plays an important role in molecular evolution and that the vertebrate genome is a result of duplications of the entire genome. He also proposed that the gene content of the mammalian species has been conserved over species not only in DNA content but also in the genes themselves—that is, nearly all mammalian species have conserved the X chromosome from their primordial X chromosome of a common ancestor (Ohno's law).

Olson, Maynard V. (1943-): Studied base-pair polymorphisms in the human genome and their significance to evolution. In 1987, with David Burke, Olson developed a new type of cloning vector, artificial chromosomes, that filled the need created by the Human Genome Project to clone very large insert DNAs (hundreds of thousands to millions of base pairs in length).

Patau, Klaus (1908-1975): German-born American geneticist who first reported a congenital disorder associated with the presence of an extra copy of chromosome 13 in 1960. The extra chromosome 13 causes numerous physical and mental abnormalities in a condition known as Patau syndrome or Bartholin-Patau syndrome, after Thomas Bartholin, a French physician who described the clinical picture associated with trisomy 13 in 1656.

Pauling, Linus (1901-1994): American chemist who received the Nobel Prize in Chemistry in 1954 for his work on the nature of the chemical bond and the 1962 Nobel Peace Prize for his antinuclear activism. His 1950's investigations of protein structure contributed to the determination of the structure of DNA.

Punnett, Reginald C. (1875-1967): English biologist who collaborated with William Bateson in a series of important breeding experiments that confirmed the principles of Mendelian inheritance. Punnett also introduced the Punnett square, the standard graphical method of depicting hybrid crosses.

Rahbar, Samuel (1929-): Iranian director at the University of Tehran, who became known as the most important immunologist in the Muslim world.

Roberts, Richard J. (1943-): Discovered that genes in eukaryotic cells and animal viruses are often discontinuous, with intervening sequences between segments of genetic material. With Phillip Sharp, Roberts received the Nobel Prize in Physiology or Medicine in 1993.

Rodbell, Martin (1925-1998): Discovered the role of membrane-bound "G" proteins in regulation of signal transduction in eukaryotic cells. With Alfred Gilman, awarded the Nobel Prize in Physiology or Medicine in 1994.

Rowley, Janet (1925-): Cytogeneticist who developed the staining techniques for observation of cell structures. She demonstrated the role of chromosomal translocation as the basis for chronic myeloid leukemia, the first example of translocation as a cause of cancer.

Rubin, Gerald M. (1950-): Major figure in developing a structure/functional relationship of genes in *Drosophila* (fruit flies) through the use of insertion mutagenesis to inactivate specific genes.

Russell, Elizabeth Shull (1913-2001): Contributed to the understanding of the role played by specific genes in creating coat variations in animals. Her later work involved the identification of genetic defects in the aging process and in the development of diseases such as muscular dystrophy.

Russell, William (1910-2003): Pioneer in the genetic effects of radiation at Oak Ridge National Laboratory whose testing of mice led to standards for acceptable levels of human exposure to radiation. Winner of the 1976 Fermi Award.

Sager, Ruth (1918-1997): During the 1950's, demonstrated the existence of nonchromosomal heredity, also known as cytoplasmic inheritence, and hence the role of cytoplasmic genes in organelle development. Later involved in study of tumor suppressor and breast cancer genes.

Sageret, Augustin (1763-1851): French botanist who discovered the ability of different traits to segregate independently in plants.

Sanger, Frederick (1918-): Determined the method for sequencing DNA. His method separated the strands of DNA and then rebuilt them in stages that allowed the terminal nucleotides to be identified. This made it possible to sequence the entire genomes of organisms. With Paul Berg and Walter Gilbert, Sanger received the 1980 Nobel Prize in Chemistry.

Sharp, Phillip A. (1944-): Discovered that genes in eukaryotic cells or animal viruses are discontinuous, with segments divided by sections separated by intervening sequences of genetic material. With Richard Roberts, received the Nobel Prize in Physiology or Medicine in 1993.

Simpson, George Gaylord (1902-1984): American paleontologist who applied population genetics to the study of the evolution of animals. Simpson was instrumental in establishing a neo-Darwinian theory of evolution (the rejection of Jean-Baptiste Lamarck's inheritance of acquired characteristics) during the early twentieth century.

Singer, Maxine (1931-): Applied the use of the newly discovered restriction enzymes in formation of recombinant DNA. Singer is most noted as a "voice of calm" in the debate over genetic research, emphasizing the application of such research and the self-policing of scientists carrying out such work.

Smith, Hamilton Othanel (1931-): Pioneered the purification of restriction enzymes, winning the 1978 Nobel Prize in Physiology or Medicine, with Werner Arber and Daniel Nathans. Arber and the team of Nathans and Smith separately described the restriction-modification system by studying bacteria and bacteriophages; the system involves the action of site-specific endonuclease and other enzymes that cleave DNA into segments.

Smith, Michael (1932-): Won the 1993 Nobel Prize in Chemistry. Smith developed site-directed mutagenesis, a means for reconfiguring genes in order to create altered proteins with distinct properties. Smith's genetic engineering tool made it possible to treat genetic disease and cancer and to create novel plant strains.

Snell, George D. (1903-1996): Snell's discovery of the H-2 histocompatibility complex, which regulates the immune response in mice, led to the later discovery of the equivalent HLA complex in humans. Awarded the Nobel Prize in Physiology or Medicine in 1980.

Sonneborn, Tracy Morton (1905-1981): Discovered crossbreeding and mating types in paramecia, integrating the genetic principles as applied to multicellular organisms with single-celled organisms such as protozoa.

Spector, David L. (1952-): Through research on live cell imaging of gene expression and noncoding RNAs, developed a live-cell gene expression system to visualize a stably integrated regulatable genetic locus and follow in real-time transcription of that locus, including visualization of its mRNA and protein products in living cells.

Spemann, Hans (1869-1941): Won the 1935 Nobel Prize in Physiology or Medicine. By transplanting bits of one embryo into a second, viable embryo, Spemann compiled evidence that an "organizer center" directs the development of an embryo and that different parts of the organizer governed distinct portions of the embryo. His experiments provided clues to the genetic control of growth from the earliest stages of an organism.

Spencer, Herbert (1820-1903): English philosopher influenced by the work of Charles Darwin. Spencer proposed the first general theory of inheritance, postulating the existence of self-replicating units within the individual which determine the traits. Spencer is more popularly known as the source of the notion of "survival of the fittest" as applied to natural selection.

Stanley, Wendell Meredith (1904-1971): American biochemist who was the first to crystallize a virus (tobacco mosaic virus), demonstrating its protein nature. Stanley was later a member of the team that determined the amino acid sequence of the TMV protein. Stanley spent the last years of his long career studying the relationship of viruses and cancer.

Stein, William H. (1911-1980): Won the 1972 Nobel Prize in Chemistry, with Stanford Moore. Stein and Moore supplemented Christian Anfinsen's research by identifying the sequence of amino acids in ribonuclease, a clue to the structure of the gene responsible for it.

Stevens, Nettie Maria (1861-1912): Discovered the existence of the specific chromosomes that determine sex, now known as the X and Y chromosomes. Described the existence of chromosomes as paired structures within the cell.

Strobell, Ella Church (1862-1920): Developed the technique of photomicroscopy for analysis of chromosomal theory.

Sturtevant, Alfred Henry (1891-1970): Colleague of Thomas Hunt Morgan and among the pioneers in application of the fruit fly (*Drosophila*) in the study of genetics. In 1913, Sturtevant constructed the first genetic map of a fruit fly chromosome. His work became a major factor in chromosome theory. In the 1930's, his work with George Beadle led to important observations of meiosis.

Sulston, John E. (1942-): Developed the first map of cell lineages in the model nematode *Caenorhabditis*, leading to the discovery of the first gene associated with programmed cell death. Sulston was also part of the team that sequenced the worm's genome. With Sydney Brenner and H. Robert Horvitz, he was awarded the Nobel Prize in Physiology or Medicine in 2002.

Sutton, Walter Stanborough (1877-1916): Biologist and geneticist who demonstrated the role of chromosomes during meiosis in gametes, and demonstrated their relationship to Gregor Mendel's laws. Sutton observed that chromosomes form homologous pairs during meiosis, with one member of each pair appearing in gametes. The particular member of each pair was subject to Mendel's law of independent assortment.

Tammes, Jantine (1871-1947): Dutch geneticist who demonstrated that the inheritance of continuous characters, traits that have a range of expression, could be explained in a Mendelian fashion. She developed a multiple allele hypothesis that helped explain some of the data.

Tan Jiazhen (C. C. Tan; 1909-2008): Considered the father of Chinese genetics. In a career spanning more than seven decades, Tan studied genetic structure and variation in a wide range of organisms. His most important work involved the study of evolution of genetic structures in *Drosophila* (fruit flies), as well as the concept of mosaic dominance in the beetle.

Tatum, Edward Lawrie (1909-1975): Along with George Beadle, Tatum demonstrated that the function of a gene is to encode an enzyme. Beadle and Tatum were awarded the 1958 Nobel Prize in Physiology or Medicine for their one gene-one enzyme hypothesis.

Temin, Howard Martin (1934-1994): Proposed that RNA tumor viruses replicate by means of a DNA intermediate. Temin's theory, initially discounted, became instrumental in revealing the process of infection and replication by such viruses. He later isolated the replicating enzyme, the RNA-directed DNA polymerase (reverse transcriptase). He was awarded the 1975 Nobel Prize in Physiology or Medicine, along with David Baltimore, for this work.

Todd, Alexander Robertus (1907-1997): Won the 1957 Nobel Prize in Chemistry. As part of wide-ranging research in organic chemistry, Todd revealed how ribose and deoxyribose bond to the nitrogenous bases on one side of a nucleotide unit and to the phosphate group on the other side. These discoveries provided necessary background for work by others that explained the structure of the DNA molecule.

Tonegawa, Susumu (1939-): Discovered the role of genetic rearrangement of DNA in lymphocytes, which plays a key role in generation of antibody diversity. In 1987, awarded the Nobel Prize in Physiology or Medicine.

Varmus, Harold Elliot (1939-): Elucidated the molecular mechanisms by which retroviruses (RNA tumor viruses) transform cells. Varmus was awarded the 1989 Nobel Prize in Physiology or Medicine.

Waelsch, Salome Gluecksohn (1907-2007): Studied the role genes play in abnormal cell differentiation and congenital abnormalities. Born in Germany, she fled to the United States from the Nazis in 1933. Her 1938 publication of the role of genes in the T (tailless) phenotype in mice is considered a genetic classic.

Waterston, Robert (1943-): Identified many of the genes that regulate muscle development in the nematode *Caenorhabditis*, as well as contributing to the sequence of the genome. His sequencing work was also applied in the Human Genome Project led by Francis Collins.

Watson, James Dewey (1928-): Along with Francis Crick, determined the double-helix structure of DNA. Together with Crick and Maurice Wilkins, Watson was awarded the 1962 Nobel Prize in Physiology or Medicine for their work in determining the structure of DNA.

Weinberg, Robert Allan (1942-): Molecular biologist who isolated the first human oncogene, the *ras* gene, associated with a variety of cancers, including those of the colon and brain. Weinberg later isolated the first tumor-suppressor gene, the retinoblastoma gene. Weinberg is considered one of the leading researchers in understanding the role played by oncogenes in development of cancer.

Weinberg, Wilhelm (1862-1937): German obstetrician who demonstrated that hereditary characteristics of humans such as multiple births and genetic diseases were subject to Mendel's laws of heredity. The mathematical application of such characteristics, published simultaneously (and independently) by Godfrey Hardy, became known as the Hardy-Weinberg equilibrium. The equation demonstrates that dominant genes do not replace recessive genes in a population; gene frequencies would not change from one generation to the next if certain criteria such as random mating and lack of natural selection were met.

Weismann, August (1834-1914): German zoologist noted for his chromosome theory of heredity. Weismann proposed that the source of heredity is in the nucleus only and that inheritance is based on transmission of a chemical or molecular substance from one generation to the next. Weismann's theory, which rejected the inheritance of acquired characteristics, came to be called neo-Darwinism. Though portions of Weismann's theory were later disproved, the nature of the chromosome was subsequently demonstrated by Thomas Hunt Morgan and his colleagues.

Wieschaus, Eric F. (1947-): Wieschaus's studies of genetic control in *Drosophila* (fruit flies) led to the discovery of genes that regulate cell patterns and shape in the embryo. Along with Edward Lewis and Christiane Nüsslein-Volhard, he was awarded the Nobel Prize in Physiology or Medicine in 1995.

Wilkins, Maurice Hugh Frederick (1916-2004): Studies on the X-ray diffraction patterns exhibited by DNA confirmed the double-helix structure of the molecule. Wilkins was a colleague of Rosalind Franklin, and it was their work that confirmed the nature of DNA as proposed by James Watson and Francis Crick. Wilkins was awarded the Nobel Prize for Physiology and Medicine in 1962, along with Watson and Crick.

Wilmut, Ian (1944-): Scottish embryologist and leader of a research team at the Roslin Institute near Edinburgh. In 1996, Wilmut and his colleagues succeeded in cloning an adult sheep, Dolly, the first adult mammal to be successfully produced by cloning.

Wilson, Edmund Beecher (1856-1939): His study of chromosomes in collaboration with Nettie Stevens led to the discovery of the X and Y chromosomes, playing a key role in the foundation of modern genetics. His later work involved the study of development and differentiation of the fertilized egg.

Witkin, Evelyn Maisel (1921-): Through her studies of induced or spontaneous mutations in bacterial DNA, discovered processes of enzymatic repair of DNA.

Woese, Carl R. (1928-   ): Based on his studies of ribosomal RNA differences in prokaryotes and eukaryotes, proposed that all life-forms exist in one of three domains: Bacteria, Archaea ("ancient" bacteria), and Eukarya (eukaryotic organisms, from microscopic plants to large animals). Woese expanded his theory in arguing that the Archaea represent the earliest form of life on Earth and that they later formed a branch which became the eukaryotes.

Wright, Sewall (1889-1988): Discovered genetic drift of genetic traits. The "Sewall Wright" effect, the random drift of characters in small populations, was explained by the random loss of genes, even in the absence of natural selection.

Yanofsky, Charles (1925-   ): Confirmed that the genetic code involved groups of three bases by demonstrating colinearity of the bases and amino acid sequences. He applied this work in demonstrating similar colinearity of mutations in the tryptophan operon and changes in amino acid sequences.

Zinder, Norton (1928-   ): With Joshua Lederberg, discovered the role of bacteriophage in transduction, the movement of genes from one host to another by means of viruses. Zinder was also noted for his discovery of RNA bacteriophage and his work on the molecular genetics of such agents.

Zinkernagel, Rolf Martin (1944-   ): Swiss geneticist who, with Peter C. Doherty, described the mechanism of the cell-mediated immune defense system of the body. They were awarded the 1996 Nobel Prize in Physiology or Medicine for their discoveries concerning the specificity of the cell-mediated immune defense.

Richard Adler, Ph.D.;
updated by Oladayo Oyelola, Ph.D., S.C. (ASCP)

NOBEL PRIZES FOR DISCOVERIES IN GENETICS

Physiology or Medicine

1910

Albrecht Kossel (German). Isolated and described molecular constituents of the cell's nucleus, notably cytosine, thymine, and uracil. These molecules later proved to be constituents of the codons in deoxyribonucleic acid (DNA) and ribonucleic acid (RNA). Thus, Kossel's research prepared the way for understanding the biochemistry of genetics.

1933

Thomas Hunt Morgan (American). Experimenting with the fruit fly *Drosophila melanogaster*, Morgan discovered that the mechanism for the Mendelian laws of heredity lies in the chromosomes inside the nucleus of cells and that specific genes on the chromosomes govern specific somatic traits in the flies. He confirmed the accuracy of Gregor Mendel's laws and ended a controversy over their physiological source.

1935

Hans Spemann (German). By transplanting bits of one embryo into a second, viable embryo, Spemann compiled evidence that an "organizer center" directs the development of an embryo and that different parts of the organizer governed distinct portions of the embryo. His experiments provided clues to the genetic control of growth from the earliest stages of an organism.

1946

Hermann J. Muller (American). Muller proved that X rays damage genes by altering their structure: radiation-induced mutation. Consequently, X rays also modify the structure of chromosomes. The mutations most often produce recessive and harmful traits in the irradiated organism.

1958

George Beadle and Edward Tatum (both American). In research on the fungus *Neurospora crassa*, Beadle and Tatum found that biotin was essential to cultivating certain mutant strains of the fungus; this fact demonstrated that genes regulate the synthesis of specific cellular chemicals, one or more of these genes being mutated in the biotin-dependent strain.

1958

Joshua Lederberg (American). Lederberg showed that the bacterium *Escherichia coli*, although not able to reproduce sexually, is capable of genetic recombination between chromosomes from different cells through a process called conjugation.

1959

Severo Ochoa (Spanish) and Arthur Kornberg (American). Ochoa and Kornberg isolated enzymes involved in the synthesis of DNA and RNA, representing the first steps in decoding the biochemical instructions preserved in the structure of genes.

1960

Frank Macfarlane Burnet (Australian) and Peter Brian Medawar (British). Burnet proposed a theory of clonal selection, confirmed by Medawar, to explain regulation of the immune response.

1962

Francis Crick (British), James Watson (American), and Maurice Wilkins (British). Using X-ray diffraction analysis and molecular modeling, Wilkins, Crick, and Watson found that DNA is structured in a double helix. They were able to identify the specific three-dimensional structure that is the basis for the ability of DNA to be replicated and transcribed.

1965

François Jacob and Jacques Monod (both French). Studying enzyme action, Jacob and Monod proved that messenger ribonucleic acid (mRNA) carries instructions from the nucleus to ribosomes, where molecules are assembled for use in the body, and they distinguished structural genes from regulatory genes.

1965

André Lwoff (French). Lwoff proposed that viral DNA can become active after invading cells and cause the cells to divide out of control, producing cancerous tumors.

1968

Robert W. Holley (American), Har Gobind Khorana (Indian), and Marshall W. Nirenberg (American). Working separately, Holley, Khorana, and Nirenberg deciphered the genetic code in RNA and DNA. Their work anticipated DNA sequencing and genetic engineering.

1969

Max Delbrück (German), Alfred D. Hershey (American), and Salvador E. Luria (Italian). In joint studies of bacteriophages and their bacterial hosts, Delbrück, Hershey, and Luria described the conformation of bacteriophage DNA, showed that different strains exchange genetic information, and proved that bacterial DNA mutated to confer protection from attack, demonstrating that bacterial heredity is based on genetic exchange. The discovery explained why bacteria gradually become resistant to pharmaceuticals.

1975

David Baltimore and Howard M. Temin (both American). Working separately, Baltimore and Temin discovered reverse transcriptase, the enzyme that inserts viral DNA into cellular DNA, which can cause cancer, They also identified retroviruses, a class of virus that includes the human immunodeficiency virus (HIV) that causes acquired immunodeficiency syndrome (AIDS).

1975

Renato Dulbecco (Italian). Dulbecco described how tumor viruses cause cellular transformation in somatic cells by suppressing the regulatory system that controls division; the cells then divide out of control.

1978

Werner Arber (Swiss), Daniel Nathans (American), and Hamilton O. Smith (American). Arber and the team of Nathans and Smith separately described the restriction-modification system by studying bacteria and bacteriophages; the system involves the action of site-specific endonuclease and other enzymes that cleave DNA into segments.

1980

Baruj Benacerraf (Venezuelan), Jean Dausset (French), and George D. Snell (American). Benacerraf, Dausset, and Snell each explained the genetic components of the major histocompatibility complex (MHC), the key to a person's immune system, and how the system produces antibodies to such a wide variety of foreign molecules and pathogens, such as viruses, fungi, and bacteria.

1983

Barbara McClintock (American). McClintock investigated the genetics of maize (corn) and discovered a new mechanism of gene modification: Some "jumping genes" (now called transposable elements or transposons) move to new sites on chromosomes and either suppress nearby structural genes or inactivate suppressor genes. The discovery was a major breakthrough in understanding novel, non-Mendelian types of genetic variation.

1985

Michael S. Brown and Joseph L. Goldstein (both American). Brown and Goldstein conducted extensive research in the regulation of cholesterol in humans. They showed that in families with a history of high cholesterol, individuals who carry two copies of a mutant gene (homozygotes) have cholesterol levels several times higher than normal and those who have one mutant gene (heterozygotes) have levels about double normal. Their discoveries proved invaluable in managing heart disease and other cholesterol-related ailments.

1987

Susumu Tonegawa (Japanese). Tonegawa explained the diversity of antibodies by showing that the antigen-sensitive part of each antibody is created by segments of three genes; since the segments from each gene can vary in length, the possible combinations from three genes can produce billions of distinct antibodies.

1989

J. Michael Bishop and Harold E. Varmus (both American). Bishop and Varmus discovered that oncogenes (genes that play a role in initiating cancer) originate in normal cells and control cellular growth and are not solely derived from retroviruses, as previously thought. Their work greatly influenced subsequent studies of tumor development.

1993

Richard J. Roberts (British) and Phillip A. Sharp (American). Roberts and Sharp separately studied the relationship between DNA and RNA. They discovered that portions of a human gene can be divided among several DNA segments, called introns, separated by noncoding segments called exons. This discovery became important to genetic engineering and to understanding the mechanism for hereditary diseases.

1994

Alfred G. Gilman and Martin Rodbell (both American). Gilman and Rodbell discovered the role of "G" proteins in regulating signal transduction in eukaryotic cells.

1995

Edward B. Lewis (American). Lewis found that an array of master genes governs embryo development.

1995

Christiane Nüsslein-Volhard (German) and Eric F. Wieschaus (American). Nüsslein-Volhard and Wieschaus worked together to extend Lewis's investigations into the genetic control of embryo development through studies of fruit flies. They isolated more than five thousand participating genes and distinguished four types of "master" control genes: gap, pair-rule, segment polarity, and even-skipped.

1996

Peter C. Doherty (Australian) and Rolf Zinkernagel (Swiss). Doherty and Zinkernagel described the mechanism of the cell-mediated immune defense.

2001

Leland H. Hartwell (American), R. Timothy Hunt (British), and Paul M. Nurse (British). Hartwell, Hunt, and Nurse conducted research on the regulation of cell cycles. Hartwell identified a class of genes that controls the cycle, including a gene that initiates it. Nurse cloned and described the genetic model of a key regulator, cyclin-dependent kinase, while Hunt discovered cyclins, a class of regulatory proteins.

2002

Sydney Brenner (American). Brenner used the transparent nematode *Caenorhabditis elegans* to establish a simple model organism for studying how genes control the development of organs.

2002

John E. Sulston (British). Sulston studied cell division and cell lineages in *Caenorhabditis elegans* following Brenner's methods. He demonstrated that genetic control of specific lineages includes programmed cell death, called apoptosis, as part of the regulatory process, and he isolated the protein that degrades the DNA of dead cells.

2002

H. Robert Horvitz (American). Using Brenner's *Caenorhabditis elegans* model, Horvitz discovered the first two "death genes" which instigate cell death. He further found that another gene helps protect cells from cell death.

2004

Richard Axel and Linda B. Buck (both American). Working as collaborators in Axel's laboratory, the two researchers discovered the extensive gene family which encodes odorant receptors in rats. They initially determined that more than one hundred related gene products are involved in detection of odors. Ultimately, Axel and Buck were able to identify approximately 1,000 different genes, constituting nearly 3 percent of the total genetic material in the cell, which encode odor receptors on the surface of cells in the nasal epithelium.

2005

Barry J. Marshall and J. Robin Warren (both Australian). Historically, peptic ulcers were successfully treated using standard procedures that inhibited acid production but routinely returned in more than 50 percent of patients. Warren, a pathologist, had observed the presence of an unusual curve-shaped bacterium in the stomach of these patients. Marshall began a clinical collaboration with Warren in which they demonstrated the relationship between a newly discovered bacterium, *Helicobacter pylori*, and inflammation leading to ulcers. Infection may lead to stomach cancer as well. Warren and Marshall developed ulcer treatment that included the use of antibiotics to eliminate the infectious agents. For most patients, antibiotics prevented recurrence. Because *Helicobacter* is found in some 50 percent of humans and not all persons develop ulcers, it is believed that genetic factors regulate the response to colonization by these bacteria.

2006

Andrew Z. Fire and Craig C. Mello (both American). Earlier work by researchers had demonstrated that gene expression could be silenced in cells following the insertion of RNA sequences corresponding to messenger RNA (mRNA) transcribed from genes. Mello termed this process RNA interference (RNAi). The mechanism was unknown. Fire and Mello discovered that the process utilized double-stranded RNA (dsRNA), which caused the degradation of mRNA and resulted in what they called post-translational gene silencing. Using the nematode *Caenorhabditis elegans* as a test organism, they observed that injection of specific dsRNA into cells was followed by cleavage of the double-stranded RNA into short small interfering RNA (siRNA) fragments, which led to the degradation of the mRNA. The process was subsequently shown to be a common method of gene regulation in both plants and animals.

2007

Mario Capecchi (American), Sir Martin Evans (British), and Oliver Smithies (American). Working independently, the laureates developed the methodology used for isolating and growing embryonic stem cells and for introducing genetic changes within those cells. Evans isolated cells from the blastocyst obtained from pregnant mice, subsequently growing those cells in laboratory cultures. Capecchi and Smithies developed a mechanism to insert specific genes into such embryonic cells, creating genetically modified mice. Their work led to the use of "knockout" mice, genetically modified animals in which specific sites on DNA could be inactivated in order to better understand the function of those genes.

2008

Harald zur Hausen (German). Zur Hausen demonstrated that certain variants of human papillomaviruses (HPV) were the etiological agents of cervical cancer and confirmed his suspicions by isolating HPV DNA from cervical cancer tissue. His work ultimately led to a better understanding of how cell division is regulated, as well as development of HPV vaccines.

2008

Françoise Barre-Sinoussi and Luc Montagnier (both French). Barre-Sinoussi and Montagnier isolated HIV, the etiological agent of AIDS. Their work led to methods for the detection of the virus in blood and lymph nodes, as well as a better understanding of how viral replication results in depression of the immune system.

CHEMISTRY

1957

Alexander Robertus Todd (British). As part of wide-ranging research in organic chemistry, Todd revealed how ribose and deoxyribose bond to the nitrogenous bases on one side of a nucleotide unit and to the phosphate group on the other side. These discoveries provided necessary background for work by others that explained the structure of the DNA molecule.

1972

Christian B. Anfinsen (American). Anfinsen, studying the three-dimensional structure of the enzyme ribonuclease, proved that its conformation was determined by the sequence of its amino acids and that to construct a complete enzyme molecule, no separate structural information was passed on from the DNA in the cell's nucleus.

1972

Stanford Moore and William H. Stein (both American). Moore and Stein supplemented Anfinsen's research by identifying the sequence of amino acids in ribonuclease, a clue to the structure of the gene responsible for it.

1980

Paul Berg (American). Berg invented procedures for removing a gene from a chromosome of one species and inserting it into the chromosome of an entirely different species, enabling him to study how the genetic information of the contributing organism interacts with the host's DNA. The recombinant DNA technology, sometimes called gene splicing, became fundamental to the genetic engineering of transgenic species.

1980

Walter Gilbert (American) and Frederick Sanger (British). Gilbert and Sanger independently developed methods for determining the sequence of nucleic acids in DNA, thus decoding the genetic information. Gilbert's method cuts DNA into small units that reveal their structure when exposed to specific chemicals; Sanger's method separates the strands of DNA and then rebuilds them in stages that allow the terminal nucleotides to be identified. Their methods later made it possible to sequence the entire genomes of organisms.

1982

Aaron Klug (British). Klug used X-ray crystallography to investigate biochemical structures, especially those of viruses. He was able to link the assembly of viral protein subunits with specific sites on viral RNA, which helped in fighting viruses that cause disease in plants and, more basically, in understanding the mechanism of RNA transfer of genetic information. He also determined the structure of transfer RNA (tRNA), which has a shape similar to that of a bent hairpin.

1989

Sidney Altman (Canadian) and Thomas R. Cech (American). Working independently, Altman and Cech discovered that RNA, like proteins, can act as a catalyst; moreover, Cech found that when ribosomal RNA participates in translation of mRNA and the synthesis of polypeptides, it acts as a catalyst in some steps.

1993

Kary B. Mullis (American). Mullis invented polymerase chain reaction (PCR), a method for swiftly making millions of copies of DNA. PCR soon became an important tool in genetic engineering, DNA fingerprinting, and medicine.

1993

Michael Smith (British). Smith developed site-directed mutagenesis, a means for reconfiguring genes in order to create altered proteins with distinct properties. Smith's genetic engineering tool made it possible to treat genetic disease and cancer and to create novel plant strains.

2004

Aaron Ciechanover (Israeli), Avram Hershko (Israeli), and Irwin Rose (American). The laureates were awarded for their work toward a better understanding of how cells regulate the hydrolysis (breakdown) of proteins. The researchers observed that proteins are "tagged" with a molecule called ubiquitin, which signals protein degradation complexes (proteosomes). The proteosomes in turn carry out the degradation process. In this manner, analogous to quality control, defective proteins are removed. A variety of cell processes, including DNA repair and cell division, are regulated in this manner. Genetic defects in the degradation process may lead to certain forms of cancer.

2006

Roger D. Kornberg (American). Kornberg was honored for his contribution toward the understanding of transcription regulation in eukaryotic cells such as yeast, plant, or animal cells. His work began with developing a system using extracts from the yeast *Saccharomyces* to study cell components involved in DNA expression. He subsequently discovered a complex of mediator proteins involved in regulating transcription in these cells. Combining the technique of X-ray crystallography with electron microscopy, Kornberg was able to produce a molecular model of the interactions. He was following in the footsteps of his father, Arthur Kornberg, a Nobel laureate in 1959 for his work in molecular biology.

2008

Osamu Shimomura, Martin Chalfie, and Roger Y. Tsien (all American). The laureates were honored for their discovery and development of green fluorescent proteins (GFP). During the early 1960's, Shimomura isolated and purified GFPs from jellyfish, molecules he named aequorin, which had the property of fluorescing in the presence of ultraviolet light. He subsequently isolated the chromophore, the fluorescent component from the molecule. During the 1980's, Chalfie received a cloned copy of the GFP gene, which he inserted into the cells of the nematode *Caenorhabditis*, resulting in the organism acquiring the ability to fluoresce. Tsien further applied the procedures originating with Chalfie to use GFPs and similar fluorescent molecules attached to proteins in the brains of mice, creating, in effect, molecular tags. With this procedure, it became possible to identify and follow gene products within cells by observing the color patterns in regions of the brain, termed "brainbow" by Tsien and colleagues.

Peace

1970

Norman Borlaug (American). Borlaug was a key figure in the "green revolution" of agriculture. Working as a geneticist and plant physiologist in a joint Mexican-American program, he developed strains of high-yield, short-strawed, disease-resistant wheat. His goal was to increase crop production and alleviate world hunger.

Roger Smith, Ph.D.; updated by Richard Adler, Ph.D.

TIME LINE OF MAJOR DEVELOPMENTS IN GENETICS

12,000 B.C.E.

With the transition from hunter-gatherer societies to agrarian civilizations, humans begin domesticating plants and animals through artificial selection, in some cases coupled with intentional hybridization for certain traits. Domesticated plants and animals become the basis that allows advanced civilizations to develop.

c. 350-322 B.C.E.

Aristotle writes and teaches on nature of living organisms, including their reproduction, hybridization, and classification.

c. 1283

Zakariya al-Qazwini (or Kazwini) writes *'Aja'ib al-makhluqat wa ghara'ib al-mawjudat* (on the marvels of nature and of the singularities of created things), in which he refers to sexual reproduction in the date palm, comparing it to sexual reproduction in humans.

1651

William Harvey publishes *Exercitationes de Generatione Animalium* (*Anatomical Exercitations, Concerning the Generation of Living Creatures,* 1653), in which he suggests that animals must originate in an egg.

1677

Antoni van Leeuwenhoek describes sperm and eggs and collects evidence that helps refute the theory of spontaneous generation.

1694

German botanist Rudolph Jakob Camerer (Camerarius) publishes *De Sexu Plantarum Epistola* (a letter on sex in plants), in which he presents experimental evidence on sexual reproduction in flowering plants.

1751

Carl von Linné (Carolus Linnaeus) publishes *Plantae Hybridae* (plant hybrids), in which he records his observation on plant hybrids, among them interspecific hybrids.

1759

In his doctoral dissertation, Kaspar Friedrich Wolff revives the hypothesis of epigenesis, which states that the complex structures of chickens develop from initially homogeneous, structureless areas of the embryo, and not from a pre-formed homunculus.

1761-1766

Joseph Gottlieb Kölreuter publishes a series of three papers titled "Vorläufige Nachricht von einigen das Geschlecht der Pflanzen betreffenden Versuchen und Bebachtungen" (preliminary report of some experiments and observations concerning sex in plants), which documents studies on fertilization and hybridization in a wide range of plant species.

1798

Thomas Robert Malthus publishes *An Essay on the Principle of Population,* in which he analyzes population growth and relates it to the struggle for existence, setting the stage for Charles Darwin's theory of natural selection.

1798

Edward Jenner publishes his findings on vaccination against smallpox in "Inquiry into the Cause and Effects of the Variolae Vaccinae." Jenner used the cowpox virus as a vaccine to induce immunity against the genetically and structurally similar, but lethal, virus that causes smallpox in humans.

1809

Jean-Baptiste Lamarck publishes *Philosophie Zoologique* (*Zoological Philosophy,* 1914), in which he sets forth his ideas on evolution, particularly his hypothesis of the inheritance of acquired characteristics. Although his notion that acquired traits alter the inherited constitution and are transmitted to the next generation was later dismissed in favor of Darwinian natural selection, Lamarck's book makes the link between evolution and inherited traits that lays a foundation for later evolutionary theory. Lamarck would have a significant influence on Darwin's development of the theory of natural selection.

1823

Thomas Andrew Knight presents "Some Remarks on the Supposed Influences of the Pollen, in Cross-Breeding, on the Colour of the Seed-Coats of Plants and the Qualities of their Fruits" to the Royal Horticultural Society. His experiments with the garden pea include hybrids, testcrosses, and reciprocal crosses. He documents dominance, parental equivalence, and the true-breeding nature of plants with a recessive phenotype but does notice mathematical patterns of the type Gregor Mendel would later observe.

1838-1839

Theodor Schwann (1838) and Matthias Jakob Schleiden (1839) propose the cell theory that all organisms are composed of cells.

1839

Gerardus Johannes Mulder publishes "On the Composition of Some Animal Substances," which provides chemical analysis of a fibrous material precipitated from cells. He calls this material "protein" (which means "of first importance") and believes it is the most important of the known components of living matter.

1839

Carl Friedrich von Gärtner publishes *Experiments and Observations on Hybrid Production in the Plant Kingdom*, documenting nearly ten thousand plant hybridization experiments. He noted repeated observations of dominance and uniformity in first-generation hybrids followed by variability in subsequent generations. Gärtner's work strongly influenced Darwin, who referred to it in *On the Origin of Species by Means of Natural Selection*, and Mendel, who cited it seventeen times in his 1866 paper and familiarized himself with Knight's work with the garden pea through Gärtner's citations of it.

1852

Franz Unger publishes *Botanical Letters*, in which he proposes that new species can arise through new combinations of inherited characteristics. As a professor at the University of Vienna, he is nearly dismissed for teaching evolution. Mendel is one of his students.

1854

Louis Pasteur begins research into fermentation. His "pasteurization" process is originally proposed as a means of preserving beer and wines. Through his work, Pasteur makes the important discovery that "life must be derived from life."

1855

Rudolph Virchow extends the cell theory of Schwann and Schleiden, proposing that all cells arise from previous cells (*Omnis cellula e cellula*).

1855-1857

Alfred Russel Wallace publishes *On the Law Which Has Regulated the Introduction of New Species*; later, in 1857, he sends Darwin a manuscript, "On the Tendency of Varieties to Depart Indefinitely from the Original Type." Today Wallace is recognized as having developed the theory of natural selection along with Darwin.

1859

Charles Darwin publishes *On the Origin of Species by Means of Natural Selection: Or, The Preservation of Favoured Races in the Struggle for Life* (usually referred to as *On the Origin of Species*) in 1859, in which he sets forth his theory of natural selection. Inheritance is a key element of Darwin's theory but is not fully understood at the time. Once genetics was studied as a discipline, it became clear that genetics is the foundation for explaining evolution through natural selection.

1862

The Organic Act establishes the U.S. Department of Agriculture (USDA). As one of its functions, the USDA is responsible for the collection of new and valuable seeds and plants and the distribution of them to agriculturists. The preservation and dissemination of agriculturally important plants is a necessity for maintaining and increasing the world's food supply.

1863

Gregor Mendel reads a German translation of Darwin's *On the Origin of Species* and makes marginal notes in it.

1865-1866

Gregor Mendel presents a paper titled "Versuche über Pflanzen-Hybriden" (experiments in plant hybrids) in 1865 to the Natural History Society of Brünn and publishes it in the society's *Proceedings* in 1866. His paper summarizes his experimental data on hybridization and his interpretations of those data to establish the theory of inheritance. Major components of this theory are now known as the principles of segregation, independent assortment, dominance, and parental equivalence. The first two are unique to Mendel's work, the latter two having been recognized by previous hybridists but explained by Mendel's theory. Although his work was cited several times following its publication, its importance was largely unrecognized until it was rediscovered in 1900 and became the foundation for the science of genetics.

1866

Ernst Haeckel develops the hypothesis that hereditary information is transmitted by the cell nucleus.

1866-1873

Gregor Mendel corresponds with Karl von Nägeli in a series of ten letters regarding plant hybridization experiments. These letters were preserved and later became an invaluable resource documenting how Mendel approached experimental design.

1868-1871

Charles Darwin publishes "the provisional hypothesis of pangenesis" as the mechanism of inheritance in his two-volume work *Plants and Animals Under Domestication.* Darwin's cousin, Francis Galton, publishes a paper in 1871, "Experiments in Pangenesis," which, on the basis of blood transfusion experiments in rabbits, offers strong experimental evidence refuting pangenesis.

1869

Francis Galton publishes *Hereditary Genius*, on the heredity of intelligence, which lays the foundation for the eugenics movement.

1869

Friedrich Miescher isolates "nuclein" from the nuclei of white blood cells. This substance is later found to consist of the nucleic acids DNA and RNA.

1870

Gregor Mendel publishes a paper on hybridization in hawkweed (*Hieracium*), an apomict, and concludes that principles of inheritance he discovered in the pea and the bean are not universal in the plant kingdom.

1876

Oskar Hertwig, a student of Ernst Haeckel, demonstrates the fertilization of an ovum in a sea urchin, thus establishing one of the basic principles of sexual reproduction: the union of egg and sperm cells.

1878-1882

In research published in 1878 and 1880, Walther Flemming describes cell division and the partitioning of chromosomes and names the process "mitosis." In 1882, he determines that chromosomes are split longitudinally during mitosis.

1883

Francis Galton founds the field of eugenics with the publication of *Inquiries into Human Faculty and Its Development.* The notion that the human species can be genetically improved by artificial selection perpetuates racism and provides a supposed scientific rationale for subsequent eugenic programs.

1883

Wilhelm Roux proposes, on the basis of experimental evidence, that mitosis must result in equal sharing of chromosomal particles by the daughter cells and describes the process.

1883

Edouard van Beneden studies the processes of meiosis and fertilization in the parasitic worm *Ascaris.* Van Beneden observes that meiosis halves the chromosome number in the resulting gametes and that fertilization restores the original chromosome number in the zygote.

1883

Emil Hansen develops a method for isolating and propagating pure yeast culture from a single cell, and it is adopted at Denmark's Carlsberg Brewery. The ability to propagate and maintain pure strains of organisms—genetically identical strains, or clones—will prove pivotal to future genetic research.

1885

August Weismann publishes *Die Continuitat des Keimplasmas als Grundlage einer Theorie der Vererbung* (the continuity of germ-plasm as the foundation of a theory of heredity), in which he maintains that only the germ cells, not somatic cells, can transmit hereditary information from one generation to the next; his experiments refuted the Larmarckian notion of inheritance of acquired characteristics.

1887-1889

In a series of articles, Émile Maupas describes the relationship of sexual conjugation and cellular senescence in protozoans.

1887-1890

Theodor Boveri investigates and describes chromosomes and their behavior, noting that they are preserved through the process of cell division and that sperm and egg contribute equal numbers of chromosomes to the zygote.

1889

Richard Altmann renames "nuclein" (isolated by Miescher in 1869) "nucleic acid."

1889

Francis Galton publishes a book titled *Natural Inheritance* in which he rejects blending inheritance and coins the term "particulate inheritance." He proposes that traits displaying continuous variation are governed by "a fine mosaic too minute for its elements to be distinguished in a general view."

1894

Albrecht Kossel identifies thymine, completing his identification of all four bases in DNA. He named it after the thymus gland from which it was isolated.

1896

Edmund B. Wilson publishes *The Cell in Development and Heredity*, in which he discusses the role of cells and chromosomes in inherited traits.

1897

Eduard Buchner shows that biochemical reactions that normally take place within living cells can proceed in cell extracts. He discovers that yeast extracts can convert glucose to ethyl alcohol. Buchner's was one of the first in vitro experiments on biochemical reactions. Performing such experiments outside the body allowed researchers to control conditions and to observe the effects of individual variables.

1899

The Royal Horticultural Society holds a meeting in Chiswick, London, in which William Bateson calls for research on discontinuous variations. The meeting is renamed the First International Conference of Genetics, later renamed the International Congress of Genetics, currently held every five years.

1900

Carl Correns, Hugo de Vries, and Erich Tschermak von Seysenegg independently rediscover Mendelian principles with experiments similar to Mendel's but not as extensive. Although de Vries soon abandons Mendelism, Correns and Tschermak continue experimentation on heredity, confirming Mendelian principles. Mendelian theory soon provides a framework for observations at the time in cytology, cellular biology, plant hybridization, and biochemistry.

1901

Karl Landsteiner discovers human blood groups.

1901

Clarence McClung describes the role of the X chromosome in determining sex in insects.

1902

W. F. R. Weldon publishes a paper titled "Mendel's Law of Alternative Inheritance in Peas," in which he subjects some of Mendel's data to Karl Pearson's recently developed (1900) chi-square statistical test. Weldon concludes that Mendel's data are exceptionally close to expectation. William Bateson responds with a booklet titled *Mendel's Principles of Heredity: A Defence.* A bitter debate ensues between Bateson and Weldon over the relevance of Mendelian inheritance.

1902

William Bateson, Reginald Punnett, Edith Saunders, Carl Correns, Erich Tschermak, Lucien Cuénot, and others begin to confirm Mendelian inheritance in numerous plant and animal species.

1902

Austrian botanist Gottlieb Haberlandt advances the cell theory with his idea of totipotentiality (later called totipotency): Cells contain all the genetic information necessary to develop into an entire, multicellular organism. Therefore, an individual plant cell is capable of developing into an entire plant.

1902

In an article titled "The Incidence of Alkaptonuria: A Study in Chemical Individuality," Sir Archibald Garrod proposes that alkaptonuria is an "inborn error of metabolism," due to the absence of a "ferment" (enzyme), and that it is inherited as a recessive trait in accordance with Mendel's principles. In this article, Garrod provides the first evidence of a specific relationship between genes and enzymes as well as Mendelian inheritance in humans.

1902

Theodor Boveri recognizes the correlation between Mendel's principles of inheritance and current studies of cellular biology; he deduces the haploid nature of sperm and egg cells (that each has equal numbers of chromosomes) and determines, by experimenting with sea urchin sperm and egg cells, that each must contribute half the total number of chromosomes to offspring for their normal development.

1903

Working independently of Boveri, Walter Sutton comes to similar conclusions using grasshoppers. Both Boveri and Sutton have formed the chromosomal theory of heredity. Mendel's principles of segregation and independent assortment coincided with Sutton's observations of how chromosomes segregated during cell division. This provided a cellular explanation for Mendel's observations.

1903

William Ernest Castle is appointed an assistant professor at Harvard University, where he directs one of the most ambitious genetics programs in the United States. He was the first to conduct genetic experiments on the fruit fly but is best known for his research on rodent genetics. In 1909, he is appointed director of the Bussey Institute at Harvard. His laboratory produces some of the most influential mammalian geneticists of the twentieth century, including L. C. Dunn, Clarence Little, Sewall Wright, and George Snell.

1903

Phoebus A. T. Levene discovers a distinction between some nucleic acid preparations in the amounts of thymine and uracil that they contain. Research will eventually show that thymine is a component of DNA and uracil a component of RNA.

1905

William Bateson coins the term "genetics" to describe the science of heredity in a letter to Adam Sedgwick.

1905

Nettie Stevens and Edmund Wilson independently describe the inheritance of sex chromosomes in relation to sex determination in insects. Their observations enhance Boveri and Sutton's evidence supporting the chromosomal theory of heredity.

1907

The first mandatory sterilization law intended to enforce eugenics is passed in Indiana. Mandatory sterilization and antimiscegenation laws will be passed throughout the United States and other countries over the next several decades as scientists, dignitaries, government leaders, the wealthy, and well-educated members of society embrace the eugenics movement. Outspoken proponents of eugenics will include Alexander Graham Bell, Theodore Roosevelt, Charles Davenport, Leonard Darwin, Ronald Fisher, and Winston Churchill, among many others.

1908

William Bateson and Reginald C. Punnett discover a case of gene linkage but fail to interpret it correctly when they observe a violation of the Mendelian principle of independent assortment, noting two traits in sweet pea that do not assort independently.

1908

George Shull self-pollinates corn plants for several generations to produce true-breeding inbred lines, notes the effect of heterosis in hybrid plants, and offers a genetic explanation for inbreeding depression.

1908

Godfrey Hardy and Wilhelm Weinberg independently discover mathematical relationships between genotypic and phenotypic frequencies in populations. Known as the Hardy-Weinberg law, the rules governing these mathematical relationships help researchers understand the dynamics of population genetics and the evolution of species.

1909

Wilhelm Johannsen, working on the statistical analysis of continuous variation, expands the modern genetic vocabulary, coining the terms "gene," "genotype," and "phenotype."

1909

Hermann Nilsson-Ehle's studies of kernel color in wheat indicate this is a polygenic trait. This was one of the first demonstrations that several genes could influence a single trait. His experiments showed that each gene contributes to the trait in an additive fashion.

1909

Carl Correns discovers a class of exceptions to Mendelian inheritance, one of the first examples of extranuclear inheritance. The idea that other cellular organelles besides the nucleus carry DNA was not recognized for decades. However, Correns's experiments in the plant *Mirabilis jalapa* showed inheritance of leaf color via the DNA in the chloroplasts.

1909

Phoebus A. T. Levene determines that the sugar in RNA is ribose.

1910

Thomas Hunt Morgan further establishes the chromosomal theory of heredity after investigating a white-eyed fruit fly and finding that the trait does not segregate exactly according to Mendelian principles but rather is influenced by the sex of the fly. This fly experiment becomes the cornerstone upon which theories of Mendelian, chromosomal, and sexual inheritance are built into a cohesive whole. Morgan also establishes the "Fly Room" at Columbia University, where he and his students will conduct groundbreaking experiments using *Drosophila* for the next quarter century. He will win the Nobel Prize in Physiology or Medicine in 1933.

1910

Albrecht Kossel wins the Nobel Prize in Physiology or Medicine for earlier work isolating and describing molecular constituents of the cell's nucleus, notably guanine (which had earlier been described by Piccard, but not as a constituent of nucleic acids), adenine, cytosine, thymine, and uracil.

1910

The Eugenics Records Office is established at Cold Spring Harbor Laboratory under the direction of geneticist Charles B. Davenport.

1911

Peyton Rous produces cellfree extracts from chicken tumors that, when injected, can induce tumors in other chickens. The tumor-producing agent in the extract is later found to be a virus. Thus, Rous has discovered a link between cancer and viruses. He wins the Nobel Prize in Physiology or Medicine in 1966.

1912-1913

The first International Eugenics Conference is held in London in 1912. Some of the earliest books on genetics, including *Heredity and Eugenics* by William Ernest Castle and others (1912), *Mendelism* by Reginald C. Punnett (1912), and *Mendel's Principles of Heredity* by William Bateson (1913), each conclude praising eugenics as a bright future for human society.

1913

Alfred H. Sturtevant, an undergraduate student studying under Thomas Hunt Morgan, publishes the first gene linkage map: five linked genes on the X chromosome of *Drosophila melanogaster.* He develops a method of gene mapping that would standardize genetic mapping of genes and genetic markers relative to one another on chromosomes.

1913

Eleanor Carothers reports her discovery of the chromosomal basis of independent assortment. By examining grasshopper chromosomes, Carothers observed the segregation of the X chromosome relative to a heteromorphic pair of autosomes during meiosis. This observation was consistent with Mendel's principle of independent assortment.

1913

Rollins A. Emerson and Edward M. East demonstrate the multiple genes in corn can influence a single trait and that their pattern of inheritance can be explained by Mendelian principles.

1914

Calvin Blackman Bridges uses the phenomenon of primary nondisjunction (a fault in cell division resulting in the failure of chromosomes to separate during metaphase I) to confirm that genes are carried on chromosomes.

1914-1919

In 1914, William Castle publishes a study on artificial selection in hooded rats, demonstrating the ability of selection to genetically alter phenotypic expression of a recessive trait. By 1919, he interprets his results as the effect of modifier genes, each of which has a minor effect on the phenotype conferred by a single gene with a major effect.

1915

The Mechanism of Mendelian Heredity, by Thomas Hunt Morgan, Alfred Sturtevant, Hermann J. Muller, and Calvin Bridges, is published, establishing *Drosophila* as a model organism for genetics research and describing fundamentals of gene mapping.

1915

In a letter to Charles Davenport, editor of the *Journal of Heredity*, Thomas Hunt Morgan privately expresses his doubts about the eugenics movement and refuses to take part in it. He delays his public opposition, however, until his Nobel speech in 1934.

1916

Edward M. East publishes an extensive analysis of controlled experiments on the inheritance of corolla length in tobacco and concludes that his observations are best explained by Mendelian inheritance of multiple genes with alleles that do not display dominance.

1916

Research on the major histocompatibility complex begins with Clarence Little and E. E. Tyzzer's experiments transplanting tumors between mice.

1917

Félix d'Herelle discovers bacteriophages, viruses that infect bacteria. Bacteriophages later played important roles in early genetics research, including confirmation that DNA is the hereditary material and their use as vectors in recombinant DNA applications.

1917

Ojvind Winge publishes "The Chromosomes: Their Number and General Importance," which describes the relationship between chromosome doubling and allopolyploidy in plants.

1917

Eleanor Carothers extends her 1913 observations of independent assortment of chromosomes to more than two pairs of chromosomes.

1917

In an article titled "Elimination of Feeblemindedness," Reginald Punnett mathematically demonstrates the futility of mandatory sterilization for reducing the frequency of recessive genetic disorders. Nonetheless, mandatory sterilization of individuals possessing genetic disorders thought to be recessive, such as "feeblemindedness," continues for decades to come.

1917

Calvin Bridges determines that the ratio of X chromosomes to autosomes determines sex in *Drosophila*, with no influence of the Y chromosome on sex determination in this species.

1921

The second International Eugenics Conference is held in New York under the theme "Eugenics is the self-direction of human evolution."

1922

Lillian Vaughan Sampson Morgan publishes her research on attached-X females in *Drosophila*, which carry a Y chromosome, confirming Bridges's proposed mechanism of sex determination. The attached-X chromosome becomes an important tool for *Drosophila* research.

1924

Donald Jones publishes the results of experiments on inbred-line development and F_1 hybrid production in corn, establishing a genetic basis for inbreeding depression and heterosis.

1924

J. B. S. Haldane publishes the first of a series of articles on the genetic basis of natural selection, initiating the merger of Mendelian genetics with Darwinian evolution, which would later be called the "modern synthesis."

1925

Alfred Sturtevant documents unequal crossing over between duplicated chromosomal segments at the *Bar* locus in *Drosophila.*

1925-1926

Sturtevant discovers position effect: An inversion may place a gene in another location in the chromosome, removing the gene from its regulatory elements and altering its expression. He also provides genetic data best explained by chromosomal inversion.

1926

Nicolai Vavilov, a Soviet geneticist who studied under Bateson, publishes his theory of centers of origin, which proposes that the centers of origin for domesticated plant species are the regions where the genetic diversity for those species is the greatest. From 1917 to 1940, he carries out a series of more than one hundred expeditions throughout the world to collect and preserve the genetic diversity of the world's food plants. These extraordinary efforts are the beginning of the worldwide establishment of gene banks to conserve and utilize genetic diversity for plant and animal breeding.

1927

The *Buck v. Bell* eugenics case reaches the U.S. Supreme Court. A woman from Virginia, Carrie Bell, challenges her legally mandated sterilization. In an 8-1 decision, the Supreme Court upholds the lower court rulings that Bell must be sterilized. In the majority opinion, Chief Justice Oliver Wendell Holmes, Jr., writes, "The principle that sustains compulsory vaccination is broad enough to cover cutting the Fallopian tubes." This case establishes precedent in support of eugenic mandatory sterilization laws.

1927-1928

Hermann J. Muller and Lewis Stadler independently discover that X rays induce inherited mutations, demonstrating the mutagenic effect of ionizing radiation. Muller works with *Drosophila* and publishes his work in 1927, whereas Stadler works with barley and publishes his work in 1928. Muller receives the Nobel Prize in Physiology or Medicine for this work in 1946.

1928

Frederick Griffith discovers the "transforming principle" in experiments targeted at developing a vaccine for pneumonia. The hereditary material has not yet been identified, but later experiments by Oswald Avery et al. in 1944 indicate that the transforming principle discovered by Griffith is DNA.

1929

Phoebus Levene determines the chemical structure of the sugar deoxyribose in the nucleic acid that would be named DNA on the basis of this discovery.

1929

Clarence Little helps found the Jackson Memorial Laboratory in Bar Harbor, Maine, which will become one of the most influential genetics research institutions in North America, particularly in mouse (mammalian) genetics, and cancer research.

1930

Ronald Aylmer Fisher publishes *The Genetical Theory of Natural Selection*, which will become one of the most influential works of the modern synthesis.

1931

Harriet Creighton and Barbara McClintock discover that a physical exchange between chromosomes in corn is associated with recombination of genes mapped to that chromosome, confirming that crossing over is responsible for recombination of linked genes. Curt Stern uses a similar approach to document a physical exchange in the X chromosome of *Drosophila.*

1931

Sewall Wright publishes a lengthy article titled "Evolution in Mendelian Populations," which establishes genetic drift as an important factor in evolution, and is another key publication in the modern synthesis.

1932-1934

The third and final International Eugenics Conference is held in New York in 1932. Nazi implementation of extreme eugenic measures in 1933 and the subsequent atrocities of the Holocaust are the start of the eugenics movement's downfall. Thomas Hunt Morgan, who has quietly opposed eugenics since 1915, publicly renounces it in his 1934 Nobel Prize acceptance speech.

1933

Theophilus Painter discovers polytene chromosomes in *Drosophila* salivary glands. These special chromosomes, resulting from numerous rounds of DNA replications without separation, are large, with distinct banding patterns. They are used extensively in mapping genes to specific regions of the chromosome.

1933

Theodosius Dobzhansky publishes an article on chromosomal evolution in *Drosophila pseudoobscura.* Dobzhansky and Alfred Sturtevant will, over a forty-year period, publish 43 articles under the collective title "Genetics of Natural Populations," firmly establishing the modern synthesis.

1934

John Desmond Bernal examines protein structure by using X-ray crystallography.

1935

Phoebus Levene correctly determines the chemical structure of single-stranded DNA, building on his previous discoveries of deoxyribose and the structure of the nucleotide. However, he incorrectly presumes that the molecule consists of repeating tetranucleotides, with equal proportions of T, C, A, and G.

1935

Hans Spemann wins the Nobel Prize in Physiology or Medicine. By transplanting bits of one embryo into a second, viable embryo, Spemann compiled evidence that an "organizer center" directs the development of an embryo and that different parts of the organizer governed distinct portions of the embryo. His experiments provided clues to the genetic control of growth from the earliest stages of an organism.

1935-1964

Trofim Denisovich Lysenko and Isaak Izrailevich Prezent initiate in 1935 a campaign against traditional geneticists in the Soviet Union, branding them as "kulak-wreckers and saboteurs." Lysenko resurrects Larmarckism as the "true" theory of genetics. The movement becomes known as Lysenkoism and receives Joseph Stalin's approval. Vavilov leads the opposition, for which he is arrested and imprisoned in 1941. He dies of maltreatment in prison in 1943. Several Soviet geneticists are also imprisoned for their refusal to support Lysenkoism, and they also die in prison or are executed. Lysenkoism is forced on Soviet geneticists until Nikita Khrushchev's forced resignation in 1964.

1936

Ronald Aylmer Fisher publishes an exhaustive analyses of Mendel's work, including reconstruction of Mendel's experiments and subjection of Mendel's data to chi-square analysis. Fisher concludes that Mendel's experiments are especially well described, literal, and conducted exactly as Mendel presented them, and that Mendel is a supporter of Darwin. Fisher points out, however, that Mendel's data are exceptionally close to expectation, and speculates that they may have been cooked by an assistant. This latter conclusion sparked the Mendel-Fisher controversy in the 1960's, which has continued to the present.

1937

Theodosius Dobzhansky publishes *Genetics and the Origin of Species.* He shows that, in natural and experimental populations of *Drosophila* species, frequency changes and geographic patterns of variation in chromosome variants are consistent with the effects of natural selection. This book is another cornerstone of the modern synthesis.

1937

George Beadle and Boris Ephrussi, on the basis of mutational analysis in two genes and tissue transplantation experiments, determine the order of two enzymes in a biochemical pathway in *Drosophila.*

1939

R. J. Gautheret demonstrates the first successful culture of isolated plant tissues as a continuously dividing callus tissue.

1940

Clarence P. Oliver discovers in *Drosophila* that a gene is not a point on a chromosome but rather a segment divisible by crossing over.

1940

Karl Landsteiner and A. S. Wiener describe the Rh blood groups.

1941

George Wells Beadle and Edward Tatum, working with a bread mold, *Neurospora*, publish results indicating that genes mediate cellular chemistry through the production of specific enzymes: the "one gene-one enzyme" experiment. This establishes the use of fungi as model systems to study genetics. Beadle and Tatum will receive the Nobel Prize in Physiology or Medicine in 1958.

1941-1942

The siege of Leningrad places Nicolai Vavilov's seed collections in peril. The curators of the collection, although they are starving along with tens of thousands of people, make a pact that they will not eat the seeds in the collections. Nine of them voluntarily starve to death, sacrificing their lives to preserve the genetic diversity collected by Vavilov.

1943-1944

The Rockefeller Foundation, in collaboration with the Mexican government, initiates the Mexican Agricultural Program. This program relies extensively on plant breeding, coupled with other improved agricultural practices, to increase Mexico's wheat, corn, and bean production. It is one of a series of projects that over a period of decades result in the Green Revolution, a dramatic increase in staple food production in less developed countries in Latin America and Asia. Norman Borlaug, who founded the wheat breeding programs in Mexico, India, and Pakistan, will receive the Nobel Peace Prize in 1970 for his contributions to the Green Revolution.

1944

Oswald T. Avery, Colin MacLeod, and Maclyn McCarty purify DNA and identify it as the "transforming principle" in the pneumococcal bacteria of Frederick Griffith's work. Although this experiment provides solid evidence that DNA is the hereditary material, most scientists still do not suspect that the conclusion may apply to other organisms.

1945

R. D. Owen conducts studies with two sets of cattle twins which demonstrate that their blood antigens could have come only from the opposite sires. These findings suggest the reciprocal passage of ancestral red blood cells. Owen's work has significant implications for immunology.

1945

Max Delbrück, Salvador Luria, and Alfred Hershey work on bacteriophage as a model system to study the mechanism of heredity. Delbrück organizes a course at Cold Spring Harbor, New York, to introduce researchers to the methods of working with bacteriophage. His course will be taught for twenty-six years, helping countless researchers to understand the use of bacteria and bacteriophages as model organisms in genetic investigations. Delbrück, Luria, and Hershey will later share the 1969 Nobel Prize in Physiology or Medicine.

1946

Joshua Lederberg and Edward Tatum discover genetic recombination through conjugation in bacteria. This discovery demonstrates that the genetic material may be transferred from one bacterial cell to another, and may recombine in the recipient cell, albeit by different mechanisms than those employed by the eukaryotes. Lederberg will win the Nobel Prize in Physiology or Medicine in 1958; Tatum and Beadle will also share in the 1958 prize for their work with cellular chemistry, enzymes, and genetics.

1946

Theodosius Dobzhansky publishes *Heredity, Race, and Society*, a book refuting the eugenics movement on the basis of genetic principles.

1949

Linus Pauling determines that sickle-cell anemia is the result of a change in the structure and chemical properties of hemoglobin. His investigations into protein structure contribute to determination of the structure of DNA. He will receive the Nobel Prize in Chemistry in 1954.

1949

Murray Barr and E. G. Bertram identify a condensed body of chromatin present in the nucleus of female mammalian cells. It will later be identified as an inactivated X chromosome and named the Barr body.

1950

Barbara McClintock publishes her conclusions that certain chromosomal elements in corn can be excised and inserted elsewhere in the chromosome. The name of this phenomenon is transposition, and the mobile elements were later termed "transposable elements," or transposons. McClintock's ideas were far ahead of her time. Her work will not be widely recognized until more evidence of transposition surfaces decades later. She will win the Nobel Prize in Physiology or Medicine in 1983.

1950

Bernard Davis shows that cell-to-cell contact is required for bacterial conjugation, ruling out transformation, of the type studied by Frederick Griffith and Oswald Avery et al., as an explanation of the Lederberg and Tatum experiments. Conjugation and transformation are two separate processes that allow bacterial genetic recombination.

1950-1951

Erwin Chargaff discovers consistent one-to-one ratios of adenine to thymine and of guanine to cytosine in DNA, suggesting a possible chemical relationship between these base pairs. Chargaff's observations become an important clue in determining the double-helical structure of DNA.

1950-1951

Maurice Wilkins obtains an X-ray diffraction image of the A form of DNA in 1950, which indicates that the crystalline molecule is symmetrical and helical. Rosalind Franklin obtains in 1951 a clear X-ray diffraction image of the B form (the form assumed by the vast majority of DNA in cells), which reveals important aspects of cellular DNA structure, among them its dimensions, helical shape, regular repeating subunits, and positioning of the sugar-phosphate backbones on the outside and bases on the inside of the molecule. Franklin's image, dubbed Photo 51, would be a key piece of evidence for deciphering the structure of DNA.

1952

Joshua Lederberg, Esther Lederberg, and Norton Zinder discover transduction, the transfer of genetic information between bacterial cells via a viral vector. Using *Escherichia coli* and a bacteriophage called *P1*, they show that transduction can be used to map genes on the bacterial chromosome.

1952

Alfred Hershey and Martha Chase use radioactive isotopes and bacteriophage infection of bacterial cells to demonstrate that bacteriophage's genetic material is DNA and not protein. This experiment is another key piece of evidence confirming that DNA is the hereditary material.

1952

Investigations into bacteriophage by Salvador Luria and Mary L. Human, and independently by Jean J. Weigle, lay the groundwork for the discovery of restriction endonucleases.

1953

James Watson and Francis Crick propose a model for the double-helical structure of DNA in a brief manuscript published in *Nature,* "Molecular Structure of Nucleic Acids: A Structure for Deoxyribose Nucleic Acid." This elegant and concise paper describes the structure of the B form of DNA and suggests a mode of replication. Watson and Crick, along with Maurice Wilkins, will receive the Nobel Prize in Physiology or Medicine in 1962. Rosalind Franklin, whose X-ray diffraction work was a major component in the discovery, will not share in the prize, having died several years earlier of cancer, possibly a result of her work with X rays.

1955

Seymour Benzer maps more than 2,400 mutations in a gene by observing bacteriophage recombination. He discovers a lower limit for the units of recombination, which he shows correspond to the individual nucleotides in DNA, a powerful correlation between DNA structure and a genetic phenomenon.

1956

Joe Hin Tjio and Albert Levan determine the chromosome number in humans to be forty-six. Until that time, the chromosome number was thought to be forty-eight. The advances that Tjio and Levan pioneered were instrumental in obtaining good chromosome preparations, allowing for significant advances in the field of cytogenetics.

1956

Elie Wollman and François Jacob map seven genes in the *E. coli* genome based on the relative times at which genes enter a recipient bacterial cell during conjugation. Time-of-entry mapping becomes the standard method for mapping genes in *E. coli*, the most researched bacterial model organism.

1956-1958

Arthur Kornberg purifies the enzyme DNA polymerase from *Escherichia coli*. This is the enzyme responsible for DNA replication, making it possible to synthesize DNA in vitro. Kornberg, along with Severo Ochoa, will win the Nobel Prize in Physiology or Medicine in 1959.

1957

Vernon M. Ingram determines that a single amino acid substitution in the beta subunit of hemoglobin is the cause of sickle-cell anemia, clarifying the alteration in protein structure discovered by Pauling in 1949.

1957

Heinz Fraenkel-Conrat and B. Singer show that tobacco mosaic virus contains RNA—the first concrete evidence that RNA, in addition to DNA, serves as the genetic material.

1957

In a landmark address to the British Society of Experimental Biology titled "On Protein Synthesis," Francis Crick articulates both the sequence hypothesis (the order of bases on a section of DNA encodes the amino acid sequence in a protein) and the "central dogma" of molecular genetics (genetic information moves from DNA to RNA to proteins, but not from proteins back to DNA).

1957

Alexander Robertus Todd wins the Nobel Prize in Chemistry. As part of wide-ranging research in organic chemistry, Todd revealed how ribose and deoxyribose bond to the nitrogenous bases on one side of a nucleotide unit and to the phosphate group on the other side. These discoveries provide a foundation for work by others that explains the structure of the DNA molecule.

1958

Matthew Meselson and Franklin Stahl, in what has been dubbed "the most elegant experiment in biology," use light and heavy isotopes of nitrogen to demonstrate that bacterial DNA replicates in a semiconservative manner: Each strand of the molecule serves as a template for the synthesis of a new, complementary strand, a hypothesis proposed by Watson and Crick in their seminal 1953 paper.

1959

Jérôme Lejeune discovers that Down syndrome is the result of an extra chromosome, evidence that genetic disorders in humans could be the result of changes in chromosome number.

1961

Sol Spiegelman and Benjamin Hall discover that single-stranded DNA will hydrogen bond to its complementary RNA. The discovery of the ability of DNA and RNA to form an association contributed to determining how genes are expressed.

1961

In a series of experiments, Jacques Monod, François Jacob, Sydney Brenner, Matthew Meselson, François Gros, and Francis Crick discover messenger RNA (mRNA), discovering that it is the molecule that carries information from DNA to the ribosome for protein synthesis. This discovery was critical to the understanding of protein synthesis and hence gene expression. Monod and Jacob will win the 1965 Nobel Prize in Physiology or Medicine for this work.

1961-1966

Marshall Nirenberg and Heinrich Matthaei discover the first sequence of three bases of RNA that encode an amino acid using RNA homopolymers: UUU encodes phenylalanine. Severo Ochoa used random RNA heteropolymers to narrow the possibilities of the genetic code. By 1963, Nirenberg have identified the nucleotide composition, albeit not the nucleotide order, of thirty-five codons, using data from his and other laboratories. In 1965, Nirenberg and Philip Leder develop a three-nucleotide "mini-RNA" method that allowed them to identify fifty of the sixty-four codons. In 1966, H. Gobind Khorana identifies the final fourteen codons using repeating RNA heteropolymers. Robert W. Holley, in 1965, sequences a yeast transfer RNA (tRNA), which led to the discovery of the tRNA function in transferring the genetic-code information in mRNA into a sequence of amino acids. Nirenberg, Khorana, and Holley will receive the Nobel Prize in Physiology or Medicine in 1968 for their contributions to understanding protein synthesis.

1962

Mary Lyon hypothesizes that during development, one of the two X chromosomes in normal mammalian females is inactivated at random, clarifying the function of the Barr body, discovered by Murray Barr in 1949. Her hypothesis is known as the Lyon hypothesis.

1962

Werner Arber finds bacteria that are resistant to infection by bacteriophage. It appears that some cellular enzymes destroy phage DNA, while others modify the bacterial DNA to prevent self-destruction. Several years later, Arber, Stuart Linn, Matthew Meselson, and Robert Yuan isolate the first restriction endonuclease and identify the modification of bacterial DNA as methylation. By this time, scientists are looking at how DNA regulates, and is regulated by, cellular activities in a new discipline, molecular genetics. Arber will win the Nobel Prize in Physiology or Medicine in 1978.

1963-1964

Evidence published in 1963 by Margit Nass and Sylvan Nass, who used chemical methods coupled with electron microscopy, points to the presence of DNA in chick embryo mitochondria. In 1964, Gottfried Schatz, E. Halsbrunner, and H. Tuppy purify yeast mitochondria and show that they contain significant amounts of DNA. These discoveries offer a plausible explanation for extranuclear inheritance observed in fungi, plants, and animals.

1964

Robin Holliday proposes a model for the recombination of DNA. Although recombination, or crossing over, is not a new idea, the molecular mechanism underlying the exchange between two homologous molecules of DNA was not known. Holliday's model explains the phenomenon and will become the basis for more detailed models in later years.

1964

John B. Gurdon transfers nuclei from toad tadpole intestinal cells into toad eggs whose native nuclei had been ablated. A small percentage of these eggs develop into adults, genetic clones of the original tadpole. This same year, F. C. Steward and coworkers regenerate whole plants from undifferentiated carrot callus cell culture. These experiments introduce animal and plant cloning.

1965

André Lwoff shares the Nobel Prize in Physiology or Medicine with Monod and Jacob. Lwoff earlier demonstrated that the genetic material of bacteriophage can become part of the host bacterium's DNA, a process known as lysogeny.

1965-1967

Sydney Brenner and colleagues identify the three stop codons.

1966

Victor McKusick publishes the first catalog of single genes that govern variation for traits in humans, *Mendelian Inheritance in Man*, which continues to be published in subsequent editions, including an online version.

1966

The International Rice Research Institute (IRRI) releases IR8, a short-stature rice variety derived from IRRI's rice breeding program. This new rice outperforms all previous varieties of rice and is widely adopted in rice-growing regions throughout the world. This event is one in a series known collectively as the Green Revolution.

1967

Mary Weiss and Howard Green introduce a method for physical chromosome mapping of human genes using somatic cell hybridization of human and mouse cells.

1967

Scientists in several laboratories isolate DNA ligase, the enzyme that joins DNA molecules.

1968

Reiji Okazaki reports the discovery of short fragments of DNA, later known as Okazaki fragments, evidence of discontinuous synthesis of the lagging DNA strand.

1970

Morton Mandel and Akiko Higa discover a method to increase the efficiency of bacterial transformation. They make the cells "competent" to take up DNA by treating bacteria with calcium chloride and then heat-shocking the cells. Introducing foreign DNA into cells was a key to the success of recombinant DNA methods.

1970

H. Gobind Khorana and twelve associates synthesize the first artificial gene: a gene encoding alanine transfer RNA in yeast.

1970

David Baltimore and Howard Temin independently discover reverse transcriptase, an enzyme used by viruses to generate a complementary DNA copy of their RNA. The reverse transcriptase enzyme becomes a key tool in genetic engineering, for which Baltimore and Temin will win the 1975 Nobel Prize in Physiology or Medicine.

1970

Hamilton O. Smith isolates the first restriction endonuclease, HindII, that cuts at a specific DNA sequence—the first "site-specific restriction enzyme." Daniel Nathans uses this enzyme to create a restriction map of the virus SV40. The use of restriction enzymes, those that cut DNA, allowed for the detailed mapping and analysis of genes. It also was pivotal for recombinant DNA techniques, including the production of transgenic organisms. Nathans and Smith will win the Nobel Prize in Physiology or Medicine in 1978 for their work on restriction enzymes.

1972

Paul Berg is the first to create a recombinant DNA molecule. He shows that restriction enzymes can be used to cut DNA in a predictable manner and that these DNA fragments can be joined together with fragments from different organisms. He will be awarded the Nobel Prize in Physiology or Medicine in 1980.

1972

Stanford Moore and William H. Stein win the Nobel Prize in Chemistry for earlier work identifying the sequence of amino acids in ribonuclease, a clue to the structure of the gene that encodes it.

1973

Joseph Sambrook and other researchers at Cold Spring Harbor improve the method of separating DNA fragments based on size, a technique called agarose gel electrophoresis. This method makes it possible to achieve an accurate interpretation of the information in DNA.

1973

Stanley Cohen and Herbert Boyer develop recombinant DNA technology by producing the first recombinant plasmid in bacteria. Plasmids—small, circular pieces of DNA—occur naturally in bacteria. Using tools of molecular biology, Cohen and Boyer inserted a new, or "foreign," piece of DNA into a genetically engineered plasmid, which then propagated itself, along with the foreign DNA, in bacterial cells. The process of replicating foreign DNA in bacteria is eventually named molecular cloning.

1975

The Asilomar Conference is held in response to increasing concerns over safety and ethics of recombinant DNA technology. Convening in Pacific Grove, California, under the auspices of the National Institutes of Health, 140 prominent international researchers and academicians air their opinions about recombinant DNA experimentation and advocate adoption of ethical guidelines. NIH later issues guidelines for recombinant DNA research to minimize potential hazards if genetically altered bacteria were released into the environment. The guidelines will be relaxed by 1981.

1975

Mary-Claire King and Allan Wilson report, based on results of a survey of protein and nucleic acid studies, that the average human protein is more than 99 percent identical to that of chimpanzees, which is confirmed by later research.

1975

Edward Southern develops a method for transferring DNA from an agarose gel to a solid membrane. This technique, known as Southern blotting, becomes one of the most important methods in recombinant DNA research.

1975

Renato Dulbecco, David Baltimore, and Howard Temin receive the Nobel Prize in Physiology or Medicine for their work on the interaction between tumor viruses and the genetic material of the cell.

1975

David Pribnow compares the sequences in the promoter regions of several bacterial genes and finds a conserved DNA sequence in them, which is later named the Pribnow box.

1976

Herbert Boyer and Robert Swanson form Genentech, a company devoted to the development and promotion of biotechnology and applications of genetic engineering.

1976

Susumu Tonegawa discovers the genetic mechanism for generation of antibody diversity. The DNA of antibody-encoding genes of immune cells is rearranged in a countless array of different combinations. This explains how millions of different antibodies can be produced from a very small number of genes. Tonegawa will win the Nobel Prize in Physiology or Medicine in 1987.

1977

Allan Maxam and Walter Gilbert develop a method to determine the nucleotide sequence of a segment of DNA. This same year, Frederick Sanger develops a different method, chain termination (dideoxy) sequencing. Although both the Gilbert-Maxam and the Sanger methods are effective, the Sanger method predominates because it is less complex and does not require toxic chemicals. Gilbert and Sanger will both receive the Nobel Prize in Physiology or Medicine in 1980.

1977

James Alwine develops Northern blotting, a variant of Southern blotting but with RNA instead of DNA.

1977

The U.S. Court of Customs and Patent Appeals rules that an inventor can patent genetically engineered microorganisms. The first patent granted for a recombinant organism, an oil-degrading bacterium, will be awarded in 1980. The legality and ethics of patenting recombinant organisms and other biological systems are highly controversial.

1977

Herbert Boyer synthesizes the human hormone somatostatin in *Escherichia coli*—the first successful use of recombinant DNA to produce a protein encoded by a eukaryotic gene in bacteria. Before recombinant DNA methods became available, isolation of mammalian somatostatin required a half million sheep brains to produce 5 milligrams of the hormone. Now, with the use of recombinant DNA, only two gallons of bacterial culture are required to produce the same amount.

1977

Phillip A. Sharp and Richard Roberts independently discover that genes in adenovirus contain intervening sequences that are not present in the mRNAs encoded by those genes. Later, the DNA segments that persist in the mRNA are named exons, and the intervening sequences removed during processing of the mRNA are named introns. Sharp and Roberts will win the 1993 Nobel Prize in Physiology or Medicine for this discovery.

1978

M. L. Goldberg and D. S. Hogness and collaborators discover a conserved promoter sequence about twenty-five nucleotide pairs upstream of the transcription startpoint in genes encoding histones in

Drosophila melanogaster. This sequence is later named the TATA box and is a transcription factor binding site present in the promoters of most eukaryotic genes.

1978

Herbert Boyer generates a synthetic version of the human insulin gene and inserts it into *Escherichia coli* cells, which produce human insulin. Development of this technology leads to commercial production of recombinant human insulin for treatment of diabetes.

1978

P. C. Steptoe and R. G. Edwards successfully use in vitro fertilization and artificial implantation in humans. Louise Brown, the first "test-tube baby," is born July 25. The process gives hope to many childless couples who, prior to this development, have been unable to conceive. It also raises concerns from ethicists and others over both the potential effects on the individual child's long-term health and social implications.

1980

A team headed by David Botstein discovers variation among individuals for the lengths of DNA fragments generated by digestion of DNA with restriction enzymes. These fragments that vary in size are named "restriction fragment length polymorphisms" (RFLPs), and they become an important tool for genetic analysis, paternity identification, DNA fingerprinting, and genetic testing for mutant alleles that cause genetic disorders.

1980

Jon W. Gordon produces the first transgenic mouse.

1980

The U.S. Supreme Court votes 5-4 that living organisms can be patented under federal law, and Ananda M. Chakrabarty receives the first patent for a genetically engineered organism, a form of bacteria, *Pseudomona originosa,* that can decompose crude oil for use in cleaning up oil spills.

1980

George Snell, Baruj Benacerraf, and Jean Dausset receive the Nobel Prize in Physiology or Medicine for their research on the genes encoding proteins of the major histocompatibility complex (MHC), a key component of the mammalian immune system.

1981

J. Michael Bishop and Harold Varmus discover that oncogenes (genes that play a role in initiating cancer) originate in normal cells as genes that control cellular growth and are not solely derived from retroviruses, as previously speculated. Their work influences subsequent studies of tumor development. Varmus and Bishop will win the Nobel Prize in Physiology or Medicine in 1989.

1982

Jorge Yunis and Om Prakash publish a comparison of G-banded human, chimpanzee, gorilla, and orangutan chromosomes. Although the chromosomes of all four species align well with one another, analysis reveals major structural differences among the aligned chromosomes which distinguish the species. These include evidence of a fusion that formed human chromosome 2, inversions, and translocations. The human and chimpanzee karyotypes are the most similar, differing by a fusion and nine inversions.

1982

The U.S. government approves commercialization of the first genetically engineered product, human insulin, dubbed Humulin. The production of pharmaceuticals through recombinant DNA technology is becoming a driving force behind both the drug industry and agriculture.

1982

Aaron Klug wins the Nobel Prize in Chemistry. Klug used X-ray crystallography to investigate biochemical structures, especially those of viruses. He linked the assembly of viral protein subunits with specific sites on viral RNA, which helped in fighting viruses that cause disease in plants and, more basically, in understanding the mechanism of RNA transfer of genetic information. He also determined the structure of transfer RNA (tRNA), which has a shape similar to that of a bent hairpin.

1983

Nancy Wexler, Michael Conneally, and James Gusella determine the chromosomal region of the gene that is mutated in people who have Huntington's disease. They are unable to locate the gene itself; it will be discovered ten years later.

1983

Thomas Cech and Sidney Altman independently discover catalytic RNA. The idea that RNA can have an enzymatic function changes researchers' views on the role of this molecule, leading to important new theories about the evolution of life. Cech and Altman will win the 1989 Nobel Prize in Chemistry.

1983

Bruce Cattanach provides evidence of genomic imprinting in mice. The phenomenon of imprinting is the differential targeting of genes for regulation in male and female gametes. This leads to differential expression of these genes during embryonic development.

1983

John Sulston, Sydney Brenner, and H. Robert Horvitz describe the cell lineage of the nematode *Caenorhabditis elegans.* The fixed developmental pattern of this small worm provides researchers with insights into how cell fate is determined during development and how cells influence the fates of neighboring cells. Sulston, Brenner, and Horvitz will win the 2002 Nobel Prize in Physiology or Medicine.

1983

Nina Fedoroff leads a team that clones the Ac and Ds elements in corn, transposable elements studied in depth more than thirty years earlier by Barbara McClintock. Characterization of these elements at the molecular level explains McClintock's observations.

1983-1984

William Bender's laboratory isolates and characterizes the molecular details of *Drosophila* homeotic genes. William McGinnis and J. Weiner discover that the base sequences of the homeotic genes they examined contain a highly similar sequence in a 180-base region. They term the conserved 180-base sequence a "homeobox." These regulatory genes direct the differentiation of body parts in animals.

1983-1985

Kary B. Mullis invents the polymerase chain reaction (PCR). This revolutionary method of copying DNA from extremely small amounts of material changes the way molecular research is done in only a few short years. It also becomes important in medical diagnostics and forensic analysis. Mullis will win the Nobel Prize in Chemistry in 1995.

1984

The Plant Gene Expression Center, a collaborative effort between academia and the U.S. Department of Agriculture, is established to research plant molecular biology, sequence plant genomes, and develop genetically modified plants.

1985

Michael S. Brown and Joseph L. Goldstein win the Nobel Prize in Physiology or Medicine for their work on the regulation of cholesterol in humans. They showed that in families with a history of high cholesterol, individuals who are homozygous for a mutant allele have cholesterol levels several times higher than normal, and those who are heterozygous for the mutant allele have levels about double normal.

1985-1987

Alec Jeffreys publishes methods for DNA fingerprinting as a way to positively link a sample of DNA with that of an individual. In a high-profile rape-murder case in England, DNA fingerprinting excludes a suspect who had been coerced into confessing and later identifies the actual perpetrator.

1985-1987

Robert Sinsheimer, Renato Dulbecco, and Charles DeLisi begin investigating the possibility of sequencing the entire human genome. DeLisi, head of the Department of Energy's Office of Health and Environmental Research, seeks federal funding. After the invention of automated sequencing, the National Research Council and later the Office of Technology Assessment support the idea.

1986

Leroy Hood, a biologist at the California Institute of Technology, constructs a prototype of the automated DNA sequencer, the most important advance in DNA sequencing technology since Gilbert and

Maxam, and Sanger, developed their sequencing methods in the 1970's. Automated sequencing replaces radioactive labels for identifying the four DNA bases with colored fluorescent dyes. Each of the four DNA bases is coded with a different dye to eliminate the need to run separate reactions. Laser and computer technology are integrated at the end stage to gather data. The result is safer, more accurate, and more rapid DNA sequencing.

1986

R. M. Myers, K. Tilly, and T. Maniatis use a saturation mutagenesis procedure to test the effect of mutations in the promoter region of the mouse beta-major globin gene. Mutations in the conserved TATA and CAAT boxes significantly alter transcription, whereas most mutations elsewhere do not, evidence that that these conserved regions have an important functional role in the initiation of transcription.

1987

Rebecca L. Cann, Mark Stoneking, and Allan C. Wilson propose on the basis of RFLP analysis of mitochondrial DNA from humans of diverse geographic origins that a single female, who lived in Africa about 200,000 years ago, is the common ancestor of the mitochondrial DNA of all living humans, the so-called "mitochondrial Eve" hypothesis.

1987

Frostban, a genetically engineered bacterium designed to prevent freezing, is tested on strawberries in California. These bacteria are released under controlled conditions outdoors, where it is hoped they will grow on the strawberries and prevent the fruit from being damaged by frost early in the growing season. The environmental release of recombinant organisms is an important and controversial step in the application of genetic engineering.

1987

Calgene receives a patent for a DNA sequence that extends the shelf life of genetically modified tomatoes.

1987

Carol Greider and Elizabeth Blackburn, using the model organism *Tetrahymena* (a protozoan), report evidence that telomeres are regenerated by an enzyme called telomerase with an RNA component. Based on the action of DNA polymerase, telomeres (located at the tips of chromosomes) should become shorter during each round of cell division. Telomerase, however, is found to extend telomere DNA, restoring the telomeres. Research in this field sparks interest in the possibility that declining levels of telomerase may contribute to aging and that the expression of this enzyme in tumor cells may be a factor in cancer.

1988

The Human Genome Organization (HUGO) is founded to coordinate and collect data from international efforts to sequence the human genome.

1988

The U.S. Congress establishes the National Center for Biotechnology Information (NCBI) as a division of the National Library of Medicine at the National Institutes of Health as a clearinghouse for molecular genetic information.

1988

The Food and Drug Administration (FDA) approves the sale of recombinant TPA (tissue plasminogen activator) as a treatment for blood clots. TPA shows promise in helping victims recovering from heart attack and stroke.

1989

Francis Collins, Lap-Chee Tsui, and researchers at Toronto's Hospital for Sick Children use chromosome walking and jumping to locate and isolate the *CF* gene, which codes the cystic fibrosis transmembrane conductance regulator (CFTR) protein. Mutations in this gene are responsible for cystic fibrosis.

1990

Maynard Olsen, Leroy Hood, Charles Cantor, and David Botstein publish "A Common Language for Physical Mapping of the Human Genome," in which they propose the use of "sequence tagged sites" to coordinate human genome sequencing efforts across laboratories. Their proposal is widely adopted and contributes to collaboration in the genome sequencing and assembly effort.

1990

Madan K. Bhattacharyya, Alison M. Smith, T. H. Noel Ellis, Cliff Hedley, and Cathie Martin publish the first molecular characterization of a gene Mendel studied, the gene that governs seed shape in garden pea. They find that the mutation that causes wrinkled seeds is a transposable element insertion.

1990

Gene therapy for severe combined immunodeficiency (SCID) is tested in clinical trials and becomes one of the first successful applications of somatic cell gene therapy. The first patient is a four-year-old girl who was born without a functioning immune system as a result of a faulty gene that makes an enzyme called ADA (adenosine deaminase).

1990

The Human Genome Project officially begins, headed by James Watson, under the auspices of the National Institutes of Health, National Center for Human Genome Research. The project is to be completed by the year 2005. The ambitious project is designed to sequence and assemble the entire euchromatic portion of the human genome and annotate it with identification of its genes and other features. Also included as part of the Human Genome Project is genome sequencing of several model organisms.

1990

At the Plant Gene Expression Center, biologist Michael Fromm announces the use of a high-speed "gene gun" to transform corn. Gene guns are used to inject genetic material directly into cells via DNA-coated microparticles.

1991

J. Craig Venter of the National Institutes of Health demonstrates the use of automated sequencing and expressed sequence tags (ESTs)—cloned sequences of complementary DNA (cDNA) molecules stored in "libraries"—to identify genes and their functions rapidly and accurately.

1991

Steven T. Warren and coworkers discover that most mutations responsible for fragile X syndrome are trinucleotide repeat expansions. Several other genetic disorders, including Huntington's disease, are found to also be a consequence of trinucleotide expansions.

1991

Scientists at the Howard Hughes Medical Institute, Yale University School of Medicine determine the DNA sequence of the ancient chromosome fusion site in human chromosome 2 and find that it was a head-to-head fusion within the telomeres of the two chromosomes.

1992

One of the first major accomplishments of the Human Genome Project is publication of an RFLP linkage map of human chromosomes with 1,416 mapped markers spanning 92 percent of the genome. Also, by 1992, physical maps of chromosome 21 and the Y chromosome had been completed with overlapping contigs in the euchromatic regions.

1992

The National Research Council in the United States publishes standards for the use of DNA technology in forensic science.

1992

The GenBank database, which will become the principal centralized database for deposition of DNA sequences, is transferred from the Department of Energy to the National Center for Biotechnology Information (NCBI).

1993

The gene that is mutated in people who have Huntington's disease is isolated and sequenced, ten years after its chromosomal region was first identified. Fifty-eight scientists collaborated on the project. The mutations responsible for Huntington's disease are trinucleotide expansion mutations within the protein-encoding portion of the gene.

1993

Michael Smith wins the Nobel Prize in Chemistry for developing "site-directed mutagenesis," a means of reconfiguring genes to produce altered proteins with distinct properties.

1994

The FDA approves the bovine hormone known as BST or BGH. The hormone is made from recombinant bacteria containing a genetically engineered version of the gene encoding BST. When injected into cows, the hormone increases milk production by up to 20 percent. Many supermarkets and manufacturers of dairy products refuse to carry or use milk from BST-injected cows.

1994

The FDA gives approval for the marketing of the Flavr Savr tomato. This genetically altered tomato can be ripened on the vine before being picked and transported. Because the spoilage process is prolonged, the tomatoes have longer shelf lives. The product is not successful, due in part to public concerns over genetically modified foods and in part to the mediocre quality of the tomato variety used as the recipient for the modified gene.

1994

Alfred G. Gilman and Martin Rodbell receive the Nobel Prize in Physiology or Medicine for discovering the role of "G" proteins in regulating signal transduction in eukaryotic cells.

1995

A mutation in the gene *BRCA1,* found by Mark Skolnick and others, is implicated in breast cancer. More than any other gene previously identified, this discovery has potential for assessing inherited breast cancer risk.

1995

J. Craig Venter of The Institute for Genome Research (TIGR) announces completion of the first DNA sequence of a nonviral, self-replicating, free-living organism, the bacterium *Haemophilus influenzae,* using "whole-genome random sequencing," nicknamed "shotgun" sequencing. This method, which precludes the need for a preliminary physical map of the genome, speeds the sequencing of other organisms significantly.

1995

Sequencing of the bacterium with the smallest known genome, *Mycoplasma genitalium,* is completed.

1995

Edward B. Lewis, Christiane Nüsslein-Volhard, and Eric Wieschaus win the Nobel Prize in Physiology or Medicine for their work on the genetic control of early development in *Drosophila.*

1995

Nüsslein-Volhard completes a genetic mutation project in zebrafish. Following her earlier work with *Drosophila,* Nüsslein-Volhard used similar techniques to begin an intensive study of development in a vertebrate system. This involved screening thousands of mutant individuals to determine if any had developmental defects.

1996

Kristen L. Kroll and Enrique Amaya develop a technique to produce stable transgenic *Xenopus* (frog) embryos.

1996

A group of more than six hundred researchers sequences the DNA of the yeast *Saccharomyces cerevisеae,* the first eukaryotic organism to have its genome sequenced.

1996

Monsanto introduces genetically engineered soybeans that are resistant to the herbicide glyphosate (Roundup). So-called Roundup-Ready soybeans become one of the most successful genetically engineered agricultural products. Roundup-Ready corn, cotton, and canola soon follow.

1997

Ian Wilmut at the Roslin Institute in Scotland announces the successful cloning of a sheep. The clone is named Dolly, the first mammal cloned from a nucleus extracted from a cell of an adult animal.

1997

The United Nations Educational, Scientific, and Cultural Organization (UNESCO) adopts the Universal Declaration on the Human Genome and Human Rights.

1997

The genomic sequence of the bacterium *Escherichia coli* is reported by Frederick Blattner and colleagues. Although *E. coli* is not the first complete bacterial sequence reported, because of the importance of *E. coli,* this event is a key step in genomic science.

1997

An eight-year court battle between the University of California and Eli Lilly & Co. over the ownership of rights to genetically engineered human insulin is settled on appeal in Lilly's favor. The litigants have spent more than $30 million in the dispute.

1998

The genome of the bacterium *Mycobacterium tuberculosis* is sequenced.

1998

Celera Genomics is founded by former National Institutes of Health researcher J. Craig Venter. Its mission is to sequence the human genome in the private sector, relying on high-throughput automated sequencers and shotgun sequencing.

1998

The genome of the nematode *Caenorhabditis elegans* is the first genome of a multicellular organism to be sequenced.

1998

The Human Genome Project completes physical mapping of human chromosomes.

1999

Laboratory tests suggest that the pollen of corn genetically engineered to contain a protein derived from the bacterium *Bacillus thuringiensis* (*Bt*) endangers monarch butterfly caterpillars. Although later evidence calls the finding into question, it prompts controversy over the environmental impact of transgenic plants.

1999

The first human death attributable to gene therapy during clinical trial is reported when an eighteen-year-old participant in a trial on gene therapy for hereditary ornithine transcarbamylase (OTC) deficiency dies of multiorgan failure caused by a severe immunological reaction to the disarmed adenovirus vector used in the trial.

1999

The publicly funded Berkeley *Drosophila* Genome Project and privately funded Celera Genomics have separately sequenced the *Drosophila melanogaster* genome using different methods. The two projects join forces to annotate the genome, saving the federal government approximately $10 million. They jointly report the results the following year in the May 24 issue of *Science.* Of the fly's 13,601 genes, a significant number are similar in sequence to human genes.

1999

The first human chromosome, chromosome 22, is completely sequenced.

2000

At a meeting in Montreal, Canada, the United Nations Convention on Biological Diversity approves the Cartegena Protocol on Biosafety, which sets the criterion internationally for patenting genetically modified organisms, including agricultural products.

2000

Chromosome 21, the smallest human chromosome, is completely sequenced; it is the second human chromosome to be completed.

2000

It is estimated that more than two-thirds of the processed foods in U.S. markets contain genetically modified ingredients, mostly products derived from soybeans or corn.

2000

The environmental organization Friends of the Earth reveals that materials derived from StarLink, a genetically engineered corn variety meant only for animal fodder, are present in food products sold in the United States. The news ignites public debate over the use of genetically modified food crops.

2000

At a press conference, a team of more than three hundred scientists from throughout the world announces that they have sequenced the genome of a plant, the model organism *Arabidopsis thaliana.*

2000

U.S. president Bill Clinton and British prime minister Tony Blair, via videolink, announce completion of the first draft of the human genome. Accompanying Clinton are Francis Collins, representing the publicly funded Human Genome Project, and J. Craig Venter representing privately funded Celera Genomics. A tenuous and limited agreement for collaboration has been reached between the public and private projects.

2001

Complete sequencing of human chromosomes 20 and 14 is published.

2001

The first working drafts of the human genome sequence are published in *Science* (which reports the results from the private company Celera Genomics, headed by J. Craig Venter) and in *Nature* (reporting the results from the publicly funded Human Genome Project). The number of protein-encoding genes is estimated to be about 30,000, which is less than previously thought but is later shown to be a significant overestimate in more refined drafts of the genome. Other key features include the observation that 43 percent of the human genome consists of transposable elements, many genes reside in clusters of similar genes, derived from duplication of preexisting genes, mutations occur more often in males than in females, and a few hundred of the estimated 30,000 human genes originated as horizontally transferred bacterial genes.

2001

Researchers complete the genomic sequence for rice, *Oryza sativa*, the first food crop to have its genome sequenced.

2001

Scientists report that genetic material from transgenic corn has mysteriously turned up in the genome of native corn species near Oaxaca, Mexico. Mexico banned transgenic crops three years earlier, and the closest known crop was located beyond the range of windborne pollen, an indication that transgenic corn is illegally produced in Mexico.

2001

Leland H. Hartwell, R. Timothy Hunt, and Paul M. Nurse win the Nobel Prize in Physiology or Medicine for their research on the regulation of cell cycles. Hartwell identified a class of genes that controls the cycle, including a gene that initiates it. Nurse cloned and described the genetic model of a key regulator, cyclin dependent kinase (cdk), and Hunt discovered cyclins, a class of regulatory proteins.

2002

The mouse genome sequence is completed, using the shotgun method; it is compared with the draft of the human genome and found to have a high degree of synteny (segments with groups of similar genes in the same spatial arrangement).

2003

Dolly, the first mammal cloned from an adult cell, is euthanized after suffering advanced arthritis and lung disease. Researchers speculate about whether cloned animals age prematurely.

2003

D. Torrents, M. Suyama, E. Zdobnov, and P. Bork identify the total number of pseudogenes (nonfunctional copies of genes) in the human genome as 19,724, almost as many as the number of genes. Approximately 70 percent are processed pseudogenes derived from reverse transcription of mRNAs. They consider their methods to be conservative and suspect that the actual number of pseudogenes is much higher.

2003-2004

The Human Genome Project announces a completed draft of the euchromatic portion of the human genome in April, 2003, two years ahead of schedule. An article describing this draft is published in 2004. Among its most notable findings is a more definitive estimate of the number of protein-encoding genes at "19,238 known genes and 2,188 predicted genes," considerably lower than the estimate of 30,000 genes in the 2001 draft.

2005

The first draft of the chimpanzee genome is published. Among the many discoveries associated with this accomplishment, is the observation that the human and chimpanzee genomes are more than 98 percent similar in aligned regions (more than 99 percent within protein-coding sequences). There is evidence of positive and negative selection, and the evolution of genes in both species is highly correlated. This genome serves as an outgroup for examination of human population genetics.

2005

454 Life Sciences introduces next-generation high-throughput DNA sequencing based on a new technology that allows researchers to sequence 400-600 million nucleotides with 400-500 nucleotide read lengths at high accuracy in a single run. The technology is especially beneficial for genomic projects because it allows rapid large-scale sequencing at a fraction of the time and cost of previous automated sequencing methods.

2006

Andrew Fire and Craig Mello receive the Nobel Prize in Physiology or Medicine for their research on RNA interference (RNAi). RNAi is hailed as one of the most significant discoveries in molecular biology, and this Nobel Prize is considered to be an unusually early award because of its significance. In an attempt beginning in 1986 to produce transgenic petunia plants with overly pigmented purple flowers, Richard Jorgensen had found that the transgenic plants have white flowers with no pigment. In 1998, he explained the phenomenon as a consequence of RNA molecules that interfere with gene expression. This same year, Fire and Mello unravel the causes of this same phenomenon in *C. elegans* and name the mechanism RNAi. Over the ensuing years, RNAi becomes a promising tool for studying gene expression and for treatment of a wide array of genetically based diseases, disorders, and cancers.

2007

The first draft of the rhesus macaque genome is published, offering genome-wide comparison of three primate genomes (human, chimpanzee, and rhesus macaque).

2007

Mario R. Capecchi, Martin J. Evans, and Oliver Smithies receive the Nobel Prize in Physiology or Medicine "for their discoveries of principles for introducing specific gene modifications in mice by the use of embryonic stem cells." This technology, dubbed gene knockout technology, allows researchers to target specific genes and deactivate them as a way of studying their functions, and has revealed much about mammalian gene expression and function.

2008

A team led by Svante Pääbo publishes the sequence of the Neanderthal mitochondrial genome, noting that the sequence has no evidence of human-Neanderthal hybridization.

2009

Completion of Neanderthal nuclear genome sequencing and annotation is announced.

Nancy Morvillo, Ph.D., and Christina J. Moose; updated by Daniel J. Fairbanks, Ph.D.

GLOSSARY

A: the abbreviation for adenine, a purine nitrogenous base found in the structure of both DNA and RNA.

acentric chromosome: a chromosome that does not have a centromere and that is unable to participate properly in cell division; often the result of a chromosomal mutation during recombination.

acquired characteristic: a change in an individual organism brought about by its interaction with its environment.

acrocentric chromosome: a chromosome with its centromere near one end. *See also* metacentric chromosome; telocentric chromosome.

activator: a protein that binds to DNA, thus increasing the expression of a nearby gene.

active site: the region of an enzyme that interacts with a substrate molecule; any alteration in the three-dimensional shape of the active site usually has an adverse effect on the enzyme's activity.

adaptation: the evolution of a trait by natural selection, or a trait that has evolved as a result of natural selection.

adaptive advantage: increased reproductive potential in offspring as a result of passing on favorable genetic information.

adenine (A): a purine nitrogenous base found in the structure of both DNA and RNA.

adenosine deaminase deficiency: a fatal immunodeficiency disorder caused by an insufficiency of the enzyme adenosine deaminase; death commonly occurs when the infant is a few months old.

adenosine triphosphate (ATP): the major energy molecule of cells, produced either through the process of cellular respiration or fermentation; it is also a component of DNA and RNA.

adenovirus: a group of viruses that cause respiratory illnesses, such as the common cold; they have been genetically altered and used for treatment of cancer and cystic fibrosis.

adult stem cell: an undifferentiated cell found among differentiated cells in a tissue or organ of an adult organism.

agarose: a chemical substance derived from algae and used to create gels for the electrophoresis of nucleic acids.

aggression: behavior directed toward causing harm to others.

***Agrobacterium tumefaciens*:** a species of bacteria that causes disease in some plants and is able to transfer genetic information, in the form of Ti plasmids, into plant cells; modified Ti plasmids can be used to produce transgenic plants.

Alagille syndrome: a genetic disorder of the liver of infants and young children whereby bile builds up in the liver because the bile ducts are abnormally narrow or absent; the child may present with jaundice, fatty deposits under the skin, facial deformities, growth retardation, and heart, kidney, eye, and vertebrae abnormalities.

albinism: the absence of pigment such as melanin in eyes, skin, hair, scales, or feathers or of chlorophyll in plant leaves and stems.

albino: a genetic condition in which an individual does not produce the pigment melanin in the skin; other manifestations of the trait may be seen in the pigmentation of the hair or eyes; albino individuals occur in many animals and plants and are due to the absence of a variety of different pigments.

alcoholism: a medical diagnosis given when there is repeated use of alcohol over the course of at least a year, despite the presence of negative consequences; tolerance, withdrawal, uncontrolled use, unsuccessful efforts to quit, considerable time spent getting or using the drug, and a decrease in other important activities because of the use are part of this condition.

algorithm: a mathematical rule or procedure for solving a specific problem.

alkaptonuria: a genetic disorder, first characterized by geneticist Archibald Garrod, in which a compound called homogentisic acid accumulates in the cartilage and is excreted in the urine of affected individuals, turning both the cartilage and the urine black (the name of the disorder literally means "black urine"); the specific genetic defect involves an inability to process by-products of phenylalanine and tyrosine metabolism.

allele: a form of a gene at a locus; each locus in an individual's chromosomes has two alleles, which may be the same or different.

allele frequency: the proportion of all the genetic variants at a locus within a population of organisms.

allergy: an abnormal immune response to a substance that does not normally provoke an immune response or that is not inherently dangerous to the body (such as plant pollens, dust, or animal dander).

allopatric speciation: a model of speciation in which parts of a population may become geographically isolated, effectively preventing interbreeding, and over time may develop differences that lead to reproductive isolation and the development of a new species.

allopolyploid: a type of polyploid species that contains genomes from more than one ancestral species.

altruism: behavior that benefits others at the evolutionary (reproductive) cost of the altruist.

Alu sequence: a repetitive DNA sequence of unknown function, approximately three hundred nucleotides long, scattered throughout the genome of primates; the name comes from the presence of recognition sites for the restriction endonuclease Alu I in these sequences.

Alzheimer's disease: a degenerative brain disorder, usually found among the elderly, characterized by brain lesions leading to loss of memory, personality changes, and deterioration of higher mental functions.

amber codon: a stop codon (UAG) found in messenger RNA (mRNA) molecules that signals termination of translation.

ambiguous genitalia: external sexual organs that are not clearly male or female.

Ames test: a test devised by molecular biologist Bruce Ames for determining the mutagenic or carcinogenic properties of various compounds based on their ability to affect the nutritional characteristics of the bacterium *Salmonella typhimurium.*

amino acid: a nitrogen-containing compound used as the building block of proteins (polypeptides); in nature, there are twenty amino acids that can be used to build proteins.

aminoacyl tRNA: a transfer RNA (tRNA) molecule with an appropriate amino acid molecule attached; in this form, the tRNA molecule is ready to participate in translation.

amniocentesis: a procedure in which a small amount of amniotic fluid containing fetal cells is withdrawn from the amniotic sac surrounding a fetus; fetal cells, found in the fluid, are then tested for the presence of genetic abnormalities.

amniotic fluid: the fluid in which the fetus is immersed during pregnancy.

amyloid plaques: protein deposits in the brain formed by fragments from amyloid precursor proteins; amyloid plaques are characteristic of Alzheimer's disease.

anabolic steroids: drugs derived from androgens and inappropriately used to enhance performance in sports.

anabolism: the part of the cell's metabolism concerned with synthesis of complex molecules and cell structures.

anaphase: the third phase in the process of mitosis; in anaphase, sister chromatids separate at the centromere and migrate toward the poles of the cell.

anaphylaxis: a severe, sometimes fatal allergic reaction often characterized by swelling of the air passages, leading to inability to breathe.

ancient DNA: DNA isolated from archaeological artifacts or fossils; it is typically extensively degraded.

androgen receptors: molecules in the cytoplasm of cells that join with circulating male hormones.

androgens: steroid hormones that cause masculinization.

anencephalus: a neural tube defect characterized by the failure of the cerebral hemispheres of the brain and the cranium to develop normally.

aneuploid: a cell or individual with one or a few missing or extra chromosomes.

angstrom: a unit of measurement equal to one ten-millionth of a millimeter; a DNA molecule is 20 angstroms wide.

animal cloning: the process of generating a genetic duplicate of an animal starting with one of its differentiated cells.

animal model: an animal that has a disorder that is similar to a human disorder; scientists can then study the disorder and transfer genes into the animals to test therapies.

annealing: the process by which two single-stranded nucleic acid molecules are converted into a double-stranded molecule through hydrogen bonding between complementary base pairs.

anthrax: an acute bacterial disease caused by *Bacillus anthracis* that affects animals and humans and that is especially deadly in its pulmonary form.

antibiotic: any substance produced naturally by a microorganism that inhibits the growth of other microorganisms; antibiotics are important in the treatment of bacterial infections.

antibody: an immune protein (immunoglobulin) that specifically recognizes an antigen; produced by B cells of the immune system.

anticodon: the portion of a transfer RNA (tRNA) molecule that is complementary in sequence to a codon in a messenger RNA (mRNA) molecule; because of this complementarity, the tRNA molecule can bind briefly to mRNA during translation and direct the placement of amino acids in a polypeptide chain.

antigen: any molecule that is capable of being recognized by an antibody molecule or of provoking an immune response.

antigenic drift or shift: minor changes in the H and N proteins of the influenza virus that enable the virus to evade the immune system of a potential host.

antioxidant: a molecule that preferentially reacts with free radicals, thus keeping them from reacting with other molecules and causing cellular damage.

antiparallel: a characteristic of the Watson-Crick double-helix model of DNA, in which the two strands of the molecule can be visualized as oriented in opposite directions; this characteristic is based on the orientation of the deoxyribose molecules in the sugar-phosphate backbone of the double helix.

antirejection medication: drugs developed to counteract the human body's natural immune system's reaction to transplanted organs.

antisense: a term referring to any strand of DNA or RNA that is complementary to a coding or regulatory sequence, for example, the strand opposite the coding strand (the sense strand) in DNA is called the antisense strand.

antisense RNA: a small RNA molecule that is complementary to the coding region of a messenger RNA (mRNA) and when bound to the mRNA prevents it from being translated.

antitoxin: a vaccine containing antibodies against a specific toxin.

Apo-B: a protein essential for cholesterol transport.

apoptosis: cell "suicide" occurring after a cell is too old to function properly, as a response to irreparable genetic damage, or as a function of genetic programming; apoptosis prevents cells from developing into a cancerous state and is a natural event during many parts of organismal development.

Archaea: the domain of life that includes diverse prokaryotic organisms distinct from the historically familiar bacteria and which often require severe conditions for growth, such as high temperatures, high salinity, or lack of oxygen.

artificial selection: selective breeding of desirable traits, typically in domesticated organisms.

ascomycetes: organisms of the phylum Ascomycota, a group of fungi known as the sac fungi, which are characterized by a saclike structure, the ascus.

ascospore: a haploid spore produced by meiosis in ascomycete fungi.

ascus: a reproductive structure, found in ascomycete fungi, that contains ascospores.

asexual reproduction: reproduction of cells or organisms without the transfer or reassortment of genetic material; results in offspring that are genetically identical to the parent.

assortative mating: mating that occurs when individuals make specific mate choices based on the phenotype or appearance of others.

ataxia telangiectasia: a genetic disorder caused by the ataxia-telangiectasia mutated gene; the portion of the brain that controls motor movement and speech deteriorates. The affected child commonly has normal or above normal intelligence, develops slurred speech and poor balance, and has delayed motor development . Telangiectasias develop in the corners of the eyes and on the surfaces of the cheeks and ears. These children are also at risk for cancer, such as acute lymphocytic leukemia and lymphoma. Many children also have a weakened immune system making them prone to infection. Ataxia telangiectasia is commonly fatal before age twenty.

ATP: *See* adenosine triphosphate.

ATP synthase: the enzyme that synthesizes ATP.

autoimmune disorders: chronic diseases that arise from a breakdown of the immune system's ability to distinguish between the body's own cells (self) and foreign substances, leading to an individual's immune system attacking the body's own organs or tissues.

autoimmune response: an immune response of an organism against its own cells.

automated fluorescent sequencing: a modification of dideoxy termination sequencing that uses fluorescent markers to identify the terminal nucleotides, allowing the automation of sequencing.

autopolyploid: a type of polyploid species that contains more than two sets of chromosomes from the same species.

autosomal dominant allele: an allele of a gene (locus) on one of the nonsex chromosomes that is always expressed, regardless of the form of the other allele at the same locus.

autosomal recessive allele: an allele of a gene (locus) that will be expressed only if there are two identical copies at the same locus.

autosomal trait: a trait that typically appears just as frequently in either sex because an autosomal chromosome, rather than a sex chromosome, carries the gene.

autosomes: non-sex chromosomes; humans have forty-four autosomes.

auxotrophic strain: a mutant strain of an organism that cannot synthesize a substance required for growth and therefore must have the substance supplied in the growth medium.

azoospermia: the absence of spermatozoa in the semen.

B cells: a class of white blood cells (lymphocytes) derived from bone marrow and responsible for antibody-directed immunity.

B-DNA: the predominant form of DNA in solution and in the cell; a right-handed double helix most similar to the Watson-Crick model. *See also* Z-DNA.

B lymphocytes: *See* B cells.

B memory cells: descendants of activated B cells that are long-lived and that synthesize large amounts of antibodies in response to a subsequent exposure to the antigen, thus playing an important role in secondary immunity.

***Bacillus thuringiensis* (*Bt*):** a species of bacteria that produces a toxin deadly to caterpillars, moths, beetles, and certain flies.

backcross: a cross involving offspring crossed with one of the parents. *See also* cross.

bacterial artificial chromosomes (BACs): cloning vectors used to clone large DNA fragments (up to 500 kb) that can be readily inserted in a bacterium, such as *Escherichia coli*.

bacteriophage: a virus that infects bacterial cells; often simply called a phage.

baculovirus: a type of virus that is capable of causing disease in a variety of insects.

Barr body: a darkly staining structure primarily present in female cells, believed to be an inactive X chromosome; used as a demonstration of the Lyon hypothesis.

base: a chemical subunit of DNA or RNA that encodes genetic information; in DNA, the bases are adenine (A), cytosine (C), guanine (G), and thymine (T); in RNA, thymine is replaced by uracil (U).

base pair (bp): often used as a measure of the size of a DNA fragment or the distance along a DNA molecule between markers; both the singular and plural are abbreviated bp.

base pairing: the process by which bases link up by hydrogen bonding to form double-stranded molecules of DNA or loops in RNA; in DNA, adenine (A) always pairs with thymine (T), and cytosine (C) pairs with guanine (G); in RNA, uracil (U) replaces thymine.

beta-amyloid peptide: the main constituent of the neuritic plaques in the brains of Alzheimer's patients.

bidirectional replication: a characteristic of DNA replication involving synthesis of DNA in both directions away from an origin of replication.

binary fission: cell division in prokaryotes in which the plasma membrane and cell wall grow inward and divide the cell in two.

biochemical pathway: the steps in the production or breakdown of biological chemicals in cells; each step usually requires a specific enzyme.

bioethics: the study of human actions and goals in a framework of moral standards relating to use and abuse of biological systems.

bioinformatics: the application of information technology to the management of biological information to organize data and extract meaning; a hybrid discipline that combines elements of computer science, information technology, mathematics, statistics, and molecular genetics.

biological clocks: genetically and biochemically based systems that regulate the timing and/or duration of biological events in an organism; examples of processes controlled by biological clocks include circadian rhythms, cell cycles, and migratory restlessness.

biological determinism: the concept that all characteristics of organisms, including behavior, are determined by the genes the organism possesses; it is now generally accepted that the characteristics of organisms are determined both by genes and environment.

biological weapon: a delivery system or "weaponization" of such pathological organisms as bacteria and viruses to cause disease and death in people, animals, or plants.

biometry: the measurement of biological and psychological variables.

biopesticides: chemicals or other agents derived from or involving living organisms that can be used to control the population of a pest species.

bioremediation: biologic treatment methods to clean up contaminated water and soils.

biotechnology: the use of biological molecules or organisms in industrial or commercial products and techniques.

bioterrorism: use of organisms as instruments or weapons of terror; for example, deliberate introduction of smallpox, anthrax, or other diseases in civilian populations.

blastocyst: a preimplantation embryo consisting of a hollow ball of two layers of cells.

blind proficiency test: a competency assessment in which laboratory personnel are unaware that it is taking place.

blood type: one of the several groups into which blood can be classified based on the presence or absence of certain molecules called antigens on the red blood cells.

blotting: the transfer of nucleic acids or proteins separated by gel electrophoresis onto a filter paper, which allows access by molecules that will interact with only one specific sequence or molecule.

***BRCA1* and *BRCA2* genes:** the best known examples of genes associated with inherited breast cancers.

***Bt* toxin:** a toxic compound naturally synthesized by bacterium *Bacillus thuringiensis*, which kills insects.

C: the abbreviation for cytosine, a pyrimidine nitrogenous base found in the structure of both DNA and RNA.

C terminus: the end of a polypeptide with an amino acid that has a free carboxyl group.

C-value: the characteristic genome size for a species.

CAG expansion: a mutation-induced increase in the number of consecutive CAG nucleotide triplets in the coding region of a gene.

callus: a group of undifferentiated plant cells growing in a clump.

cAMP: *See* cyclic adenosine monophosphate.

cancer: a disease in which there is unrestrained growth and reproduction of cells, loss of contact inhibition, and, eventually, metastasis (the wandering of cancer cells from a primary tumor to other parts of the body); invasion of various tissues and organs by cancer cells typically leads to death.

candidate gene: a gene suspected of causing a specific disease.

capsid: the protective protein coating of a virus particle.

carcinogen: any physical or chemical cancer-causing agent.

carrier: a healthy individual who has one normal allele and one defective allele for a recessive genetic disease.

carrier screening: specialized testing to see whether an individual carries a mutated gene for a specific condition.

catabolism: the part of the cell's metabolism concerned with the breakdown of complex molecules, usually as an energy-generating mechanism.

catabolite repression: a mechanism of operon regulation involving an enzyme reaction's product used as a regulatory molecule for the operon that encodes the enzyme; a kind of feedback inhibition.

cDNA: *See* complementary DNA.

cDNA library: a collection of clones produced from all the RNA molecules in the cells of a particular organism, often from a single tissue. *See also* complementary DNA.

ceiling principle: a method for estimating the frequency in which a DNA's profile occurs in a population containing a variety of ethnic groups.

cell culture: growth and maintenance of cells or tissues in laboratory vessels containing a precise mixture of nutrients and hormones.

cell cycle: the various growth phases of a cell, which include (in order) G1 (gap phase 1), S (DNA synthesis), G2 (gap phase 2), and M (mitosis).

cell differentiation: a process during which a cell becomes specialized as a specific type of cell, such as a neuron, or undergoes programmed cell death (apoptosis).

cell line: a cell culture maintained for an indeterminate time.

cell signaling: communication between cells that occurs most commonly when one cell releases a specific "signaling" molecule that is received and recognized by another cell.

centiMorgan (cM): a unit of genetic distance between genes on the same chromosome, equal to a recombination frequency of 1 percent; also called a map unit, since these distances can be used to construct genetic maps of chromosomes.

central dogma: a foundational concept in modern genetics stating that genetic information present in the form of DNA can be converted to the form of messenger RNA (or other types of RNA) through transcription and that the information in the form of mRNA can be converted into the form of a protein through translation.

centriole: a eukaryotic cell structure involved in cell division, possibly with the assembly or disassembly of the spindle apparatus during mitosis and meiosis; another name for this organelle is the microtubule organizing center (MTOC).

centromere: a central region where a pair of chromatids are joined before being separated during anaphase of mitosis or meiosis; also, the region of the chromatids where the microtubules of the spindle apparatus attach.

checkpoint: the time in the cell cycle when molecular signals control entry to the next phase.

chemical mutagens: chemicals that can directly or indirectly create mutations in DNA.

chiasma: the point at which two homologous chromosomes exchange genetic material during the process of recombination; the word literally means "crosses," which refers to the appearance of these structures when viewed with a microscope.

chi-square analysis: a nonparametric statistical analysis of data from an experiment to determine how well the observed data correlate with the expected data.

chloroplast: the cellular organelle in plants responsible for photosynthesis.

chloroplast DNA (cpDNA): circular DNA molecules found in multiple copies in chloroplasts; they contain some of the genes required for chloroplast functions.

cholera: an intestinal disease caused by the bacteria *Vibrio cholerae*, which is often spread by water contaminated with human waste.

chorion: an embryonic structure that develops into the placenta; cells from this structure have the same genetic composition as the fetus and are sampled during chorionic villus testing.

chorionic villus sampling: a procedure in which fetal cells are obtained from an embryonic structure called the chorion and analyzed for the presence of genetic abnormalities in the fetus.

chromatid: one half of a chromosome that has been duplicated in preparation for mitosis or meiosis; each chromatid is connected to its sister chromatid by a centromere.

chromatin: the form chromosomes take when not undergoing cell division; a complex of fibers composed of DNA, histone proteins, and nonhistone proteins.

chromatography: a separation technique involving a mobile solvent and a stationary, adsorbent phase.

chromosome: the form in which genetic material is found in the nucleus of a cell; composed of a single DNA molecule that is extremely tightly coiled, usually visible only during the processes of mitosis and meiosis.

chromosome jumping: similar to chromosome walking, but involving larger fragments of DNA and thus resulting in faster analysis of longer regions of DNA. *See also* chromosome walking.

chromosome map: a diagram showing the locations of genes on a particular chromosome; generated through analysis of linkage experiments involving those genes.

chromosome mutation: a change in chromosome structure caused by chromosome breakage followed by improper rejoining; examples include deletions, insertions, inversions, and translocations.

chromosome puff: an extremely unwound or uncoiled region of a chromosome indicative of a transcriptionally active region of the chromosome.

chromosome theory of inheritance: a concept, first proposed by geneticists Walter Sutton and Theodor Boveri, that genes are located on chromosomes and that the inheritance and movement of chromosomes during meiosis explain Mendelian principles on the cellular level.

chromosome walking: a molecular genetics technique used for analysis of long DNA fragments; the name comes from the technique of using previously cloned and characterized fragments of DNA to "walk" into uncharacterized regions of the chromosome that overlap with these fragments. *See also* chromosome jumping.

circadian rhythm: a cycle of behavior, approximately twenty-four hours long, that is expressed independent of environmental changes.

cirrhosis: a disease of the liver, marked by the development of scar tissue that interferes with organ function, that can result from chronic alcohol consumption.

cistron: a unit of DNA that is equivalent to a gene; it encodes a single polypeptide.

clinical trial: an experimental research study used to determine the safety and effectiveness of a medical treatment or drug.

clone: a molecule, cell, or organism that is a perfect genetic copy of another.

cloning: the technique of making a perfect genetic copy of an item such as a DNA molecule, a cell, or an entire organism.

cloning vector: a DNA molecule that can be used to transport genes of interest into cells, where these genes can then be copied.

codominance: a genetic condition involving two alleles at a locus in a heterozygous organism; each of these alleles is fully expressed in the phenotype of the organism.

codon: a group of three nucleotides in messenger RNA (mRNA) that represent a single amino acid in the genetic code; this is mediated through binding of a transfer RNA (tRNA) anticodon to the codon during translation.

color blindness: an inherited condition in people whose eyes lack one or more of the three color receptors.

complementary base pairing: hydrogen bond formation in DNA and RNA that occurs only between cytosine and guanine (in both DNA and RNA) or between adenine and thymine (in DNA) or adenine and uracil (in RNA).

complementary DNA (cDNA): a DNA molecule that is synthesized using messenger RNA (mRNA) as a template and that is catalyzed by the enzyme reverse transcriptase.

complementation testing: performing a cross between two individuals with the same phenotype to determine whether or not the mutations occur within the same gene.

composite transposon: a transposable element that contains genes other than those required for transposition.

concerted evolution: a process in which the members of a gene family evolve together.

concordance: the presence of a trait in both members of a pair of twins.

cones: the light-sensitive structures in the retina that are the basis for color vision.

congenital defect: a defect or disorder that occurs during prenatal development.

conjugation: a form of genetic transfer among bacterial cells involving the F pilus.

consanguine: of the same blood or origin; in genetics, the term implies the sharing of genetic traits or characteristics from the same ancestors (as cousins, for example).

consensus sequence: a sequence with no or only slight differences commonly found in DNA molecules from various sources, implying that the sequence has been actively conserved and plays an important role in some genetic process.

contig: A map that shows regions on a chromosome where neighboring portions of DNA overlap; the map provides information about a large segment of a genome by looking at a series of overlapping clones .

cordocentesis: removing a sample of blood for prenatal testing from the umbilical cord using ultrasound guidance.

cosmid: a cloning vector partially derived from genetic sequences of lambda, a bacteriophage; cosmids are useful in cloning relatively large fragments of DNA.

craniosynostosis: a congenital defect involving the skull bones of an infant; the bones are fused together at birth requiring surgical intervention to allow for normal brain development.

cross: the mating of individuals to produce offspring by sexual reproduction.

crossing over: the exchange of genetic material between two homologous chromosomes during prophase I of meiosis, providing an important source of genetic variation; also called "recombination" or crossover.

cultivar: a variety of plant developed through controlled breeding techniques.

cyclic adenosine monophosphate (cAMP): an important cellular molecule involved in cell signaling and regulation pathways.

cyclins: a group of eukaryotic proteins with characteristic patterns of synthesis and degradation during the cell cycle; part of an elaborate mechanism of cell cycle regulation, and a key to the understanding of cancer.

cystic fibrosis: the most common recessive lethal inherited disease among Caucasians in the United States and the United Kingdom.

cytogenetics: the study of chromosome number and

structure, including identification of abnormalities.

cytokines: soluble intercellular molecules produced by cells such as lymphocytes that can influence the immune response.

cytokinesis: the division of the cytoplasm, typically occurring in concert with nuclear division (mitosis or meiosis).

cytoplasmic inheritance: *See* extranuclear inheritance.

cytosine (C): a pyrimidine nitrogenous base found in the structure of both DNA and RNA.

cytoskeleton: the structure, composed of microtubules and microfilaments, that gives shape to a eukaryotic cell, enables some cells to move, and assists in such processes as cell division.

dalton: a unit of molecular weight equal to the mass of a hydrogen atom; cellular molecules such as proteins are often measured in terms of a kilodalton, equal to 1,000 daltons.

daughter cells: cells that result from cell division.

deamination: the removal of an amino group from an organic molecule.

degenerate: refers to a property of the genetic code via which two or more codons can code for the same amino acid.

degradation: the act of chemically or physically breaking down DNA.

deletion: a type of chromosomal mutation in which a genetic sequence is lost from a chromosome, usually through an error in recombination.

denaturation: changes in the physical shape of a molecule caused by changes in the immediate environment, such as temperature or pH level; denaturation usually involves the alteration or breaking of various bonds within the molecule and is important in DNA and protein molecules.

deoxyribonucleic acid (DNA): the genetic material found in all cells; DNA consists of nitrogenous bases (adenine, guanine, cytosine, and thymine), sugar (deoxyribose), and phosphate.

deoxyribose: a five-carbon sugar used in the structure of DNA.

diabetes: a syndrome in which the body cannot metabolize glucose appropriately.

diakinesis: a subphase of prophase I in meiosis in which chromosomes are completely condensed and position themselves in preparation for metaphase.

dicentric chromosome: a chromosome with two centromeres, usually resulting from an error of recombination.

dideoxy termination sequencing: *See* Sanger sequencing.

differentiation: the series of changes necessary to convert an embryonic cell into its final adult form, usually with highly specialized structures and functions.

dihybrid: an organism that is hybrid for each of two genes—for example, *AaBb*; when two dihybrid organisms are mated, the offspring will appear in a 9:3:3:1 ratio with respect to the traits controlled by the two genes.

diphtheria: an acute bacterial disease caused by *Corynebacterium diphtheriae*; symptoms are primarily the result of a toxin released by the bacteria.

diploid: a cell or organism with two complete sets of chromosomes, usually represented as 2*N*, where *N* stands for one set of chromosomes; for example, humans have two sets of twenty-three chromosomes in their somatic cells, making them diploid.

diplotene: a subphase of prophase I in meiosis in which synapsed chromosomes begin to move apart and the chiasmata are clearly visible.

discontinuous replication: replication on the lagging strand of a DNA molecule, resulting in the formation of Okazaki fragments. *See also* Okazaki fragments.

discontinuous variation: refers to a set of related phenotypes that are distinct from one another, with no overlapping.

discordant: a condition where a set of twins does not display the same characteristics.

disjunction: the normal division of chromosomes that occurs during meiosis or mitosis; the related term "nondisjunction" refers to problems with this process.

disomy: a case in which both copies of a chromosome come from a single parent, rather than (as is usual) one being maternal and one being paternal.

dizygotic: developed from two separate zygotes; fraternal twins are dizygotic because they develop from two separate fertilized ova (eggs).

DNA: *See* deoxyribonucleic acid.

DNA fingerprinting: a DNA test used by forensic scientists to aid in the identification of criminals or to resolve paternity disputes, which involves looking at known, highly variable DNA sequences; more correctly called DNA genotyping.

DNA footprinting: a molecular biology technique involving DNA-binding proteins that are allowed to bind to DNA; the DNA is then degraded by DNases, and the binding sites of the proteins are revealed by the nucleotide sequences protected from degradation.

DNA gyrase: a bacterial enzyme that reduces tension in DNA molecules that are being unwound during replication; a type of cellular enzyme called a topoisomerase.

DNA library: a collection of cloned DNA fragments from a single source, such as a genome, chromosome, or set of messenger RNA (mRNA) molecules; most common examples are genomic and cDNA libraries.

DNA ligase: a cellular enzyme used to connect pieces of DNA together; important in genetic engineering procedures.

DNA polymerase: the cellular enzyme responsible for making new copies of DNA molecules through replication of single-stranded DNA template molecules or, more rarely, using an RNA template molecule as in the case of RNA-dependent DNA polymerase or reverse transcriptase.

DNA replication: synthesis of new DNA strands complementary to template strands resulting in new double-stranded DNA molecules comprising the old template and the newly synthesized strand joined by hydrogen bonds; described as a semiconservative process in that half (one strand) of the original template is retained and passed on.

DNase: refers to a class of enzymes, deoxyribonucleases, which specifically degrade DNA molecules.

domain: the highest-level division of life, sometimes called a superkingdom.

dominant: an allele or a trait that will mask the presence of a recessive allele or trait.

dosage compensation: an equalization of gene products that can occur whenever there are more or fewer genes for specific traits than normal.

double helix: a model of DNA structure proposed by molecular biologists James Watson and Francis Crick; the major features of this model are two strands of DNA wound around each other and connected by hydrogen bonds between complementary base pairs.

down-regulation: generally used in reference to gene expression and refers to reducing the amount that a gene is transcribed and/or translated; up-regulation is the opposite.

Down syndrome: a genetic defect caused by possession of an extra copy of chromosome 21; symptoms include mental retardation, mongoloid facial features, and premature aging.

downstream: in relation to the left-to-right direction of DNA whose nucleotides are arranged in sequence with the 5′ carbon on the left and the 3′ on the right, downstream is to the right.

drug resistance: a phenomenon in which pathogens no longer respond to drug therapies that once controlled them; resistance can arise by recombination, by mutation, or by several methods of gene transfer, and is made worse by misuse of existing drugs.

duplication: a type of chromosomal mutation in which a chromosome region is duplicated because of an error in recombination during prophase I of meiosis; thought to play an important role in gene evolution.

dwarfism: the condition of adults of short stature who are less than 50 inches in height, which can be caused by genetic factors, endocrine malfunction, acquired conditions, or growth hormone deficiency; many dwarfs prefer to be called "little people."

dye blobs: an artifact that may occur during short tandem repeats testing.

dysplasia: abnormal cell or tissue growth or development.

***E. coli*:** See *Escherichia coli.*

Edwards syndrome: a genetic disorder in which there is an extra copy of chromosome 18; infants with this disorder commonly have small eyes, deformed ears, heart defects, and severe intellectual impairment. These children rarely live longer than 1 year; also known as trisomy 18.

electron transport chain: a series of protein complexes that use high-energy electrons to do work such as pumping H^+ ions out of the mitochondrial matrix into the intermembrane space as a way of storing energy that is then used by ATP synthase to make ATP.

Electrophoresis: *See* gel electrophoresis.

embryo: the term for a complex organism (particularly humans) during its earliest period of development, the stage of development that begins at fertilization and ends with the eighth week of development, after which the embryo is called a fetus.

embryology: the study of developing embryos.

embryonic stem cell: a cell derived from an early embryo that can replicate indefinitely in vitro and can differentiate into other cells of the developing embryo.

emerging disease: a disease whose incidence in humans or other target organisms has increased.

empiric risk: estimated risk for an individual developing a certain condition based on observations of other families affected by the condition.

endemic: prevalent and recurring in a particular geographic region; for example, an organism that is specific to a particular region is characterized as endemic to that region.

endocrine gland: a gland that secretes hormones into the circulatory system.

endonuclease: an enzyme that degrades a nucleic acid molecule by breaking phosphodiester bonds within the molecule.

endosymbiotic hypothesis: a hypothesis stating that mitochondria and chloroplasts were once free-living bacteria that entered into a symbiotic relationship with early pre-eukaryotic cells; structural and genetic similarities between these organelles and bacteria provide support for this hypothesis.

enhancer: a region of a DNA molecule that facilitates the transcription of a gene, usually by stimulating the interaction of RNA polymerase with the gene's promoter.

enzyme: a protein that acts as a catalyst to speed up or facilitate a specific biochemical reaction in a cell.

epigenesis: the formation of differentiated cell types and specialized organs from a single, homogeneous fertilized egg cell without any preexisting structural elements.

epistasis: a genetic phenomenon in which a gene at one locus influences the expression of a second gene at another locus, usually by masking the effect of the second gene; however, only one trait is being controlled by these two genes, so epistasis is characterized by modified dihybrid ratios.

equational division: refers to meiosis II, in which the basic number of chromosome types remains the same although sister chromatids are separated from one another; after equational division occurs, functional haploid gametes are present.

Escherichia coli: a bacterium widely studied in genetics research and extensively used in biotechnological applications.

estrogens: steroid hormones or chemicals that stimulate the development of female sexual characteristics and control the female reproductive cycle.

ethidium bromide: a chemical substance that inserts itself (intercalates) into the DNA double helix; when exposed to ultraviolet light, ethidium bromide fluoresces, making it useful for the visualization of DNA molecules in molecular biology techniques.

etiology: the cause or causes of a disease or disorder.

euchromatin: chromatin that is loosely coiled during interphase; thought to contain transcriptionally active genes.

eugenics: a largely discredited field of genetics that seeks to improve humankind by selective breeding; can be positive eugenics, in which individuals with desirable traits are encouraged or forced to breed, or negative eugenics, in which individuals with undesirable traits are discouraged or prevented from breeding.

eukaryote: a cell with a nuclear membrane surrounding its genetic material (a characteristic of a true nucleus) and a variety of subcellular, membrane-bound organelles; eukaryotic organisms include all known organisms except bacteria, which are prokaryotic. *See also* prokaryote.

euploid: the normal number of chromosomes for a cell or organism.

eusociality: an extreme form of altruism and kin selection in which most members of the society do not reproduce but rather feed and protect their relatives; bees, for example, are eusocial.

euthanasia: the killing of suffering individuals; sometimes referred to as "mercy" killing.

exogenous gene: a gene produced or originating from outside an organism.

exon: a protein-coding sequence in eukaryotic genes, usually flanked by introns.

exonuclease: an enzyme that degrades a nucleic acid molecule by breaking phosphodiester bonds at either end of the molecule.

expressed sequence tags (ESTs): an STS (sequence tagged site) that has been derived from a cDNA library.

expression vector: a DNA cloning vector designed to allow genetic expression of inserted genes via promoters engineered into the vector sequence.

expressivity: the degree to which a genotype is expressed as a phenotype.

extranuclear inheritance: inheritance involving genetic material located in the mitochondria or chloroplasts of a eukaryotic cell; also known as maternal inheritance (because these organelles are generally inherited from the mother) and cytoplasmic inheritance (because the organelles are found in the cell's cytoplasm rather than its nucleus).

extreme halophiles: microorganisms that require extremely high salt concentrations for optimal growth.

F pilus: also called the fertility pilus; a reproductive structure found on the surface of some bacterial cells that allows the cells to exchange plasmids or other DNA during the process of conjugation.

F_1 generation: first filial generation; offspring produced from a mating of P (parental) generation individuals.

F_2 generation: second filial generation; offspring produced from a mating of F_1 generation individuals.

familial Mediterranean fever: a genetic inflammatory disorder of the abdominal cavity, affecting people of Mediterranean descent, that causes abdominal pain and recurrent fevers; also known as recurrent polyseroserositis.

fate map: a description of the adult fate of embryonic cells.

faulty gene: a gene that does not function properly.

fertilization: the fusion of two cells (egg and sperm) in sexual reproduction.

fitness: a measure of the ability of a genotype or individual to survive and reproduce; when fitness is compared to other genotypes or individuals it is called relative fitness.

fluorescence in situ hybridization (FISH): an extremely sensitive assay for determining the presence of deletions on chromosomes, which uses a fluorescence-tagged segment of DNA that binds to the DNA region being studied.

foreign DNA: DNA taken from a source other than the host cell that is joined to the DNA of the cloning vector; also known as "insert DNA."

forensic genetics: the use of genetic tests and principles to resolve legal questions.

formylmethionine (fMet): the amino acid used to start all bacterial proteins; it is attached to the initiator transfer RNA (tRNA) molecule.

frame-shift mutation: a DNA mutation involving the insertion or deletion of one of several nucleotides that are not in multiples of three, resulting in a shift of the codon reading frame; usually produces nonfunctional proteins. *See also* open reading frame; reading frame.

fraternal twins: twins that develop and are born simultaneously but are genetically unique, being produced from the fertilization of two separate eggs; a synonymous term is "dizygotic twins."

free radical: *See* oxygen free radical.

G: the abbreviation for guanine, a purine nitrogenous base found in the structure of both DNA and RNA.

G_0: a point in the cell cycle at which a cell is no longer progressing toward cell division; can be considered a "resting" stage.

G_1 checkpoint: a point in the cell cycle at which a cell commits either to progressing toward cell division (by replicating its DNA and eventually engaging in mitosis) or to entering the G_0 phase, thereby withdrawing from the cell cycle either temporarily or permanently.

gamete: a sex cell, either sperm or egg, containing half the genetic material of a normal cell.

gel electrophoresis: a technique of molecular biology in which biological molecules are placed into a gel-like matrix (such as agarose or polyacrylamide) and then subjected to an electric current; using this technique, researchers can separate molecules of varying sizes and properties.

GenBank: a comprehensive, annotated collection of publicly available DNA sequences maintained by the National Center for Biotechnology Information and available through its Web site.

gene: a portion of a DNA molecule containing the genetic information necessary to produce a molecule of messenger RNA (via the process of transcription) that can then be used to produce a protein (via the process of translation); also includes regions of DNA that are transcribed to RNA that does not get translated, but carries out other roles in the cell.

gene amplification: a rise in the number of copies of a DNA sequence for a particular protein without a proportional rise in others; this process takes place in tumor cells.

gene expression: the combined biochemical processes, called "transcription" and "translation," that convert the linearly encoded information in the bases of DNA into the three-dimensional structures of proteins.

gene families: multiple copies of the same or similar genes in the same genome; the copies can be identical and tandemly repeated, or they may differ slightly and be scattered on the same or different chromosomes.

gene flow: movement of alleles from one population to another by the movement of individuals or gametes.

gene frequency: the occurrence of a particular allele present in a population, expressed as a percentage of the total number of alleles present for the locus.

gene pool: the complete assortment of genes present in the gametes of the members of a population that are eligible to reproduce.

gene silencing: any form of genetic regulation in which the expression of a gene is completely repressed, either by preventing transcription (pretranscriptional gene silencing) or after a messenger RNA (mRNA) has been transcribed (posttranscriptional gene silencing).

gene therapy: any procedure to alleviate or treat the symptoms of a disease or condition by genetically altering the cells of the patient.

gene transfer: the movement of fragments of genetic information, whole genes, or groups of genes between organisms.

genetic code: the correspondence between the sequence of nucleotides in DNA or messenger RNA (mRNA) molecules and the amino acids in the polypeptide a gene codes for.

genetic counseling: a discipline concerned with analyzing the inheritance patterns of a particular genetic defect within a given family, including the determination of the risk associated with the presence of the genetic defect in future generations and options for treatment of existing genetic defects.

genetic drift: chance fluctuations in allele frequencies within a population, resulting from random variation in the number and genotypes of offspring produced by different individuals.

genetic engineering: a term encompassing a wide variety of molecular biology techniques, all concerned with the modification of genetic characteristics of cells or organisms to accomplish a desired effect.

genetic load: the average number of the recessive deleterious (lethal or sublethal) alleles in individuals in a population.

genetic map: a "map" showing distances between genes in terms of recombination frequency; using DNA sequence data, a physical map with distance in base pairs can also be produced.

genetic marker: a distinctive DNA sequence that shows variation in the population and can therefore potentially be used for identification of individuals and for discovery of disease genes.

genetic screening: the testing of individuals for disease-causing genes or genetic disease.

genetic testing: the use of the techniques of genetics research to determine a person's risk of developing, or status as a carrier of, a disease or other disorder.

genetically modified (GM) foods: foods produced through the application of recombinant DNA technology, whereby genes from the same or different species are transferred and expressed in crops that do not naturally harbor those genes.

genetically modified organism (GMO): an organism produced by using biotechnology to introduce a new gene or genes, or new regulatory sequences for genes, into it for the purpose of giving the organism a new trait, usually to adapt the organism to a new environment, provide resistance to pest species, or enable the production of new products from the organism. *See also* transgenic organism.

genetics: an area of biology involving the scientific study of heredity.

genome: all of the DNA in the nucleus or in one of the organelles, such as a chloroplast or mitochondrion.

genomic imprinting: a genetic phenomenon in which the phenotype associated with a particular allele depends on which parent donated the allele.

genomic library: a collection of clones that includes the entire genome of a single species as fragments ligated to vector DNA.

genomics: that branch of genetics dealing with the study of genetic sequences, including their structure and arrangement.

genotype: the genetic characteristics of a cell or organism, expressed as a set of symbols representing the alleles present at one or more loci.

germ cells: reproductive cells such as eggs and sperm.

germ-line gene therapy: a genetic modification in gametes or fertilized ova so all cells in the organism will have the change, which potentially can be passed on to offspring.

germ-line mutation: a heritable change in the genes of an individual's reproductive cells, often linked to hereditary diseases.

gonad: an organ that produces reproductive cells and sex hormones; termed ovaries in females and testes in males.

Green Revolution: the introduction of scientifically bred or selected varieties of grain (such as rice, wheat, and maize) that, with high enough inputs of fertilizer and water, can greatly increase crop yields.

guanine (G): a purine nitrogenous base found in the structure of both DNA and RNA.

H substance: a carbohydrate molecule on the surface of red blood cells; when modified by certain monosaccharides, this molecule provides the basis of the ABO blood groups.

haplodiploidy: a system of sex determination in which males are haploid (developing from unfertilized eggs) and females are diploid.

haploid: refers to a cell or an organism with one set of chromosomes; usually represented as the *N* number of chromosomes, with 2*N* standing for the diploid number of chromosomes.

haploinsufficiency: protein produced by a single gene copy is insufficient for normal function.

haplotype: a sequential set of genes on a single chromosome inherited together from one parent; the other parent provides a matching chromosome with a different set of genes.

Hardy-Weinberg law: a concept in population genetics stating that, given an infinitely large population that experiences random mating without mutation or any other such affecting factor, the frequency of particular alleles will reach a state of equilibrium, after which their frequency will not change from one generation to the next.

HeLa cells: the first human tumor cells shown to form a continuous cell line; they were derived from a cervical cancer tumor removed from a woman known as Henrietta Lacks.

helicase: a cellular enzyme that breaks hydrogen bonds between the strands of the DNA double helix, thus unwinding the helix and facilitating DNA replication.

helper T cells: a class of white blood cells (lymphocytes) derived from bone marrow that prompts the production of antibodies by B cells in the presence of an antigen.

hemizygous: characterized by having a gene present in a single copy, such as any gene on the X chromosome in a human male.

hemoglobin: a molecule made up of two alpha and two beta amino acid chains whose precise chemical and structural properties normally allow it to bind with oxygen in the lungs and transport it to other parts of the body.

hemophilia: an X-linked recessive disorder in which an individual's blood does not clot properly because of a lack of blood-clotting factors; as in all X-linked recessive traits, the disease is most common in males, the allele for the disease being passed from mother to son.

heredity: the overall mechanism by which characteristics or traits are passed from one generation of organisms to the next; genetics is the scientific study of heredity.

heritability: a proportional measure of the extent to which differences among organisms within a population for a particular characteristic result from genetic rather than environmental causes (a measure of nature versus nurture).

hermaphrodite: an individual who has both male and female sex organs.

heterochromatin: a highly condensed form of chromatin, usually transcriptionally inactive.

heterochrony: a change in the timing or rate of development of characters in an organism relative to those same events in its evolutionary ancestors.

heteroduplex: a double-stranded molecule of nucleic acid with each strand from a different source, formed either through natural means such as recombination or through artificial means in the laboratory.

heterogametic sex: the particular sex of an organism that produces gametes containing two types of sex chromosome; in humans, males are the heterogametic sex, producing sperm that can carry either an X chromosome or a Y chromosome.

heterogeneous nuclear RNA (hnRNA): an assortment of RNA molecules of various types found in the nucleus of the cell and in various stages of processing prior to their export to the cytoplasm.

heterozygote: an individual with two different alleles at a gene locus.

heterozygous: composed of two alleles that are different, for example *Aa*; synonymous with "hybrid."

highly conserved sequence: A DNA sequence that is similar among different organisms.

Hirschsprung's disease: genetic disorder in which the nerve cells in the colon do not develop properly leading to constipation and abdominal distention; neonates suffering from this disorder may not have a bowel movement and may vomit bile. Surgical intervention is the treatment of choice.

histones: specialized proteins in eukaryotic cells that bind to DNA molecules and cause them to become more compact; thought to be involved in regulation of gene expression as well.

HLA: *See* human leukocyte antigens.

hnRNA: *See* heterogeneous nuclear RNA.

holandric: refers to a trait passed from father to son via a sex chromosome such as the Y chromosome in human males.

homeobox: a DNA sequence encoding a highly basic protein known as a homeodomain; a homeodomain functions as a transcription factor and is thought to help regulate major events in the embryonic development of higher organisms.

homeotic gene: a gene that helps determine body plan early in development; the products of homeotic genes are transcription factors that control the expression of other genes.

homogametic sex: the particular sex of an organism that produces gametes containing only one type of sex chromosome; in humans, females are the homogametic sex, producing eggs with X chromosomes.

homologous: refers to chromosomes that are identical in terms of types of genes present and the location of the centromere; because of their high degree of similarity, homologous chromosomes can synapse and recombine during prophase I of meiosis.

homology: similarity resulting from descent from a common evolutionary ancestor.

homozygote: an individual with two identical alleles at a gene locus.

homozygous: characterized by a genotype composed of two alleles at the same locus that are the same, for example *AA* or *aa*; synonymous with "purebred."

human artificial chromosome: an agent constructed similarly to a human chromosome, that is used to transfer large segments of human DNA.

Human Genome Project: a multiyear genetic research endeavor to sequence the entire human genome, as well as the genomes of related organisms; the human genome sequence was officially completed in 2003.

human leukocyte antigens (HLAs): molecules found on the surface of cells that allow the immune system to differentiate between foreign, invading cells and the body's own cells.

hybrid: any cell or organism with genetic material from two different sources, through either natural processes such as sexual reproduction or more artificial processes such as genetic engineering.

hybridization: a process of base pairing involving two single-stranded nucleic acid molecules with complementary sequences; the extent to which two unrelated nucleic acid molecules will hybridize is often used as a way to determine the amount of similarity between the sequences of the two molecules.

hybridoma: a type of hybrid cancer cell created by artificially joining a cancer cell with an antibody-producing cell; hybridomas have useful applications in immunological research.

hydrogen bond: a bond formed between molecules containing hydrogen atoms with positive charges and molecules containing atoms such as nitrogen or oxygen that can possess a negative charge; a relatively weak but important bond in nature that, among other things, connects water molecules, allows DNA strands to base-pair, and contributes to the three-dimensional shape of proteins.

identical twins: a pair of genetically identical offspring that develop from a single fertilized egg; also known as monozygotic twins.

immune system: the system in the body that normally responds to foreign agents by producing antibodies and stimulating antigen-specific lymphocytes, leading to destruction of these agents.

in situ hybridization: the process of base pairing a sequence of DNA to metaphase chromosomes on a microscope slide.

in vitro: literally, "in glass"; an event occurring in an artificial setting such as in a test tube, as opposed to inside a living organism.

in vitro fertilization: a fertilization that occurs unnaturally in a laboratory or other controlled environment.

in vivo: literally, "in the living"; an event occurring in a living organism, as opposed to an artificial setting.

inborn error of metabolism: a genetic defect in one of a cell's metabolic pathways, usually at the level of an enzyme, that causes the pathway to malfunction; results in phenotypic alterations at the cellular or organismal level.

inbreeding: mating between genetically related individuals.

inbreeding depression: a reduction in the health and vigor of offspring from closely related individuals, a common and widespread phenomenon among nonhuman organisms.

incidence: the frequency in which a disorder appears in a given population.

inclusive fitness: an individual's total genetic contribution to future generations, comprising both direct fitness, which results from individual reproduction, and indirect fitness, which results from the reproduction of close relatives.

incomplete dominance: a phenomenon involving two alleles, neither of which masks the expression of the other; instead, the combination of the alleles in the heterozygous state produces a new phenotype that is usually intermediate to the phenotypes produced by either allele alone in the homozygous state.

independent assortment: a characteristic of standard Mendelian genetics referring to the random assortment or shuffling of alleles and chromosomes that occurs during meiosis I; independent assortment is responsible for the offspring ratios observed in Mendelian genetics.

inducer: a molecule that activates some bacterial operons, usually by interacting with regulatory proteins bound to the operator region.

induction: a process in which a cell or group of cells signals an adjacent cell or group of cells to pursue a different developmental pathway and so become differentiated from neighboring cells.

informed consent: the right of patients to know the risks of medical treatment and to determine what is done to their bodies, including the right to accept or refuse treatment based on this information.

inherited: genetic transfer from parents to children.

initiation codon: also called the "start codon," a codon, composed of the nucleotides AUG, that signals the beginning of a protein-coding sequence in a messenger RNA (mRNA) molecule; in the genetic code, AUG always represents the amino acid methionine.

insert DNA: *See* foreign DNA.

insertion: a process whereby a DNA sequence is abnormally inserted into a gene, disrupting the normal function and structure of the gene and leading to a chromosomal abnormality.

insertion sequence: a small, independently transposable genetic element.

intelligence quotient (IQ): the most common measure of intelligence; it is based on the view that there is a single capacity for complex mental work and that this capacity can be measured by testing.

intercalary deletion: a type of chromosome deletion in which DNA has been lost from within the chromosome (as opposed to a terminal deletion involving a region of DNA lost from the end of the chromosome).

interference: in genetic linkage, a mathematical expression that represents the difference between the expected and the observed number of double recombinant offspring; this can be a clue to the physical location of linked genes on the chromosome.

interphase: the period of the cell cycle in which the cell is preparing to divide, consisting of two distinct growth phases (G_1 and G_2) separated by a period of DNA replication (S phase).

introgression: the transfer of genes from one species to another or the movement of genes between species (or other well-marked genetic populations) mediated by backcrossing.

intron: an intervening sequence within eukaryotic DNA, transcribed as part of a messenger RNA (mRNA) precursor but then removed by splicing before the mRNA molecule is translated; introns are thought to play an important role in the evolution of genes.

inversion: a chromosomal abnormality resulting in a region of the chromosome where the normal order of genes is reversed.

isotope: an alternative form of an element with a variant number of neutrons in its atomic nucleus; isotopes are frequently radioactive and are important tools for numerous molecular biology techniques.

jumping: *See* chromosome jumping.

junk DNA: a disparaging (and now known to be inaccurate) characterization of the noncoding DNA content of a genome.

karyokinesis: division of a cell's nuclear contents, as opposed to cytokinesis (division of the cytoplasm). *See also* cytokinesis.

karyotype: the complete set of chromosomes possessed by an individual, usually isolated during metaphase and arranged by size and type as a method of detecting chromosomal abnormalities.

kilobase (kb): a unit of measurement for nucleic acid molecules, equal to 1,000 bases or nucleotides.

kinase: an enzyme that catalyzes phosphate addition to molecules.

kinetochore: a chromosome structure found in the region of the centromere and used as an attachment point for the microtubules of the spindle apparatus during cell division.

Klinefelter syndrome: a human genetic disorder in males who possess an extra X chromosome; Klinefelter males have forty-seven chromosomes instead of the normal forty-six and suffer from abnormalities such as sterility, body feminization, and mental retardation.

knockout: the inactivation of a specific gene within a cell (or whole organism, as in the case of knockout mice) to determine the effects of loss of function of that gene.

lactose: a disaccharide that is an important part of the metabolism of many bacterial species; lactose metabolism in these species is genetically regulated via the *lac* operon.

lagging strand: in DNA replication, the strand of DNA being synthesized in a direction opposite to that of replication fork movement; this strand is synthesized in a discontinuous fashion as a series of Okazaki fragments later joined together. *See also* Okazaki fragments.

Lamarckianism: the theory, originally proposed by Jean-Baptiste Lamarck, that traits acquired by an organism during its lifetime can be passed on to offspring.

lambda (λ) phage: a bacteriophage that infects bacteria and then makes multiple copies of itself by taking over the infected bacteria's cellular machinery.

lateral gene transfer: the movement of genes between organisms; also called horizontal gene transfer.

leading strand: in DNA replication, the strand of DNA being synthesized in the same direction as the movement of the replication fork; this strand is synthesized in a continuous fashion.

leptotene: a subphase of prophase I of meiosis in which chromosomes begin to condense and become visible.

lethal allele: an allele capable of causing the death of an organism; a lethal allele can be recessive (two copies of the allele are required before death results) or dominant (one copy of the allele produces death).

leucine zipper: an amino acid sequence, found in some DNA-binding proteins, characterized by leucine residues separated by sets of seven amino acids; two molecules of this amino acid sequence can combine via the leucine residues and "zip" together, creating a structure that can then bind to a specific DNA sequence.

library: a cloned collection of DNA created from a specific organism.

linkage: a genetic phenomenon involving two or more genes inherited together because they are physically located on the same chromosome; Gregor Mendel's principle of independent assortment does not apply to linked genes, but genotypic and phenotypic variation is possible through crossing over.

linkage mapping: a form of genetic mapping that uses recombination frequencies to estimate the relative distances between linked genes.

locus (*pl.* loci): the specific location of a particular gene on a chromosome.

LOD score: statistical calculation of whether two gene sites (loci) are likely to be located near one another on a chromosome and be genetically linked; also known as logarithm of the odds score.

lymphocytes: sensitized cells of the immune system that recognize and destroy harmful agents via antibody and cell-mediated responses that include B lymphocytes from the bone marrow and T lymphocytes from the thymus.

Lyon hypothesis: a hypothesis stating that one X chromosome of the pair found in all female cells must be inactivated in order for those cells to be normal; the inactivated X chromosome is visible by light microscopy and stains as a Barr body.

lysis: the breaking open of a cell.
lysogeny: a viral process involving repression and integration of the viral genome into the genome of the host bacterial cell.

major histocompatibility complex (MHC): a group of molecules found on the surface of cells, allowing the immune system to differentiate between foreign, invading cells and the body's own cells; in humans, this group of molecules is called HLAs (human leukocyte antigens). *See also* human leukocyte antigens.
malformation: an inherited organ defect that occurs during fetal development.
map unit: *See* centiMorgan.
maternal inheritance: *See* extranuclear inheritance.
Maxam-Gilbert sequencing: a method of base-specific chemical degradation to determine DNA sequence; this method has largely been supplanted by the Sanger method. *See also* Sanger sequencing.
megakaryocyte: a large bone marrow cell that has a nucleus containing many lobes; the cell functions to produce platelets.
meiosis: a process of cell division in which the cell's genetic material is reduced by half and sex cells called gametes are produced; important as the basis of sexual reproduction.
melanism: the opposite of albinism, a condition that leads to the overproduction of melanin.
melting: a term sometimes used to describe the denaturation of a DNA molecule as it is heated in solution; as the temperature rises, hydrogen bonds between the DNA strands are broken until the double-strand molecule has been completely converted into two single-strand molecules.
Mendelian genetics: the genetics of traits that show simple inheritance patterns; based on the work of Gregor Mendel, a nineteenth century monk who studied the genetics of pea plants.
messenger RNA (mRNA): a type of RNA molecule containing the genetic information necessary to produce a protein through the process of translation; produced from the DNA sequence of a gene in the process of transcription.
metabolic pathway: a series of enzyme-catalyzed reactions leading to the breakdown or synthesis of a particular biological molecule.
metabolism: the collection of biochemical reactions occurring in an organism.
metacentric chromosome: a chromosome with the centromere located at or near the middle of the chromosome. *See also* acrocentric chromosome; telocentric chromosome.
metafemale: a term used to describe *Drosophila* (fruit fly) females that have more X chromosomes than sets of autosomes (for example, a female that has two sets of autosomes and three X chromosomes); also used in reference to human females with more than two X chromosomes.
metaphase: the second phase in the process of mitosis, involving chromosomes lined up in the middle of the cell on a line known as the equator.
methylation: the process of adding a methyl chemical group (one carbon atom and three hydrogen atoms) to a particular molecule, such as to the base portion of a nucleotide in a DNA nucleotide.
metric trait: *See* quantitative trait.
microarray: a flat surface on which 10,000 to 100,000 tiny spots of short DNA molecules (oligonucleotides) are fixed and are used to detect the presence of DNA or RNA molecules that are homologous to the oligonucleotides.
microsatellite DNA: a type of variable number tandem repeat (VNTR) in which the repeated motif is 1 to 6 base pairs; also called a simple sequence repeat (SSR) or a short tandem repeat (STR).
microtubule: a cell structure involved in the movement and division of chromosomes during mitosis and meiosis; part of the cell's cytoskeleton, microtubules can be rapidly assembled and disassembled.
microtubule organizing center (MTOC): *See* centriole.
minimal media: an environment that contains the simplest set of ingredients that a microorganism can use to produce all the substances required for reproduction and growth.
minisatellite DNA: a type of variable number tandem repeat (VNTR) in which the repeated motif is 12 to 500 base pairs in length.
miscegenation: sexual activity or marriage between members of two different human races.
mismatch repair: a cellular DNA repair process in which improperly base-paired nucleotides are enzymatically removed and replaced with the proper nucleotides.
missense mutation: a DNA mutation that changes an existing amino acid codon in a gene to some

other amino acid codon; depending on the nature of the change, this can be a harmless or a serious mutation (for example, sickle-cell disease in humans is the result of a missense mutation).

mitochondrial genome (mtDNA): DNA found in mitochondria, which contains some of the genes that code for proteins involved in energy metabolism; it is a circular molecule similar in structure to the genome of bacteria.

mitochondrion: the organelle responsible for production of ATP through the process of cellular respiration in a eukaryotic cell; sometimes referred to as the "powerhouse of the cell."

mitosis: a process of cell division in which a cell's duplicated genetic material is evenly divided between two daughter cells, so that each daughter cell is genetically identical to the original parent cell.

model organism: an organism well suited for genetic research because it has a well-known genetic history, a short life cycle, and genetic variation between individuals in the population.

modern synthesis: the merging of the Darwinian mechanisms for evolution with Mendelian genetics to form the modern fields of population genetics and evolutionary biology; also called the neo-Darwinian synthesis.

molecular clock hypothesis: a hypothesis that predicts that amino acid changes in proteins and nucleotide changes in DNA are approximately constant over time.

molecular cloning: the process of splicing a piece of DNA into a plasmid, virus, or phage vector to obtain many identical copies of that DNA.

molecular genetics: the branch of genetics concerned with the central role that molecules, particularly the nucleic acids DNA and RNA, play in heredity.

monoclonal antibodies: identical antibodies (having specificity for the same antigen) produced by a single type of antibody-producing cell, either a B cell or a hybridoma cell line; important in various types of immunology research techniques.

monoculture: the agricultural practice of growing the same cultivar on large tracts of land.

monohybrid: an organism that is hybrid with respect to a single gene (for example, *Aa*); when two monohybrid organisms are mated, the offspring will generally appear in a 3:1 ratio involving the trait controlled by the gene in question.

monosomy: a genetic condition in which one chromosome from a homologous chromosome pair is missing, producing a 2n–1 genotype; usually causes significant problems in the phenotype of the organism.

monozygotic: developed from a single zygote; identical twins are monozygotic because they develop from a single fertilized ovum that splits in two.

morphogen: a protein or other molecule made by cells in an egg that creates a concentration gradient affecting the developmental fate of surrounding cells by altering their gene expression or their ability to respond to other morphogens.

morphogenesis: the induction and formation of organized body parts or organs.

mosaicism: a condition in which an individual has two or more cell populations derived from the same fertilized ovum, or zygote, as in sex chromosome mosaics, in which some cells contain the usual XY chromosome pattern and others contain extra X chromosomes.

mRNA: *See* messenger RNA.

mtDNA: *See* mitochondrial genome.

MTOC: *See* centriole.

multifactorial: characterized by a complex interaction of genetic and environmental factors.

multiple alleles: a genetic phenomenon in which a particular gene locus is represented by more than two alleles in a population; the greater the number of alleles, the greater the genetic diversity.

mutagen: any chemical or physical substance capable of increasing mutations in a DNA sequence.

mutant: a trait or organism different from the normal, or wild-type, trait or organism seen commonly in nature; mutants can arise either through expression of particular alleles in the organism or through spontaneous or intentional mutations in the genome.

mutation: a change in the genetic sequence of an organism, usually leading to an altered phenotype.

N terminus: the end of a polypeptide with an amino acid that has a free amino group.

natural selection: a process involving genetic variation on the genotypic and phenotypic levels that contributes to the success or failure of various species in reproduction; thought to be the primary force behind evolution.

negative eugenics: improving human stocks through the restriction of reproduction by individuals with inferior traits or who are known to carry alleles for inferior traits.

neural tube: the embryonic precursor to the spinal cord and brain, which normally closes at small openings, or neuropores, by the twenty-eighth day of gestation in humans.

neurofibromatosis: a genetic disorder whereby tumors develop along peripheral nerves. The tumors may cause disfiguration, hearing and vision loss, and learning disabilities.

neurotransmitter: a chemical that carries messages between nerve cells.

neutral mutation: a mutation in a gene, or some other portion of the genome, that is considered to have no effect on the fitness of the organism.

neutral theory of evolution: Motoo Kimura's theory that nucleotide substitutions in the DNA often have no effect on fitness, and thus changes in allele frequencies in populations are caused primarily by genetic drift.

Niemann-Pick disease, Type C: a fatal genetic disorder involving the inability to metabolize cholesterol and other lipids within the cell, causing cholesterol to accumulate in the spleen and liver and excessive amounts of other lipids to accumulate in the brain.

noncoding DNA: a strand of DNA that is unable to make protein because it lacks the necessary information.

nondisjunction: the improper division of chromosomes during anaphase of mitosis or meiosis, resulting in cells with abnormal numbers of chromosomes and sometimes seriously altered phenotypes.

nonhistone proteins: a heterogeneous group of acidic or neutral proteins found in chromatin that may be involved with chromosome structure, chromatin packaging, or the control of gene expression.

nonsense codon: another term for a termination or stop codon (UAA, UAG, or UGA).

nonsense mutation: a DNA mutation that changes an existing amino acid codon in a message to one of the three termination, or stop, codons; this results in an abnormally short protein that is usually nonfunctional.

Northern blot: a molecular biology procedure in which a labeled single-stranded DNA probe is exposed to cellular RNA immobilized on a filter; under the proper conditions, the DNA probe will seek out and bind to its complementary sequence in the RNA molecules if such a sequence is present.

nuclease: an enzyme that degrades nucleic acids by breaking the phosphodiester bond that connects nucleosides.

nucleic acid: the genetic material of cells, found in two forms: deoxyribonucleic acid (DNA) and ribonucleic acid (RNA); composed of repeating subunits called nucleotides.

nucleocapsid: a viral structure including the capsid, or outer protein coat, and the nucleic acid of the virus.

nucleoid: a region of a prokaryotic cell containing the cell's genetic material.

nucleolus: a eukaryotic organelle located in the nucleus of the cell; the site of ribosomal RNA (rRNA) synthesis.

nucleoside: a building block of nucleic acids, composed of a sugar (deoxyribose or ribose) and one of the nitrogenous bases: adenine (A), cytosine (C), guanine (G), thymine (T), or uracil (U).

nucleosome: the basic unit molecule of chromatin, composed of a segment of a DNA molecule that is bound to and wound around histone molecules; DNA with nucleosomes appears as beads on a string when viewed by electron microscopy.

nucleotide: a building block of nucleic acids, composed of a sugar (deoxyribose or ribose), one of the nitrogenous bases (adenine, cytosine, guanine, thymine, or uracil) and one or more phosphate groups.

nucleus: the "control center" of eukaryotic cells, where the genetic material is separated from the rest of the cell by a membrane; site of DNA replication and transcription.

nullisomy: a genetic condition in which both members of a homologous chromosome pair are absent; usually, embryos with this type of genetic defect are not viable.

ochre codon: a stop codon (UAA) found in messenger RNA (mRNA) molecules that signals termination of translation.

Okazaki fragments: short DNA fragments, approximately two thousand or fewer bases in length, produced during discontinuous replication of the "lagging" strand of a DNA molecule.

oligonucleotide: a short molecule of DNA, generally fewer than twenty bases long and usually synthesized artificially; an important tool for numerous molecular biology procedures, including site-directed mutagenesis.

oncogene: any gene capable of stimulating cell division, thereby being a potential cause of cancer if unregulated; found in all cells and in many cancer-causing viruses.

oncovirus: a group of retroviruses that form tumors; these viruses have been associated with leukemia and sarcoma in animals and may also cause them in humans.

oogenesis: the process of producing eggs in a sexually mature female organism; another term for meiosis in females.

opal codon: a stop codon (UGA) found in messenger RNA (mRNA) molecules; signals termination of translation.

open reading frame (ORF): a putative protein-coding DNA sequence, marked by a start codon at one end and a stop codon at the other end.

operator: a region of a bacterial operon serving as a control point for transcription of the operon; a regulatory protein of some type usually binds to the operator.

operon: a genetic structure found only in bacteria, whereby a set of genes are controlled together by the same control elements; usually these genes have a common function, such as the genes of the lactose operon in *Escherichia coli* for the metabolism of lactose.

oxygen free radical: a highly reactive form of oxygen in which a single oxygen atom has a free, unpaired electron; free radicals are common by-products of chemical reactions.

***p53* gene:** a gene that typically controls the cell cycle, protecting the cell from damage; cancer cells may develop if this gene mutates and no longer protects the cell.

P generation: parental generation; the original individuals mated in a genetic cross.

pachytene: a subphase of prophase I in meiosis in which tetrads become visible.

palindrome: in general, a word that reads the same forward and backward (such as the words "noon" and "racecar"); in genetics, a DNA sequence that reads the same on each strand of the DNA molecule, although in opposite directions because of the antiparallel nature of the double helix; most DNA palindromes serve as recognition sites for restriction endonucleases.

pandemic: a worldwide outbreak of a disease.

paracentric inversion: an inversion of a chromosome's sequence that does not involve the centromere, taking place on a single arm of the chromosome.

parthenogenesis: production of an organism from an unfertilized egg.

paternal: coming from the father.

Patau syndrome: genetic disorder in which there is an extra copy of chromosome 13; individuals born with this condition commonly have malformed, low set ears, severe intellectual impairment, kidney and heart defects, and microcephaly. Also known as trisomy 13.

pedigree: a diagram of a particular family, showing the relationships between all members of the family and the inheritance pattern of a particular trait or genetic defect; especially useful for research into human traits that may otherwise be difficult to study.

penetrance: a quantitative term referring to the percentage of individuals with a certain genotype that also exhibit the associated phenotype.

peptide bond: a bond found in proteins; occurs between the carboxyl group of one amino acid and the amino group of the next, linking them together.

pericentric inversion: an inversion of a chromosome's sequence involving the centromere.

pharmacogenomics: the branch of human medical genetics that evaluates how an individual's genetic makeup influences his or her response to drugs.

phenotype: the physical appearance or biochemical and physiological characteristics of an individual, which is determined by both heredity and environment.

phenotypic plasticity: the ability of a genotype to produce different phenotypes when exposed to different environments.

phosphodiester bond: in DNA, the phosphate group connecting one nucleoside to the next in the polynucleotide chain.

photoreactivation repair: a cellular enzyme system responsible for repairing DNA damage caused by ultraviolet light; the system is activated by light.

phylogeny: often called an evolutionary tree, the

branching patterns that show evolutionary relationships, with the taxa on the ends of the branches.

pilus: a hairlike reproductive structure possessed by some species of bacterial cells that allows them to engage in a transfer of genetic material known as conjugation.

plasmid: a small, circular DNA molecule commonly found in bacteria and responsible for carrying various genes, such as antibiotic resistance genes; important as a cloning vector for genetic engineering.

pleiotropy: a genetic phenomenon in which a single gene has an effect on two or more traits.

-ploid, -ploidy: a suffix that refers to a chromosome set; humans have two sets of chromosomes and are referred to as being diploid, whereas some plants may have four sets, called tetraploid. Other terms include "autoploidy" and "polyploidy." *See also* allopolyploid; aneuploid; autopolyploid; diploid; euploid; haplodiploidy; haploid; polyploid; triploid.

pluripotency: the ability of a cell to give rise to all the differentiated cell types in an embryo.

point mutation: a DNA mutation involving a single nucleotide.

polar body: a by-product of oogenesis used to dispose of extra, unnecessary chromosomes while preserving the cytoplasm of the developing ovum.

polycistronic: characterizing messenger RNA (mRNA) molecules that contain coding sequences for more than one protein, common in prokaryotic cells.

polydactyly: a genetic disorder where the embryo develops extra fingers or toes.

polygenic inheritance: expression of a trait depending on the cumulative effect of multiple genes; human traits such as skin color, obesity, and intelligence are thought to be examples of polygenic inheritance.

polymerase: a cellular enzyme capable of creating a phosphodiester bond between two nucleotides, producing a polynucleotide chain complementary to a single-stranded nucleic acid template; the enzyme DNA polymerase is important for DNA replication, and the enzyme RNA polymerase is involved in transcription.

polymerase chain reaction (PCR): a technique of molecular biology in which millions of copies of a single DNA sequence can be artificially produced in a relatively short period of time; important for a wide variety of applications when the source of DNA to be copied is either scarce or impure.

polymorphism: the presence of many different alleles for a particular locus in individuals of the same species.

polypeptide: a single chain of amino acids connected to one another by peptide bonds; all proteins are polypeptides, but a protein may comprise one or more polypeptide molecules.

polyploid: a cell or organism that possesses multiple sets of chromosomes, usually more than two.

polysome: a group of ribosomes attached to the same messenger RNA (mRNA) molecule and producing the same protein product in varying stages of completion.

population: a group of organisms of the same species in the same place at the same time and thus potentially able to mate; populations are the basic unit of speciation.

population genetics: the study of how genes behave in populations; often a highly mathematical branch of genetics in which evolutionary processes are modeled.

positional cloning: a gene mapping technique used to locate a specific gene that causes a disorder when evidence is lacking about the biochemical nature of the disorder; also known as reverse genetics.

positive eugenics: selecting individuals to reproduce who have desirable genetic traits, as seen by those in control.

post-translational modification: chemical alterations to proteins after they have been produced at a ribosome that alters their properties.

prenatal testing: testing that is done during pregnancy to examine the chromosomes or genes of a fetus to detect the presence or absence of a genetic disorder.

prevalence: the percentage of people in a given population that are affected by a disorder at that particular time.

primary immunodeficiency: a genetic disorder involving a defect in the immune system, leaving the affected individual prone to infection; the defect may involve a single cell or one or more components of the immune system; more than 150 primary immunodeficiency disorders have been identified.

primer: a short nucleic acid molecule used as a beginning point for the enzyme DNA polymerase as it replicates a single-stranded template.

prion: an infectious agent composed solely of protein; thought to be the cause of various human and animal diseases characterized by neurological degeneration, including scrapie in sheep, mad cow disease in cattle, and Creutzfeldt-Jakob disease in humans.

probe: in genetics research, typically a single-stranded nucleic acid molecule or antibody that has been labeled in some way, either with radioactive isotopes or fluorescent dyes; this molecule is then used to seek out its complementary nucleic acid molecule or protein target in a variety of molecular biology techniques such as Southern, Northern, or Western blotting.

product rule: a rule of probability stating that the probability associated with two simultaneous yet independent events is the product of the events' individual probabilities.

prokaryote: a cell that lacks a nuclear membrane (and therefore has no true nucleus) and membrane-bound organelles; bacteria are the only known prokaryotic organisms.

promoter: a region of a gene that controls transcription of that gene; a physical binding site for RNA polymerase.

pronucleus: nucleus of a sperm or an egg before fertilization; when fertilization occurs the pronucleus of the egg and sperm join together to form the nucleus of the embryo that contains chromosomes from both the egg and the sperm.

prophase: the first phase in the process of mitosis or meiosis, in which the nuclear membrane disappears, the spindle apparatus begins to form, and chromatin takes on the form of chromosomes by becoming shorter and thicker.

propositus: the individual in a human pedigree who is the focus of the pedigree, usually by being the first person who came to the attention of the geneticist.

protease: enzyme required to digest protein.

protein: a biological molecule composed of amino acids linked together by peptide bonds; used as structural components of the cell or as enzymes; the term "protein" can refer to a single chain of amino acids or to multiple chains of amino acids functioning in a concerted way, as in the molecule hemoglobin.

proteomics: the study of which proteins are expressed in different types of cells, tissues, and organs during normal and abnormal conditions.

proto-oncogene: a gene, found in eukaryotic cells, that stimulates cell division; ordinarily, expression of this type of gene is tightly controlled by the cell, but in cancer cells, proto-oncogenes have been converted into oncogenes through alteration or elimination of controlled gene expression.

pseudodominance: a genetic phenomenon involving a recessive allele on one chromosome that is automatically expressed because of the deletion of its corresponding dominant allele on the other chromosome of the homologous pair.

pseudogenes: DNA sequences derived from partial copies, mutated complete copies, or normal copies of functional genes that have lost their control sequences and therefore cannot be transcribed; may originate by gene duplication or retrotransposition and are apparently nonfunctional regions of the genome that may evolve at a maximum rate, free from the evolutionary constraints of natural selection.

pseudohermaphrodite: individual born with either ambiguous genitalia or external genitalia that are the opposite of the chromosomal sex.

punctuated equilibrium: a model of evolutionary change in which new species originate abruptly and then exist through a long period of stasis; important as an explanation of the stepwise pattern of species change seen in the fossil record.

purine: either of the nitrogenous bases adenine or guanine; used in the structure of nucleic acids.

pyrimidine: any of the nitrogenous bases cytosine, thymine, or uracil; used in the structure of nucleic acids.

quantitative trait: a trait, such as human height or weight, that shows continuous variation in a population and can be measured; also called a metric trait.

quantitative trait loci (QTLs): genomic regions that affect a quantitative trait, generally identified via DNA-based markers.

reaction norm: the relationship between environment and phenotype for a given genotype.

reading frame: refers to the manner in which a messenger RNA (mRNA) sequence is interpreted as a series of amino acid codons by the ribosome;

because of the triplet nature of the genetic code, a typical messenger RNA (mRNA) molecule has three possible reading frames, although usually only one of these will actually code for a functional protein.

receptors: molecules to which signaling molecules bind in target cells.

recessive: a term referring to an allele or trait that will only be expressed if another, dominant, trait or allele is not also present.

reciprocal cross: a mating that is the reverse of another with respect to the sex of the organisms that possess certain traits; for example, if a particular cross were tall male H short female, then the reciprocal cross would be short male H tall female.

reciprocal translocation: a two-way exchange of genetic material between two nonhomologous chromosomes, resulting in a wide variety of genetic problems depending on which chromosomes are involved in the translocation.

recombinant DNA: DNA molecules that are the products of artificial recombination between DNA molecules from two different sources; important as a foundation of genetic engineering.

recombination: an exchange of genetic material, usually between two homologous chromosomes; provides one of the foundations for the genetic reassortment observed during sexual reproduction.

reductional division: refers to meiosis I, in which the amount of genetic material in the cell is reduced by half through nuclear division; it is at this stage that the diploid cell is converted to an essentially haploid state.

reductionism: the explanation of a complex system or phenomenon as merely the sum of its parts.

replication: the process by which a DNA or RNA molecule is enzymatically copied.

replicon: a region of a chromosome under control of a single origin of replication.

replisome: a multiprotein complex that functions at the replication fork during DNA replication; it contains all the enzymes and other proteins necessary for replication, including DNA polymerase.

repressor: a protein molecule capable of preventing transcription of a gene, usually by binding to a regulatory region close to the gene.

resistance plasmid (R plasmid): a small, circular DNA molecule that replicates independently of the bacterial host chromosome and encodes a gene for antibiotic resistance.

restriction endonuclease: a bacterial enzyme that cuts DNA molecules at specific sites; part of a bacterial cell's built-in protection against infection by viruses; an important tool of genetic engineering.

restriction enzyme: *See* restriction endonuclease.

restriction fragment length polymorphism (RFLP): a genetic marker, consisting of variations in the length of restriction fragments in DNA from individuals being tested, allowing researchers to compare genetic sequences from various sources; used in a variety of fields, including forensics and the Human Genome Project.

retrotransposon (retroposon): a DNA sequence that is transcribed to RNA and reverse transcribed to a DNA copy able to insert itself at another location in the genome.

retrovirus: a virus that carries reverse transcriptase that converts its RNA genome into a DNA copy that integrates into the host chromosome.

reverse transcriptase: a form of DNA polymerase, discovered in retroviruses, that uses an RNA template to produce a DNA molecule; the name indicates that this process is the reverse of the transcription process occurring naturally in the cell.

reverse transcription polymerase chain reaction (RT-PCR): a technique, requiring isolated RNA, for quickly determining if a gene or a small set of genes are transcribed in a population of cells.

RFLP analysis: *See* restriction fragment length polymorphism.

Rh factor: a human red blood cell antigen, first characterized in rhesus monkeys, that contributes to blood typing; individuals can be either Rh positive (possessing the antigen on their red blood cells) or Rh negative (lacking the antigen).

ribonucleic acid (RNA): a form of nucleic acid in the cell used primarily for genetic expression through transcription and translation; in structure, it is virtually identical to DNA, except that ribose is used as the sugar in each nucleotide and the nitrogenous base thymine is replaced by uracil; present in three major forms in the cell: messenger RNA (mRNA), transfer RNA (tRNA), and ribosomal RNA (rRNA).

ribose: a five-carbon sugar used in the structure of ribonucleic acid (RNA).

ribosomal RNA (rRNA): a type of ribonucleic acid in the cell that constitutes some of the structure of the ribosome and participates in the process of translation.

ribosome: a cellular structure, composed of ribosomal RNA (rRNA) and proteins, that is the site of translation.

ribozyme: an RNA molecule that can function catalytically as an enzyme.

RNA: *See* ribonucleic acid.

RNA interference (RNAi): an artificial technique using small, interfering RNAs that cause gene silencing by binding to the part of a messenger RNA (mRNA) to which they are complementary, thus blocking translation.

RNA polymerase: the cellular enzyme required for making an RNA copy of genetic information contained in a gene; an integral part of transcription.

RNase: refers to a group of enzymes, ribonucleases, capable of specifically degrading RNA molecules.

rRNA: *See* ribosomal RNA.

Sanger sequencing: also known as dideoxy termination sequencing, a method using nucleotides that are missing the 3′ hydroxyl group in order to terminate the polymerization of new DNA at a specific nucleotide; the most common sequencing method, used almost exclusively.

segregation: a characteristic of Mendelian genetics, resulting in the division of homologous chromosomes into separate gametes during the process of meiosis.

semiconservative replication: a characteristic of DNA replication, in which every new DNA molecule is actually a hybrid molecule, being composed of a parental, preexisting strand and a newly synthesized strand.

sequence-tagged sites: a short sequence of DNA that occurs once in the genome; the location and sequence of this DNA segment are known and can be easily found using polymerase chain reaction when particular primers are used.

severe immunodeficiency: a primary disorder of the immune system characterized by a severe defect in the T- and B-lymphocyte systems; infants born with this disorder commonly develop a severe sometimes life-threatening infection, such as meningitis, pneumonia, or bloodstream infection within the first few months of life.

sex chromosome: a chromosome carrying genes responsible for determination of an organism's sex; in humans, the sex chromosomes are designated X and Y.

sex-influenced inheritance: inheritance in which the expression of autosomal traits is influenced or altered relative to the sex of the individual possessing the trait; pattern baldness is an example of this type of inheritance in humans.

sex-limited inheritance: inheritance of traits expressed in only one sex, although these traits are usually produced by non-sex-linked genes (that is, they are genes located on autosomes instead of sex chromosomes).

sex-linked: disorder that is inherited through one of the sex chromosomes, the X or Y chromosomes.

sexual reproduction: reproduction of cells or organisms involving the transfer and reassortment of genetic information, resulting in offspring that can be phenotypically and genotypically distinct from either of the parents; mediated by the fusion of gametes produced during meiosis.

Shine-Dalgarno sequence: a short sequence in prokaryotic messenger RNA (mRNA) molecules complementary to a sequence in the prokaryotic ribosome; important for proper positioning of the start codon of the mRNA relative to the P site of the ribosome.

short interspersed sequences (SINES): short repeats of DNA sequences scattered throughout a genome.

shotgun cloning: a technique by which random DNA fragments from an organism's genome are inserted into a collection of vectors to produce a library of clones, which can then be used in a variety of molecular biology procedures.

sigma factor: a molecule that is part of RNA polymerase molecules in bacterial cells; allows RNA polymerase to select the genes that will be transcribed.

signal transduction: all of the molecular events that occur between the arrival of a signaling molecule at a target cell and its response; typically involves a cascading series of reactions that can eventually determine expression of many dozens of genes.

single nucleotide polymorphism (SNP): differences at the individual nucleotide level among individuals.

site-directed mutagenesis: a molecular genetics procedure in which synthetic oligonucleotide molecules are used to induce carefully planned mutations in a cloned DNA molecule.

small nuclear RNA (snRNA): small, numerous RNA molecules found in the nuclei of eukaryotic cells and involved in splicing of messenger RNA (mRNA) precursors to prepare them for translation.

snRNA: *See* small nuclear RNA.

sociobiology: the study of social structures, organizations, and actions in terms of underlying biological principles.

solenoid: a complex, highly compacted DNA structure consisting of many nucleosomes packed together in a bundle.

somatic cell: any cell in the body, excluding the reproductive cells.

somatic mutation: a mutation occurring in a somatic, or nonsex, cell; because of this, somatic mutations cannot be passed to the next generation.

Southern blot: a molecular biology technique in which a labeled single-stranded DNA probe is exposed to denatured cellular DNA immobilized on a filter; under the proper conditions, the DNA probe will seek out and bind to its complementary sequence among the cellular DNA molecules, if such a sequence is present.

speciation: the process of evolutionary change that leads to the formation of new species.

species: a group of organisms that can interbreed with one another but not with organisms outside the group; generally, members of a particular species share the same gene pool; defining a species is still controversial and remains a debated concept.

spermatogenesis: the process of producing sperm in a sexually mature male organism; another term for meiosis in males.

spindle apparatus: a structure, composed of microtubules and microfilaments, important for the proper orientation and movement of chromosomes during mitosis and meiosis; appears during prophase and begins to disappear during anaphase.

spliceosome: a complex of nuclear RNA and protein molecules responsible for the excision of introns from messenger RNA (mRNA) precursors before they are translated.

***SRY* gene:** the sex-determining region of the Y chromosome; a gene encoding a protein product called testis determining factor (TDF), responsible for conversion of a female embryo to a male embryo through the development of the testes.

stem cell: an undifferentiated cell that retains the ability to give rise to other, more specialized cells.

stutter: polymerase chain reaction (PCR) amplification product that arises as a result of strand slippage during PCR ; it is one or more repeat units less than the sample's allele.

suicide gene: a gene that when expressed in a cell becomes lethal to the cell; suicide genes make cancer cells vulnerable to chemotherapy.

sum rule: a rule of probability theory stating that the probability of either of two mutually exclusive events occurring is the sum of the events' individual probabilities.

supercoil: a complex DNA structure in which the DNA double helix is itself coiled into a helix; usually observed in circular DNA molecules such as bacterial plasmids.

sympatric speciation: the genetic divergence of populations, not separated geographically, that eventually results in formation of new species.

synapsis: the close association of homologous chromosomes occurring during early prophase I of meiosis; during synapsis, recombination between these chromosomes can occur.

T: the abbreviation for thymine, a pyrimidine nitrogenous base found in the structure of DNA; in RNA, thymine is replaced by uracil.

Taq polymerase: DNA polymerase from the bacterium *Thermus aquaticus*; an integral component of polymerase chain reaction.

tautomerization: a spontaneous internal rearrangement of atoms in a complex biological molecule that often causes the molecule to change its shape or its chemical properties.

taxon (*pl.* taxa): a general term used by evolutionists to refer to a type of organism at any level in a classification of organisms.

telocentric chromosome: a chromosome with a centromere at the end. *See also* acrocentric chromosome; metacentric chromosome.

telomere: the end of a eukaryotic chromosome, protected and replaced by the cellular enzyme telomerase.

telophase: the final phase in the process of mitosis or meiosis, in which division of the cell's nuclear contents has been completed and division of the cell itself occurs.

template: a single-stranded DNA molecule (or RNA molecule) used to create a complementary strand of nucleic acid through the activity of a polymerase.

teratogen: any chemical or physical substance, such as thalidomide, that creates birth defects in offspring.

testcross: a mating involving an organism with a recessive genotype for desired traits crossed with an organism that has an incompletely determined genotype; the types and ratio of offspring produced allow geneticists to determine the genotype of the second organism.

tetrad: a group of four chromosomes formed as a result of the synapsis of homologous chromosomes that takes place early in meiosis.

tetranucleotide hypothesis: a disproven hypothesis, formulated by geneticist P. A. Levene, stating that DNA is a structurally simple molecule composed of a repeating unit known as a tetranucleotide (composed, in turn, of equal amounts of the bases adenine, cytosine, guanine, and thymine).

thermal cycler: a machine that can rapidly heat and cool reaction tubes; used for performing PCR reactions.

theta structure: an intermediate structure in the bidirectional replication of a circular DNA molecule; the name comes from the resemblance of this structure to the Greek letter theta.

thymine (T): a pyrimidine nitrogenous base found in the structure of DNA; in RNA, thymine is replaced by uracil.

thymine dimer: a pair of thymine bases in a DNA molecule connected by an abnormal chemical bond induced by ultraviolet light; prevents DNA replication in the cell unless it is removed by specialized enzymes.

topoisomerases: cellular enzymes that relieve tension in replicating DNA molecules by introducing single- or double-stranded breaks into the DNA molecule; without these enzymes, replicating DNA becomes progressively more supercoiled until it can no longer unwind, and DNA replication is halted.

totipotent: the ability of a cell to produce an entire adult organism through successive cell divisions and development; as cells become progressively differentiated, they lose this characteristic.

trait: a phenotypic characteristic that is heritable.

transcription: the cellular process by which genetic information in the form of a gene in a DNA molecule is converted into the form of a messenger RNA (mRNA) molecule; dependent on the enzyme RNA polymerase.

transcription factor: a protein that is involved in initiation of transcription but is not part of the RNA polymerase.

transduction: DNA transfer between cells, with a virus serving as the genetic vector.

transfer RNA (tRNA): a type of RNA molecule necessary for translation to occur properly; provides the basis of the genetic code, in which codons in a messenger RNA (mRNA) molecule are used to direct the sequence of amino acids in a polypeptide; contains a binding site for a particular amino acid and a region complementary to a messenger RNA (mRNA) codon (an anticodon).

transformation: the process by which a normal cell is converted into a cancer cell; also refers to the change in phenotype accompanying entry of foreign DNA into a cell, such as in bacterial cells being used in recombinant DNA procedures.

transgenic organism: an organism possessing one or more genes from another organism, such as mice that possess human genes; important for the study of genes in a living organism, especially in the study of mutations within these genes. *See also* genetically modified organism (GMO).

transition mutation: a DNA mutation in which one pyrimidine (cytosine or thymine) takes the place of another, or a purine (adenine or guanine) takes the place of another.

translation: the cellular process by which genetic information in the form of a messenger RNA (mRNA) molecule is converted into the amino acid sequence of a protein, using ribosomes and RNA molecules as accessory molecules.

translocation: the movement of a chromosome segment to a nonhomologous chromosome as a result of an error in recombination; also refers to the movement of a messenger RNA (mRNA) codon from the A site of the ribosome to the P site during translation.

transposable element: *See* transposon.

transposon: a DNA sequence capable of moving to various places in a chromosome, discovered by geneticist Barbara McClintock; transposons are thought to be important as mediators of genetic variability in both prokaryotes and eukaryotes.

transversion: a DNA mutation in which a pyrimidine (cytosine or thymine) takes the place of a purine (adenine or guanine), or vice versa.

triploid: possessing three complete sets of chromosomes, or 3*N*; important in the development of desirable characteristics in the flowers or fruit of some plants; triploids are often sterile.

trisomy: a genetic condition involving one chromosome of a homologous chromosome pair that has been duplicated in some way, giving rise to a $2N + 1$ genotype and causing serious phenotypic abnormalities; a well-known example is trisomy 21, or Down syndrome, in which the individual possesses three copies of chromosome 21 instead of the normal two copies.

tRNA: *See* transfer RNA.

tumor-suppressor genes: any of a number of genes that limit or halt cell division under certain circumstances, thereby preventing the formation of tumors in an organism; two well-studied examples are the retinoblastoma gene and the *p53* gene; mutations in tumor-suppressor genes can lead to cancer.

Turner syndrome: a human genetic defect in which an individual has only forty-five chromosomes, lacking one sex chromosome; the sex chromosome present is an X chromosome, making these individuals phenotypically female, although with serious abnormalities such as sterility and anatomical defects.

uracil (U): a pyrimidine nitrogenous base found in the structure of RNA; in DNA, uracil is replaced by thymine.

variable number tandem repeat (VNTR): a repetitive DNA sequence of approximately fifty to one hundred nucleotides; important in the process of forensic identification known as DNA fingerprinting.

vector: a DNA molecule, such as a bacterial plasmid, into which foreign DNA can be inserted and then transported into a cell for further manipulation; important in a wide variety of recombinant DNA techniques.

virions: mature infectious virus particles.

viroids: naked strands of RNA, 270 to 380 nucleotides long, that are circular and do not code for any proteins that are able to cause disease in susceptible plants, many of them economically important. *See also* virusoids.

virus: a microscopic infectious particle composed primarily of protein and nucleic acid; bacterial viruses, or bacteriophages, have been important tools of study in the history of molecular genetics.

virusoids: similar to viroids, microscopic infectious particles composed primarily of protein and nucleic acid; unlike viroids, virusoids are packaged in the protein coat of other plant viruses, referred to as helpers, and are therefore dependent on the other virus. *See also* viroids.

VNTR: *See* variable number tandem repeat.

walking: *See* chromosome walking.

Western blot: a molecular biology technique involving labeled antibodies exposed to cellular proteins immobilized on a filter; under the proper conditions, the antibodies will seek out and bind to the proteins for which they are specific, if such proteins are present.

wild-type: interacting with more than one messenger RNA (mRNA) codon by virtue of the inherent flexibility present in the third base of the anticodon; first proposed by molecular biologist Francis Crick.

Wolfram syndrome: an autosomal recessive disorder, caused by a mutation in a gene responsible for the production of the wolfram protein, resulting in the functional loss of the protein. Patients who suffer from this neurodegenerative disorder typically develop diabetes insipidus, diabetes mellitus, blindness, and deafness. Also known as DIDMOAD, an acronym for diabetes insipidus, diabetes mellitus, optic atrophy, and deafness.

X linkage: a genetic phenomenon involving a gene located on the X chromosome; the typical pattern of X linkage involves recessive alleles, such as that for hemophilia, which exert their effects when passed from mother to son and are more likely to be exhibited by males than females.

X-ray diffraction: a method for determining the structure of molecules that infers structure by the way crystals of molecules scatter X rays as they pass through.

xenotransplants: transplants of organs or cellular tissue between different species of animals, such as between pigs and humans.

Y linkage: a genetic phenomenon involving a gene located on the Y chromosome; as a result, such a condition can be passed only from father to son.

yeast artificial chromosome (YAC): a cloning vector that has been engineered with all of the major genetic characteristics of a eukaryotic chromosome so that it will behave as such during cell division; YACs are used to clone extremely large DNA fragments from eukaryotic cells and are an integral part of the Human Genome Project.

Z-DNA: a zigzag form of DNA in which the strands form a left-handed helix instead of the normal right-handed helix of B-DNA; Z-DNA is known to be present in cells and is thought to be involved in genetic regulation. *See also* B-DNA.

zinc finger: an amino acid sequence, found in some DNA-binding proteins, that complexes with zinc ions to create polypeptide "fingers" that can then wrap around a specific portion of a DNA molecule.

zygote: a diploid cell produced by the union of a male gamete (sperm) with a female gamete (egg); through successive cell divisions, the zygote will eventually give rise to the adult form of the organism.

zygotene: a subphase of prophase I of meiosis involving synapsis between homologous chromosomes.

Randall K. Harris, Ph.D., and Bryan Ness, Ph.D.;
updated by Collette Bishop Hendler, R.N., M.S.

BIBLIOGRAPHY

GENERAL

Andreasen, Nancy C., ed. *Research Advances in Genetics and Genomics: Implications for Psychiatry.* Washington, D.C.: American Psychiatric Publishing, 2005.

Audesirk, Teresa, Gerald Audesirk, and Bruce E. Myers. *Biology: Life on Earth.* 6th ed. Upper Saddle River, N.J.: Prentice Hall, 2001.

Bailey, Philip S., Jr., Christina A. Bailey, and Robert J. Ouellette. *Organic Chemistry.* New York: Prentice Hall, 1997.

Banaszak, Leonard J. *Foundations of Structural Biology.* San Diego: Academic Press, 2000.

Barash, David P. *Revolutionary Biology: The New, Gene-Centered View of Life.* New Brunswick, N.J.: Transaction Publishers, 2003.

Berg, Paul, and Maxine Singer. *Dealing with Genes: The Language of Heredity.* Mill Valley, Calif.: University Science Books, 1992.

Brenner, Sydney, and Jeffrey K. Miller, eds. *Encyclopedia of Genetics.* New York: Elsevier, 2001.

Campbell, Neil A., and Jane B. Reece. *Biology.* 6th ed. San Francisco: Benjamin Cummings, 2002.

Carey, Gregory. *Human Genetics for the Social Sciences.* Thousand Oaks, Calif.: Sage Publications, 2003.

Clark, David P., and Lonnie D. Russell. *Molecular Biology Made Simple and Fun.* 2d ed. Vienna, Ill.: Cache River Press, 2000.

Compston, Alastair, et al. *McAlpine's Multiple Sclerosis.* Sydney: Elsevier/Churchill Livingstone, 2005.

Cummings, Michael. *Human Heredity: Principles and Issues.* Australia: Brooks/Cole Cengage Learning, 2009.

DeBusk, Ruth M. *Genetics: The Nutrition Connection.* Chicago: American Dietetic Association, 2003.

Diagram Group. *Genetics and Cell Biology on File.* Rev. ed. New York: Facts On File, 2003.

Entine, Jon. *Abraham's Children: Race, Identity, and the DNA of the Chosen People.* New York: Grand Central, 2007.

Fairbanks, Daniel J., and W. Ralph Anderson. *Genetics: The Continuity of Life.* New York: Brooks/Cole, 1999.

Finch, Caleb Ellicott. *Longevity, Senescence, and the Genome.* Reprint. Chicago: University of Chicago Press, 1994.

Goldstein, David B. *Jacob's Legacy: A Genetic View of Jewish History.* New Haven, Conn.: Yale University Press, 2008.

Hartl, Daniel L. *Genetics: Analysis of Genes and Genomes.* 5th ed. Boston: Jones and Bartlett, 2001.

Hartwell, L. H., et al. *Genetics: From Genes to Genomes.* 2d ed. New York: McGraw-Hill, 2003.

Hill, John, et al. *Chemistry and Life: An Introduction to General, Organic, and Biological Chemistry.* 6th ed. New York: Prentice Hall, 2000.

Jones, Steve Y. *The Descent of Men.* Boston: Houghton Mifflin, 2003.

King, Robert C., and William D. Stansfield. *A Dictionary of Genetics.* 6th ed. New York: Oxford University Press, 2001.

Klug, William S. *Essentials of Genetics.* 3d ed. Upper Saddle River, N.J.: Prentice Hall, 1999.

Lerner, K. Lee. *World of Genetics.* Detroit: Gale Research, 2001.

Lewin, Benjamin. *Genes VII.* New York: Oxford University Press, 2001.

Madigan, Michael T., John M. Martinko, and Jack Parker. *Brock Biology of Micro-organisms.* 10th ed. Upper Saddle River, N.J.: Prentice Hall, 2003.

Marks, Jonathan. *What It Means to Be 98 Percent Chimpanzee: Apes, People, and Their Genes.* Berkeley: University of California Press, 2002.

Micklos, David A., Greg A. Freyer, and David A. Crotty. *DNA Science: A First Course.* Cold Spring Harbor, N.Y.: Cold Spring Harbor Laboratory Press, 2003.

Moore, David S. *Dependent Gene: The Fallacy of Nature vs. Nurture.* New York: W. H. Freeman, 2001.

Moore, John A. *Science as a Way of Knowing.* 1993. Reprint. Cambridge, Mass.: Harvard University Press, 1999.

Raven, Peter H., and George B. Johnson. *Biology.* 6th ed. New York: W. H. Freeman/Worth, 1999.

Reeve, Eric C. R. *Encyclopedia of Genetics.* Chicago: Fitzroy Dearborn, 2001.

Relethford, John H., and John Relethford. *Reflections of Our Past: How Human History Is Revealed in Our Genes.* Boulder, Colo.: Westview Press, 2003.

Ridley, Matt. *Nature Via Nurture: Genes, Experience, and What Makes Us Human.* London: Fourth Estate, 2003.

Robinson, Richard, ed. *Genetics.* New York: Macmillan, 2002.

Russell, Peter J. *Genetics.* San Francisco: Benjamin Cummings, 2002.

Salter, Frank. *On Genetic Interests: Family, Ethnicity, and Humanity in an Age of Mass Migration.* Frankfurt am Main, Germany: Peter Lang, 2003.

Simon, Anne. *The Real Science Behind "The X-Files": Microbes, Meteorites, and Mutants.* New York: Simon & Schuster, 1999.

Singer, Maxine, and Paul Berg, eds. *Exploring Genetic Mechanisms.* Sausalito, Calif.: University Science Books, 1997.

Starr, Cecie. *Biology: Concepts and Applications.* 5th ed. New York: Brooks/Cole, 2003.

Sykes, Brian. *Adam's Curse: A Future Without Men.* London: Bantam, 2003.

Tamarin, Robert H. *Principles of Genetics.* Boston: McGraw-Hill, 2002.

Tortora, Gerard. *Microbiology: An Introduction.* 7th ed. San Francisco: Benjamin Cummings, 2001.

Weaver, Robert F., and Philip W. Hedrick. *Genetics.* 3d ed. New York: McGraw-Hill, 1997.

Wells, Spencer. *The Journey of Man: A Genetic Odyssey.* London: Penguin, 2003.

AGRICULTURE AND GENETICALLY MODIFIED FOODS

Altieri, Miguel A. *Genetic Engineering in Agriculture.* Chicago: LPC Group, 2001.

Anderson, Luke. *Genetic Engineering, Food, and Our Environment.* White River Junction, Vt.: Chelsea Green, 1999.

Avery, Dennis T. *Saving the Planet with Pesticides and Plastic: The Environmental Triumph of High-Yield Farming.* 2d ed. Indianapolis: Hudson Institute, 2000.

Carozzi, Nadine, and Michael Koziel, eds. *Advances in Insect Control: The Role of Transgenic Plants.* Bristol, Pa.: Taylor & Francis, 1997.

Chrispeels, Maarten J., and David E. Sadava. *Plants, Genes, and Crop Biotechnology.* Boston: Jones and Bartlett, 2003.

Engel, Karl-Heinz, et al. *Genetically Modified Foods: Safety Aspects.* Washington, D.C.: American Chemical Society, 1995.

Entwhistle, Philip F., Jenny S. Cory, Mark J. Bailey, and Steven R. Higgs. *"Bacillus thuringiensis," an*

Environmental Biopesticide: Theory and Practice. New York: Wiley, 1994.

Fedoroff, Nina, and Nancy Marie Brown. *Mendel in the Kitchen: A Scientist's View of Genetically Modified Foods.* Washington, D.C.: Joseph Henry Press, 2004.

Galun, Esra, and Adina Breiman. *Transgenic Plants.* London: Imperial College Press, 1997.

Goodman, David, et al. *From Farming to Biotechnology: A Theory of Agro-Industrial Development.* New York: Basil Blackwell, 1987.

Hall, Franklin R., and Julius J. Menn, eds. *Biopesticides: Use and Delivery.* Totowa, N.J.: Humana Press, 1999.

Henry, Robert J. *Practical Applications of Plant Molecular Biology.* New York: Chapman & Hall, 1997.

Hindmarsh, Richard, and Geoffrey Lawrence, eds. *Recoding Nature: Critical Perspectives on Genetic Engineering.* Sydney: UNSW Press, 2004.

Koul, Opender, and G. S. Dhaliwal, eds. *Microbial Biopesticides.* New York: Taylor & Francis, 2002.

Krimsky, Sheldon, et al., eds. *Agricultural Biotechnology and the Environment: Science, Policy, and Social Issues.* Urbana: University of Illinois Press, 1996.

Lambrecht, Bill. *Dinner at the New Gene Cafe: How Genetic Engineering Is Changing What We Eat, How We Live, and the Global Politics of Food.* New York: Thomas Dunne Books, 2001.

Lurquin, Paul F. *The Green Phoenix: A History of Genetically Modified Plants.* New York: Columbia University Press, 2001.

Rissler, Jane, and Margaret Mellon. *The Ecological Risks of Engineered Crops.* Cambridge, Mass.: MIT Press, 1996.

Winston, Mark L. *Travels in the Genetically Modified Zone.* Cambridge, Mass.: Harvard University Press, 2002.

Zohary, Daniel, and Maria Hopf. *Domestication of Plants in the Old World: The Origin and Spread of Cultivated Plants in West Asia, Europe, and the Nile Valley.* 3d ed. New York: Oxford University Press, 2001.

BACTERIAL GENETICS

Adelberg, Edward A., ed. *Papers on Bacterial Genetics.* Boston: Little, Brown, 1966.

Birge, Edward A. *Bacterial and Bacteriophage Genetics.* 5th ed. New York: Springer, 2006.

Brock, Thomas D. *The Emergence of Bacterial Genetics.* Cold Spring Harbor, N.Y.: Cold Spring Harbor Laboratory Press, 1990.

Carlberg, David M. *Essentials of Bacterial and Viral Genetics.* Springfield, Ill.: Charles C Thomas, 1976.

Dale, Jeremy. *Molecular Genetics of Bacteria.* New York: John Wiley & Sons, 1994.

Day, Martin J. *Plasmids.* London: Edward Arnold, 1982.

Dean, Alastair Campbell Ross, and Sir Cyril Hinshelwood. *Growth, Function, and Regulation in Bacterial Cells.* Oxford, England: Clarendon Press, 1966.

De Bruijn, Frans J., et al., eds. *Bacterial Genomes: Physical Structure and Analysis.* New York: Chapman & Hall, 1998.

Dorman, Charles J. *Genetics of Bacterial Virulence.* Oxford, England: Blackwell Scientific, 1994.

Drlica, Karl, and Monica Riley, eds. *The Bacterial Chromosome.* Washington, D.C.: American Society for Microbiology, 1990.

Fry, John C., and Martin J. Day, eds. *Bacterial Genetics in Natural Environments.* London: Chapman & Hall, 1990.

Goldberg, Joanna B., ed. *Genetics of Bacterial Polysaccharides.* Boca Raton, Fla.: CRC Press, 1999.

Jacob, François, and Elie L. Wollman. *Sexuality and the Genetics of Bacteria.* New York: Academic Press, 1961.

Joset, Françoise, et al. *Prokaryotic Genetics: Genome Organization, Transfer, and Plasticity.* Boston: Blackwell Scientific Publications, 1993.

Miller, Jeffrey H. *A Short Course in Bacterial Genetics: A Laboratory Manual and Handbook for Escherichia coli and Related Bacteria.* Cold Spring Harbor, N.Y.: Cold Spring Harbor Laboratory Press, 1999.

Murray, Patrick, ed. *Manual of Clinical Microbiology.* Washington, D.C.: ASM Press, 2003.

Park, Simon. *Molecular Genetics of Bacteria.* Chichester, England: John Wiley & Sons, 2005.

Schumann, Wolfgang, S. Dusko Ehrlich, and Naotake Ogasawara, eds. *Functional Analysis of Bacterial Genes: A Practical Manual.* New York: Wiley, 2001.

Siezen, Roland J., et al., eds. *Lactic Acid Bacteria: Genetics, Metabolism, and Applications.* 7th ed. Boston: Kluwer Academic, 2002.

Snyder, Larry, and Wendy Champness. *Molecular Genetics of Bacteria.* Washington, D.C.: ASM Press, 1997.

Summers, David K. *The Biology of Plasmids.* Malden, Mass.: Blackwell, 1996.

Thomas, Christopher M., ed. *The Horizontal Gene*

Pool: Bacterial Plasmids and Gene Spread. Amsterdam: Harwood Academic, 2000.

U.S. Congress. Office of Technology Assessment. *New Developments in Biotechnology—Field-Testing Engineered Organisms: Genetic and Ecological Issues.* Washington, D.C.: National Technical Information Service, 1988.

Vaughan, Pat, ed. *DNA Repair Protocols: Prokaryotic Systems.* Totowa, N.J.: Humana Press, 2000.

BIOETHICS AND SOCIAL POLICY

Becker, Gerhold K., and James P. Buchanan, eds. *Changing Nature's Course: The Ethical Challenge of Biotechnology.* Hong Kong: Hong Kong University Press, 1996.

Bonnicksen, Andrea L. *Crafting a Cloning Policy: From Dolly to Stem Cells.* Washington, D.C.: Georgetown University Press, 2002.

Boylan, Michael, and Kevin E. Brown. *Genetic Engineering: Science and Ethics on the New Frontier.* Upper Saddle River, N.J.: Prentice Hall, 2001.

Brannigan, Michael C., ed. *Ethical Issues in Human Cloning: Cross-Disciplinary Perspectives.* New York: Seven Bridges Press, 2001.

Bulger, Ruth Ellen, Elizabeth Heitman, and Stanley Joel Reiser, eds. *The Ethical Dimensions of the Biological and Health Sciences.* 2d ed. New York: Cambridge University Press, 2002.

Caplan, Arthur. *Due Consideration: Controversy in the Age of Medical Miracles.* New York: Wiley, 1997.

Chadwick, Ruth, et al., eds. *The Ethics of Genetic Screening.* Boston: Kluwer Academic, 1999.

Chapman, Audrey R., ed. *Perspectives on Genetic Patenting: Religion, Science, and Industry in Dialogue.* Washington, D.C.: American Association for the Advancement of Science, 1999.

Charon, Rita, and Martha Montello, eds. *Stories Matter: The Role of Narrative in Medical Ethics.* New York: Routledge, 2002.

Comstock, Gary L., ed. *Life Science Ethics.* Ames: Iowa State Press, 2002.

Danis, Marion, Carolyn Clancy, and Larry R. Churchill, eds. *Ethical Dimensions of Health Policy.* New York: Oxford University Press, 2002.

Davis, Bernard D., ed. *The Genetic Revolution: Scientific Prospects and Public Perceptions.* Baltimore: Johns Hopkins University Press, 1991.

De Waal, Franz. *Good Natured: The Origins of Right and Wrong in Humans and Other Animals.* Cambridge, Mass.: Harvard University Press, 1996.

Doherty, Peter, and Agneta Sutton, eds. *Man-Made Man: Ethical and Legal Issues in Genetics.* Dublin: Four Courts Press, 1997.

Espejo, Roman, ed. *Biomedical Ethics: Opposing Viewpoints.* San Diego: Greenhaven Press, 2003.

Evans, John Hyde. *Playing God? Human Genetic Engineering and the Rationalization of Public Bioethical Debate.* Chicago: University of Chicago Press, 2002.

Gonder, Janet C., Ernest D. Prentice, and Lilly-Marlene Russow, eds. *Genetic Engineering and Animal Welfare: Preparing for the Twenty-first Century.* Greenbelt, Md.: Scientists Center for Animal Welfare, 1999.

Grace, Eric S. *Biotechnology Unzipped: Promises and Reality.* Washington, D.C.: National Academy Press, 1997.

Harpignies, J. P. *Double Helix Hubris: Against Designer Genes.* Brooklyn, N.Y.: Cool Grove Press, 1996.

Holland, Suzanne, Karen Lebacqz, and Laurie Zoloth, eds. *The Human Embryonic Stem Cell Debate: Science, Ethics, and Public Policy (Basic Bioethics).* Cambridge, Mass.: MIT Press, 2001.

Hubbard, Ruth, and Elijah Wald. *Exploding the Gene Myth: How Genetic Information Is Produced and Manipulated by Scientists, Physicians, Employers, Insurance Companies, Educators, and Law Enforcers.* Boston: Beacon Press, 1999.

Kass, Leon R. *Life, Liberty, and the Defense of Dignity: The Challenge for Bioethics.* San Francisco: Encounter Books, 2002.

Kevles, Daniel J. *In the Name of Eugenics: Genetics and the Uses of Human Heredity.* Cambridge, Mass.: Harvard University Press, 1995.

Kristol, William, and Eric Cohen, eds. *The Future Is Now: America Confronts the New Genetics.* Lanham, Md.: Rowman & Littlefield, 2002.

Leroy, Bonnie, Dianne M. Bartels, and Arthur L. Caplan, eds. *Prescribing Our Future: Ethical Challenges in Genetic Counseling.* New York: Aldine de Gruyter, 1993.

Long, Clarisa, ed. *Genetic Testing and the Use of Information.* Washington, D.C.: AEI Press, 1999.

MacKinnon, Barbara, ed. *Human Cloning: Science, Ethics, and Public Policy.* Urbana: University of Illinois Press, 2000.

May, Thomas. *Bioethics in a Liberal Society: The Political Framework of Bioethics Decision Making.* Baltimore: Johns Hopkins University Press, 2002.

O'Neill, Onora. *Autonomy and Trust in Bioethics.* New York: Cambridge University Press, 2002.

Post, Stephen G., ed. *Encyclopedia of Bioethics.* New York: Macmillan Reference USA, 2003.

Rantala, M. L., and Arthur J. Milgram, eds. *Cloning: For and Against.* Chicago: Open Court, 1999.

Real, Leslie A., ed. *Ecological Genetics.* Princeton, N.J.: Princeton University Press, 1994.

Reilly, Philip R. *Abraham Lincoln's DNA and Other Adventures in Genetics.* Cold Spring Harbor, N.Y.: Cold Spring Harbor Laboratory Press, 2000.

Reiss, Michael J., and Roger Straughan, eds. *Improving Nature? The Science and Ethics of Genetic Engineering.* New York: Cambridge University Press, 2001.

Resnik, David B., Holly B. Steinkraus, and Pamela J. Langer. *Human Germline Gene Therapy: Scientific, Moral, and Political Issues.* Austin, Tex.: R. G. Landes, 1999.

Singer, Peter. *Unsanctifying Human Life: Essays on Ethics.* Edited by Helga Kuhse. Malden, Mass.: Blackwell, 2002.

Singer, Peter, and Deane Wells. *Making Babies: The New Science and Ethics of Conception.* New York: Charles Scribner's Sons, 1985.

U.S. Congress. Senate. Committee on Health, Education, Labor, and Pensions. *Fulfilling the Promise of Genetics Research: Ensuring Nondiscrimination in Health Insurance and Employment.* Washington, D.C.: Government Printing Office, 2001.

_______. *Protecting Against Genetic Discrimination: The Limits of Existing Laws.* Washington, D.C.: Government Printing Office, 2002.

Veatch, Robert M. *The Basics of Bioethics.* 2d ed. Upper Saddle River, N.J.: Prentice Hall, 2003.

Vogel, Fredrich, and Reinhard Grunwald, eds. *Patenting of Human Genes and Living Organisms.* New York: Springer, 1994.

Wailoo, Keith. *Dying in the City of the Blues: Sickle Cell Anemia and the Politics of Race and Health.* Chapel Hill: University of North Carolina Press, 2001.

Walters, LeRoy, and Julie Gage Palmer. *The Ethics of Human Gene Therapy.* Illustrated by Natalie C. Johnson. New York: Oxford University Press, 1997.

Yount, Lisa, ed. *The Ethics of Genetic Engineering.* San Diego: Greenhaven Press, 2002.

BIOINFORMATICS. *See also* **GENOMICS AND PROTEOMICS**

Barnes, Michael R. *Bioinformatics for Geneticists.* Hoboken, N.J.: Wiley, 2003.

Baxevanis, Andreas D., and B. F. Francis Ouellette. *Bioinformatics: A Practical Guide to the Analysis of Genes and Proteins.* 2d ed. Hoboken, N.J.: John Wiley & Sons, 2003.

Bergeron, Bryan P. *Bioinformatics Computing.* Upper Saddle River, N.J.: Prentice Hall, 2002.

Bird, R. Curtis, and Bruce F. Smith, eds. *Genetic Library Construction and Screening: Advanced Techniques and Applications.* New York: Springer, 2002.

Bishop, Martin J., ed. *Guide to Human Genome Computing.* 2d ed. San Diego: Academic Press, 1998.

Campbell, A. Malcolm. *Discovering Genomics, Proteomics, and Bioinformatics.* San Francisco: Benjamin Cummings, 2003.

Claverie, Jean-Michel, and Cedric Notredame. *Bioinformatics for Dummies.* Hoboken, N.J.: Wiley, 2003.

Clote, Peter. *Computational Molecular Biology: An Introduction.* New York: John Wiley, 2000.

Dwyer, Rex A. *Genomic Perl: From Bioinformatics Basics to Working Code.* Cambridge, England: Cambridge University Press, 2003.

Kohane, Isaac S. *Microarrays for an Integrative Genomics.* Cambridge, Mass.: MIT Press, 2003.

Krane, Dan E. *Fundamental Concepts of Bioinformatics.* San Francisco: Benjamin Cummings, 2003.

Krawetz, Stephen A., and David D. Womble. *Introduction to Bioinformatics: A Theoretical and Practical Approach.* Totowa, N.J.: Humana Press, 2003.

Lesk, Arthur M. *Introduction to Bioinformatics.* New York: Oxford University Press, 2002.

Lim, Hwa A. *Genetically Yours: Bioinforming, Biopharming, Biofarming.* River Edge, N.J.: World Scientific, 2002.

Meller, Jaroslaw, and Wieslaw Nowak, eds. *Applications of Statistical and Machine Learning Methods in Bioinformatics.* Frankfurt: Peter Lang, 2007.

Mount, David W. *Bioinformatics: Sequence and Genome Analysis.* Cold Spring Harbor, N.Y.: Cold Spring Harbor Laboratory Press, 2001.

Rashidi, Hooman H. *Bioinformatics Basics: Applications in Biological Science and Medicine.* Boca Raton, Fla.: CRC Press, 2000.

Westhead, David R. *Bioinformatics.* Oxford, England: BIOS, 2002.

CELLULAR BIOLOGY

Attardi, Giuseppe M., and Anne Chomyn, eds. *Methods in Enzymology: Mitochondrial Biogenesis and Genetics.* Vols. 260, 264. San Diego: Academic Press, 1995.

Becker, Peter B. *Chromatin Protocols.* Totowa, N.J.: Humana Press, 1999.

Becker, W. M., L. J. Kleinsmith, and J. Hardin. *The World of the Cell.* 5th ed. San Francisco: Benjamin Cummings, 2003.

Bell, John I., et al., eds. *T Cell Receptors.* New York: Oxford University Press, 1995.

Bickmore, Wendy A. *Chromosome Structural Analysis: A Practical Approach.* New York: Oxford University Press, 1999.

Blackburn, Elizabeth H., and Carol W. Greider, eds. *Telomeres.* Cold Spring Harbor, N.Y.: Cold Spring Harbor Laboratory Press, 1995.

Broach, J., J. Pringle, and E. Jones, eds. *The Molecular and Cellular Biology of the Yeast Saccharomyces.* 3 vols. Cold Spring Harbor, N.Y.: Cold Spring Harbor Laboratory Press, 1991-1997.

Carey, M., and S. T. Smale. *Transcriptional Regulation in Eukaryotes: Concepts, Strategies and Techniques.* Cold Spring Harbor, N.Y.: Cold Spring Harbor Laboratory Press, 2000.

Darnell, James, et al. *Molecular Cell Biology.* 4th ed. New York: W. H. Freeman, 2000.

Franklin, T. J., and G. A. Snow. *Biochemistry of Antimicrobial Action.* New York: Chapman and Hall, 2001.

Freshney, R. Ian. *Culture of Animal Cells.* New York: Wiley-Liss, 2000.

Gilchrest, Barbara A., and Vilhelm A. Bohr, eds. *The Role of DNA Damage and Repair in Cell Aging.* New York: Elsevier, 2001.

Gillham, Nicholas W. *Organelle Genes and Genomes.* London: Oxford University Press, 1997.

John, Bernard. *Meiosis.* New York: Cambridge University Press, 1990.

Kipling, David, ed. *The Telomere.* New York: Oxford University Press, 1995.

Lewis, Claire E., and James O'D. McGee. *The Macrophage: The Natural Immune System.* Oxford, England: IRL Press at Oxford University Press, 1992.

Lodish, Harvey, David Baltimore, and Arnold Berk. *Molecular Cell Biology.* 4th ed. New York: W. H. Freeman, 2000.

Marshak, Daniel R., Richard L. Gardner, and David Gottlieb, eds. *Stem Cell Biology.* Cold Spring Harbor, N.Y.: Cold Spring Harbor Laboratory Press, 2002.

Murray, A. W., and Tim Hunt. *The Cell Cycle: An Introduction.* New York: W. H. Freeman, 1993.

Pon, Liza, and Eric A. Schon, eds. *Mitochondria.* San Diego: Academic Press, 2001.

Scheffler, Immo E. *Mitochondria.* New York: John Wiley & Sons, 1999.

Sharma, Archana, and Sumitra Sen. *Chromosome Botany.* Enfield, N.H.: Science Publishers, 2002.

Vig, Baldev K., ed. *Chromosome Segregation and Aneuploidy.* New York: Springer-Verlag, 1993.

CLASSICAL TRANSMISSION GENETICS. *See also* **HISTORY OF GENETICS**

Berg, Paul, and Maxine Singer. *Dealing with Genes: The Language of Heredity.* Mill Valley, Calif.: University Science Books, 1992.

Cittadino, E. *Nature as the Laboratory.* New York: Columbia University Press, 1990.

Ford, E. B. *Understanding Genetics.* London: Faber & Faber, 1979.

Gardner, Eldon J. *Principles of Genetics.* 5th ed. New York: John Wiley & Sons, 1975.

Goodenough, Ursula. *Genetics.* 2d ed. New York: Holt, Rinehart and Winston, 1978.

Klug, William, and Michael Cummings. *Concepts of Genetics.* 4th ed. New York: Macmillan College, 1994.

Lewin, Benjamin. *Genes VII.* New York: Oxford University Press, 2001.

Mendel, Gregor. *Experiments in Plant-Hybridization.* Foreword by Paul C. Mangelsdorf. Cambridge, Mass.: Harvard University Press, 1965.

Rothwell, Norman V. *Understanding Genetics.* 4th ed. New York: Oxford University Press, 1988.

Russell, Peter. *Genetics.* 2d ed. Boston: Scott, Foresman, 1990.

Singer, Sam. *Human Genetics: An Introduction to the Principles of Heredity.* 2d ed. New York: W. H. Freeman, 1985.

Suzuki, David T., et al., eds. *An Introduction to Genetic Analysis.* 2d ed. San Francisco: W. H. Freeman, 1981.

Weaver, Robert F., and Philip W. Hendrick. *Genetics.* 3d ed. Dubuque, Iowa: Wm. C. Brown, 1997.

Wilson, Edward O. *The Diversity of Life.* New York: W. W. Norton, 1992.

CLONING. *See also* **GENETIC ENGINEERING**

Brannigan, Michael C., ed. *Ethical Issues in Human Cloning: Cross-Disciplinary Perspectives.* New York: Seven Bridges Press, 2001.

Chen, Bing-Yuan, and Harry W. Janes, eds. *PCR*

Cloning Protocols. Rev. 2d ed. Totowa, N.J.: Humana Press, 2002.

DeSalle, Robert, and David Lindley. *The Science of Jurassic Park and the Lost World.* New York: BasicBooks, 1997.

Drlica, Karl. *Understanding DNA and Gene Cloning: A Guide for the Curious.* Rev. ed. New York: Wiley, 2003.

Jones, P., and D. Ramji. *Vectors: Cloning Applications and Essential Techniques.* New York: J. Wiley, 1998.

Klotzko, Arlene Judith. *A Clone of Your Own? The Science and Ethics of Cloning.* New York: Cambridge University Press, 2006.

_______, ed. *The Cloning Sourcebook.* New York: Oxford University Press, 2001.

Lauritzen, Paul, ed. *Cloning and the Future of Human Embryo Research.* New York: Oxford University Press, 2001.

Lu, Quinn, and Michael P. Weiner, eds. *Cloning and Expression Vectors for Gene Function Analysis.* Natick, Mass.: Eaton, 2001.

MacKinnon, Barbara, ed. *Human Cloning: Science, Ethics, and Public Policy.* Urbana: University of Illinois Press, 2000.

Prentice, David A. *Stem Cells and Cloning.* New York: Benjamin Cummings, 2003.

Rantala, M. L., and Arthur J. Milgram, eds. *Cloning: For and Against.* Chicago: Open Court, 1999.

Sambrook, Joseph, and David W. Russell, eds. *Molecular Cloning: A Laboratory Manual.* 3d ed. 3 vols. Cold Spring Harbor, N.Y.: Cold Spring Harbor Laboratory Press, 2001.

Shostak, Stanley. *Becoming Immortal: Combining Cloning and Stem-Cell Therapy.* Albany: State University of New York Press, 2002.

Wilmut, Ian, Keith Campbell, and Colin Tudge. *The Second Creation: The Age of Biological Control by the Scientists That Cloned Dolly.* London: Headline, 2000.

DEVELOPMENTAL GENETICS

Beurton, Peter, Raphael Falk, and Hans-Jorg Rheinberger, eds. *The Concept of the Gene in Development and Evolution: Historical and Epistemological Perspectives.* New York: Cambridge University Press, 2000.

Bier, Ethan. *The Coiled Spring: How Life Begins.* Cold Spring Harbor, N.Y.: Cold Spring Harbor Laboratory Press, 2000.

Bowman, John L. *Arabidopsis: An Atlas of Morphology and Development.* New York: Springer-Verlag, 1993.

Carroll, Sean B., Jennifer K. Grenier, and Scott D. Weatherbee. *From DNA to Diversity: Molecular Genetics and the Evolution of Animal Design.* Malden, Mass.: Blackwell, 2001.

Cronk, Quentin C. B., Richard M. Bateman, and Julie A. Hawkins, eds. *Developmental Genetics and Plant Evolution.* New York: Taylor & Francis, 2002.

Davidson, Eric H. *Gene Activity in Early Development.* Orlando, Fla.: Academic Press, 1986.

DePamphilis, Melvin L., ed. *Gene Expression at the Beginning of Animal Development.* New York: Elsevier, 2002.

DePomerai, David. *From Gene to Animal: An Introduction to the Molecular Biology of Animal Development.* New York: Cambridge University Press, 1985.

Dyban, A. P., and V. S. Baranov. *Cytogenetics of Mammalian Embryonic Development.* Translated by V. S. Baranov. New York: Oxford University Press, 1987.

Gilbert, Scott F. *Developmental Biology.* Sunderland, Mass.: Sinauer Associates, 2003.

Gottlieb, Frederick J. *Developmental Genetics.* New York: Reinhold, 1966.

Gurdon, John B. *The Control of Gene Expression in Animal Development.* Cambridge, Mass.: Harvard University Press, 1974.

Hahn, Martin E., et al., eds. *Developmental Behavior Genetics: Neural, Biometrical, and Evolutionary Approaches.* New York: Oxford University Press, 1990.

Harvey, Richard P., and Nadia Rosenthal, eds. *Heart Development.* San Diego: Academic Press, 1999.

Hennig, W., ed. *Early Embryonic Development of Animals.* New York: Springer-Verlag, 1992.

Hsia, David Yi-Yung. *Human Developmental Genetics.* Chicago: Year Book Medical Publishers, 1968.

Hunter, R. H. F. *Sex Determination, Differentiation, and Intersexuality in Placental Mammals.* New York: Cambridge University Press, 1995.

Leighton, Terrance, and William F. Loomis, Jr. *The Molecular Genetics of Development.* New York: Academic Press, 1980.

McKinney, M. L., and K. J. McNamara. *Heterochrony: The Evolution of Ontogeny.* New York: Plenum Press, 1991.

Malacinski, George M., ed. *Developmental Genetics of Higher Organisms: A Primer in Developmental Biology.* New York: Macmillan, 1988.

Massaro, Edward J., and John M. Rogers, eds. *Folate and Human Development.* Totowa, N.J.: Humana Press, 2002.

Moore, Keith. *The Developing Human: Clinically Ori-*

ented Embryology. 7th ed. Amsterdam: Elsevier Science, 2003.

Nüsslein-Volhard, Christiane, and J. Kratzschmar, eds. *Of Fish, Fly, Worm, and Man: Lessons from Developmental Biology for Human Gene Function and Disease.* New York: Springer, 2000.

Piontelli, Alessandra. *Twins: From Fetus to Child.* New York: Routledge, 2002.

Prescott, John, and Beverly J. Tepper, eds. *Genetic Variation in Taste Sensitivity.* New York: Marcel Dekker, 2004.

Pritchard, Dorian J. *Foundations of Developmental Genetics.* Philadelphia: Taylor & Francis, 1986.

Raff, Rudolf. *The Shape of Life: Genes, Development, and the Evolution of Animal Form.* Chicago: University of Chicago Press, 1996.

Ranke, M., and G. Gilli. *Growth Standards, Bone Maturation, and Idiopathic Short Stature.* Farmington, Conn.: S. Karger, 1996.

Rao, Mahendra S., ed. *Stem Cells and CNS Development.* Totowa, N.J.: Humana Press, 2001.

Sang, James H. *Genetics and Development.* New York: Longman, 1984.

Saunders, John Warren, Jr. *Patterns and Principles of Animal Development.* New York: Macmillan, 1970.

Seidman, S., and H. Soreq. *Transgenic Xenopus: Microinjection Methods and Developmental Neurobiology.* Totowa, N.J.: Humana Press, 1997.

Stewart, Alistair D., and David M. Hunt. *The Genetic Basis of Development.* New York: John Wiley & Sons, 1982.

Tomanek, Robert J., and Raymond B. Runyan, eds. *Formation of the Heart and Its Regulation.* Foreword by Edward B. Clark. Boston: Birkhauser, 2001.

Ulijaszek, S. J., Francis E. Johnston, and Michael A. Preece. *Cambridge Encyclopedia of Human Growth and Development.* New York: Cambridge University Press, 1998.

Vijg, Jan. *Aging of the Genome: The Dual Role of the DNA in Life and Death.* New York: Oxford University Press, 2007.

Wilkins, Adam S. *Genetic Analysis of Animal Development.* New York: Wiley-Liss, 1993.

DISEASES AND SYNDROMES

GENERAL

American Psychiatric Association. *Diagnostic and Statistical Manual of Mental Disorders: DSM-IV-TR.* Rev. 4th ed. Washington, D.C.: Author, 2000.

Bennett, Robin L. *The Practical Guide to the Genetic Family History.* New York: Wiley-Liss, 1999.

Bianchi, Diana W., Timothy M. Crombleholme, and Mary E. D'Alton. *Fetology: Diagnosis and Management of the Fetal Patient.* New York: McGraw-Hill, 2000.

Brooks, G. F., J. S. Butel, and S. A. Morse. *Medical Microbiology.* 21st ed. Stamford, Conn.: Appleton and Lange, 1998.

Browne, M. J., and P. L. Thurlby, eds. *Genomes, Molecular Biology, and Drug Discovery.* San Diego: Academic Press, 1996.

Cotran, R. S., et al. *Robbins Pathologic Basis of Disease.* 6th ed. Philadelphia: Saunders, 1999.

Epstein, Richard J. *Human Molecular Biology; An Introduction to the Molecular Basis of Health and Disease.* Cambridge, England: Cambridge University Press, 2003.

Gallo, Robert C., and Flossie Wong-Staal, eds. *Retrovirus Biology and Human Disease.* New York: Marcel Dekker, 1990.

Gelehrter, Thomas D., Francis S. Collins, and David Ginsburg. *Principles of Medical Genetics.* 2d ed. Baltimore: Williams & Wilkins, 1998.

Gilbert, Patricia. *Dictionary of Syndromes and Inherited Disorders.* 3d ed. Chicago: Fitzroy Dearborn, 2000.

Hogenboom, Marga. *Living with Genetic Syndromes Associated with Intellectual Disability.* Philadelphia: Jessica Kingsley, 2001.

Jorde, L. B., J. C. Carey, M. J. Bamshad, and R. L. White. *Medical Genetics.* 2d ed. St. Louis, Mo.: Mosby, 2000.

Massimini, Kathy, ed. *Genetic Disorders Sourcebook: Basic Consumer Information About Hereditary Diseases and Disorders.* 2d ed. Detroit: Omnigraphics, 2001.

Neumann, David, et al. *Human Variability in Response to Chemical Exposures: Measures, Modeling, and Risk Assessment.* Boca Raton, Fla.: CRC Press, 1998.

New, Maria I., ed. *Diagnosis and Treatment of the Unborn Child.* Reddick, Fla.: Idelson-Gnocchi, 1999.

O'Rahilly, S., and D. B. Dunger, eds. *Genetic Insights in Paediatric Endocrinology and Metabolism.* Bristol, England: BioScientifica, 1999.

Pai, G. Shashidhar, Raymond C. Lewandowski, and Digamber S. Borgaonkar. *Handbook of Chromosomal Syndromes.* New York: John Wiley & Sons, 2002.

Pasternak, Jack J. *An Introduction to Human Molecular Genetics: Mechanisms of Inherited Diseases.* Bethesda, Md.: Fitzgerald Science Press, 1999.

Petrikovsky, Boris M., ed. *Fetal Disorders: Diagnosis and Management.* New York: Wiley-Liss, 1999.

Pilu, Gianluigi, and Kypros H. Nicolaides. *Diagnosis of Fetal Abnormalities: The 18-23-Week Scan.* New York: Parthenon Group, 1999.

Rakel, Robert E., et al., eds. *Conn's Current Therapy.* Philadelphia: W. B. Saunders, 2003.

Sasaki, Mutsuo, et al., eds. *New Directions for Cellular and Organ Transplantation.* New York: Elsevier Science, 2000.

Stephens, Trent D., and Rock Brynner. *Dark Remedy: The Impact of Thalidomide and Its Revival as a Vital Medicine.* Cambridge, Mass.: Perseus, 2001.

Twining, Peter, Josephine M. McHugo, and David W. Pilling, eds. *Textbook of Fetal Abnormalities.* New York: Churchill Livingstone, 2000.

Weaver, David D., with the assistance of Ira K. Brandt. *Catalog of Prenatally Diagnosed Conditions.* 3d ed. Baltimore: Johns Hopkins University Press, 1999.

Weiss, Kenneth M. *Genetic Variation and Human Disease: Principles and Evolutionary Approaches.* New York: Cambridge University Press, 1993.

Wynbrandt, James, and Mark D. Ludman. *The Encyclopedia of Genetic Disorders and Birth Defects.* 2d ed. New York: Facts On File, 2000.

Zallen, Doris Teichler. *Does It Run in the Family? A Consumer's Guide to DNA Testing for Genetic Disorders.* New Brunswick, N.J.: Rutgers University Press, 1997.

ALZHEIMER'S DISEASE

Gauthier, S., ed. *Clinical Diagnosis and Management of Alzheimer's Disease.* 2d ed. London: Martin Dunitz, 2001.

Hamdy, Ronald, James Turnball, and Joellyn Edwards. *Alzheimer's Disease: A Handbook for Caregivers.* New York: Mosby, 1998.

Mace, M., and P. Rabins. *The Thirty-six-Hour Day: A Family Guide to Caring for Persons with Alzheimer Disease, Related Dementing Illnesses, and Memory Loss in Later Life.* Baltimore: Johns Hopkins University Press, 1999.

Nelson, James Lindemann, and Hilde Lindemann Nelson. *Alzheimer's: Answers to Hard Questions for Families.* New York: Main Street Books, 1996.

Powell, L., and K. Courtice. *Alzheimer's Disease: A Guide for Families and Caregivers.* Cambridge, Mass.: Perseus, 2001.

Richter, Ralph W., and Brigitte Zoeller Richter, eds. *Alzheimer's Disease: A Physician's Guide to Practical Management.* Totowa, N.J.: Humana Press, 2004.

Sadowski, Marcin, and Thomas M. Wisniewski. *100 Questions and Answers About Alzheimer's Disease.* Boston: Jones and Bartlett, 2004.

Terry, R., R. Katzman, K. Bick, and S. Sisodia. *Alzheimer Disease.* 2d ed. Philadelphia: Lippincott Williams & Wilkins, 1999.

CANCER

Angier, Natalie. *Natural Obsessions: Striving to Unlock the Deepest Secrets of the Cancer Cell.* Boston: Mariner Books/Houghton Mifflin, 1999.

Bowcock, Anne M., ed. *Breast Cancer: Molecular Genetics, Pathogenesis, and Therapeutics.* Totowa, N.J.: Humana Press, 1999.

Cannon, L Bishop, et al. "Genetic Epidemiology of Prostate Cancer in the Utah Mormon Genealogy." In *Inheritance of Susceptibility to Cancer in Man,* edited by W. F. Bodmer. New York: Oxford University Press, 1983.

Coleman, William B., and Gregory J. Tsongalis, eds. *The Molecular Basis of Human Cancer.* Totowa, N.J.: Humana Press, 2002.

Cooper, Geoffrey M. *Oncogenes.* 2d ed. Boston: Jones and Bartlett, 1995.

Cowell, J. K., ed. *Molecular Genetics of Cancer.* 2d ed. San Diego: Academic Press, 2001.

Davies, Kevin, and Michael White. *Breakthrough: The Race to Find the Breast Cancer Gene.* New York: John Wiley, 1996.

DeVita, V. T., S. Hellman, and S. A. Rosenberg, eds. *Principles and Practice of Oncology.* Philadelphia: Lippincott-Raven, 1997.

Dickson, Robert B., and Marc E. Lipman, eds. *Genes, Oncogenes, and Hormones: Advances in Cellular and Molecular Biology of Breast Cancer.* Boston: Kluwer Academic, 1992.

Dunbier, Anita K., and Parry J. Guilford. "Gastric Cancer: Inherited Predisposition." In *Encyclopedia of Cancer.* 2d ed. New York: Elsevier Science, 2002.

Ehrlich, Melanie, ed. *DNA Alterations in Cancer: Genetic and Epigenetic Changes.* Natick, Mass.: Eaton, 2000.

Fisher, David E., ed. *Tumor Suppressor Genes in Human Cancer.* Totowa, N.J.: Humana Press, 2001.

Goode, Ellen L., and Gail P. Jarvik. "Genetics Predisposition to Prostate Cancer." In *Encyclopedia of Cancer.* 2d ed. New York: Elsevier Science, 2002.

Greaves, Mel F. *Cancer: The Evolutionary Legacy.* New York: Oxford University Press, 2000.

Habib, Nagy A., ed. *Cancer Gene Therapy: Past Achievements and Future Challenges.* New York: Kluwer Academic/Plenum, 2000.

Hanski, C., H. Scherübl, and B. Mann, eds. *Colorectal Cancer: New Aspects of Molecular Biology and Immunology and Their Clinical Applications.* New York: New York Academy of Sciences, 2000.

Heim, S., and Felix Mitelman. *Cancer Cytogenetics.* 2d ed. New York: J. Wiley, 1995.

Hodgson, Shirley V., and Eamonn R. Maher. *A Practical Guide to Human Cancer Genetics.* 2d ed. New York: Cambridge University Press, 1999.

Kemeny, Mary Margaret, and Paula Dranov. *Beating the Odds Against Breast and Ovarian Cancer: Reducing Your Hereditary Risk.* Reading, Mass.: Addison-Wesley, 1992.

Krupp, Guido, and Reza Parwaresch, eds. *Telomerases, Telomeres, and Cancer.* New York: Kluwer Academic/Plenum, 2003.

La Thangue, Nicholas B., and Lasantha R. Bandara, eds. *Targets for Cancer Chemotherapy: Transcription Factors and Other Nuclear Proteins.* Totowa, N.J.: Humana Press, 2002.

Lattime, Edmund C., and Stanton L. Gerson, eds. *Gene Therapy of Cancer: Translational Approaches from Preclinical Studies to Clinical Implementation.* 2d ed. San Diego: Academic Press, 2002.

Maruta, Hiroshi, ed. *Tumor-Suppressing Viruses, Genes, and Drugs: Innovative Cancer Therapy Approaches.* San Diego: Academic Press, 2002.

Mendelsohn, John, et al. *The Molecular Basis of Cancer.* 2d ed. Philadelphia: Saunders, 2001.

Mulvihill, John J. *Catalog of Human Cancer Genes: McKusick's Mendelian Inheritance in Man for Clinical and Research Oncologists.* Foreword by Victor A. McKusick. Baltimore: Johns Hopkins University Press, 1999.

National Cancer Institute. *Genetic Testing for Breast Cancer: It's Your Choice.* Bethesda, Md.: Author, 1997.

Ruddon, Raymond. *Cancer Biology.* 3d ed. New York: Oxford University Press, 1995.

Schneider, Katherine A. *Counseling About Cancer: Strategies for Genetic Counseling.* 2d ed. New York: Wiley-Liss, 2002.

Varmus, Harold, and Robert Weinberg. *Genes and the Biology of Cancer.* New York: W. H. Freeman, 1993.

Vogelstein, Bert, and Kenneth W. Kinzler, eds. *The Genetic Basis of Human Cancer.* 2d ed. New York: McGraw-Hill, 2002.

Wilson, Samuel, et al. *Cancer and the Environment: Gene-Environment Interaction.* Washington, D.C.: National Academy Press, 2002.

Cholera

Coleman, William, and I. Edward Alcamo, eds. *Cholera.* Philadelphia: Chelsea House, 2003.

Keusch, Gerald, and Masanobu Kawakami, eds. *Cytokines, Cholera, and the Gut.* Amsterdam: IOS Press, 1997.

Wachsmuth, Kate, et al., eds. *Vibrio Cholerae and Cholera: Molecular to Global Perspectives.* Washington, D.C.: ASM Press, 1994.

Color Blindness

Rosenthal, Odeda, and Robert H. Phillips. *Coping with Color Blindness.* Garden City Park, N.Y.: Avery, 1997.

Congenital Adrenal Hyperplasia

Speiser, Phyllis W., ed. *Congenital Adrenal Hyperplasia.* Philadelphia: W. B. Saunders, 2001.

Cystic Fibrosis

Hodson, Margaret E., and Duncan M. Geddes, eds. *Cystic Fibrosis.* 2d ed. New York: Oxford University Press, 2000.

Orenstein, David M., Beryl J. Rosenstein, and Robert C. Stern. *Cystic Fibrosis: Medical Care.* Philadelphia: Lippincott Williams & Wilkins, 2000.

Shale, Dennis. *Cystic Fibrosis.* London: British Medical Association, 2002.

Tsui, Lap-Chee, et al., eds. *The Identification of the CF (Cystic Fibrosis) Gene: Recent Progress and New Research Strategies.* New York: Plenum Press, 1991.

Yankaskas, James R., and Michael R. Knowles, eds. *Cystic Fibrosis in Adults.* Philadelphia: Lippincott-Raven, 1999.

Diabetes

American Diabetes Association. *American Diabetes Association Complete Guide to Diabetes: The Ultimate Home Reference from the Diabetes Experts.* New York: McGraw-Hill, 2002.

Becker, Gretchen. *The First Year: Type 2 Diabetes, an Essential Guide for the Newly Diagnosed.* New York: Marlowe, 2001.

Flyvbjerg, Allan, Hans Orskov, and George Alberti, eds. *Growth Hormone and Insulin-like Growth Factor I in Human and Experimental Diabetes.* New York: John Wiley & Sons, 1993.

Kahn, C. R., et al. *Joslin's Diabetes Mellitus.* 14th ed. Philadelphia: Lippincott Williams and Wilkins, 2003.

Lowe, William L., Jr., ed. *Genetics of Diabetes Mellitus.* Boston: Kluwer Academic, 2001.

Milchovich, Sue K., and Barbara Dunn-Long. *Diabetes Mellitus: A Practical Handbook.* 8th ed. Boulder, Colo.: Bull, 2003.

Pinette, Gilles. *Diabetes and Diet: Ivan's Story.* Owen Sound, Ont.: Ningwakwe Learning Press, 2002.

Roith, Derek Le, and Haim Werner. *Insulin Growth Factors.* In *Encyclopedia of Cancer.* 2d ed. New York: Elsevier Science, 2002.

Down Syndrome and Mental Retardation

Broman, Sarah H., and Jordan Grafman, eds. *Atypical Cognitive Deficits in Developmental Disorders: Implications for Brain Function.* Hillsdale, N.J.: Lawrence Erlbaum, 1994.

Cohen, William I., Lynn Nadel, and Myra E. Madnick, eds. *Down Syndrome: Visions for the Twenty-first Century.* New York: Wiley-Liss, 2002.

Cunningham, Cliff. *Understanding Down Syndrome: An Introduction for Parents.* 1988. Reprint. Cambridge, Mass.: Brookline Books, 1999.

Dykens, Elisabeth M., Robert M. Hodapp, and Brenda M. Finucane. *Genetics and Mental Retardation Syndromes: A New Look at Behavior and Interventions.* Baltimore: Paul H. Brookes, 2000.

Faraone, Stephen V., Ming T. Tsuang, and Debby W. Tsuang. *Genetics of Mental Disorders: A Guide for Students, Clinicians, and Researchers.* New York: Guilford Press, 1999.

Jobling, Anne, and Naznin Virji-Babul. *Down Syndrome: Play, Move and Grow.* Burnaby, B.C.: Down Syndrome Research Foundation, 2004.

Lubec, G. *Protein Expression in Down Syndrome Brain.* New York: Springer, 2001.

Newton, Richard. *The Down's Syndrome Handbook: A Practical Guide for Parents and Caregivers.* New York: Arrow Books, 1997.

Selikowitz, Mark. *Down Syndrome: The Facts.* 2d ed. New York: Oxford University Press, 1997.

Shannon, Joyce Brennfleck, ed. *Mental Retardation Sourcebook: Basic Consumer Health Information About Mental Retardation and Its Causes, Including Down Syndrome, Fetal Alcohol Syndrome, Fragile X Syndrome, Genetic Conditions, Injury, and Environmental Sources.* Detroit: Omnigraphics, 2000.

Fragile X Syndrome

Hagerman, Paul J. *Fragile X Syndrome: Diagnosis, Treatment, and Research.* 3d ed. Baltimore: Johns Hopkins University Press, 2002.

Hagerman, Randi Jenssen, and Paul J. Hagerman, eds. *Fragile X Syndrome: Diagnosis, Treatment, and Research.* Baltimore: Johns Hopkins University Press, 2002.

Parker, James N., and Philip M. Parker, eds. *The 2002 Official Parent's Sourcebook on Fragile X Syndrome.* San Diego: Icon Health, 2002.

Gender Identity and Sex Errors

Berch, Daniel B., and Bruce G. Bender, eds. *Sex Chromosome Abnormalities and Human Behavior.* Boulder, Colo.: Westview Press, 1990.

Dreger, Alice Domurat. *Hermaphrodites and the Medical Invention of Sex.* Cambridge, Mass.: Harvard University Press, 1998.

Gilbert, Ruth. *Early Modern Hermaphrodites: Sex and Other Stories.* New York: Palgrave, 2002.

Money, John. *Sex Errors of the Body and Related Syndromes: A Guide to Counseling Children, Adolescents, and Their Families.* 2d ed. Baltimore: Paul H. Brookes, 1994.

Heart Diseases and Defects

Braunwald, Eugene, Douglas P. Zipes, Peter Libby, and Douglas D. Zipes. *Heart Disease: A Textbook of Cardiovascular Medicine.* 6th ed. Philadelphia: W. B. Saunders, 2001.

Edwards, Jesse E. *Jesse E. Edwards' Synopsis of Congenital Heart Disease.* Edited by Brooks S. Edwards. Armonk, N.Y.: Futura, 2000.

Goldbourt, Uri, Kare Berg, and Ulf de Faire, eds. *Genetic Factors in Coronary Heart Disease.* Boston: Kluwer Academic, 1994.

Kramer, Gerri Freid, and Shari Maurer. *The Parent's Guide to Children's Congenital Heart Defects: What They Are, How to Treat Them, How to Cope with Them.* Foreword by Sylvester Stallone and Jennifer Flavin-Stallone. New York: Three Rivers Press, 2001.

Marian, Ali J. *Genetics for Cardiologists: The Molecular Genetic Basis of Cardiovascular Disorders.* London: ReMEDICA, 2000.

Hemophilia

Buzzard, Brenda, and Karen Beeton, eds. *Physiotherapy Management of Haemophilia.* Malden, Mass.: Blackwell, 2000.

Jones, Peter. *Living with Haemophilia.* 5th ed. New York: Oxford University Press, 2002.

Monroe, Dougald M., et al., eds. *Hemophilia Care in the New Millennium.* New York: Kluwer Academic/Plenum, 2001.

Resnik, Susan. *Blood Saga: Hemophilia, AIDS, and the Survival of a Community.* Berkeley: University of California Press, 1999.

Steinberg, Martin H., et al., eds. *Disorders of Hemoglobin: Genetics, Pathophysiology, and Clinical Management.* Foreword by H. Franklin Bunn. New York: Cambridge University Press, 2001.

IMMUNE DISORDERS AND ALLERGIES

Abbas, Abul K., and Richard A. Flavell, eds. *Genetic Models of Immune and Inflammatory Diseases.* New York: Springer, 1996.

Bona, Constantin A., et al., eds. *The Molecular Pathology of Autoimmune Diseases.* 2d ed. New York: Taylor and Francis, 2002.

Clark, William R. *At War Within: The Double-Edged Sword of Immunity.* New York: Oxford University Press, 1995.

Cutler, Ellen W. *Winning the War Against Asthma and Allergies.* Albany, N.Y.: Delmar, 1998.

Hadley, Andrew G., and Peter Soothill, eds. *Alloimmune Disorders of Pregnancy: Anaemia, Thrombocytopenia, and Neutropenia in the Fetus and Newborn.* New York: Cambridge University Press, 2002.

Shomon, Mary J. *Living Well with Chronic Fatigue Syndrome and Fibromyalgia: What Your Doctor Doesn't Tell You That You Need to Know.* New York: Quill, 2004.

Walsh, William. *The Food Allergy Book.* New York: J. Wiley, 2000.

INFECTIOUS AND EMERGING DISEASES

Anderson, R. M., and R. M. May. *Infectious Diseases of Humans: Dynamics and Control.* Oxford, England: Oxford University Press, 1992.

DeSalle, Rob, ed. *Epidemic! The World of Infectious Disease.* New York: The New Press, 1999.

Drexler, Madeline. *Secret Agents: The Menace of Emerging Infections.* Washington, D.C.: Joseph Henry Press, 2002.

Garrett, Laurie. *The Coming Plague: Newly Emerging Diseases in a World Out of Balance.* New York: Penguin, 1995.

Hacker, J., and J. B. Kaper, eds. *Pathogenicity Islands and the Evolution of Pathogenic Microbes.* 2 vols. New York: Springer, 2002.

Kolata, Gina. *Flu: The Story of the Great Influenza Pandemic of 1918 and the Search for the Virus That Caused It.* New York: Simon and Schuster, 2001.

Lappe, Marc. *Breakout: The Evolving Threat of Drug-Resistant Disease.* San Francisco: Sierra Club Books, 1996.

Levy, Stuart B. *The Antibiotic Paradox: How Miracle Drugs Are Destroying the Miracle.* New York: Plenum Press, 1992.

McNeill, William H. *Plagues and Peoples.* New York: Anchor Books, 1998.

Parker, James N., and Philip M. Parker, eds. *The Official Patient's Sourcebook on "E. coli."* San Diego: Icon Health, 2002.

World Health Organization. *Future Research on Smallpox Virus Recommended.* Geneva, Switzerland: World Health Organization Press, 1999.

KLINEFELTER SYNDROME

Bock, Robert. *Understanding Klinefelter Syndrome: A Guide for XXY Males and Their Families.* Bethesda, Md.: Department of Health and Human Services, Public Health Service, National Institutes of Health, National Institute of Child Health and Human Development, 1997.

Parker, James N., and Philip M. Parker, eds. *The Official Parent's Sourcebook on Klinefelter Syndrome: A Revised and Updated Directory for the Internet Age.* San Diego: Icon Health, 2002.

Probasco, Terri, and Gretchen A. Gibbs. *Klinefelter Syndrome: Personal and Professional Guide.* Richmond, Ind.: Prinit Press, 1999.

LACTOSE INTOLERANCE

Auricchio, Salvatore, and G. Semenza, eds. *Common Food Intolerances 2: Milk in Human Nutrition and Adult-Type Hypolactasia.* New York: Karger, 1993.

METABOLIC DISORDERS

Econs, Michael J., ed. *The Genetics of Osteoporosis and Metabolic Bone Disease.* Totowa, N.J.: Humana Press, 2000.

Evans, Mark I., ed. *Metabolic and Genetic Screening.* Philadelphia: W. B. Saunders, 2001.

Pacifici, G. M., and G. N. Fracchia, eds. *Advances in Drug Metabolism in Man.* Brussels: European Commission, 1995.

Scriver, Charles, et al., eds. *The Metabolic and Molecular Bases of Inherited Disease.* 8th ed. 4 vols. New York: McGraw-Hill, 2001.

Tew, K. D., et al., eds. *Structure and Function of Glutathione Transferases.* Boca Raton, Fla.: CRC Press, 1993.

Mitochondrial Diseases

Lestienne, Patrick, ed. *Mitochondrial Diseases: Models and Methods.* New York: Springer, 1999.

Neural Defects

Bock, Gregory, and Joan Marsh, eds. *Neural Tube Defects.* New York: Wiley, 1994.

Goldstein, Sam, and Cecil R. Reynolds, eds. *Handbook of Neurodevelopmental and Genetic Disorders in Children.* New York: Guilford Press, 1999.

Phenylketonuria

Koch, Jean Holt. *Robert Guthrie, the PKU Story: A Crusade Against Mental Retardation.* Pasadena, Calif.: Hope, 1997.

Parker, James N., ed. *The Official Parent's Sourcebook on Phenylketonuria.* San Diego: Icon Health, 2002.

Prion Diseases

Baker, Harry F., ed. *Molecular Pathology of the Prions.* Totowa, N.J.: Humana Press, 2001.

Groschup, Martin H., and Hans A. Kretzschmar, eds. *Prion Diseases: Diagnosis and Pathogenesis.* New York: Springer, 2000.

Klitzman, Robert. *The Trembling Mountain: A Personal Account of Kuru, Cannibals, and Mad Cow Disease.* New York: Plenum Trade, 1998.

_______, ed. *Prion Biology and Diseases.* Cold Spring Harbor, N.Y.: Cold Spring Harbor Laboratory Press, 1999.

Rabenau, Holger F., Jindrich Cinatl, and Hans Wilhelm Doerr, eds. *Prions: A Challenge for Science, Medicine, and Public Health System.* New York: Karger, 2001.

Ratzan, Scott C., ed. *The Mad Cow Crisis: Health and the Public Good.* New York: New York University Press, 1998.

Prader-Willi Syndrome

Eiholzer, Urs. *Prader-Willi Syndrome: Effects of Human Growth Hormone Treatment.* New York: Karger, 2001.

Sickle-Cell Disease

Anionwu, Elizabeth N., and Karl Atkin. *The Politics of Sickle Cell and Thalassaemia.* Philadelphia: Open University Press, 2001.

Gordon, Melanie Appel. *Let's Talk About Sickle Cell Anemia.* New York: PowerKids Press, 2000.

Harris, Jacqueline. *Sickle Cell Disease.* Brookfield, Conn.: Twenty-First Century Books, 2001.

Parker, James N., and Philip M. Parker, eds. *The 2002 Official Patient's Sourcebook on Sickle Cell Anemia.* San Diego: Icon Health, 2002.

Platt, Allan F., and Alan Sacerdote. *Hope and Destiny: The Patient's and Parent's Guide to Sickle Cell Disease and Sickle Cell Trait.* Roscoe, Ill.: Hilton, 2002.

Serjeant, Graham R., and Beryl E. Serjeant. *Sickle Cell Disease.* 3d ed. New York: Oxford University Press, 2001.

Simon, Dyson M. *Ethnicity and Screening for Sickle Cell/Thalassaemia: Lessons for Practice from the Voices of Experience.* Edinburgh: Elsevier Churchill Livingstone, 2005.

Tapper, Melbourne. *In the Blood: Sickle Cell Anemia and the Politics of Race.* Philadelphia: University of Pennsylvania Press, 1999.

Wailoo, Keith, and Stephen Pemberton. *The Troubled Dream of Genetic Medicine: Ethnicity and Innovation in Tay-Sachs, Cystic Fibrosis, and Sickle Cell Disease.* Baltimore: Johns Hopkins University Press, 2006.

Tay-Sachs Disease

Desnick, Robert J., and Michael M. Kaback, eds. *Tay-Sachs Disease.* San Diego: Academic Press, 2001.

National Tay-Sachs and Allied Diseases Association. *A Genetics Primer for Understanding Tay-Sachs and the Allied Diseases.* Brookline, Mass.: Author, 1995.

Parker, James N., and Philip M. Parker, eds. *The Official Parent's Sourcebook on Tay-Sachs Disease: A Revised and Updated Directory for the Internet Age.* San Diego: Icon Press, 2002.

Turner Syndrome

Albertsson-Wikland, Kerstin, and Michael B. Ranke, eds. *Turner Syndrome in a Life Span Perspective: Research and Clinical Aspects.* New York: Elsevier, 1995.

Rieser, Patricia A., and Marsha Davenport. *Turner Syndrome: A Guide for Families.* Houston: Turner Syndrome Society, 1992.

Rosenfeld, Ron G. *Turner Syndrome: A Guide for Physicians.* 2d ed. Houston: Turner Syndrome Society, 1992.

EVOLUTION

Avital, Eytan, and Eva Jablonka. *Animal Traditions: Behavioural Inheritance in Evolution.* New York: Cambridge University Press, 2000.

Baxter, Brian. *A Darwinian Worldview: Sociobiology, Environmental Ethics and the Work of Edward O. Wilson.* Aldershot, England: Ashgate, 2007.

Beurton, Peter, Raphael Falk, and Hans-Jorg Rheinberger, eds. *The Concept of the Gene in Development and Evolution: Historical and Epistemological Perspectives.* New York: Cambridge University Press, 2000.

Calos, Michele. *Molecular Evolution of Chromosomes.* New York: Oxford University Press, 2003.

Calvin, William H. *A Brain for All Seasons: Human Evolution and Abrupt Climate Change.* Chicago: University of Chicago Press, 2002.

_______. *A Brief History of the Mind: From Apes to Intellect and Beyond.* New York: Oxford University Press, 2004.

Campbell, Bernard G., ed. *Sexual Selection and the Descent of Man: The Darwinian Pivot.* New Brunswick, N.J.: AldineTransaction, 2006.

Cronk, Lee, Napoleon Chagnon, and William Irons, eds. *Adaptation and Human Behavior: An Anthropological Perspective.* New York: Aldine de Gruyter, 2000.

Darwin, Charles. *The Descent of Man and Selection in Relation to Sex.* London: John Murray, 1871. Reprint. Princeton, N.J.: Princeton University Press, 1981.

_______. *On the Origin of Species by Means of Natural Selection: Or, The Preservation of Favored Races in the Struggle for Life.* London: John Murray, 1859. Reprint. New York: Random House, 1999.

Davidson, Iain. *Sex, Politics, and Religion: Archaeology for a Polite Society.* Armidale, N.S.W.: University of New England, 1999.

Dawkins, Richard. *The Blind Watchmaker: Why the Evidence of Evolution Reveals a Universe Without Design.* New York: W. W. Norton, 1996.

_______. *Climbing Mount Improbable.* New York: W. W. Norton, 1997.

_______. *Extended Phenotype: The Long Reach of the Gene.* Rev. 2d ed. Afterword by Daniel Dennett. New York: Oxford University Press, 1999.

_______. *The Selfish Gene.* 2d ed. New York: Oxford University Press, 1990.

Depew, David, and Bruce Weber. *Darwinism Evolving: Systems Dynamics and the Genealogy of Natural Selection.* Boston: MIT Press, 1995.

Dessalles, Jean-Louis. *Why We Talk: The Evolutionary Origins of Language.* Translated by James Grieve. Oxford, England: Oxford University Press, 2007.

Dickens, Peter. *Society and Nature: Changing Our Environment, Changing Ourselves.* Cambridge, England: Polity Press, 2004.

Dobzhansky, Theodosius G. *Genetics and the Origin of Species.* New York: Columbia University Press, 1937.

Edey, Maitland A., and Donald C. Johnson. *Blueprints: Solving the Mystery of Evolution.* Reprint. New York: Viking, 1990.

Fisher, Helen. *Why We Love: The Nature and Chemistry of Romantic Love.* New York: Henry Holt, 2004.

Fisher, Ronald Aylmer. *The Genetical Theory of Natural Selection: A Complete Variorum Edition.* 1958. Edited with a foreword and notes by J. H. Bennett. New York: Oxford University Press, 1999.

Foley, R. A. *Unknown Boundaries: Exploring Human Evolutionary Studies.* Cambridge, England: Cambridge University Press, 2006.

Freeman, Scott, and Jon C. Herron. *Evolutionary Analysis.* 2d ed. Upper Saddle River, N.J.: Prentice Hall, 2000.

Gesteland, Raymond F., Thomas R. Cech, and John F. Atkins, eds. *The RNA World: The Nature of Modern RNA Suggests a Prebiotic RNA.* 2d ed. Cold Spring Harbor, N.Y.: Cold Spring Harbor Laboratory Press, 1999.

Gould, Stephen Jay. *Eight Little Piggies.* New York: W. W. Norton, 1994.

_______. *The Panda's Thumb.* New York: W. W. Norton, 1980.

_______. *The Structure of Evolutionary Theory.* Cambridge, Mass.: Harvard University Press, 2002.

Graur, Dan, and Wen-Hsiung Li. *Fundamentals of Molecular Evolution.* 2d ed. Sunderland, Mass.: Sinauer Associates, 1999.

Heinrich, Bernd. *Why We Run: A Natural History.* New York: Ecco, 2002.

Herrmann, Bernd, and Susanne Hummel, eds. *Ancient DNA: Recovery and Analysis of Genetic Material from Paleographic, Archaeological, Museum, Medical, and Forensic Speciments.* New York: Springer-Verlag, 1994.

Jones, Martin. *The Molecule Hunt: Archaeology and the Search for Ancient DNA.* New York: Arcade, 2002.

Keller, Laurent, ed. *Levels of Selection in Evolution.* Princeton, N.J.: Princeton University Press, 1999.

Kimura, Motoo, and Naoyuki Takahata, eds. *New As-*

pects of the Genetics of Molecular Evolution. New York: Springer-Verlag, 1991.

Landweber, Laura F., and Andrew P. Dobson, eds. *Genetics and the Extinction of Species: DNA and the Conservation of Biodiversity.* Princeton, N.J.: Princeton University Press, 1999.

Langridge, John. *Molecular Genetics and Comparative Evolution.* New York: John Wiley & Sons, 1991.

Levy, Charles K. *Evolutionary Wars, a Three-Billion-Year Arms Race: The Battle of Species on Land, at Sea, and in the Air.* Illustrations by Trudy Nicholson. New York: W. H. Freeman, 1999.

Lewontin, R. *The Genetic Basis of Evolutionary Change.* New York: Columbia University Press, 1974.

Li, Wen-Hsiung. *Molecular Evolution.* Sunderland, Mass.: Sinauer Associates, 1997.

Lloyd, E. *The Structure of Evolutionary Theory.* Westport, Conn.: Greenwood Press, 1987.

Lynch, John M., ed. *Darwin's Theory of Natural Selection: British Responses, 1859-1871.* 4 vols. Bristol, England: Thoemmes Press, 2002.

Magurran, Anne E., and Robert M. May, eds. *Evolution of Biological Diversity: From Population Differentiation to Speciation.* New York: Oxford University Press, 1999.

Mayr, Ernst. *One Long Argument: Charles Darwin and the Genesis of Modern Evolutionary Thought.* Cambridge, Mass.: Harvard University Press, 1991.

Michod, Richard E. *Darwinian Dynamics: Evolutionary Transitions in Fitness and Individuality.* Princeton, N.J.: Princeton University Press, 1999.

Miller, Stanley L. *From the Primitive Atmosphere to the Prebiotic Soup to the Pre-RNA World.* Washington, D.C.: National Aeronautics and Space Administration, 1996.

Nei, Masatoshi, and Sudhir Kumar. *Molecular Evolution and Phylogenetics.* New York: Oxford University Press, 2000.

Persell, Stuart Michael. *Neo-Lamarckism and the Evolution Controversy in France, 1870-1920.* Lewiston, N.Y.: Edwin Mellen Press, 1999.

Pollard, Tessa M. *Western Diseases: An Evolutionary Perspective.* Cambridge, England: Cambridge University Press, 2008.

Prothero, Donald R. *Bringing Fossils to Life: An Introduction to Paleobiology.* New York: McGraw-Hill, 2003.

Quammen, David. *Song of the Dodo.* New York: Simon & Schuster, 1997.

Raff, Rudolf. *The Shape of Life: Genes, Development, and the Evolution of Animal Form.* Chicago: University of Chicago Press, 1996.

Ranzi, Carlo, with illustrations by the author. *Seventy Million Years of Man.* New York: Greenwich House, 1983.

Ryan, Frank. *Darwin's Blind Spot: Evolution Beyond Natural Selection.* Boston: Houghton Mifflin, 2002.

Selander, Robert K., et al., eds. *Evolution at the Molecular Level.* Sunderland, Mass.: Sinauer Associates, 1991.

Shlain, Leonard. *Sex, Time, and Power: How Women's Sexuality Shaped Human Evolution.* New York: Penguin, 2004.

Singh, Rama S., and Costas B. Krimbas, eds. *Evolutionary Genetics: From Molecules to Morphology.* New York: Cambridge University Press, 2000.

Somit, Albert, and Steven A. Peterson, eds. *The Dynamics of Evolution: The Punctuated Equilibrium Debate in the Natural and Social Sciences.* Ithaca, N.Y.: Cornell University Press, 1992.

Stanford, Craig. *Upright: The Evolutionary Key to Becoming Human.* Boston: Houghton Mifflin, 2003.

Steele, Edward J., Robyn A. Lindley, and Robert V. Blanden. *Lamarck's Signature: How Retrogenes Are Changing Darwin's Natural Selection Paradigm.* Reading, Mass.: Perseus Books, 1998.

Wessen, K. P. *Simulating Human Origins and Evolution.* New York: Cambridge University Press, 2005.

Williams, George C. *Adaptation and Natural Selection: A Critique of Some Current Evolutionary Thought.* 1966. Reprint. Princeton, N.J.: Princeton University Press, 1996.

Wolf, Jason B., Edmund D. Brodie III, and Michael J. Wade. *Epistasis and the Evolutionary Process.* New York: Oxford University Press, 2000.

Zimmer, Carl. *Smithsonian Intimate Guide to Human Origins.* Washington, D.C.: Smithsonian Books, 2005.

FORENSIC GENETICS

Ballantyne, Jack, George Sensabaugh, and Jan Witkowski, eds. *DNA Technology and Forensic Science.* Cold Spring Harbor, N.Y.: Cold Spring Harbor Laboratory Press, 1989.

Budowle, Bruce, et al. *DNA Typing Protocols: Molecular Biology and Forensic Analysis.* Natick, Mass.: Eaton, 2000.

Buckleton, John, Christopher M. Triggs, and Simon J. Walsh. *Forensic DNA Evidence Interpretation.* Boca Raton, Fla.: CRC Press, 2005.

Burke, Terry, R. Wolf, G. Dolf, and A. Jeffreys, eds. *DNA Fingerprinting: Approaches and Applications.* Boston: Birkhauser, 2001.

Butler, John M. *Forensic DNA Typing: Biology and Technology Behind STR Markers.* London: Academic Press, 2001.

Butzel, Henry M. *Genetics in the Courts.* Lewiston, N.Y.: Edwin Mellen Press, 1987.

Carracedo, Angel. *Forensic DNA Typing Protocols.* Totowa, N.J.: Humana Press, 2005.

Coleman, Howard, and Eric Swenson. *DNA in the Courtroom: A Trial Watcher's Guide.* Seattle: GeneLex Press, 1994.

Connors, Edward, et al. *Convicted by Juries, Exonerated by Science: Case Studies in the Use of DNA Evidence to Establish Innocence After Trial.* Washington, D.C.: U.S. Department of Justice, Office of Justice Programs, National Institute of Justice, 1996.

Fridell, Ron. *DNA Fingerprinting: The Ultimate Identity.* New York: Scholastic, 2001.

Goodman, Christi. *Paternity, Marriage, and DNA.* Denver, Colo.: National Conference of State Legislatures, 2001.

Goodwin, William, Adrian Linacre, and Sibte Hadi. *An Introduction to Forensic Genetics.* Chichester, England: John Wiley & Sons, 2007.

Hummel, Susanne. *Fingerprinting the Past: Research on Highly Degraded DNA and Its Applications.* New York: Springer-Verlag, 2002.

Jarman, Keith, and Norah Rudin. *An Introduction to Forensic DNA Analysis.* 2d ed. Boca Raton, Fla.: CRC Press, 2001.

Kirby, L. T. *DNA Fingerprinting: An Introduction.* New York: Stockton Press, 1990.

Kobilinsky, Lawrence, Louis Levine, and Henrietta Margolis-Nunno. *Forensic DNA Analysis.* New York: Chelsea House, 2006.

Lincoln, Patrick J., and Jim Thompson, eds. *Forensic DNA Profiling Protocols.* Vol. 98. Totowa, N.J.: Humana Press, 1998.

National Research Council. *DNA Technology in Forensic Science.* Washington, D.C.: National Academy Press, 1992.

Rose, David, and Lisa Goos. *DNA: A Practical Guide.* Toronto: Thomson Carswell, 2004.

Rudin, Norah, and Keith Inman. *An Introduction to Forensic DNA Analysis.* Boca Raton, Fla.: CRC Press, 2002.

Sonenstein, Freya L., Pamela A. Holcomb, and Kristin S. Seefeldt. *Promising Approaches to Improving Paternity Establishment Rates at the Local Level.* Washington, D.C.: Urban Institute, 1993.

United States National Research Council. *The Evaluation of Forensic DNA Evidence.* Rev. ed. Washington, D.C.: National Academy Press, 1996.

GENETIC ENGINEERING AND BIOTECHNOLOGY. *See also* **AGRICULTURE AND GENETICALLY MODIFIED FOODS; CLONING**

Aldridge, Susan. *The Thread of Life: The Story of Genes and Genetic Engineering.* New York: Cambridge University Press, 1996.

Bassett, Pamela. *Emerging Markets in Tissue Engineering: Angiogenesis, Soft and Hard Tissue Regeneration, Xenotransplant, Wound Healing, Biomaterials, and Cell Therapy.* Southborough, Mass.: D & MD Reports, 1999.

Boylan, Michael, and Kevin E. Brown. *Genetic Engineering: Science and Ethics on the New Frontier.* Upper Saddle River, N.J.: Prentice Hall, 2001.

Burley, Justine, and John Harris, eds. *A Companion to Genetics.* Malden, Mass.: Blackwell, 2002.

Chrispeels, Maarten J., and David E. Sadava. *Plants, Genes, and Crop Biotechnology.* Boston: Jones and Bartlett, 2003.

Crocomo, O. J., ed. *Biotechnology of Plants and Microorganisms.* Columbus: Ohio State University Press, 1986.

Curran, Brendan, ed. *A Terrible Beauty Is Born: Clones, Genes and the Future of Mankind.* London: Taylor & Francis, 2003.

Dale, Jeremy. *Molecular Genetics of Bacteria.* New York: John Wiley & Sons, 1994.

Davis, Bernard D., ed. *The Genetic Revolution: Scientific Prospects and Public Perceptions.* Baltimore: Johns Hopkins University Press, 1991.

Emery, Alan E., and H. Emery. *An Introduction to Recombinant DNA.* Chichester, Sussex, England: Wiley, 1984.

Ettore, Elizabeth. *Reproductive Genetics, Gender and the Body.* London: Routledge, 2002.

Evans, Gareth M. *Environmental Biotechnology: Theory and Application.* Hoboken, N.J.: Wiley, 2003.

Fincham, J. R. S. *Genetically Engineered Organisms: Benefits and Risks.* Toronto: University of Toronto Press, 1991.

Gaillardin, Claude, and Henri Heslot. *Molecular Biology and Genetic Engineering of Yeasts.* Boca Raton, Fla.: CRC Press, 1992.

Glick, Bernard R. *Molecular Biotechnology: Principles and Applications of Recombinant DNA*. Washington, D.C.: ASM Press, 2003.

Goodnough, David. *The Debate over Human Cloning*. Berkeley Heights, N.J.: Enslow, 2003.

Grange, J. M., et al., eds. *Genetic Manipulation: Techniques and Applications*. Boston: Blackwell Scientific, 1991.

Hill, Walter E. *Genetic Engineering: A Primer*. Newark, N.J.: Harwood Academic, 2000.

Jacobson, G. K., and S. O. Jolly. *Gene Technology*. New York: VCH Verlagsgesellschaft, 1989.

Joyner, Alexandra L., ed. *Gene Targeting: A Practical Approach*. New York: Oxford University Press, 1993.

Kiessling, Ann, and Scott C. Anderson. *Human Embryonic Stem Cells: An Introduction to the Science and Therapeutic Potential*. Boston: Jones and Bartlett, 2003.

Kontermann, Roland, and Stefan Dübel, eds. *Antibody Engineering*. New York: Springer, 2001.

Kowalski, Kathiann M. *The Debate over Genetically Engineered Food: Healthy or Harmful?* Berkeley Heights, N.J.: Enslow, 2002.

Kreuzer, Helen, and Adrianne Massey. *Recombinant DNA and Biotechnology: A Guide for Teachers*. Washington, D.C.: ASM Press, 2001.

Krimsky, Sheldon. *Biotechnics and Society: The Rise of Industrial Genetics*. New York: Praeger, 1991.

Lappe, Marc. *Broken Code: The Exploitation of DNA*. San Francisco: Sierra Club Books, 1984.

Le Vine, Harry, III. *Genetic Engineering: A Reference Handbook*. Santa Barbara, Calif.: ABC-CLIO, 1999.

McGee, Glenn. *The Perfect Baby: A Pragmatic Approach to Genetics*. Lanham, Md.: Rowman & Littlefield, 1997.

McKay, David, and Mark Walker. *Unravelling Genes: A Layperson's Guide to Genetic Engineering*. Frenchs Forest, N.S.W.: Pearson Prentice Hall, 2006.

McKelvey, Maureen D. *Evolutionary Innovations: The Business of Biotechnology*. New York: Oxford University Press, 1996.

Mak, Tak W., et al., eds. *The Gene Knockout Factsbook*. 2 vols. San Diego: Academic Press, 1998.

Mayforth, Ruth D. *Designing Antibodies*. San Diego: Academic Press, 1993.

Nicholl, Desmond S. T. *An Introduction to Genetic Engineering*. 3d ed. New York: Cambridge University Press, 2008.

Nossal, Gustav J. V., and Ross L. Coppel. *Reshaping Life: Key Issues in Genetic Engineering*. New York: Cambridge University Press, 1985.

Nottingham, Stephen. *Genescapes: The Ecology of Genetic Engineering*. New York: Zed Books, 2002.

Old, R. W., and S. B. Primrose. *Principles of Gene Manipulation: An Introduction to Genetic Engineering*. Boston: Blackwell Scientific, 1994.

Olson, Steve. *Biotechnology: An Industry Comes of Age*. Washington, D.C.: National Academy Press, 1986.

Oxender, Dale L., and C. Fred Fox, eds. *Protein Engineering*. New York: Liss, 1987.

Prokop, Ales, and Rakesh K. Bajpai. *Recombinant DNA Technology I*. New York: New York Academy of Sciences, 1991.

Ronald, Pamela C., and R. W. Adamchak. *Tomorrow's Table: Organic Farming, Genetics, and the Future of Food*. Oxford, England: Oxford University Press, 2008.

Rothman, Barbara Katz. *The Book of Life: A Personal and Ethical Guide to Race, Normality, and the Implications of the Human Genome Project*. Boston: Beacon Press, 2001.

Russo, V. E. A., and David Cove. *Genetic Engineering: Dreams and Nightmares*. New York: W. H. Freeman, 1995.

Shannon, Thomas A., ed. *Genetic Engineering: A Documentary History*. Westport, Conn.: Greenwood Press, 1999.

Singer, Maxine, and Paul Berg, eds. *Exploring Genetic Mechanisms*. Sausalito, Calif.: University Science Books, 1997.

Sofer, William. *Introduction to Genetic Engineering*. Boston: Butterworth-Heinemann, 1991.

Spallone, Patricia. *Generation Games: Genetic Engineering and the Future for Our Lives*. Philadelphia: Temple University Press, 1992.

Steinberg, Mark, and Sharon D. Cosloy, eds. *The Facts On File Dictionary of Biotechnology and Genetic Engineering*. New ed. New York: Checkmark Books, 2001.

Thomas, John, et al., eds. *Biotechnology and Safety Assessment*. 2d ed. Philadelphia: Taylor & Francis, 1999.

Vega, Manuel A., ed. *Gene Targeting*. Boca Raton, Fla.: CRC Press, 1995.

Wade, Nicholas. *The Ultimate Experiment: Man-Made Evolution*. New York: Walker, 1977.

Walker, Mark, and David McKay. *Unravelling Genes: A*

Layperson's Guide to Genetic Engineering. St. Leonards, N.S.W.: Allen & Unwin, 2000.

Walker, Matthew R., with Ralph Rapley. *Route Maps in Gene Technology.* Oxford, England: Blackwell Scientific, 1997.

Wang, Henry Y., and Tadayuki Imanaka, eds. *Antibody Expression and Engineering.* Washington, D.C.: American Chemical Society, 1995.

Warr, J. Roger. *Genetic Engineering in Higher Organisms.* Baltimore: E. Arnold, 1984.

Watson, James D., et al. *Recombinant DNA.* 2d ed. New York: W. H. Freeman, 1992.

Williams, J. G., A. Ceccarelli, and A. Wallace. *Genetic Engineering.* 2d ed. New York: Springer, 2001.

Wu-Pong, S., and Y. Rojanasakul. *Biopharmaceutical Drug Design and Development.* Totowa, N.J.: Humana Press, 1999.

GENOMICS AND PROTEOMICS. *See also* **BIOINFORMATICS**

Bradbury, E. Morton, and Sandor Pongor, eds. *Structural Biology and Functional Genomics.* Boston: Kluwer Academic, 1999.

Dunn, Michael J., ed. *From Genome to Proteome: Advances in the Practice and Application of Proteomics.* New York: Wiley-VCH, 2000.

Gibson, Greg. *A Primer of Genome Science.* Sunderland, Mass.: Sinauer, 2002.

Innis, Michael A., David H. Gelfand, and John J. Sninsky, eds. *PCR Applications: Protocols for Functional Genomics.* San Diego: Academic Press, 1999.

Jolles, P., and H. Jornvall, eds. *Proteomics in Functional Genomics: Protein Structure Analysis.* Boston: Birkhauser, 2000.

Kang, Manjit S. *Quantitative Genetics, Genomics, and Plant Breeding.* Wallingford, Oxon, England: CABI, 2002.

Liebler, David G. *Introduction to Proteomics: Tools for the New Biology.* Totowa, N.J.: Humana Press, 2001.

Link, Andrew J., ed. *2-D Proteome Analysis Protocols.* Totowa, N.J.: Humana Press, 1999.

Liu, Ben-Hui. *Statistical Genomics: Linkage, Mapping, and QTL Analysis.* Boca Raton, Fla.: CRC Press, 1998.

Pennington, S. R., and M. J. Dunn, eds. *Proteomics: From Protein Sequence to Function.* New York: Springer, 2001.

Rabilloud, Thierry, ed. *Proteome Research: Two-Dimensional Gel Electrophoresis and Identification Methods.* New York: Springer, 2000.

HISTORY OF GENETICS

Alibeck, Ken, with Stephen Handelman. *Biohazard: The Chilling True Story of the Largest Covert Biological Weapons Program in the World, Told from the Inside by the Man Who Ran It.* New York: Random House, 1999.

Ayala, Francisco J., and Walter M. Fitch, eds. *Genetics and the Origin of Species: From Darwin to Molecular Biology Sixty Years After Dobzhansky.* Washington, D.C.: National Academies Press, 1997.

Beighton, Peter. *The Person Behind the Syndrome.* Rev. ed. New York: Springer-Verlag, 1997.

Bowler, Peter J. *Evolution: The History of an Idea.* Berkeley: University of California Press, 1990.

Brookes, Martin. *Fly: The Unsung Hero of Twentieth-Century Science.* San Francisco: HarperCollins, 2001.

Corcos, A., and F. Monaghan. *Mendel's Experiments on Plant Hybrids: A Guided Study.* New Brunswick, N.J.: Rutgers University Press, 1993.

Darwin, Charles. *Charles Darwin's Notebooks, 1836-1844.* Edited by P. H. Barrett et al. Ithaca, N.Y.: Cornell University Press, 1987.

_______. *The Correspondence of Charles Darwin.* Cambridge, England: Cambridge University Press, 1994.

Dover, Gabriel A. *Dear Mr. Darwin: Letters on the Evolution of Life and Human Nature.* Berkeley: University of California Press, 2000.

Dunn, L. C. *A Short History of Genetics: The Development of Some of the Main Lines of Thought, 1864-1939.* New York: McGraw-Hill, 1965.

Edelson, Edward. *Gregor Mendel and the Roots of Genetics.* New York: Oxford University Press, 1999.

Fast, Julius. *Blueprint for Life: The Story of Modern Genetics.* New York: St. Martin's Press, 1965.

Fitzgerald, Patrick J. *From Demons and Evil Spirits to Cancer Genes: The Development of Concepts Concerning the Causes of Cancer and Carcinogenesis.* Washington, D.C.: American Registry of Pathology, Armed Forces Institute of Pathology, 2000.

Fredrickson, Donald S. *The Recombinant DNA Controversy, a Memoir: Science, Politics, and the Public Interest, 1974-1981.* Washington, D.C.: ASM Press, 2001.

Fujimura, Joan H. *Crafting Science: A Sociohistory of the Quest for the Genetics of Cancer.* Cambridge, Mass.: Harvard University Press, 1996.

Gillham, Nicholas Wright. *A Life of Sir Francis Galton: From African Exploration to the Birth of Eugenics.* New York: Oxford University Press, 2001.

Haldane, J. B. S. *Selected Genetic Papers of J. B. S. Haldane.* Edited with an introduction by Krishna R. Dronamraju, foreword by James F. Crow. New York: Garland, 1990.

Harper, Peter S. *A Short History of Medical Genetics.* New York: Oxford University Press, 2008.

Henig, Robin Marantz. *The Monk in the Garden: The Lost and Found Genius of Gregor Mendel, the Father of Genetics.* Boston: Houghton Mifflin, 2000.

Hubbell, Sue. *Shrinking the Cat: Genetic Engineering Before We Knew About Genes.* Illustrations by Liddy Hubbell. Boston: Houghton Mifflin, 2001.

Iltis, Hugo. *Life of Mendel.* Translated by Eden Paul and Cedar Paul. London: Allen & Unwin, 1932.

Jacob, François. *The Logic of Life: A History of Heredity.* New York: Pantheon, 1973.

Johnson, George B. *How Scientists Think: Twenty-one Experiments That Have Shaped Our Understanding of Genetics and Molecular Biology.* Dubuque, Iowa: Wm. C. Brown, 1996.

Judson, Horace Freeland. *The Eighth Day of Creation: Makers of the Revolution in Biology.* Rev. ed. Cold Spring Harbor, N.Y.: Cold Spring Harbor Laboratory Press, 1997.

Kay, Lily E. *Who Wrote the Book of Life? A History of the Genetic Code.* Stanford, Calif.: Stanford University Press, 2000.

Keller, Evelyn Fox. *A Feeling for the Organism: The Life and Work of Barbara McClintock.* 10th anniversary ed. New York: W. H. Freeman, 1993.

Kornberg, Arthur. *For the Love of Enzymes: The Odyssey of a Biochemist.* Reprint. Cambridge, Mass.: Harvard University Press, 1991.

Kühl, Stefan. *The Nazi Connection: Eugenics, American Racism, and German National Socialism.* New York: Oxford University Press, 2002.

Laffin, John. *Hitler Warned Us: The Nazis' Master Plan for a Master Race.* Totowa, N.J.: Barnes and Noble, 1998.

McCarty, Maclyn. *The Transforming Principle: Discovering That Genes Are Made of DNA.* New York: W. W. Norton, 1994.

Maddox, Brenda. *Rosalind Franklin: The Dark Lady of DNA.* New York: HarperCollins, 2002.

Mayr, Ernst. *One Long Argument: Charles Darwin and the Genesis of Modern Evolutionary Thought.* Cambridge, Mass.: Harvard University Press, 1991.

Mendel, Gregor. *Experiments in Plant-Hybridization.* Foreword by Paul C. Mangelsdorf. Cambridge, Mass.: Harvard University Press, 1965.

Olby, Robert C. *Origins of Mendelism.* New York: Schocken Books, 1966.

Orel, Vítezslav. *Gregor Mendel: The First Geneticist.* Translated by Stephen Finn. New York: Oxford University Press, 1996.

Palladino, Paolo. *Plants, Patients, and the Historians: On (RE)Membering in the Age of Genetic Engineering.* New Brunswick, N.J.: Rutgers University Press, 2003.

Persell, Stuart Michael. *Neo-Lamarckism and the Evolution Controversy in France, 1870-1920.* Lewiston, N.Y.: Edwin Mellen Press, 1999.

Potts, D. M., and W. T. W. Potts. *Queen Victoria's Gene: Haemophilia and the Royal Family.* Stroud, Gloucestershire, England: Sutton, 1999.

Rutter, Michael Sir, ed. *Genetic Effects on Environmental Vulnerability to Disease.* New York: Wiley, 2008.

Shannon, Thomas A., ed. *Genetic Engineering: A Documentary History.* Westport, Conn.: Greenwood Press, 1999.

Shermer, Michael. *In Darwin's Shadow: The Life and Science of Alfred Russel Wallace, a Biographical Study on the Psychology of History.* New York: Oxford University Press, 2002.

Stubbe, H. *A History of Genetics.* Cambridge, Mass.: MIT Press, 1968.

Sturtevant, A. H. *A History of Genetics.* 1965. Reprint. Introduction by Edward B. Lewis. Cold Spring Harbor, N.Y.: Cold Spring Harbor Laboratory Press, 2001.

Sulston, John, and Georgina Ferry. *The Common Thread: A Story of Science, Politics, Ethics, and the Human Genome.* Washington, D.C.: Joseph Henry Press, 2002.

Tudge, Colin. *In Mendel's Footnotes: An Introduction to the Science and Technologies of Genes and Genetics from the Nineteenth Century to the Twenty-Second.* London: Jonathan Cape, 2000.

Watson, James. *The Double Helix.* New York: Simon and Schuster, 2001.

Weiner, Jonathan. *The Beak of the Finch: A Story of Evolution in Our Time.* New York: Random House, 1995.

Wood, Roger J., and Vitezslav Orel. *Genetic Prehistory in Selective Breeding: A Prelude to Mendel.* New York: Oxford University Press, 2001.

HUMAN GENETICS

Adolph, Kenneth W., ed. *Human Molecular Genetics.* New York: Academic Press, 1996.

Alcock, John. *The Triumph of Sociobiology.* Reprint. New York: Oxford University Press, 2003.

Andreasen, Nancy C. *Brave New Brain: Conquering Mental Illness in the Era of the Genome.* New York: Oxford University Press, 2001.

Arking, Robert, ed. *Biology of Aging: Observations and Principles.* 2d ed. Sunderland, Mass.: Sinauer, 2001.

Austad, Steven N. *Why We Age: What Science Is Discovering About the Body's Journey Throughout Life.* New York: John Wiley & Sons, 1997.

Badcock, C. R. *Evolutionary Psychology: A Critical Introduction.* Malden, Mass.: Polity Press in association with Blackwell, 2000.

Baudrillard, Jean. *The Vital Illusion.* Edited by Julia Witwer. New York: Columbia University Press, 2000.

Blackmore, Susan J. *The Meme Machine.* Foreword by Richard Dawkins. New York: Oxford University Press, 1999.

Bock, Gregory R., and Jamie A. Goode. *Genetics of Criminal and Antisocial Behaviour.* New York: John Wiley & Sons, 1996.

Bock, Gregory R., Jamie A. Goode, and Kate Webb, eds. *The Nature of Intelligence.* New York: John Wiley & Sons, 2001.

Boyer, Samuel, ed. *Papers on Human Genetics.* Englewood Cliffs, N.J.: Prentice-Hall, 1963.

Brierley, John Keith. *The Thinking Machine: Genes, Brain, Endocrines, and Human Nature.* Rutherford, N.J.: Fairleigh Dickinson University Press, 1973.

Briley, Mike, and Fridolin Sulser, eds. *Molecular Genetics of Mental Disorders: The Place of Molecular Genetics in Basic Mechanisms and Clinical Applications in Mental Disorders.* Malden, Mass.: Blackwell, 2001.

British Medical Association. *Biotechnology, Weapons, and Humanity.* Amsterdam, Netherlands: Harwood Academic, 1999.

Burnet, Sir Frank Macfarlane. *Endurance of Life: The Implications of Genetics for Human Life.* New York: Cambridge University Press, 1978.

Burnham, Terry, and Jay Phelan. *Mean Genes: From Sex to Money to Food—Taming Our Primal Instincts.* Cambridge, Mass.: Perseus, 2000.

Carson, Ronald A., and Mark A. Rothstein. *Behavioral Genetics: The Clash of Culture and Biology.* Baltimore: Johns Hopkins University Press, 1999.

Carter, Cedric O. *Human Heredity.* Baltimore: Penguin Books, 1962.

Cartwright, John. *Evolution and Human Behavior: Darwinian Perspectives on Human Nature.* Cambridge, Mass.: MIT Press, 2000.

Cavalli-Sforza, L. L., and Francesco Cavalli-Sforza. *The Great Human Diasporas: The History of Diversity and Evolution.* Translated by Sarah Thorne. Reading, Mass.: Addison-Wesley, 1995.

Centers for Disease Control and Prevention. "Bioterrorism-Related Anthrax." *Emerging Infectious Diseases* 8 (October, 2002): 1013-1183.

Clark, William R., and Michael Grunstein. *Are We Hardwired? The Role of Genes in Human Behavior.* New York: Oxford University Press, 2000.

Clegg, Edward J. *The Study of Man: An Introduction to Human Biology.* London: English Universities Press, 1968.

Cole, Leonard A. *The Eleventh Plague: The Politics of Biological and Chemical Warfare.* New York: W. H. Freeman, 1996.

Cooper, Colin. *Intelligence and Abilities.* New York: Routledge, 1999.

Cooper, Necia Grant, ed. *The Human Genome Project: Deciphering the Blueprint of Heredity.* Foreword by Paul Berg. Mill Valley, Calif.: University Science Books, 1994.

Cronk, Lee. *That Complex Whole: Culture and the Evolution of Human Behavior.* Boulder, Colo.: Westview Press, 1999.

Crow, Tim J., ed. *The Speciation of Modern Homo Sapiens.* Oxford, England: Oxford University Press, 2002.

Cummings, Michael J. *Human Heredity: Principles and Issues.* 5th ed. Pacific Grove, Calif.: Brooks/Cole, 2000.

Curran, Charles E. *Politics, Medicine, and Christian Ethics: A Dialogue with Paul Ramsey.* Philadelphia: Fortress Press, 1973.

Cziko, Gary. *The Things We Do: Using the Lessons of Bernard and Darwin to Understand the What, How, and Why of Our Behavior.* Cambridge, Mass.: MIT Press, 2000.

DeMoss, Robert T. *Brain Waves Through Time: Twelve Principles for Understanding the Evolution of the Human Brain and Man's Behavior.* New York: Plenum Trade, 1999.

Dennis, Carina, and Richard Gallagher. *The Human Genome.* London: Palgrave Macmillan, 2002.

Diamant, L., and R. McAnuity, eds. *The Psychology of Sexual Orientation, Behavior, and Identity: A Handbook.* Westport, Conn.: Greenwood Press, 1995.

Dobzhansky, Theodosius G. *Mankind Evolving: The Evolution of the Human Species.* New Haven, Conn.: Yale University Press, 1967.

Edlin, Gordon. *Human Genetics: A Modern Synthesis.* Boston: Jones & Bartlett, 1990.

Fish, Jefferson M., ed. *Race and Intelligence: Separating Science from Myth.* Mahwah, N.J.: Lawrence Erlbaum, 2002.

Fishbein, Diana H., ed. *The Science, Treatment, and Prevention of Antisocial Behaviors: Application to the Criminal Justice System.* Kingston, N.J.: Civic Research Institute, 2000.

Fooden, Myra, et al., eds. *The Second X and Women's Health.* New York: Gordian Press, 1983.

Gallagher, Nancy L. *Breeding Better Vermonters: The Eugenics Program in the Green Mountain State.* Hanover, N.H.: University Press of New England, 2000.

Gardner, Howard. *Frames of Mind: The Theory of Multiple Intelligences.* 10th anniversary ed. New York: Basic Books, 1993.

Glassy, Mark C. *The Biology of Science Fiction Cinema.* Jefferson, N.C.: McFarland, 2001.

Goldhagen, Daniel J. *Hitler's Willing Executioners: Ordinary Germans and the Holocaust.* New York: Random House, 1996.

Gould, Stephen Jay. *The Mismeasure of Man.* Rev. ed. New York: W. W. Norton, 1996.

Graves, Joseph L., Jr. *The Emperor's New Clothes: Biological Theories of Race at the Millennium.* New Brunswick, N.J.: Rutgers University Press, 2001.

Hahn, Martin E., et al., eds. *Developmental Behavior Genetics: Neural, Biometrical, and Evolutionary Approaches.* New York: Oxford University Press, 1990.

Haldane, J. B. S. *Selected Genetic Papers of J. B. S. Haldane.* Edited with an introduction by Krishna R. Dronamraju, foreword by James F. Crow. New York: Garland, 1990.

Hamer, Dean, and Peter Copeland. *Living with Our Genes: Why They Matter More than You Think.* New York: Doubleday, 1998.

Harris, Harry. *The Principles of Human Biochemical Genetics.* New York: Elsevier/North-Holland Biomedical Press, 1980.

Herrnstein, Richard J., and Charles Murray. *The Bell Curve: Intelligence and Class Structure in America.* New York: Free Press, 1994.

Heschl, Adolf. *The Intelligent Genome: On the Origin of the Human Mind by Mutation and Selection.* Drawings by Herbert Loserl. New York: Springer, 2002.

Hsia, David Yi-Yung. *Human Developmental Genetics.* Chicago: Year Book Medical Publishers, 1968.

Jacquard, Albert. *In Praise of Difference: Genetics and Human Affairs.* Translated by Margaret M. Moriarty. New York: Columbia University Press, 1984.

Jensen, Arthur Robert. *Genetics and Education.* New York: Harper & Row, 1972.

Karlsson, Jon L. *Inheritance of Creative Intelligence.* Chicago: Nelson-Hall, 1978.

Kevles, Daniel J. *In the Name of Eugenics: Genetics and the Uses of Human Heredity.* New York: Alfred A. Knopf, 1985.

Korn, Noel, and Harry Reece Smith, eds. *Human Evolution: Readings in Physical Anthropology.* New York: Holt, 1959.

Krebs, J., and N. Davies. *An Introduction to Behavioral Ecology.* Malden, Mass.: Blackwell, 1991.

Lasker, Gabriel W., ed. *The Processes of Ongoing Human Evolution.* Detroit: Wayne State University Press, 1960.

Levitan, Max. *Textbook of Human Genetics.* New York: Oxford University Press, 1988.

Lewis, Ricki. *Human Genetics: Concepts and Applications.* 5th ed. New York: McGraw-Hill, 2003.

Lewontin, Richard C. *Human Diversity.* New York: Scientific American Library, 1995.

Ludmerer, Kenneth M. *Genetics and American Society: A Historical Appraisal.* Baltimore: Johns Hopkins University Press, 1972.

Lynn, Richard. *Dysgenics: Genetic Deterioration in Modern Populations.* Westport, Conn.: Praeger, 1996.

McConkey, Edwin H. *Human Genetics: The Molecular Revolution.* Boston: Jones & Bartlett, 1993.

Macieira-Coelho, Alvaro. *Biology of Aging.* New York: Springer, 2002.

McKusick, Victor A., comp. *Mendelian Inheritance in Man: A Catalog of Human Genes and Genetic Disorders.* 12th ed. Baltimore: Johns Hopkins University Press, 1998.

McWhirter, David P., et al. *Homosexuality/Heterosexuality: Concepts of Sexual Orientation.* New York: Oxford University Press, 1990.

Mange, Elaine Johansen, and Arthur P. Mange. *Basic Human Genetics.* 2d ed. Sunderland, Mass.: Sinauer Associates, 1999.

Manuck, Stephen B., et al., eds. *Behavior, Health, and Aging.* Mahwah, N.J.: Lawrence Erlbaum, 2000.

Marks, Jonathan M. *Human Biodiversity: Genes, Race, and History.* New York: Aldine de Gruyter, 1995.

Marteau, Theresa, and Martin Richards, eds. *The Troubled Helix: Social and Psychological Implications of the New Human Genetics.* New York: Cambridge University Press, 1999.

Medina, John J. *The Clock of Ages: Why We Age, How We Age—Winding Back the Clock.* New York: Cambridge University Press, 1996.

Mielke, James H., and Michael H. Crawford, eds. *Current Developments in Anthropological Genetics.* New York: Plenum Press, 1980.

Miller, Judith, Stephen Engelberg, and William Broad. *Germs: Biological Weapons and America's Secret War.* New York: Simon & Schuster, 2001.

Miller, Orlando J., and Eeva Therman. *Human Chromosomes.* 4th ed. New York: Springer Verlag, 2001.

Moffitt, Terrie E., Avshalom Caspi, Michael Rutter, and Phil A. Silva. *Sex Differences in Antisocial Behaviour: Conduct Disorder, Delinquency, and Violence in the Dunedin Longitudinal Study.* New York: Cambridge University Press, 2001.

Moody, Paul Amos. *Genetics of Man.* New York: W. W. Norton, 1967.

Nelkin, Dorothy, and Laurence Tancredi. *Dangerous Diagnostics: The Social Power of Biological Information.* New York: Basic Books, 1989.

Ostrer, Harry. *Non-Mendelian Genetics in Humans.* New York: Oxford University Press, 1998.

Ott, Jurg. *Analysis of Human Genetic Linkage.* Baltimore: Johns Hopkins University Press, 1991.

Pearson, Roger. *Eugenics and Race.* Los Angeles: Noontide Press, 1966.

Pierce, Benjamin A. *The Family Genetic Sourcebook.* New York: John Wiley & Sons, 1990.

Plomin, Robert, et al. *Behavioral Genetics.* 4th ed. New York: Worth, 2001.

Puterbaugh, Geoff. *Twins and Homosexuality: A Casebook.* New York: Garland, 1990.

Resta, Robert G., ed. *Psyche and Helix: Psychological Aspects of Genetic Counseling.* New York: Wiley-Liss, 2000.

Ricklefs, Robert E., and Caleb E. Finch. *Aging: A Natural History.* New York: W. H. Freeman, 1995.

Rifkin, Jeremy. *The Biotech Century: Harnessing the Gene and Remaking the World.* New York: Putnam, 1998.

Roderick, Gordon Wynne. *Man and Heredity.* New York: St. Martin's Press, 1968.

Rosenberg, Charles, ed. *The History of Hereditarian Thought: A Thirty-two Volume Reprint Series Presenting Some of the Classic Books in This Intellectual Tradition.* New York: Garland, 1984.

Ruse, M. *Sociobiology: Sense or Nonsense?* Boston: D. Riedel, 1979.

Santos, Miguel A. *Genetics and Man's Future: Legal, Social, and Moral Implications of Genetic Engineering.* Springfield, Ill.: Thomas, 1981.

Segerstråle, Ullica. *Defenders of the Truth: The Battle for Science in the Sociobiology Debate and Beyond.* New York: Oxford University Press, 2000.

Singh, Jai Rup, ed. *Current Concepts in Human Genetics.* Amritsar, India: Guru Nanak Dev University, 1996.

Steen, R. Grant. *DNA and Destiny: Nurture and Nature in Human Behavior.* New York: Plenum, 1996.

Strachan, Tom, and Andrew P. Read. *Human Molecular Genetics.* 2d ed. New York: Wiley-Liss, 1999.

Sussman, Robert, ed. *The Biological Basis of Human Behavior: A Critical Review.* 2d ed. New York: Simon and Schuster, 1998.

Suzuki, D., and P. Knudtson. *Genethics: The Clash Between the New Genetics and Human Value.* Cambridge, Mass.: Harvard University Press, 1989.

Terwilliger, Joseph Douglas, and Jurg Ott. *Handbook of Human Genetic Linkage.* Baltimore: Johns Hopkins University Press, 1994.

Thorner, M., and R. Smith. *Human Growth Hormone: Research and Clinical Practice.* Vol. 19. Totowa, N.J.: Humana Press, 1999.

Timiras, Paola S. *Physiological Basis of Aging and Geriatrics.* 3d ed. Boca Raton, Fla.: CRC Press, 2003.

Toussaint, Olivier, et al., eds. *Molecular and Cellular Gerontology.* New York: New York Academy of Sciences, 2000.

Turney, Jon. *Frankenstein's Footsteps: Science, Genetics, and Popular Culture.* New Haven, Conn.: Yale University Press, 1998.

Underwood, Jane H. *Human Variation and Human Microevolution.* Englewood Cliffs, N.J.: Prentice-Hall, 1979.

U.S. Congress. Office of Technology Assessment. *Mapping Our Genes: Genome Projects—How Big, How Fast?* Baltimore: Johns Hopkins University Press, 1988.

Vandenberg, Steven G., ed. *Methods and Goals in Human Behavior Genetics.* New York: Academic Press, 1965.

Van der Dennen, Johan M. G., David Smillie, and Daniel R. Wilson, eds. *The Darwinian Heritage and Sociobiology.* Westport, Conn.: Praeger, 1999.

Varmus, Harold, and Robert Weinberg. *Genes and the Biology of Cancer.* New York: W. H. Freeman, 1993.

Vogel, Friedrich, and A. G. Motulsky. *Human Genet-*

ics: Problems and Approaches. New York: Springer-Verlag, 1997.

Wasserman, David, and Robert Wachbroit, eds. *Genetics and Criminal Behavior.* New York: Cambridge University Press, 2001.

Weil, Jon. *Psychosocial Genetic Counseling.* New York: Oxford University Press, 2000.

Weir, Bruce S. *Human Identification: The Use of DNA Markers.* New York: Kluwer Academic, 1995.

Weiss, Kenneth M. *Genetic Variation and Human Disease: Principles and Evolutionary Approaches.* New York: Cambridge University Press, 1993.

Wexler, Alice. *Mapping Fate: A Memoir of Family, Risk, and Genetic Research.* Berkeley: University of California Press, 1996.

Whittinghill, Maurice. *Human Genetics and Its Foundations.* New York: Reinhold, 1965.

Wright, Lawrence. *Twins: And What They Tell Us About Who We Are.* New York: John Wiley & Sons, 1997.

Wright, William. *Born That Way: Genes, Behavior, Personality.* New York: Knopf, 1998.

Young, Ian D. *Introduction to Risk Calculation in Genetic Counseling.* 2d ed. New York: Oxford University Press, 1999.

Yu, Byung Pal, ed. *Free Radicals in Aging.* Boca Raton, Fla.: CRC Press, 1993.

IMMUNOGENETICS

Bell, John I., et al., eds. *T Cell Receptors.* New York: Oxford University Press, 1995.

Bernal, J. E. *Human Immunogenetics: Principles and Clinical Applications.* Translated by Derek Roberts. London: Taylor & Francis, 1986.

Bibel, Debra Jan. *Milestones in Immunology.* New York: Springer-Verlag, 1988.

Bona, Constantin A., et al., eds. *The Molecular Pathology of Autoimmune Diseases.* 2d ed. New York: Taylor and Francis, 2002.

Clark, William R. *At War Within: The Double-Edged Sword of Immunity.* New York: Oxford University Press, 1995.

Coleman, Robert M., et al. *Fundamental Immunology.* Dubuque, Iowa: Wm. C. Brown, 1989.

Dwyer, John M. *The Body at War: The Miracle of the Immune System.* New York: New American Library, 1988.

Falus, Andras, ed. *Immunogenomics and Human Disease.* Chichester, West Sussex, England: John Wiley, 2006.

Frank, Steven A. *Immunology and Evolution of Infectious Disease.* Princeton, N.J.: Princeton University Press, 2002.

Fudenberg, H. Hugh, et al. *Basic Immunogenetics.* New York: Oxford University Press, 1984.

Gallo, Robert C., and Flossie Wong-Staal, eds. *Retrovirus Biology and Human Disease.* New York: Marcel Dekker, 1990.

Holland, J. J., ed. *Current Topics in Microbiology and Immunology: Genetic Diversity of RNA Viruses.* New York: Springer-Verlag, 1992.

Janeway, Charles A., Paul Travers, et al. *Immunobiology: The Immune System in Health and Disease.* 5th rev. ed. Philadelphia: Taylor & Francis, 2001.

Joneja, Janice M. V., and Leonard Bielory. *Understanding Allergy, Sensitivity, and Immunity.* New Brunswick, N.J.: Rutgers University Press, 1990.

Kimball, John W. *Introduction to Immunology.* 3d ed. New York: Macmillan, 1990.

Kreier, Julius P., and Richard F. Mortensen. *Infection, Resistance, and Immunity.* New York: Harper & Row, 1990.

Kuby, Janis. *Immunology.* 4th ed. New York: W. H. Freeman, 2000.

Litwin, Stephen D., David W. Scott, et al., eds. *Human Immunogenetics: Basic Principles and Clinical Relevance.* New York: Dekker, 1989.

Mak, Tak W., and John J. L. Simard. *Handbook of Immune Response Genes.* New York: Plenum Press, 1998.

Mizel, Steven B., and Peter Jaret. *In Self-Defense.* San Diego: Harcourt Brace Jovanovich, 1985.

Panayi, Gabriel S., and Chella S. David, eds. *Human Immunogenetics.* London: Butterworths, 1984.

Pines, Maya, ed. *Arousing the Fury of the Immune System.* Chevy Chase, Md.: Howard Hughes Medical Institute, 1998.

Oksenberg, Jorge, and David Brassat, eds. *Immunogenetics of Autoimmune Disease.* Georgetown, Tex.: Landes Bioscience/Eurekah.com, 2006.

Roitt, Ivan, Jonathan Brostoff, and David Male, eds. *Immunology.* New York: Mosby, 2001.

Samter, Max. *Immunological Diseases.* 4th ed. Boston: Little, Brown, 1988.

Sasazuki, Takehiko, and Tomio Tada, eds. *Immunogenetics: Its Application to Clinical Medicine.* Orlando, Fla.: Academic Press, 1984.

Silverstein, Arthur M. *A History of Immunology.* San Diego: Academic Press, 1989.

Smith, George P. *The Variation and Adaptive Expres-*

sion of Antibodies. Cambridge, Mass.: Harvard University Press, 1973.

Stewart, John. *The Primordial VRM System and the Evolution of Vertebrate Immunity.* Austin, Tex.: R. G. Landes, 1994.

Theofilopoulos, A. N., ed. *Genes and Genetics of Autoimmunity.* New York: Karger, 1999.

Tizard, Ian R. *Immunology: An Introduction.* 2d ed. New York: W. B. Saunders, 1988.

MOLECULAR GENETICS

Adolph, Kenneth W., ed. *Gene and Chromosome Analysis.* San Diego: Academic Press, 1993.

_______. *Human Molecular Genetics.* San Diego: Academic Press, 1996.

Alberts, Bruce, Dennis Bray, Julian Lewis, Martin Raff, Keith Roberts, and James D. Watson. *Molecular Biology of the Cell.* 4th ed. New York: Garland, 2002.

Ausubel, Fredrick, Roger Brent, Robert Kingston, David Moore, J. Seidman, and K. Struhl. *Current Protocols in Molecular Biology.* Hoboken, N.J.: John Wiley & Sons, 1998.

Baltimore, David, ed. *Nobel Lectures in Molecular Biology, 1933-1975.* New York: Elsevier North-Holland, 1977.

Barry, John Michael, and E. M. Barry. *Molecular Biology: An Introduction to Chemical Genetics.* Englewood Cliffs, N.J.: Prentice-Hall, 1973.

Benjamin, Jonathan, Richard P. Ebstein, and Robert H. Belmaker, eds. *Molecular Genetics and the Human Personality.* Washington, D.C.: American Psychiatric Association, 2002.

Berul, Charles I., and Jeffrey A. Towbin, eds. *Molecular Genetics of Cardiac Electrophysiology.* Boston: Kluwer Academic, 2000.

Bishop, T., P. Sham, Pak Sham, and D. Timothy Bishop. *Analysis of Multifactorial Disease.* Oxford, England: BIOS, 2000.

Brändén, Carl-Ivar, and John Tooze. *Introduction to Protein Structure.* 2d ed. New York: Garland, 1999.

Broach, J., J. Pringle, and E. Jones, eds. *The Molecular and Cellular Biology of the Yeast Saccharomyces.* 3 vols. Cold Spring Harbor, N.Y.: Cold Spring Harbor Laboratory Press, 1991-1997.

Brown, Terence A. *Genetics: A Molecular Approach.* 3d ed. New York: Chapman & Hall, 1998.

Browning, Michael, and Andrew McMichael, eds. *HLA and MHC: Genes, Molecules, and Function.* New York: Academic Press, 1999.

Capy, Pierre, et al. *Dynamics and Evolution of Transposable Elements.* New York: Chapman & Hall, 1998.

Cotterill, Sue, ed. *Eukaryotic DNA Replication: A Practical Approach.* New York: Oxford University Press, 1999.

Crick, Francis. *Life Itself: Its Origin and Nature.* New York: Simon & Schuster, 1981.

_______. *What Mad Pursuit.* New York: Basic Books, 1988.

DePamphilis, Melvin L., ed. *Concepts in Eukaryotic DNA Replication.* Cold Spring Harbor, N.Y.: Cold Spring Harbor Laboratory Press, 1999.

De Pomerai, David. *From Gene to Animal: An Introduction to the Molecular Biology of Animal Development.* New York: Cambridge University Press, 1985.

Dillon, Lawrence S. *The Gene: Its Structure, Function, and Evolution.* New York: Plenum Press, 1987.

Dizdaroglu, Miral, and Ali Esat Karakaya, eds. *Advances in DNA Damage and Repair: Oxygen Radical Effects, Cellular Protection, and Biological Consequences.* New York: Plenum Press, 1999.

Eckstein, Fritz, and David M. J. Lilley, eds. *Catalytic RNA.* New York: Springer, 1996.

Elgin, Sarah C. R., and Jerry L. Workman, eds. *Chromatin Structure and Gene Expression.* 2d ed. New York: Oxford University Press, 2000.

Erickson, Robert P., and Jonathan G. Izant, eds. *Gene Regulation: Biology of Antisense RNA and DNA.* New York: Raven Press, 1992.

Frank-Kamenetskii, Maxim D. *Unraveling DNA.* Reading, Mass.: Addison-Wesley, 1997.

Freedman, Leonard P., and M. Karin, eds. *Molecular Biology of Steroid and Nuclear Hormone Receptors.* Boston: Birkhauser, 1999.

Friedberg, Errol C., et al., eds. *DNA Repair and Mutagenesis.* Washington, D.C.: ASM Press, 1995.

Gomperts, Kramer, et al. *Signal Transduction.* San Diego, Calif.: Academic Press, 2002.

Gros, François. *The Gene Civilization.* New York: McGraw-Hill, 1992.

Gwatkin, Ralph B. L., ed. *Genes in Mammalian Reproduction.* New York: Wiley-Liss, 1993.

Hancock, John T. *Molecular Genetics.* Boston: Butterworth-Heinemann, 1999.

Harris, David A., ed. *Prions: Molecular and Cellular Biology.* Portland, Oreg.: Horizon Scientific Press, 1999.

Hatch, Randolph T., ed. *Expression Systems and Processes for rDNA Products.* Washington, D.C.: American Chemical Society, 1991.

Hatfull, Graham F., and William R. Jacobs, Jr., eds. *Molecular Genetics of Mycobacteria.* Washington, D.C.: ASM Press, 2000.

Hawkins, John D. *Gene Structure and Expression.* New York: Cambridge University Press, 1996.

Hekimi, Siegfried, ed. *The Molecular Genetics of Aging.* New York: Springer, 2000.

Henderson, Daryl S., ed. *DNA Repair Protocols: Eukaryotic Systems.* Totowa, N.J.: Humana Press, 1999.

Hoch, James A., and Thomas J. Silhavy, eds. *Two-Component Signal Transduction.* Washington, D.C.: ASM Press, 1995.

Holmes, Roger S., and Hwa A. Lim, eds. *Gene Families: Structure, Function, Genetics and Evolution.* River Edge, N.J.: World Scientific, 1996.

Joklik, Wolfgang K., ed. *Microbiology: A Centenary Perspective.* Washington, D.C.: ASM Press, 1999.

Kimura, Motoo, and Naoyuki Takahata, eds. *New Aspects of the Genetics of Molecular Evolution.* New York: Springer-Verlag, 1991.

Langridge, John. *Molecular Genetics and Comparative Evolution.* New York: John Wiley & Sons, 1991.

Lewin, Benjamin M. *Gene Expression.* New York: John Wiley & Sons, 1980.

Litvack, Simon. *Retroviral Reverse Transcriptases.* Austin, Tex.: R. G. Landes, 1996.

McDonald, John F., ed. *Transposable Elements and Genome Evolution.* London: Kluwer Academic, 2000.

McGuffin, Peter, Michael J. Owen, and Irving I. Gottesman, eds. *Psychiatric Genetics and Genomics.* Oxford, England: Oxford University Press, 2002.

MacIntyre, Ross J., ed. *Molecular Evolutionary Genetics.* New York: Plenum Press, 1985.

Maraia, Richard J., ed. *The Impact of Short Interspersed Elements (SINEs) on the Host Genome.* Austin, Tex.: R. G. Landes, 1995.

Miesfeld, Roger L. *Applied Molecular Genetics.* New York: John Wiley, 1999.

Müller-Hill, Benno. *The Lac Operon: A Short History of a Genetic Paradigm.* New York: Walter de Gruyter, 1996.

Murphy, Kenneth P. *Protein Structure, Stability, and Folding.* Totowa, N.J.: Humana Press, 2001.

Murray, James A. H., ed. *Antisense RNA and DNA.* New York: Wiley-Liss, 1992.

Nei, Masatoshi. *Molecular Evolutionary Genetics.* New York: Columbia University Press, 1987.

Pasternak, Jack J. *An Introduction to Human Molecular Genetics: Mechanisms of Inherited Diseases.* Hoboken, N.J.: Wiley, 2005.

Pfeifer, John D., ed., with Daniel A. Arber et al. *Molecular Genetic Testing in Surgical Pathology.* Philadelphia: Lippincott, Williams & Wilkins, 2006.

Pollack, Robert. *Signs of Life: The Language and Meanings of DNA.* New York: Houghton Mifflin, 1994.

Ptashne, Mark, and Alexander Gann. *Genes and Signals.* Cold Spring Harbor, N.Y.: Cold Spring Harbor Press, 2002.

Roe, Bruce A., Judy S. Crabtree, and Akbar S. Khan, eds. *DNA Isolation and Sequencing.* New York: John Wiley & Sons, 1996.

Sarma, Ramaswamy H., and M. H. Sarma. *DNA Double Helix and the Chemistry of Cancer.* Schenectady, N.Y.: Adenine Press, 1988.

Schleif, Robert F. *Genetics and Molecular Biology.* Reading, Mass.: Addison-Wesley, 1986.

Selander, Robert K., et al., eds. *Evolution at the Molecular Level.* Sunderland, Mass.: Sinauer Associates, 1991.

Simons, Robert W., and Marianne Grunberg-Manago, eds. *RNA Structure and Function.* Cold Spring Harbor, N.Y.: Cold Spring Harbor Laboratory Press, 1997.

Smith, Paul J., and Christopher J. Jones, eds. *DNA Recombination and Repair.* New York: Oxford University Press, 2000.

Smith, Thomas B., and Robert K. Wayne. *Molecular Genetic Approaches in Conservation.* New York: Oxford University Press, 1996.

Snyder, Larry, and Wendy Champness. *Molecular Genetics of Bacteria.* Washington, D.C.: ASM Press, 1997.

Stone, Edwin M., and Robert J. Schwartz. *Intervening Sequences in Evolution and Development.* New York: Oxford University Press, 1990.

Strachan, Tom, and Andrew P. Read. *Human Molecular Genetics.* 2d ed. New York: Garland Press, 2004.

Trainor, Lynn E. H. *The Triplet Genetic Code: The Key to Molecular Biology.* River Edge, N.J.: World Scientific, 2001.

Turner, Bryan. *Chromatin and Gene Regulation: Mechanisms in Epigenetics.* Malden, Mass.: Blackwell, 2001.

Watson, James D. *The Double Helix.* New York: Atheneum, 1968.

_______, et al. *Molecular Biology of the Gene.* 5th ed. 2 vols. Menlo Park, Calif.: Benjamin Cummings, 2003.

Weaver, Robert F. *Molecular Biology.* 2d ed. New York: McGraw-Hill, 2002.

White, Robert J. *Gene Transcription: Mechanisms and Control.* Malden, Mass.: Blackwell, 2001.

POPULATION GENETICS

Avise, John, and James Hamrick, eds. *Conservation Genetics: Case Histories from Nature.* New York: Chapman and Hall, 1996.

Ayala, Francisco J. *Population and Evolutionary Genetics: A Primer.* Menlo Park, Calif.: Benjamin/Cummings, 1982.

Boorman, Scott A., and Paul R. Levitt. *The Genetics of Altruism.* New York: Academic Press, 1980.

Bushman, Frederick. *Lateral Gene Transfer: Mechanisms and Consequences.* Cold Spring Harbor, N.Y.: Cold Spring Harbor Laboratory Press, 2001.

Charlesworth, Brian. *Evolution in Age-Structured Populations.* New York: Cambridge University Press, 1980.

Christiansen, Freddy B. *Population Genetics of Multiple Loci.* New York: Wiley, 2000.

Conner, Jeffrey K., and Daniel L. Hartl. *A Primer of Ecological Genetics.* Sunderland, Mass.: Sinauer Associates, 2004.

Costantino, Robert F., and Robert A. Desharnais. *Population Dynamics and the Tribolium Model: Genetics and Demography.* New York: Springer-Verlag, 1991.

Crow, James F. *Basic Concepts in Population, Quantitative, and Evolutionary Genetics.* New York: W. H. Freeman, 1986.

Dawson, Peter S., and Charles E. King, eds. *Readings in Population Biology.* Englewood Cliffs, N.J.: Prentice-Hall, 1971.

De Waal, Franz. *Good Natured: The Origins of Right and Wrong in Humans and Other Animals.* Cambridge, Mass.: Harvard University Press, 1996.

Falconer, D. S., and Trudy F. MacKay. *Introduction to Quantitative Genetics.* 4th ed. Reading, Mass.: Addison-Wesley, 1996.

Gale, J. S. *Population Genetics.* New York: John Wiley & Sons, 1980.

Gavrilets, Sergey. *Fitness Landscapes and the Origin of Species.* Princeton, N.J.: Princeton University Press, 2004.

Gillespie, John H. *Population Genetics: A Concise Guide.* Baltimore: Johns Hopkins University Press, 2004.

Gould, Steven Jay. *Ontogeny and Phylogeny.* Cambridge, Mass.: Belknap Press, 1977.

Halliburton, Richard. *Introduction to Population Genetics.* Upper Saddle River, N.J.: Pearson/Prentice Hall, 2004.

Harper, J. L. *Population Biology of Plants.* New York: Academic Press, 1977.

Hartl, Daniel L. *A Primer of Population Genetics.* Rev. 3d ed. Sunderland, Mass.: Sinauer Associates, 2000.

Hartl, Daniel L., and Andrew G. Clark. *Principles of Population Genetics.* Sunderland, Mass.: Sinauer Associates, 2007.

Hedrick, Philip W. *Genetics of Populations.* 3d ed. Boston: Jones and Bartlett, 2005.

Hoelzel, A. R., ed. *Molecular Genetic Analysis of Populations: A Practical Approach.* New York: IRL Press at Oxford University Press, 1992.

Kang, Manjit S., and Hugh G. Gauch, Jr. *Genotype-by-Environment Interaction.* Boca Raton, Fla.: CRC Press, 1996.

Kingsland, S. E. *Modeling Nature: Episodes in the History of Population Ecology.* Chicago: University of Chicago Press, 1985.

Lack, D. *The Natural Regulation of Animal Numbers.* Oxford, England: Clarendon Press, 1954.

Laikre, Linda. *Genetic Processes in Small Populations: Conservation and Management Considerations with Particular Focus on Inbreeding and Its Effects.* Stockholm: Division of Population Genetics, Stockholm University, 1996.

Lynch, Michael. *The Origins of Genome Architecture.* Sunderland, Mass.: Sinauer Associates, 2007.

Lynn, Richard. *Dysgenics: Genetic Deterioration in Modern Populations.* Westport, Conn.: Praeger, 1996.

McKinney, M. L., and K. J. McNamara. *Heterochrony: The Evolution of Ontogeny.* New York: Plenum Press, 1991.

Magurran, Anne E., and Robert M. May, eds. *Evolution of Biological Diversity: From Population Differentiation to Speciation.* New York: Oxford University Press, 1999.

Papiha, Surinder S., Ranjan Deka, and Ranajit Chakraborty, eds. *Genomic Diversity: Applications in Human Population Genetics.* New York: Kluwer Academic/Plenum, 1999.

Provine, William B. *The Origins of Theoretical Population Genetics.* 2d ed. Chicago: University of Chicago Press, 2001.

Real, Leslie A., ed. *Ecological Genetics.* Princeton, N.J.: Princeton University Press, 1994.

Rousset, François. *Genetic Structure and Selection in*

Subdivided Populations. Princeton, N.J.: Princeton University Press, 2004.

Ruse, M. *Sociobiology: Sense or Nonsense?* Boston: D. Riedel, 1979.

Schmitt, L. H., Leonard Freedman, and Rayma Pervan. *Genes, Ethnicity, and Ageing*. Nedlands: Centre for Human Biology, University of Western Australia, 1995.

Schonewald-Cox, Christine M., et al., eds. *Genetics and Conservation: A Reference for Managing Wild Animal and Plant Populations*. Menlo Park, Calif.: Benjamin/Cummings, 1983.

Slatkin, Montgomery, and Michel Veuille, eds. *Modern Developments in Theoretical Population Genetics: The Legacy of Gustave Malécot*. New York: Oxford University Press, 2002.

Syvanen, Michael, and Clarence Kado. *Horizontal Gene Transfer*. 2d ed. Burlington, Mass.: Academic Press, 2002.

Templeton, Alan R. *Population Genetics and Microevolutionary Theory*. Hoboken, N.J.: Wiley-Liss, 2006.

Thornhill, Nancy Wilmsen, ed. *The Natural History of Inbreeding and Outbreeding: Theoretical and Empirical Perspectives*. Chicago: University of Chicago Press, 1993.

Wilson, E. O. *Sociobiology: The New Synthesis*. Cambridge, Mass.: Belknap Press, 1975.

REPRODUCTIVE TECHNOLOGY

Andrews, Lori B. *The Clone Age: Adventures in the New World of Reproductive Technology*. New York: Henry Holt, 1999.

Bentley, Gillian R., and C. G. Nicholas Mascie-Taylor. *Infertility in the Modern World: Present and Future Prospects*. New York: Cambridge University Press, 2000.

Bonnicksen, Andrea L. *In Vitro Fertilization: Building Policy from Laboratories to Legislature*. Reprint. New York: Columbia University Press, 1991.

Brinsden, Peter R., ed. *A Textbook of In Vitro Fertilization and Assisted Reproduction: The Bourn Hall Guide to Clinical and Laboratory Practice*. 2d ed. New York: Parthenon, 1999.

Campbell, Annily. *Childfree and Sterilized: Women's Decisions and Medical Responses*. New York: Cassell, 1999.

Elder, Kay, and Brian Dale. *In Vitro Fertilization*. 2d ed. New York: Cambridge University Press, 2000.

Ettore, Elizabeth. *Reproductive Genetics, Gender, and the Body*. New York: Routledge, 2002.

Heyman, Bob, and Mette Henriksen. *Risk, Age, and Pregnancy: Case Study of Prenatal Genetic Screening and Testing*. New York: Palgrave, 2001.

Jansen, Robert, and D. Mortimer, eds. *Towards Reproductive Certainty: Fertility and Genetics Beyond 1999*. Boca Raton, Fla.: CRC Press, 1999.

McElreavey, Ken, ed. *The Genetic Basis of Male Infertility*. New York: Springer, 2000.

McGee, Glenn. *The Perfect Baby: A Pragmatic Approach to Genetics*. Lanham, Md.: Rowman & Littlefield, 1997.

Mader, Sylvia S. *Human Reproductive Biology*. 3d ed. New York: McGraw-Hill, 2000.

Marrs, Richard, et al. *Dr. Richard Marrs' Fertility Book*. New York: Dell, 1997.

Parry, Vivienne. *Antenatal Testing Handbook: The Complete Guide to Testing in Pregnancy*. Collingdale, Pa.: DIANE, 1998.

Rapp, Rayna. *Testing Women, Testing the Fetus: The Social Impact of Amniocentesis in America*. New York: Routledge, 1999.

Rosenthal, M. Sara. *The Fertility Sourcebook: Everything You Need to Know*. 2d ed. Los Angeles: Lowell House, 1998.

Rothman, Barbara Katz. *The Tentative Pregnancy: How Amniocentesis Changes the Experience of Motherhood*. Rev. ed. New York: Norton, 1993.

Seibel, Machelle M., and Susan L. Crockin, eds. *Family Building Through Egg and Sperm Donation*. Boston: Jones and Bartlett, 1996.

Trounson, Alan O., and David K. Gardner, eds. *Handbook of In Vitro Fertilization*. 2d ed. Boca Raton, Fla.: CRC Press, 2000.

TECHNIQUES AND METHODOLOGIES. *See also* **BIOINFORMATICS**

Adolph, Kenneth W., ed. *Gene and Chromosome Analysis*. San Diego: Academic Press, 1993.

Braman, Jeff, ed. *In Vitro Mutagenesis Protocols*. 2d ed. Totowa, N.J.: Humana Press, 2002.

Budowle, Bruce, et al. *DNA Typing Protocols: Molecular Biology and Forensic Analysis*. Natick, Mass.: Eaton, 2000.

Butler, John M. *Forensic DNA Typing: Biology, Technology, and Genetics of STR Markers*. Amsterdam: Elsevier Academic Press, 2005.

Chen, Bing-Yuan, and Harry W. Janes, eds. *PCR Cloning Protocols*. Rev. 2d ed. Totowa, N.J.: Humana Press, 2002.

Crawley, Jacqueline N. *What's Wrong with My Mouse? Behavioral Phenotyping of Transgenic and Knockout Mice.* New York: Wiley-Liss, 2000.

Davis, Rowland H. *Neurospora: Contributions of a Model Organism.* New York: Oxford University Press, 2000.

Double, John A., and Michael J. Thompson, eds. *Telomeres and Telomerase: Methods and Protocols.* Totowa, N.J.: Humana Press, 2002.

Farrell, Robert. *RNA Methodologies.* 2d ed. San Diego, Calif.: Academic Press, 1998.

Gjerde, Douglas T., Christopher P. Hanna, and David Hornby. *DNA Chromatography.* Weinheim, Germany: Wiley-VCH, 2002.

Grange, J. M., et al., eds. *Genetic Manipulation: Techniques and Applications.* Boston: Blackwell Scientific, 1991.

Guilfoile, P. *A Photographic Atlas for the Molecular Biology Laboratory.* Englewood, Colo.: Morton, 2000.

Hall, Walter A., ed. *Methods in Molecular Biology: Immunotoxin Methods and Protocols.* Vol. 166. Totowa, N.J.: Humana Press, 2000.

Hames, B. D., and D. Rickwood, eds. *Gel Electrophoresis of Nucleic Acids: A Practical Approach.* 2d ed. New York: Oxford University Press, 1990.

Harlow, Ed, and David Lane, eds. *Using Antibodies: A Laboratory Manual.* Rev. ed. Cold Spring Harbor, N.Y.: Cold Spring Harbor Laboratory Press, 1999.

Jackson, J. F., H. F. Linskens, and R. B. Inman, eds. *Testing for Genetic Manipulation in Plants.* New York: Springer, 2002.

Kochanowski, Bernd, and Udo Reischl, eds. *Quantitative PCR Protocols.* Methods in Molecular Medicine 26. Totowa, N.J.: Humana Press, 1999.

Lai, Eric, and Bruce W. Birren, eds. *Electrophoresis of Large DNA Molecules: Theory and Applications.* Cold Spring Harbor, N.Y.: Cold Spring Harbor Laboratory, 1990.

Lewis, Ricki. *Human Genetics: Concepts and Applications.* Boston: McGraw-Hill Higher Education, 2007.

Lloyd, Ricardo V., ed. *Morphology Methods: Cell and Molecular Biology Techniques.* Totowa, N.J.: Humana Press, 2001.

McPherson, M. J., and S. G. Møller. *PCR Basics.* Oxford, England: BIOS Scientific, 2000.

McRee, Duncan Everett. *Practical Protein Crystallography.* 2d ed. San Diego: Academic Press, 1999.

O'Connell, Joe, ed. *RT-PCR Protocols.* Totowa, N.J.: Humana Press, 2002.

Pacifici, O. G. M., Julio Collado-Vides, and Ralf Hofestadt, eds. *Gene Regulation and Metabolism: Postgenomic Computational Approaches.* Cambridge, Mass.: MIT Press, 2002.

Perlmann, P., and H. Wigzell, eds. *Handbook of Experimental Pharmacology.* Heidelberg: Springer, 1999.

Pretlow, T. G., and T. P. Pretlow, eds. *Biochemical and Molecular Aspects of Selected Cancers.* Vol. 1. San Diego: Academic Press, 1991.

Sandor, Suhai, ed. *Theoretical and Computational Methods in Genome Research.* New York: Plenum Press, 1997.

Silver, Lee. *Mouse Genetics: Concepts and Applications.* New York: Oxford University Press, 1995.

Suzuki, D. T., et al. *An Introduction to Genetic Analysis.* 7th ed. New York: W. H. Freeman, 2000.

Wilson, Zoe A. *Arabidopsis: A Practical Approach.* New York: Oxford University Press, 2000.

VIRAL GENETICS

Becker, Yechiel, and Gholamreza Darai, eds. *Molecular Evolution of Viruses: Past and Present.* Boston: Kluwer Academic, 2000.

Cann, Alan J. *DNA Virus Replication.* New York: Oxford University Press, 2000.

Carlberg, David M. *Essentials of Bacterial and Viral Genetics.* Springfield, Ill.: Charles C Thomas, 1976.

Domingo, Esteban, Robert Webster, and John Holland, eds. *Origin and Evolution of Viruses.* New York: Academic Press, 1999.

Gallo, Robert C., and Flossie Wong-Staal, eds. *Retrovirus Biology and Human Disease.* New York: Marcel Dekker, 1990.

Holland, J. J., ed. *Current Topics in Microbiology and Immunology: Genetic Diversity of RNA Viruses.* New York: Springer-Verlag, 1992.

Kolata, Gina. *The Story of the Great Influenza Pandemic of 1918 and the Search for the Virus That Caused It.* New York: Simon and Schuster, 2001.

World Health Organization. *Future Research on Smallpox Virus Recommended.* Geneva, Switzerland: World Health Organization Press, 1999.

Updated by Oladayo Oyelola, Ph.D., S.C. (ASCP)

WEB SITES

The sites listed below were visited by the editors of Salem Press in June of 2009. Because URLs frequently change or are moved, their accuracy cannot be guaranteed; however, long-standing sites—such as those of university departments, national organizations, and government agencies—generally maintain links when sites move or otherwise may upgrade their offerings and hence remain useful.

General Genetics

Dolan DNA Learning Center, Cold Spring Harbor Laboratory
Gene Almanac
http://www.dnalc.org

This is the best entry point to the Web for newcomers to genetics. An online science center devoted to public education in genetics at high school and college levels, this site provides information on DNA science, genetics and medicine, and biotechnology through interactive features, animated tutorials, downloads, and extensive links.

Emory University and the University of Alabama at Birmingham
Ask the Geneticist
http://www.askthegen.org

This site provides answers to site visitors' questions about genetic concepts and genetic disorders. Selected questions and answers are posted within three weeks after submission and are archived in a searchable database. Genetics fact sheets and other general information are also posted.

Genetic Science Learning Center, The University of Utah
Learn.Genetics
http://learn.genetics.utah.edu

A repository of educational resources on genetics, bioscience, and health topics for students, teachers, and the public. Information about DNA, genetic disorders, cloning, stem cells, gene therapy, and genetic testing is presented through articles, animations, and interactive features. The site also describes simple experiments via a virtual lab. Comprehensive teacher resources, professional development programs, and lesson plans are available on the subsite Teach.Genetics (http://teach.genetics.utah.edu).

Genetics Society of America
Home Page
http://www.genetics-gsa.org

Although dedicated to subscribers who are professional geneticists, scientists, teachers, and engineers, this Web site supports genetics education for all ages and offers a history of the organization and short position statements on evolution and genetically modified organisms. There are also links to databases and related Web sites.

Johns Hopkins University
Genetics and Public Policy Center
http://www.dnapolicy.org

This site provides information and analysis of human genetics public policy. The site includes articles about genetics and health, genetic testing, assisted reproduction, cloning and stem cells, and genetic modification. It also provides a bibliography, reports, testimony, and statements from the Center and its staff. A video library and a searchable database of international laws are also provided.

Kimball's Biology Pages
Home Page
http://biology-pages.info

A reliable place to start for those new to genetics, this online biology textbook features clear, well-illustrated explanations of topics in biology by university professor John W. Kimball, including all aspects of genetics and biotechnology, with news updates.

MedBioWorld
Home Page
http://www.medbioworld.com

A portal providing links to several databases for medical and biotechnology professionals. Site features include Reuters Health News Explorer article database, MedBio Access full-access scholarly research collections, links to journals ranked by Thomson Scientifics' Journal Citation Reports (JCR), and the Post Genomics Blog that includes information about upcoming conferences and meetings, funding resources, and research tools. Also features job postings and links to national health observances.

National Public Radio, SoundVision Productions
The DNA Files
http://www.dnafiles.org
This Web site features radio documentaries and in-depth features that provide an informal education for nonscientists about the genomics revolution, including the science of genetics and its ethical, social, and legal implications. The third series of the Peabody Award-winning audio programs from 2007 features topics including the human genome, DNA and behavior, DNA and evolution, genetics and ecology, gene therapy, genetic medicine, biotechnology, predictive genetic testing, and more.

Nature Publishing Group
genetics@nature.com
http://www.nature.com/genetics
Part of the Web site for *Nature*, Britain's premier science journal. Written for educated general readers, the sections offer news and recently published articles, commentary, and an encyclopedia of life sciences. There are links to specialty journals concerning topics in genetics and biotechnology. Some articles are accessible by the public, but full use of the site requires a subscription.

Netspace
MendelWeb
http://www.mendelweb.org
An educational site for teachers and students about the origins of classical genetics, introductory data analysis, elementary plant science, and the history and literature of science. It reproduces early publications by such pioneers as Gregor Mendel and William Bateson, accompanied by commentaries and reference resources.

Rutgers University
Morgan
http://morgan.rutgers.edu/MorganWebFrames/How_to_use/HTU_intro.html
A multimedia tutorial for advanced high school students or beginning college students. Its six levels review basic principles in genetics, with particular attention to molecular interactions.

University of Massachusetts
DNA Structure
http://molvis.sdsc.edu/dna/index.htm
An interactive, animated tutorial on the molecular composition and structure of DNA for high school students and undergraduates. It can be downloaded and is available in Spanish, German, French, and Portuguese.

U.S. Department of Energy
Gene Gateway
http://genomics.energy.gov/genegateway
This Web site introduces various Internet tools that anyone can use to investigate genetic disorders, chromosomes, genome maps, genes, sequence data, genetic variants, and molecular structures. The downloadable Gene Gateway Workbook features a collection of guides and tutorials to help students and other novice users get started with some of the publicly available genetics resources.

U.S. Department of Energy Office of Science
Virtual Library on Genetics
http://www.ornl.gov/TechResources/Human_Genome/genetics.html
A comprehensive catalog of Web site links, arranged by subject, pertaining to the Human Genome Project. The links lead to gene and chromosome databases and specific information on genetics, bioinformatics, and genetic disorders.

BIOINFORMATICS

Bioinformatics Organization
Home Page
http://www.bioinformatics.org
This international organization's site is dedicated to the exchange of genetic information, serving the scientific and educational needs of bioinformatic practitioners and the public. The organization has about 25,000 members and more than three hundred projects. Membership is free to the public and is fee-based for professionals, with additional discounts and services offered. It includes free online databases and analysis tools, group hosting services, software, professional bioinformatics education, forums, and a newsletter. Fee-based data analysis and software development services are provided through the organization's Bioinformatics Core Facility.

European Bioinformatics Institute
The Path to Knowledge
http://www.ebi.ac.uk

A bioinformatics research center, this institute maintains databases concerning nucleic acids, protein sequences, and macromolecular structures. Also posts downloadable fact sheets, archived publications, news, events, training opportunities, and descriptions of ongoing scientific projects.

Technical University of Denmark DTU
Center for Biological Sequence Analysis
http://www.cbs.dtu.dk

The center conducts basic research in bioinformatics and systems biology. The site, primarily for researchers and university students, offers sequencing data sets, prediction servers, software, bioinformatics tools, information about its research groups and education programs, events, and news.

Biotechnology

Bio-Link
Educating the Biotechnology Workforce
http://bio-link.org

Intended for technicians, this site offers information and instruction covering recent advances in biotechnology, as well as a virtual laboratory and library, news, discussion forums, job links, and internship postings. A U.S. map also links to directories of regional biotechnology education centers, organizations, and industries.

Biotechnology Industry Organization (BIO)
Home Page
http://bio.org

BIO provides advocacy, business development, and communications services for more than 1,200 international members involved in the research and development of health care, agricultural, industrial, and environmental biotechnology technologies. The site provides free access to articles on national, state, and local biotechnology issues with links to testimonies, statements, and fact sheets. Additional site features are provided to members.

Carolina Biological Supply Company
Biotechnology and Genetics
http://www.carolina.com/biotech

This supplier of biological, science, and math supplies and teaching materials provides a rich educational resource for elementary and high school teachers and students. Resources include classroom activities, articles, a newsletter, workshops, and videos, all focusing on the use of biotechnology and laboratory techniques.

Dow AgroSciences LLC
Plant Genetics and Biotechnology
http://www.dowagro.com

This corporate Web site concerns the marketing of its agricultural biotechnology. Given that bias, it contains news releases, descriptions of products, labels and safety data sheets, articles, publication links, and a media kit of interest to general readers.

The Hastings Center
Home Page
http://www.thehastingscenter.org

The Hastings Center is an independent, nonpartisan, and nonprofit bioethics research institute that addresses ethical issues in health, medicine, and the environment. Its site contains news postings, articles on bioethics and aspects of genetics science, research project overviews, and announcements of events and publications.

National Institutes of Health
National Center for Biotechnology Information
http://www.ncbi.nlm.nih.gov

This Web site for the main health agency of the United States contains links to the various specialized institutes under its umbrella as well as public databases in genomics and sequencing, articles and handbooks on a wide range of biotechnology topics, and more than a dozen types of free software for analyzing genetic data. Primarily intended for researchers, but also includes a comprehensive science primer and other resources for educators, students, and other nonspecialists.

Genomics

American Institute of Biological Sciences (AIBS)
ActionBioscience.org
http://www.actionbioscience.org/genomic

Includes a growing collection of articles by distinguished scientists, educators, and other writers on a wide range of biological subjects for K-12 and undergraduate educators, as well as students and the public. The site's genomics subsection features articles on genomic mapping, ethics, genetic information, genetic manipulation, and genomes.

BioMed Central Ltd.
Genome Biology
http://genomebiology.com

A print journal that serves as an international forum for the biological research community to aid the dissemination, discussion, and critical review of information. The journal articles are accessible online to subscribers, and a sampling of full-text articles is posted for public access, including select reviews, reports, and research.

Göteborg University, Sweden
RatMap, the Rat Genome Database
http://ratmap.gen.gu.se

A professional database of information on approximately six thousand rat genes, their positions on chromosomes, pertinent nomenclature, and gene functions.

Massachusetts Institute of Technology and Harvard University
The Broad Institute
http://www.broad.mit.edu/

Intended for scientists and university students, the site provides access to genomics research news, software, and sequencing databases. It also features press releases and news archives, organized by research areas. The Outreach section features programs and events to teach high school students and the public about genomics.

National Center for Biotechnology Information
Human Genome Resources
http://www.ncbi.nlm.nih.gov/genome/guide/human

An integrated, one-stop genomic resource for biomedical researchers who can use the site's search and analysis tools to find a gene's location in the genome, find other genes in the same region, correlate diseases to genes, determine if similar genes exist in other organisms, and view genetic variations.

National Center for Biotechnology Information
A New Gene Map of the Human Genome: The International RH Mapping Consortium
http://www.ncbi.nlm.nih.gov/genemap99

An electronic data supplement for the Gene Map paper, published by P. Deloukas et al. in *Science* 282 (1998): 744-746.

National Human Genome Research Institute
Home Page
http://www.genome.gov

In addition to information for researchers and health professionals, this site contains public information including a comprehensive introduction to the Human Genome Project, a glossary of genetic terms, fact sheets, multimedia education kits, and links to online education resources.

PHG Foundation (Foundation for Genomics and Population Health)
Home Page
http://www.phgfoundation.org

This independent not-for-profit public health organization focuses on science and biomedical innovation to improve health, especially genome-based science and technologies. The foundation's Web site, with free registration, provides access to a database of policy literature, interactive educational resources, rapid responses to genetics news stories, full-text reports, key papers, and links.

Sanger Institute, Wellcome Trust
Home Page
http://www.sanger.ac.uk

This genomics research institute's site, targeted to scientists, offers news updates, a searchable database, explanations of gene sequencing and computer software aids, and descriptions of genomics research projects. There is also a section with resources for students, teachers, and the community, including a glossary of terms, links, and information about scheduling site visits or video conferences with Sanger Institute researchers.

U.S. Department of Energy Office of Science
Genomics.Energy.gov
http://genomics.energy.gov

This gateway provides links to several subsites of the DOE including Human Genome Project Information, the Genomic Science Program, and the Microbial Genome Program. Site features include a press room with image gallery, resources for educators, a genomics primer, and articles about biofuels, medicine and genetics, as well as ethical, legal, and social issues.

U.S. Department of Energy Office of Science
Genomic Science Program
http://genomicscience.energy.gov

This research program aims to provide sufficient

scientific understanding of plants and microbes to develop new strategies to produce biofuels, clean up waste, or sequester carbon. The site provides research and technology summaries, as well as links to teaching tools for educators to provide instruction on genomics, systems biology, and microbes. Includes an image library.

U.S. Department of Energy Office of Science
Joint Genome Institute (JGI)
http://www.jgi.doe.gov

The JGI unites the expertise and resources in DNA sequencing, informatics, and technology development at five national laboratories, including Lawrence Berkeley, Lawrence Livermore, Los Alamos, Oak Ridge, and Pacific Northwest, along with the HudsonAlpha Institute for Biotechnology. Primarily for researchers, the site includes a Genomic Encyclopedia of Bacteria and Archaea, sequencing information and protocols, video links, news items and meeting notices, software tools, educational opportunities, current genome project lists, research group lists, and job postings.

U.S. Department of Energy Office of Science, Office of Biological and Environmental Research, Human Genome Program
Human Genome Project Information
http://www.ornl.gov/hgmis/home.shtml

A comprehensive site about the Human Genome Project, the international thirteen-year effort to discover all of the human genes and make them accessible for further study. Although the project concluded in 2003, data analyses continue, and the site provides project information, research archives, publication links, and information and article links about ethical, legal, and social issues; gene testing; gene therapy; genetic counseling; and pharmacogenomics. A section for young students and science teachers also is provided.

World Health Organization (WHO)
The Genomic Resource Centre
http://www.who.int/genomics/en

Provides information to raise awareness on human genetics and human genomics. The site provides an A-to-Z listing of health topics and WHO initiatives, as well as a repository of articles for patients, the public, health professionals, and policy makers.

Medicine and Genetics

American College of Medical Genetics
Home Page
http://www.acmg.net

Designed for physicians, this site contains some information for patients about medical genetics, selecting a geneticist, and genetic testing, along with news articles, practice guidelines, and reference materials.

Centers for Disease Control
Office of Public Health Genomics
http://www.cdc.gov/genomics/default.htm

Intended for the public, the site provides information about human genomic discoveries and how they can be used to improve health and prevent disease. It includes a glossary, frequently asked questions, fact sheets, news articles, student and teacher resources, and state activities. It also includes information about CDC public health activities for disease prevention. Available in Spanish.

Genetic Alliance, Inc.
Home Page
http://www.geneticalliance.org

A nonprofit coalition of advocacy, research, and health care organizations that represent millions of individuals with genetic conditions. The open network connects members of parent and family groups, community organizations, disease-specific advocacy organizations, professional societies, educational institutions, corporations, and government agencies.

National Center for Biotechnology Information
Online Mendelian Inheritance in Man
http://www.ncbi.nlm.nih.gov/sites/entrez?db=omim

This catalog of human genes and genetic disorders is intended for use by physicians, genetics researchers, and advanced science and medicine students. The site's full-text, referenced overviews contain information on all known hereditary disorders and more than 12,000 genes, focusing on the relationship between phenotype and genotype. It is updated daily, and the entries contain copious links to similar sites.

The National Fragile X Foundation
Home Page
http://www.nfxf.org

Provides extensive general information about fragile X syndrome, a cause of inherited mental impairments, and advises caregivers on testing, medical treatment, education, and life planning. With links to other resources.

National Organization for Rare Disorders (NORD)
Home Page
http://www.rarediseases.org
Offers a useful and long index of genetic disorders and diseases. The rare disease database links to fact sheets on more than 1,150 indexed conditions. The organizational database, searchable by topic, provides links to more than 2,000 resources and support organizations.

National Society of Genetic Counselors
Society Home Page
http://www.nsgc.org
Although much of this site is devoted to society members, it has a search engine for locating genetic counselors in the United States, general information about genetic testing, and a newsroom with press releases and fact sheets about genetic counseling.

The University of Nottingham
INTUTE: Health and Life Sciences
http://www.intute.ac.uk/healthandlifesciences/medicine
This gateway to health and life sciences information provides education and research resources that are evaluated and selected by a network of subject specialists.

University of Washington, Seattle, and National Center for Biotechnology Information
GeneTests
http://www.ncbi.nlm.nih.gov/sites/GeneTests/?db=GeneTests
This NIH-funded site provides reliable, up-to-date genetic counseling and testing information for physicians and researchers. The site includes educational materials; expert-authored, peer-reviewed disease descriptions; a voluntary directory of U.S. and international laboratories offering genetic testing; and a voluntary directory of clinics providing genetic evaluation and genetic counseling.

TRANSGENICS

National Academy of Sciences
Transgenic Plants and World Agriculture
http://www.nap.edu/openbook.php?record_id=9889
A downloadable report published in July, 2000, by an international consortium of leading research societies. It assesses the need to modify crops genetically in order to feed the increasing world population and then discusses examples of the technology, its safety, environmental effects, funding sources, and intellectual property issues.

Syngenta Foundation for Sustainable Agriculture
Home Page
http://www.syngentafoundation.org
The Foundation provides resources to small farmers in developing countries by extending science-based knowledge, facilitating access to quality inputs, and linking them to markets in profitable ways. The site provides resource articles, press releases, project summaries, and stories.

University of Michigan
Transgenic Animal Model Core
http://www.med.umich.edu/tamc
A professional Web site for researchers seeking a host animal to test transgenes. However, it contains useful general information about transgenics (especially transgenic rats), vectors, and laboratory procedures. With links and a photo gallery.

Roger Smith, Ph.D.; updated by Angela Costello

Indexes

CATEGORY INDEX

BACTERIAL GENETICS

BIOETHICS

BIOINFORMATICS

CELLULAR BIOLOGY

CLASSICAL TRANSMISSION GENETICS

DEVELOPMENTAL GENETICS

DISEASES AND SYNDROMES

EVOLUTIONARY BIOLOGY

GENETIC ENGINEERING AND BIOTECHNOLOGY

HISTORY OF GENETICS

HUMAN GENETICS AND SOCIAL ISSUES

IMMUNOGENETICS

MOLECULAR GENETICS

POPULATION GENETICS

TECHNIQUES AND METHODOLOGIES

VIRAL GENETICS

PERSONAGES INDEX

SUBJECT INDEX

A page number or range in boldface type indicates that an entire entry devoted to that topic appears in the set.